ORDER IN the COMPLEXITY of Evolving Worlds · Searching for

ACKNOWLEDGMENTS

The SFI Press is supported by William H. Miller and the Miller Omega Program.

◊ ◊ ◊

To produce a mighty book, you must choose a mighty theme. No great and enduring volume can ever be written on the flea, though many there be who have tried it.
—HERMAN MELVILLE
Moby-Dick (1851)

These four volumes are a product of collective intelligence. They have come into existence through the coordinated insights of a global network of complexity scientists. We thank every one of them for their insights and efforts.

We thank our generous Board of Trustees, research foundations, and federal agencies for their support of science, reason, and debate.

We thank our colleagues Kate Joyce, Tim Taylor, Renée Tursi, Katherine Mast, and Bronwynn Woodsworth for reading, commenting, adding to, and improving on the project.

We dedicate these four volumes to the friends and colleagues we have lost during their making: Phil Anderson, Dan Dennett, Herb Gintis, James Hartle, Erica Jen, Richard Lewontin, Dan Lynch, Robert May, Cormac McCarthy, David Padwa, James Pelkey, William Sick, Chuck Stevens, and Douglas White.

David C. Krakauer
Laura Egley Taylor
Sienna Latham
Zato Hebbert
Ellis Wylie

FOUNDATIONAL PAPERS IN COMPLEXITY SCIENCE

Volume Four

1989–2000

DAVID C. KRAKAUER

editor

© 2024 Santa Fe Institute
All rights reserved.

1399 Hyde Park Road
Santa Fe, New Mexico 87501

Foundational Papers in Complexity Science, Vol. 4
ISBN (HARDCOVER): 978-1-947864-55-9
Library of Congress Control Number: 2024938011

The SFI Press is generously supported by
the Miller Omega Program.

A SCIENTIFIC REVOLUTION IS AN ACT OF MOTION. The mind leaves one major door of perception, one high window, and turns to another. The landscape is seen in a fresh perspective, under different lights and shadows, in new contours and foreshortenings. Features that were salient now appear to be secondary or are recognized as elements in a more comprehensive form. Details hitherto unobserved or casually grouped assume a dominant focus. The grid of the world alters ...

—GEORGE STEINER
"Lifelines," *Extraterritorial* (1972)

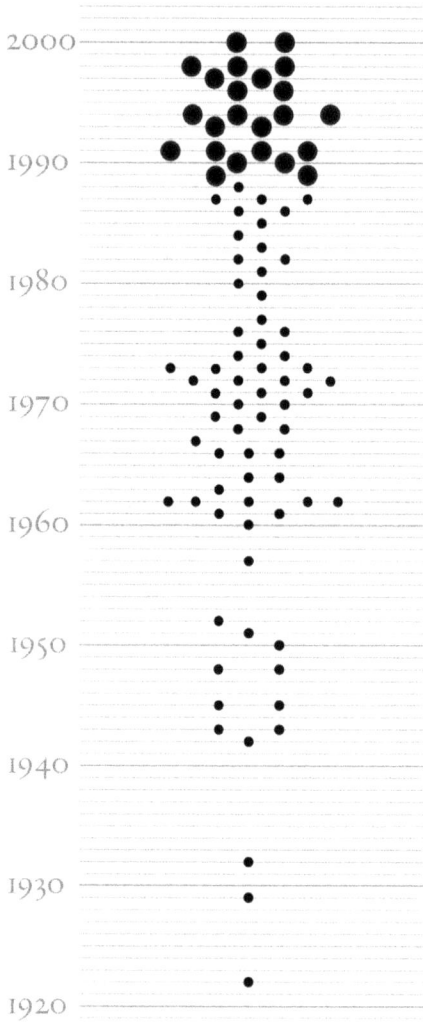

THE PAPERS

Large dots represent papers included in this volume; small dots are the remaining papers in the collection. Dots are positioned from bottom to top according to the year the paper was published. Volume 4 begins with Arthur (1989) and ends with Ostrom (2000).

TABLE OF CONTENTS
— Volume Four —

Listed chronologically, with the introduction to each paper followed by the (☞) annotated paper

67: *Scott E. Page, "Competing Technologies, Increasing Returns, and Lock-In"* 2069
☞ W. B. Arthur, "Competing Technologies, Increasing Returns, and Lock-In by Historical Events" (1989) 2073

68: *Rob J. de Boer, "Immune Network Theory: A Retrospective"* 2097
☞ A. S. Perelson, "Immune Network Theory" (1989) 2104

69: *Stephanie Forrest, "Connecting Connectionist Models"* 2143
☞ J. D. Farmer, "A Rosetta Stone for Connectionism" (1990) 2151

70: *Jessica C. Flack, "And It Is All One to Me"* 2211
☞ J. A. Wheeler, "Information, Physics, Quantum: The Search for Links" (1989) .. 2219

71: *Vijay Balasubramanian, "What a Neuron Says and How It Says It"* 2251
☞ W. Bialek, F. Rieke, R. de Ruyter van Stevenink, and D. Warland, "Reading a Neural Code" (1991) 2255

72: *Richard Bookstaber, "Economics for Humans"* 2269
☞ J. H. Holland and J. H. Miller, "Artificial Adaptive Agents in Economic Theory" (1991) 2273

73: *W. Brian Arthur, "Simple Dynamics with Rich Implications"* 2285
☞ K. Lindgren, "Evolutionary Phenomena in Simple Dynamics" (1991) 2290

74: *Suresh Naidu, "What Simon Says: Introducing Organizations and Markets"* .. 2311
☞ H. A. Simon, "Organizations and Markets" (1991) 2315

75: *Mirta Galesic, "Evolving Centralized Institutions from the Bottom Up"* 2341
☞ J. S. Lansing and J. N. Kremer, "Emergent Properties of Balinese Water Temple Networks: Coadaptation on a Rugged Fitness Landscape" (1993) ... 2346

76: *David H. Ackley, "Life and Computation: From Statistical Physics to Emergent Physics"* 2373
☞ M. Mitchell, P. T. Hraber, and J. P. Crutchfield, "Revisiting the Edge of Chaos: Evolving Cellular Automata to Perform Computations" (1993) 2378

77: *Willemien Kets, "The Costs of Miscoordination"* 2429
☞ W. B. Arthur, "Inductive Reasoning and Bounded Rationality" (1994) 2435

78: *Peter M. A. Sloot, Rick Quax, and Mile Gu, "Natural Information Processing"* . . 2447
- J. P. Crutchfield, "The Calculi of Emergence: Computation: Dynamics, and Induction" (1994) 2457

79: *Anil Somayaji, "Learning from the Immune System"* 2535
- S. Forrest, A. S. Perelson, L. Allen, and R. Cherukuri, "Self–Nonself Discrimination in a Computer" (1994) 2540

80: *Evandro Ferrada, "The Architecture of Genotype–Phenotype Maps"* 2563
- P. Schuster, W. Fontana, P. F. Stadler, and I. L. Hofacker, "From Sequences to Shapes and Back: A Case Study in RNA Secondary Structures" (1994) 2567

81: *Miguel Fuentes, "Characterization of Complex Systems and the Evolution of Scientific Theories"* 2581
- M. Gell-Mann and S. Lloyd, "Information Measures, Effective Complexity, and Total Information" (1996) 2585

82: *David C. Krakauer, "The Evolution of Selection"* 2609
- F. J. Odling-Smee, K. N. Laland, and M. W. Feldman, "Niche Construction" (1996) 2614

83: *Pablo Marquet, "Quarter-Power Allometric Scaling and Life Dynamics"* 2625
- G. B. West, J. H. Brown, and B. J. Enquist, "A General Model for the Origin of Allometric Scaling Laws in Biology" (1997) 2632

84: *Charlie Strauss and Vijay Balasubramanian, "No Free Lunch"* 2649
- D. H. Wolpert and W. G. Macready, "No Free Lunch Theorems for Optimization" (1997) 2653

85: *Nihat Ay, "From the Euclidean to the Natural"* 2699
- S.-I. Amari, "Natural Gradient Works Efficiently in Learning (1998) 2705

86: *Rajiv Sethi, "Sam Bowles on Endogenous Preferences"* 2733
- S. Bowles, "Endogenous Preferences: The Cultural Consequences of Markets and Other Economic Institutions" (1998) 2737

87: *Michelle Girvan, "How the Small-World Model Transformed How We Think about Connectivity"* 2795
- D. J. Watts and S. H. Strogatz, "Collective Dynamics of 'Small-World' Networks" (1998) 2798

88: *Christopher P. Kempes, "Mesocosmos: In Search of Universal Laws of Living Material"* 2809
- R. B. Laughlin, D. Pines, J. Schmalian, B. P. Stojković, and P. Wolynes, "The Middle Way" (2000) 2815

89: *John M. Anderies, "Collective Action & the Dynamics of Institutions"* 2835
- E. Ostrom, "Collective Action and the Evolution of Social Norms" (2000) . . . 2839

Biographical Oddities, Volume 4 2868

— FOUNDATIONAL PAPERS: Volume One —

Available for purchase. Visit www.sfipress.org *to learn more.*

1: *Luís Bettencourt, "Maximum Power as a Physical Principle of Evolution"*
☞ A. J. Lotka, "Contribution to the Energetics of Evolution" (1922)

2: *Susanne Still, "To Work is to Think"*
☞ L. Szilárd, "On the Decrease of Entropy in a Thermodynamic System by the Intervention of Intelligent Beings" (1929)

3: *Sergey Gavrilets, "Navigating the Landscape of Complexity: Sewall Wright's Seminal Contributions and their Far-Reaching Impact"*
☞ S. Wright, "The Roles of Mutation, Inbreeding, Crossbreeding, and Selection in Evolution" (1932)

4: *Walter Fontana, "The Shadow of a Mechanism"*
☞ C. H. Waddington, "Canalization of Development and the Inheritance of Acquired Characters" (1942)

5: *Bruno Olshausen and Christopher Hillar, "The Genesis of Modern Computing and AI"*
☞ W. S. McCulloch and W. Pitts, "A Logical Calculus of the Ideas Immanent in Nervous Activity" (1943)

6: *Andrew Pickering, "The Birth of Cybernetics"*
☞ A. Rosenblueth, N. Wiener, and J. Bigelow, "Behavior, Purpose, and Teleology" (1943)

7: *Samuel Bowles, "Friedrich Hayek's Economy of Knowledge"*
☞ F. A. Hayek, "The Use of Knowledge in Society" (1945)

8: *Cosma Shalizi, "Opening a Closed Box"*
☞ A. Rosenblueth and N. Wiener, "The Role of Models in Science" (1945)

9: *Seth Lloyd, "From the Analog to the Digital"*
☞ C. E. Shannon, "A Mathematical Theory of Communication" (1948)

10: *Stuart Kauffman, "If, and, If So, How?"*
☞ W. Weaver, "Science and Complexity" (1948)

11: *Daniel Dennett, "Turing's Brainchild: Artificial Intelligence"*
☞ A. M. Turing, "Computing Machinery and Intelligence" (1950)

12: *Robert Axtell, "The Nash Equilibrium"*
☞ J. Nash, "Non-Cooperative Games" (1951)

13: *Karen Page, "Do Turing Patterns Provide a Chemical Basis of Morphogenesis?"*
☛ A. M. Turing, "The Chemical Basis of Morphogenesis" (1952)

14: *Dawn Holmes, "Jaynes and the Principle of Maximum Entropy"*
☛ E. T. Jaynes, "Information Theory and Statistical Mechanics" (1957)

15: *Maxim Raginsky, "The State–Space Revolution in the Study of Complex Systems"*
☛ R. E. Kálmán, "Contributions to the Theory of Optimal Control" (1960)

16: *David Wolpert, "The Relationship between Physics and Computation: The Minimal Thermodynamic Cost of to Erase a Bit"*
☛ R. Landauer, "Irreversibility and Heat Generation in the Computing Process" (1961)

17: *Melanie Mitchell, "Symbols versus Cybernetics: Marvin Minsky on the Prospects for Artificial Intelligence"*
☛ M. Minsky, "Steps Toward Artificial Intelligence" (1961)

18: *John Geanakoplos, "Arrow's 'Learning by Doing' and Complexity Economics"*
☛ K. J. Arrow, "The Economic Implications of Learning by Doing" (1962)

19: *David C. Krakauer, "The Unstable Foundations of Simplicity"*
☛ M. Bunge, "The Complexity of Simplicity" (1962)

20: *John H. Miller, "Mastering the Glass Bead Game"*
☛ J. H. Holland, "Outline for a Logical Theory of Adaptive Systems" (1962)

— FOUNDATIONAL PAPERS: Volume Two —

Available for purchase. Visit www.sfipress.org to learn more.

21: *Simon A. Levin, "The Architect of Complexity"*
☞ H. A. Simon, "The Architecture of Complexity" (1962)

22: *Erica Jen, "Café Roots & Fruits"*
☞ S. Ulam, "On Some Mathematical Problems Connected with Patterns of Growth in Figures" (1962)

23: *J. Doyne Farmer, "The Revolutionary Discovery of Chaos"*
☞ E. N. Lorenz, "Deterministic Nonperiodic Flow" (1963)

24: *Cristopher Moore, "Easy vs. Hard: The Dawn of Computational Complexity*
☞ A. Cobham, "The Intrinsic Computational Difficulty of Functions" (1964)

25: *Paul M.B. Vitányi, "A General System for Machine Learning"*
☞ R. J. Solomonoff, "A Formal Theory of Inductive Inference, Part 1" (1964)

26: *Simon DeDeo, "Portrait of the Artist as a Young Complexity Hacker"*
☞ G. J. Chaitin, "On the Length of Programs for Computing Finite Binary Sequences" (1966)

27: *Ricard Solé, "Morphospaces, the Possible, and the Actual"*
☞ D. M. Raup, "Geometric Analysis of Shell Coiling; General Problems" (1966)

28: *Neil Gershenfeld, "Self-Reproducing Ideas"*
☞ J. von Neumann, "Theory of Self-Reproducing Automata" (1966)

29: *Geoffrey B. West, "Fractals, Self-Similarity, and Power Laws"*
☞ B. B. Mandelbrot, "How Long is the Coast of Britain? Statistical Self-Similarity and Fractional Dimension" (1967)

30: *Carl T. Bergstrom and Michael Lachmann, "Origins of the Neutral Theory"*
☞ M. Kimura, "Evolutionary Rate at the Molecular Level" (1968)

31: *Simon DeDeo, "Inspiring, Enigmatic, Incomplete"*
☞ A. N. Kolmogorov, "Three Approaches to the Quantitative Definition of Information" (1968)

32: *Daniel P. Schrag, "A Faulty Foundation Supports a Powerful Idea: Mikhail Budyko and His Work on the Ice–Albedo Feedback"*
☞ M. I. Budyko, "The Effect of Solar Radiation Variations on the Climate of the Earth" (1969)

33: *Sanjay Jain, "Attractors of a Random Networked Dynamical System Have Something to Say about Life"*
☞ S. Kauffman, "Metabolic Stability and Epigenesis in Randomly Constructed Genetic Nets" (1969)

34: *James P. Crutchfield, "Requisite Complexity—A Contemporary View of Cybernetic Control and Communication"*
☞ R. C. Conant and W. R. Ashby, "Every Good Regulator of a System Must Be a Model of That System" (1970)

35: *Michael Lachmann, "The Price Equation: The Mathematical Basis of Evolutionary Theory?"*
☞ G. R. Price, "Selection and Covariance" (1970)

36: *Philipp Honegger and Walter Fontana, "The Wheels of Chemistry"*
☞ O. E. Rössler, "A System–Theoretic Model of Biogenesis (Ein systemtheoretisches Modell zur Biogenese)" (1971)

37: *H. Peyton Young, "Tipping Points: Schelling's Account of Sorting and Segregation"*
☞ T. C. Schelling, "Dynamic Models of Segregation" (1971)

38: *Timothy A. Kohler, "Prehistory in the Land of Malthus"*
☞ E. B. W. Zubrow, "Carrying Capacity and Dynamic Equilibrium in the Prehistoric Southwest" (1971)

39: *Tanmoy Bhattacharya, "Rise of Complexity Sciences in a Reductionist World"*
☞ P. W. Anderson, "More Is Different" (1972)

40: *Cristopher Moore, "Gardens of Forking Paths: Exponential Hardness and Exhaustive Search"*
☞ R. M. Karp, "Reducibility Among Combinatorial Problems" (1972)

41: *Jennifer A. Dunne, "Diversity/Complexity/Stability"*
☞ R. M. May, "Will a Large Complex System Be Stable?" (1972)

42: *Manfred Laubichler, "The Deep Co-Evolutionary Roots of Complexity"*
☞ H. von Foerster, "Notes on an Epistemology for Living Things" (1972)

43: *Jon Machta, "The Energy Cost of Computing"*
☞ C. H. Bennett, "Logical Reversibility of Computation" (1973)

44: *Duncan Watts, "Weak Ties and the Origins of Network Science"*
☞ M. S. Granovetter, "The Strength of Weak Ties" (1973)

— FOUNDATIONAL PAPERS: Volume Three —

Available for purchase. Visit www.sfipress.org *to learn more.*

45: *Van Savage, "Resilience Redounding"*
☛ C. S. Holling, "Resilience and Stability of Ecological Systems" (1973)

46: *John Kaag, "The Organization of the Organization of Complex Systems"*
☛ H. A. Simon, "The Organization of Complex Systems" (1973)

47: *Karl Sigmund, "The Emergence of Evolutionary Stability"*
☛ J. Maynard Smith, "The Theory of Games and the Evolution of Animal Conflicts" (1974)

48: *Randall D. Beer, "A Factory that Makes Itself"*
☛ F. G. Varela, H. R. Maturana, and R. Uribe, "Autopoiesis: The Organization of Living Systems, Its Characterization, and a Model" (1974)

49: *Sid Redner, "On the Simplest (and Perhaps Too Simple) Solvable Model of a Spin Glass"*
☛ D. Sherrington and S. Kirkpatrick, "Solvable Model of a Spin-Glass" (1975)

50: *Axel Hutt, "Synergetics: The Doctrine of Interaction"*
☛ H. Haken, "Synergetics" (1976)

51: *Nicholas W. Watkins "Brownian Motion as Mathematical Superstructure to Organize the Science of Climate and Weather"*
☛ K. Hasselmann, "Stochastic Climate Models Part I. Theory" (1976)

52: *Douglas H. Erwin, "Punctuation and Discontinuities: A New View of History (?)"*
☛ S. J. Gould and N. Eldredge, "Punctuated Equilibria: The Tempo and Mode of Evolution Reconsidered" (1977)

53: *Daniel L. Stein, "Spin Glasses, Disordered Systems, and Complexity"*
☛ G. Parisi, "Infinite Number of Order Parameters for Spin-Glasses" (1979)

54: *Neil Gershenfeld, "Finding Hidden Meaning"*
☛ N. H. Packard, J. P. Crutchfield, J. D. Farmer, and R. S. Shaw, "Geometry from a Time Series" (1980)

55: *Michael E. Hochberg, "Complexity in the Evolution of Cooperation"*
☛ R. Axelrod and W. D. Hamilton, "The Evolution of Cooperation" (1981)

56: *D. Eric Smith, "Small Molecule Self-Organization before Genes"*
☞ F. J. Dyson, "A Model for the Origin of Life" (1982)

57: *David Sherrington, "Attractor Neural Networks"*
☞ J. J. Hopfield, "Neural Networks and Physical Systems with Emergent Collective Computational Abilities" (1982)

58: *Elizabeth Bradley, "Universality in Nonlinear Dynamical Systems"*
☞ M. J. Feigenbaum, "Universal Behavior in Nonlinear Systems" (1983)

59: *Hector Zenil, "The Emergent Behavior of Computer Programs of Short Description Length in Discrete Time and Discrete Space"*
☞ S. Wolfram, "Universality and Complexity in Cellular Automata" (1984)

60: *Stefan Thurner, "Irreversible, Nonlinear, Stochastic Systems"*
☞ I. Prigogine and G. Nicolis, "Self-Organisation in Nonequilibrium Systems: Towards a Dynamics of Complexity" (1985)

61: *Sara Imari Walker, "Simulating Life"*
☞ C. G. Langton, "Studying Artificial Life with Cellular Automata" (1986)

62: *David Kinney, "Causal Structures for Evidential Reasoning"*
☞ J. Pearl, "Fusion, Propagation, and Structuring in Belief Networks" (1986)

63: *Gunnar Pruessner, "The Physics of Fractals"*
☞ P. Bak, C. Tang, and K. Wiesenfeld, "Self-Organized Criticality: An Explanation of the 1/f Noise" (1987)

64: *Christoph Adami, "Tripping the Walks Fantastic"*
☞ S. Kauffman and S. Levin, "Towards a General Theory of Adaptive Walks on Rugged Landscapes" (1987)

65: *Robert L. Axtell, "Simple Agents and the Emergence of Complex Behaviors"*
☞ C. Reynolds, "Flocks, Herds, and Schools: A Distributed Behavioral Model" (1987)

66: *David C. Krakauer, "The Evolutionary Cloud in Sequence Space"*
☞ M. Eigen, J. McCaskill, and P. Schuster, "Molecular Quasi-Species" (1988)

VOLUME 4 CONTRIBUTORS

David H. Ackley *Living Computation Foundation; University of New Mexico*

John M. Anderies *Arizona State University*

W. Brian Arthur *SRI International; Santa Fe Institute*

Nihat Ay *Hamburg University of Technology*

Vijay Balasubramanian *University of Pennsylvania; Santa Fe Institute*

Richard Bookstaber

Rob J. de Boer *Utrecht University*

Evandro Ferrada *Universidad de Valparaíso*

Jessica C. Flack *Santa Fe Institute*

Stephanie Forrest *Arizona State University; Santa Fe Institute*

Miguel Fuentes *Argentine Society of Philosophical Analysis; Santa Fe Institute*

Mirta Galesic *Santa Fe Institute; Complexity Science Hub Vienna*

Michelle Girvan *University of Maryland; Santa Fe Institute*

Mile Gu *Nanyang Technological University*

Christopher P. Kempes *Santa Fe Institute*

Willemien Kets *Utrecht University; Santa Fe Institute*

David C. Krakauer *Santa Fe Institute*

Pablo Marquet *Pontificia Universidad Católica de Chile; Santa Fe Institute*

Suresh Naidu *Columbia University; Santa Fe Institute*

Scott E. Page *University of Michigan, Ann Arbor; Santa Fe Institute*

Rick Quax *University of Amsterdam*

Rajiv Sethi *Columbia University; Santa Fe Institute*

Peter M. A. Sloot *University of Amsterdam*

Anil Somayaji *Carleton University*

Charlie Strauss *Los Alamos National Laboratory*

HOW TO CITE

When citing *Foundational Papers of Complexity Science* **in full**, please use the following approach:

BIBLIOGRAPHY:

Foundational Papers in Complexity Science. Edited by David C. Krakauer. 4 volumes. Santa Fe, NM: SFI Press, 2024.

LATEX:

```
@book{Krakauer_FP_2024,
       editor = {Krakauer, David C.},
       title = {Foundational Papers in Complexity Science},
       year = {2024},
       publisher = {SFI Press},
       location = {Santa Fe, NM}}
```

When citing a particular **volume** of *Foundational Papers in Complexity Science,* please use the following approach (*volume 1 serves as an example here*):

BIBLIOGRAPHY:

Foundational Papers in Complexity Science. Edited by David C. Krakauer. Volume 1. Santa Fe, NM: SFI Press, 2024.

LATEX:

```
@book{Krakauer_FP_1_2024,
       editor = {Krakauer, David C.},
       title = {Foundational Papers in Complexity Science},
       volume = {1},
       year = {2024},
       publisher = {SFI Press},
       location = {Santa Fe, NM}}
```

When citing a specific **chapter** from this project, please treat the introduction and the annotated paper as a single unit, as follows:

BIBLIOGRAPHY:

Bettencourt, Luís M.A. "Maximum Power as a Physical Principle of Evolution." In *Foundational Papers in Complexity Science*. Edited by David C. Krakauer. Volume 1. Santa Fe, NM: SFI Press, 2024, 1–15.

LATEX:
```
@incollection{Bettencourt_Lotka_2024,
    author = {Bettencourt, Luís M. A.},
    title = {Maximum Power as a Physical Principle of Evolution},
    year = {2024},
    editor = {Krakauer, David C.},
    booktitle = {Foundational Papers in Complexity Science},
    volume = {1},
    publisher = {SFI Press},
    location = {Santa Fe, NM}
    pages = {1–15}}
```

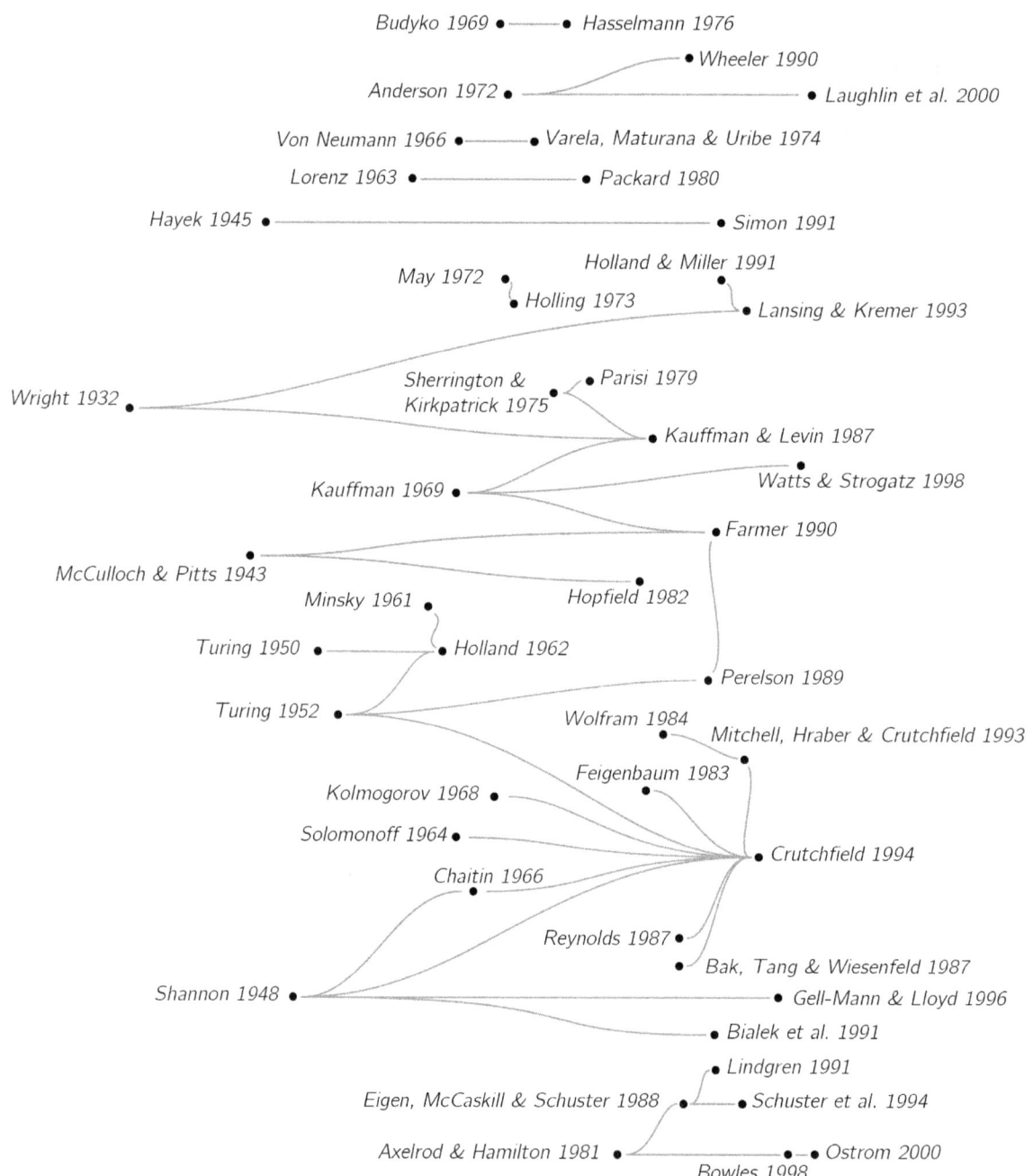

FOUNDATIONAL PAPERS: CITATION CONNECTIVITY

This visualization illustrates citation connections between foundational papers. Only papers that cite or are cited by other foundational papers are included. The papers are arranged left to right according to publication date. The gray connection lines signify that the later paper cites the earlier. *Examples*: Hasselmann 1976 cites Budyko 1969. Both Wheeler 1989 and Laughlin et al. 2000 reference Anderson 1972.

RATIONALE

1. *Foundational Papers in Complexity Science* is a project to discern the unity of an evolving inquiry.

2. After polling members of the extended network of Santa Fe Institute complexity researchers, we selected eighty-nine papers spanning just under a century that chart the formation of the field.

3. These papers—some classics, others cultish—collectively investigate the principles governing open, out-of-equilibrium systems that are self-organizing or selected, in the natural and cultural world.

4. The papers are ordered chronologically to establish patterns of influence and an emerging consensus.

5. Each paper has a unique introduction by a complexity researcher placing its ideas in historical context, and highlighting its perdurable contributions, and the new ideas that it has spawned.

6. Each paper is annotated by a researcher to underscore points where critical insights are made.

7. The year 2000 was established as the cut-off year for the *Foundational Papers*, recognizing that not enough time has elapsed to label subsequent work foundational.

8. These papers are being made available as a print-only four-volume set. The decision in favor of print is based on the prohibitive cost of licensing these papers as online materials.

9. A great deal of effort has gone into making these four volumes as beautiful and practical as possible in order to engage the senses and the mind.

10. *Searching for Order in the Complexity of Evolving Worlds*

—DAVID C. KRAKAUER AND THE SFI PRESS TEAM

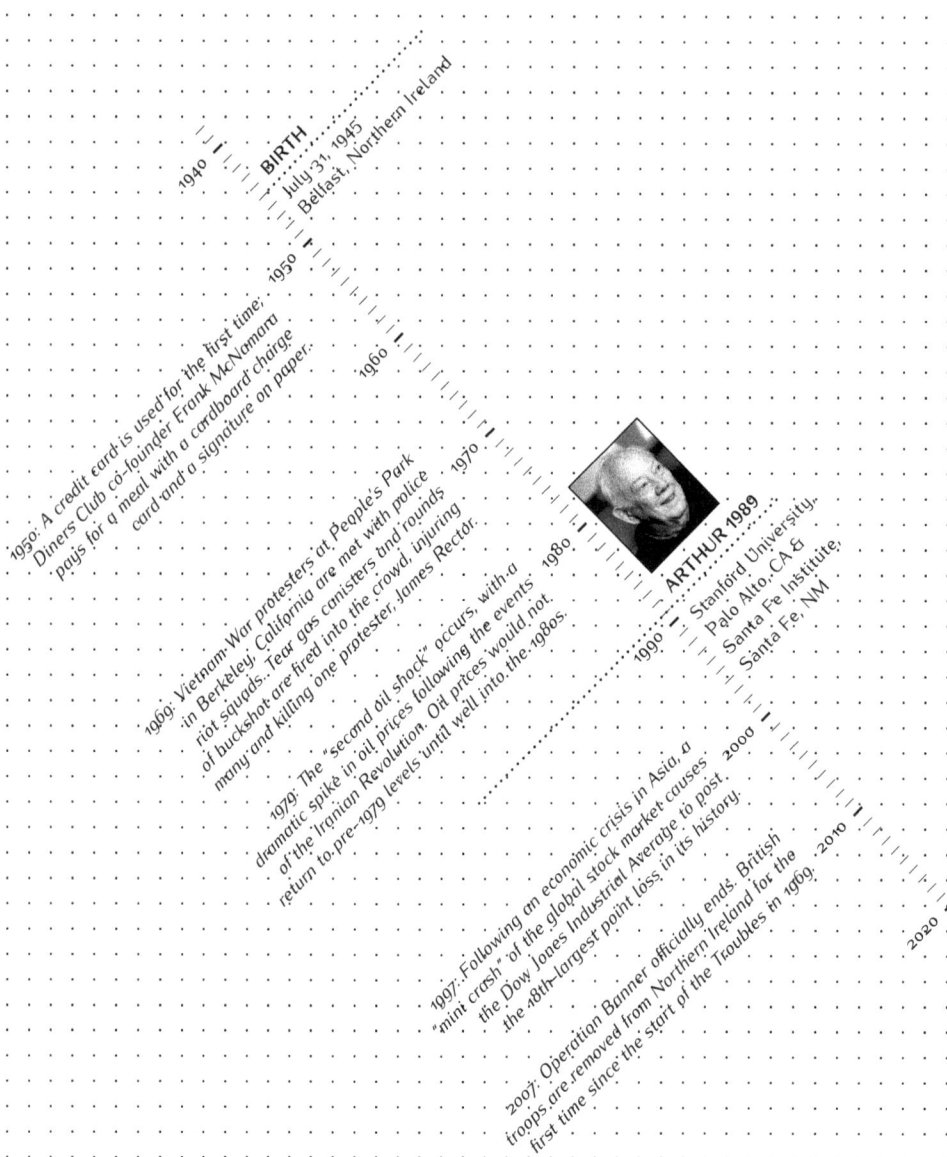

BIRTH
1945
July 31, 1945
Belfast, Northern Ireland

1950: A credit card is used for the first time; Diners Club co-founder Frank McNamara pays for a meal with a cardboard charge card and a signature on paper.

1969: Vietnam-War protesters at People's Park in Berkeley, California are met with police riot squads. Tear gas canisters and rounds of buckshot are fired into the crowd, injuring many and killing one protester, James Rector.

1979: The "second oil shock" occurs, with a dramatic spike in oil prices following the events of the Iranian Revolution. Oil prices would not return to pre-1979 levels until well into the 1980s.

ARTHUR 1989
Stanford University, Palo Alto, CA &
Santa Fe Institute, Santa Fe, NM

1997: Following an economic crisis in Asia, a "mini crash" of the global stock market causes the Dow Jones Industrial Average to post the 18th-largest point loss in its history.

2007: Operation Banner officially ends. British troops are removed from Northern Ireland for the first time since the start of the Troubles in 1969.

WILLIAM BRIAN ARTHUR

[67]

COMPETING TECHNOLOGIES, INCREASING RETURNS, AND LOCK-IN

Scott E. Page, University of Michigan, Ann Arbor
and Santa Fe Institute

As measured by academic criteria, Brian Arthur's pathbreaking and agenda-defining paper "Competing Technologies, Increasing Returns, and Lock-In by Historical Events" lies at the far end of the long tail. It has garnered over 13,000 citations, can be found on any number of syllabi in economics and business, receives deep engagement from leading scholars, and, most impressive, has spawned scores of extensions and elaborations.

It also brought attention to the prevalence of positive feedbacks in the economy. They arise in decision-making, investment strategies, network dynamics, and institutional designs. Systems with path dependence can have multiple attractors with different levels of efficiency, path dependence, which differs from sensitivity to initial conditions and therefore unpredictability. All are properties associated with the science of complexity.

In the paper, Arthur unpacks and communicates the then-unfamiliar logic of increasing returns in a model of technological choice. In doing so, he made conditional the prevailing economic logic that markets chose the best technology. He showed instead that history *does* matter, and how some paths may lock in to inferior technologies.

At the moment of publication, while both increasing returns and lock-in were well-known concepts of growing importance, path dependence, which in the title Arthur wisely refers to as "historical events" to reduce jargon, was mere background noise in the world of ideas. Since the publication of Arthur's paper and work by Paul David (1985, 1986), path dependence has been on an exorable rise. More relevant, path dependence has become a core concept in disciplines beyond economics, including but not limited to political science, sociology, history, biology, and ecology.

W. B. Arthur, "Competing Technologies, Increasing Returns, and Lock-In by Historical Events," *The Economic Journal* 99 (394), 116–131 (1989).

Reprinted by permission of Oxford University Press.

While Arthur notes that path dependence means that we may not be able to predict market shares, other scholars quickly realized that the same can be said about species distributions, features of species, institutional choices, legal rulings, ideologies, and so on. His ideas traveled effortlessly across disciplinary boundaries.

The academic influence of this paper rests in no small part on the brilliantly conceived model at its core. I like to imagine Arthur having a flash of insight that economic contexts with increasing returns were fundamentally different along the lines described: contingent, inefficient, unpredictable—that is, *complex*.

He might have constructed an abstract framework and derived necessary conditions for inefficient lock-in and path dependence to occur. Arthur, in fact, does this in the penultimate section of the paper, and while that powerful mathematical result makes a larger scientific contribution than any single model, the abstract theorem does not reveal the interplay between increasing returns and path-dependence results, nor does it show how path dependent nor how inefficient choices might be. Unlike a simple model, a general theorem does not invite extensions and elaborations. Most important, the general theorem is not user-friendly; it does not likely yield straightforward strategic insights. To borrow an idea, the deep, general model does not produce increasing returns.

To produce a pivot in economic thinking, Arthur needed a clean, intuitive model. If, as economic theorist Ariel Rubenstein proposes, we should see models as works of art, then this is a cubist Picasso: spare, geometric, and allowing for multiple views of a single imagine. Arthur's model achieves both *simplicity*—two technologies and two types of consumers designed so that each type experiences different direct benefits along with returns to scale that can be decreasing, constant, or increasing— and *realism*—people and societies choose between technologies, and the payoffs we receive depend on how many others have made the same choice.

From such few moving parts, he demonstrates the key differences between decreasing, constant, and increasing returns to scale and how historical contingencies matter for last case. And do they matter! He shows not just that market choices may miss efficiency's bull's-eye in marginal cases—that we might be typing from a slightly inefficient keyboard configuration. In fact, we might choose a wildly inefficient technology—

the wrong design of nuclear reactors, the wrong propulsion technology for cars (gas rather than steam or electric), Microsoft Word rather than, well, just about anything.

Given the paper's immediate and resounding academic echo, it is all the more amazing that Arthur's paper has had a greater effect on society writ large. It did so because the paper redefined strategic thinking, particularly in the tech sector. Arthur's model pointed to an ensemble of strategies that would become defining features of Silicon Valley.

Arthur showed that entrepreneurs should temper a singular focus on building the best mouse trap. Instead, innovators must anticipate a market opportunity and jump in; however, as the model makes clear, the first entrant need not win. Successful firms have to understand and leverage the causes of increasing return consciously and constantly—outcomes are path dependent, not five-year-plan dependent. Firms must monitor and manage that path: learn, improve, repeat. Relatedly, firms should build webs: webs of users that produce positive feedbacks and webs of supporting products and services that amplify increasing returns and lower the lock-in threshold.

Not lost to entrepreneurs was the payoff in Arthur's model. Winning in a market with increasing returns produces a pot of gold; a market capitalization followed by ten, eleven, or even twelve zeros.

Look about you. The economy is dominated by technomonopolists in markets with increasing returns: Apple, Microsoft, Google, Facebook, Amazon, Netflix, Uber, Airbnb. "Siri, what video should I watch?" "Alexa, do I have to buy this from Amazon?"

You are living in the world Arthur anticipated.

And yet, not so fast. Google supplanted Yahoo!, and Lyft entered after Uber had a solid foothold. Arthur would surely add, with a twinkle in his eye, that Google was a far sight better than Yahoo!, the car replaced the horse, and Lyft did have to burn over a billion in marketing to become competitive.

Arthur's paper focuses primarily on the economics of increasing returns, leaving others to contemplate the negative consequences implied by his analysis, among them potentially massive wealth inequality and the accumulation of information by a few corporations.

As important as the moment-in-time impacts of this paper—Silicon Valley and all that—those may pale in the long run to the formal demonstration that the economic is complex and contingent. Most important, the accumulation of small events need not average away, or, as put in Arthur's more technical terms, *dynamics have a memory*. Small efforts and tiny events contribute to the ultimate outcomes, which may be for better or worse.

We must, therefore, keep front of mind that ideas, like technologies, exhibit increasing returns and that lives spent producing those ideas matter. They can shape history.

REFERENCES

David, P. A. 1985. "Clio and the Economics of QWERTY." *American Economic Review* 75 (2): 332–337. https://www.jstor.org/stable/1805621.

———. 1986. "Understanding the Economics of QWERTY: The Necessity of History." In *Economic History and the Modern Economist,* edited by W. N. Parker, 30–49. Oxford, UK: Basil Blackwell.

COMPETING TECHNOLOGIES, INCREASING RETURNS, AND LOCK-IN BY HISTORICAL EVENTS

W. Brian Arthur, Stanford University

This paper explores the dynamics of allocation under increasing returns in a context where increasing returns arise naturally: agents choosing between technologies competing for adoption.[*]

Modern, complex technologies often display increasing returns to adoption in that the more they are adopted, the more experience is gained with them, and the more they are improved.[1] When two or more increasing-return technologies 'compete' then, for a 'market' of potential adopters, insignificant events may by chance give one of them an initial advantage in adoptions. This technology may then improve more than the others, so it may appeal to a wider proportion of potential adopters. It may therefore become further adopted and further improved. Thus a technology that by chance gains an early lead in adoption may eventually 'corner the market' of potential adopters, with the other technologies becoming locked out. Of course, under different 'insignificant events'—unexpected successes in the performance of prototypes, whims of early developers, political circumstances—a different technology might achieve sufficient adoption and improvement to come to dominate. Competitions between technologies may have multiple potential outcomes.

Arthur describes how learning by doing, a well-established phenomenon in economics, produces increasing returns to scale, putting foundations on his core assumption.

---

[*] I thank Robin Cowan, Paul David, Joseph Farrell, Ward Hanson, Charles Kindleberger, Richard Nelson, Nathan Rosenberg, Paul Samuelson, Martin Shubik, and Gavin Wright for useful suggestions and criticisms. An earlier version of part of this paper appeared in 1983 as Working Paper 83-90 at the International Institute for Applied Systems Analysis, Laxenburg, Austria. Support from the Centre for Economic Policy Research, Stanford, and from the Guggenheim Foundation is acknowledged.

[1] Rosenberg (1982) calls this 'Learning by Using' (see also Atkinson and Stiglitz 1969). Jet aircraft designs like the Boeing 727, for example, undergo constant modification and they improve significantly in structural soundness, wing design, payload capacity and engine efficiency as they accumulate actual airline adoption and use.

> The idea of increasing returns and its connection to multiple equilibria was *not* a new one. The novelty of the paper lies in introducing dynamics and showing that historical contingencies contribute to the equilibrium outcome.

It is well known that allocation problems with increasing returns tend to exhibit multiple equilibria, and so it is not surprising that multiple outcomes should appear here. Static analysis can typically locate these multiple equilibria, but usually it cannot tell us *which* one will be 'selected'. A dynamic approach might be able to say more. By allowing the possibility of 'random events' occurring during adoption, it might examine how these influence 'selection' of the outcome—how some sets of random 'historical events' might cumulate to drive the process towards one market-share outcome, others to drive it towards another. It might also reveal how the two familiar increasing-returns properties of *non-predictability* and *potential inefficiency* come about: how increasing returns act to magnify chance events as adoptions take place, so that *ex-ante* knowledge of adopters' preferences and the technologies' possibilities may not suffice to predict the 'market outcome'; and how increasing returns might drive the adoption process into developing a technology that has inferior long-run potential. A dynamic approach might also point up two new properties: *inflexibility* in that once an outcome (a dominant technology) begins to emerge it becomes progressively more 'locked in'; and *non-ergodicity* in that historical 'small events' are not averaged away and 'forgotten' by the dynamics—they may decide the outcome.

> A strength of the model is that it can accommodate decreasing, constant, and increasing returns by changing a parameter The phenomena the model produces change discontinuously as the parameter changes. We might think of those changes as phase transitions.

This paper contrasts the dynamics of technologies' 'market shares' under conditions of increasing, diminishing and constant returns. It pays special attention to how returns affect predictability, efficiency, flexibility, and ergodicity; and to the circumstances under which the economy might become locked-in by 'historical events' to the monopoly of an inferior technology.

I. A Simple Model

Nuclear power can be generated by light-water, or gas-cooled, or heavy-water, or sodium-cooled reactors. Solar energy can be generated by crystalline-silicon or amorphous-silicon technologies. I abstract from cases like this and assume in an initial, simple model that two new technologies, A and B, 'compete' for adoption by a large number of economic agents. The technologies are not sponsored[2] or strategically manipulated by any firm; they are open to all. Agents are simple consumers of the technologies who act directly or indirectly as developers of them.

Agent i comes into the market at time t_i; at this time he chooses the latest version of either technology A or technology B; and he uses this version thereafter.[3] Agents are of two types, R and S, with equal numbers in each, the two types independent of the times of choice but differing in their preferences, perhaps because of the use to which they will put their choice. The version of A or B each agent chooses is fixed or frozen in design at his time of choice, so that his payoff is affected only by past adoptions of his chosen technology. (Later I examine the expectations case where payoffs are also affected by future adoptions.)

Not all technologies enjoy increasing returns with adoption. Sometimes factor inputs are bid upward in price so that diminishing returns accompany adoption. Hydro-electric power, for example, becomes more costly as dam sites become scarcer and less suitable. And some technologies are unaffected by adoption—their returns are constant. I include these cases by assuming that the returns to choosing A or B realised by any agent (the net present value of the version of the technology available to him) depend upon the number of previous adopters, n_A and n_B, at the time of his choice (as in Table 1[4]) with increasing, diminishing, or constant returns to adoption given by r and s simultaneously positive, negative, or zero. I also assume $a_R > b_R$ and $a_S < b_S$ so that R-agents have a natural preference for A, and S-agents have a natural preference for B.

To complete this model, I want to define carefully what I mean by 'chance' or 'historical events'. Were we to have infinitely

That his examples involve technological choices on billion-dollar investments rather than the QWERTY typewriter adds gravitas to the subsequent analysis.

This comment appeases economists who assume agents are forward-looking. This creates some difficult for Arthur in that no one can rationally anticipate an uncertain outcome.

---

[2] Following terminology introduced in Arthur (1983), *sponsored* technologies are proprietary and capable of being priced and strategically manipulated; *unsponsored* technologies are generic and not open to manipulation or pricing.

[3] Where technologies are improving, it may pay adopters under certain conditions to wait; so that no adoptions take place (Balcer and Lippman 1984; Mamer and McCardle 1987). We can avoid this problem by assuming adopters need to replace an obsolete technology that breaks down at times $\{t_i\}$.

[4] More realistically, where the technologies have uncertain monetary returns we can assume von Neumann-Morgenstern agents, with Table 1 interpreted as the resulting determinate expected-utility payoffs.

	TECHNOLOGY A	TECHNOLOGY B
R-agent	$a_R + rn_A$	$b_R + rn_B$
S-agent	$a_S + sn_A$	$b_s + sn_B$

Table 1. Returns to Choosing A or B given Previous Adoptions

detailed prior knowledge of events and circumstances that might affect technology choices—political interests, the prior experience of developers, timing of contracts, decisions at key meetings—the outcome or adoption market-share gained by each technology would presumably be determinable in advance. We can conclude that our limited discerning power, or more precisely the limited discerning power of an implicit *observer*, may cause indeterminacy of outcome. I therefore define 'historical small events' to be those events or conditions that are outside the *ex-ante* knowledge of the observer—beyond the resolving power of his 'model' or abstraction of the situation.

Building on my previous comment, here Arthur explicitly details why historical events may be unpredictable. He could have chosen a single reason but makes the wise choice to include many.

To return to *our* model, let us assume an observer who has full knowledge of all the conditions and returns functions, except the set of events that determines the times of entry and choice $\{t_i\}$ of the agents. The observer thus 'sees' the choice order as a binary sequence of R and S types with the property that an R or an S comes nth in the adoption line with equal likelihood, that is, with probability one half.

We now have a simple neoclassical allocation model where two types of agents choose between A and B, each agent choosing his preferred alternative when his time comes. The supply (or returns) functions are known, as is the demand (each agent demands one unit inelastically). Only one small element is left open, and that is the set of historical events that determine the sequence in which the agents make their choice. Of interest is the adoption-share outcome in the different cases of constant, diminishing, and increasing returns, and whether the fluctuations in the order of choices these small events introduce make a difference to adoption shares.

We will need some properties. I will say that the process is: *predictable* if the small degree of uncertainty built in 'averages away' so that the observer has enough information to pre-determine market shares accurately in the long-run; *flexible* if a subsidy or tax adjustment to one of the technologies' returns can always influence future market choices; *ergodic* (not path-dependent) if different sequences of historical events lead to the same market outcome with probability one. In this allocation problem choices define a 'path' or sequence of *A*- and *B*-technology versions that become adopted or 'developed', with early adopters possibly steering the process onto a development path that is right for them, but one that may be regretted by later adopters. Accordingly, and in line with other sequential-choice problems, I will adopt a 'no-regret' criterion and say that the process is *path-efficient* if at all times equal development (equal adoption) of the technology that is behind in adoption would not have paid off better.[5] (These informal definitions are made precise in the Appendix.)

Arthur's decision to place the precise definitions in the appendix made the paper more readable and surely increased its impact.

ALLOCATION IN THE THREE REGIMES

Before examining the outcome of choices in our R and S agent model, it is instructive to look at how the dynamics would run in a trivial example with increasing-returns where agents are of one type only (Table 2). Here choice order does not matter; agents are all the same; and unknown events can make no difference so that ergodicity is not an issue. The first agent chooses the more favourable technology, A say. This enhances the returns to adopting A. The next agent *a-fortiori* chooses A too. This continues, with A chosen each time, and B incapable of 'getting started'. The end result is that A 'corners the market' and B is excluded. This outcome is trivially predictable, and path-efficient if returns rise at the same rate. Notice though that if returns increase at different rates, the adoption process may easily become path-inefficient, as Table 2 shows. In this case after thirty choices in the adoption process, all of which are A, equivalent adoption of B would have delivered higher returns. But if the process has gone far enough, a given subsidy-adjustment g to B can no longer close the gap between the returns to A and the returns to B at the starting point. Flexibility is not present here; the market becomes increasingly 'locked-in' to an inferior choice.

---

[5] An alternative efficiency criterion might be total or aggregate payoff (after n choices). But in this problem we have two agent types with different preferences operating under the 'greedy algorithm' of each agent taking the best choice at hand for himself; it is easy to show that under any returns regime maximisation of total payoffs is never guaranteed.

NUMBER OF PREVIOUS ADOPTIONS	0	10	20	30	40	50	60	70	80	90	100
TECHNOLOGY A	10	11	12	13	14	15	16	17	18	19	20
TECHNOLOGY B	4	7	10	13	16	19	22	25	28	31	34

Table 2. An Example: Adoption Payoffs for Homogeneous Agents

The example clearly makes his point—the process could be stacked against the better technology—but a graph would have been better than the table.

Now let us return to the case of interest, where the unknown choice-sequence of two types of agents allows us to include some notion of historical 'small events'. Begin with the constant-returns case, and let $n_A(n)$ and $n_B(n)$ be the number of choices of A and B respectively, when n choices in total have been made. We can describe the process by x_n, the market share of A at stage n, when n choices in total have been made. We will write the difference in adoption, $n_A(n) - n_B(n)$ as d_n. The market share of A is then expressible as

$$x_n = 0 \cdot 5 + d_n/2n. \tag{1}$$

Note that through the variables d_n and n—the difference and total—we can fully describe the dynamics of adoption of A versus B. In this constant-returns situation R-agents always choose A and S-agents always choose B, regardless of the number of adopters of either technology. Thus the way in which adoption of A and B cumulates is determined simply by the sequence in which R- and S-agents 'line up' to make their choice, $n_A(n)$ increasing by one unit if the next agent in line is an R, with $n_B(n)$ increasing by one unit if the next agent in line is an S, and with the difference in adoption, d_n, moving upward by one unit or downward one unit accordingly. To our observer, the choice-order is random, with agent types equally likely. Hence to him, the state d_n appears to perform a simple coin-toss gambler's random walk with each 'move' having equal probability $0 \cdot 5$.

In the increasing-returns case, these simple dynamics are modified. New R-agents, who have a natural preference for A, will switch allegiance if by chance adoption pushes B far enough *ahead* of A in numbers and in payoff. That is, new R-agents will 'switch' if

$$d_n = n_A(n) - n_B(n) < \Delta_R = \frac{(b_R - a_R)}{r}. \tag{2}$$

Arthur (1989)

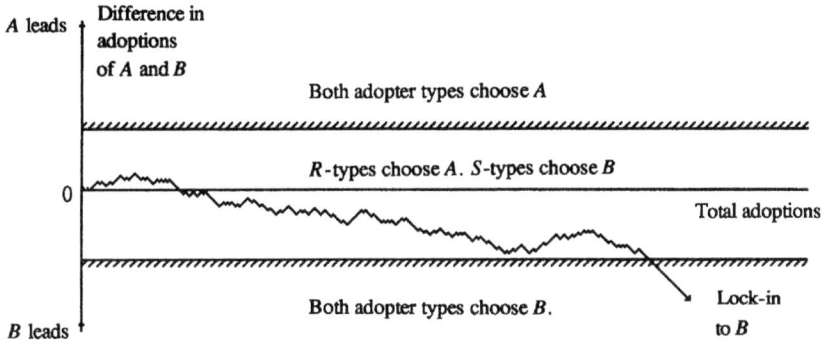

 This graph is spectacular. It shows the random walk, the lock-in thresholds, and the behavior that distinguishes the two regimes.

Figure 1. Increasing returns adoption: a random walk with absorbing barriers

Similarly new S-agents will switch preference to A if numbers adopting A become sufficiently ahead of the numbers adopting B, that is, if

$$d_n = n_A(n) - n_B(n) > \Delta_S = \frac{(b_S - a_S)}{s}. \quad (3)$$

Regions of choice now appear in the d_n, n plane (see Fig. 1), with boundaries between them given by (2) and (3). Once one of the outer regions is entered, both agent types choose the same technology, with the result that this technology further increases its lead. Thus in the d_n, n plane (2) and (3) describe barriers that 'absorb' the process. Once either is reached by random movement of d_n, the process ceases to involve both technologies—it is 'locked-in' to one technology only. Under increasing returns then, the adoption process becomes a random walk with absorbing barriers. I leave it to the reader to show that the allocation process with diminishing returns appears to our observer as a random walk with *reflecting* barriers given by expressions similar to (2) and (3).

Several of Arthur's papers include tables with columns labeled "equilibrium economics" and "complexity economics." Here, he has rows. The top two are old-school thinking. The bottom (which I would have put in bold) captures the new complexity economics.

	PREDICTABLE	FLEXIBLE	ERGODIC	NECESSARILY PATH-EFFICIENT
CONSTANT RETURNS	Yes	No	Yes	Yes
DIMINISHING RETURNS	Yes	Yes	Yes	Yes
INCREASING RETURNS	No	No	No	No

Table 3. Properties of the Three Regimes

PROPERTIES OF THE THREE REGIMES

We can now use the elementary theory of random walks to derive the properties of this choice process under the different linear returns regimes. For convenient reference the results are summarised in Table 3. To prove these properties, we need first to examine long-term adoption shares. Under constant returns, the market is shared. In this case the random walk ranges free, but we know from random walk theory that the standard deviation of d_n increases with $\sqrt{n}$. It follows that the $d_n/2n$ term in equation (1) disappears and that x_n tends to $0 \cdot 5$ (with probability one), so that the market is split 50-50. In the diminishing returns case, again the adoption market is shared. The difference-in-adoption, d_n, is trapped between finite constants; hence $d_n/2n$ tends to zero as n goes to infinity, and x_n must approach $0 \cdot 5$. (Here the 50–50 market split results from the returns falling at the same rate.) In the increasing-returns-absorbing-barrier case, by contrast, the adoption share of A *must* eventually become zero or one. This is because in an absorbing random walk d_n eventually crosses a barrier with probability one. Therefore the two technologies cannot coexist indefinitely: one *must* exclude the other.

Nice—the power of a simple declarative sentence. Cormac McCarthy would be proud!

Predictability is therefore guaranteed where the returns are constant, or diminishing: in both cases a forecast that the market will settle to 50-50 will be correct, with probability one. In the increasing returns case, however, for accuracy the observer must predict A's eventual share either as 0 or 100%. But either choice will be wrong with probability one-half. Predictability is lost. Notice though that the observer *can* predict that one technology will take the market; theoretically he can also predict that it will be A with probability $s\left(a_R - b_R\right) / \left[s\left(a_R - b_R\right) + r\left(b_s - a_S\right)\right]$; but he cannot predict the actual market-share outcome with any accuracy—in spite of his knowledge of supply and demand conditions.

Flexibility in the constant-returns case is at best partial. Policy adjustments to the returns can affect choices at all times, but only if they are large enough to bridge the gap in preferences between technologies. In the two other regimes adjustments correspond to a shift of one or both of the barriers. In the diminishing-returns case, an adjustment g can always affect future choices (in absolute numbers, if not in market shares), because reflecting barriers continue to influence the process (with probability one) at times in the future. Therefore diminishing returns are flexible. Under increasing returns however, once the process is absorbed into A and B, the subsidy or tax adjustment necessary to shift the barriers enough to influence choices (a precise index of the degree to which the system is 'locked-in') increases without bound. Flexibility does not hold.

Ergodicity can be shown easily in the constant and diminishing returns cases. With constant returns only extraordinary line-ups (for example, twice as many R-agents as S-agents appearing indefinitely) with associated probability zero can cause deviation from fifty–fifty. With diminishing returns, any sequence of historical events—any line-up of the agents—must still cause the process to remain between the reflecting barriers and drive the market to fifty–fifty. Both cases forget their small-event history. In the increasing returns case the situation is quite different. Some proportion of agent sequences causes the market outcome to 'tip' towards A, the remaining proportion causes it to 'tip' towards B. (Extraordinary line-ups—say S followed by R followed by S followed by R and so on indefinitely—that could cause market sharing, have probability or measure zero.) Thus, the small events that determine $\{t_i\}$ *decide* the path of market shares; the process is non-ergodic or path-dependent—it is determined by its small-event history.

Policymakers like to believe that they can intervene through taxes and subsidies. What he is calling "flexibility" means policy efficacy.

Path-efficiency is easy to prove in the constant- and diminishing-returns cases. Under constant-returns, previous adoptions do not affect pay-off. Each agent-type chooses its preferred technology and there is no gain foregone by the failure of the lagging technology to receive further development (further adoption). Under diminishing returns, if an agent chooses the technology that is ahead, he must prefer it to the available version of the lagging one. But further adoption of the lagging technology by definition lowers its payoff. Therefore there is no possibility of choices leading the adoption process down an inferior development path. Under increasing returns, by contrast, development of an inferior option can result. Suppose the market locks in to technology A. R-agents do not lose; but S-agents would each gain $(b_S - a_S)$ if their favoured technology B had been equally developed and available for choice. There is regret, at least for one agent type.

Inefficiency can be exacerbated if the technologies improve at different rates. An early run of agent-types who prefer an initially attractive but slow-to-improve technology can lock the market in to this inferior option; equal development of the excluded technology in the long run would pay off better to both types.

EXTENSIONS, AND THE RATIONAL EXPECTATIONS CASE

It is not difficult to extend this basic model in various directions. The same qualitative results hold for M technologies in competition, and for agent types in unequal proportions (here the random walk 'drifts'). And if the technologies arrive in the market at different times, once again the dynamics go through as before, with the process now starting with initial n_A or n_B not at zero. Thus in practice an early-start technology may already be locked in, so that a new potentially-superior arrival cannot gain a footing.

Where agent numbers are finite, and not expanding indefinitely, absorption or reflection and the properties that depend on them still assert themselves providing agent numbers are large relative to the numerical width of the gap between switching barriers.

For technologies *sponsored* by firms, would the possibility of strategic action alter the outcomes just described? A complete answer is not yet known. Hanson (1985) shows in a model based on the one above that again market exclusion goes through: firms engage in penetration pricing, taking losses early on in exchange for potential monopoly profits later, and all but one firm exit with probability one. Under strong discounting, however, firms may be more interested in immediate sales than in shutting rivals out, and market sharing can reappear.[6]

Perhaps the most interesting extension is the expectations case where agents' returns are affected by the choices of future agents. This happens for example with *standards*, where it matters greatly whether later users fall in with one's own choice. Katz and Shapiro (1985, 1986) have shown, in a two-period case with strategic interaction, that agents' expectations about these future choices act to destabilise the market. We can extend their findings to our stochastic-dynamic model.

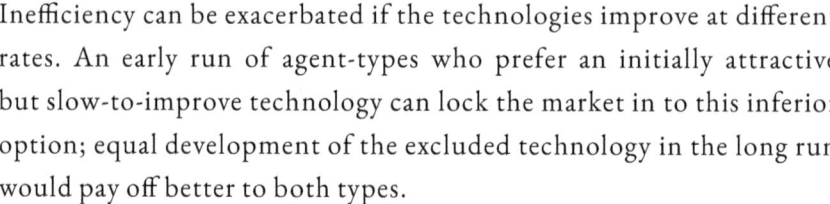

As Shakespeare wrote, "There is a tide in the affairs of men." S-agents are stuck in the shallows and miseries of technology A because of technological lock-in caused by historical contingencies.

We still feel the implications of this today as giant firms dominate the information and technology landscapes. Arthur may have missed an opportunity here to point out the negative consequence of monopolization.

---

[6]For similar findings see the literature on the dynamics of commodity competition under increasing returns (e.g. Spence 1981; Fudenberg and Tirole 1983).

Assume agents form expectations in the shape of beliefs about the type of stochastic process they find themselves in. When the *actual* stochastic process that results from these beliefs is identical with the *believed* stochastic process, we have a rational-expectations fulfilled-equilibrium process. In the Appendix, I show that under increasing returns, rational expectations also yield an absorbing random walk, but one where expectations of lock-in hasten lock-in, narrowing the absorption barriers and worsening the fundamental market instability.

II. A General Framework

It would be useful to have an analytical framework that could accommodate sequential-choice problems with more general assumptions and returns mechanisms than the basic model above. In particular it would be useful to know under what circumstances a competing-technologies adoption market must end up dominated by a single technology.

In designing a general framework it seems important to preserve two properties: *(i)* That choices between alternative technologies may be affected by the numbers of each adopted at the time of choice; *(ii)* That small events 'outside the model' may influence adoptions, so that randomness must be allowed for. Thus adoption market shares may determine not the next technology chosen directly but rather the *probability* of each technology's being chosen.

Consider then a dynamical system where one of K technologies is adopted each time an adoption choice is made, with probabilities $p_1(\mathbf{x}), p_2(\mathbf{x}), \ldots, p_K(\mathbf{x})$, respectively. This vector of probabilities $\mathbf{p}$ is a function of the vector $\mathbf{x}$, the adoption-shares of technologies 1 to K, out of the total number n of adoptions so far. The initial vector of proportions is given as $\mathbf{x}_0$. I will call $\mathbf{p}(\mathbf{x})$ the *adoption function*.

We may now ask what happens to the long run proportions or adoption shares in such a dynamical system. Consider the two different adoption functions in Fig. 2, where $K = 2$. Now, where the probability of adoption of A is higher than its market share, in the adoption process A tends to increase in proportion; and where it is lower, A tends to decrease. If the proportions or adoption-shares settle down as total adoptions increase, we would conjecture that they settle down at a fixed point of the adoption function.

Wonderful, clear explanation of how the insights produced by his simple models can be embedded within a general framework.

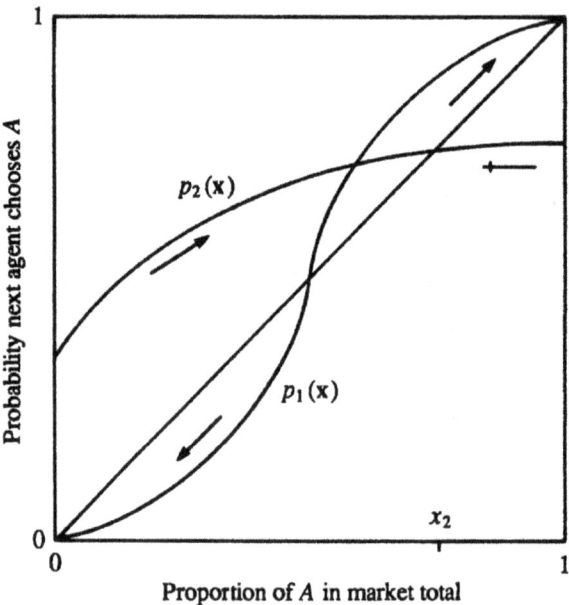

Figure 2. Two illustrative adoption functions.

In 1983 Arthur, Ermoliev, and Kaniovski proved that under certain technical conditions (see the Appendix) this conjecture is true. A stochastic process of this type converges with probability one to one of the fixed points of the mapping from proportions (adoption shares) to the probability of adoption. Not all fixed points are eligible. Only 'attracting' or stable fixed points (ones that expected motions of the process lead towards) can emerge as the long run outcomes. And where the adoption function varies with time n, but tends to a limiting function **p**, the process converges to an attracting fixed point of **p**.

Thus in Fig. 2 the possible long-run shares are 0 and 1 for the function p_1 and x_2 for the function p_2. Of course, where there are multiple fixed points, *which* one is chosen depends on the path taken by the process: it depends on the cumulation of random events that occur as the process unfolds.

We now have a general framework that immediately yields two useful theorems on path-dependence and single-technology dominance.

Theorem I. *An adoption process is non-ergodic and non-predictable if and only if its adoption function* **p** *possesses multiple stable fixed points.*

Theorem II. *An adoption process converges with probability one to the dominance of a single technology if and only if its adoption function* **p** *possesses stable fixed points only where* **x** *is a unit vector.*

Both theorems have straightforward proofs. Neither has much wow or surprise factor: Once we know the assumptions, the conclusion follows. This does not make them less economically relevant, but it does mean that there was no excitement about the novel method of proof. That is not true of Arthur's work on the Pólya process, which garnered a lot of attention.

These theorems follow as simple corollaries of the basic theorem above. Thus where two technologies compete, the adoption process will be path-dependent (multiple fixed points must exist) as long as there exists at least one unstable 'watershed' point in adoption shares, above which adoption of the technology with this share becomes self-reinforcing in that it tends to increase its share, below which it is self-negating in that it tends to lose its share. It is therefore not sufficient that a technology gain advantage with adoption; the advantage must (at some market share) be self-reinforcing (see Arthur, Ermoliev, and Kaniovski 1988).

NON-LINEAR INCREASING RETURNS WITH A CONTINUUM OF ADOPTER TYPES

Consider, as an example, a more general version of the basic model above, with a continuum of adopter types rather than just two, choosing between K technologies, with possibly non-linear improvements in payoffs. Assume that if n_j previous adopters have chosen technology j previously, the next agent's payoff to adopting j is $\Pi_j(n_j) = a_j + r(n_j)$ where a_j represents the agent's 'natural preference' for technology j and the monotonically increasing function r represents the technological improvement that comes with previous adoptions. Each adopter has a vector of natural preferences $\mathbf{a} = (a_1, a_2, \ldots, a_K)$ for the K alternatives, and we can think of the continuum of agents as a distribution of points **a** (with bounded support) on the positive orthant. We assume an adopter is drawn at random from this probability distribution each time a choice occurs. Dominance of a single technology j corresponds to positive probability of the distribution of payoffs Π being driven by adoptions to a point where Π_j exceeds Π_i for all $i \neq j$.

The Arthur–Ermoliev–Kaniovski theorem above allows us to derive:

Theorem III. *If the improvement function r increases at least at rate ϵ as n_j increases, the adoption process converges to the dominance of a single technology, with probability one.*

Theorem III is the most subtle: You only need a little increasing returns to go a long way. This result was foreshadowed by his simple model in which any increasing returns produce path dependence.

Proof. In this case, the adoption function varies with total adoptions n. (We do not need to derive it explicitly however.) It is not difficult to establish that as n becomes large: (i) At any point in the neighbourhood of any unit vector of adoption shares, unbounded increasing returns cause the corresponding technology to dominate *all* choices; therefore the unit-vector shares are stable fixed points. (ii) The equal-share point is also a fixed point, but unstable. (iii) No other point is a fixed point. Therefore, by the general theorem, since the limiting adoption function has stable fixed points only at unit vectors the process converges to one of these with probability one. Long-run dominance by a single technology is assured. ∎

Dominance by a single technology is no longer inevitable, however, if the improvement function r is bounded, as when learning effects become exhausted. This is because certain sequences of adopter types could bid improvements for two or more technologies upward more or less in concert. These technologies could then reach the upper bound of r together, so that none of these would dominate and the market would remain shared from then on. Under other adopter sequences, by contrast, one of the technologies may reach the upper bound sufficiently fast to shut the others out. Thus, in the bounded case, some event histories dynamically lead to a shared market; other event histories lead to dominance. Increasing returns, if they are bounded, are in general *not* sufficient to guarantee eventual monopoly by a single technology.

This conditionality of his result, like the possibility of prior information about values undermining the revenue equivalence theorem of Myerson, should have been written as a theorem given its eventual significance.

III. Remarks

(1) To what degree might the actual economy be locked-in to inferior technology paths? As yet we do not know. Certainly it is easy to find cases where an early-established technology becomes dominant, so that later, superior alternatives cannot gain a footing.[7] Two important studies of historical events leading to lock-ins have now been carried out: on the QWERTY typewriter keyboard (David 1985);

---

[7] Examples might be the narrow gauge of British railways (Kindleberger 1983); the US colour television system; the 1950s programming language FORTRAN; and of course the QWERTY keyboard (Arthur 1984; David 1985; Hartwick 1985). In these particular cases the source of increasing returns is network externalities however rather than learning effects. Breaking out of locked-in technological standards has been investigated by Farrell and Saloner (1985, 1986)

and on alternating current (David and Bunn 1987). (In both cases increasing returns arise mainly from coordination externalities.)

Promising empirical cases that may reflect lock-in through learning are the nuclear-reactor technology competition of the 1950s and 1960s and the US steam-versus-petrol car competition in the 1890s. The US nuclear industry is practically 100% dominated by light-water reactors. These reactors were originally adapted from a highly compact unit designed to propel the first nuclear submarine, the U.S.S. *Nautilus*, launched in 1954. A series of circumstances—among them the Navy's role in early construction contracts, political expediency, the Euratom programme, and the behaviour of key personages—acted to favour light water. Learning and construction experience gained early on appear to have locked the industry in to dominance of light water and shut other reactor types out (Bupp and Derian 1978; Cowan 1987). Yet much of the engineering literature contends that, given equal development, the gas-cooled reactor would have been superior (see Agnew 1981).

Oh, by the way, QWERTY! But why not the English System of Weights and Measures? Feet, inches, cups, pints—what a horrible lock-in, though not in England!

In the petrol-versus-steam car case, two different developer types with predilections toward steam or petrol depending on their previous mechanical experience, entered the industry at varying times and built upon the best available versions of each technology. Initially petrol was held to be the less promising option: it was explosive, noisy, hard to obtain in the right grade, and it required complicated new parts.[8] But in the United States a series of trivial circumstances (McLaughlin 1954; Arthur 1984) pushed several key developers into petrol just before the turn of the century and by 1920 had acted to shut steam out. Whether steam might have been superior given equal development is still in dispute among engineers (see Burton 1976; Strack 1970).

Perhaps we do not live in the best of all possible worlds, and we should reconsider some of our past technological choices.

(2) The argument of this paper suggests that the interpretation of economic history should be different in different returns regimes. Under constant and diminishing returns, the evolution

---

[8] Amusingly, Fletcher ([1904]1973) writes: '… unless the objectionable features of the petrol carriage can be removed, it is bound to be driven from the road by its less objectionable rival, the steam-driven vehicle of the day.'

of the market reflects only *a-priori* endowments, preferences, and transformation possibilities; small events cannot sway the outcome. But while this is comforting, it reduces history to the status of mere carrier—the deliverer of the inevitable. Under increasing returns, by contrast many outcomes are possible. Insignificant circumstances become magnified by positive feedbacks to 'tip' the system into the actual outcome 'selected'. The small events of history become important.[9] Where we observe the predominance of one technology or one economic outcome over its competitors we should thus be cautious of any exercise that seeks the means by which the winner's innate 'superiority' came to be translated into adoption.

(3) The usual policy of letting the superior technology reveal itself in the outcome that dominates is appropriate in the constant and diminishing-returns cases. But in the increasing returns case laissez-faire gives no guarantee that the 'superior' technology (in the long-run sense) will be the one that survives. Effective policy in the (unsponsored) increasing-returns case would be predicated on the nature of the market breakdown: in our model early adopters impose externalities on later ones by rationally choosing technologies to suit only themselves; missing is an inter-agent market to induce them to explore promising but costly infant technologies that might pay off handsomely to later adopters.[10] The standard remedy of assigning to early developers (patent) rights of compensation by later users would be effective here only to the degree that early developers can appropriate later payoffs. As an alternative, a central authority could underwrite adoption and exploration along promising but less popular technological paths. But where eventual returns to a technology are hard to ascertain—as in the U.S. Strategic Defence Initiative case for example—the

A double complexity whammy! When the technological development process is complex, the rate of increasing returns will be more uncertain (recall: economists would be prone to assume we can know future returns), so the likelihood of choosing the wrong technology increases.

---

[9]For earlier recognition of the significance of both non-convexity and path-dependence for economic history see David (1975).

[10]Competition between sponsored technologies suffers less from this missing market. Sponsoring firms can more easily appropriate later payoffs, so they have an incentive to develop initially costly, but promising technologies. And financial markets for sponsoring investors together with insurance markets for adopters who may make the 'wrong' choice, mitigate losses for the risk-averse. Of course, if a product succeeds and locks-in the market, monopoly-pricing problems may arise. For further remarks on policy see David (1987).

authority then faces a classic multi-arm bandit problem of choosing which technologies to bet on. An early run of disappointing results (low 'jackpots') from a potentially superior technology may cause it perfectly rationally to abandon this technology in favour of other possibilities. The fundamental problem of possibly locking-in a regrettable course of development remains (Cowan 1987).

IV. Conclusion

This paper has attempted to go beyond the usual static analysis of increasing-returns problems by examining the dynamical process that 'selects' an equilibrium from multiple candidates, by the interaction of economic forces and random 'historical events'. It shows how dynamically, increasing returns can cause the economy gradually to lock itself in to an outcome not necessarily superior to alternatives, not easily altered, and not entirely predictable in advance.

Welcome to the world of complexity economics!

Under increasing returns, competition between economic objects—in this case technologies—takes on an evolutionary character, with a 'founder effect' mechanism akin to that in genetics.[11] 'History' becomes important. To the degree that the technological development of the economy depends upon small events beneath the resolution of an observer's model, it may become impossible to predict market shares with any degree of certainty. This suggests that there may be theoretical limits, as well as practical ones, to the predictability of the economic future.

Always the provocateur, Arthur ends with a shot across the bow.

Appendix

A. DEFINITIONS OF THE PROPERTIES

Here I define precisely the properties used above. Denote the market share of A after n choices as x_n. The allocation process is:

(i) *predictable* if the observer can *ex-ante* construct a forecasting sequence $\{x_n^*\}$ with the property that $|x_n - x_n^*| \to 0$, with probability one, as $n \to \infty$;

---

[11]For other selection mechanisms affecting technologies see Dosi (1988), Dosi et al. (1988), and Metcalfe (1985)

(ii) *flexible* if a given marginal adjustment g to the technologies' returns can alter future choices;

(iii) *ergodic* if, given two samples from the observer's set of possible historical events, $\{t_i\}$ and $\{t'_i\}$, with corresponding time-paths $\{x_n\}$ and $\{x'_n\}$, then $|x'_n - x_n| \to 0$, with probability one, as $n \to \infty$;

(iv) *path-efficient* if, whenever an agent chooses the more-adopted technology α, versions of the lagging technology β would not have delivered more had they been developed and available for adoption. That is, path-efficiency holds if returns Π remain such that $\Pi_\alpha(m) \geqslant \mathrm{Max}_j \{\Pi_\beta(j)\}$ for $k \leqslant j \leqslant m$, where there have been m previous choices of the leading technology and k of the lagging one.

B. THE EXPECTATIONS CASE

Consider here the competing standards case where adopters are affected by *future* choices as well as past choices. Assume in our earlier model that R-agents receive additional net benefits of Π_A^R, Π_B^R, if the process locks-in to their choice, A or B respectively; similarly S-agents receive Π_A^S, Π_B^S. (Technologies improve with adoption as before.) Assume that agents know the state of the market (n_A, n_B) when choosing and that they have expectations or beliefs that adoptions follow a stochastic process Ω. They choose rationally under these expectations, so that actual adoptions follow the process $\Gamma(\Omega)$. This actual process is a *rational expectations equilibrium process* when it bears out the expected process, that is, when $\Gamma(\Omega) \equiv \Omega$.

We can distinguish two cases, corresponding to the degree of heterogeneity of preferences in the market.

Case (i). Suppose initially that $a_R - b_R > \Pi_B^R$ and $b_S - a_S > \Pi_A^S$ and that R and S-types have beliefs that the adoption process is a random walk Ω with absorption barriers at Δ'_R, Δ'_S, with associated probabilities of lock-in to A, $P(n_A, n_B)$ and lock-in to B, $1 - P(n_A, n_B)$. Under these beliefs, R-type expected payoffs for choosing A or B are, respectively:

$$a_R + rn_A + P(n_A, n_B)\Pi_A^R \tag{4}$$

$$b_R + rn_B + [1 - P(n_A, n_B)]\Pi_B^R. \tag{5}$$

S-type payoffs may be written similarly. In the actual process R-types will switch to B when n_A and n_B are such that these two expressions become equal. Both types choose B from then on. The actual probability of lock-in to A is zero here; so that if the expected process is fulfilled, P is also zero here and we have n_A and n_B such that

$$a_R + rn_A = b_R + rn_B + \Pi_B^R$$

with associated barrier given by

$$\Delta_R = n_A - n_B = -\left(a_R - b_R - \Pi_B^R\right)/r. \qquad (6)$$

Similarly S-types switch to A at boundary position given by

$$\Delta_S = n_A - n_B = \left(b_S - a_S - \Pi_A^S\right)/s. \qquad (7)$$

It is easy to confirm that beyond these barriers the actual process is indeed locked in to A or to B and that within them R-agents prefer A, and S-agents prefer B. Thus if agents believe the adoption process is a random walk with absorbing barriers Δ_R', Δ_S' given by (68) and (6), these beliefs will be fulfilled, and this random walk will be a rational expectations equilibrium.

Case (ii). Suppose now that $a_R - b_R < \Pi_B^R$ and $b_S - a_S < \Pi_A^S$. Then (4) and (5) show that switching will occur immediately if agents hold expectations that the system will definitely lock-in to A or to B. These expectations become self-fulfilling and the absorbing barriers narrow to zero. Similarly, when non-improving standards compete, so that r and s are zero, in this case again beliefs that A or B will definitely lock-in become self-fulfilling.

Taking cases (i) and (ii) together, expectations either narrow or collapse the switching boundaries. They exacerbate the fundamental market instability.

C. THE PATH-DEPENDENT STRONG-LAW THEOREM

Consider a dependent-increment stochastic process that starts with an initial vector of units $\mathbf{b}_0$, in the K categories, 1 through K. At each event-time a unit is added to one of the categories 1 through K, with probabilities $\mathbf{p} = [p_1(x), p_2(x), \ldots, p_K(x)]$, respectively. (The Borel function $\mathbf{p}$ maps the unit simplex of proportions S^K into the unit simplex of probabilities S^K.) The process is iterated to yield the vectors of proportions $\mathbf{X}_1, \mathbf{X}_2, \mathbf{X}_3, \ldots$.

Theorem. Arthur, Ermoliev, and Kaniovski (1983, 1986)

1. Suppose $p: S^K \to S^K$ is continuous, and suppose the function $p(x) - x$ possesses a Lyapunov function (that is, a positive, twice-differentiable function V with inner product $\{[p(x) - x], V_x\}$ negative). Suppose also that the set of fixed points of p, $B = \{x : p(x) = x\}$ has a finite number of connected components. Then the vector of proportions $\{X_n\}$ converges, with probability one, to a point z in the set of fixed points B, or to the border of a connected component.

2. Suppose p maps the interior of the unit simplex into itself, and that z is a stable point (as defined in the conventional way). Then the process has limit point z with positive probability.

3. Suppose z is a non-vertex unstable point of p. Then the process cannot converge to z with positive probability.

4. Suppose probabilities of addition vary with time n, and the sequence $\{p_n\}$ converges to a limiting function p faster than $1/n$ converges to zero. Then the above statements hold for the limiting function p. That is, if the above conditions are fulfilled, the process converges with probability one to one of the stable fixed points of the limiting function p.

The theorem is extended to non-continuous functions p and to non-unit and random increments in Arthur, Ermoliev and Kaniovski (1987b). For the case $K = 2$ with p stationary see the elegant analysis of Hill *et al.* (1980).

REFERENCES

Agnew, H. 1981. "Gas-Cooled Nuclear Power Reactors." *Scientific American* 244:55–63.

Arthur, W. B. 1983. "Competing Technologies, Increasing Returns, and Lock-In by Historical Events." *The Economic Journal* 99 (394): 116–131.

———. 1984. "Competing Technologies and Economic Prediction." *Options, International Institute for Applied Systems Analysis* (Laxenburg, Austria), nos. 1984/2, 10–3.

Arthur, W. B., Y. M. Ermoliev, and Y. M. Kaniovski. 1983. "On Generalized Urn Schemes of the Polya Kind." *Cybernetics* 19:61–71.

———. 1986. "Strong Laws for a Class of Path-Dependent Stochastic Processes with Applications." In *Stochastic Optimization: Proceedings of the International Conference, Kiev, 1984*, 81:14. Lecture Notes in Control and Information Sciences. Berlin, Germany: Springer.

———. 1988. "Self-Reinforcing Mechanisms in Economics." In *The Economy as an Evolving Complex System*, edited by P. Anderson, K. Arrow, and D. Pines. Reading, MA: Addison-Wesley.

———. 1987b. *Non-Linear Urn Processes: Asymptotic Behavior and Applications.* Technical report WP-87-85. Laxenburg, Austria: International Institute for Applied Systems Analysis.

———. 1987a. "Path-Dependent Processes and the Emergence of Macro-Structure." *European Journal of Operational Research* 30:294–303.

Atkinson, A., and J. Stiglitz. 1969. "A New View of Technical Change." *The Economic Journal* 79:573–80.

Balcer, Y., and S. Lippman. 1984. "Technological Expectations and the Adoption of Improved Technology." *Journal of Economic Theory* 34:292–318.

Bupp, I., and J. Derian. 1978. *Light Water: How The Nuclear Dream Dissolved.* New York, NY: Basic Books.

Burton, R. 1976. *Recent Advances in Vehicular Steam Engine Efficiency.* Technical report 760340. Society of Automotive Engineers.

Cowan, R. 1987. "Backing the Wrong Horse: Sequential Choice Among Technologies of Unknown Merit." PhD diss., Stanford University.

David, P. 1975. *Technical Choice, Innovation, and Economic Growth.* Cambridge, UK: Cambridge University Press.

———. 1985. "Clio and the Economics of QWERTY." *The American Economic Review* 75 (2): 332–337.

———. 1987. "Some New Standards for the Economics of Standardization in the Information Age." In *Economic Policy and Technological Performance,* edited by P. Dasgupta and P. Stoneman, 206–239. Cambridge, UK: Cambridge University Press.

David, P., and J. Bunn. 1987. "The Economics of Gateway Technologies and Network Evolution: Lessons from Electricity Supply History." *Information Economics and Policy* 3 (2): 165–202.

Dosi, G. 1988. "Sources, Procedures and Microeconomic Effects of Innovation." *Journal of Economic Literature* 26:1120–71.

Dosi, G., C. Freeman, R. Nelson, G. Silverberg, and L. Soete, eds. 1988. *Technical Change and Economic Theory.* London, UK: Pinter.

Farrell, J., and G. Saloner. 1985. "Standardization, Compatibility, and Innovation." *Rand Journal of Economics* 16:70–83.

———. 1986. "Installed Base and Compatibility: Innovation, Product Preannouncements and Predation." *American Economic Review* 76:940–55.

Fletcher, W. [1904]1973. *English and American Steam Carriages and Traction Engines.* Newton Abbot, UK: David and Charles.

Fudenberg, D., and J. Tirole. 1983. "Learning by Doing and Market Performance." *Bell Journal of Economics* 14:522–30.

Hanson, W. 1985. "Bandwagons and Orphans: Dynamic Pricing of Competing Systems Subject to Decreasing Costs." PhD diss., Stanford University.

Hartwick, J. 1985. *The Persistence of QWERTY and Analogous Seemingly Suboptimal Standards.* Mimeo 592. Queen's University, Kingston, Ontario. https://ideas.repec.org/p/qed/wpaper/592.html.

Hill, B., D. Lane, and W. Sudderth. 1980. "A Strong Law for Some Generalized Urn Processes." *Annals of Probability* 8:214–26.

Katz, M., and C. Shapiro. 1985. "Network Externalities, Competition, and Compatibility." *American Economic Review* 75:424–40.

———. 1986. "Technology Adoption in the Presence of Network Externalities." *Journal of Political Economy* 94:822–4.

Kindleberger, C. 1983. "Standards as Public, Collective and Private Goods." *Kyklos* 36:377–96.

Mamer, J., and K. McCardle. 1987. "Uncertainty, Competition and the Adoption of New Technology." *Management Science* 33:161–77.

McLaughlin, C. 1954. "The Stanley Steamer: A Study in Unsuccessful Innovation." *Explorations in Entrepreneurial History* 7:37–47.

Metcalfe, J. S. 1985. "On Technological Competition." PhD diss., University of Manchester.

Rosenberg, N. 1982. *Inside the Black Box: Technology and Economics.* Cambridge, UK: Cambridge University Press.

Spence, A. M. 1981. "The Learning Curve and Competition." *Bell Journal of Economics* 12:49–70.

Strack, W. 1970. *Condensers and Boilers for Steam-Powered Cars.* Technical report TN D-5813. Washington, DC: NASA.

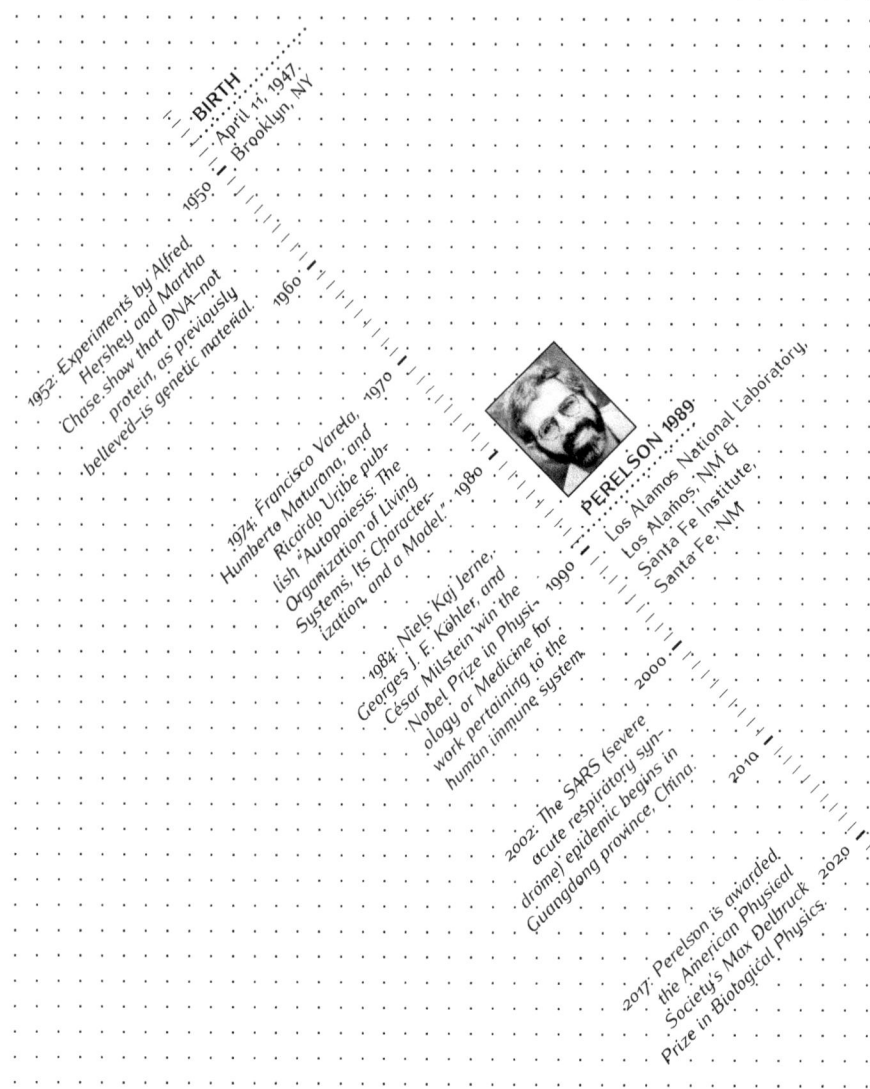

BIRTH
April 11, 1947
Brooklyn, NY

1952: Experiments by Alfred Hershey and Martha Chase show that DNA—not protein, as previously believed—is genetic material.

1974: Francisco Varela, Humberto Maturana, and Ricardo Uribe publish "Autopoiesis: The Organization of Living Systems, Its Characterization, and a Model."

1984: Niels Kaj Jerne, Georges J. F. Köhler, and César Milstein win the Nobel Prize in Physiology or Medicine for work pertaining to the human immune system.

PERELSON 1989:
Los Alamos National Laboratory, Los Alamos, NM &
Santa Fe Institute, Santa Fe, NM

2002: The SARS (severe acute respiratory syndrome) epidemic begins in Guangdong province, China.

2017: Perelson is awarded the American Physical Society's Max Delbruck Prize in Biological Physics.

ALAN STUART PERELSON

[68]

IMMUNE NETWORK THEORY: A RETROSPECTIVE

Rob J. de Boer, Utrecht University

The immune system is a distributed complex adaptive system that makes decisions on a daily basis about how to respond to intruders, to commensals, and to itself. These decisions are typically remembered for life. Lifelong immunity typically protects us from a large variety of pathogens in our environment. The absence of immunity to the commensal bacteria in our microbiomes is essential for our physiological health, and the absence of immunity to our own tissues prevents us from suffering from autoimmune disease. These "cognitive" properties of deciding and remembering are typically attributed to the adaptive immune system, that is, the lymphocytes, which are circulating white blood cells, expressing a randomly made receptor, with which they can potentially bind proteins that they encounter while patroling the body. The proteins (or peptides) that lymphocytes bind to with high affinity are called antigens (or epitopes), and lymphocytes tend to be "specific" for a particular antigen because their random receptor only binds to a very small fraction (typically $1:10^5$ to $1:10^6$) of the proteins in their environment. Lymphocytes specifically binding to a novel antigen—that is, those involved in a high-affinity interaction—trigger a process of cell division and differentiation, and expand into a large clone of "effector cells," all expressing the same receptor. This process is called "clonal expansion." Together these effector cells are capable of binding and clearing the antigen.

There are two major classes of lymphocytes, B cells and T cells. B cells express a random receptor that can bind native proteins, and they release this receptor into the circulation after clonal expansion as so-called "antibodies." These antibodies can bind their cognate antigen in the circulation, forming "immune complexes" that are cleared. Thus, circulating antibodies clear viruses and bacteria from circulation;

A. S. Perelson, "Immune Network Theory," *Immunological Reviews* 110 (1), 5–36 (1989).

Published by Munksgaard, Copenhagen, Denmark. Reprinted by permission of John Wiley and Sons.

this is the major mechanism underlying the protection induced by vaccination. T cells express a random receptor that binds short peptides presented on the surface of other cells. After clonal expansion, T cells can mature into cytotoxic cells that can kill cells infected with bacteria or viruses, whenever these cells express short peptides sampled from these intracellular pathogens on their surface. Importantly, different antigens require different immune responses, ranging from tolerance to the "self proteins" in our tissues and the commensal bacteria in our microbiomes, to antibodies clearing circulating pathogens and the toxins they may secrete and cytotoxic T cells killing cells harboring intracellular pathogens. Somehow the immune system has to decide on the most appropriate immune response to this large variety of antigens.

The random receptor of lymphocytes is made by a process called "somatic recombination." Each receptor is composed of several gene segments, and because our genome codes for multiple variants of each gene segment, the combinatorics of sampling variants in a semi-random manner allow for a very diverse repertoire of receptors. True randomness is added because at the ends of the segments several nucleotides can be deleted and inserted. Moreover, during B-cell responses the random receptors are mutated in a process called "affinity maturation," allowing for an evolutionary process selecting the antibodies with the highest affinity for the cognate antigen. Nowadays, we can measure the diversity of the B- and T-cell repertoires by next-generation sequencing. Using ecological concepts to account for the species missed in small samples taken from the blood (Chao and Bunge 2002), we currently estimate that T-cell repertoires are composed of at least 10^9 different receptors (Qi, Liu, and Cheng 2014). Importantly, because lymphocyte receptors are proteins with a binding site that is at least partly random, these receptors themselves also form novel antigens that other lymphocytes can bind and respond to. This is the basis of idiotypic network theory that was originally formulated by the Nobel Prize winner Niels Jerne (1974) and forms the basis of Alan Perelson's (1989) "Immune Network Theory." Interestingly, Giorgio Parisi (1990), who shared the Nobel Prize in physics for his work in complexity theory, also worked on idiotypic networks around that time, also focusing on immune memory and the similarity of the theory with spin-glass models.

According to idiotypic network theory, the immune system functions as a very diverse network of antibodies binding to each other in solution and on the surface of cells, leading to stimulatory and inhibitory interactions among the millions to billions of different "clones" of B cells in a human immune system (Jerne 1974). The idiotypic network is a powerful and attractive theory, because such large and highly connected networks would readily account for cognitive properties like decision making and memory. Moreover, idiotypic interactions seem inevitable in these very diverse repertoires of random lymphocyte receptors. For instance, for a conservative estimate based on a probability of binding of one in 10^6 antigens, and a repertoire of about 10^8 receptors, clones are expected to be connected to about a hundred other clones in the network. Idiotypic network theory therefore attracted a lot of attention from theoreticians interested in immunology, and several mathematical models have been proposed as mechanisms underlying the regulation of immune responses and the maintenance of immune memory. In "Immune Network Theory" Perelson provides a general mathematical basis of the theory while reviewing several of the models of B-cell networks. T-cell receptors were also shown to be involved in idiotypic interactions, which lead to the counterintuitive idea of "T-cell vaccination," that is, a vaccination against autoimmune disease with the very T cells that are actually responsible for the disease (Cohen 1986). We have had several workshops at the Santa Fe institute covering both B- and T-cell networks.

In his review paper, Perelson employed his general concept of a "shape space" representing both receptors and antigens (Perelson and Oster 1979) to argue that immune repertoires are expected to be "complete": in other words, almost all antigens are expected to be bound by at least one receptor in the repertoire. The idiotypic network follows as a necessary consequence of this completeness, because lymphocyte receptors are also shapes expected to be bound (Perelson 1989). Another innovation at the time was to represent antibody shapes as binary strings (which resulted from Perelson's collaboration with Doyne Farmer and Norman Packard: Farmer, Packard, and Perelson 1986) and to use the Hamming distance as a model for affinity with which antibodies interact. This opened up the way for large-scale computer simulations of idiotypic networks involving many species of B cells. An important realization was that networks with more than a few edges per node tend to be fully connected (De Boer 1989; Perelson 1989), and hence that the first perturbation of such a network could percolate throughout the entire network. To prevent a single stimulus, for example, an antigen from a pathogen, from affecting the entire network, specific network topologies "distributing" the signal were introduced (Weisbuch, De Boer, and Perelson 1990). Perelson's review finishes with a fairly complete model defining both B cells, their antibodies, and the immune complexes these can form. Interestingly, this model builds on earlier classic models for receptors binding ligands and forming crosslinks on the cell surface (Perelson and DeLisi 1980). This mechanistic derivation revealed that the activation function is bell-shaped, with optimal stimulation at intermediate antibody levels, and inhibitory interactions at high concentrations (Sulzer, De Boer, and Perelson 1996)—that is, a highly nonlinear interaction function.

However, interest in idiotypic networks waned at the end of the last century (although the theory was never falsified), and most textbooks no longer discuss it. This is due to a lack of experimental evidence that the immune system truly functions as an idiotypic network (the original evidence demonstrating that anti-idiotypic antibodies can be produced under favorable experimental conditions does not prove that this is also prevalent during normal circumstances; see below). The immunologist Klaus Eichmann (2008) published a book analyzing the rise and fall of idiotypic network theory that included interviews with colleagues who also studied it. Only a few novel papers on the theory continue to appear in the immunological literature, some by immunologists, such as two recent SARS-CoV-2 papers (Arthur *et al.* 2021; Murphy and Longo 2022), and some by theoreticians continuing to work on modeling idiotypic interactions (and citing Perelson's classic review paper). Importantly, since we have argued that idiotypic interactions are inevitable due to the completeness of the repertoire, we need to discuss if and how they could remain rare, overlooked, and/or nonfunctional during normal immune reactions.

One likely explanation for this is the more recent insight that the lymphocytes forming the adaptive immune system require signals from the innate immune system, composed of many different cell types (mostly phagocytes). Cells in the innate immune system have "pattern-recognition receptors" (PRRs) recognizing "pathogen-associated molecular patterns" (PAMPs), and can be activated by damage and inflammation in a tissue (Takeuchi and Akira 2010). This requirement has been called the "dirty little secret" (Janeway, Jr. 1989) to account for the fact that to trigger PRRs we have to add adjuvants to our vaccines. In the absence of an innate response—that is, if we were to provide antigen only—there would be hardly any adaptive response.

This dependence would restrict the development of anti-idiotypic responses to the early time period of inflammation that is associated with infection by a pathogen or with the adjuvant in a vaccination experiment (with antigen or antibody). Such a requirement on the early activation of the innate immune system could still allow for early anti-idiotypic responses but would perhaps no longer allow for later anti-anti-idiotypic responses to these anti-idiotypic antibodies. Potentially anti-idiotypic (Arthur *et al.* 2021; Murphy and Longo 2022) autoantibodies binding the cell-surface protein that SARS-CoV-2 uses to enter cells were indeed identified in the vast majority of patients suffering from severe COVID-19 and in only a few SARS-CoV-2-infected people experiencing little inflammation and mild symptoms (Arthur *et al.* 2021). Thus, the existence of a first layer of anti-idiotypic interactions does not necessarily imply that the immune system functions as a deep network.

Interestingly, the adaptive immune system's dependence on innate signals provides a new insight on the decisions taken by the immune system. Since the PRRs are evolutionarily conserved, and because they bind specifically to particular PAMPs, the innate signals carry information on the type of pathogen that is being encountered and the amount of damage that is being done. Thus, lymphocytes that are becoming activated into clonal expansion by a novel antigen can be informed by the innate immune system on what type of immune response would be most appropriate. Lymphocytes can "remember" this information by changing the expression of the genes determining their effector phenotype, and thus provide an appropriate response to their cognate antigen for the rest of their life. In this scenario, decisions tend to be taken by the innate immune system (based upon an evolutionary learning process). Because these decisions are subsequently remembered by the adaptive immune system, we learn from every initial challenge by a pathogen in our environment how to respond rapidly, specifically, and appropriately to the ongoing intrusion and to subsequent challenges by the same or related pathogens (Borghans and De Boer 2002).

Acknowledgments

I thank Alan Perelson and Peter de Greef for helpful suggestions.

REFERENCES

Arthur, J. M., J. C. Forrest, K. W. Boehme, J. L. Kennedy, S. Owens, C. Herzog, J. Liu, and T. O. Harville. 2021. "Development of ACE2 Autoantibodies after SARS-CoV-2 Infection." *PLOS One*, https://doi.org/10.1371/journal.pone.0257016j.

Borghans, J. A. M., and R. J. De Boer. 2002. "Memorizing Innate Instructions Requires a Sufficiently Specific Adaptive Immune System." *International Immunology* 14:525–32. https://doi.org/10.1093/intimm/14.5.525.

Chao, A., and J. Bunge. 2002. "Estimating the Number of Species in a Stochastic Abundance Model." *Biometrics* 58 (3): 531–539. https://doi.org/10.1111/j.0006-341X.2002.00531.x.

Chao, D. L., M. P. Davenport, S. Forrest, and A. S. Perelson. 2004. "A Stochastic Model of Cytotoxic T Cell Responses." *Journal of Theoretical Biology* 228:227–40. https://doi.org/10.1016/j.jtbi.2003.12.011.

Cohen, I. R. 1986. "Regulation of Autoimmune Disease Physiological and Therapeutic." *Immunological Reviews* 94:5–21. https://doi.org/10.1111/j.1600-065x.1986.tb01161.x.

De Boer, R. J. 1989. In *Theories of Immune Networks*, edited by H. Atlan and I. Cohen. Berlin, Germany: Springer-Verlag.

De Boer, R. J., and A. S. Perelson. 1993. "How Diverse Should the Immune System Be?" *Proceedings of the Royal Society B: Biological Sciences* 252:171–75. https://doi.org/10.1098/rspb.1993.0062.

Detours, V., R. Mehr, and A. S. Perelson. 2000. "Deriving Quantitative Constraints on T Cell Selection from Data on the Mature T Cell Repertoire." *Journal of Immunology* 164:121–8. https://doi.org/10.4049/jimmunol.164.1.121.

Eichmann, K. 2008. *The Network Collective: Rise and Fall of a Scientific Paradigm*. Basel, Switzerland: Birkhauser.

Farmer, J., N. H. Packard, and A. S. Perelson. 1986. "The Immune System, Adaptation, and Machine Learning." *Physica D: Nonlinear Phenomena* 22:187–204. https://doi.org/10.1016/0167-2789(86)90240-X.

Giorgetti, O. B., P. Shingate, C. P. O'Meara, V. Ravi, N. E. Pillai, B.-H. Tay, A. Prasad, *et al.* 2021. "Antigen Receptor Repertoires of One of the Smallest Known Vertebrates." *Science Advances* 7 (1). https://doi.org/10.1126/sciadv.abd8180.

Janeway, Jr., C. A. 1989. "Approaching the Asymptote? Evolution and Revolution in Immunology." *Cold Spring Harbor Symposia on Quantitative Biology* 54:1–13. https://doi.org/10.1101/sqb.1989.054.01.003.

Jerne, N. K. 1974. "Towards a Network Theory of the Immune System." *Annals of Allergy, Asthma & Immunology* (Paris, France) 125:373–89.

Macken, C. A., and A. S. Perelson. 1989. "Protein Evolution on Rugged Landscapes." *Proceedings of the National Academy of Sciences* 86 (16): 6191–5. https://doi.org/10.1073/pnas.86.16.6191.

Murphy, W. J., and D. L. Longo. 2022. "A Possible Role for Anti-Idiotype Antibodies in SARS-CoV-2 Infection and Vaccination." *The New England Journal of Medicine* 386:394–6. https://doi.org/10.1056/NEJMcibr2113694.

Parisi, G. 1990. "A Simple Model for the Immune Network." *Proceedings of the National Academy of Sciences* 87:429–33. https://doi.org/10.1073/pnas.87.1.429.

Perelson, A. S. 1989. "Immune Network Theory." *Immunological Reviews* 110:5–36. https://doi.org/10.1111/j.1600-065x.1989.tb00025.x.

Perelson, A. S., and C. DeLisi. 1980. "Receptor Clustering on a Cell Surface. I. Theory of Receptor Cross-Linking by Ligands Bearing Two Chemically Identical Functional Groups." *Mathematical Biosciences* 48:71–110. https://doi.org/10.1016/0025-5564(80)90017-6.

Perelson, A. S., and G. F. Oster. 1979. "Theoretical Studies of Clonal Selection: Minimal Antibody Repertoire Size and Reliability of Self–Non-Self Discrimination." *Journal of Theoretical Biology* 81:645–70. https://doi.org/10.1016/0022-5193(79)90275-3.

Qi, Q., Y. Liu, and Y. Cheng. 2014. "Diversity and Clonal Selection in the Human T-Cell Repertoire." *Proceedings of the National Academy of Sciences* 111:13139–44. https://doi.org/10.1073/pnas.1409155111.

Segel, L. A., and A. S. Perelson. 1988. "Computations in Shape Space: A New Approach to Immune Network Theory." Edited by A. S. Perelson. (Reading, MA) II:321–343.

Sulzer, B., R. J. De Boer, and A. S. Perelson. 1996. "Cross-Linking Reconsidered: Binding and Cross-Linking Fields and the Cellular Response." *Biophysical Journal* 70:1154–68. https://doi.org/10.1016/S0006-3495(96)79676-5.

Takeuchi, O., and S. Akira. 2010. "Pattern Recognition Receptors and Inflammation." *Cell* 140:805–20. https://doi.org/10.1016/j.cell.2010.01.022.

Weisbuch, G., R. J. De Boer, and A. S. Perelson. 1990. "Localized Memories in Idiotypic Networks." *Journal of Theoretical Biology* 146:483–499. https://doi.org/10.1016/s0022-5193(05)80374-1.

Wortel, I. M. N., C. Kesmir, R. J. De Boer, Mandl J. N., and J. Textor. 2020. "Is T Cell Negative Selection a Learning Algorithm?" *Cells* 9:690. https://doi.org/10.3390/cells9030690.

IMMUNE NETWORK THEORY

Alan S. Perelson

Introduction

The theoretical development of idiotypic networks was initiated by Jerne (1974a). In a review, Jerne (1974b) attempted to put the network proposal into mathematical terms. Before describing Jerne's and more modern efforts in detail it is worth remembering what Jerne said about this endeavor. "Although this task is a challenge to immunologists, I suspect that several would rather turn away from such ambitious and, for the present, unrealistic exercises in order to contemplate what they would call real experimental facts. To those I would reply that as long as the quantitation of the immune response remains elusive, immunology will remain a phenomenology, an ever accumulating catalogue of such phenomena as are at present our daily bread."

Jerne constructed a differential equation to describe the dynamics of a set of identical lymphocytes. By identical Jerne meant cells that were indistinguishable with respect to their state of differentiation as well as to their receptors and to the antibody molecules they produce. To illustrate Jerne's idea, call these identical lymphocytes, lymphocytes of type i, and let L_i denote the number of lymphocytes of type i. Lymphocytes of type i interact with lymphocytes of other types, e.g. lymphocytes of type j, and antibody of type j via idiotopes and combining sites. The interactions can be either excitatory or inhibitory. The lymphocytes of type j of course interact with other lymphocyte types and these with others and so on. Jerne suggested that the rate at which lymphocytes of a particular type increase or decrease in number is given by:

$$\frac{dL_i}{dt} = \alpha - \beta L_i + L_i \sum_{j=1}^{m} \varphi(E_j, K_j, t) - L_i \sum_{j=1}^{n} \psi(I_j, K_j, t). \quad (1)$$

In this equation, α is the rate at which lymphocytes enter set i from other compartments in the immune system and β is the rate (per lymphocyte) at which the lymphocytes die or leave the set. The functions ϕ and ψ keep track of the excitatory and inhibitory signals. The first sum is over all excitatory signals generated by idiotopes in the sets E_j that are recognized with association constants K_j by the combining sites on lymphocytes of type i. The second sum is over the inhibitory interactions generated by lymphocytes in sets I_j whose combining sites recognize idiotopes on cells in L_i. A further term would be needed to deal with the effects of an external antigen. Recognizing that the number of elements in the inhibitory and excitatory sets could change in time and that a differential equation of this type would be needed for each element of the network, Jerne concluded that there existed no satisfactory mathematical method of treating network problems. In the subsequent 15 years methods have been developed that allow us to treat networks of this level of complexity. In this paper I shall review these new approaches and discuss some of the theoretical insights that result.

Completeness of the Repertoire

Coutinho (1980) has postulated that the immune system in its ability to recognize antigen is "complete" (also see Forni and Coutinho 1981). If the antibody repertoire is complete then it follows that antibody molecules that have immunogenic idiotopes will be recognized by other antibody molecules and an idiotypic network will be created (Jerne 1984). Attempting to prove that the repertoire is complete is difficult. Perelson and Oster (1979) presented a simple theoretical argument showing that a complete repertoire is attainable within the known parameters of immune recognition. This argument, reproduced below, is based on the idea of "shape space".

Binding between a receptor and a ligand or a paratope and an idiotope generally involves short-range non-covalent interactions based on electrostatic charge, hydrogen binding, van der Waals interactions, etc. In order for the molecules to approach each other over an appreciable portion of their surfaces, there must be extensive regions of complementarity. In some cases, the complementary regions may be planar, while in others they more closely resemble a bump and a groove. Both shape and charge distributions, as well as the existence in the appropriate complementary positions of chemical groups that can form

FOUNDATIONAL PAPERS IN COMPLEXITY SCIENCE

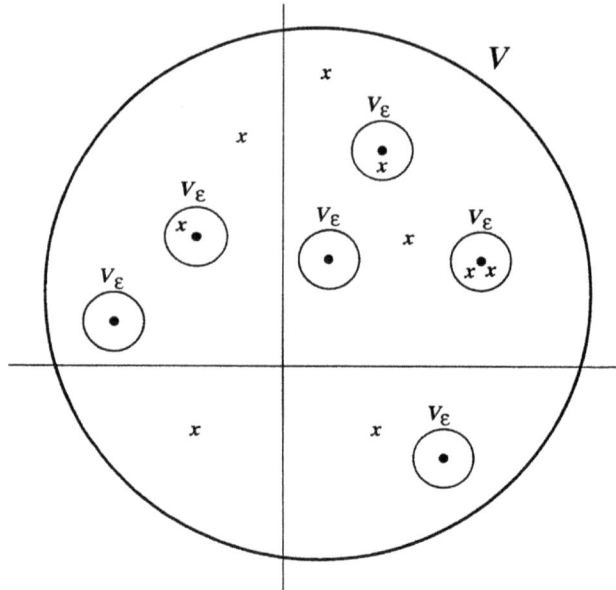

Figure 1. Diagramatic representation of shape space. Within the space there is a volume V in which paratope (•) and epitope (×) shapes are located. An antibody is assumed to recognize with affinity greater than or equal to K all epitopes within a volume $V_{\varepsilon(K)}$ surrounding it.

hydrogen bonds and interact in other ways are properties of antigens and antibodies that are important in determining the interactions between these molecules. We call this constellation of features the *generalized shape* of a molecule. Suppose that one can adequately describe the generalized shape of an antibody combining site (paratope) by N parameters: the length, width and height of any bump or groove in the combining site, its charge, etc. The precise number of parameters or their values is not important for the argument that follows, only that a finite number of them is needed. Then a point in an N-dimensional space, "shape space", specifies the generalized shape of a paratope with regard to its antigen binding properties. If an animal has a repertoire of size N_{Ab}, then the shape space for that animal would contain N_{Ab} different points. One would expect these points to lie in some finite volume V of the space since there is only a restricted range of widths, lengths, charges, etc. that an antibody combining site can assume (Fig. 1). For example, one would never find an antibody with a combining site dimension of one centimeter!

Antigenic determinants (epitopes or idiotopes) are also characterized by generalized shapes which should lie within V. For example, a combining site with a length of 2 nm cannot be expected to recognize a determinant 10 nm long. In order to estimate how well an animal with a repertoire of size N_{Ab} can recognize molecular determinants, let us assume that a paratope and epitope fit together perfectly if they have the same shape, i.e. lie at the same point in V. (This is clearly a fiction because epitopes and paratopes must have complementary shapes. However, this is a useful mathematical fiction because it simplifies the argument and yields the same result as a more precise treatment which includes the notion of complementarity). If the paratope and epitope shapes are not quite complementary then the two molecules may still bind but with lower affinity. At some low level of affinity, e.g. $10^4 M^{-1}$, we say the interaction is not specific and that the epitope and paratope are not complementary. To describe this we assume that each paratope specifically interacts with all epitopes that are within a small surrounding region in shape space, called a "recognition ball" (see Fig. 1). Let $V_{\varepsilon(K)}$ (or for simplicity V_ε) be the volume of a recognition ball when the threshold affinity is set as K. (If one wanted to be more precise one could assign an explicit affinity to each paratope-epitope pair depending upon their location in shape space. This approach will be taken with regard to more restricted models in subsequent sections). Because each antibody can recognize all epitopes within a recognition ball, a finite number of antibodies can recognize an infinite number of epitopes, i.e. one can put an infinite number of points into the volume V_ε. If antibodies are multispecific then each paratope might be viewed as having multiple shapes, say corresponding to different pockets and grooves in the combining region. This case was analyzed by Perelson and Oster (1979) but will not be pursued here. Suffice it to say that with multispecific antibodies completeness of the repertoire is easier to attain than the analysis below shows.

To complete the argument, let us assume that antibodies are made with random shapes. Thus the N_{Ab} antibodies lie scattered at random in the shape space. If each antibody has roughly the same recognition volume V_ε, then the total volume covered by all of the antibodies in the repertoire is $N_{Ab} V_\varepsilon$. If this volume is large compared with the total volume of shape space V, then one would expect that the various antibodies would have recognition regions that overlap and completely cover shape space. In fact, each epitope would on average be recognized by $N_{Ab}V_\varepsilon/V$ different antibodies, and the probability, P, that an epitope is *not* recognized by some antibody is (Perelson and Oster 1979)

$$P = e^{-N_{Ab}V_\varepsilon/V}. \qquad (2)$$

We can use Eq. (2) to quantify the completeness of the repertoire. Typically, of order 1 in 100 000 B cells responds to an epitope (cf. Klinman and Press 1975). We will use this value as an estimate of $p(K)$, the probability that an antibody recognizes a random antigenic determinant with an affinity above the threshold value K. To interpret $p(K)$ within the context of shape space theory notice that, if one randomly places an epitope in shape space, the probability that it lands in the volume V_e surrounding any given antibody is V_e/V, the fraction of the shape space volume covered by a single antibody. (An easy way to see this is to consider throwing darts at the two-dimensional version of shape space depicted in Fig. 1. Assume that the darts are thrown at random and all hit the board. The probability that a dart will land in a recognition ball is then the area of the ball divided by the area of the dart board, V.) Thus, if the readout of immune recognition is B-cell stimulation

$$p(K) = V_\varepsilon/V \simeq 10^{-5}. \qquad (3)$$

With this rough estimate, Eq. (2) predicts that animals with a repertoire of 10^5 antibodies will only be marginal, i.e. e^{-1} or 37% of epitopes will escape detection. However, if $N_{Ab} = 5 \times 10^5$ then P falls to 6.7×10^{-3} and less than 1% of epitopes escape detection. If $N_{Ab} = 10^6$ then $P = 4.5 \times 10^{-5}$ and essentially all epitopes will be recognized. Thus a repertoire of order 10^6, composed of antibodies with random shapes, will be complete. This is interesting because the smallest known immune system, that of a young tadpole, is estimated to have 10^6 lymphocytes and thus a repertoire of order 10^5 to 10^6 (Du Pasquier 1973; Du Pasquier and Haimovitch 1976). Smaller immune systems do not exist and the "back of the envelope" calculation given above suggests that this is the case because such immune systems would recognize antigen so infrequently that they would provide little, if any, protective advantage.

To summarize, we have argued that the repertoire will be complete if three hypotheses are satisfied: 1) Each antibody can recognize a set of

Interestingly, a recent paper described the immune system of one of the smallest vertebrates, the fish species *Paedocypris*, which is estimated to have about 37,000 T cells (Giorgetti *et al.* 2021), more than an order of magnitude lower than the 10^6 lymphocytes in a young tadpole. Apparently, such a small immune system can be functional. The reason could be that most pathogens are typically presented as many different epitopes and that the fish would already be protected if it mounts an immune response to least one of the epitopes (De Boer and Perelson 1993).

related epitopes, each of which differs slightly in shape. The strength of binding may differ for different epitopes and is accounted for by differences in affinity. 2) The antibodies in the repertoire have shapes that are randomly distributed throughout shape. 3) The repertoire size is of order 10^6 or greater.

Less stringent forms of hypotheses 2) and 3) should also lead to complete repertoires. For example, antibodies need not be randomly distributed in shape space; some regions could have substantially higher density than others but empty regions larger than V_ε in volume need to be rare. Under such conditions a slightly larger repertoire might be needed. The distribution of antibodies in shape space is unknown. Antibody V-regions are not made at random but derive from recombination of germline V, J, and D gene segments, with added junctional and mutational diversity. Depending on gene usage one might expect regions of different density in shape space, and the existence of holes.

Completeness is not absolute. As our quantitative examples have shown there is a small probability that an epitope will not be recognized. However, when the completeness conditions are met, or even nearly met with some holes being present, the chances of antibodies recognizing idiotopes on other antibodies are so overwhelming that, as Jerne (1984) has said, "the idiotypic network idea is unavoidable".

Repertoire completeness only ensures that at least one antibody will recognize an idiotope. Whether that recognition will lead to a successful regulatory interaction depends upon many other factors, such as concentrations and affinities of the interacting molecules. In the physical chemical theory of antigen-antibody reactions one typically finds that it is the product KC of antibody affinity, K, and antigen concentration, C, that is important in determining the fraction of antibody combining sites bound by antigen at equilibrium. Thus at high concentrations low affinity interactions are significant, whereas at low concentrations only high affinity interactions are important. A repertoire which is complete at one concentration level need not be complete at lower concentrations that require higher affinity interactions. For example, if one required an affinity level of $10^8 M^{-1}$ for a response, then one might find only 10^{-6} or 10^{-7} of B cells responding. With $p(K) = V_{\varepsilon(K)}/V = 10^{-7}$, repertoire sizes of order 10^8 would be needed for completeness. Because of the interplay

This argument that a repertoire of more than 10^6 receptors would be complete, and hence protective, has recently been confirmed by next-generation sequencing of naive T cells in volunteers. The naive T cell repertoire size in young human adults was estimated to be at least 10^9 receptors (Qi et al. 2014), which is even a thousand-fold larger. Interestingly, in elderly people this diversity declined less than five-fold (Qi et al. 2014), which would argue that the repertoires of naive T cells in healthy aged people remain complete. The erosion of immunity due to aging would therefore require a different mechanism (if the erosion is random).

between affinity and concentration and the degree of interconnection in a network, dynamic models of the type proposed by Jerne in Eq. 1 become important. After discussing recent advances in modeling we shall return to the question of repertoire completeness and network connectivity.

Antibody Representations

One of the major stumbling blocks in formulating a mathematical model of an idiotypic network is determining which antibodies react with which other antibodies. In a system with a repertoire of say 10^7 elements how can one ever determine all of the possible interactions? Three approaches have been taken. In the first, no attempt is made and one simply assumes some simple relationship among idiotypes and anti-idiotypes. For example, Hoffmann's plus-minus network theory (Hoffmann 1975, 1979, 1980; Gunther and Hoffmann 1982) only deals with two specificities, an antigen-specific population and its anti-idiotypic partner. Richter (1975, 1978) dealt with a linear idiotypic network (Fig. 2) in which antibodies and/or cells (a distinction between the two has generally not been made in models) at idiotypic level i interacts with populations at levels $i - 1$ and $i + 1$, for $i = 0, 1, \ldots, l$, and antigen is considered level 0. Although l, the number of levels in the network, could be very high in simulation studies l is generally chosen to be rather small, i.e. $l < 10$. Hiernaux (1977) dealt with a small cyclic network (Fig. 2). By prescribing the topology of the network one has, in essence, determined that all the antibodies (cells) in the repertoire can be assigned to the few classes in the network diagram.

The second approach assumes that the network is so large and complex that it is impossible to determine the relationships between the elements and one simply assigns them at random (cf. De Boer 1988; Hoffmann 1982; Hoffmann *et al.* 1988; Parisi 1990). This approach begins to confront the complexity that inherently can be in a network.

The third approach, which is one that I have pursued, assumes that the interactions in a network are determined by the specific chemical interactions between the various cells and molecules in the immune system. The basis of these interactions is what we previously called generalized shape. Thus if one knew the shape of each molecule one could predict which molecules would react and the affinity of their interaction. Even though we do not know the actual shapes of molecules one can find simple

Giorgio Parisi formulated a very simple symmetric network model to study the memory capacity of the idiotypic network and pointed out that this model coincides with models for spin glasses. Based on his groundbreaking work on spin glasses and the theory of complex systems, he was awarded a (shared) Nobel Prize for physics in 2021.

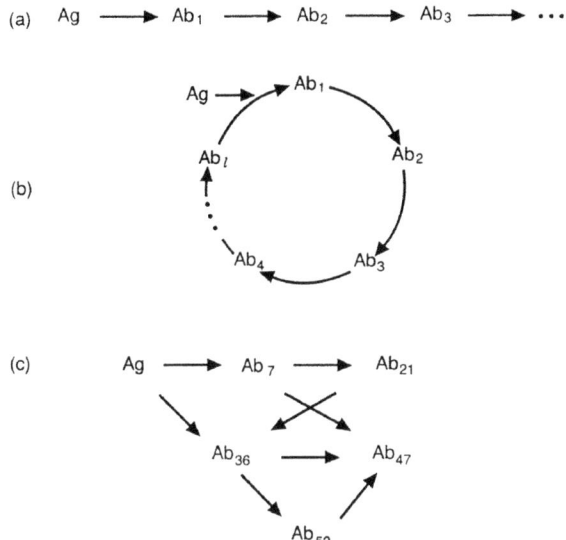

Figure 2. (a) Linear idiotypic network. (b) Cyclic idiotypic network. (c) A random network.

mathematical representations of antibodies that allow us to compute the degree of complementarity between molecules and even assign an affinity to their interaction. Various rules will give slightly different results but this type of formulation makes it possible to begin asking detailed questions about the topology of idiotypic networks and how that topology might vary due to structural constraints among antibody V regions. The representation introduced by (Farmer, Packard, and Perelson 1986) assigns to each antibody two binary strings of length n, one representing the paratope and the other an epitope or idiotope (see Fig. 3). An antigen containing a single epitope is represented by a single binary string, whereas antigens with multiple epitopes are represented by multiple strings. With this representation shape space is a hypercube of dimension n. If one chooses $n = 32$, then one can represent $2^{32} \approx 4 \times 10^9$ different determinants in this shape space. Thus with a 32-bit computer, such as a VAX or SUN workstation, one can represent systems with diversity comparable to that expressed in the mammalian immune system. One can view the binary strings representing an antibody as being related to the string of nucleic acids or amino acids that code for the V region.

Complementarity can be defined by any of a number of rules. Fig. 4 illustrates the simplest rule: two determinants are complementary if some threshold number of bits in their binary string representations

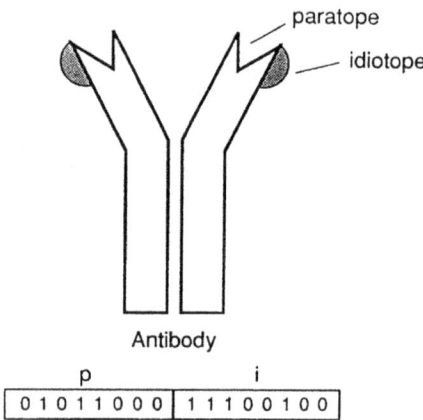

Figure 3. Antibody and its binary string representation.

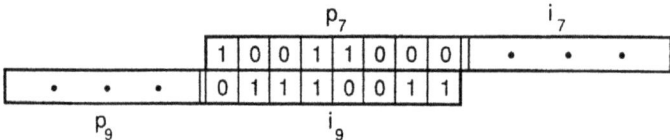

Figure 4. Complementarity between paratopes and idiotopes can be assessed by aligning the paratope and idiotope strings and then summing the number of positions at which a 1 is matched by 0. Here 6 out of 8 bits of the paratope of Ab_7 are complementary to the idiotope on Ab_9.

are complementary. The number of complementary bits can be used to assign an affinity to the interaction (Farmer, Packard, and Perelson 1986). Although computing complementarity by comparing binary strings was chosen for convenience, there is evidence suggesting that in some cases complementarity at the level of DNA may imply complementarity at the protein level. If a peptide is transcribed from one strand of a double stranded DNA molecule and a "complementary peptide" is synthesized by reading from the complementary DNA strand, then the peptide and complementary peptide may bind specifically and with high affinity (Bost, Smith, and Blalock, 1985a, 1985b; Smith, Bost, and Blalock 1987; Shai, Flashner, and Chaiken 1987). Further, in the case of the hormone ACTH, antibodies against ACTH and antibodies against the complementary peptide seem to be an idiotypic-anti-idiotypic pair, leading to the speculation that idiotopes and anti-idiotopes may represent complementary sequences in the hypervariable regions of such immunoglobulin pairs (Smith, Bost, and Blalock 1987).

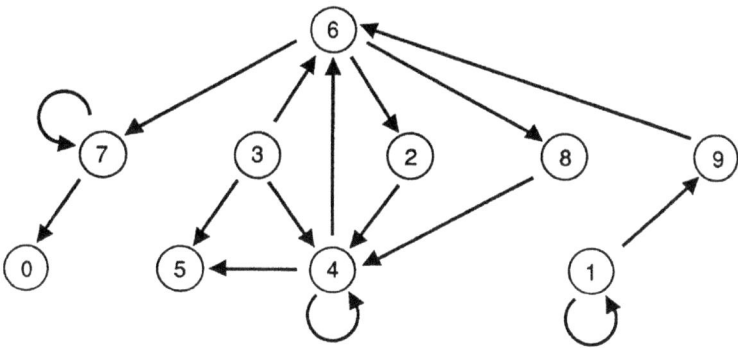

Figure 5. An example of an idiotypic network generated using the complementary match rule of Fig. 4. Idiotopes and paratopes are each 8 bits long and have been chosen at random. A line has been drawn connecting Ab_i to Ab_j if the paratope on Ab_i is complementary to 6 or more bits of the idiotope on Ab_j. Paratope-paratope matches can also be allowed but they are not shown here.

Other rules can also be used for determining complementarity. For example, since the strings represent molecules they need not be aligned when they interact. Thus the number of complementary bits can be summed or chosen to be the maximum over all possible alignments (cf. Farmer, Packard, and Perelson 1986). Molecules generally do not interact over their entire length, but rather interactions tend to be localized. Thus one can use a complementarity rule in which the number of adjacent complementary bits is important (cf. Stadnyk 1987).

The bit string representation is very powerful. Paratopes and idiotopes need not be the same size nor do they have to be distinct. The paratope and idiotope can overlap or even be chosen to be identical as in the symmetric network theory of Hoffmann. Multiple idiotopes can also be represented and framework regions introduced.

An example of an idiotypic network generated from the bit string representation is shown in Fig. 5. To make the diagram easy to visualize only 10 antibodies are in the network. Here epitopes and paratopes have been chosen as random sequences of 8 bits. A line is drawn connecting two antibodies if their paratope and idiotope are complementary at six or more bits. The lines are oriented in the direction of paratope-idiotope recognition. Thus the paratope on antibody 6 recognizes the idiotope on antibody 7. The network does not resemble either the linear or cyclic network of Fig. 2; rather it represents a complex interrelationship between the antibodies in the network. To a certain extent the network resembles

Mathematical models at the time could be based on symmetric interactions, treating idiotype and anti-idiotype as complementary shapes, or could be asymmetric when the binding site (i.e., the idiotype) and the recognition size (i.e., the paratope) were nonoverlapping.

a map of the antibody interactions connecting the immune responses to the acetylcholine receptor and $a - 1, 3$-dextran mapped by Dwyer et al. (1986). The network diagram can be decomposed into levels so that it appears analogous to the classical linear picture with antigen being level 0, Ab_1 being the set of antibodies directed against the antigen, Ab_2 being the anti-idiotypic antibodies, etc. The Ab_2 population can be decomposed into anti-idiotypic antibodies whose paratopes recognize the idiotope of Ab_1 ($Ab_{2\alpha}$), and internal image antibodies that have idiotopes which are recognized by the paratope of Ab_1 ($Ab_{2\beta}$) populations (Jerne, Roland, and Cazenave 1982). For example, if antibody 6 were an Ab_1, then antibodies 3, 4, and 9 would be anti-idiotypic, i.e. $Ab_{2\alpha}$, and antibodies 2, 7, and 8 would be internal images, $Ab_{2\beta}$. Because our complementarity rule does not require all bits to be complementary, $Ab_{2\beta}$ antibodies need not be exact images. In this example, antibodies 2 and 7 are only complementary at 6 out of the 8 bits and hence would be poor internal images, whereas antibody 8 is complementary at 7 bit positions and hence would be a better internal image. Other possibilities also arise. The paratope of an antibody can recognize its own idiotope. Self-recognizing antibodies have been found by Kohler and called "autobodies" (Kang and Kohler 1986). In the figure, three antibodies are autobodies, a surprisingly large number. Antibodies, termed epibodies or $Ab_{2\varepsilon}$, have been found (Bona et al. 1982) which are complementary to both idiotopes on Ab_1 and the antigen. In this example, if antigen is introduced the paratopes on antibodies 6, 7 and 9 recognize the antigen (not shown). However antibody 6 also recognizes the idiotope on antibody 7 and hence would be an epibody. The antibodies in the network have different connectivities. Antibody 6 interacts with six other antibodies, and antibody 4 interacts with five other antibodies and itself. These antibodies mimic the high connectivity self antibodies found by Holmberg et al. (1986) in newborn mice.

The percolation threshold predicted that almost all B cells in a fully connected network would become activated by the pathogen evoking a B cell response leading to the production of antibodies. This seemed unrealistic as most B cells in the repertoire tend to remain quiescent until they are activated by their cognate foreign antigen.

Phase Transitions in Idiotype Networks

Analyzing the topology of idiotypic networks in detail, I uncovered what appears to be a phase transition in the structure of the network (Perelson 1989). De Boer (1989) and De Boer and Hogeweg (1989c) have made similar findings. Consider a system containing N_{Ab} antibodies and a single antigen. Let each paratope, idiotope, and epitope be exactly n bits long. Use the simple complementarity rule in Fig. 4 in which paratope and idiotope or epitope are aligned and the number of complementary bits counted. The number of bits must be above a threshold, denoted n_θ, for recognition. Starting with the antigen, determine all antibodies that have paratopes that are complementary to the antigen. These are labeled as being in level 1, i.e. belong to Ab_1. All antibodies that have paratopes that recognize idiotopes on antibodies in Ab_1 or which have idiotopes recognized by the paratopes of Ab_1 antibodies are placed in level 2, i.e. Ab_2, as long as they are not in Ab_1. We continue in this manner, assigning an antibody to level i if it matches an antibody in level $i-1$, and if it has not already been assigned to some previous level. This latter condition ensures that each antibody is assigned to a unique level. The network can then be represented as a tree with the antigen as its root.

Fig. 6 illustrates a network diagram drawn in this way. Notice all N_{Ab} antibodies in the system need not appear in the diagram. For example, if no antibody matches the antigen the diagram will contain only the antigen root. There is nothing special about the antigen, the diagram can "die out" at any level because of a lack of matches. Thus, beginning with any root all of the antibodies in the system need not be assigned to an idiotypic level, only those that are somehow connected to the antigen appear. When different antigens are presented different subsets of the system may appear in such diagrams. Also, if a different set of N_{Ab} antibodies is constructed the diagram will most likely look different. For example, the number of levels need not be the same.

Both the probability of the tree lacking some of the antibodies in the system and the number of levels in the tree depend on the complementarity rule being used and the recognition threshold n_θ. For example, consider a system with 100 antibodies each represented by a single string 32 bits long (i.e. $N_{Ab} = 100$ and $n = 32$) and the complementarity rule of Fig. 4. If we choose $n_\theta = 32$, so that all bits must match, then the tree will be trivial, almost assuredly containing only the antigen. The probability of a match is $1/2^{32} \approx 2^{-10}$ and hence the system would need to contain 5×10^9 antibodies in order to expect even a single match. On the other hand, if $n_\theta = 10$, all antibodies will most likely match the antigen. Because bits are chosen at random with 0 and 1 having equal probability, on average half of the bits will be complementary. Thus if $n_\theta \leq n/2$ we expect all

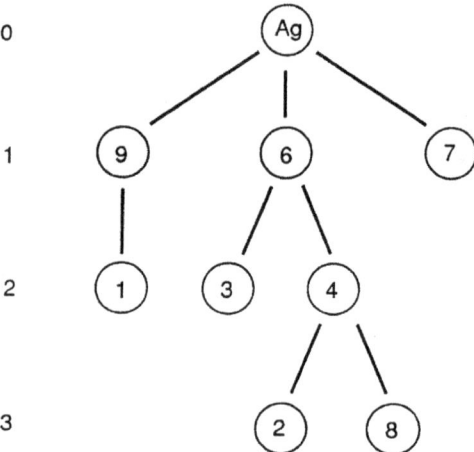

Figure 6. A network of 10 antibodies drawn as a rooted tree.

molecules to match each other, and all of the N_{Ab} antibodies to be in level 1. This is not a very realistic case. At values of $n_\theta > n/2$ there should be some matching between network elements, but all elements will no longer match the antigen. Thus one would expect trees with increasing numbers of levels. As n_θ approaches n matches become rare, and both the number of levels and the number of antibodies in the tree will decrease. As reasoned above, when n_θ reaches n we expect no antibodies in the tree if $N_{Ab} \ll 2^n$.

To summarize, if one varies n_θ or, equivalently, the probability of an antibody recognizing an epitope or idiotope, one expects the graph of the maximum level reached in the tree versus n_θ to approximate a smooth curve starting at one, rising, going through a maximum, and declining toward zero. Not surprising, this is precisely what we find via Monte Carlo simulations as we scan all possible values of n_θ. However, if one examines the curves that result for different values of N_{Ab} one finds a surprising feature: the graphs approach a curve with a singularity at a critical value of n_θ as N_{Ab} becomes large. In Fig. 7 I illustrate this for systems of size 100, 300 and 500. The behavior shown in the figure is typical for a system with a *phase transition*. There is a critical value of n_θ, which I call n_c, at which the number of levels in an idiotypic network rises very sharply. For the system with 500 antibodies a maximum of 29 layers are encountered at n_c. For systems containing 10^7 antibody types, I imagine that at the critical point thousands of levels may be present. Outside the phase transition region the network may be minimal. Simulations with $N_{Ab} = 10^7$ are not feasible

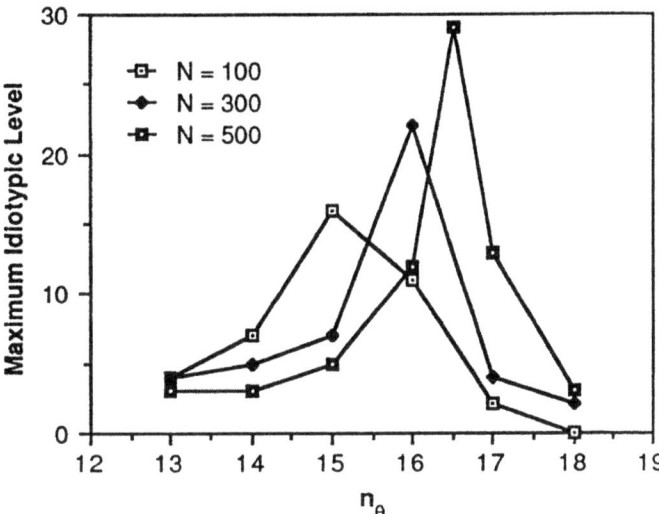

Figure 7. Phase transition in an idiotypic network. Shown are the results of 50 ($N_{Ab} = 100$) or 100 ($N_{Ab} = 300, 500$) Monte Carlo runs. In each run a network graph of the form shown in Fig. 5 was generated, and the maximum level reached recorded. The maximum level attained in all of the runs is plotted for systems with 100, 300, and 500 antibodies. The number of bits in each string, $n = 20$. In the graph for $N_{Ab} = 500$ a set of runs was done with $n = 32$ to mimic a connectance that could be attained with $n_\theta = 16.5$.

and thus it is important to understand how the phase-transition scales with system size.

A simple argument based on percolation theory (cf. Stauffer 1985) on a Bethe lattice can be used to explain the phase transition in idiotypic networks. A Bethe lattice is an infinite lattice in which each node is connected to at most z other nodes; z is called the coordination number of the lattice. The lattice can be drawn in levels (or generations) in the same manner as an idiotypic network (Fig. 8). Because the lattice is infinite some modification of the following argument will be needed to make it rigorous for a finite system but the general idea should remain valid. Let $E(i)$ be the expected number of antibodies on the ith level, and let p be the probability that two antibodies are connected. Then, the expected number of antibodies on level $i + 1$,

$$E(i+1) = p(z-1)E(i).$$

The factor $z - 1$ arises because each antibody on level i, $i \geq 1$, is connected to an antibody on level $i - 1$ and thus can be connected to at most $z - 1$ antibodies on level $i + 1$. Thus, on average, the total number of possible

FOUNDATIONAL PAPERS IN COMPLEXITY SCIENCE

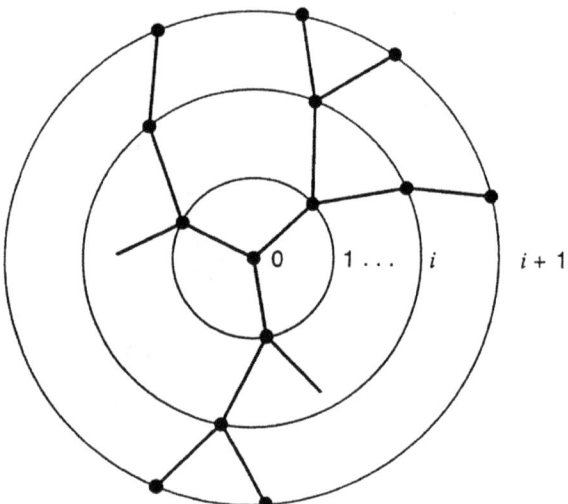

Figure 8. Percolation on a Bethe lattice with coordination number z. Each antibody on the ith level is connected to 1 antibody on level $i-1$ and at most $z-1$ antibodies on level $i+1$. If each such connection occurs with probability p, then the expected number of antibodies in level $i+1$ is equal to the number in level i times $(z-1)p$. If $(z-1)p > 1$ there is a positive probability that the lattice will grow to infinite size. When this occurs a disturbance created by an antigen at level 0 can affect an infinite number of antibodies.

connections from level i to level $i+1$ is $(z-1)E(i)$, and p is the fraction of these that are actually made. When

$$p > p_c = 1/(z-1),$$

$E(i+1) > E(i)$, and the network will grow without bound as i increases. When $p < p_c$, $E(i+1) < E(i)$ and the graph will terminate after a finite number of levels. p_c is the threshold between these two qualitatively different behaviors and is called the critical percolation threshold. Precisely the same argument for a percolation threshold has been used to explain the formation of antibody-antigen precipitates and underlies the quantitative theory of the precipitin curve and the formation of microscopically visible cell surface receptor-ligand aggregates called "patches" (cf. Goldberg 1952; DeLisi 1974; DeLisi and Perelson 1976).

For idiotypic networks each antibody can, in principle, be connected to all others. Hence $z = N_{Ab} - 1$ and

$$p_c \simeq 1/N_{Ab}. \tag{4}$$

Further, from the complementarity rule one can easily compute the probability that a paratope and idiotope are complementary. Because 0 and

PRE-CRITICAL REGION $p < p_c$	TRANSITION REGION $p \simeq p_c$	POST-CRITICAL REGION $p > p_c$
network loosely connected	network connected	network highly connected
many independent small components	few components: one large, many small	fewer components: one large, few small
almost no cycles	few cycles	many cycles
antigen connected via network to few antibodies	antigen connected via network to many antibodies	antigen connected via network to almost all antibodies

Table 1. Phase transitions in idiotypic networks

1 are equally probable,

$$p = \sum_{i=n_\theta}^{n} \binom{n}{i} \left(\frac{1}{2}\right)^i \left(\frac{1}{2}\right)^{n-i} = 2^{-n} \sum_{i=n_\theta}^{n} \binom{n}{i}. \quad (5)$$

Hence each value of n_θ corresponds to a different value of p. Using this correspondence one finds (Perelson 1989) n_c, the critical value of n_θ that corresponds to $p = p_c$ to be approximately 15 for $N_{Ab} = 100$, approximately 16 for $N_{Ab} = 300$, and to be between 16 and 17 for $N_{Ab} = 500$. Fig. 7 shows the validity of these three predictions.

The properties of the immune system on the two sides of the phase transition, and in the transition region, are summarized in Table 1. In the "pre-critical region", where $p < p_c$, i.e. $n_\theta > n_c$, the network is very sparse and composed of many unconnected components. Each antibody, on average, is connected to less than one antibody on the next level since $p(z - 1) < 1$. Including the connection to the previous level, each antibody is connected to less than two antibodies in the network. (De Boer (1989), discusses an alternative criterion for the phase transition derived from the theory of random graphs by Erdös and Rényi and based on the average connectance of each node). In the pre-critical region, no network *per se* exists. Rather, there are many small, discrete, non-interacting subnetworks, or network "components". Each antigen is connected to a small number of antibodies and these in turn are connected to rather few other antibodies. With one antigen one can not excite the entire network, only the component in which the antigen lies. In the "post-critical region" $p > p_c$, i.e. $n_\theta < n_c$, the network is highly connected. Although there still may be more than one non-interacting subnetwork, there is a non-zero probability that all antibodies are part of a single component. In general, the post-critical region is characterized by a single large component and many small ones. This is analogous to a large molecular network, i.e. a gel, being in equilibrium with many small molecular aggregates, i.e. a sol. As p is increased or n_θ decreased the probability of a single global network increases. Concomitantly, the observed connectivity of the network increases. The number of antibodies at each level increases and in a finite system fewer and fewer levels are needed to account for all antibodies. Ultimately, when $p = 1$, all antibodies recognize the antigen and are on level 1.

Why is the existence of this phase-transition interesting? Among immunologists the relevance of idiotypic networks to the functioning of the immune system is controversial (cf. Cohn 1986). Langman and Cohn (1986) have argued that a complete idiotypic network is an absurd immune system. Some immunologists believe that idiotypic networks are an epiphenomenon and of no functional relevance, whereas others believe that idiotypic networks are the core of the immune system, accounting for much of the normal activity of the immune system in times of health and controlling the system in times of disease (cf. Coutinho *et al.* 1984). The phase-transition is a marker. On one side of it the network is so sparse that signals can not propagate through many idiotypic levels. Some idiotypic anti-idiotypic interactions will be present but the topology of the network will prevent a cascade of antibodies against antibodies from occurring. On the other side the network is highly connected and idiotypic interactions can lead to communication among all clones in the immune system. Signals have pathways by which they can propagate very deeply into the idiotypic network and network interactions may dominate any response. Such systems will require control or else a small perturbation could trigger the entire immune system. Antibody affinities and concentrations become important in determining whether or not deep penetration of the network will in fact occur. Based on our knowledge of neural networks one can hypothesize that in the post-critical region the immune system will have

These citations prove that the role of idiotypic networks in regulating the immune system were also heavily debated. Whether or the system functioned as well-connected network or as small disconnected pairs of idiotypic and anti-idiotypic interactions was an important element of this debate.

the potential for very many different modes of activity, a richness that one might expect of a system that is to learn, have memory, and react in different ways to different antigens.

In order to assess the importance of this phase transition one must estimate parameters characterizing the immune system. From the percolation result, $p_c \simeq N_{Ab}^{-1}$, one sees that it is crucial to estimate N_{Ab}, the number of different antibodies in the repertoire, and p, the probability that two randomly chosen antibodies recognize each other with an affinity above that required to activate B cells. The size of the repertoire that is expressed in a given individual is not precisely known. A mouse contains approximately 10^8 small B lymphocytes. Average clone sizes are thought to be between 10 and 100 Jerne (1984), and thus the expressed repertoire, i.e. the repertoire of immunoglobulins carried on the surfaces of B cells, would be between 10^6 and 10^7, values that agree with other estimates (Holmberg et al. 1986). Repertoires of 10^7 or larger are also quoted (cf. Klinman and Press 1975; Sigal and Klinman 1978). Using these values for N_{Ab}, $p_c \leq 10^{-6}$. Since typical values for p are 10^{-5} it seems that, at the level of the expressed repertoire, not only is a phase transition possible but the immune system is in the post-critical regime, i.e. $p > p_c$. In this regime the immune network is highly connected and deep penetration is possible.

If one considers the actual repertoire (Holmberg et al. 1986) represented by serum antibody, the situation changes. Not all B cells in the expressed repertoire are expected to be active. If 10% or fewer of B-cell clones are secreting antibody then N_{Ab} may be of order 10^5. If this is the case then p and p_c may be comparable and it is a delicate matter as to which side of the phase transition the immune system lies.

Topology is only one aspect of immune networks. It places bounds on what is possible. In the pre-critical region the network is very sparse and it is impossible for antigen or any internal perturbation of the network to spread throughout the immune system. In the post-critical region, the immune network is highly connected and it is possible for global excitation to occur. However, whether it does occur or not is now determined by the dynamical interactions between the elements of the immune system. In a dynamical model of idiotypic networks (Segel and Perelson 1988) described below, we have shown that the tuning of a single parameter can cause the behavior of the model to switch from network-like, in which

Novel technologies like next-generation sequencing have markedly increased our estimates for the diversity of T-cell repertoires to about 10^9 different receptors (Qi et al. 2014). Such estimates are more difficult to obtain for B-cell repertoires, partly because they contain mutated sequences generated by somatic recombination, but there is no reason to think that B cell repertoires would be less diverse. However, most of the clones in the T- and B-cell repertoires will be naive. Hence, if only a very small fraction of the B-cell clones is secreting antibody, this estimate of 10^5 "active" clones could still be realistic.

many clones and their anti-idiotypic clones are excited, to a model in which only a single clone that recognizes antigen is excited. For random networks made of elements which can either be "on" or "off" conditions have been derived (cf. Weisbuch 1989) which ensure that the system will have rich dynamical behaviors that remain localized in the network. In the real immune system, T cells, antigen-presenting cells, various cytokines and growth factors must play an important, if not crucial, role in regulating the activity of the immune network. These controls seem to prevent global activation of the network, still leaving the question of whether they allow substantial network activity.

Shape Space Analysis of Immune Networks

Large complex networks such as those generated by matching antibody bit strings can only be studied by computer simulation. Farmer et al. (1986, 1987) and De Boer and Hogeweg (cf. De Boer 1988, 1989; De Boer and Hogeweg, 1989a, 1989b, 1989c, 1989d) have pioneered this approach. While insights can be gained by this approach, one is uncertain if the dynamical equations being simulated are correct (cf. Perelson 1988). Modeling the dynamics of the immune system mathematically seems to be at the same stage as was modeling in neurophysiology before Hodgkin and Huxley; there are many mathematical models but none clearly summarizes a wide body of experimental findings. In fact, much of De Boer's work has been aimed at pointing out that a variety of network models are unrealistic. A further concern in all modeling, but especially when numerical solutions or computer simulations are employed, is whether or not the parameter values chosen are the most relevant ones. Dynamic behavior can change drastically with small adjustments in parameter values (cf. Kevrekidis, Zecha, and Perelson 1988).

Segel and Perelson (1988, 1989a, 1989b, 1989c) have developed a class of simple models that captures the cross-reactivity of immune networks and which are amenable to mathematical analysis. With these models we have tried to emphasize certain principles of stability and design of the immune system and we have begun to explore the ability of the immune system to generate organized patterns of behavior.

The basis of this work is the creation of a one-dimensional shape space. We describe the shape of a molecule by a single continuous variable

x which, for example, one can view as the height of a bump in the combining site. If x is negative then it describes an indentation rather than a bump. Hence we assume that molecules of shape x and $-x$ are exactly complementary. Other molecules y, i.e. with shape y, may also fit x but not as well as when $y = -x$. To describe this we assume the affinity of y with x is a gaussian centered at $y = -x$.

Using this one-dimensional shape space we (Segel and Perelson 1988) formulated a "toy model" embodying the dynamics of interactions in shape space. In this model, the fundamental unknown function is $b(x,t)$, which can be regarded for definiteness as the number (or concentration) of lymphocytes with receptors of shape x that constitute the immune system at time t. If one were to plot $b(x,t)$ versus x at a fixed time, the graph would show the distribution of clones in shape space. If the graph were a horizontal line, i.e. $b(x,t) = \bar{b} =$ constant, then all clones would be present at the same concentration. (Fig. 9). If some clones were present in high concentration then the distribution would have peaks (Fig. 9c). The distribution of clones should evolve in time. When antigen-expressing epitopes of shape x are present the system should respond with a high concentration of complementary clones with shape $-x$. Memory in this system would correspond to the peak at $-x$ remaining for long periods of time. Thus patterns in shape space should reflect both the internal activities of the immune system and its antigenic history. Consequently, an important feature of an immune system model is that it should be able to develop patterns in shape space.

The general problem of how biological systems develop pattern has been most intensively studied in developmental biology beginning with the pioneering work of Turing (1952). From the work of Gierer and Meinhardt (1972), Ermentrout, Campbell, and Oster (1986), Ermentrout and Cowan (1979), Levin and Segel (1982) and others covering applications in fields as diverse as developmental biology, neurobiology and ecology, one finds that a general principle for the formation of pattern is the presence of influences which activate over short distances and which inhibit over long distances (cf. Meinhardt 1982; Oster 1988). This principle is sometimes called "short-range activation and long-range inhibition". To bring this principle to bear, Segel and I developed the following model:

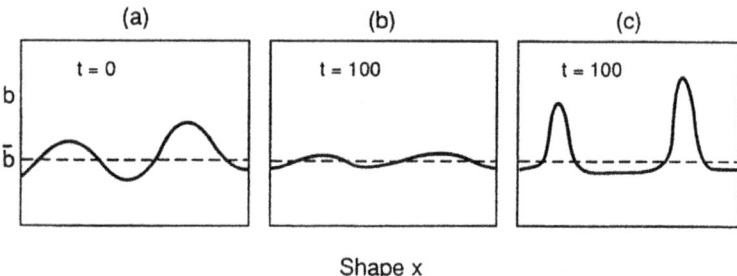

Figure 9. The distribution of clones in shape space. The dashed line indicates a uniform distribution in which each clone is present at the same population level. This is a potential steady state of the system, i.e. $b(x,t) = \bar{b}$. The stability of this uniform state can be examined by perturbing it (solid line). If the perturbed distribution (a) returns toward a uniform distribution (b) then the uniform distribution is said to be stable. In the case of instability, (a) evolves toward a less uniform distribution (c).

The lymphocyte population is divided into two classes, stimulated and unstimulated. The structure of the model can be represented as

$$\begin{array}{c}\text{rate of}\\\text{population}\\\text{increase}\end{array} = \begin{array}{c}\text{influx}\\\text{from bone}\\\text{marrow}\end{array} - \begin{array}{c}\text{death of}\\\text{unstimulated}\\\text{cells}\end{array} + \begin{array}{c}\text{reproduction}\\\text{of stimulated}\\\text{cells.}\end{array}$$

In mathematical form this equation becomes

$$\partial b/\partial t = m - db[1-a] + r(B,b)a. \qquad (6)$$

Here m is the rate of supply of new cells from the bone marrow, d is the death rate of unstimulated cells and $r(B,b)$ is the rate of growth of stimulated cells. This growth rate is assumed to depend upon the total number of lymphocytes in the system, B, as well as the size of the clone with receptors of shape x, i.e. $b(x)$. Perhaps the most important point in the formulation is that the fraction of cells of shape x which are stimulated to grow, a, depends on both activating and suppressing signals. If shapes range between L and $-L$, then the strength of the activation signal

$$A(b) = \begin{array}{l}\text{fraction of } x\text{-cells' activating}\\\text{receptors bound when cell}\\\text{distribution in shape space is}\\\text{given by } b(y,t)\end{array} = \int_{-L}^{L} a(x,y)b(y,t)dy.$$

The function

$$a(x,y) = a_M \left(2\pi\sigma_a^2\right)^{-1/2} e^{-(x+y)^2/2\sigma_a^2}$$

which is a gaussian centered at $y = -x$ can be viewed as the association constant for y binding to activating receptors on x cells in the limit that only a small fraction of the receptors are bound. The constants a_M and σ_a represent the amplitude and the width of the gaussian.

The strength of suppressive signals, $S(b)$, is defined in a similar way using a gaussian kernel $s(x,y)$ of width σ_s representing the affinity of "suppressive receptors". The fraction of cells activated, a, is then a function of A and S that increases (decreases) if more activator (suppressor) receptors are bound. For example,

$$a = \frac{A}{p + qS + A}$$

has these properties when p and q are chosen as positive constants. The principle of short-range activation and long-range inhibition is incorporated into the model by choosing the width of the suppressive gaussian greater than the width of the activation gaussian $\sigma_s > \sigma_a$.

The suppressive receptors are not explicitly defined in the model. On B cells they may be Fc$_\gamma$ receptors (cf. Uher and Dickler 1986), receptors for interferon-γ (cf. Reynolds, Boom, and Abbas 1987), transforming growth factor-β receptors (Kehrl et al. 1986, a), or the immunoglobulin receptor itself, which when cross-linked can give a suppressive signal to B cells in some stages of development (cf. Teale and Klinman 1980, 1984; Dintzis, Middleton, and Dintzis 1983, 1985). On basophils, cross-linking surface IgE can give both activation and desensitization signals (cf. Goldstein 1988).

With this model we were able to show that patterns form if the short-range activation and long-range inhibition principle is followed (Segel and Perelson 1988, 1989a, 1989b). However, unlike other systems in biology we were able to establish that patterns can also form if the inhibition is short-range and the activation is long-range (Segel and Perelson 1989c). Range in shape space is equivalent to specificity. Thus, short-range activation and long-range inhibition translates into specific activation and less specific inhibition. A signal which is non-specific has infinite range in shape space, i.e. it affects all cells. We have not been successful in generating a pattern with infinite range inhibition.

Specificity and range also correlate with the affinity threshold used to define a recognition ball in shape space. Thus in Fig. 1, as the affinity

Interestingly, this shape space model of the idiotypic network is able to form patterns when the ranges of activation and inhibition are sufficiently different (Segel and Perelson 1988). Perelson developed this model with Lee Segel, during Segel's yearly summer visits to the Santa Fe Institute at that time.

threshold is decreased the recognition ball will get larger and interactions can occur over a longer range. The pattern formation rules, short-range activation and long-range inhibition or vice versa, thus translates into having different affinity thresholds for activation and tolerance induction in B cells. This may relate to the difference in affinity thresholds for mature and immature B cells (Riley and Klinman 1986).

In addition to illustrating that patterns can be formed in shape space, one can learn several things from this highly simplified model. I shall discuss only one: the trade-off between good stability properties and good controllability properties. In designing a control system for an aircraft or an automobile one attempts to create a system that remains largely unaffected by small random disturbances, but one which can modify its course in response to a "purposeful" command. A system whose state more or less remains unaffected by small disturbances is called *stable*. If a system is too stable, it will be relatively insensitive to commands. Thus an airplane that remains virtually unbuffeted by even quite strong gusts of wind will also respond very sluggishly to a deflection of its rudder. Because of this, modern fighter planes are designed to be slightly unstable and computers constantly adjust their course.

The immune system is constrained by the same stability–controllability trade-off. Thus, as a general principle, we believe the immune system should be *stable but not too stable*. If this is the case, the immune system can remain insensitive to small random disturbances but yet be responsive to antigenic challenge. The concept of stability is further illustrated in Fig. 9.

Motivated by these considerations, we analyzed the stability of an immune system in a "virgin" state in which all clones have a population level determined by a balance between influx from the bone marrow and unperturbed death. The distribution in shape space was assumed to be uniform (Fig. 9a). The stability of other, non-uniform, distributions can also be analyzed, but this is more difficult mathematically. Because of the stability-controllability trade-off we suggested that the immune system has evolved so that its parameters are in the stable domain, but not too far from the borders of this domain. We then calculated relations between parameters (Segel and Perelson 1988) that corresponded to the boundary between stability and instability and showed that our model immune system was more responsive to antigenic challenge near this border than when far away from it. Further, we were able to show that the system had the capability of responding either as a network, with idiotypic and anti-idiotypic responses, or in a non-network manner with only the clones recognizing the antigen responding, the differences between these classes of responses having to do with the details of the clonal growth law $r(B, b)$. Even in network responses, we found that, because of subtle feedbacks in the system due to the range of the activating and suppressive interactions, clones which were neither complementary to the antigen or Ab_1 antibodies were excited. Thus the system spontaneously established background activity.

Memory in a Network Environment

Theorists have two views of memory. One view is dynamic in which clones are maintained at an elevated level due to stimulation via network interactions or retained antigen. In this view memory cells should be activated cells. The other, more classical, view is one in which memory is static and maintained by resting, long-lived memory cells. The views are not contradictory and memory may be carried by both static and dynamic means. Using the one-dimensional shape space model, Segel and Perelson (1989b) examined how long-lived memory cells would fare in a network environment. In this study memory cells were assumed to be identical to virgin cells except that they live longer, i.e. have a lower death rate d. The question asked was, in the absence of antigen would network interactions allow memory clones to be maintained at elevated levels or would the network try to self-regulate in such a way that memory clones would be reduced in population size so as to approach a common stable background level?

To study this question the parameters in the immune system model were set in a regime where a uniform distribution of clones in shape space would be stable. The death rate of one clone was then lowered in order to mimic its being a memory clone. From Eq. 6 it is easy to see that lowering the death rate of a clone increases its steady state population size. Segel & Perelson (1988) also showed that lowering the death rate tends to shift the parameters characterizing a clone toward a region of instability. They thus hypothesized that the memory clone might maintain an elevated population level by being an "unstable" island in a stable "sea". Numerical solution of Eq. 6 showed that the memory clone persisted when its parameters were set in the unstable range, as was expected; but, surprisingly, the memory clone also persisted with its parameters set somewhat in the stable range. Further, the existence of the memory clone had an effect on the network and a set of clones centered around its anti-idiotypic partner had population levels that were elevated (depressed) compared with background when activation was shorter (longer) range than inhibition. This analysis does not argue for or against static or dynamic memory. It only shows that the classical notion of a memory cell is compatible with a simple network model of the immune system. It remains to be seen whether these conclusions remain valid with more realistic network models (cf. Perelson 1988; Segel and Perelson, 1989a).

Network Models with Both Chemical and Cellular Components

Idiotypic network models that have been formulated mathematically generally have not distinguished between cells and molecules; models by Hiernaux and Bona (1979), Varela *et al.* (1988) and De Boer and Hogeweg (1989c) being exceptions. Since idiotypic determinants are found on both cells and antibodies this was not considered an important issue for early models. This clearly needs to change and we need to come to grips with the different properties of antibodies, B cells and T cells. Idiotypic interactions can either stimulate or suppress. In the model of Jerne's given by Eq. 1 idiotypes were broken down into stimulatory and inhibitory subsets. Although both types of interactions may be necessary to control an idiotypic network they need not be built in to the properties of specific cell types. For example, it is well known that if cell triggering is assumed to be proportional to the fraction of cell surface receptors that are cross-linked, then little or no stimulation will occur in antigen excess (cf. Perelson and DeLisi 1980). Also various feedback signals that may be important in regulation can also be incorporated into idiotypic networks once chemistry is allowed. Here I would like to illustrate some of these points with what I consider to be one of the simplest idiotypic networks with chemistry. This model

Perelson (1989)

is derived from a one-dimensional shape space model that incorporates antibody—receptor binding but no cross-linking (Segel and Perelson, 1989a). Possible inhibitory effects mediated by antibody-dependent cellular cytotoxicity or the binding of immunoglobulin complexes to Fc receptors are ignored here but can easily be included in later models.

Consider a system composed of two B cell populations, B_1 and B_2 (see Fig. 10). Let B_1 recognize antigen G and assume that B_2 has receptors that are complementary to those on B_1. For example the receptors on B_2 may have an idiotope that is an internal image of the antigen. At a minimum, the following events seem important to model: Antigen binds to receptors on B_1, cross-links them and possibly leads to proliferation and the production of antibody, A_1. Antibody A_1 binds antigen and leads to its elimination. It also binds receptors on B_2, cross-links them and possibly leads to proliferation and the production of antibody A_2. Antibody A_2 can bind A_1 in solution to form immune complexes that are eliminated, or A_2 can bind receptors on B_1 and either stimulate or inhibit the cell depending on the level of cross-linking induced. Note that the formation of A_1–A_2 immune complexes eliminates A_1, the signal stimulating B_2. Thus the response of B_2 can turn off the signal that is stimulating it if immune complex formation dominates the chemistry or it can lead to the production of more A_1 and possibly greater stimulation through receptor cross-linking and stimulation of B_1. Predicting the outcome of such events is difficult and a mathematical model can help determine the various possibilities.

The basic chemical reactions underlying the model are the binding of A_1 to free receptor sites (S_2) on B_2:

$$A_1 + S_2 B_2 \underset{k_-}{\overset{k_+}{\rightleftarrows}} A_1^s B_2. \tag{7a}$$

The concentration of free immunoglobulin receptor sites per B_2 cell is denoted S_2. Because the concentration of B cells can change in the model their concentration is indicated in the reaction. This reaction leads to the formation of an antibody-receptor complex in which A_1 is bound by one Fab. We call this species a singly bound antibody and denote its concentration per cell A_{1s}. A singly bound antibody still has one free Fab which can bind to another free receptor site (S_2) on B_2 to

Anti-idiotypic antibodies, being complementary to an antibody that is complementary to antigen, were also called an "internal image" of the antigen. A recent paper on SARS-CoV-2 speculated about the pathological role that anti-idiotypic antibodies, mimicking (part of) the spike protein of the virus, may play by also binding angiotensin-converting-enzyme 2 (ACE2) (Murphy and Longo 2022).

It is very elegant that the interaction function of nodes in the idiotypic network was derived mechanistically by Perelson's own pioneering work on the theory of bivalent ligands forming crosslinks of the receptors on the surface of the B cells. Since the concentration of crosslinks is a bell-shaped function of the ligand concentration, this delivered a nonlinear interaction function, allowing for activation at intermediate concentrations and inhibition at high concentrations of ligand.

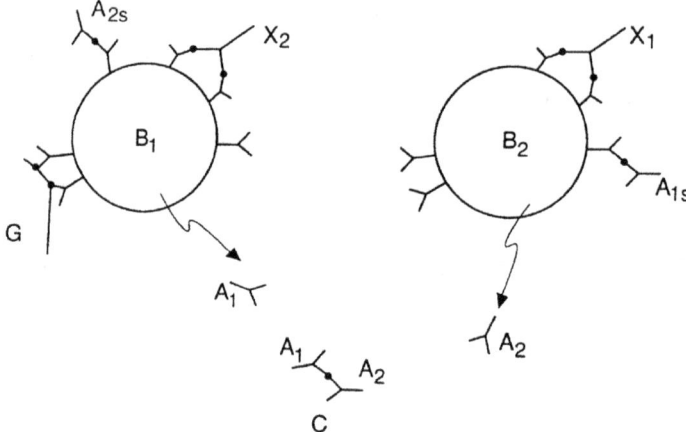

Figure 10. A simple idiotypic network with chemistry. The network contains two populations of B cells, B_1 and B_2. The receptors on B_1 are complementary to the antigen G. The receptors on B_2 are complementary to those on B_1. When stimulated, via receptor cross-linking, B_1 and B_2 proliferate and secrete bivalent antibodies A_1 and A_2, which react in solution to form a complex C, or which bind to complementary cell surface immunoglobulin. Antibody 1 bound to surface immunoglobulin by one Fab arm is called singly bound (A_{1s}), whereas it is called a cross-link (X_1) when bound by both Fab arms to two different cell surface immunoglobulin molecules.

form a receptor cross-link (X_1) on the surface of B_2:

$$A_{1s}B_2 + S_2B_2 \underset{k_{x-}}{\overset{k_{x+}}{\rightleftarrows}} X_1 B_2. \tag{7b}$$

A similar set of reactions occurs between A_2 and receptors on B_1, i.e.

$$A_2 + S_1 B_1 \underset{k_-}{\overset{k_+}{\rightleftarrows}} A_{2s} B_1, \tag{7c}$$

$$A_{2s}B_1 + S_1B_1 \underset{k_{x-}}{\overset{k_{x+}}{\rightleftarrows}} X_2 B_1. \tag{7d}$$

In solution, the formation of antibody-antibody complexes (C) occurs:

$$A_1 + A_2 \underset{k_{c-}}{\overset{k_{c+}}{\rightleftarrows}} C. \tag{7e}$$

For simplicity, consider A_2 to be the antigen which stimulates B_1. Also, assume that the rate constants k_{x+} and k_{x-} characterizing the reaction of A_1 with S_2 are the same as those characterizing the reaction of A_2 with S_1. A similar symmetry condition is applied to the cross-linking reactions 7b and 7d.

The total number of immunoglobulin receptors per cell is assumed to be constant. In later models this restriction can be relaxed and receptor synthesis and internalization modeled. Since receptors are bivalent, the total number of receptor sites per cell S_0 obeys the following conservation law (Perelson and DeLisi 1980):

$$S_0 = S_i + A_{is} + 2X_i, i = 1, 2. \tag{8}$$

These chemical reactions can now be incorporated into a model for the growth and decay of the B-cell populations:

$$\frac{dB_1}{dt} = m + r(X_2, B_1, s_1) B_1 - \mu_b B_1, \tag{9a}$$

$$\frac{dB_2}{dt} = m + r(X_1, B_2, s_2) B_2 - \mu_b B_2, \tag{9b}$$

where m, is the rate of generation of new B cells in the bone marrow and μ_b is their specific death rate. The stimulation and subsequent growth of B cells is modeled by r, the specific growth rate function. I assume that r depends on the degree of receptor cross-linking, X, the B-cell density, and the rate of antibody secretion s. Receptor cross-linking is included to model the degree of cell stimulation. If the B cells are growing in a lymphoid organ then space, nutrients and growth factors may limit the size of the clone. The density dependence models these effects. Lastly, as B cells differentiate into plasma cells their antibody secretion rate increases but their rate of growth slows or even stops. We model this by making the B-cell growth rate a decreasing function of the antibody secretion rate. A specific choice for r is:

$$r(X_2, B_1) = \left(\frac{r_0 X_2}{f_x R + X_2}\right) e^{-\lambda B_1} e^{-\eta s}, \tag{10}$$

where R is the total number of immunoglobulin receptors on a B cell and f_x is a parameter controlling the fraction of receptors that need to be cross-linked to give a stimulatory signal. When X is equal to $f_x R$ the term in parentheses is half the maximal growth rate r_0. The constant λ controls the density dependence so that growth slows only when $\lambda B_1 \gg 1$. Likewise, the constant η controls the decrease in growth with increases in the rate of antibody production s_i.

The rate of antibody secretion s is assumed to be determined by the degree of B-cell activation. As a simple model I use the fraction of receptor sites cross-linked as an indicator of the state of stimulation. Thus for cell i,

$$\frac{ds_i}{dt} = k_s q_A \left(X_i/S_0\right) - k_s s_i, s_i(0) = 0, i = 1, 2. \qquad (11)$$

This equation has the property that the secretion rate starts at zero and ultimately reaches a steady value on a time scale determined by k_s. The function q_A describes the dependence of the steady state secretion rate on the extent of receptor cross-linking. In the example given below (Fig. 11), $q_A\left(X_i/S_0\right) = qX_i/S_0$ was chosen for simplicity, where the constant q is the maximal antibody secretion rate. However, a function which rises monotonically and then saturates would also be a reasonable choice for q_A. By using a differential equation to determine s, we account for the delays involved in lymphocyte activation and the "gearing up" for large amounts of antibody secretion that occurs during B-cell differentiation into a plasma cell.

The concentration of antibody in solution is determined by a balance between secretion, degradation and binding to cell surface receptors and anti-idiotypic antibodies. Thus

$$\begin{aligned}\frac{dA_1}{dt} = s_1 B_1 - \mu_A A_1 - 2k_+ A_1 S_2 B_2 \\ + k_- A_{2s} B_2 - 4k_c + A_1 A_2 - k_{c-} C,\end{aligned} \qquad (12a)$$

where μ_A is the rate of antibody loss through catabolism. The factors 2 and 4 are statistical factors introduced because the rate constants are defined per site. An antibody in solution has two free sites and thus there are two ways in which it can combine with a free receptor site. Similarly, there are four ways in which a free site on A_1 can combine with a free site on A_2. The factor B_2 enters because S_2 and A_{1s} are concentrations of free and singly bound receptors per cell and antibody can bind receptors on any cell.

The concentrations of singly bound antibody and cross-links change according to

$$\frac{dA_{1s}}{dt} = 2k_+ A_1 S_2 - k_- A_{1s} - k_{x+} A_{1s} S_2 + 2k_{x-} X_1, \qquad (12b)$$

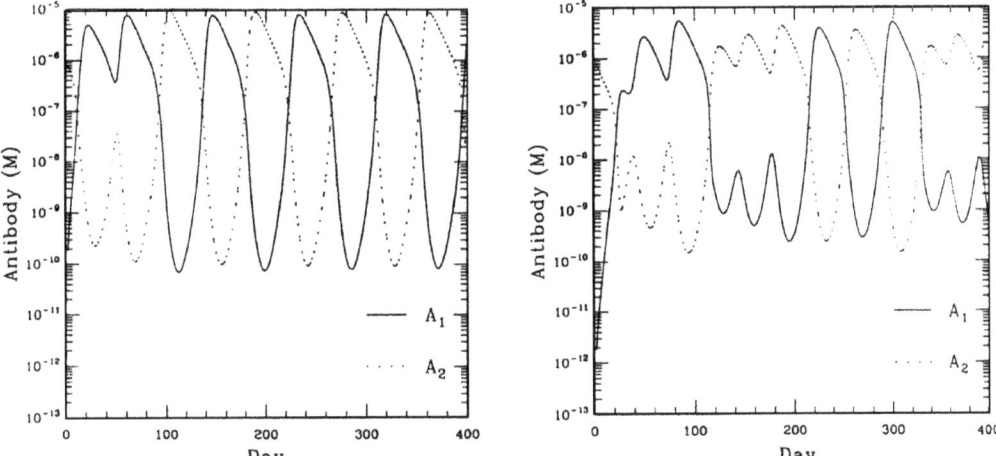

Figure 11. Dynamics of the two B-cell network. (a) The system oscillates with the A_1 and A_2 populations having an inverse relationship. The period is approximately 80 d. The parameters used in the numerical solution correspond to receptor affinities of $10^7 M^{-1}$. The antibody-antibody association constant $k_{c+}/k_{c-} = 10^7 M^{-1}$. The number of receptors per cell, $R = S_0/2$, was taken as 10^5 or 1.67×10^{-10} cm^{-2}. The cross-linking equilibrium constant $k_{x+}/k_{x-} = 10^{10}$, so that when multiplied by R it was of order one. The decay rates $\mu_b = 0.3$ d^{-1}, $\mu_A = 0.1$ d^{-1}, $\mu_c = 2$. d^{-1}, correspond to 3 d B cell, 10 d antibody and 12 h complex lifetimes. An average of 5 d was chosen for "gearing up" to full antibody production, i.e. plasma cell formation. Thus, $k_s = 0.2$ d^{-1}. The maximum antibody secretion rate $q = 10^4$ molecules per cell per second, or 1.44×10^{-15} mol d^{-1}. The maximal B cell growth rate $r_0 = 1.0$ d^{-1}, and $f_x = 0.1$ so that 10% of receptors need to be cross-linked for half-maximal stimulation. The parameters $\lambda = 0$ and $\eta = 0.5$. The unit of volume was chosen as 1 cm^3, and the cell influx rate $m = 2.4 \times 10^4$ cells d^{-1}. The initial conditions were chosen such that B_1 and B_2 were in their virgin state, i.e. $B_1 = B_2 = m/\mu_b$, $A_1 = 0$, and the "antigen" $A_2 = 10^{-6}$ M. All other variables were initially zero. (b) Parameters same as (a) except the influx rate $m = 2.4 \times 10^2$.

$$\frac{dX_1}{dt} = k_{x+} A_{1s} S_2 - 2 k_{x-} X_1. \tag{12c}$$

The factor 2 multiplying k_{x-} is also a statistical factor—either of the bound Fab arms of a cross-linked antibody can dissociate. Precisely the same equations hold for A_2, A_{2s} and X_2 if the subscripts 1 and 2 are interchanged in Eqs. (12a)-(12c). Lastly, the concentration of antibody complexes is governed by

$$\frac{dC}{dt} = 4 k_{c+} A_1 A_2 - k_{c-} C - \mu_c C, \tag{13}$$

where μ_c is the rate of elimination of antibody complexes.

Introducing chemistry has clearly complicated matters. This "simple" system of an idiotype and its anti-idiotype is described by a system of 11 ordinary differential equations for the variables: B_1, B_2, s_1, s_2, A_1, A_2, A_{1s}, A_{2s}, X_1, X_2, and C. The conservation equations (8) can be used to eliminate two of the equations. Also, because the chemical reactions typically take place on a time scale of seconds to minutes, whereas cell growth and death and elimination of antibody and immune complexes typically occurs over a period of hours and days, one can further simplify the model by assuming the chemical reactions are at equilibrium on the time scale relevant to cellular changes. This can be done formally using the "method of multiple scales" (cf. Lin and Segel 1974) as shown in Segel and Perelson (1989a), and results in a mixed system of ordinary differential and nonlinear algebraic equations. This procedure is not the same as the quasi-steady-state assumption used in enzyme kinetics, and differential equations for the total amount of antibody and complex in the system remain.

Analyzing the dynamics of this system for different parameter values shows that the system can exhibit quite complex behavior: approach virgin or immune steady states, undergo sustained or damped oscillations, and exhibit aperiodic (i.e. chaotic) behavior. Figure 11a shows that under quite reasonable parameter choices the system can undergo sustained oscillations with the concentrations of idiotypic and anti-idiotypic antibodies fluctuating inversely with a period of approximately 80 days. These oscillations are reminiscent of those found by Rodkey and Adler (1983) in a year-long study of naturally induced auto-anti-idiotypic antibodies in the rabbit, in which one could also identify peaks separated by 80 d. The oscillations found by Rodkey & Adler were much more irregular than those in Fig. 11a. Choosing slightly different parameters gives rise to irregular oscillations with peaks in A_1 at d 50, 90, 220 and 300 (Fig. 11b). Figs 2 and 4A of Rodkey and Adler (1983) show similar peaks at similar times (d 120 and 260, and 200 and 280, respectively). Inverse idiotypic and anti-idiotypic fluctuations have also been found by Kelso and Cerny (1979) and Bona (1982).

Summary

Theoretical ideas have played a profound role in the development of idiotypic network theory. Mathematical models can help in the precise translation of speculative ideas into quantitative predictions. They can also help establish general principles and frameworks for

Idiotypic network theory attracted both immunologists and theoreticians. They were brought together during many collaborative workshops, and they slowly learned to speak each others' language and to view the immune system as a complex adaptive system rather than as a large collection of individual clones independently responding to antigens. Although interest in the idiotypic network has waned, nowadays we realize that the immune system functions as a large complex network of interacting cell types and cytokines that is heterogeneously distributed over many different tissues.

thinking. Using the idea of shape space, criteria were introduced for evaluating the completeness and overlap in the antibody repertoire. Thinking about the distribution of clones in shape space naturally leads to considerations of stability and controllability. An immune system which is too stable will be sluggish and unresponsive to antigenic challenge; one which is unstable will be driven into immense activity by internal fluctuations. This led us to postulate that the immune system should be stable but not too stable.

In many biological contexts the development of pattern requires both activation and inhibition but on different spatial scales. Similar ideas can be applied to shape space. The principle of short-range activation and long-range inhibition translates into specific activation and less specific inhibition. Application of this principle in model immune systems can lead to the stable maintenance of non-uniform distributions of clones in shape space. Thus clones which are useful and recognize antigen or internal images of antigen can be maintained at high population levels whereas less useful clones can be maintained at lower population levels. Pattern in shape space is a minimal requirement for a model. Learning and memory correspond to the development and maintenance of particular patterns in shape space.

Representing antibodies by binary strings allows one to develop models in which the binary string acts as a tag for a specific molecule or clone. Thus models with huge numbers of cells and molecules can be developed and analyzed using computers. Using parallel computers or finite state models it should soon be feasible to study model immune systems with 10^5 or more elements. Although idiotypic networks were the focus of this paper, these modeling strategies are general and apply equally well to non-idiotypic models.

Using bit string or geometric models of antibody combining sites, the affinity of interaction between any two molecules, and hence the connections in a model idiotypic network, can be determined. This approach leads to the prediction of a phase transition in the structure of idiotypic networks. On one side of the transition networks are small localized structures much as might be predicted by clonal selection and circuit ideas. On the other side of the transition profound idiotypic networks become possible, signals can percolate throughout

The representation of receptors and ligands as strings of bits or characters, for which complementary matching rules are defined, has frequently been adopted in modeling various processes in the immune system, ranging from selection and MHC restriction (Detours, Mehr, and Perelson 2000; Wortel et al. 2020), to affinity maturation (Macken and Perelson 1989), to T cell–mediated immune responses to viruses (Chao et al. 2004).

the network, and antigen stimulation can affect large portions of the network. Whether signals actually have such profound effects or just stimulate local portions of the network depends crucially on antibody affinities and concentrations, molecules such as lymphokines which play roles in controlling the proliferation and differentiation of lymphocytes, and the dynamical laws governing the interactions among the cells and molecules of the immune system. These dynamical laws still need to be uncovered and provide a challenge to both experimentalists and theorists.

Acknowledgments

Portions of this work were performed under the auspices of the U. S. Department of Energy. I thank the Santa Fe Institute for providing continuing support for Theoretical Immunology and the US-Israel Binational Science Foundation (Grant 86-00107) for travel support. My collaborators, Lee Segel, J. Doyne Farmer, Norman Packard and George Oster played major roles in the development of the ideas and models presented here. I also thank Rob De Boer for his assistance with the development of the model and the computer simulations underlying Fig. 11.

REFERENCES

Bona, C., S. Finley, S. Waters, and H. G. Kunkel. 1982. "Anti-Immunoglobulin Antibodies. III. Properties of Sequential Anti-Idiotypic Antibodies to Heterologous Anti-Gamma Globulins. Detection of Reactivity of Anti-Idiotype Antibodies with Epitopes of Fc Fragments (Homobodies) and with Epitopes and Idiotopes (Epibodies)." *Journal of Experimental Medicine* 156 (4): 986–99.

Bona, C. A. 1982. "Inverse Fluctuations of Idiotypes and Anti-Idiotypes during the Immune Response." In *Regulation of Immune Response Dynamics,* edited by C. DeLisi and J. R. J. Hiernaux, I:75. Boca Raton, FL: CRC Press.

Bost, K. L., E. M. Smith, and J. E. Blalock. 1985b. "Regions of Complementarity Between the Messenger RNAs for Epidermal Growth Factor, Transferrin, Interleukin- 2 and Their Respective Receptors." *Biochemical and Biophysical Research Communications* 128 (3): 1373–80.

———. 1985a. "Similarity Between the Corticotropin (ACTH) Receptor and a Peptide Encoded by an RNA that is Complementary to ACTH mRNA." *Proceedings of the National Academy of Sciences* 82:1372.

Cohn, M. 1986. "The Concept of a Functional Idiotypic Network for Immune Regulation Mocks All and Comforts None." *Annales De Linstitut Pasteur Immunologie* 137C (1): 64–76.

Coutinho, A. 1980. "The Self–Nonself Discrimination and the Nature and Acquisition of the Antibody Repertoire." *Annales De Linstitut Pasteur Immunologie* 131D (3): 235–53.

Coutinho, A., L. Forni, D. Holmberg, F. Ivars, and N Vaz. 1984. "From an Antigen Centered Clonal Perspective of Immune Responses to an Organism Centered Network Perspective of Autonomous Activity in a Self-Referential Immune System." *Immunological Reviews* 79.

De Boer, R. J. 1988. "Symmetric Idiotypic Networks: Connectance and Switching, Stability and Suppression." In *Theoretical Immunology, Part Two*, edited by A. S. Perelson, 265. Redwood City, CA: Addison-Wesley.

———. 1989. "Extensive Percolation in Reasonable Idiotypic Networks." In *Theories of Immune Networks*. Berlin, Germany: Springer-Verlag.

De Boer, R. J., and P. Hogeweg. 1989d. "Idiotypic Networks Incorporating T–B Cell Cooperation. The Conditions for Percolation." *Journal of Theoretical Biology* 139 (1): 17–38.

———. 1989b. "Memory but No Suppression in Low-Dimensional Symmetric Idiotypic Networks." *The Bulletin of Mathematical Biology* 51:223.

———. 1989a. "Stability of Symmetric Idiotypic Networks: A Critique of Hoffmann's Analysis." *The Bulletin of Mathematical Biology* 51 (2): 217–22.

———. 1989c. "Unreasonable Implications of Reasonable Idiotypic Network Assumptions." *The Bulletin of Mathematical Biology* 51:381.

DeLisi, C. 1974. "A Theory of Precipitation and Agglutination Reactions in Immunological Systems." *Journal of Theoretical Biology* 45 (2): 555–575.

DeLisi, C., and A. S. Perelson. 1976. "The Kinetics of Aggregation Phenomena. I. Minimal Models for Patch Formation on Lymphocyte Membranes." *Journal of Theoretical Biology* 62:159.

Dintzis, R. Z., M. H. Middleton, and H. M. Dintzis. 1983. "Studies on the Immunogenicity and Tolerogenicity of T-independent Antigens." *Journal of Immunology* 131 (5): 2196–2203.

———. 1985. "Inhibition of Anti-DNP Antibody Formation by High Doses of DNP-Polyacrylamide Molecules; Effects of Hapten Density and Hapten Valence." *Journal of Immunology* 135:423.

Du Pasquier, L. 1973. "Ontogeny of the Immune Response in Cold Blooded Vertebrates." *Current Topics in Microbiology and Immunology* 61:37.

Du Pasquier, L., and J. Haimovitch. 1976. "The Antibody Response During Amphibian Ontogeny." *Immunogenetics* 3:381.

Ermentrout, B., J. Campbell, and G. Oster. 1986. "A Model for Shell Patterns Based on Neural Activity." *The Veliger* 28:369.

Ermentrout, G. B., and J. D. Cowan. 1979. "A Mathematical Theory of Visual Hallucination Patterns." *Biological Cybernetics* 34:137.

Farmer, J. D., S. A. Kauffman, N. H. Packard, and A. S. Perelson. 1987. "Adaptive Dynamic Networks as Models for the Immune System and Autocatalytic Sets." *Annals of the New York Academy of Sciences* 504.

Farmer, J. D., N. H. Packard, and A. S. Perelson. 1986. "The Immune System, Adaptation, and Machine Learning." *Physica D* 22:187.

Forni, L., and A. Coutinho. 1981. "Individuality of Immune Systems: The Thousand Ways and One Way of Being Complete." In *The Immune System, Vol. 1,* edited by C. M. Steinberg and I. Lefkovits. Basel, Switzerland: Karger.

Gierer, A., and H. Meinhardt. 1972. "A Theory of Biological Pattern Formation." *Kybernetik* 12:30.

Goldberg, R. J. 1952. "A Theory of Antibody-Antigen Reactions. I. Theory for Reactions of Multivalent Antigen with Bivalent and Univalent Antibody." *Journal of the American Chemical Society* 74:5715.

Goldstein, B. 1988. "Desensitization, Histamine Release and the Desensitization of IgE on Human Basophils." In *Theoretical Immunology, Part One,* edited by A. S. Perelson, 3. Redwood City, CA: Addison-Wesley.

Gunther, N., and G. W. Hoffmann. 1982. "Qualitative Dynamics of a Network Model of Regulation of the Immune System: A Rationale for the IgM to IgG Switch." *Journal of Theoretical Biology* 94:815.

Hiernaux, J. 1977. "Some Remarks on the Stability of the Idiotype Network." *Immunochemistry* 14:733–739.

Hiernaux, J., and C. Bona. 1979. "Network Regulatory Mechanisms of the Immune Response." In *Regulation of Function of Lymphocytes by Antibody,* edited by C. Bona and P. A. Cazenave. New York, NY: John Wiley and Sons.

Hoffmann, G. W. 1975. "A Theory of Regulation and Self–Nonself Discrimination in an Immune Network." *European Journal of Immunology* 5:638.

———. 1979. "A Mathematical Model of the Stable States of a Network Theory of Self-Regulation." In *Lecture Notes in Biomathematics,* vol. 32. Berlin, Germany: Springer.

———. 1980. "On Network Theory and H-2 Restriction." *Contemporary Topics in Molecular Immunology* 11.

———. 1982. "The Application of Stability Criteria in Evaluating Network Regulation Models." In *Regulation of Immune Response Dynamics, Vol. I,* edited by C. DeLisi and J. R. J. Hiernaux. Boca Raton, FL: CRC Press.

Hoffmann, G. W., T. A. Kion, R. B. Forsyth, K. G. Soga, and A. Cooper-Willis. 1988. "The N-dimensional Network." In *Theoretical Immunology, Part Two,* edited by A. S. Perelson. Redwood, CA: Addison-Wesley.

Holmberg, D., A. A. Freitas, D. Portnoi, F. Jacquemart, S. Avrameas, and A. Coutinho. 1986. "Antibody Repertoires of Normal BALB/c mice: B Lymphocytes Populations Defined by State of Activation." *Immunological Reviews* 93:147.

Jerne, N. K. 1984. "Idiotypic Networks and Other Preconceived Ideas." *Immunological Reviews* 79:5.

———. 1974b. "Clonal Selection in a Lymphocyte Network." Edited by G. M. Edelman. (New York, NY).

———. 1974a. "Towards a Network Theory of the Immune System." *Annales d'immunologie* 125C:373–389.

Jerne, N. K., J. Roland, and P.-A. Cazenave. 1982. "Recurrent Idiotopes and Internal Images." *The EMBO Journal* 1 (2): 243–7.

Kang, C.-Y., and H. Kohler. 1986. "A Novel Chimeric Antibody with Circular Network Characteristics: Autobody." *Annals of the New York Academy of Sciences* 475:114.

Kehrl, J. H., A. B. Roberts, L. M. Wakefield, S. Jakowlew, M. B. Sporn, D. B. Burlington, H. C. Lane, and A. S. Fauci. 1986. "Transforming Growth Factor Beta is an Important Immunomodulatory Protein for Human B Lymphocytes." *Journal of Immunology* 137:3855.

Kelso, G., and J. Cerny. 1979. "Reciprocal Expressions of Idiotypic and Antiidiotypic Clones Following Antigen Stimulation." *Nature* 279 (333).

Kevrekidis, I. G., A. D. Zecha, and A. S. Perelson. 1988. "Modeling Dynamical Aspects of the Immune Response. I. T Cell Proliferation and the Effect of IL-2." In *Theoretical Immunology, Part One,* edited by A. S. Perelson, 167. Redwood City, CA: Addison-Wesley.

Klinman, N. R., and J. L. Press. 1975. "The B Cell Specificity Repertoire: Its Relationship to Definable Subpopulations." *Transplantation Reviews* 24:41–83.

Langman, R., and M. Cohn. 1986. "The "Complete" Idiotype Network is an Absurd Immune System." *Immunology Today* 7.

Levin, S. A., and L. A. Segel. 1982. "Models of the Influence of Predation on Aspect Diversity in Prey Population." *Journal of Mathematical Biology* 14:253–284.

Lin, C. C., and L. A. Segel. 1974. *Mathematics Applied to Deterministic Problems in the Natural Sciences.* New York, NY: Macmillan.

Meinhardt, H. 1982. *Models of Biological Pattern Formation.* New York, NY: Academic Press.

Oster, G. F. 1988. "Lateral Inhibition Models of Developmental Processes." *Mathematical Biosciences* 90 (1): 265–286.

Parisi, G. 1990. "A Simple Model for the Immune Network." *Proceedings of the National Academy of Sciences* 87 (1): 429–433.

Perelson, A. S. 1988. "Toward a Realistic Model of the Immune System." In *Theoretical Immunology, Part Two,* edited by A. S. Perelson, 377. Redwood, City, CA: Addison-Wesley.

———. 1989. "Immune Networks: a Topological View." In *Cell to Cell Signalling: From Experiments to Theoretical Models,* edited by A. Goldbeter. New York, NY: Academic Press.

Perelson, A. S., and C. DeLisi. 1980. "Receptor Clustering on a Cell Surface. I. Theory of Receptor Cross-Linking by Ligands Bearing Two Chemically Identical Functional Groups." *Mathematical Biosciences* 48:71–100.

Perelson, A. S., and G. F. Oster. 1979. "Theoretical Studies of Clonal Selection: Minimal Antibody Repertoire Size and Reliability of Self- non-Self Discrimination." *Journal of Theoretical Biology* 81:645.

Reynolds, D. S., W. H. Boom, and A. A. Abbas. 1987. "Inhibition of B Lymphocyte Activation by Interferon-Gamma." *Journal of Immunology* 139:767.

Richter, P. H. 1975. "A Network Theory of the Immune System." *European Journal of Immunology* 5:350.

Richter, P. H. 1978. "The Network Idea and the Immune Response." In *Theoretical Immunology,* edited by G. I. Bell, A. S. Perelson, and G. H. Jr. Pimbley. New York, NY: 539.

Riley, R. L., and N. R. Klinman. 1986. "The Affinity Threshold for Antigenic Triggering Differs for Tolerance Susceptible Immature Precursors vs Mature Primary B Cells." *Journal of Immunology* 136:3147.

Rodkey, L. S., and F. L. Adler. 1983. "Regulation of Natural Antiallotype Antibody Responses in Idiotype Network-Induced Auto-Antiidiotypic Antibodies." *Journal of Experimental Medicine* 157:1920.

Segel, L. A., and A. S. Perelson. 1988. "Computations in Shape Space. A New Approach to Immune Network Theory." In *Theoretical Immunology, Part Two,* edited by A. S. Perelson, 321. Redwood City, CA: Addison-Wesley.

———. 1989c. "A Paradoxical Instability Caused by Relatively Short-Range Inhibition." *SIAM Journal on Applied Mathematics* 50 (1): 91–107.

———. 1989a. "Shape Space Analysis of Immune Networks." In *Cell to Cell Signalling: From Experiments to Theoretical Models,* edited by A. Goldbeter. New York, NY: Academic Press.

———. 1989b. "Some Reflections on Memory in Shape Space." In *Theories of Immune Networks,* edited by H. Atlan and I. R. Cohen. Berlin, Germany: Springer-Verlag.

Shai, Y., M. Flashner, and I. M. Chaiken. 1987. "Anti-sense Peptide Recognition of Sense Peptides: Direct Quantitative Characterization with the Ribonuclease S-peptide System Using Analytical High-Performance Affinity Chromatography." *Biochemistry* 26:669.

Sigal, N. H., and N. R. Klinman. 1978. "The B Cell Clonotype Repertoire." *Advances in Immunology* 26:255.

Smith, L. R., K. L. Bost, and J. E. Blalock. 1987. "Generation of Idiotypic and Anti-idiotypic Antibodies by Immunization With Peptides Encoded by Complementary RNA: A Possible Molecular Basis jor the Network Theory." (Journal of Immunology) 138 (7).

Stadnyk, I. 1987. "Schema Recombination in Pattern Recognition Problems." In *Genetic Algorithms and Their Applications,* edited by J. J. Grefenstette. Hillsdale, NJ: Lawrence Erlbaum Associates.

Stauffer, D. 1985. *Introduction to Percolation Theory.* Philadelphia, PA: Taylor and Francis.

Teale, J. M., and N. R. Klinman. 1980. "Tolerance as an Active Process." *Nature* 288:385.

———. 1984. "Membrane and Metabolic Requirements for Tolerance Induction of Neonatal B Cells." *Journal of Immunology* 133 (4): 1811–7.

Turing, A. M. 1952. "The Chemical Basis of Morphogenesis." *Philosophical Transactions of the Royal Society B* 237:37.

Uher, F., and H. B. Dickler. 1986. "Cooperativity Between B Lymphocyte Membrane Molecules: Independent Ligand Occupancy and Cross-Linking of Antigen Receptors and Fc_y Receptors Down-regulates B lymphocyte Function." *Journal of Immunology* 137:3124.

Varela, F. J., A. Coutinho, B. Dupire, and N. N. Vaz. 1988. "Cognitive Networks: Immune, Neural, and Otherwise." In *Theoretical Immunology, Part II,* edited by A. S. Perelson. Redwood City, CA: Santa Fe Institute/Addison-Wesley.

Weisbuch, G. 1989. "Dynamical Behavior of Discrete Models of Jerne's Network." In *Theories of Immune Networks.* Berlin, Germany: Springer-Verlag.

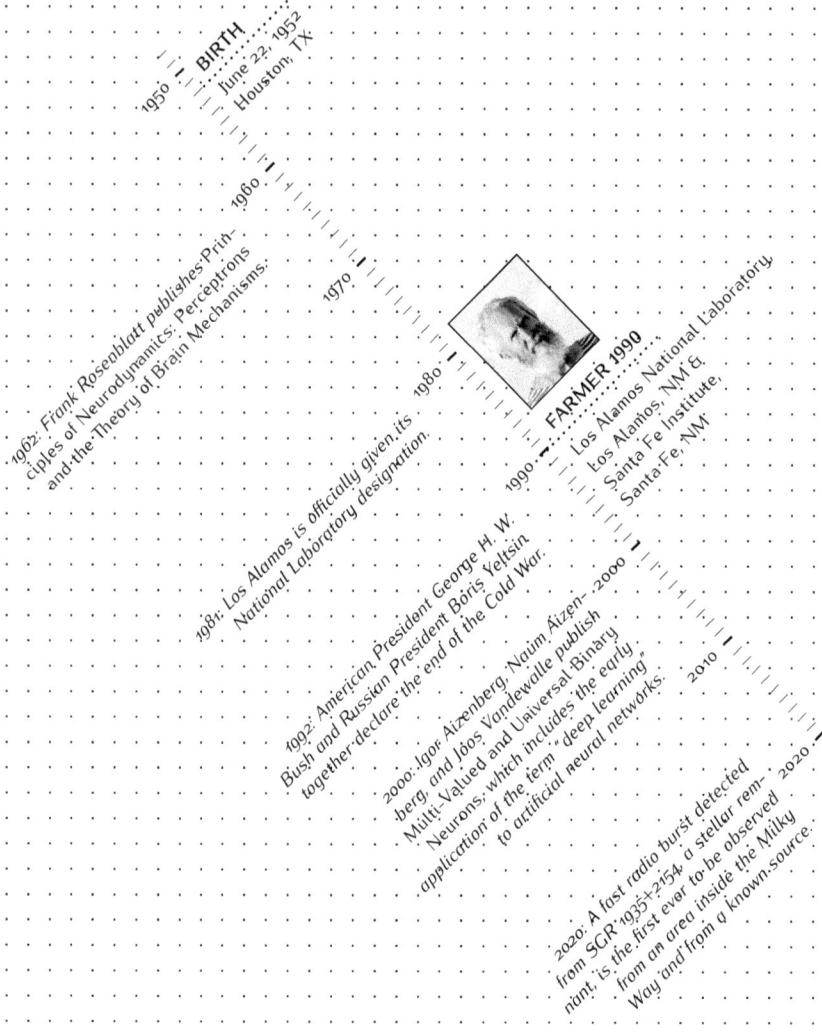

- 1950
- **BIRTH** — June 22, 1952, Houston, TX
- 1960
- 1962: Frank Rosenblatt publishes Principles of Neurodynamics: Perceptrons and the Theory of Brain Mechanisms.
- 1970
- 1980
- 1981: Los Alamos is officially given its National Laboratory designation.
- **FARMER 1990** — Los Alamos National Laboratory, Los Alamos, NM & Santa Fe Institute, Santa Fe, NM
- 1990
- 1992: American President George H. W. Bush and Russian President Boris Yeltsin together declare the end of the Cold War.
- 2000
- 2000: Igor Aizenberg, Naum Aizenberg, and Joos Vandewalle publish Multi-Valued and Universal Binary Neurons, which includes the early application of the term "deep learning" to artificial neural networks.
- 2010
- 2020
- 2020: A fast radio burst detected from SGR 1935+2154, a stellar remnant, is the first ever to be observed from an area inside the Milky Way and from a known source.

JAMES DOYNE FARMER

[69]

CONNECTING CONNECTIONIST MODELS

Stephanie Forrest, Arizona State University; Santa Fe Institute

How do systems like the brain, the immune system, ecosystems, or financial markets learn and adapt over time? From the early days of computing, scientists and engineers alike have sought to abstract the key mechanisms of adaptation, learning, and intelligence and encode them in a computer. Norbert Weiner, an early cyberneticist, hypothesized that feedback mechanisms were the key ingredient of intelligence (Wiener 1948), Ross Ashby's general theory of adaptive systems emphasized system states and the role of modeling in intelligence (Ashby 1956), and Donald O. Hebb's influential book *Organization of Behavior* emphasized synaptic learning and cell assemblies (Hebb 1949). Meanwhile, engineers developed computational realizations of different aspects of adaptive systems: neural networks (McCulloch and Pitts 1943), genetic algorithms and classifier systems (Holland 1962, 1975; Holland *et al.* 1989), reinforcement learning and minimax algorithms (Samuel 1959; Kaelbling, Littman, and Moore 1996; Russell and Norvig 2003), and symbolic artificial intelligence models (Newell and Simon 1956). For understandable reasons, these efforts focused primarily on intelligence in the brain and on human thought processes. However, over time each developed its own formalisms, terminology, and research communities specific to the type of system being studied. As computing matured, other researchers developed computational models in fields far afield from the brain, including chemical reaction networks and immunology.

J. Doyne Farmer's Rosetta Stone paper develops a common mathematical framework that highlights both similarities and differences among these models, focusing on those that emphasize networks, or connectionism, as it was known then. He uses the language of dynamical systems as his Rosetta Stone for connectionism. The paper is important because it broadens our view of adaptive systems beyond brain-centric models, a bias that remains

J. D. Farmer, "A Rosetta Stone for Connectionism," *Physica D* 42 (1–3), 153–87 (1990).

Reprinted with permission of Elsevier.

today, and it returns to the original enterprise of the early cyberneticists by providing a unified mathematical treatment of adaptive processes. Although he focuses on just four exemplar connectionist models, the paper highlights other examples that fall into the framework: Bayesian inference networks, Boolean networks, ecological models and population genetics, economics, game theory, and other models of molecular evolution.

On the face of it, dynamical systems theory is not an obvious choice for a unifying framework. Most dynamical systems models are expressed as differential equations, do not focus on learning or adaptation, and define well-mixed systems rather than systems that are structured by a network of interactions. Computational approaches, such as neural networks or reinforcement learning, on the other hand tend to be discrete, and learning is central, as is the network of connections. Bringing a physicist's lens to the problem, Farmer overcomes these challenges and recasts four representative models in a single dynamical systems framework.

He begins by defining a class of models that has two properties:

(1) Interactions between components are constrained by a graph of connections, and

(2) the connections are dynamic, both in terms of activity across connections and (optionally) in terms of the graph topology.

In standard dynamical systems, system states change quickly, system parameters change slowly (e.g., between experiments), and the interaction graph between component types is fixed. Connectionism relaxes these assumptions, requiring Farmer to define three kinds of dynamics, each with their own characteristic time scale: node dynamics that change the state of the system (referred to as the transition rule), parameter dynamics (referred to as the learning rule), and graph dynamics that change the topology of the graph. These are formalized as coupled dynamical systems living on a graph, with the possibility that the graph (or architecture of the system) can change over time. Depending on the system in question, the dynamics are defined in terms of discrete maps (transition rules), function evaluation (e.g., a learning rule for updating weights in a neural network), ordinary differential equations, etc.

Having established a general connectionist framework, the paper describes four quite different connectionist systems and places them in the framework, which is summarized in a table near the end of the paper. It is worth a word or two about his choice of systems. Neural networks were probably the earliest connectionist model, originally described by Warren McCulloch and Walter Pitts (1943). They have stood the test of time and form the core of most of today's machine learning systems, with remarkably few changes over the years to the basic idea of a connected graph with weights on the edges and a learning rule to adjust the weights over time. Learning classifier systems are not as well known but provide an important early example of reinforcement learning, inspired in part by Arthur Samuel's famous checker-playing program (Samuel 1959). In reinforcement learning, an agent seeks to maximize its reward from the environment based on actions that it takes in the environment (Kaelbling, Littman, and Moore 1996), balancing exploration of new actions against exploitation of actions that were rewarded previously. Classifier systems contain two learning mechanisms—discovery of new rules and credit assignment—which work together to provide an executable theory of induction processes in cognition. Today, other reinforcement learning methods, often based on dynamic programming, are more popular for real-world decision and control problems, but the learning classifier system certainly foreshadowed these later approaches, and was the first system to marry the rule-based paradigm of symbolic AI systems with statistical learning models like neural networks.

Immune network theory, as proposed by Niels Kaj Jerne (1973, 1974), suggested that the immune system consists of a network of interacting cells and molecules—lymphocytes (B cells and T cells), antibodies, and foreign antigen. The network was postulated as a solution to the problem of auto-reactive antibodies, the theory being that new lymphocytes would be generated to control auto-reactive antibodies, and these in turn would be controlled by other lymphocytes, leading to a complex regulatory network of interactions between lymphocytes, antigens, and antibodies. Farmer, Norman Packard, and Alan Perelson (1986) modeled immune networks, focusing on B cells, treating immune interactions as a connectionist network, and using random strings and string matching to model chemical reactions between the various components. It is this version of the immune network theory that Farmer places in the Rosetta Stone framework. Finally, Farmer considers the autocatalytic chemical network, an origin-of-life model intended to demonstrate how a "soup" of monomers could spontaneously evolve a polymer metabolism with autocatalytic properties.

More than thirty years on, what does the Rosetta Stone paper offer us today? First, it reminds us that adaptation and learning are ubiquitous processes that occur across a wide range of systems, yet operate in similar ways following basic principles. With today's fascination with brain-inspired machine learning, the paper also provides an important reminder of the many other ways beyond the brain that systems learn in nature. A recent work, "Liquid Brains, Solid Brains," picks up this theme and argues that cognitive networks come in many forms, some with well-defined static architectures (solid) like the brain and others with dynamic (liquid) architectures, like ant colonies or immune systems (Solé, Moses, and Forrest 2019).

By focusing on connectionist systems, Farmer's paper paved the way for what became network science, which also defines a common framework based on graphs and dynamics and applies to a multitude of systems. Network science, however, has succeeded in finding common properties among disparate networks such as community structure, degree distributions, and preferential attachment models. Although the Rosetta Stone paper succeeds in finding a common language for expressing connectionist models, it is less clear about what the language tells us about the different systems or what the machinery of dynamical systems theory reveals about the common principles of adaptation and learning.

As for the four connectionist systems Farmer chose to analyze in detail, immune network theory has fallen out of favor over the years, due to the lack of convincing experimental evidence to support it; Holland's learning classifier system has been supplanted by other reinforcement learning systems, many based on Markovian assumptions about the environment and instantiated with neural networks; neural networks, as we mentioned earlier, are still going strong; and since the Rosetta Stone paper was written, a number of autocatalytic experimental systems have been demonstrated (Sievers and Von Kiedrowski 1994; Kim and Joyce 2004; Ashkenasy *et al.* 2004; Vaidya *et al.* 2012), including one that is based on inorganic molecules (Haralampos *et al.* 2020). Yet, we have not seen the spontaneous emergence of these systems in the absence of explicit design by an experimenter, as would be required for solving the origin of life (Richert 2018).

Although immune network theory as originally conceived may not be so relevant today, other immune network models have had remarkable success, such as Derek Smith's influential vaccination model based on Kannerva's connectionist model (Smith 1999; Smith *et al.* 1999). Concerns over auto-antibodies and their role in disease remain relevant today (Jernbom *et al.* 2024), and scientists have yet to untangle the overall logic of the complex cytokine signalling system, which is highly complex and amenable to the connectionist framework.

There are too few serious efforts to find commonalities across disparate fields in science and engineering, but at its core this is what a science of complexity should aim to achieve. One recent effort in the field of modern artificial intelligence and reinforcement learning, where literally tens of thousands of papers are published each year, is the work of Dimitri Bertsekas (2019, 2022). In this work he distills reinforcement learning to its two core mathematical ideas, from which he develops a unifying framework for control theory (model predictive control and adaptive control), multiagent systems and decentralized control, and discrete and Bayesian optimization. This ambitious effort, using a different mathematical language from Farmer's dynamical systems approach, provides hope that Farmer's original enterprise of unifying the treatment of connectionist models is achievable, even with today's vast array of learning and optimization approaches.

REFERENCES

Ashby, W. R. 1956. *An Introduction to Cybernetics*. London, UK: Chapman and Hall Ltd.

Ashkenasy, G., R. Jagasia, M. Yadav, and M. R. Ghadiri. 2004. "Design of a Directed Molecular Network." *Proceedings of the National Academy of Sciences* 101 (30): 10872–7. https://doi.org/10.1073/pnas.0402674101.

Bertsekas, D. P. 2019. *Reinforcement Learning and Optimal Control*. Belmont, MA: Athena Scientific.

———. 2022. *Lessons from AlphaZero for Optimal, Model Predictive, and Adaptive Control*. Belmont, MA: Athena Scientific.

Dam, H. H., H. A. Abbass, C. Lokan, and X. Yao. 2008. "Neural-Based Learning Classifier Systems." *IEEE Transactions on Knowledge and Data Engineering* 20 (1): 26–39. https://doi.org/10.1109/TKDE.2007.190671.

Dick, J. M., and E. L. Shock. 2021. "The Release of Energy During Protein Synthesis at Ultramafic-Hosted Submarine Hydrothermal Ecosystems." *Journal of Geophysical Research: Biogeosciences* 126 (11): e2021JG006436. https://doi.org/10.1029/2021JG006436.

Farmer, J. D., N. H. Packard, and A. S. Perelson. 1986. "The Immune System, Adaptation, and Machine Learning." *Physica D: Nonlinear Phenomena* 22 (1–3): 187–204. https://doi.org/10.1016/0167-2789(86)90240-X.

Haralampos, N. M., C. Mathis, W. Xuan, D.-L. Long, R. Pow, and L. Cronin. 2020. "Spontaneous Formation of Autocatalytic Sets with Self-Replicating Inorganic Metal Oxide Clusters." *Proceedings of the National Academy of Sciences* 117 (20): 10699–705. https://doi.org/10.1073/pnas.1921536117.

Hebb, D. O. 1949. "The Organization of Behaviour." (New York, NY).

Hochreiter, S., and J. Schmidhuber. 1997. "Long Short-Term Memory." *Neural Computation* 9 (8): 1735–1780. https://doi.org/10.1162/neco.1997.9.8.173.

Holland, J. H. 1962. "Outline for a Logical Theory of Adaptive Systems." *Journal of the ACM* 9 (3): 297–314. https://doi.org/10.1145/321127.321128.

———. 1975. "Adaptation in Natural and Artificial Systems: An Introductory Analysis with Applications to Biology, Control, and Artiificial Intelligence." (Ann Arbor, MI).

Holland, J. H., K. J. Holyoak, R. E. Nisbett, and P. R. Thagard. 1989. *Induction: Processes of Inference, Learning, and Discovery.* Cambridge, MA: MIT Press.

Hordijk, W., L. Hasenclever, J. Gao, D. Mincheva, and J. Hein. 2014. "An Investigation into Irreducible Autocatalytic Sets and Power Law Distributed Catalysis." *Natural Computing* 13:287–296. https://doi.org/10.1007/s11047-014-9429-6.

Jernbom, A. F., L. Skoglund, E. Pin, R. Sjöberg, H. Tegel, S. Hober, E. Rostami, *et al*. 2024. "Prevalent and Persistent New-Onset Autoantibodies in Mild to Severe COVID-19." *Nature Communications* 15:8941. https://doi.org/10.1038/s41467-024-53356-5.

Jerne, N. K. 1973. "The Immune System." *Scientific American* 229 (1): 52–63. https://doi.org/10.1038/scientificamerican0773-52.

———. 1974. "Towards a Network Theory of the Immune System." *Annales D'Immunologie* 125:373–389.

Kaelbling, L. P., M. L. Littman, and A. W. Moore. 1996. "Reinforcement Learning: A Survey." *Journal of Artificial Intelligence Research* 4 (1): 237–285. https://doi.org/10.5555/1622737.1622748.

Kim, D.-E., and G. F. Joyce. 2004. "Cross-Catalytic Replication of an RNA Ligase Ribozyme." *Chemistry & Biology* 11 (11): 1505–12. https://doi.org/10.1016/j.chembiol.2004.08.021.

Kim, J.-Y., and S.-B. Cho. 2019. "Exploiting Deep Convolutional Neural Networks for a Neural-Based Learning Classifier System." *Neurocomputing* 354:61–70. https://doi.org/10.1016/j.neucom.2018.05.137.

Kitchin, J. R. 2018. "Machine Learning in Catalysis." *Nature Catalysis* 1 (4): 230–232. https://doi.org/10.1038/s41929-018-0056-y.

Matziner, P. 1994. "Tolerance, Danger, and the Extended Family." *Annual Review of Immunology* 12:991–1045. https://doi.org/10.1146/annurev.iy.12.040194.005015.

McCulloch, W. S., and W. Pitts. 1943. "A Logical Calculus of the Ideas Immanent in Nervous Activity." *The Bulletin of Mathematical Biophysics* 5 (4): 115–33. https://doi.org/10.1007/BF02478259.

Miikkulainen, R., J. Liang, E. Meyerson, A. Rawal, D. Fink, O. Francon, B. Raju, *et al.* 2019. "Evolving Deep Neural Networks." In *Artificial Intelligence in the Age of Neural Networks and Brain Computing*, edited by R. Kozma, C. Alippi, Y. Choe, and F. C. Morabito, 267–287. Amsterdam, Netherlands: Elsevier. https://doi.org/10.1016/B978-0-323-96104-2.00002-6.

Newell, A., and H. Simon. 1956. "The Logic Theory Machine: A Complex Information Processing System." *IRE Transactions on Information Theory* 2 (3): 61–79. https://doi.org/10.1109/TIT.1956.1056797.

Piovesan, A., F. Antonaros, L. Vitale, P. Strippoli, M. C. Pelleri, and M. Caracausi. 2019. "Human Protein-Coding Genes and Gene Feature Statistics in 2019." *BMC Research Notes* 12 (1): 315. https://doi.org/10.1186/s13104-019-4343-8.

Richert, C. 2018. "Prebiotic Chemistry and Human Intervention." *Nature Communications* 9 (5177). https://doi.org/10.1038/s41467-018-07219-5.

Russell, S., and P. Norvig. 2003. *Artificial Intelligence: A Modern Approach.* https://aima.cs.berkeley.edu/.

Samuel, A. L. 1959. "Some Studies in Machine Learning Using the Game of Checkers." *IBM Journal of Research and Development* 3 (3): 210–29. https://doi.org/10.1147/rd.33.0210.

Segel, L. A., and R. L. Bar-Or. 1999. "On the Role of Feedback in Promoting Conflicting Goals of the Adaptive Immune System." *Journal of Immunology* 163 (3): 1342–1349. https://doi.org/10.4049/jimmunol.163.3.1342.

Segel, L. A., and I. R. Cohen. 2001. *Design Principles for the Immune System and Other Distributed Autonomous System.* New York, NY: Oxford University Press.

Sievers, D., and G. Von Kiedrowski. 1994. "Self-replication of Complementary Nucleotide-based Oligomers." *Nature* 369 (6477): 221–4. https://doi.org/10.1038/369221a0.

Smith, D. J. 1999. *Immunological Memory is Associative,* edited by D. Dasgupta, 105–14. Berlin, Germany: Springer. https://doi.org/10.1007/978-3-642-59901-9_6.

Smith, D. J., S. Forrest, D. H. Ackley, and A. S. Perelson. 1999. *Variable Efficacy of Repeated Annual in Infuenza Vaccination.* 96:14001–6. 24. https://doi.org/10.1073/pnas.96.24.14001.

Solé, R., M. Moses, and S. Forrest. 2019. "Liquid Brains, Solid Brains." *Philosophical Transactions of the Royal Society B* 374:20190040. https://doi.org/10.1098/rstb.2019.0040.

Sompayrac, L. M. 2022. *How the Immune System Works.* Hoboken, NJ: John Wiley & Sons.

Stanley, K., and R. Miikkulainen. 2002. "Evolution of Neural Networks through Augmenting Topologies." *Evolutionary Computation* 10 (2): 99–127. https://doi.org/10.1162/106365602320169811.

Sutton, A.G., and R.S. Barto. 1990. "Time-Derivative Models of Pavlovian Reinforcement." In *Learning and Computational Neuroscience: Foundations of Adaptive Networks,* edited by M. Gabriel and J. Moore, 497–537. Cambridge, MA: MIT Press.

Vaidya, N., M. L. Manapat, I. A. Chen, R. Xulvi-Brunet, E. J. Hayden, and N. Lehman. 2012. *Spontaneous Network Formation Among Cooperative RNA Replicators.* 491:72–77. 7422. https://doi.org/10.1038/nature11549.

Vasas, V., C. Fernando, M. Santos, S. Kauffman, and E. Szathmáry. 2012. "Evolution before Genes." *Biology Direct* 7:1–14. https://doi.org/10.1186/1745-6150-7-1.

Wiedemann, G. M., E. K. Santosa, S. Grassmann, S. Sheppard, J.-B. Le Luduec, N. M. Adams, C. Dang, K. C. Hsu, J. C. Sun, and C. M. Lau. 2021. "Deconvoluting Global Cytokine Signaling Networks in Natural Killer Cells." *Nature Immunology* 22:627–638. https://doi.org/10.1038/s41590-021-00909-1.

Wiener, N. 1948. *Cybernetics: or Control and Communication in the Animal and the Machine.* New York, NY: John Wiley & Sons.

Xavier, J. C., W. Hordijk, S. Kauffman, M. Steel, and W. F. Martin. 2020. "Autocatalytic Chemical Networks at the Origin of Metabolism." *Proceedings of the Royal Society B* 287 (1922): 20192377. https://doi.org/10.1098/rspb.2019.2377.

A ROSETTA STONE FOR CONNECTIONISM

J. Doyne Farmer, Los Alamos National Laboratory; Santa Fe Institute

Abstract

The term connectionism is usually applied to neural networks. There are, however, many other models that are mathematically similar, including classifier systems, immune networks, autocatalytic chemical reaction networks, and others. In view of this similarity, it is appropriate to broaden the term connectionism. I define a connectionist model as a dynamical system with two properties: (1) The interactions between the variables at any given time are explicitly constrained to a finite list of connections. (2) The connections are fluid, in that their strength and/or pattern of connectivity can change with time.

This paper reviews the four examples listed above and maps them into a common mathematical framework, discussing their similarities and differences. It also suggests new applications of connectionist models, and poses some problems to be addressed in an eventual theory of connectionist systems.

1. Introduction

This paper has several purposes. The first is to identify a common language across several fields in order to make their similarities and differences clearer. A central goal is that practitioners in neural nets, classifier systems, immune nets, and autocatalytic nets will be able to make correspondences between work in their own field as compared to the others, more easily importing mathematical results across disciplinary boundaries. This paper attempts to provide a coherent statement of what connectionist models are and how they differ in mathematical structure and philosophy from conventional "fixed" dynamical system models. I hope that it provides a first step toward clarifying some of the mathematical issues needed for a generally applicable theory of connectionist models. Hopefully this will also provide a natural framework for connectionist models in other areas, such as ecology, economics, and game theory.

FOUNDATIONAL PAPERS IN COMPLEXITY SCIENCE

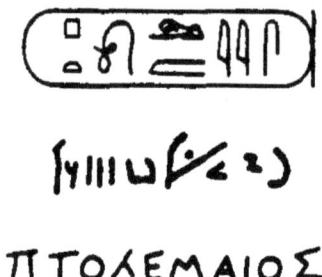

Figure 1. "Ptolemy", in hieroglyphics, Demotic, and Greek. This cartouche played a seminal role in deciphering hieroglyphics, by providing a hint that the alphabet was partially phonetic (Wallis Budge 1929). (The small box is a "p", and the half circle is a "t"—literally it reads "ptolmis".)

1.1. BREAKING THE JARGON BARRIER

Language is the medium of cultural evolution. To a large extent differences in language define culture groupings. Someone who speaks Romany, for example, is very likely a Gypsy; the existence of a common and unique language is one of the most important bonds preserving Gypsy culture. At times, however, communication between subcultures becomes essential, so that we must map one language to another.

The language of science is particularly specialized. It is also particularly fluid; words are tools onto which we map ideas, and which we invent or redefine as necessary. Our jargon evolves as science changes. Although jargon is a necessary feature of communication in science, it can also pose a barrier impeding scientific progress.

When models are based on a given class of phenomena, such as neurobiology or ecology, the terminology used in the models tends to reflect the phenomenon being modeled rather than the underlying mathematical structure. This easily obscures similarities in the mathematical structure. "Neural activation" may appear quite different from "species population", even though relative to given mathematical models the two may be identical. Differences in jargon place barriers to communication that prevent results in one field from being transparent to workers in another field. Proper nomenclature should identify similar things but distinguish those that are genuinely different.

At present this problem is particularly acute for adaptive systems. The class of mathematical models that are employed to understand adaptive systems contain subtle but nonetheless significant new features that are not easily categorized by conventional mathematical terminology. This adds to the problem of communication between disciplines, since there are no standard mathematical terms to identify the features of the models.

1.2. WHAT IS CONNECTIONISM?

Connectionism is a term that is currently applied to neural network models such as those described in (Rummelhart and McClelland 1986; Cowan and Sharp 1988). The models consist of elementary units, which can be "connected" together to form a network. The form of the resulting connection diagram is often called the *architecture* of the network. The computations performed by the network are highly dependent on the architecture. Each connection carries information in its weight, which specifies how strongly the two variables it connects interact with each other. Since the modeler has control over how the connections are made, the architecture is plastic.

Since the advent of deep learning and related systems, which consist of multiple neural networks interconnected in various ways, the term *architecture* refers to how the component networks are organized to form a single system—for example, how many layers of networks are there and what flavor of neural network is in each layer?

This contrasts with the usual approach in dynamics and bifurcation theory, where the dynamical system is a fixed object whose variability is concentrated into a few parameters. The plasticity of the connections and connection strengths means that we must think about the entire family of dynamical systems described by all possible architectures and all possible combinations of weights. Dynamics occurs on as many as three levels, that of the states of the network, the values of connection strengths, and the architecture of the connections themselves.

Mathematical models with this basic structure are by no means unique to neural networks. They occur in several other areas, including classifier systems, immune networks, and autocatalytic networks. They also have potential applications in other areas, such as economics, game-theoretic models and ecological models. I propose that the term connectionism be extended to this wider class of models.

This insight is also the foundation of what became known as *network science*, although the formalizations that evolved in that field are distinct from those proposed here.

By comparing connectionist models for different phenomena using a common nomenclature, we get a clear view of the extent to which these models are similar or different. We also get a *glimpse* of the extent to which the underlying phenomena are similar or different. I emphasize the word glimpse to make it clear that we are simplifying a complicated phenomenon when we model it in connectionist terms. Comparing two connectionist models of, for example, the nervous systems and the immune system, provides a means of extracting certain aspects of their similarities, but we must be very careful in doing this; much richness and complexity is lost at this level of description.

This tension is at the heart of all models—does the simplification provided by the model reveal important insights about the modeled system, which are not evident otherwise?

Connectionism represents a particular level of abstraction. By reducing the state of a neuron to a single number, we are collapsing its properties relative to a real neuron, or relative to those of another potentially more comprehensive mathematical formalism. For example, consider fluid dynamics. At one level of description the state of a fluid is a function whose evolution is governed by a partial differential equation. At another level we can model the fluid as a finite collection of spatial modes whose interactions are described by a set of ordinary differential equations. The partial differential equation is not a connectionist model; there are no identifiable elements to connect together; a function simply evolves in time. The ordinary differential equations are *more* connectionist; the nature of the solution depends critically on the particular set of modes, their connections, and their coupling parameters. In fluid dynamics we can sometimes calculate the correct couplings from first principles, in which case the model is just a fixed set of ordinary differential equations. In contrast, for a connectionist model there are dynamics for the couplings and/or connections. In a fully connectionist model, the connections and couplings would be allowed to change, to find the best possible model with a given degree of complexity.

Another alternative is to model the fluid on a grid with a finite difference scheme or a cellular automaton. In this case each element is "connected" to its neighbors, so there might be some justification for calling these connectionist models. However, the connections are fixed, completely regular, and have no dynamics. I will not consider them as "connectionist".

Just as there are limits to what can be described by a finite number of distinct modes, there are also limits to what can be achieved by connectionist models. For more detailed descriptions of many adaptive phenomena we may need models with explicit spatial structure, such as partial differential equations or cellular automata. Nonetheless, connectionism is a useful level of abstraction, which solves some problems efficiently.

The Rosetta Stone is a fragment of rock in which the same text is inscribed in several different languages and alphabets (fig. 1). It provides a key that greatly facilitated the decoding of these languages, but it is by no means a complete description of them. My goal is similar; by presenting several connectionist models side by side, I hope to make it clear how some aspects of the underlying phenomena compare with one another, but I offer the warning that quite a bit has been omitted in the process.

1.3. ORGANIZATION OF THIS PAPER

In section 2, I describe the basic mathematical framework that is common to connectionist models. I then discuss four different connectionist models: neural networks, classifier systems, immune networks, and autocatalytic networks. In each case I begin with a background discussion, make a correspondence to the generic framework described in section 2, and then discuss general issues. Finally, the conclusion contains the "Rosetta Stone" in table 3, which maps the jargon of each area into a common nomenclature. I also make a few suggestions for applications of connectionist models and comment on what I learned in writing this paper.

Connectionist models are ultimately dynamical systems. Readers who are not familiar with terms such as automaton, map, or lattice model may wish to refer to the appendix.

2. The General Mathematical Framework of Connectionist Models

In this section I present the mathematical framework of a "generic" connectionist model. I make some arbitrary choices about nomenclature, in order to provide a standard language, noting common synonyms whenever appropriate.

To first approximation a connectionist model is a pair of coupled dynamical systems living on a graph. In some cases the graph itself may also have dynamics. The remainder of this section explains this in more detail.

2.1. THE GRAPH

The foundation of any connectionist model is a graph consisting of *nodes* (or vertices) and *connections* (also called links or edges) between them as shown in fig. 2. The graph describes the architecture of the system and provides the channels in which the dynamics takes place. There are

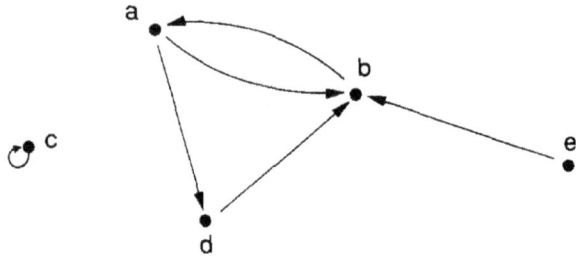

Figure 2. A directed graph.

different types of graphs; for example, the links can be either directed (with arrows), or undirected (without arrows). For some purposes, such as modeling catalysis, it is necessary to allow complicated graphs with more than one type of node or more than one type of link.

For many purposes it is important to specify the pattern of connections, with a *graph representation*. The simplest way to represent a graph is to draw a picture of it, but for many purposes a more formal description is necessary. One common graph representation is a *connection matrix*. The nodes are assigned an arbitrary order, corresponding to the rows or columns of a matrix. The row corresponding to each node contains a nonzero entry, such as "1", in the columns corresponding to the nodes to which it makes connections. For example, if we order the nodes of fig. 2 lexicographically, the connection matrix is

$$C = \begin{bmatrix} 0 & 1 & 0 & 1 & 0 \\ 1 & 0 & 0 & 0 & 0 \\ 0 & 0 & 1 & 0 & 0 \\ 0 & 1 & 0 & 0 & 0 \\ 0 & 1 & 0 & 0 & 0 \end{bmatrix} \qquad (1)$$

If the graph is undirected then the connection matrix is symmetric. It is sometimes economical to combine the representation of the graph and the connection parameters associated with it into a matrix of connection parameters.

A *connection list* is an alternative graph representation. For example, the graph of fig. 2 can also be represented as

$$a \to b, \quad a \to d, \quad b \to a, \quad c \to c, \quad d \to b, \quad e \to b. \qquad (2)$$

Note that the nodes are implicitly contained in the connection list. In some cases, if there are isolated nodes, it may be necessary to provide an additional list of nodes that do not appear on the connection list. For the connectionist models discussed here isolated nodes, if any, can be ignored.

A graph can also be represented by an algorithm. A simple example is a program that creates connections "at random" using a deterministic random number generator. The program, together with the initial speed, forms a representation of a graph.

For a *dense* graph almost every node is connected to almost every other node. For a *sparse* graph most nodes are connected to only a small fraction of the other nodes. A connection matrix is a more efficient representation for a dense graph, but a connection list is a more efficient representation for a sparse graph.

2.2. DYNAMICS

In conventional dynamical models the form of the dynamical system is fixed. The only part of the dynamical system that changes is the state, which contains all the information we need to know about the system to determine its future behavior. The possible ways the "fixed" dynamical form "might change" are encapsulated as parameters. These are usually thought of as fixed in any given experiment, but varying from experiment to experiment. Alternatively we can think of the parameters as knobs that can be slowly changed in the background. In reality the quantities that we incorporate as parameters are usually aspects of the system that change on a time scale slower than those we are modeling with the dynamical system.

Connectionist models extend this view by giving the parameters an explicit dynamics of their own, and in some cases, by giving the list of variables and their connections a dynamics of its own. Typically this also involves a separation of time scales. Although a separation of time scales is not necessary, it provides a good starting point for the discussion. The fast scale dynamics, which changes the *states* of the system, is usually associated with short-term information processing. This is the *transition rule*. The intermediate scale dynamics changes the *parameters*, and is usually associated with learning. I will call this the *parameter dynamics* or the *learning rule*. On the longest time scale, the graph itself may change. I will call this the *graph dynamics*. The graph dynamics may also be used for learning; hopefully this will not lead to confusion.

Of course, strictly speaking the states, parameters, and graph representation described above are just the states of a larger dynamical system with multiple time scales. Reserving the word state for the shortest time scale is just a convenience. The association of time scales given above is the natural generalization of "conventional" dynamical systems, in which the states change quickly, the parameters change slowly, and the graph is fixed. For some purposes, however, it might prove to be useful to relax this separation, for example, letting the graph change at a rate comparable to that of the states. Although all the models discussed here have at most three time scales, in principle this framework could be iterated to higher levels to incorporate an arbitrary number of time scales.

The information that resides on the graph typically consists of integers, real numbers, or vectors, but could in principle be any mathematical objects. The state transition and learning rules can potentially be any type of dynamical system. For systems with continuous states and continuous parameters the natural dynamics are ordinary differential equations or discrete time maps. In principle, the states or parameters could also be functions whose dynamics are partial differential equations or functional maps. This might be natural, for example, in a more realistic model of neurons where the spatio-temporal form of pulse propagation in the axon is important (Scott 1977). When the activities or parameters are integers, their dynamics are naturally automata, although it is also common to use continuous dynamics even when the underlying states are discrete.

Since the representation of the graph is intrinsically discrete, the graph dynamics usually has a different character. Often, as in classifier systems, immune networks, or autocatalytic networks, the graph dynamics contains random elements. In other cases, it may be a deterministic response to statistical properties of the node states or the connection strengths, for example, as in pruning algorithms. Dynamical systems with graph dynamics are sometimes called *metadynamical systems* (Farmer, Kauffman, and Packard 1986; Bagley et al. 1990).

In all of the models discussed here the states of the system reside on the nodes of the graph[1]. The states are denoted x_i, where i is an integer labeling the node. The parameters reside at either nodes or connections; θ_i

---

[1] It is also possible that states could be attached to connections, but this is not the case in any of the models discussed here.

refers to a *node parameter* residing at node i, and w_{ij} refers to a *connection parameter* residing at the connection between node i and node j.

The degree to which the activity at one node influences the activity at another node, or the *connection strength*, is an important property of connectionist models. Although this is often controlled largely by the connection parameters w_{ij}, the node parameters θ_i may also have an influence, and in some cases, such as B-cell immune networks, provide the *only* means of changing the average connection strength. Thus, it is misleading to assume that the connection parameters are equivalent to the connection strengths. Since the connection strength of any given instant may vary depending on the states of the system, and since the form of the dynamics may differ considerably in different models, we need to discuss connection strength in terms of a quantity that is representation-independent, which is well defined for any dynamical model.

For a continuous transition rule the natural way to discuss the connection strength is in terms of the Jacobian. When the transition rule is an ordinary differential equation, of the form

$$\frac{dx_i}{dt} = f_i(x_1, x_2, \ldots, x_N),$$

the *instantaneous connection strength* of the connection from node i to node j (where i is an input to j) is the corresponding term in the Jacobian matrix

$$J_{ji} = \frac{\partial x_j}{\partial x_i} = \frac{\partial f_j}{\partial x_i}.$$

A connection is excitatory if $J_{ji} > 0$ and inhibitory if $J_{ji} < 0$. Similarly, for discrete time dynamical systems (continuous maps), of the form

$$x_j(t+1) = f_j(x_1, x_2, \ldots, x_N),$$

a connection is excitatory if $|J_{ji}| > 1$ and inhibitory if $|J_{ji}| < 1$. In a continuous system, the average connection strength is $\langle J_{ji} \rangle$, where $\langle \ \rangle$ denotes an appropriate average; in a discrete system it is $\langle |J_{ji}| \rangle$. To make this more precise it is necessary to specify the ensemble over which the average is taken.

For automaton transition rules, since the states x_i are discrete the notion of instantaneous connection strength no longer makes sense. The average connection strength may be defined in one of many ways; for example, as the fraction of times node j changes state when node i changes state. In situations where x_i is an integer but nonetheless approximately preserves continuity, if $|\Delta x_i(t)|$ is the magnitude of the change in x_i at time t, the average connection strength can be defined as

$$\left\langle \frac{|\Delta x_j(t+1)|}{|\Delta x_i(t)|} \right\rangle_{|\Delta x_i(t)|>0}.$$

3. Neural Nets

3.1. BACKGROUND

Neural networks originated with early work of McCulloch and Pitts (1943), Rosenblatt (1958), and others. Although the form of neural networks was originally motivated by neurophysiology, their properties and behavior are not constrained by those of real neural systems, and indeed are often quite different. There are two basic applications for neural networks: one is to understand the properties of real neural systems, and the other is for machine learning. In either case, a central question for developing a theory of learning is: Which behaviors of real neurons are essential to their information processing capabilities, and which are simply irrelevant side effects?

For machine learning problems neural networks have many uses that go considerably beyond the problem of modeling real neural systems. There are several reasons for dropping the constraints of modeling real neurons:

(i) We do not understand the behavior of real neurons.

(ii) Even if we understood them, it would be computationally inefficient to implement the full behavior of real neurons.

(iii) It is unlikely that we need the full complexity of real neurons in order to solve problems in machine learning.

(iv) By experimenting with different approaches to simplified models of neurons, we can hope to extract the basic principles under which they operate, and discover which of their properties are truly essential for learning.

Because of the factors listed above, for machine learning problems there has been a movement towards simpler artificial neural networks that are less motivated by real neural networks. Such networks are often called "artificial neural networks", to distinguish them from the real thing, or from more realistic models. Similar arguments apply to all the models discussed here; it might also be appropriate to say "artificial immune networks" and "artificial autocatalytic networks". However, this is cumbersome and I will assume that the distinction between the natural and artificial worlds is taken for granted.

Neural networks are constructed with simple units, often called "neurons". Until about five years ago, there were almost as many different types of neural networks as there were active researchers in the field. In the simplest and probably currently most popular form, each neuron is a simple element that sums its inputs with respect to weights, subtracts a threshold, and applies an *activation function* to the result. If we assume that time is discrete so that we can write the dynamics as a map, then we have

$t = 1, 2, \ldots =$ time;
$x_i(t) =$ state of neuron i;
$w_{ij} =$ weight of connection from i to j;
$\theta_j =$ threshold;
$S =$ the activation function, often a sigmoidal function such as tanh.

The response of a single neuron can be characterized as

$$x_j(t+1) = S\left(\sum_i w_{ij} x_i(t) - \theta_j\right). \qquad (3)$$

We could also write the dynamics in terms of automata, differential equations, or, if we assume that the neurons have a refractory period during which they do not change their state, as delay differential equations.

The instantaneous connection strength is

$$\frac{\partial x_j(t+1)}{\partial x_i(t)} = w_{ij} S'\left(\sum_i w_{ij} x_i(t) - \theta_j\right), \qquad (4)$$

where S' is the derivative of S. If S is a sigmoid, then S' is always positive and a connection with $w_{ij} > 0$ is always excitatory and a connection with $w_{ij} < 0$ is always inhibitory.

A currently popular procedure for constructing neural networks is to line the neurons up in rows, or "layers". A standard architecture has one

These layers within a single network are not to be confused with the layers used in deep learning, which are themselves neural networks or other machine learning systems.

layer of input units, one or two layers of "hidden" units, and a layer of output units, with full connections between adjacent layers. For a *feed-forward* architecture the graph has no loops so that the fixed parameters information flows only in one direction, from the inputs to the outputs. If the graph has loops so that the activity of a neuron feeds back on itself then the network is *recurrent*.

For layered networks it is sometimes convenient to assign the neurons an extra label that indicates which layer they are in. For feed-forward networks the dynamics across layers is particularly simple, since first the input layer is active, then the first hidden layer, then the next, etc., until the output layer is reached. If, for definiteness, we choose tanh as the activation function, and let 1 refer to the input layer, 2 to the first hidden layer, etc., the dynamics can be described by eq. (5). Note that because the activity of each layer is synchronized and depends only on that of the previous layer at the previous time step, the role of time is trivial. Since each variable only changes its value once during a given feed-forward step, we can drop time labels without ambiguity:

$$x_{2j} = \tanh\left(\sum_i w_{1ji} x_{1i} - \theta_{1j}\right),$$
$$x_{3k} = \tanh\left(\sum_j w_{2kj} x_{2j} - \theta_{2k}\right), \qquad (5)$$
$$x_{4l} = \tanh\left(\sum_k w_{3lk} x_{3k} - \theta_{3l}\right).$$

From this point of view the neural network simply implements a particular family of nonlinear functions, parameterized by the weights w and the thresholds θ (Farmer and Sidorowich 1988). For feed-forward networks the transition rule dynamics is equivalent to a single (instantaneous) mapping. For a recurrent network, in contrast, the dynamics is no longer trivial; any given neuron can change state more than once during a computation. This more interesting dynamics effectively gives the network a memory, so that the set of functions that can be implemented with a given number of neurons is much larger. However, it becomes necessary to make a decision as to when the computation is completed, which complicates the learning problem.

Long short-term memory (LSTM) networks are a modern example of this idea (Hochreiter and Schmidhuber 1997), which are widely used in machine learning.

To solve a given problem we must select values of the parameters w and θ, i.e. we must select a particular member of the family of functions specified by the network. This is done by a learning rule.

The Hebbian learning rules are perhaps the simplest and most time honored. They do not require detailed knowledge of the desired outputs, and are easy to implement locally. The idea is simply to strengthen neurons with coincident activity. One simple implementation changes the weights according to the product of the activities on each connection,

$$\Delta w_{ij} = c x_i x_j. \qquad (6)$$

Hebbian rules are appealing because of their simplicity and particularly because they are local. They can be implemented under very general circumstances. However, learning with Hebbian rules can be ineffective, particularly when there is more detailed knowledge available for training. For example, in some situations we have a training set of patterns for which we know both the correct input and the correct output. Hebbian rules fail to exploit this information, and are correspondingly inefficient when compared with algorithms that do.

Given a learning set of desired input/output vectors, the parameters of the network can be determined to match these input/output vectors by minimizing an error function based on them. The back-propagation algorithm, for example, minimizes the least mean square error and is effectively a nonlinear least-squares fitting algorithm. For more on this, see Rummelhart and McClelland (1986).

Back-propagation, which can be though of as stochastic differentiation, is still widely used today in most machine learning systems.

Since there is an extensive and accessible literature on neural networks, I will not review it further (Rummelhart and McClelland 1986; Cowan and Sharp 1988).

3.2. COMPARISON TO A GENERIC NETWORK

Neural networks are the canonical example of connectionism and their mapping into generic connectionist terms is straightforward.

Nodes correspond to neurons.

Connections correspond to the axons, synapses, and dendrites of real neurons. The average connection strength is proportional to the weight of each connection.

Node dynamics. There are many possibilities. For feed-forward networks the dynamics is reduced to function evaluation. For recurrent networks the node dynamics may be an automaton, a system of coupled mappings, or a system of ordinary differential equations. The attractors of such systems can be fixed points, limit cycles, or chaotic attractors. More realistic models of the refractory periods of the neurons yield systems of delay-differential equations.

Learning rules. Again, there are many possibilities. For feed-forward networks with carefully chosen neural activation functions such as radial basis functions (Broomhead and Lowe 1988; Casdagli 1989; Poggio and Girosi 1989) where the weights can be solved through a linear algorithm, the dynamics reduces to a function evaluation. Nonlinear search algorithms such as back-propagation are nonlinear mappings which usually have fixed point attractors. Nondeterministic algorithms such as simulated annealing have stochastic dynamics.

Graph dynamics. For real neural systems this corresponds to plasticity of the synapses. There is increasing evidence that plasticity plays an important role, even in adults (Alkon 1989). As currently practiced, most neural networks do not have explicit graph dynamics; the user simply tinkers with the architecture attempting to get good results. This approach is clearly limited, particularly for large problems where the graph must be sparse and the most efficient way to restrict the architecture is not obvious from the symmetries of the problem. There is currently a great deal of interest in implementing graph dynamics for neural networks, and there are already some results in this direction (Harp, Samad, and Guha 1989; Miller, Todd, and Hegde 1989; Montana and Davis 1989; Whitley and Hanson 1989; Wilson 1990). This is likely to become a major field of interest in the future.

4. Classifier Systems

4.1. BACKGROUND

The classifier system is an approach to machine learning introduced by Holland (1986). It was inspired by many influences, including production systems in artificial intelligence (Newell 1973), population genetics, and economics. The central motivation was to avoid the problem of brittleness encountered in expert systems and conventional approaches to artificial intelligence. The classifier system learns and adapts using a low-level

> The field of neuroevolution addresses this issue by using evolutionary computation to evolve the network architecture, as in the case of NEAT (Neuro Evolution of Augmenting Topologies; see Stanley and Miikkulainen 2002) where the network topology (graph structure) is learned simultaneously with weight training, and in deep learning models to design the network architecture, known as neural architecture search (Miikkulainen *et al.* 2019).

> Holland's classifier system is a computational mechanism that illustrates how a cognitive system could iteratively learn and update models of its environment. Although the computational realization itself is not used widely today, many of the core ideas remain relevant and presaged later developments, for example, reinforcement learning, generalization, short- and long-term memory, and the need to accommodate nonstationary environments.

abstract representation that it constructs itself, rather than a high-level explicit representation constructed by a human being.

On the surface the classifier system appears quite different from a neural network, and at first glance it is not obvious that it is a connectionist system at all. On closer examination, however, classifier systems and neural networks are quite similar. In fact, by taking a sufficiently broad definition of "classifier systems" and "neural networks", any particular implementation of either one may be viewed as a special case of the other. Classifier systems and neural networks are part of the same class of models, and represent two different design philosophies for the connectionist approach to learning. The analogy between neural networks and classifier systems has been explored by Compiani *et al.* (1988), Belew and Gherrity (1989), Davis (1989). There are many different versions of classifier systems; I will generally follow the version originally introduced by Holland (1986), but with a few more recent modifications such as intensity and support (Holland *et al.* 1986).

At its core, the classifier system has a rule-based language with content addressable memories. The addressing of instructions occurs by matching of patterns or rules rather than by the position of the instructions, as it does in traditional von Neumann languages. Each rule or *classifier* consists of a condition and an action, both of which are fixed length strings. One rule invokes another when the action part of one matches the condition part of the other. This makes it possible to set up a chain of associations; when a given rule is active it may invoke a series of other rules, effecting a computation. The activity of the rules is mediated by a *message list*, which serves as a blackboard or short-term memory on which the rules post messages for each other. While many of the messages sages for each other. While many of the messages on the list are posted by other classifiers, some of them are also external messages, inputs to the program posted by activity from the outside world. In the most common implementations the message list is of fixed length, although there are applications where its length may vary. See the schematic diagram shown in fig. 3. You may also want to refer to the example in section 5.

The conditions, actions, and messages are all strings of the same fixed length. The messages are strings over the binary alphabet $\{0, 1\}$, while the conditions and actions are over the alphabet $\{0, 1, \#\}$, where # is a

One advantage of a rule-based architecture is that it is more readily interpretable than lower-level representations such as neural networks.

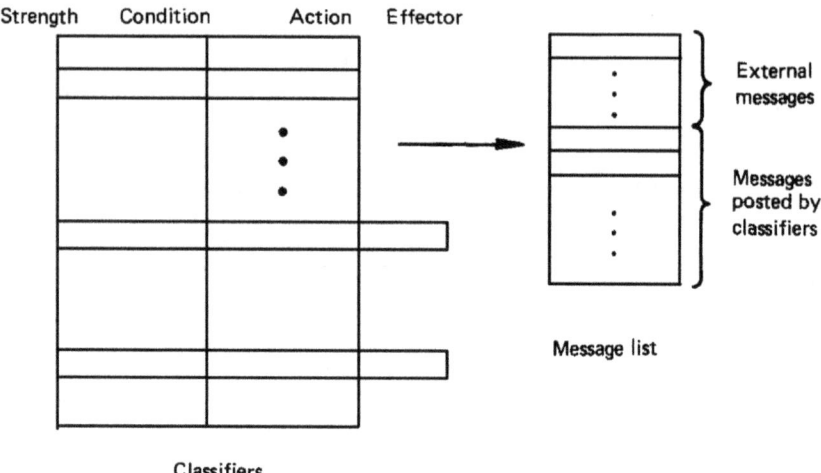

Figure 3. A schematic diagram of the classifier system.

"wildcard" or "don't care" symbol. The length of the message list controls how many messages can be active at a given time, and is typically much smaller than the total number of rules.

The way in which a classifier system "executes programs" is apparent by examining what happens during a cycle of its operation. At a given time, suppose there is a set of messages on the message list, some of which were posted by other classifiers, and some of which are inputs from the external world. The condition parts of all the rules are matched against all the messages on the message list. A match occurs if each symbol matches with the symbol in the corresponding position. The symbol # matches everything. The rules that make matches on a given time step post their actions as messages on the next time step. By going through a series of steps like this, the classifier system can perform a computation. Note that in most implementations of the classifier system each rule can have more than one condition part; a match occurs only when both conditions are satisfied.

In general, because of the # symbol, more than one rule may match a given message. The parameters of the classifier system (frequency of #, length of messages, length of message list, etc.) are usually chosen so that the number of matches typically exceeds the size of the message list. The rules then *bid* against each other to decide which of them will be allowed to post messages. The bids are used to compute a threshold, which is adjusted to keep the number of messages on the message list (that will be posted on the next step) less than or equal to the size of the message list. Only those rules whose bids exceed the threshold are allowed to post their messages on the next time step[2].

An important factor determining the size of the bid is the *strength* of a classifier, which is a real number attached to each classifier rule. The strength is a central part of the learning mechanism. If a classifier wins the bidding competition and successfully posts a message, an amount equal to the size of its bid is subtracted from its strength and divided among the classifiers that (on the previous time step) posted the messages that match the bidding classifier's condition parts on the current time step[3].

Another factor in determining the size of bids is the *specificity* of a classifier, which is defined as the percentage of characters in its condition part that are either zero or one, i.e. that are not #. The motivation is that when there are "specialists" to solve a problem, their input is more valuable than that of "generalists".

The final factor that determines the bid size is the *intensity* $x_i(t)$ associated with a given message. In older implementations of the classifier system, the intensity is a Boolean variable, whose value is one if the message is on the message list, and zero otherwise. In newer implementations the intensity is allowed to take on real values $0 \leq x_i \leq 1$. Thus, some messages on the list are "more intense" than others, which means they have more influence on subsequent activity. Under the *support rule*, the intensity of a message is computed by taking the sum over all the matching messages on the previous time step, weighted by the strength of the classifier making the match.

The size of a bid is

$$\text{bid} = \text{const} \times w \times \text{specificity} \times F(\text{intensity}). \qquad (7)$$

$F(\text{intensity})$ is a function of the intensities of the matching messages. There are many options; for example, it can be the intensity of the message generating the highest bid, or the sum of the intensities of all the matching messages (Riolo 1986).

---

[2] Some implementations allow stochastic bidding.

[3] Other variants are also used. Many authors think that this step is unnecessary, or even harmful; this is a topic of active controversy.

To produce outputs the classifier system must have a means of deciding when a computation halts. The most common method is to designate certain classifiers as outputs. When these classifiers become active the classifier system makes the output associated with that classifier's message. If more than one output classifier becomes active it is necessary to resolve the conflict. There are various means of doing this; a simple method is to simply pick the output with the largest bid.

Neglecting the learning process, the state of a classifier system is determined by the intensities of its messages (most of which may be zero). In many cases it is important to be able to pass along a particular set of information from one time step to another. This is done by a construction called *pass-through*. The # symbol in the action part of the rule has a different meaning than it has in the condition part of the rule. In the action part of the rule it is used to "pass through" information from the message list on one time step to the message list on the next time step; anywhere there is a # symbol in the action part, the message that is subsequently posted contains either a zero or a one according to whether the *message matched by the condition part* on the previous time step contained a zero or a one.

The procedure described above allows the classifier system to implement any finite function, as long as the necessary rules are present in the system with the proper strengths (so that the correct rules will be evoked). The transfer of strengths according to bid size defines a learning algorithm called the *bucket brigade*. The problem of making sure the necessary rules are present is addressed by the use of *genetic algorithms* that operate on the bit strings of the rules as though they were haploid chromosomes. For example, *point mutations* randomly changes a bit in one of the rules. *Crossover* or *recombination* mimics sexual reproduction. It is performed by selecting two rules, picking an arbitrary position, and interchanging substrings so that the left part of the first rule is concatenated to the right part of the second rule and vice versa. When the task to be performed has the appropriate structure, crossover can speed up the time required to generate a good set of rules, as compared to pure point mutation[4].

---

[4] Several specialized graph manipulation operators, for example triggered cover operators, have also been developed for classifier systems (Riolo 1986).

4.2. COMPARISON TO GENERIC NETWORK

The classifier system is rich with structure, nomenclature, and lore, and has a literature of its own that has evolved more or less independently of the neural network literature. Nonetheless, the two are quite similar, as can be seen by mapping the classifier system to standard connectionist terms.

For the purpose of this discussion we will assume that the classifiers only have one condition part. The extension to classifiers with multiple condition parts has been made by Compiani *et al.* (1988).

Nodes. The messages are labels for the nodes of the connectionist network. For a classifier system with word length N the 2^N possible messages range from $i = 0, 1, \ldots, 2^N - 1$. (In practice, for a given set of classifiers, only a small subset of these may actually occur.) The state of the ith node is the intensity x_i. The node activity also depends on a globally defined threshold $\theta(t)$, which varies in time.

Connections. The condition and action parts of the classifier rules are a connection list representation of a graph, in the form of eq. (2). Each classifier rule connects a set of nodes $\{i\}$ to a node j and can be written $\{i\} \to j$. A rule consisting entirely of ones and zeros corresponds to a single connection; a rule with n don't care symbols represents 2^n different connections. Note that if two rules share their output node j and some of their input nodes i then there are multiple connections between two nodes. The connection parameters w_{ij} are computed as the product of the classifier rule strength and the classifier rule specificity, i.e.

$$w_{ij} = \text{specificity} \times \text{strength}.$$

When the graph is sparse there are many nodes that have no rule connecting them so that implicitly $w_{ij} = 0$.

Note that only the connections are represented explicitly; the nodes are implicitly represented by the right-hand parts of the connection representations, which give all the nodes that could ever conceivably become active. Thus nodes with no inputs are not represented. This can be very efficient when the graph is sparse.

Although on the surface pass-through appears to be a means of keeping recurrent information, as first pointed out by Miller and Forrest (1989), in connectionist terms it is a mechanism for efficient graph representation. Pass-through occurs when a classifier has # symbols at the same location in both its condition and action parts. (If the # is only in the action part, then the pass-through value is always the same, and so it is irrelevant.) The net effect is that the node that is activated on the output depends on the node that was active on the input. This amounts to representing more than one connection with a single classifier. For example, consider the classifier 0# → 1#. If node 00 becomes active, then the second 0 is "passed through", so the output is 10. Similarly, if 01 becomes active, the output is 11. The net result is that two connections are represented by the same classifier. From the point of view of the network, the classifier 0# → 1# is equivalent to the two classifiers 00 → 10 and 01 → 11. The net effect is thus a more efficient graph representation, and pass-through is just a representational convenience.

Transition rule. In traditional classifier systems a node j becomes active on time step $t + 1$ if it has an input connection i on time step t such that $x_i(t)w_{ij} > \theta$. Using the support rule,

$$x_j(t+1) = \sum_i x_i(t)w_{ij}, \qquad (8)$$

where the sum is taken over all i that satisfy $x_i(t)w_{ij} > \theta$. With the support rule the dynamics is thus piecewise linear, with nonlinearity due to the effect of the threshold θ. Without the support rule the intensity is $x_j(t+1) = \max_i \{x_i(t)\}$.

There are two approaches to computing the threshold θ. The simplest approach is to simply set it to a constant value θ. A more commonly used approach in traditional classifier systems is to adjust $\theta(t)$ on each time step so that the number of messages that are active on the message list is less than or equal to a constant, which is equivalent to requiring that the number of nodes active on a given time step is less than or equal to a constant. In connectionist terms this may be visualized as adding a special thresholding unit that has input and output connections to every node.

Learning rule. The traditional learning algorithm for classifier systems is the bucket brigade, which is a particular modified Hebbian learning rule. (See eq. (6).) When a node becomes active, strength is transferred from its active output connections to its active input connections. This transfer occurs on the time step after it was active. To be more precise, consider a wave of activity $x_j(t) > 0$ propagating through node j, as shown in fig. 4.

The bucket brigade is an early example of a reinforcement learning algorithm. Its most direct descendent is the temporal-difference method proposed by Sutton and Barto (1990). Today, most reinforcement learning algorithms use a variation of these ideas.

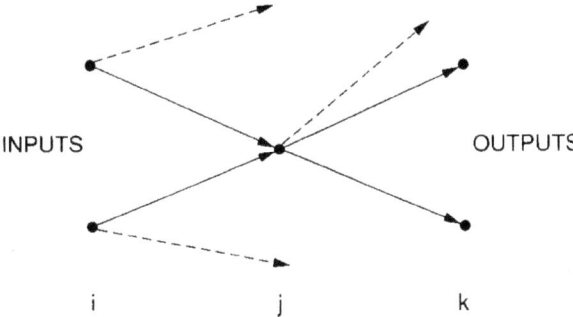

Figure 4. *The bucket brigade learning algorithm.* A wave of activity propagates from nodes $\{i\}$ at time $t-1$ through node j at time t to nodes $\{k\}$ at time $t+1$. The solid lines represent active connections, and the dashed lines represent inactive connections. Strength is transferred from the input connections of j to output connections of j according to eq. (11). The motivation is that connections "pay" the connections that activate them.

Suppose this activity is stimulated by m activities $x_i(t-1) > 0$ through input connection parameters w_{ij}, and in turn stimulates activities $x_k(t+1) > 0$ through output connection parameters w_{jk}. Letting H be the Heaviside function $H(x) = 1$ for $x > 0$, $H(x) = 0$ for $x \leq 0$, the input connections gain strength according to

$$\Delta w_{ij} = \frac{x_j}{m} \sum_k w_{jk} H\left(x_j w_{jk} - \theta\right), \tag{9}$$

$$\Delta w_{jk} = -x_j w_{jk} H\left(x_j w_{jk} - \theta\right), \tag{10}$$

where

$$\Delta w_{ij} = w_{ij}(t+1) - w_{ij}(t). \tag{11}$$

All the quantities on the right-hand side are evaluated at time t.

This is only one of several variants of the bucket brigade learning algorithm; for discussion of other possibilities see Booker, Goldberg, and Holland (1989).

In order to learn, the system must receive feedback about the quality of its performance[5]. To provide feedback about the overall performance

---

[5] It is clearly important to maintain an appropriate distribution of strength within a classifier system, which does not overly favor input or output classifiers and which can set up chains of appropriate associations. Strength is added to classifiers that participate in good outputs, and then the bucket brigade causes a local transfer of feedback, in the form of connection strength, from outputs to inputs. This is further complicated by the recursive structure of classifier systems, which corresponds to loops in the graph. Maintaining an appropriate gradient of strength from outputs to inputs has proved to be a difficult issue in classifier systems.

The bucket brigade is a clever idea. In practice it suffered from exponential decay as strength was passed back through chains of activating rules. This impeded the ability of the algorithm to assign credit to long chains of classifiers that set up an eventual reward.

of the system, the output connections of the system, or the *effectors*, are given strength according to the quality of their outputs. Judgements as to the quality must be made according to a predefined evaluation function. To prevent the system from accumulating useless classifiers, causing isolated connections, there is an activity tax which amounts to a dissipation term. Putting all of these effects together and following Farmer, Packard, and Perelson (1986) we can write the bucket brigade dynamics (the learning rule) as

$$\begin{aligned}\Delta w_{ij} =& \frac{1}{m} \sum_k x_j w_{jk} H\left(x_j w_{jk} - \theta\right) \\ & - x_i w_{ij} H\left(x_i w_{ij} \theta\right) \\ & + x_i P(t) + k w_{ij},\end{aligned} \qquad (12)$$

where k is the dissipation rate for the activity tax, and $P(t)$ is the evaluation function for outputs at time t.

Graph dynamics. The graph dynamics occurs through manipulations of the graph representation (the classifier rules) through genetic algorithms such as point mutation and crossover. These operations are stochastic and are highly nonlocal; they preserve either the input or output of each connection, but the other part can move to a very different part of the graph. The application of these operators generates new connections, which is usually accompanied by the removal of other connections.

4.3. AN EXAMPLE

An example makes the graph-theoretic view of classifier systems clearer. For example, consider the classic problem of exclusive-or. (See also Belew and Gherrity (1989).) The exclusive-or function is 0 if both inputs are the same and 1 if both inputs are different. The standard neural net solution of this problem is easily implemented with three classifiers:

(i) 0# → 10 : +1;

(ii) 0# → 11 : +1;

(iii) 10 → 11 : −2.

(The number after the colon is $w = $ strength $\times$ specificity.) Although there are only three classifiers, because of the # symbols they make five connections, as shown in fig. 5.

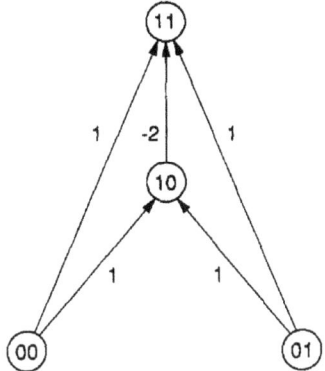

Figure 5. A classifier network implementing the exclusive-or in standard neural net fashion. The binary numbers, which in classifier terms would be messages on the message list, label the nodes of the network.

With this representation the node 00 represents one of the inputs, and 01 represents the other input; the state of each input is its intensity. If both inputs are 1, for example, then nodes 00 and 01 become active, in other words, they have intensity > 0, which is equivalent to saying that the messages 00 and 01 are placed on the message list. Assume that we use the support rule, eq. (8), that outputs occur when the activity on the message list settles to a fixed point, and that the message list is large enough to accommodate at least four messages. An example illustrating how the computation is accomplished is shown in table 1.

This example is unusual from the point of view of common classifier system practice in several respects. (1) The protocol of requiring that the system settle to a fixed point in order to make an output. A more typical practice would be to make an output whenever one of the output classifiers becomes active. (2) The message list is rather large for the number of classifiers, so the threshold is never used. (3) There are no recursive connections (loops in the graph).

There are simpler ways to implement exclusive-or with a classifier system. For example, if we change the input protocol and let the input message be simply the two inputs, then the classifier system can solve this with four classifiers whose action parts are the four possible outputs. This always solves the problem in one step with a message list of length one. Note that in network terms this corresponds to unary inputs, with the four possible input nodes representing each possible input configuration. While this is a cumbersome way to solve

Table 1. A wave of activity caused by the inputs (1, 1) is shown. The numbers from left to right are the intensities on successive iterations. Initially the two input messages have intensity 1, and the others are 0. The input messages activate messages 10 and 11, and then 10 switches 11 off. For the input (0, 0), in contrast, the network immediately settles to a fixed point with the intensities of all the nodes at zero.

NODE	INTENSITY
00	1 1 1 1
01	1 1 1 1
10	0 1 1 1
11	0 1 0 0

There is some work combining classifier systems with neural networks, for example, neural-based learning classifier systems that use a simple neural network (Dam et al. 2008) or a convolutional neural network to process the action of a rule (classifier) (Kim and Cho 2019).

Classical neural-based learning systems struggle with issues such as catastrophic forgetting (memory) or domain shift (novelty), two issues that an architecture like classifier systems can handle naturally.

the problem with a network, it is actually quite natural with a classifier system.

4.4. COMPARISON OF CLASSIFIERS AND NEURAL NETWORKS

There are many varieties of classifier systems and neural networks. Once the classifier system is described in connectionist terms, it becomes difficult to distinguish between them. In practice, however, there are significant distinctions between neural nets *as they are commonly used* and classifier systems *as they are commonly used*. The appropriate distinction is not between classifiers and neural networks, but rather between the two design philosophies represented by the typical implementations of connectionist networks within the classifier system and neural net communities. A comparison of classifier systems and neural networks in a common language illustrates their differences more clearly and suggests a natural synthesis of the two approaches.

Graph topology and representation. The connection list graph representation of the classifier system is efficient for sparse graphs, in contrast to the connection matrix representation usually favored by neural net researchers. This issue is not critical on small problems that can be solved by small networks which allow the luxury of a densely connected graph. On larger problems, use of a sparsely connected graph is essential. If a large problem cannot be solved with a sparsely connected network, then it cannot feasibly be implemented in hardware or on parallel machines where there are inevitable constraints on the number of connections to a given node.

To use a sparse network it is necessary to discover a network topology suited to a given problem. Since the number of possible network topologies is exponentially large, this can be difficult. For a classifier system the sparseness of the network is controlled by the length of each message, and by the number of classifiers and their specificity. Genetic algorithms provide a means of discovering a good network, while maintaining the sparseness of the network throughout the learning process. (Of course, there may be problems with convergence time.) For neural nets, in contrast, the most commonly used approach is to begin with a network that is fully wired across adjacent layers, train the network, and then prune connections if their weights decay to zero. This is useless for a large problem because of the dense network that must be present at the beginning.

The connection list representation of the classifier system, which can be identified with that of production systems, potentially makes it easier to incorporate prior knowledge. For example, Forrest has shown that the semantic networks of KL-One can be mapped into a classifier system (Forrest 1985). On the other hand, another common form of prior knowledge occurs in problems such as vision, when there are group invariances such as translation and rotation symmetry. In the context of neural nets, Giles, Griffin, and Maxwell (1986) have shown that such invariances can be hard-wired into the network by restricting the network weights and connectivity in the proper manner. This could also be done with a classifier system by imposing appropriate restrictions on the rules produced by the genetic algorithm.

Transition rule. Typical implementations of the classifier system apply a threshold to each input separately, before it is processed by the node, whereas in neural networks it is more common to combine the inputs and then apply thresholds and activation functions. It is not clear which of these approaches is ultimately more powerful, and more work is needed.

Most implementations of the classifier system are restricted to either linear threshold activation functions or maximum input activation functions. Neural nets, in contrast, utilize a much broader class of activation functions. The most common example is probably the sigmoid, but in recent work there has been a move to more flexible functions, such as radial basis functions (Broomhead and Lowe 1988; Casdagli 1989; Moody and Darken 1988; Poggio and Girosi 1989) and local linear functions (Farmer and Sidorowich 1988; Jones *et al.* 1989; Wolpert 1989). Some of these functions also have the significant speed advantage of linear learning rules[6]. In smooth environments, smooth activation functions allow more compact representations. Even in environments where a priori it is not obvious the smoothness plays a role, such as learning Boolean functions, smooth functions often yield better generalization results and accelerate the learning process (Wolpert 1989). Implementation of smoother activation functions may improve performance of classifier systems in some problems.

Traditionally, classifier systems use a threshold computed on each time step in order to keep the number of active nodes below a maximum value. Computation of the threshold in this way requires a global computation that is expensive from a connectionist point of view. Future work should concentrate on constant or locally defined thresholds.

From a connectionist point of view, classifiers with the # symbol correspond to multiple connections constrained to have the same strength. There is no obvious reason why their lack of specificity should give them less connection strength. This intuition seems to be borne out in numerical experiments using simplified classifier systems (Wilson 1989).

Learning rule. The classifier system traditionally employs the bucket brigade learning algorithm, whose feedback is condensed into an overall performance score. In problems where there is more detailed feedback, for example a set of known input–output pairs, the bucket-brigade algorithm fails to use this information. This, combined with the lack of smoothness in the activation function, causes it to perform poorly in problems such as learning and forecasting smooth dynamical systems (Pope). Since there are now recurrent implementations of back-propogation (Pineda 1988), it makes sense to incorporate this into a classifier system with smooth activation functions, to see whether this gives better performance on such problems (Belew and Gherrity 1989).

---

[6]Linear learning rules are sometimes criticized as "not local". Linear algorithms are, however, easily implemented in parallel by systolic arrays, and converge in logarithmic time.

Farmer (1990)

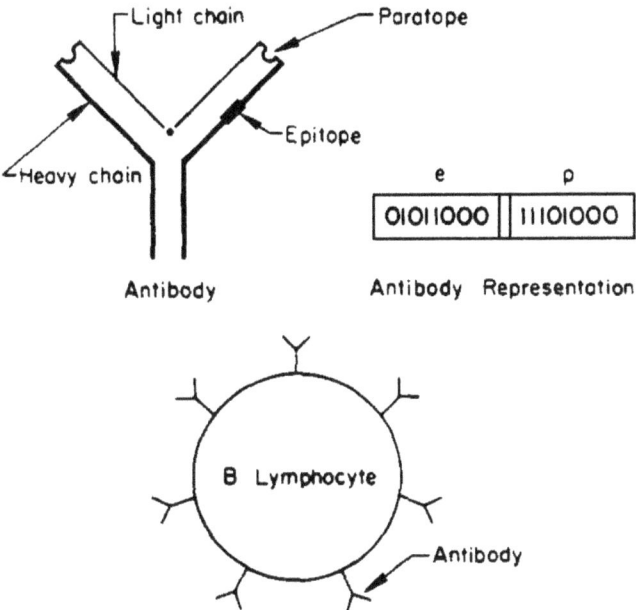

Figure 6. A schematic representation of the structure of an antibody, an antibody as we represent it in our model, and a B-lymphocyte with antibodies on its surface that function as antigen detectors.

For problems where there is only a performance score, the bucket brigade is more appropriate. Unfortunately, there have been no detailed comparisons of the bucket brigade algorithm against other algorithms that use "learning with a critic". The form of the bucket brigade algorithm is intimately related to the activation dynamics, in that the size of the connection strength transfers are proportional to the size of the input activation signal (the bid). Although coupling of the connection strength dynamics to the activation dynamics is certainly necessary for learning, it is not clear that the threshold activation level is the correct or only quantity to which the learning algorithm should be coupled. Further work is needed in this area.

5. Immune Networks

5.1. BACKGROUND

The basic task of the immune system is to distinguish between self and non-self, and to eliminate non-self. This is a problem of pattern learning and pattern recognition in the space of chemical patterns. This is a

A bit simplistic (Segel and Bar-Or 1999; Segel and Cohen 2001). At the least, the immune system needs to perform this task while minimizing harm to self and tolerating nondangerous self (Matzinger 1994).

Lymphocytes and antibodies are key players in what is known as the adaptive immune response, but there are numerous cell types associated with the innate immune response, including macrophages, mast cells, natural killer cells and many more. Sompayrac (2002) is an excellent and readable introduction to the immune system.

This paragraph has a typo, although the basic argument is correct. Current studies of the human genome estimate that there are about 20,000 genes that code for proteins, and that each gene code for two or three different proteins (Piovesan *et al.* 2019).

In 1987, Susumu Tonegawa received the Nobel Prize in Physiology or Medicine for his discovery of this principle.

difficult task, and the immune system performs it with high fidelity, with an extraordinary capacity to make subtle distinctions between molecules that are quite similar.

The basic building blocks of the immune system are *antibodies*, "y" shaped molecules that serve as identification tags for foreign material; *lymphocytes*, cells that produce antibodies and perform discrimination tasks; and *macrophages*, large cells that remove material tagged by antibodies. Lymphocytes have antibodies attached to their surface which serve as antigen detectors. (See fig. 6.) Foreign material is called *antigen*. A human contains roughly 10^{20} antibodies and 10^{12} lymphocytes, organized into roughly 10^8 distinct *types*, based on the chemical structure of the antibody. Each lymphocyte has only one type of antibody attached to it. Its type is equivalent to the type of its attached antibodies. The majority of antibodies are *free antibodies*, i.e. not attached to lymphocytes. The members of a given type form a clone, i.e. they are chemically identical.

The difficulty of the problem solved by the immune system can be estimated from the fact that mammals have roughly 10^5 genes, coding for the order of 10^5 proteins. An *antigenic determinant* is a region on the antigen that is recognizable by an antibody. The number of antigenic determinants on a protein such as myoglobin is the order of 50, with $6-8$ amino acids per region. We can compare the difficulty of telling proteins apart to a more familar task by assuming that each antigenic determinant is roughly as difficult to recognize as a face. In this case the pattern recognition task performed by the immune system is comparable to recognizing a million different faces. A central question is the means by which this is accomplished. Does the immune system function as a gigantic look up table, like a neural network with billions of "grandmother cells"? Or, does it have an associative memory with computational capabilities?

The argument given above neglects the important fact that there are 10^5 distinct proteins only *if we neglect the immune system*. Each antibody is itself a protein, and there are 10^8 distinct antibody, which appears to be a contradiction: How do we generate 10^8 antibody types with only 10^5 genes? The answer lies in combinatorics. Each antibody is chosen from seven gene segments, and each gene segment is chosen from a "family" or set of possible variants. The total number of possible antibody types is then the product of the sizes of each gene family. This is not known exactly, but is believed to be on the order of $10^7 - 10^8$. Additional diversity is created by *somatic mutation*. When the lymphocytes replicate, they do so with an unusually large error rate in their antibody genes. Although it is difficult to estimate the number of *possible* types precisely, it is probably much larger than the number of types that are actually present in a given organism.

The ability to recognize and distinguish self is *learned*. How the immune system accomplishes this task is unknown. However, it is clear that one of the main tools the immune system uses is *clonal selection*. The idea is quite simple: A particular lymphocyte can be stimulated by a particular antigen if it has a chemical reaction with it. Once stimulated it replicates, producing more lymphocytes of the same type, and also secreting free antibodies. These antibodies bind to the antigen, acting as a "tag" instructing macrophages to remove the antigen. Lymphocytes that do not recognize antigen do not replicate and are eventually removed from the system.

While clonal selection explains how the immune system recognizes and removes antigen, it does not explain how it distinguishes it from self. From both experiments and theoretical arguments, it is quite clear that this distinction is learned rather than hard-wired. Clonal selection must be suppressed for the molecules of self. How this actually happens is unknown.

A central question for self–nonself discrimination is: Where is the seat of computation? It is clear that a significant amount of computation takes place in the lymphocytes, which have a sophisticated repertoire of different behaviors. It is also clear that there are complex interactions between lymphocytes of the same type, for example, between the different varieties of T-lymphocytes and B-lymphocytes. These interactions are particularly strong during the early stages of development.

Jerne proposed that a significant component of the computational power of the immune system may come from the interactions of different types of antibodies and lymphocytes *with each other* (Jerne 1973, 1974). The argument for this is quite simple: Since antibodies are

Although immune network theory as described here is an appealing idea intellectually, there has been little experimental evidence uncovered to support it. There are many other important immune networks, however, for example cytokine signaling networks (Wiedemann *et al.* 2021).

after all just molecules, then from the point of view of a given molecule other molecules are effectively indistinguishable from antigens. He proposed that much of the power of the immune system to regulate its own behavior may come from interacting antibodies and lymphocytes of many different types[7].

There is good experimental evidence that network interactions take place, particularly in young animals. Using the nomenclature that an antibody that reacts directly with antigen AB1, an antibody that reacts directly with AB1 is AB2, etc., antibodies in categories as deep as AB4 have been observed experimentally[8]. Furthermore, rats raised in sterile environments have active immune systems, with activity between types. Nonetheless, the relevance of networks in immunology is highly controversial.

5.2. CONNECTIONIST MODELS OF THE IMMUNE SYSTEM

While Jerne proposed that the immune system could form a network similar to that of the nervous system, his proposal was not specific. Early work on immune networks put this proposal into more quantitative terms, assuming that a given AB1 type interacted only with one antigen and one other AB2 type. These interactions were modeled in terms of simple differential equations whose three variables represented antigen, AB1, and AB2 (Richter 1975; Hoffmann 1975). A model that treats immune interactions in a connectionist network[9], allowing interactions between arbitrary types, was proposed in Farmer, Packard, and Perelson (1986). The complicated network of chemical interactions between different antibody types, which are impossible to model in detail from first principles, was taken into account by constructing an artificial antibody chemistry. Each antigen and antibody type is assigned a random binary string, describing its "chemical properties". Chemical interactions are assigned based on complementary matching between strings. The strength of a chemical reaction is proportional to

> The immune system solves two problems elegantly that neural machine learning systems struggle with: continual learning and catastrophic forgetting. It would be productive to revisit Farmer's connectionist interpretation of the immune system in light of modern immunology, with an eye to proposing a connectionst algorithm that addresses these two concerns.

---

[7] Such networks are often called *idiotypic networks*.

[8] This classification of antibodies should not be confused with their type; a given type can simultaneously be AB1 and AB2 relative to different antigens, and many different types may be AB1.

[9] Another connectionist model with a somewhat different philosophy was also proposed by Hoffmann *et al.* (1988).

the length of the matching substrings, with a threshold below which no reaction occurs. Even though this artificial chemistry is unrealistic in detail, hopefully it correctly captures some essential qualitative features of real chemistry.

A model of gene shuffling provides metadynamics for the network. This is most realistically accomplished with a gene library of patterns, mimicking the gene families of real organisms. These families are randomly shuffled to produce an initial population of antibody types. This gives an initial assignment of chemical reactions, through the matching procedure described above, including rate constants and other parameters[10]. Kinetic equations implement clonal selection; some types are stimulated by their chemical reactions, while others are suppressed. Types with no reactions are slowly flushed from the system so that they perish. Through reshuffling of the gene library new types are introduced to the system. It is also possible to stimulate somatic mutation through point mutations of existing types, proportional to their rate of replication.

It is difficult to model the kinetics of the immune system realistically. There are five different classes of antibodies, with distinct interactions and properties. There are different types of lymphocytes, including helper, killer and suppressor T-cells, which perform regulatory functions, as well as B-cells, which can produce free antibodies. All of these have developmental stages, with different responses in each stage. Chemical reactions include cell–cell, antibody–antibody, and cell–antibody interactions. Furthermore, the responses of cells are complicated and often state dependent. Thus, any kinetic equations are necessarily highly approximate, and applicable to only a subset of the phenomena.

In our original model we omitted T-cells, treating only B-cells. (This can also be thought of as modeling the response to certain polymeric antigens, for which T-cells seem to be irrelevant.) We assumed that the concentration of free antibodies is in equilibrium with the concentration of lymphocytes, so that their populations can be lumped together into a single concentration variable. Since

---

[10] The genetic operations described here are more sophisticated than those actually used in Farmer, Packard, and Perelson (1986); more realistic mechanisms have been employed in subsequent work (Perelson 1989; De Boer and Hogeweg 1989; De Boer 1990).

the characteristic time scale for the production of free antibodies is minutes or hours, while that of the population of lymphocytes is days, this is a good approximation for some purposes. It turns out, however, that separating the concentration of lymphocytes and free antibodies and considering the cell–cell, antibody–antibody, and cell–antibody reactions separately give rise to new phenomena that are important for the connectionist view. In particular, this generates a more interesting repertoire of steady states, including "mildly excited" self-stimulated states suggestive of those observed in real immune systems (Perelson 1989; De Boer and Hogeweg 1989; De Boer 1990).

5.3. COMPARISON TO A GENERIC NETWORK

As with classifier systems and neural networks, there are several varieties of immune networks (Farmer, Packard, and Perelson 1986; De Boer and Hogeweg 1989; Hoffmann *et al.* 1988; Varela *et al.* 1988), and it is necessary to choose one in order to make a comparison. The model described here is based on that of Farmer, Packard, and Perelson (1986), with some modifications due to later work by Perelson (1989) and De Boer and Hogeweg (1989). Also, since this model only describes B-cells, whenever necessary I will refer to it as a B-cell network, to distinguish it from models that also incorporate the activity of T-cells.

To discuss immune networks in connectionist terms it is first necessary to make the appropriate map to nodes and connections. The most obvious mapping is to assign antibodies and antigens to nodes. However, since antibodies and antigens typically have more than one antigenic determinant, and each region has a distinct chemical shape[11], we could also make the regions (or chemical shapes) the fundamental variable. Since all the models discussed above treat the concentration of antibodies and lymphocytes as the fundamental variables, I shall make the identification at this level. This leads to the following connectionist description:

Nodes correspond to antibodies, or more accurately, to distinct antibody types. Antigens are another type of node with different dynamics; from a certain point of view the antigen concentrations

---

[11]"Chemical shape" here means all the factors that influence chemical properties, including geometry, charge, polarization, etc.

may be regarded as the input and output nodes of the network[12]. The free antibody concentrations, which can change on a rapid time scale, are the states of the nodes. They are the immediate indicators of information processing in the network. The lymphocyte concentrations, which change on an intermediate time scale, are node parameters. (Recall that there is a one-to-one correspondence between free antibody types and lymphocyte types.) Changes in lymphocyte concentration are the mechanism for learning in the network.

Connections. The physical mechanisms which cause connections between nodes are chemical reactions between antibodies, lymphocytes, and antigens. The strength of the connections depends on the strength of the chemical reactions. This is in part determined by chemical properties, which are fixed in time, and in part by the concentrations of the antibodies, lymphocytes, and antigens, which change with time. Thus the instantaneous connection strength changes in time as conditions change in the network. The precise way of representing and modeling the connections is explained in more detail in the following.

Graph representation. To model the notion of "chemical properties" we assign each antibody type a binary string. To determine the rate of the chemical reaction between type i and type j, the binary string corresponding to type i is compared to binary string corresponding to type j. A match strength matrix m_{ij} is assigned to this connection, which depends on the degrees of complementary matching between the two strings. Types whose strings have a high degree of complementary matching are assigned large reaction rates. Since the matching algorithm is symmetric[13] $m_{ij} = m_{ji}$.

There is a threshold for the length of the complementary matching region below which we assume that no reaction occurs and set $m_{ij} = 0$. Since m_{ij} is the connection matrix of the graph, setting $m_{ij} = 0$ amounts to deleting the corresponding connection from the graph. We thus neglect reactions that are so weak that they have an insignificant effect on the behavior of the network. The match threshold together

---

[12]Future models should include chemical types identified with self as yet another type of node.

[13]In our original paper (Farmer, Packard, and Perelson 1986) we also considered the case of asymmetric interactions. However, this is difficult to justify chemically, and it is probably safe to assume that the connections are symmetric (Hoffmann 1975).

with the length of the binary strings determines the sparseness of the graph. When the system is sparse the matrix m_{ij} can be represented in the form of a connection list. The match strength for a given pair of immune types does not change with time. However, as new types are added or deleted from the system, the m_{ij} that are relevant to the types *in the network* change.

The graph dynamics provides a mechanism of learning in the immune system; as new types are tested by clonal selection, the graph changes, and the system "evolves". Another mechanism for dynamical learning depends on the lymphocyte concentrations, as discussed below.

Dynamics. The m_{ij} are naturally identified as connection parameters for the network. For any given i and j, however, the m_{ij} are fixed. Thus, in B-cell immune networks the parameter dynamics, analogous to the learning rule in neural networks, occurs not by changing connection parameters, but rather by changing the lymphocyte concentration, which is a parameter node. The net reaction flux (or strength of the reaction) is a nonlinear function of the lymphocyte concentrations. Thus changing the lymphocyte concentration changes the effective connection strength. This is a fundamental difference between neural networks and B-cell immune networks; while the connection strength is changeable in both cases, in B-cell immune networks all the connection strengths to a given node change in tandem as the lymphocyte concentration varies. However, since the reaction rates are nonlinear functions, a change in lymphocyte concentration may affect each connection differently, depending on the concentration of the other nodes.

The dynamics of the real immune system are not well understood. The situation is similar to that of neural networks; we construct simplified heuristic immune dynamics based on a combination of chemical kinetics and experimental observations, attempting to recover some of the phenomena of real immune systems. The real complication arises because lymphocytes are cells, and understanding their kinetics requires understanding how they respond to stimulation and suppression by antigens, antibodies, and other cells. At this point our understanding of this is cells. At this point our understanding of this is highly approximate and comes only from experimental data.

The kinetic equations used in our original paper were highly idealized (Farmer, Packard, and Perelson 1986). The more realistic equations quoted here are due to De Boer and Hogeweg[14] (1989).

Let i label the nodes of the system, x_i the concentration of antibodies, and θ_i the concentration of lymphocytes[15]. The amount of stimulation received by lymphocytes of type i is approximated as

$$s_i = \sum_j m_{ij} x_j. \qquad (13)$$

The rate of change of antibody concentration is due to production by lymphocytes, removal from the system, and binding with other antibodies. The equations are

$$\frac{dx_i}{dt} = \theta_i f(s_i) - k x_i - c x_i s_i. \qquad (14)$$

k is a dissipation constant and c the binding constant. f is a function describing the degree of stimulation of a lymphocyte. Experimental observations show that f is bell-shaped. A function with this rough qualitative behavior can be constructed by taking the product of a sigmoid with an inverted sigmoid, for example

$$f(z) = \frac{z k_2}{(k_1 + z)(k_2 + z)}. \qquad (15)$$

The production of lymphocytes is due to replenishment by the bone marrow, cell replication, and removal from the system. The equations are

$$\frac{d\theta_i}{dt} = r + p \theta_i f(s_i) - k \theta_i. \qquad (16)$$

r is the rate of replenishment and p is a rate constant for replication.

5.4. COMPARISON TO NEURAL NETWORKS AND CLASSIFIER SYSTEMS

There are significant differences between the dynamics of immune networks and neural networks. The most obvious is in the form

---

[14] More realistic equations have also been proposed by Segel and Perelson (1989), Perelson (1988, 1989), and Varela et al. (1988).

[15] Note that I use θ to represent lymphocytes because they play the role of node parameters. However, they are not thresholds, but rather quantities whose primary function is to modify connection strength.

of the transition and learning rules. The nodes of the immune network are activated by a bell-shaped function rather than a sigmoid function. Since the bell-shaped function undergoes an inflection and its derivative changes sign, the dynamics are potentially more complicated.

B-cell immune networks differ from neural networks in that there is no variable which acts as a connection parameter. Instead, the connection strength is indirectly determined by the node parameters (concentrations and kinetic equations). The instantaneous connection strength is

$$\frac{\partial \dot{x}_i}{\partial x_j} = \left[\theta_i f'(s_i) - cx_i\right] m_{ij} - cs_i - k\delta_{ij}, \qquad (17)$$

where $\delta_{ij} = 0$ for $i \neq j, \delta_{ii} = 1$. All of the terms in this equation except for f' are greater than or equal to zero. For low values of s_i, $f'(s_i) > 0$, but for large values of s_i, $f'(s_i) < 0$. Given the structure of these equations, as s_i increases, at some point before f reaches a maximum, all the connections to a given node change from excitatory to inhibitory. The point at which this happens depends on the lymphocyte concentration of i, the antibody concentration, the concentration of the other antibodies, and on the exact form of the stimulation function. Thus, in contrast to neural networks or the classifier system, a given connection can be either excitatory or inhibitory depending on the state of the system.

Many T-lymphocytes reside in tissue, where spatial constraints are important, which these equations don't capture.

The connections in the immune system are chemical reactions. Insofar as the immune system is well stirred, this allows a potentially very large connectivity, as high as the number of different chemical types a given type can react with. In practice, the number of types that a given type reacts with can be as high as about 1000. Thus, the connectivity of real immune networks is apparently of the same order of magnitude as that of real neural networks.

One of the central differences between the B-cell immune networks and neural or classifier networks is that for the immune system there are no independent parameters on the connections. If the average strength of a connection to a given node cannot be adjusted independently of that of other nodes, the learning capabilities of the network may be much weaker or more inefficient than those of networks where the connection parameters are independent. As discussed in section 5.5, this may be altered by the inclusion of T-cells in the models.

5.5. DIRECTIONS FOR FUTURE RESEARCH

Whether immune networks are a major component of the computational machinery of the immune system is a subject of great debate. The analogy between neural networks and immune networks suggests that immune networks potentially possess powerful capabilities, such as associative memory, that could be central to the functioning of the immune system. However, before this idea can reach fruition we need more demonstrations of what immune networks can do. At this point the theory of immune networks is still in its infancy and their utility remains an open question.

The immune network may be able to perform tasks that would be impossible for individual cells. Consider, for example, a large antigen such as a bacterium with many distinct antigenic determinants. If each region is chemically distinct, a single type can interact with at most a few of them (and thus a single cell can interact with at most a few of them). Network interactions, in contrast, potentially allow different cells and cell types to communicate with each other and make a collective computation to reinforce or suppress each other's immune responses. For example, suppose A, B, C and D are active sites. It might be useful for a network to implement an associative memory rule such as: If any three of A, B, C, and D are present, then generate an immune response; otherwise do not. Such an associative memory requires the capability to implement a repertoire of Boolean functions. A useful rule might be: "Generate an immune response if active site A is present, or active site B is present, but not if both are present simultaneously". Such a rule, which is equivalent to taking the exclusive-or function of A and B, might be useful for implementing self tolerance. Such logical rules are easily implemented by networks. It is difficult to see how they could be implemented by individual cells acting on their own.

Immune memory is another task in which networks may play an essential role. Currently the prevailing belief is that immune memory comes about because of special memory cells. It is certainly true that some cells go into developmental states that are indicative of memory. Although the typical lifetime of a lymphocyte is about five days, there are some lymphocytes that have been demonstrated to persist for as long as a month. This is a far cry, however, from the eighty or more years that a human may display an immune memory. Since cells are normally flushed from the system at a steady rate, it is difficult to believe that any individual cell could last this long. It is only the type, then, that persists, but in order to achieve this individual cells must periodically replicate themselves. However, in order to hold the population stable the replication rate must be perfectly balanced against the removal rate. This is an unstable process unless there is feedback holding the population stable. It is difficult to see how feedback on the population size can be given unless there are network interactions.

In an immune network a memory can potentially be modeled by a fixed point of the network. The concentrations at the fixed point are held constant through the feedback of one type to another type. Models of the form of eqs. (14) and (16) contain fixed points that might be appropriate for immune memory. However, it is clear from experiments that T-cells are necessary for memory, and so must be added to immune networks to recover this effect.

T-cells are a key element missing from most current immune network models. T-cells play an important role in stimulating or suppressing reactions between antibodies and antigens, and are essential to immune memory. From the point of view of learning in the network, they may also indirectly act as specific connection parameters.

One of the most interesting activities of the immune system is "antigen presentation". When a B-cell or macrophage reacts with an antigen it may process it, discarding all but the antigenic determinants. It then presents the antigenic determinant on its surface (as a peptide bound to an MHC molecule). The T-cell reacts with the antigenic determinant and the B-cell, and based on this information may either stimulate or suppress the B-cell. Note that antigen presentation provides information about *both* the B-cell *and* an antigen, and thus potentially about a specific connection in the network.

In a connectionist model, this may amount to a connection strength parameter; a B-cell presenting a given active site contains information that is specific to two nodes, one for the B-cell of the same type as the T-cell, and one for the antigen whose active site is being presented (which may also be another antibody). Due to their interactions with T-cells, the B-cell populations of type i presenting antigenic determinants from type j may play the roles of the connection parameters w_{ij}.

At this point, it is not clear how strongly the absence of explicit connection parameters limits the computational and learning power of immune networks. However, it seems likely that before they can realize their full potential, connection parameters must be included, taking into account the operation of T-cells. T-cells act like catalysts, either suppressing or enhancing reactions. Since catalytic activity is one of the primary tools used to implement the internal functions of living organisms, it is not surprising that it should play a central role in the immune system as well. Autocatalytic activity is discussed in more detail in section 6.

6. Autocatalytic Networks

6.1. BACKGROUND

All the models discussed so far are designed to perform learning tasks. The autocatalytic network model of this section differs in that it is designed to solve a problem in evolutionary chemistry. Of course, evolution may also be regarded as a form of learning. Still, the form that learning takes in autocatalytic networks is significantly different from the other models discussed here.

The central goal of the autocatalytic network is to solve a classic problem in the origin of life, namely, to demonstrate an evolutionary pathway from a soup of monomers to a polymer metabolism with selected autocatalytic properties, which in turn could provide a substrate for the emergence of contemporary (or other) life forms. When Miller and Urey discovered that amino acids could be synthesized de novo from the hypothetical primordial constituents "earth, fire and water" (Miller and Urey 1959), it seemed but a small step to the synthesis of polymers built out of amino acids (polypeptides and proteins). It was hoped that RNA and DNA could be created similarly. However, under normal circumstances longer polymers are not favored at equilibrium. Living systems, in contrast, contain DNA, RNA, and proteins, specific long polymers which exist in high concentration. They are maintained in abundance by their symbiotic relationship with each other: Proteins help replicate RNA and DNA, and DNA and RNA

There is now evidence that an autocatalytic network was indeed embedded at the core of microbial metabolism (Xavier et al. 2020).

There is now evidence that under chemical conditions which are hypothesized to be critical for the origin of life, longer polymers are indeed thermodynamically favored (Dick and Shock 2021).

help synthesize proteins. Without the other, neither would exist. How did such a complex system ever get started, unless there were proteins and RNA to begin with? The question addressed in Kauffman (1986), Farmer, Kauffman, and Packard (1986), and Bagley *et al.* (1990) is: Under what circumstances can the synthesis of specific long polymers be achieved beginning with simple constituents such as monomers and dimers?

The model here applies to any situation in which unbranched polymers are built out of monomers through a network of catalytic activity. The monomers come from a fixed alphabet, a, b, c, ... They form one-dimensional chains which are represented as a string of monomers, acabbacbc... The monomer alphabet could be the twenty amino acids, or it could equally well be the four nucleotides. This changes the parameters but not the basic properties of the model. The model assumes that the polymers have catalytic properties, i.e. that they can undergo reactions in which one polymer catalyzes the formation of another. If A, B, C, and E are polymers, and H is water, then the basic reaction is:

$$A + B \overset{E}{\rightleftharpoons} C + H, \qquad (18)$$

where E is written over the arrows to indicate that it catalyzes the reaction.

Our purpose is to model a chemostat, a reaction vessel into which monomers are added at a steady rate. The chemical species that are added to the chemostat are called the *food set*. We assume that the mass in the vessel is conserved, for example, by simply letting the excess soup overflow. For convenience we assume that the soup is well stirred, so that we can model it by a system of ordinary differential equations.

In any real system it is extremely difficult to determine from first principles which reactions will be catalyzed, and with what affinity. Very few if any of the relevant properties have been measured experimentally in any detail, and the number of measurements or computations that would have to be made in order to predict all the chemical properties is hopelessly complex. Our approach is to invent an artificial chemistry and attempt to make its properties at least qualitatively similar to those of a real chemical system. Actually we use one of two different artificial chemistries, based on two different principles:

(i) Random assignment of catalytic properties.
(ii) Assignment of catalytic properties based on string matching.

This remains a central problem in computational chemistry, but machine learning methods have proved useful (Kitchin 2018).

It has now been shown that the graph can be generated under different conditions and still generate autocatalytic sets (Hordijk *et al.* 2014).

These two simple artificial chemistries lie on the borders of extreme behavior in real chemistry. In some cases, we know that changing one monomer can have a dramatic effect on the chemical properties of a polymer, either because it causes a drastic change in the configuration of the polymer or because it alters a critical site. If this were always the case, then random chemistry would be a reasonable model.

In other cases, changing a monomer has only a small effect on the chemical properties. Our string matching model is closer to this case; altering a single monomer will only change the quality of matching between two strings by an incremental amount, and should never cause a dramatic alteration in the chemical properties of the polymer.

Another difficulty of modeling real chemistry is that there is an extraordinarily large number of possible reactions. In a vessel with all polymers of length l or less, for example, the total number of polymer species is $\sum_{i=1}^{i=l} m^l$, where m is the number of distinct monomers. For example, with $m = 20$ and $l = 100$, the number of polymer species is in excess of 20^{100}, an extremely large number, and the number of possible reactions is still larger than this. To get around this problem, to first approximation, we neglect spontaneous reactions, and assume that the catalytic properties are sufficiently strong that all catalyzed reactions are much faster than spontaneous reactions[16].

Once we have assigned chemical properties, we can represent the network of catalyzed chemical reactions as a graph, or more precisely, as a polygraph with two types of nodes and two types of connections (Farmer, Kauffman, and Packard 1986). Because of catalysis the graph must be more complicated than for any of the other networks discussed so far. An example is shown in fig. 7. One type of node is labeled by ovals containing the string representation of the polymer species. The other type of node corresponds to catalyzed reactions, and is labeled by black dots. The dark black connections are undirected (because the reactions are reversible), and connect each reaction to the three polymer species that participate in it; the dotted connections are directed, and connect the reaction to its catalysts.

---

[16]In more recent work (Bagley and Farmer 1990) we make a tractable model for approximate treatment of spontaneous reactions by lumping together all the polymer species of a given length that are not in the autocatalytic network, assuming that they all have the same concentration. These can be viewed as a new type of node in the network. This allows us to include the effect of spontaneous reactions when necessary.

All the edges connect polymers to reactions, and each reaction has at least four connections, three connections for the reaction products and one or more for the catalyst(s). In this illustration we have labeled the members of the food set by double ovals.

If we use the random method of assigning chemical properties, then the graph is a random graph and can be studied using standard techniques. The probability p that a reaction selected at random will be catalyzed controls the ratio of connections to nodes. As p increases so does this ratio. As p grows the graph becomes more and more connected, i.e. more dense.

The graph-theoretic analysis only addresses the question of who reacts with whom, and begs the central (and much more difficult) question of concentrations. Numerical modeling of the kinetics *for any given catalyzed reaction* is straightforward but cumbersome. We introduced a simplified technique for treating catalyzed reactions of this type in Farmer, Kauffman, and Packard (1986) that approximates the true catalyzed kinetics fairly well.

Modeling of the complete kinetics for an entire reaction graph is impossible, since the graph is infinite and under the laws of continuous mass action, even if we initialize all but a finite number of the species to zero concentration, an instant later they will all have non-zero concentrations. From a practical point of view, however, it is possible to circumvent this problem by realizing that any chemical reaction vessel is finite, and species whose continuous concentrations are significantly below the concentration corresponding to the presence of a single molecule are unlikely to participate in any reactions. Thus, to cope with this problem we introduce a concentration threshold, and only consider reactions where all the members on either side of the reaction equation (either A, B, and E, or C and E) are above the concentration threshold. This then becomes a metadynamical system: At any given time, only a finite number of species are above the threshold, and we only consider a finite graph. As the kinetics act, species may rise above the concentration threshold, so that the graph grows, or they may drop below the threshold, so that the graph shrinks.

One of the main goals of this model is to obtain closure in the form of an *autocatalytic set*, which is a set of polymer species such that each member of the set is produced by at least one catalyzed reaction involving only other members of the set (including the catalysts). Since the reactions are (including the catalysts). Since the reactions are reversible, a species can

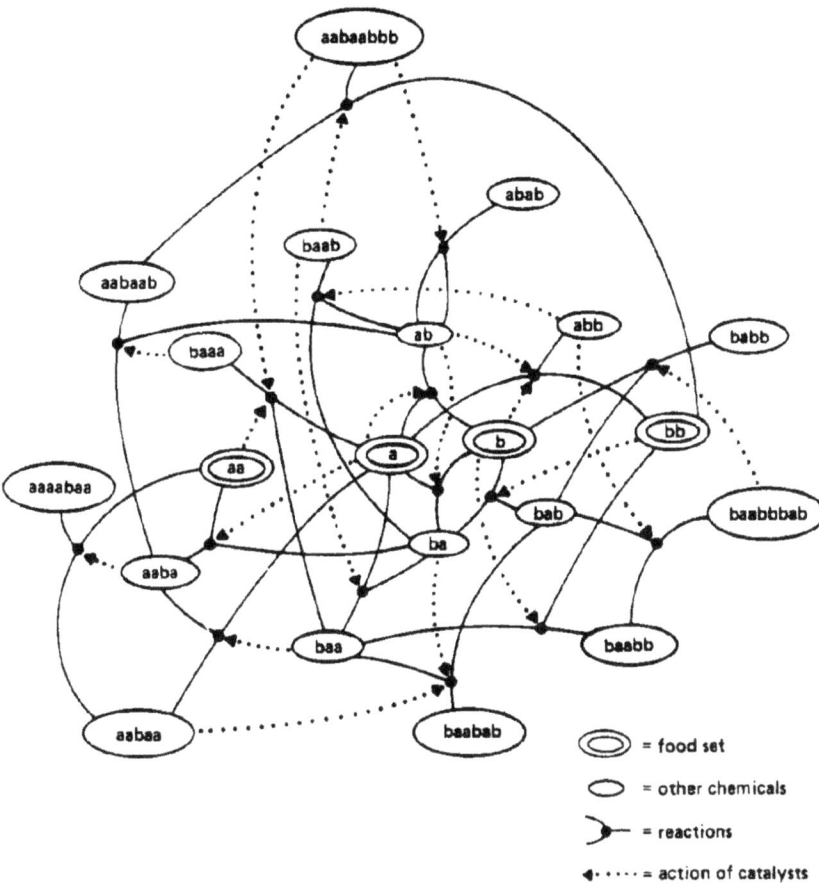

Figure 7. The graph for an autocatalytic network. The ovals represent polymer species, labeled by strings. The black dots represent reactions. The solid lines are connections from polymer nodes to the reactions in which they participate. The dotted lines go from polymer species to the reactions they catalyze. The double ovals are special polymer nodes corresponding to the elements of the food set, whose concentrations are supplied externally.

be "produced" either by cleavage or condensation, depending on which side of equilibrium it finds itself. Thus an autocatalytic set can be quite simple; for example,

$$A + B \overset{A}{\rightleftharpoons} C + H \qquad (19)$$

is an autocatalytic set, and so is

$$A + B \overset{C}{\rightleftharpoons} C + H. \qquad (20)$$

A, B, and C will be regenerated by supplying either A and B, or by supplying C. Note, however, that such simple autocatalytic sets are only likely to occur when the probability of catalysis is very high. Even for small values of p it is always possible to find autocatalytic sets as long as the food set is big enough. However, the typical autocatalytic set is more complicated than the examples given in eqs. (19) and (20). There is a critical transition from the case where graphs with autocatalytic sets are very rare to that in which they are very common, as described in Farmer, Kauffman, and Packard (1986) and Kauffman (1986). The results given there show that it is possible to create autocatalytic sets (in this graph theoretic sense) under reasonably plausible prebiotic conditions.

There are three notions of the formation of autocatalytic sets, depending on what we mean by "produced by" in the definition given above:

(i) *Graph theoretic.* The subgraph defined by the autocatalytic set is closed, so that each member is connected (by a solid connection) to at least one reaction catalyzed by another member.

(ii) *Kinetics.* Each member is produced at a level exceeding a given concentration threshold.

(iii) *Robust.* The autocatalytic set is robust under at least some changes in its food set, i.e. its members are at concentrations sufficiently large and there are enough pathways so that for some alterations of the food set it remains a kinetic autocatalytic set, capable of regenerating removed elements at concentrations above the threshold.

These notions are arranged in order of their strength, i.e. an autocatalytic set in the sense of kinetics is automatically an autocatalytic set in the graph-theoretic sense, and a robust autocatalytic set is automatically a kinetic autocatalytic set.

Describing the details of the conditions under which autocatalytic sets can be created is outside of the scope of this paper. Suffice it to say that, within our artificial chemistry we can create robust autocatalytic sets. Consider, for example, an autocatalytic set based on the monomers a and b, originally formed by a food set consisting of the species a, b, ab, and bb, as shown in fig. 8 and table 2.

We plot the concentrations of the 21 polymer species in the reactor against an index that is arbitrary except that it orders the species according to their length. We compare four different alterations of the original food set, all of which have the same rate of mass input. For two of the altered food sets the concentration of the members of the autocatalytic set remains

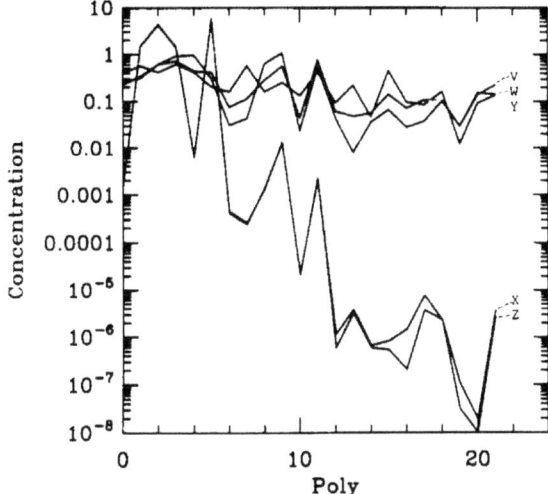

Figure 8. An experiment demonstrating the robust properties of an autocatalytic set. The food set is originally a, b, ab, and bb. The food set is altered in four different ways, as shown in table 2. For each alteration of the food set the concentrations of all 21 polymers in the autocatalytic set are plotted against the "polymer index". (The polymer index assigns a unique label to each polymer. It is ordered according to length, but is otherwise arbitrary.) Two of the alterations of the food set cause the autocatalytic set to die, while the other two hardly change it. Like a robust metabolism, the autocatalytic set can digest a variety of different foods.

almost the same; they are all maintained at high concentration. For the other two, the autocatalytic set "dies" in that some of the members of the set fall below the concentration threshold, and most of the concentrations decrease dramatically (Bagley and Farmer 1990).

Our numerical evidence suggests that any fixed reaction network always approaches a fixed point where the concentrations are constant. However, since spontaneous reactions always take place, there is the

Table 2. An experiment in varying the food set of an autocatalytic set. The table shows the four species of the food set, and the concentration of each that is supplied externally per unit time. Case v is used to "grow" the autocatalytic set, and cases $w - z$ are four changes made once the autocatalytic set is established. x and z kill the autocatalytic set, while w and y sustain it with only minimal alteration, as shown in fig. 8.

	a	b	ab	bb
v	5	5	5	5
w	5	0	5	7.5
x	0	0	10	5
y	10	20	0	0
z	0	10	10	0

possibility that a new species will be created that is on the graph of the autocatalytic set, but which the kinetics did not yet reach. If the catalyzed pathway is sufficiently strong, then the new species may be regenerated and added to the (kinetic) autocatalytic set. This is the way the autocatalytic sets evolve; spontaneous reactions provide natural variation, and kinetics provides selection.

Autocatalytic networks create a rich, focused set of enzymes at high concentration. They form simple metabolisms, which might have provided a substrate for contemporary life.

The results discussed here, as well as many others, will be described in more detail in a future paper (Bagley and Farmer 1990). We intend to study the evolution of autocatalytic sets, and to make a closer correspondence to experimental parameter values.

6.2. COMPARISON TO GENERIC NETWORK

(i) *Nodes* correspond to both polymer species and to reactions. The states are determined by the concentrations of the polymers.

(ii) *Connections*. The graph connections are quite different in this system, in that there are no direct reaction connections to the same types of nodes. Each reaction node is connected by undirected links to exactly three polymer nodes, and contains one (directed) catalytic link to one or more polymer nodes. A polymer node can be connected by a solid link to any number of reaction nodes, and can have any number of catalytic links to reaction nodes.

(iii) *Dynamics*. The dynamics is based on the laws of mass action. The equations are physically realistic, and are considerably more complicated than those of the other networks we have discussed. Arbitrarily label all the polymer species by an index i, and let x_i represent the concentration of the ith species. Assume that all the forward reactions in eq. (18) have the same rate constant k_f, all the backward reactions have the same rate constant k_r, and that all catalyzed reactions have the same velocity ν. Let the quantity m_{ijke} represent the connections in the two graphs, where i and j refer to the two species that join together to form k under enzyme e. $m_{ijke} = 1$ when there is a catalyzed reaction, and $m_{ijke} = 0$ otherwise. $m_{ijke} = m_{jike}$. Let the dissipation constant be k, let the rate at which elements are added to the foodset be d, and let h be the concentration of water. Neglecting the effects of enzyme saturation, the equations can be written

$$\frac{\mathrm{d}x_k}{\mathrm{d}t} = \sum_{e,i,j} m_{ijke} \left(1 + \nu x_e\right) \left(k_\mathrm{f} x_i x_j - k_\mathrm{r} h x_k\right)$$
$$+ 2 \sum_{l,m,e} m_{klme} \left(1 + \nu x_e\right) \left(k_\mathrm{r} h x_m - k_\mathrm{f} x_k x_l\right) \quad (21)$$
$$- k x_k + d f\left(x_k\right).$$

f is a function whose value is one if x_k is in the food set, and is zero otherwise. More accurate equations incorporating the effect of enzyme saturation are given in Farmer, Kauffman, and Packard (1986).

An effective instantaneous connection strength can be computed by evaluating $\partial \dot{x}_k / \partial x_p$. The resulting expression is too complicated to write here. Like the immune network, the instantaneous connection strength can be either excitatory or inhibitory depending on where the network is relative to its steady state value. In contrast to the other networks we have studied, there are no special variables in eq. (21) that explicitly play the role of either node or connection parameters. The concentration of the enzymes x_e that catalyze a given reaction is suggestive of the connection parameters in other connectionist networks. However, since any species can be a reactant in one equation and an enzyme in another, there is no explicit separation of time scales between x_e and the other variables.

(iv) *Graph dynamics.* The separation of time scales usually associated with learning occurs entirely through modification of the graph. The deterministic behavior for any given graph apparently goes to a fixed point. However, in a real autocatalytic system there are always spontaneous reactions creating new species not contained in the catalytic reaction graph. It occasionally happens that one of the new species catalyzes a pathway that feeds back to create that species. Such a fluctuation can be amplified enormously, altering the part of the catalyzed graph that is above the concentration threshold. This provides a mechanism for the evolution of autocatalytic networks.

It is debated whether or not autocatalytic networks can undergo evolution in a Darwinian sense, at the least it seems to impose further constraints on the topology of the graph itself (Vasas et al. 2012).

Autocatalytic networks are interesting from a connectionist point of view because of their rich graph structure and because of the possibilities opened up by catalytic activity. Catalytic activity is analogous to amplification in electronic circuits; it results in multiplicative terms that either amplify or suppress the activity of a given node. The fixed points of the network may be thought of as self-sustaining memories, caused by the feedback of catalytic activity. The dynamical equations that we use here are based on reversible chemical reactions, and lead to unique fixed points. However, other chemical reaction networks can have multiple fixed points, and it seems likely that when we alter the model to study irreversible reactions such as those observed in contemporary metabolisms, we will see multiple fixed points. In this case the computational possibilities of such networks become much more complex.

7. Other Potential Examples and Applications

The four examples discussed here are by no means the only ones where connectionist models have been used, or could be used. Limitations of space and time prevent a detailed examination of all the possibilities, but a few deserve at least cursory mention.

Bayesian inference networks, Markov networks, and constraint networks are procedures used in artificial intelligence and decision theory for organizing and codifying casual relationships in complex systems (Pearl 1988). Each variable corresponds to a node of the network. Each node is connected to the other variables on which it depends. Bayesian networks are based on conditional probability distributions, and use directed graphs; Markov networks are based on joint probability and have undirected graphs; constraint networks assume deterministic constraints between variables. These networks are most commonly used to incorporate prior knowledge, make predictions and test hypotheses. Learning good graph representations is an interesting problem where further work is needed.

Boolean networks. A neural network whose transition rule is a binary automaton is an example of a Boolean network. In general there is no need to restrict the dynamics to the sum and threshold rules usually used in neural nets (other than the fact that this may make the learning problem simpler). Instead, the nodes can implement arbitrary logical (Boolean) functions. Kauffman studied the emergent properties of networks in which each node implements a random Boolean function (Kauffman 1969, 1984). (The functions are fixed in time, but each node implements a different function.) More recently, Miller and Forrest (1989) have shown that the dynamics of classifier systems can be mapped into Boolean networks. This allows them to describe the emergent properties of classifier systems. Their work implicitly maps Boolean networks to the generic connectionist framework. The formulation of learning rules for general Boolean networks is an interesting problem that deserves further study. Kauffman has done some work using point mutation to modify the graph (Kauffman 1990).

Ecological models and population genetics are a natural area for the application of connectionism. There is a large body of work modeling plant and animal populations and their interactions with their environment in terms of differential equations. In these models it is necessary to explicitly state how the populations interact, and translate this into mathematical form. An alternative is to let these interactions *evolve*. A natural framework for such models is provided by the work of Maynard Smith in the application of game-theoretic models to population genetics and ethology (Maynard Smith 1986). The interactions of the populations with each other are modeled as game-theoretic strategies. In these models, however, it is necessary to state in advance what these strategies are. A natural alternative is to let the strategies evolve. Some aspects of this have been addressed in the fledgling theory of evolutionary games (Friedman 1989). A connectionist approach is a natural extension of this work. The immune networks discussed here are very similar to predator-prey models. The strings encoding chemical properties are analogous to genotypes of a given population, and the matrix of interactions are analogous to phenotypes.

Economics is another natural area of application. Again, existing game-theoretic work suggests a natural avenue for a connectionist approach, which could be implemented along the lines of the immune model. The binary strings can be viewed as encoding simple strategies, specifying the interactions of economic agents. Indeed, there are already investigations of models of this type based on classifier systems (Arthur 1989a, 1989b; Marimon, McGrattan, and Sargeant 1989).

Game theory is a natural area of application. For example, Axelrod (1986) has studied the game of iterated prisoner's dilemma. His approach was to encode recent past moves as binary variables, and encode the strategy of the player as a Boolean function. He demonstrated that genetic algorithms can be used to evolve Boolean functions that correspond to good strategies. An alternative approach would be to distribute the strategy over many nodes, and use a connectionist model instead of a look-up table. Such models may have applications in many different problems where evolutionary games are relevant, such as economics and ethology.

Molecular evolution models. The autocatalytic model discussed in detail here is by no means the only connectionist model for molecular evolution. Perhaps one of the earliest example is the hypercycle model of Eigen and Schuster (1979), which has recently been compared to the Hopfield neural network models (Hopfield and Tank 1985; Pichler, Keeler, and Ross 1989). For a review see Hofbauer and Sigmund (1988).

8. Conclusions

I hope that presenting four different connectionist systems in a common framework and notation will make it easier to transfer results from one field to another. This should be particularly useful in areas such as immune networks, where connectionist models are not as well developed as they are in other areas, such as neural networks. By showing how similar mathematical structure manifests itself in quite different contexts, I hope that I have conveyed the broad applicability of connectionism. Finally, I hope that these mathematical analogies make the underlying phenomena clearer. For example, comparing the role of the lymphocyte in these models to the role of neurons may give more insight into the construction of immune networks with more computational power.

8.1. OPEN QUESTIONS

Hopefully the framework for connectionist models presented here will aid the development of a broader mathematical theory of connectionist systems. From an engineering point of view, the central question is: What is the most effective way to construct good connectionist networks? Questions that remain unclear include:

(i) In some systems, such as neural networks and classifier systems, a connection is always either inhibitory or excitatory. In others, such as immune networks and autocatalytic networks, a connection can be either inhibitory or excitatory, depending on the state of the system. Does the latter more flexible approach complicate learning? Does it give the network any useful additional computational power?

(ii) Is it essential to have independent parameters for each connection? In neural nets, each connection has its own parameter. In classifier systems, the use of the "don't care" symbol means that many connections are represented by one classifier, and thus share a common connection parameter. This decreases the flexibility of the network, but at the same time gives an efficient graph representation, and aids the genetic algorithms in finding good graphs. In B-cell immune networks the parameters reside entirely in the nodes, and thus as a single parameter changes many different connections are effected. Does this make it impossible to implement certain functions? How does this effect learning and evolution? (It is conceivable that the reduction of parameters may actually cause some improvements.)

(iii) What is the optimal level of complexity for the transition rule? Some neural nets and classifier systems employ simple activation functions, such as linear threshold rules. Somewhat more complicated nonlinear functions, such as sigmoids, have the advantage of being smooth; immune networks have even more complicated activation functions. An alternative is to make each node a flexible function approximation box, for example, with its own set of local linear functions, so that the node can approximate functions with more general shapes (Farmer and Sidorowich 1988; Wolpert 1989). However, complexity also increases the number of free parameters and potentially increases the amount of data needed for learning.

(iv) A related question concerns the role of catalysis. In autocatalytic networks, a node can be switched on or off by another node through *multiplicative* coupling terms. In contrast to networks in which inputs can only be summed, this allows a single unit to exert over-riding control over another. A similar approach has been suggested in Σ-Π neural networks (Rummelhart and McClelland 1986); T-cells and neurotransmitters may play a similar role in real biological systems. How valuable is specific catalysis to a network? How difficult is the learning problem when it is employed?

(v) What are the optimal approaches to evolving good graph representations? Most of the work in this area has been done for classifier systems, although even here many important issues remain to be clarified. All known algorithms that can *create* connections and nodes, such as the genetic algorithms, are stochastic; there are deterministic pruning algorithms that can only destroy connections, such as orthogonal projection. Are there efficient deterministic algorithms for *creating* new graph connections?

(vi) What are the best learning algorithms? A great deal of effort has been devoted to answering this question, but the answer is still obscure. A perusal of the literature suggests certain general conclusions. For example, in problems with detailed feedback, e.g. a list of known input–output pairs, deterministic function fitting algorithms such as least-squares minimization (of which back-propagation is an example) can be quite effective. However, if the search space is not smooth, for example because the samples are too small to be statistically stable, stochastic algorithms such as crossover are often more effective (Ackley 1987). In more general situations where there is no detailed feedback, there seems to be no general consensus as to which learning algorithms are superior.

Thus far, very few connectionist networks make use of nontrivial computational capabilities. In typical applications most connectionist networks end up functioning as stimulus–response systems, simply mapping inputs to outputs without making use of conditional looping, subroutines, or any of the power we take for granted in computer programs. Even in systems that clearly have a great deal of computational power *in principle*, such as classifier systems, the solutions actually learned are usually close to look-up tables. It seems to be much easier to implement effective learning rules in simpler architectures that sacrifice computational complexity, such as feed-forward networks.

It may be that there is an inherent trade-off between the complexity of learning and the complexity of computation, so that the difficulty of learning increases with computational power. At one end of the spectrum is a look-up table. Learning is trivial; examples are simply inserted as they occur. Unfortunately, all too often neural network applications have not been compared to this simple approach. In the infamous NET-talk problem (Sejnowski and Rosenberg 1987), for example, a simple look-up table gives better performance than a sum/sigmoid back-propagation network (An *et al.* 1987). Simple function approximation is one level above a look-up table in computational complexity; functions can at least attempt to interpolate between examples, and generalize to examples that are not in the learning data set. Learning is still *fairly* simple, although already the subtleties of probability and statistics begin to complicate the matter. However, simple function approximation has less computational capability than a finite state machine. At present, there are no good learning algorithms for finite state machines. Without counting, conditional looping, etc., many problems will simply remain insoluble.

It is probably more likely that learning *is possible* with more sophisticated computational power, and that we simply do not yet know how to accomplish it. I suspect that the connectionist networks of the future will be full of loops.

Connectionist models are a useful tool for solving problems in learning and adaptation. They make it possible to deal with situations in which there are an infinite number of possible variables, but in which only a finite number are active at any given time. The connections are explicit but changeable. We have only recently begun to acquire the computational capabilities to realize their potential. I suspect that the next decade will witness an enormous explosion in the application of the connectionist methodology.

However, connectionism represents a level of abstraction that is ultimately limited by such factors as the need to specify connections explicitly, and the lack of built-in spatial structure. Many problems in adaptive systems ultimately require models such as partial differential equations or cellular automata with spatial structure (Langton 1989). The molecular evolution models of Fontana et al., for example, explicitly model the spatial structure of individual polymers in an artificial chemistry. As a

Table 3. A Rosetta Stone for connectionism.

GENERIC	NEURAL NET	CLASSIFIER SYSTEM	IMMUNE NET	AUTOCATALYTIC NET
node	neuron	message	antibody type	polymer species
state	activation level	intensity	free antibody/ antigen concentration	polymer concentration
connection	axon/synapse/ dendrite	classifier	chemical reaction of antibodies	catalyzed chemical reaction
parameters	connection weight	strength and specificity	reaction affinity lymphocyte concentration	catalytic velocity
interaction rule	sum/sigmoid	linear threshold and maximum	bell-shaped	mass action
learning algorithm	Hebb, back-propagation	bucket brigade (gen. Hebb)	clonal selection (gen. Hebb)	approach to attractor
graph dynamics	synaptic plasticity	genetic algorithms	genetic algorithms	artificial chemistry rules, spontaneous reactions

result the phenotypes emerge more naturally than in the artificial chemistry in the autocatalytic network model discussed here. On the other hand, the approach of Fontana et al. requires more computational resources. For many problems connectionism may provide a good compromise between accurate modeling and tractability, appropriate to the study of adaptive phenomena during the last decade of this millennium.

8.2. ROSETTA STONE

This paper is a modest start toward creating a common vocabulary for connectionist systems, and unifying work on adaptive systems. Like the Rosetta Stone, it contains only a small fragment of knowledge. I hope it will nonetheless lead to a deeper understanding in the future. Table 3 summarizes the analogies developed in this paper.

Acknowledgements

I would like to thank Rob De Boer, Walter Fontana, Stephanie Forrest, André Longtin, Steve Omohundro, Norman Packard, Alan Perelson, and Paul Stolorz for valuable discussions, and Ann and Bill Beyer for lending valuable references on the Rosetta Stone.

I urge the reader to use these results for peaceful purposes.

Appendix. A Superficial Taxonomy of Dynamical Systems

Dynamical systems can be trivially classified according to the continuity or locality of the underlying variables. A variable either can be discrete, i.e. describable by a finite integer, or continuous. There are three essential properties:

(i) *Time.* All dynamical systems contain time as either a discrete or continuous variable.

(ii) *State.* The state can either be a vector of real numbers, as in an ordinary differential equation, or integers, as for an automaton.

(iii) *Space* plays a special role in dynamical systems. Some dynamical models, such as automata or ordinary differential equations, do not contain the notion of space. Other models, such as lattice maps or cellular automata, contain a notion of locality and therefore space even though they are not fully continuous. Partial differential equations or functional maps have continuous spatial variables.

This is summarized in table 4.

Table 4. Types of dynamical systems, characterized by the nature of time, space, and state. "Local" means that while this property is discrete, there is typically some degree of continuity and a clear notion of neighborhood.

TYPE OF DYNAMICAL SYSTEM	SPACE	TIME	REPRESENTATION
partial differential equations	continuous	continuous	continuous
computer representation of a PDE	local	local	local
functional maps	continuous	discrete	continuous
ordinary differential equations	none	continuous	continuous
lattice models	local	discrete or continuous	continuous
maps (difference equations)	none	discrete	continuous
cellular automata	local	discrete	discrete
automata	none	discrete	discrete

REFERENCES

Ackley, D. H. 1987. "An Empirical Study of Bit Vector Function Optimization." In *Genetic Algorithms and Simulated Annealing,* edited by L. Davis. Los Altos, CA: Morgan Kaufmann.

Alkon, D. L. 1989. "Memory Storage and Neural Systems." *Scientific American* 261:26–34.

An, Z. G., S. M. Mniszewski, Y. C. Lee, G. Papcun, and G. D. Doolen. 1987. "HI-ERtalker: A Default Hierarchy of High Order Neural Networks That Learns To Read English Aloud." *Center for Nonlinear Studies Newsletter* (Los Alamos National Laboratory).

Arthur, W. B. 1989a. *Nash-Discovery Automata for Finite-Action Games.* Technical report. Santa Fe Institute.

———. 1989b. *On the Use of Classifier Systems on Economics Problems.* Technical report. Santa Fe Institute.

Axelrod, R. 1986. "An Evolutionary Approach to Norms." *American Political Science Review* 80:1095–1111.

Bagley, R. J., and J. D. Farmer. 1990. *Robust Autocatalytic Sets.* In progress.

Bagley, R. J., J. D. Farmer, S. A. Kauffman, N. H. Packard, A. S. Perelson, and I. M. Stadnyk. 1990. "Modeling Adaptive Biological Systems." *Biocybernetics* 23:113–137.

Belew, R. K., and M. Gherrity. 1989. *Back Propagation for the Classifier System.* Technical report. University of California San Diego.

Booker, L. B., D. E. Goldberg, and J. H. Holland. 1989. "Classifier Systems and Genetic Algorithms." *Artificial Intelligence* 40:235–282.

Broomhead, D., and D. Lowe. 1988. *Radial Basis Functions, Multivalued Functional Interpolation and Adaptive Networks.* Technical report Memorandum 4148. Royal Signals and Radar Establishment.

Casdagli, M. 1989. "Nonlinear Prediction of Chaotic Time Series." *Physica D* 35:335–356.

Compiani, M., D. Montanari, R. Serra, and G. Valastro. 1988. "Classifier Systems and Neural Networks." In *Proceedings of the Second Workshop on Parallel Architectures and Neural Networks,* edited by E. Caianiello. Singapore: World Scientific.

Cowan, J. D., and D. H. Sharp. 1988. "Neural Nets." *Quart. Rev. Biophys.* 21:365–427.

Davis, L. 1989. "Mapping Classifier Systems into Neural Networks." In *Neural Information Processing Systems 1,* edited by D.S. Touretzsky. Los Altos, CA: Kaufmann.

De Boer, R. J. 1990. *Dynamical and Topological Patterns in Developing Idiotypic Networks.* Technical report. Los Alamos National Laboratory.

De Boer, R. J., and P. Hogeweg. 1989. "Unreasonable Implications of Reasonable Idiotypic Network Assumptions." *Bulletin of Mathematical Biology* 51:381–408.

Eigen, M., and P. Schuster. 1979. *The Hypercycle: A Principle of Natural Self-Organization.* Berlin, Germany: Springer-Verlag.

Farmer, J. D., S. A. Kauffman, and N. H. Packard. 1986. "Autocatalytic Replication of Polymers." *Physica D* 22:50–67.

Farmer, J. D., N. H. Packard, and A. S. Perelson. 1986. "The Immune System, Adaptation, and Machine Learning." *Physica D* 22:187–204.

Farmer, J. D., and J. J. Sidorowich. 1988. "Exploiting Chaos to Predict the Future and Reduce Noise." In *Evolution, Learning and Cognition,* edited by Y. C. Lee. Singapore: World Scientific.

Forrest, S. 1985. "Implementing Semantic Network Structures using the Classifier System." In *Proceedings of the First International Conference on Genetic Algorithms and their Applications,* edited by John J. Grefenstette. Hillsdale, N.J: Lawrence Erlbaum Associates.

Friedman, D. 1989. *Evolutionary Games in Economics.* Technical report. Stanford University.

Giles, C. L., R. D. Griffin, and T. Maxwell. 1986. "Encoding Geometric Invariances in Higher Order Neural Networks." In *Neural Networks for Computing,* edited by J.S. Denker. New York, NY: AIP.

Harp, S. A., T. Samad, and A. Guha. 1989. "Towards the Genetic Synthesis of Neural Networks," in *Proceedings of the Third International Conference on Genetic Algorithms,* edited by J.D. Schaffer. Los Altos, CA: Kaufmann.

Hofbauer, J., and K. Sigmund. 1988. *The Theory of Evolution and Dynamical Systems.* Cambridge, UK: Cambridge Univ. Press.

Hoffmann, G. W. 1975. "A Theory of Regulation and Self-Nonself Discrimination in an Immune Network." *European Journal of Immunology* 5:638–647.

Hoffmann, G. W., T. A. Kion, R. B. Forsyth, K. G. Soga, and A. Cooper-Willis. 1988. "The n-dimensional network." In *Theoretical Immunology, Part II,* edited by A. S. Perelson. Redwood City, CA: Santa Fe Institute/Addison-Wesley.

Holland, J. H. 1986. "Escaping Brittleness: The Possibilities of General Purpose Machine Learning Algorithms applied to Parallel Rule-based Systems." In *Machine Learning II.* Los Altos, CA: Kaufmann.

Holland, J. H., K. J. Holyoak, R. E Nisbett, and P. R. Thagard. 1986. *Induction: Processes of Inference, Learning, and Discovery.* Cambridge, MA: MIT Press.

Hopfield, J. J., and D. W. Tank. 1985. ""Neural" Computation of Decisions in Optimization Problems." *Biological Cybernetics* 52:141–152.

Jerne, N. K. 1973. "The Immune System." *Scientific American* 229:52–60.

———. 1974. "Towards A Network Theory of the Immune System." *Ann. Immunology (Inst. Pasteur)* 125C:373–389.

Jones, R., C. Barnes, Y. C. Lee, and K. Lee. 1989. "Fast Algorithm for Localized Prediction." Private communication.

Kauffman, S. A. 1969. "Metabolic Stability and Epigenesis in Randomly Constructed Genetic Nets." *Journal of Theoretical Biology* 22:437.

———. 1984. "Emergent Properties in Random Complex Automata." *Physica D* 10:145–156.

———. 1986. "Autocatalytic Sets of Proteins." *Journal of Theoretical Biology* 119:1–24.

———. 1990. *Origins of Order, Self-Organization, and Selection in Evolution.* Oxford, UK: Oxford University Press.

Langton, C. G., ed. 1989. *Artificial Life.* Redwood City, CA: Addison-Wesley.

Marimon, R., E. McGrattan, and T. J. Sargeant. 1989. *Money as a Medium of Exchange in an Economy with Artificially Intelligent Agents.* Technical report 89-004. Santa Fe, NM: Santa Fe Institute.

Maynard Smith, J. 1986. "Evolutionary Game Theory." *Physica D* 22:43–49.

McCulloch, W. S., and W. Pitts. 1943. "A Logical Calculus of the Ideas Immanent in Nervous Activity." *Bulletin of Mathematical Biophysics* 5:115–133.

Miller, G. F., P. M. Todd, and S. U. Hegde. 1989. "Designing Neural Networks Using Genetic Algorithms." Edited by J.D. Schaffer. *Kaufmann* (Los Altos, CA).

Miller, J. H., and S. Forrest. 1989. "The Dynamical Behavior of Classifier Systems." In *Proceedings of the Third International Conference on Genetic Algorithms,* edited by J. D. Schaffer. Los Altos, CA: Morgan Kaufmann Publishers Inc.

Miller, S. L., and H. C. Urey. 1959. "Organic Compound Synthesis on the Primitive Earth." *Science* 130:245–251.

Montana, D., and L. Davis. 1989. "Training Feedforward Neural Networks Using Genetic Algorithms." In *Proceedings of the Eleventh International Joint Conference on Artificial Intelligence.*

Moody, J., and C. Darken. 1988. *Learning With Localized Receptive Fields.* Technical report. Department of Computer Science, Yale University.

Newell, A. 1973. "Production systems: models of control structures." In *Visual Information Processing,* edited by W. G. Chase. New York: Academic Press.

Pearl, J. 1988. *Probabilistic Reasoning in Intelligent Systems: Networks of Plausible Inference.* Los Altos, CA: Kaufmann.

Perelson, A. S. 1988. "Toward a Realistic Model of the Immune System." In *Theoretical Immunology, Part II,* edited by A. S. Perelson. Redwood City, CA: Santa Fe Institute/Addison-Wesley.

———. 1989. "Immune Network Theory." *Immunological Reviews* 110:5–36.

Pichler, E. S., J. D. Keeler, and J. Ross. 1989. *Comparison of Self-Organization and Optimization in Evolution and Neural Network Models.* Technical report. Chemistry Department, Stanford University.

Pineda, F. J. 1988. "Generalization of Backpropagation to Recurrent and Higher Order Neural Networks." In *Neural Information Processing Systems,* edited by D. Z. Anderson. New York, NY: AIP.

Poggio, T., and F. Girosi. 1989. *A Theory of Networks for Approximation and Learning.* Technical report. MIT.

Pope, S. Unpublished research.

Richter, P. H. 1975. "A Network Theory of the Immune System." *European Journal of Immunology* 5:350–354.

Riolo, R. 1986. *CFS-C: A Package of Domain Independent Subroutines for Implementing Classifier Systems in Arbitrary, User-Defined Environments.* Technical report. Logic of Computers Group, University of Michigan.

Rosenblatt, F. 1958. "The Perceptron: A Probabilistic Model for Information Storage and Organization in the Brain." *Psychological Review* 65:386.

Rummelhart, D., and J. McClelland. 1986. *Parallel Distributed Processing, Vol 1.* Cambridge, MA: MIT Press.

Scott, A. C. 1977. *Neurophysics.* New York, NY: Wiley.

Segel, L. A., and A. S. Perelson. 1989. "Shape Space Analysis of Immune Networks." In *Cell to Cell Signalling: From Experiments to Theoretical Models,* edited by A. Goldbeter. New York, NY: Academic Press.

Sejnowski, T. J., and C. R. Rosenberg. 1987. "Parallel Networks that Learn to Pronounce English Text." *Complex Systems* 1:145–168.

Varela, F. J., A. Coutinho, B. Dupire, and N. N. Vaz. 1988. "Cognitive Networks: Immune, Neural, and Otherwise." In *Theoretical Immunology, Part II,* edited by A. S. Perelson. Redwood City, CA: Santa Fe Institute/Addison-Wesley.

Wallis Budge, E. A. 1929. *The Rosetta Stone.* London, UK: The Religious Tract Society.

Whitley, D., and T. Hanson. 1989. "Optimizing Neural Networks Using Faster, More Accurate Genetic Search." In *Proceedings of the Third International Conference on Genetic Algorithms,* edited by J. D. Schaffer. Los Altos, CA, Kaufmann.

Wilson, S. W. 1989. "Bid Competition and Specificity Reconsidered." *Complex Systems* 2:705–723.

———. 1990. "Perceptron Redux: Emergence of Structure." *Physica D* 42:249–256.

Wolpert, D. H. 1989. "A Benchmark for How Well Neural Nets Generalize." *Biological Cybernetics* 61:303–313.

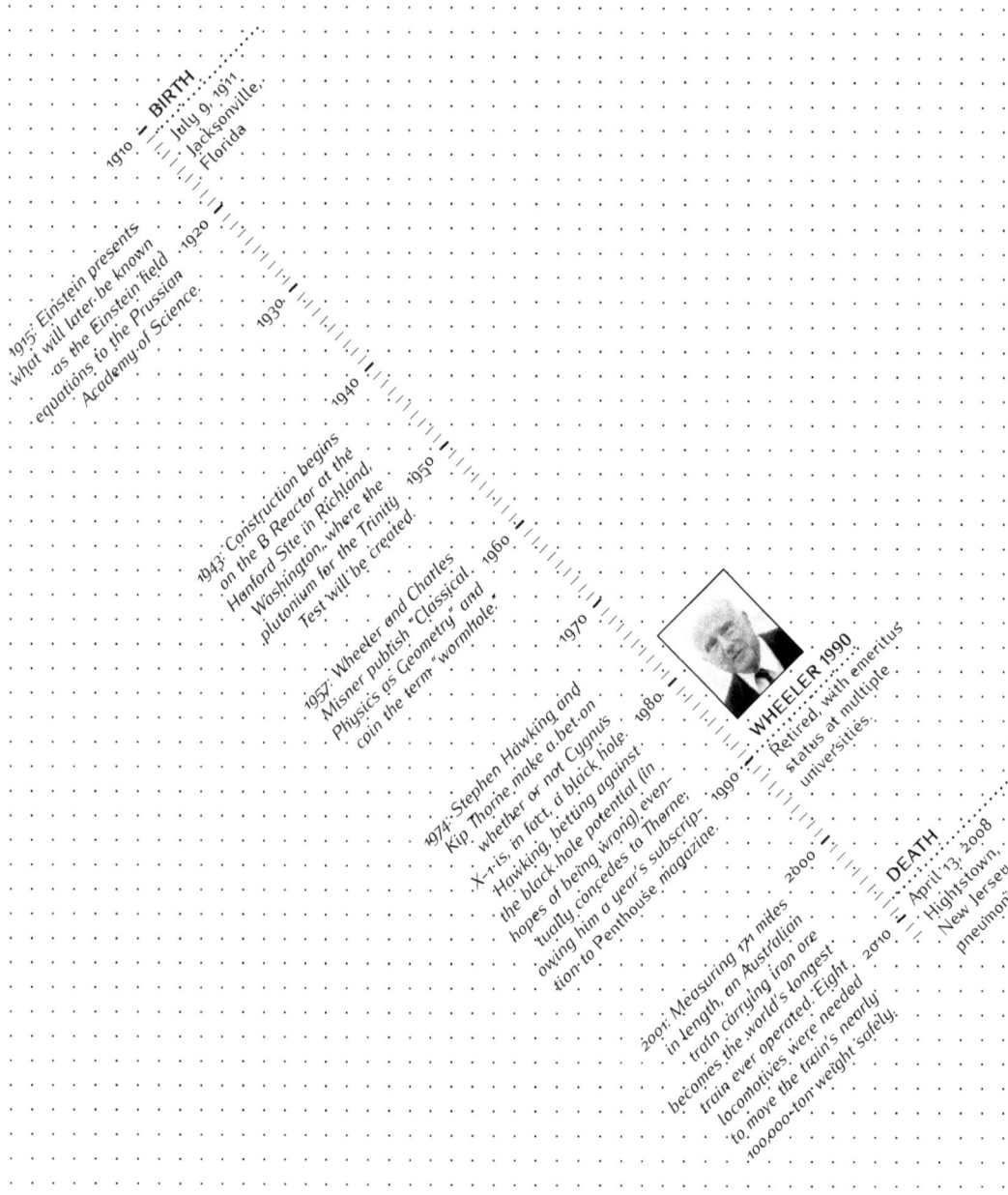

BIRTH — July 9, 1911, Jacksonville, Florida

1915: Einstein presents what will later be known as the Einstein field equations to the Prussian Academy of Science.

1943: Construction begins on the B Reactor at the Hanford Site in Richland, Washington, where the plutonium for the Trinity Test will be created.

1957: Wheeler and Charles Misner publish "Classical Physics as Geometry" and coin the term "wormhole."

1974: Stephen Hawking and Kip Thorne make a bet on whether or not Cygnus X-1 is, in fact, a black hole. Hawking, betting against the black hole potential (in hopes of being wrong), eventually concedes to Thorne, owing him a year's subscription to Penthouse magazine.

WHEELER 1990 — Retired, with emeritus status at multiple universities.

2001: Measuring 171 miles in length, an Australian train carrying iron ore becomes the world's longest train ever operated. Eight locomotives were needed to move the train's nearly 100,000-ton weight safely.

DEATH — April 13, 2008, Hightstown, New Jersey, pneumonia.

JOHN ARCHIBALD WHEELER

[70]

AND IT IS ALL ONE TO ME

Jessica C. Flack, Santa Fe Institute

And it is all one to me
Where I am to begin; for I shall return there again.
—PARMENIDES, *Fragment 5* (trans. David Gallop)

In the contents of consciousness we recognize three sorts of elements, Firstness, Secondness, **Thirdness**. *[—] What a Third is depends on two other things between which it mediates. Firstness is feeling-quality; secondness is brute reaction;* **thirdness** *is mediation.*
—C. S. PEIRCE, *The Logic Notebook* (1865–1909)

John Archibald Wheeler, in this little understood but widely read and referenced essay, coins the phrase "it from bit" and argues that information is the natural language of the universe. Bits are its building blocks, preceding objects and matter. Order, meaning, and causality itself result from the simple act of registering a choice—an idea Wheeler encapsulates with the term "observer-participancy" (OP). Making a binary choice—yes or no—reduces uncertainty from an information-theoretic standpoint as the number of options (Haji-Akbari *et al.* 2009; Walchover 2017) before the question is answered—and hence the entropy—goes from two (one bit) to zero upon answering. The decrease in uncertainty also allows the OP to allocate its investments more efficiently, giving it access to new options and increasing entropy again (Flack 2017).

In my read of Wheeler, out of this cycle of symmetry-breaking come the sensations of space and time and, ultimately, with the concordant global increase in entropy, an arrow of time. The sensation that time passes is amplified by the mesoscale emergence of hierarchically structured physical and living systems. Further partitioning cause from effect by creating a separation of a scales, this hierarchy allows for feedback, reifying the loop created by the simple act of asking

J. A. Wheeler, "Information, Physics, Quantum: The Search for Links," *Complexity, Entropy, and the Physics of Information*, Santa Fe Institute Studies in the Sciences of Complexity, Redwood City, CA: Addison–Wesley, 354–368 (1989).

and answering questions, and also creating temporal and spatial heterogeneity and clustering. Reality—the universe—is a loop, but with complex internal structure. Wheeler's essay is mostly an argument for the former with allusions to the latter.

To state this another way and connect with more familiar debates within complexity science, the determinism of fundamental physics—and to a lesser extent the determinism seemingly implied by the emergent hierarchy of screened off organizational levels in biology—is an illusion in the deepest sense, although not necessarily in a proximal sense. Reality, Wheeler writes, is fundamentally circular: "physics gives rise to observer-participancy; observer-participancy gives rise to information; and information gives rise to physics." If there is any inception point it is in the interaction of particles—initially the massless fundamental particles of quantum phenomena—and registration of these interactions by the changes they induce. In this simple Peircean quantum triad of particle interaction and registration, which Wheeler refers to as elementary observer-participancy, existence, the universe, and consciousness, have their origins.

There's a tad more to unpack but for the moment it is worth highlighting that many essays have been written about Wheeler's seductive yet initially perplexing idea, generally with the same conceptual formula. Most begin by reiterating the meaning of "it from bit," discuss the double-slit and delayed-choice particle physics experiments—the surprising results of which helped Wheeler reach his conclusions—then wander on to the loosely woven connections he makes with adaptive systems and consciousness, and finally settle reluctantly into the conclusion that, while provocative, the paper is a failure, having not provided a satisfactory account of how observer-participants themselves arise. This formula follows Wheeler's own conceptual line through his essay but only as it strikes surface.

I have now read the paper many times and must concede to having had these same thoughts as I struggled (always with enjoyment) through it. However, the more attention I give it the more convinced I am that although Wheeler characteristically creates many idea portals to untrodden intellectual territory that we readers could spend lifetimes exploring, he also creates efficient and creative paths through heavily trodden bogs, even if he also takes a few semantic and didactic liberties to get us to follow him.

Wheeler's most intoxicating conclusion is that information precedes matter, rather than the other way around, as is typically the case in grand unified theories. Wheeler's story turns not on objects like strings but on information-theoretic entities—decoders or registrants—his so-called observer-participants or choice-makers, as I call them.

Readers presumably assume, as I initially did, that an observer-participant needs to be an object with mass and, more egregiously, the capacity for computation. In other words, the universe could not come into existence without complex objects to start. This is both confusing (how could complexity be there from the beginning?) and a confusion. The confusion stems in part from human (perhaps Western) biases about what is required to perceive or register or pose a question. It also stems from Wheeler's strategic decision to use several experiments in particle physics and cosmology to ground OP. In stressing the role of the measuring apparatus in these experiments as the registrant—the observer-participant—he immediately makes the concept of OP graspable, but the anthropomorphism also obscures.

The experiments that best distill the OP concept concern the nature of photons, whether they exist, where they are located, and whether they travel as particles or waves. To know a photon exists we must make measurements. Specifically, we can register the existence of a photon when it strikes and is absorbed by an electron in a photomultiplier that accepts its energy. The photon does not exist prior to emission or after detection—as it is destroyed when it is absorbed, meaning, importantly, it must be destroyed to confirm its existence, a point I will return to shortly.

For the moment, the key takeaway is the Peircean triadic nature of this experiment—there is a photon, an electron, and an observer-participant. In Wheeler's telling, and in most descriptions of these experiments by others, the observer is the measurement device, or, if one backs up a step or two, the experiment-performing scientist. This gives the impression that observer-participants must be complex, but this is not generally assumed by physicists. And in this tiny modification to Wheeler's presentation lies a key to averting the essay's frequent but superficial dismissal—a third object with mass is not necessary—the triad can be simplified into two objects and two actions—the photon, electron, their collision, and the registration of the absorption of energy by the electron itself. In other words, the observer can be a third object with mass like a scientist, or it can be one of the interacting partners that in experiencing the event effectively registers it. The jump from first-person participancy to third-person participancy presumably entails an increase in complexity. Wheeler does not discuss this distinction or its implications but it potentially relates to the distinction between first- and third-person probabilities of what is observed vs. what occurs in quantum cosmology, as discussed in Hartle and Hertog (2017). I leave that thought to be explored by others.

The participancy part of observer-participancy is easier to understand. It comes from the fact that to register the photon an electron must absorb and consequently destroy it. The act of registering, or measuring, changes the nature of reality. Interestingly, when the electron absorbs the photon entropy in one sense decreases and in another increases. It decreases from the point of view of the photon's existence—the absorption confirms that answer as "yes," the photon existed. On the other hand, the absorption also pushes the electron into a higher energy state, increasing its options. Wheeler does not discuss this information transition, but recent work in biology on the information-theoretic conditions supporting macroscopic expansions suggests that it is an important factor in emergence (Flack 2017).

Wheeler follows his explanation of observer-participants and photons with a discussion of black holes, arguing that bits—via yes–no questions—not only bring objects into existence but give them their properties. In the case of black holes, their size—an "it"—is defined by the number of bits of information they contain. Through this example Wheeler gives substance to information as the language of the universe and bits as its building blocks. He is deriving physics from information—writing, "The quantum, $\hbar$, in whatever correct physics formula it appears thus serves as a lamp. It lets us see the horizon area as information lost [measured as entropy], understand wave number of light as photon momentum, and think of field flux as bit-registered fringe shift."

At this juncture, Wheeler begins to consider the implications of this new way of thinking for understanding how the universe is organized and what, besides the quantum, could be fundamental. Nothing, it turns out. He concludes there is no tower of turtles—existence is not in its most essential description a hierarchy but a loop. Laws and any hierarchical structure are consequences of self-synthesis that ultimately contribute to the loop. Space and time are emergent and additionally relative, with fluctuations at distances of the Planck length so great that space becomes lumpy and time nonsensical, emerging more meaningfully instead from convenient coarse-grainings over fluctuations that give rise to discreet rather than continuous objects.

Among the most important statements in the paper is, "Reality is a theory; the past has no evidence except as it is recorded in the present." Registration is an act of amplification that screens off the quantum level but also produces bits that observer-participants use to do work (the electron enters a higher-energy state). This sounds very much like biology and like arguments I have made in my own papers! As the number of observer-participants grows, their joint interventions enrich the universe, the objective reality of which is defined by the degree of agreement among observer-participants in the answers to their yes–no questions. The objective reality resulting from this collective computation of consensus—from this collective intelligence—creates meaning, and observer-participants use the consolidated meaning map to communicate and work more efficiently, keeping the loop looping.

Wheeler is proposing that the universe comes into existence with a quantum triad of particle interaction and registration. This loop has no micro- and macroscale, no intrinsic spacetime, and no continuum. All of that is emergent—or, in the case of the continuum, an illusion, contributing to the loop lumpy, complex mesoscale phenomena that in cycling back on itself through feedback keeps the loop looping. OP requires no complexity to start. No mass and no resource intensive computation. Only quantum phenomena.

Abruptly with this insight comes the possibility that Wheeler's essay is not, in fact, as so many have argued, ultimately a failure. Rather there is tremendous foresight in its reframing and refocusing, as well as actionable research directions. To understand the origins of the loop, we need to understand the origins of the quantum (perhaps the hardest part?). To understand looping, we need a theory of information (perhaps the most open-ended?). And for that, Wheeler writes, we should look to biology and the study of inference and perception.

A natural retort to this claim is that biology is measure zero given the scale of the universe and hence inconsequential. But that view, Wheeler wants us to understand, is surface. It privileges space and time, which, like biological systems, are emergent, not fundamental, and—more significantly—are to some extent a consequence of emergence, and, specifically, of coarse-graining by biological systems as they process the environment and distill regularities. The vastness of space and time is an illusion or at least a red herring; the vastness neither trivializes biology nor ensures biology's contribution is negligible—as Wheeler points out, much can come from almost nothing (the boundary of a boundary is zero).

This concession to biology is a massive frameshift. Biological systems are not a consequence of physics, existing "above it" and largely screened off, but a core, causal set of sub-processes. Not in terms of their influence on particle physics, but in terms of their contribution to loop dynamics. Biological information processing contributes the mesoscale dynamics that keep the loop looping. Carlo Rovelli and colleagues (2018) make a similar point when they suggest that time may not be unitary, but a complex phenomenon with many layers, united only in their connection to entropy, and with each contributing unique properties to reality.

Wheeler, having opened an idea portal, leaves it to descendants to explore. Once one sits with this reframing for a while, the shock of the idea that biology could "contribute to physics" subsides and the natural teleological property of biology comes to the fore as holding the explanation of how.

Where Wheeler leaves off, Jim Hartle picks up. Jim was a student of Wheeler's at Princeton and a collaborator of Santa Fe Institute cofounder and discoverer of the quark, Murray Gell-Mann Gell-Mann. During Jim's visits to SFI we discussed overlaps in our work—overlaps that Jim first perceived even as he worked largely in cosmology and I on computation in adaptive systems—an ideal Institute interaction. It was not, unfortunately, until just after he died this past year in 2023, upon re-reading some of his papers on time, coarse-graining, and information processing, that I realized just how extensively our ideas and interests overlapped.

Jim's work further develops the idea that information should be central in any theory of the universe's origins and structure. It also connects Wheeler's incipient physics of information more concretely to fundamental physics as well as to the biology of information processing, contextualizing Wheeler and his essay within complexity science and pushing "it from bit" towards its conceptual center.

Hartle (2005) gives a form to Wheeler's observer-participants that a biologist easily recognizes. Jim reformulates OPs as IGUSs (information gathering and utilizing systems)—a term Jim borrowed from Murray, who introduced it in *The Quark & the Jaguar*—and uses this more developed concept of information processing to show how human notions of past, present, and future can be sensible within the four-dimensional world of fundamental physics. Perhaps most significantly, Jim outlines the spacetime conditions supporting a common now—that is the capacity of entities separated in space to collectively experience approximately the same present. With this insight Jim provides another central detail to the "it from bit" story, as the capacity to experience or perceive approximately the same present is a key condition for the emergence of hierarchy and levels of organization in biology. I think Wheeler would like this generative trajectory. There is much to do! 🖝

REFERENCES

Flack, J. C. 2017. "Coarse-Graining as a Downward Causation Mechanism." *Philosophical Transactions of the Royal Society A* 375 (2109): 20160338. https://doi.org/10.1098/rsta.2016.0338.

Gell-Mann, M. 1994. *The Quark and the Jaguar: Adventures in the Simple and the Complex*. New York, NY: W. H. Freeman.

Haji-Akbari, A., M. Engel, A. S. Keys, X. Zheng, R. G. Petschek, P. Palffy-Muhoray, and S. C. Glotzer. 2009. "Disordered, Quasicrystalline, and Crystalline Phases of Densely Packed Tetrahedra." *Nature* 462:773–777. https://doi.org/10.1038/nature08641.

Hartle, J. B. 2005. "The Physics of Now." *American Journal of Physics* 73 (2): 101–109. https://doi.org/10.1119/1.1783900.

Hartle, J. B., and T. Hertog. 2017. "The Observer Strikes Back." In *The Philosophy of Cosmology*, edited by K. Chamcham, J. Silk, J. D. Barrow, and S. Saunders. Cambridge, UK: Cambridge University Press.

Parmenides. 1991. *Parmenides of Elea: A Text and Translation with an Introduction*. Translated by D. Gallop. Toronto, Canada: University of Toronto Press.

FOUNDATIONAL PAPERS IN COMPLEXITY SCIENCE

Peirce, Charles S. 1865-1909. "The Logic Notebook. MS [R] 339." http://www.commens.org/bibliography/manuscript/peirce-charles-s-1865-1909-logic-notebook-ms-r-339.

Rovelli, C. 2018. *The Order of Time.* Translated by E. Segre and S. Carnell. New York, NY: Riverhead Books.

Walchover, N. 2017. "'Digital Alchemist' Seeks Rules of Emergence." *Quanta Magazine,* https://www.quantamagazine.org/digital-alchemist-sharon-glotzer-seeks-rules-of-emergence-20170308/.

INFORMATION, PHYSICS, QUANTUM: THE SEARCH FOR LINKS

John Archibald Wheeler
Princeton University and University of Texas, Austin

This report reviews what quantum physics and information theory have to tell us about the age-old question, How come existence? No escape is evident from four conclusions: (1) The world cannot be a giant machine, ruled by any pre-established continuum physical law. (2) There is no such thing at the microscopic level as space or time or spacetime continuum. (3) The familiar probability function or functional, and wave equation or functional wave equation, of standard quantum theory provide mere continuum idealizations and by reason of this circumstance conceal the information-theoretic source from which they derive. (4) No element in the description of physics shows itself as closer to primordial than the elementary quantum phenomenon, that is, the elementary device-intermediated act of posing a yes–no physical question and eliciting an answer or, in brief, the elementary act of observer-participancy. Otherwise stated, every physical quantity, every it, derives its ultimate significance from bits, binary yes-or-no indications, a conclusion which we epitomize in the phrase *it from bit*.

1 Quantum Physics Requires a New View of Reality

Revolution in outlook though Kepler (1619), Newton (1687), and Einstein (1915, 1916, 1917) brought us,[1] and still more startling the story of life (Mendel 1866; Darwin 1859; Watson and Crick 1953)

Order, meaning, and causality itself result from the simple act of registering a choice—an idea that Wheeler encapsulates with the term "observer-participancy."

---

[1] The appendix of Kepler's Book 5 contains one side, and the publications of the English physician and thinker Robert Fludd (1574–1637) the other side, of a great debate that was analyzed by Wolfgang Pauli (1952). Totally in contrast to Fludd's concept of intervention from on high was Kepler's guiding principle, *Ubi materia, ibi geometria*— where there is matter, there is geometry (Kepler 1621). It was not directly from Kepler's writings, however, that Newton learned of Kepler's three great geometry-driven findings about the motions of the planets in space and in time, but from the distillation of Kepler's work offered by Thomas Streete (1661).

that evolution forced upon an unwilling world, the ultimate shock to preconceived ideas lies ahead, be it a decade hence, a century or a millennium. The overarching principle of twentieth-century physics, the quantum (Planck 1900)—and the principle of complementarity[2] that is the central idea of the quantum—leaves us no escape, Niels Bohr (1935) tells us, from "a radical revision of our attitude as regards physical reality" and a "fundamental modification of all ideas regarding the absolute character of physical phenomena." Transcending Einstein's summons of 1908 (see Einstein 1995), "This quantum business is so incredibly important and difficult that everyone should busy himself with it," Bohr's modest words direct us to the supreme goal: *Deduce the quantum* from an understanding of *existence*.

How do we make headway toward a goal so great against difficulties so large? The search for understanding presents to us three questions, four no's, and five clues:

Three **questions**,

- How come existence?

- How come the quantum?

- How come "one world" out of many observer-participants?

Four **no's**,

- No tower of turtles

- No laws

- No continuum

- No space, no time

Five **clues**,

- The boundary of a boundary is zero.

Wheeler (1990) provides a brief and accessible summary of Einstein's 1915 and still standard geometrodynamics, which capitalizes on Élie Cartan's appreciation of the central idea of the theory: the boundary of a boundary is zero.

[2] See Bohr (1928). The mathematics of complementarity I have not been able to discover stated anywhere more sharply, more generally, and earlier than in Weyl (1928), in the statement that the totality of operators for all the physical quantities of the system in question forms an irreducible set.

- No question? No answer!

- The super-Copernican principle.

- "Consciousness"

- More is different.

2 "It from Bit" as a Guide in the Search for a Link Connecting Physics, Quantum, and Information

In default of a tentative idea or working hypothesis, these questions, no's, and clues—yet to be discussed—do not move us ahead. Nor will any abundance of clues assist a detective who is unwilling to theorize how the crime was committed! A wrong theory? The policy of the engine inventor, John Kris, reassures us, "Start her up and see why she don't go!" In this spirit, I, like other searchers,[3] attempt formulation after formulation of the central issues, and here present a wider overview, taking for a working hypothesis the most effective one that has survived this winnowing: **It from bit**. Otherwise put, every **it**—every particle, every field of force, even the spacetime continuum itself—derives its function, its meaning, its very existence entirely—even if in some contexts indirectly—from the apparatus-elicited answers to yes or no questions, binary choices (see Tukey 1947), *bits*.

The most famous lines and idea in the paper: *it from bit*. Wheeler is making a very controversial claim: that bits are the building blocks of the universe, preceding objects and matter.

It from bit symbolizes the idea that every item of the physical world has at bottom—at a very deep bottom, in most instances—an immaterial source and explanation; that which we call reality arises in the last analysis from the posing of yes–no questions and the registering of equipment-evoked responses; in short, that all things physical are information-theoretic in origin and this is a **participatory universe**.

Wheeler is arguing that the natural language of the universe is information not energy or matter.

Three examples may illustrate the theme of it from bit. First, the photon. With a polarizer over the distant source and analyzer of polarization over the photodetector under watch, we ask the yes or no question, "Did the counter register a click during the specified second?" If yes, we often say, "A photon did it." We know perfectly well that the photon existed neither before the emission nor after the detection.

---

[3]Please refer to the author's publications listed in the bibliography. Others include Bell (1987), d'Espagnat (1989), Greenberger (1986), and Mittelstaedt and Stachow (1985).

Observer-participancy has two properties: observation registers an event. Participancy captures the fact that reality is altered by registering the event. In the case of the photon, it must collide with an electron to be registered. In colliding, it is absorbed and destroyed.

However, we also have to recognize that any talk of the photon "existing" during the intermediate period is only a blown-up version of the raw fact, a count.

The yes or no that is recorded constitutes an unsplittable bit of information. A photon, Wootters and Zurek (1982, 1983) demonstrate, cannot be cloned.

As a second example of it from bit, we recall the Aharonov and Bohm (1959) scheme to measure a magnetic flux. Electron counters stationed off to the right of a doubly-slit screen give yes-or-no indications of the arrival of an electron from the source located off to the left of the screen, both before the flux is turned on and afterward. That flux of magnetic lines of force finds itself embraced between—but untouched by—the two electron beams that fan out from the two slits. The beams interfere. The shift in interference fringes between field off and field on reveals the magnitude of the flux,

(phase change around perimeter of the included area)

$= 2\pi \times$ (shift of interference pattern, measured in number of fringes)

$=$ (electron charge) $\times$ (magnetic flux embraced)$/(\hbar c)$.

Here $\hbar = 1.0546 \times 10^{-27}$ gcm^2/s is the quantum in conventional units, or in geometric units (Misner, Thorne, and Wheeler 1973; Wheeler 1990)—where both time and mass are measured in the units of length—$\hbar = \hbar c = 2.612 \times 10^{-66}$ cm^2 = the square of the Planck length, 1.616×10^{-33} = what we hereafter term the *Planck area*.

Not only in electrodynamics but also in geometrodynamics and in every other gauge-field theory, as Anandan, Aharonov,[4] and others point out, the difference around a circuit in the phase of an appropriately chosen quantum-mechanical probability amplitude provides a measure of the field. Here again the concept of it from bit applies (Wheeler 1988b). Field strength or spacetime curvature reveals itself through a shift of interference fringes, fringes that stand for nothing but a statistical pattern of yes-or-no registrations.

When a magnetometer reads that *it* which we call a magnetic field, no reference at all to a bit seems to show itself. Therefore we look closer. The idea behind the operation of the instrument is simple. A

---

[4]See Anandan (1988) and Anandan and Aharonov (1988), for example.

wire of length l carries a current i through a magnetic field B that runs perpendicular to it. In consequence the piece of copper receives in the time t a transfer of momentum p in a direction z perpendicular to the directions of the wire and of the field,

$$\begin{aligned}p &= Blit \\ &= \text{(flux per unit } z\text{)} \\ &\quad \times \text{(charge, } e\text{, of the elementary carrier of current)} \\ &\quad \times \text{(number, } N\text{, of carriers that pass in the time } t\text{)}.\end{aligned} \quad (1)$$

This impulse is the source of the force that displaces the indicator needle of the magnetometer and gives us an instrument reading. We deal with bits wholesale rather than bits retail when we run the fiducial current through the magnetometer coil, but the definition of field founds itself no less decisively on bits.

As a third and final example of it from bit we recall the wonderful quantum finding of Bekenstein (1972, 1973, 1980)—a totally unexpected denouement of earlier classical work of Penrose (1969), Christodoulou (1970), and Christodoulou and Ruffini (1971)—refined by Hawking (1975, 1976), that the *surface area* of the horizon of a black hole, rotating or not, *measures* the *entropy* of the black hole. Thus this surface area, partitioned in the imagination into domains (see figure 1) each of size $4\hbar\log_e 2$, that is, $2.77\ldots$ times the Planck area, yields the *Bekenstein number*, N; and the Bekenstein number, so Zurek and Thorne (1985) explain, tells us the number of binary digits, the number of bits, that would be required to specify in all detail the configuration of the constituents out of which the black hole was put together. Entropy is a measure of lost information. To no community of newborn outside observers can the black hole be made to reveal out of which particular one of the 2^N configurations it was put together. Its size, an *it*, is fixed by the number, N, of *bits* of information hidden within it.

The quantum, $\hbar$, in whatever correct physics formula it appears, thus serves as a lamp. It lets us see the horizon area as information lost, understand the wave number of light as photon momentum and think of the field flux as bit-registered fringe shift.

Information as bits also makes objects by giving them properties; the number of bits describing a black hole gives the black hole its size.

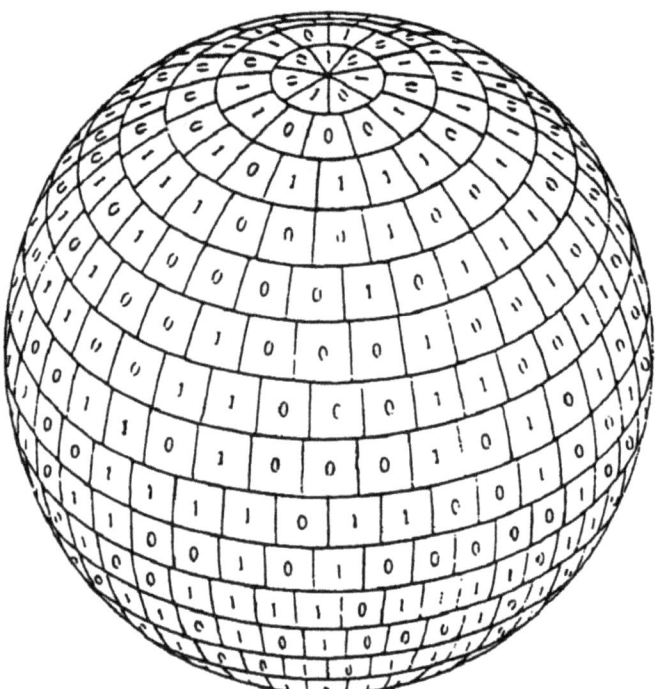

Figure 1. Symbolic representation of the "telephone number" of the particular one of the 2^N conceivable, but by now indistinguishable, configurations out of which this particular black hole, of Bekenstein number N and horizon area $4N\hbar \log_e 2$, was put together. Symbol, also, in a broader sense, of the theme that *every* physical entity, every *it*, derives from bits. Reproduced from Wheeler (1990, 220); reprinted by permission of Freeman Pub. Co.

By now it should start to be apparent that Wheeler's argument is less mysterious than his introduction makes it seem: OP does not require complex objects to register changes and events. Initially only quantum phenomena are required.

Giving us its as bits, the quantum presents us with physics as information.

How come a value for the quantum so small as $\hbar = 2.612 \times 10^{-66}\,\text{cm}^2$? As well as ask why the speed of light is so great as $c = 3 \times 10^{10}\,\text{cm/s}$! No such constant as the speed of light ever makes an appearance in a truly fundamental account of special relativity or Einstein geometrodynamics, and for a simple reason: Time and space are both tools to measure interval. We only then properly conceive them when we measure them in the same units (Misner, Thorne, and Wheeler 1973; Wheeler 1990). The numerical value of the ratio between the second and the centimeter totally lacks teaching power. It is an historical accident. Its occurrence in equations obscured for decades one of Nature's great simplicities. Likewise with $\hbar$! Every equation that contains an $\hbar$ floats a banner, "It from bit." The formula displays a piece of physics that we have learned to translate into information-theoretic terms. Tomorrow we will have learned to understand and express *all* of physics in the language of information. At that point we will revalue $\hbar = 2.612 \times 10^{-66}$ cm^2—as we downgrade $c = 3 \times 10^{10}$ cm/s today—from constant of Nature to artifact of history, and from foundation of truth to enemy of understanding.

3 Four 'No's

To the question "How come the quantum?" we thus answer, "Because what we call existence is an information-theoretic entity." But how come existence? Its as bits, yes; and physics as information, yes; but *whose* information? How does the vision of one world arise out of the information-gathering activities of many observer-participants? In the consideration of these issues we adopt for guidelines four no's.

FIRST NO

"No tower of turtles," advised William James. Existence is not a globe supported by an elephant, supported by a turtle, supported by yet another turtle, and so on. In other words, no infinite regress. No structure, no plan of organization, no framework of ideas underlaid by another structure or level of ideas, underlaid by yet another level, by yet another, *ad infinitum*, down to a bottomless night. To endlessness no alternative is evident but a loop,[5] such a loop as this: Physics gives rise to observer-participancy; observer-participancy gives rise to information; and information gives rise to physics.

In addition to the idea information precedes matter, Wheeler's other major and provocative point is that the universe is a loop, not a hierarchy.

Existence thus built on "insubstantial nothingness"? Rutherford and Bohr made a table no less solid when they told us it was 99.9 . . . percent emptiness. Thomas Mann (1937) may exaggerate when he suggests that ". . . we are actually bringing about what seems to be happening to us," but Leibniz reassures us, "Although the whole of this life were said to be nothing but a dream and the physical world nothing but a phantasm, I should call this dream or phantasm real enough if, using reason well, we were never deceived by it" (Newman 1956).

---

[5]See Misner, Thorne, and Wheeler (1973, 1217) and Wheeler (1988b).

SECOND NO

Wheeler is not saying no laws at all. He is saying no fundamental law that underlies everything. He accepts emergent, level-specific laws.

No laws: "So far as we can see today, the laws of physics cannot have existed from everlasting to everlasting. They must have come into being at the big bang. There were no gears and pinions, no Swiss watchmakers to put things together, not even a pre-existing plan ... Only a principle of organization which is no organization at all would seem to offer itself. In all of mathematics, nothing of this kind more obviously offers itself than the principle that 'the boundary of boundary is zero.' Moreover, all three great field theories of physics use this principle twice over ... This circumstance would seem to give us some reassurance that we are talking sense when we think of ... physics being" (Wheeler 1982c) as foundation-free as a logic loop, the closed circuit of ideas in a self-referential deductive axiomatic system (Steenrod 1962; Ehresmann 1965; Lohmer 1989; Weil 1979).

Universe as machine? This universe one among a great ensemble of machine universes, each differing from the others in the values of the dimensionless constants of physics? Our own selected from this ensemble by an anthropic principle of one or another form (Barrow and Tipler 1986)? We reject here the concept of universe as machine not least because it "has to postulate explicitly or implicitly, a supermachine, a scheme, a device, a miracle, which will turn out universes in infinite variety and infinite number" (Wheeler 1988b).

Directly opposite to the concept of universe as machine built on law is the vision of a world self-synthesized. On this view, the notes struck out on a piano by the observer-participants of all places and all times, bits though they are, in and by themselves constitute the great wide world of space and time and things.

THIRD NO

No continuum. No continuum in mathematics and therefore no continuum in physics. A half-century of development in the sphere of mathematical logic[6] has made it clear that there is no evidence supporting the belief in the existential character of the number continuum. "Belief in this transcendental world," Hermann Weyl tells

---

[6] See for example the survey by S. Feferman, "Turing in the Land of O(z)," pages 113–47, and related papers on mathematical logic in Herken (1988).

us, "taxes the strength of our faith hardly less than the doctrines of the early Fathers of the Church or of the scholastic philosophers of the Middle Ages" (Weyl 1946). This lesson out of mathematics applies with equal strength to physics. "Just as the introduction of the irrational numbers ... is a convenient myth [which] simplifies the laws of arithmetic ... so physical objects," Willard Van Orman Quine (1980) tells us, "are postulated entities which round out and simplify our account of the flux of existence ... The conceptual scheme of physical objects is a convenient myth, simpler than the literal truth and yet containing that literal truth as a scattered part."

Nothing so much distinguishes physics as conceived today from mathematics as the difference between the continuum character of the one and the discrete character of the other. Nothing does so much to extinguish this gap as the elementary quantum phenomenon "brought to a close," as Bohr (1935) puts it, by "an irreversible act of amplification," such as the click of a photodetector or the blackening of a grain of photographic emulsion. Irreversible? More than one idealized experiment (Wheeler and Zurek 1984) illustrates how hard it is, even today, to give an all-inclusive definition of the term irreversible. Those difficulties supply pressure, however, not to retreat to old ground, but to advance to new insight. In brief, continuum-based physics, no; information-based physics, yes.

FOURTH AND LAST NO

No space, no time: Heaven did not hand down the word "time." Man invented it, perhaps positing hopefully as he did that "Time is Nature's way to keep everything from happening all at once."[7] If there are problems with the concept of time, they are of our own creation! As Leibniz (1951) tells us, "... time and space are not things, but orders of things ...;" or as Einstein put it, "Time and space are modes by which we think, and not conditions in which we live" (Forsee 1963).

What are we to say about that weld of space and time into spacetime which Einstein gave us in his 1915 and still standard classical geometrodynamics? On this geometry quantum theory, we know,

Again, the point here is that space and time are emergent, not fundamental, and might arise more from perception by entities than directly from the underlying physics.

---

[7]Discovered among the graffiti in the men's room of the Pecan Street Cafe, Austin, Texas.

imposes fluctuations.[8] Moreover, the predicted fluctuations grow so great at distances of the order of the Planck length that in that domain they put into question the connectivity of space and deprive the very concepts of "before" and "after" of all meaning (Wheeler 1979a, 411). This circumstance reminds us anew that no account of existence can ever hope to rate as fundamental which does not translate all of continuum physics into the language of bits.

We will not feed time into any deep-reaching account of existence. We must derive time—and time only in the continuum idealization—out of it. Likewise with space.

4 Five Clues

FIRST CLUE

The boundary of a boundary is zero. This central principle of algebraic topology (Spanier 1966), identity, triviality, tautology, though it is, is also the unifying theme of Maxwell electrodynamics, Einstein geometrodynamics, and almost every version of modern field theory.[9] That one can get so much from so little, almost everything from almost nothing, inspires hope that we will someday complete the mathematization of physics and derive everything from nothing, all law from no law.

SECOND CLUE

No question, no answer. Better put, no bit-level question, no bit-level answer. So it is in the game of twenty questions in its surprise version.[10] And so it is for the electron circulating within the atom or a field within a space. To neither field nor particle can we attribute a coordinate or momentum until a device operates to measure the one or the other. Moreover any apparatus that *accurately* (Wootters and Zurek 1979) measures the one quantity inescapably rules out then and there the operation of equipment to measure the other (Bohr 1928; Bohr and Rosenfeld 1933; Heisenberg 1927). In brief, the choice of question

---

[8] See Wheeler (1957a, 1957b) and Misner, Thorne, and Wheeler (1973, sec. 43.4).

[9] See Misner, Thorne, and Wheeler (1973, Chap. 15), Atiyah (1988), Cartan (1925a, 1925b), and Kheyfets and Wheeler (1986).

[10] See Wheeler (1978, 41–42) and Wheeler (1979a, 397–98).

asked, and choice of when it's asked, play a part—not the whole part, but a part—in deciding what we have the right to say (Wheeler and Zurek 1984; Wheeler 1986a).

Bit-registration of a chosen property of the electron, a bit-registration of the arrival of a photon, Aharonov-Bohm bit-based determination of the magnitude of a field flux, bulk-based count of bits bound in a black hole: All are examples of physics expressed in the language of information. However, into a bit count that one might have thought to be a private matter, the rest of the nearby world irresistibly thrusts itself. Thus the atom-to-atom distance in a ruler—basis for a bit count of distance—evidently has no invariant status, depending as it does on the temperature and pressure of the environment. Likewise the shift of fringes in the Aharonov-Bohm experiment depends not only upon the magnetic flux itself, but also on the charge of the electron. But this electron charge—when we take the quantum itself to be Nature's fundamental measuring unit—is governed by the square root of the quantity $e^2/\hbar c = 1/137.036...$, a "constant" which—for extreme conditions—is as dependent on the local environment (Gross 1989) as is a dielectric "constant" or the atom-to-atom spacing in the ruler.

The contribution of the environment becomes overwhelmingly evident when we turn from length of bar or flux of field to the motion of alpha particle through cloud chamber, dust particle through 3°K-background radiation, or Moon through space. This we know from the analyses of Bohr and Mott (1929), Zeh (1970, 1989), Joos and Zeh (1985), Zurek (1981, 1982, 1983), and Unruh and Zurek (1989). It from bit, yes; but the rest of the world also makes a contribution, a contribution that suitable experimental design can minimize but not eliminate. Unimportant nuisance? No. Evidence that the whole show is wired up together? Yes. Objection to the concept of every *it* from *bits*? No.

Build physics, with its false face of continuity, on bits of information! What this enterprise is we perhaps see more clearly when we examine for a moment a thoughtful, careful, wide-reaching exposition (Hartle 1989) of the directly opposite thesis, that physics at bottom is *continuous*; that the bit of information is *not* the basic entity. Rate as false the claim that the bit of information is the basic entity.

Instead, attempt to build everything on the foundation of some "grand unified field theory" such as string theory (Brink and Henneaux 1988; Green, Schwarz, and Witten 1987)—or in default of that, on Einstein's 1915 and still standard geometrodynamics. Hope to derive that theory by way of one or another plausible line of reasoning. But don't try to derive quantum theory. Treat it as supplied free of charge from on high. Treat quantum theory as a magic sausage grinder which takes in as raw meat this theory, that theory or the other theory and turns out a "wave equation," one solution of which is "the" wave function for the universe (Hartle and Hawking 1983; Hartle 1989; Hawking 1982; Vilenkin 1982; Wheeler 1957b). From start to finish accept continuity as right and natural: Continuity in the manifold, continuity in the wave equation, continuity in its solution, continuity in the features that it predicts. Among conceivable solutions of this wave equation select as reasonable one which "maximally decoheres," one which exhibits "maximal classicity"—maximal classicity by reason, not of "something external to the framework of wave function and Schrödinger equation," but something in "the initial conditions of the universe specified within quantum theory itself."

Here Wheeler is distinguishing it from bit from so-called grand unified theories that usually start with objects like strings.

How do we compare the opposite outlooks of decoherence and it-from-bit? Remove the casing that surrounds the workings of a giant computer. Examine the bundles of wires that run here and there. What is the status of an individual wire? Mathematical limit of the bundle? Or building block of bundle? The one outlook regards the wave equation and wave function to be primordial and precise and built on continuity, and the bit to be idealization. The other outlook regards the bit to be the primordial entity, and wave equation and wave function to be secondary and approximate—and derived from bits via information theory.

Derived, yes; but how? No one has done more than William Wootters (1980, 1981) towards opening up a pathway from information to quantum theory. He puts into connection two findings, long known, but little known. Already before the advent of wave mechanics, he notes, the analyst of population statistics R. A. Fisher (1922, 1956) proved that the proper tool to distinguish one population from another is not the probability of this gene, that gene and the third gene (for example), but the square roots of these probabilities; that is to say, the two probability amplitudes, each probability amplitude being a vector with three components. More precisely, Wootters proves, the *distinguishability* between the two populations is measured by the angle in Hilbert space between the two state vectors, both real. Fisher, however, was dealing with information that sits "out there." In microphysics, however, the information does not sit out there. Instead, Nature in the small confronts us with a revolutionary pistol, "No question, no answer." Complementarity rules. And complementarity as E. C. G. Stueckelberg (1952, 1960) proved as long ago as 1952, and as Saxon (1964) made more readily understandable in 1964, demands that the probability amplitudes of quantum physics must be complex. Thus Wootters *derives* familiar Hilbert space with its familiar complex probability amplitudes from the twin demands of complementarity and measure of distinguishability.

Try to go on from Wootters's finding to *deduce* the full blown machinery of quantum field theory? Exactly not to try to do so—except as an idealization—is the demand laid on us by the concept of it from bit. How come?

Probabilities exist "out there" no more than do space or time or the position of the atomic electron. Probability, like time, is a concept invented by humans, and humans have to bear the responsibility for the obscurities that attend it. Obscurities there are whether we consider probability defined as frequency (Larson 1974) or defined à la Bayes (Jaynes 1986; Rosenkrantz 1989; Schrödinger 1947; Viertl 1987). Probability in the sense of frequency has no meaning as applied to the spontaneous fission of the particular plutonium nucleus that triggered the November 1, 1952 H-bomb blast.

What about probabilities of a Bayesian cast, probabilities "interpreted not as frequencies observable through experiments, but as degrees of plausibility one assigns to each hypothesis based on the data and on one's assessment of the plausibility of the hypotheses prior to seeing the data" (Denning 1989). Belief-dependent probabilities, different probabilities assigned to the same proposition by different people (Berger and Berry 1988)? Probabilities associated (Burke 1985) with the view that "objective reality is simply an interpretation of data agreed to by large numbers of people?"

Heisenberg directs us to the experiences of the early nuclear-reaction-rate theorist Fritz Houtermans, imprisoned in Kharkov during the time of the Stalin terror: "... the whole cell would get together to produce an adequate confession ... [and] helped them [the prisoners] to compose their 'legends' and phrase them properly, implicating as few others as possible" (Beck and Godin 1951).

Existence as confession? Myopic but in some ways illuminating formulation of the demand for intercommunication implicit in the theme of it from bit!

So much for "No question, no answer."

THIRD CLUE

The super-Copernican principle (Wheeler 1988b). This principle rejects now-centeredness in any account of existence as firmly as Copernicus repudiated here-centeredness. It repudiates most of all any tacit adoption of here-centeredness in assessing observer-participants and their number.

What is an observer-participant? One who operates an observing device and participates in the making of meaning, meaning in the sense of Føllesdal (1975), "Meaning is the joint product of all the evidence that is available to those who communicate." Evidence that is available? The investigator slices a rock and photographs the evidence for the heavy nucleus that arrived in the cosmic radiation of a billion years ago (Wheeler and Zurek 1984). Before he can communicate his findings, however, an asteroid atomizes his laboratory, his records, his rocks and him. No contribution to meaning! Or at least no contribution then. A forensic investigation of sufficient detail and wit to reconstruct the evidence of the arrival of that nucleus is difficult to imagine. What about the famous tree that fell in the forest with no one around (Berkeley 1734)? It leaves a fallout of physical evidence so near at hand and so rich that a team of up-to-date investigators can establish what happened beyond all doubt. Their findings contribute to the establishment of meaning.

"Measurements and observations," it has been said, "cannot be fundamental notions in a theory which seeks to discuss the early universe when neither existed" (Herken 1988). On this view the past has a status beyond all questions of observer-participancy. It from bit offers us a different vision: "Reality is theory";[11] "the past has no evidence except as it is recorded in the present" (Wheeler 1978, 41). The photon that we are going to register tonight from that four-billion-year-old quasar cannot be said to have had an existence "out there" three billion years ago, or two (when it passed an intervening gravitational lens) or one, or even a day ago. Not until we have fixed arrangements at our telescope do we register tonight's quantum as having passed to the left (or right) of the lens or by both routes (as in a double slit experiment). This registration, like every delayed-choice experiment (Miller and Wheeler 1984; Wheeler 1978), reminds us that no elementary quantum phenomenon is a phenomenon until, in Bohr's words, "It has been brought to a close" by "an irreversible act of amplification" (Bohr 1935). What we call the past is built on bits.

Irreversible amplification is related to emergence and akin to the concept of screening off to describe the separation of levels in biological systems.

Enough bits to structure a universe so rich in features as we know this world to be. Preposterous! Mice and men and all on Earth who may ever come to rank as intercommunicating meaning-establishing observer-participants will never mount a bit count sufficient to bear so great a burden.

The count of bits needed, huge though it may be, nevertheless, so far as we can judge, does not reach infinity. In default of a better estimate, we follow familiar reasoning (Zel'dovich and Novikov 1971) and translate into the language of the bits the entropy of the primordial cosmic fireball as deduced from the entropy of the present 2.735°K (uncertainty < 0.05°K) microwave relic radiation (Mather *et al.* 1990) totaled over a 3-sphere of radius 13.2×10^9 light years (uncertainty < 35%)[12] or 1.25×10^{28} cm and of volume $2\pi^2$ radius^3,

$$\begin{aligned}
(\text{number of bits}) &= (\log_2 e) \times (\text{number of nats}) \\
&= (\log_2 e) \times (\text{entropy/Boltzmann's constant}, k) \\
&= 1.44\ldots \times \left[(8\pi^4/45)(radius \cdot kT/\hbar c)^3\right] \\
&= 8 \times 10^{88}
\end{aligned}$$

(2)

It would be totally out of place to compare this overpowering number with the number of bits of information elicited up to date by observer-

---

[11] See T. Segerstedt, as quoted in Wheeler (1979a, 415).

[12] See Misner, Thorne, and Wheeler (1973, 738, Box 27.4) or Wheeler (1990, Chap. 13, 242).

participancy. So warns the super-Copernican principle. We today, to be sure, through our registering devices, give a tangible meaning to the history of the photon that started on its way from a distant quasar long before there was any observer-participancy anywhere. However, the far more numerous establishers of meaning of time to come have a like inescapable part—by device-elicited question and registration of answer—in generating the "reality" of today. For this purpose, moreover, there are billions of years yet to come, billions on billions of sites of observer-participancy yet to be occupied. How far foot and ferry have carried meaning-making communication in fifty thousand years gives faint feel for how far interstellar propagation is destined (O'Neill 1989; Jastrow 1989) to carry it in fifty billion years.

Do bits needed balance bits achievable? They must, declares the concept of "world as system self-synthesized by quantum networking" (Wheeler 1988b). By no prediction does this concept more clearly expose itself to destruction, in the sense of Popper (1962).

FOURTH CLUE

"Consciousness." We have traveled what may seem a dizzying path. First, elementary quantum phenomenon brought to a close by an irreversible act of amplification. Second, the resulting information expressed in the form of bits. Third, this information used by observer-participants—via communication—to establish meaning. Fourth, from the past through the billeniums to come, so many observer-participants, so many bits, so much exchange of information, as to build what we call existence.

Doesn't this it-from-bit view of existence seek to elucidate the physical world, about which we know something, in terms of an entity about which we know almost nothing, consciousness (Calvin 1990; Edelman 1987; Fuller and Putnam 1966; Fuller 1967)? And doesn't Marie Sklodowska Curie tell us, "Physics deals with things, not people?" Using such and such equipment, making such and such a measurement, I get such and such a number. Who I am has nothing to do with this finding. Or does it? Am I sleepwalking (Collins 1868; Hobson 1989)? Or am I one of those poor souls without the critical power to save himself from pathological science (Hetherington 1988; Sheehan 1988; Langmuir 1989)?

Under such circumstances any claim to have "measured" something falls flat until it can be checked out with one's fellows. Checked how?

Wheeler emphasizes the importance of consensus and collectivity throughout the essay. Both are key to overcoming subjectivity and computing a shared reality. This is quite an unusual take for physics, but has much in common with the study of collective phenomena in biology.

Morton White (1972) reminds us how the community applies its tests of credibility, and in this connection quotes analyses by Chauncey Wright, Josiah Royce and Charles Saunders Peirce.[13] Parmenides of Elea may tell us that "What is . . . is identical with the thought that recognizes it" (Parmenides of Elea [ca. 515–450 BCE] 1959). We, however, steer clear of the issues connected with *"consciousness"*. The line between the unconscious and the conscious begins to fade (Pugh 1976) in our day as computers evolve and develop—as mathematics has—level upon level upon level of logical structure. We may someday have to enlarge the scope of what we mean by a "who." This granted, we continue to accept—as essential part of the concept of it from bit—Føllesdal's guideline, "Meaning is the joint product of all the evidence that is available to those who *communicate*" (Føllesdal 1975). What shall we say of a view of existence[14] that appears, if not anthropomorphic in its use of the word "who," still overly centered on life and consciousness? It would seem more reasonable to dismiss for the present the semantic overtones of "who" and explore and exploit the insights to be won from the phrases, "communication" and "communication employed to establish meaning."

Føllesdal's statement supplies, not an answer, but the doorway to new questions. For example, man has not yet learned how to communicate with an ant. When he does, will the questions put to the world around by the ant and the answers that he elicits contribute their share, too, to the establishment of meaning? As another issue associated with communication, we have yet to learn how to draw the line between a communication network that is closed, or parochial, and one that is open. And how to use that difference to distinguish between reality and poker— or another game (von Neumann and Morgenstern 1944; Wang 1988)—so

---

[13] See Peirce (1940), especially passages from pages 335–37, 353, and 358. Peirce's position on the forces of nature, "May they not have naturally grown up," foreshadowing though it does the concept of the world as a self-synthesized system, differs from it in one decisive point, in that it tacitly takes time as a primordial category supplied free of charge from outside.

[14] See von Schelling (1958–1959), especially volume 5, 428–30, as kindly summarized for me by B. Kanitscheider: "daß das Universum von vornherein ein ihm immanentes Ziel, eine teleologische Struktur, besitzt und in allen seinen Produkten auf evolutionäre Stadien ausgerichtet ist, die schliesslich die Hervorbringung von Selbet-bewußtsein einschliessen, welches dann aber wiederum den Entstehungsprozess reflektiert und diese Reflexion ist die notwendige Bedingung für die Konstitution der Gegenstände des Bewusstseins."

intense as to appear more real than reality. No term in Føllesdal's statement poses a greater challenge to reflection than "communication," descriptor of a domain of investigation (Pierce 1961; Schwartz 1987; Roden 1988) that enlarges in sophistication with each passing year.

FIFTH AND FINAL CLUE

More is different (Anderson 1972). Not by plan but by inner necessity a sufficiently large number of H_2O molecules collected in a box will manifest solid, liquid and gas phases. Phase changes, superfluidity and superconductivity all bear witness to Anderson's pithy point, more is different.

We do not have to turn to objects so material as electrons, atoms and molecules to see big numbers generating new features. The evolution from small to large has already in a few decades forced on the computer a structure (Mead and Conway 1980; Schneck 1987) reminiscent of biology by reason of its segregation of different activities into distinct organs. Distinct organs, too, the giant telecommunications system of today finds itself inescapably evolving (Schwartz 1987; Roden 1988). Will we someday understand time and space and all the other features that distinguish physics—and existence itself—as the similarly self-generated organs of a self-synthesized information system (Yates 1987; Haken 1979; Kohonen 1989)?

5 Conclusion

The spacetime continuum? Even continuum existence itself? Except as idealization neither one entity nor the other can make any claim to be a primordial category in the description of Nature. It is wrong, moreover, to regard this or that physical quantity as sitting "out there" with this or that numerical value in default of the question asked and the answer obtained by way of an appropriate observing device. The information thus solicited makes physics and comes in bits. The count of bits drowned in the dark night of a black hole displays itself as horizon area, expressed in the language of the Bekenstein number. The bit count of the cosmos, however it is figured, is ten raised to a very large power. So also is the number of elementary acts of observer-participancy over any time of the order of fifty billion years. And, except via those time-leaping quantum phenomena that we rate as elementary acts of observer-participancy, no way has ever offered itself to construct what we call "reality." That's why we take seriously the theme of it from bit.

6 Agenda

Intimidating though the problem of existence continues to be, the theme of it from bit breaks it down into six issues that invite exploration:

1. Go beyond Wootters and determine what, if anything, has to be added to distinguishability and complementarity to obtain *all* of standard quantum theory.

2. Translate the quantum versions of string theory and of Einstein's geometrodynamics from the language of continuum to the language of bits.

3. Sharpen the concept of bit. Determine whether "an elementary quantum phenomenon brought to a close by an irreversible act of amplification" has at bottom (1) the 0-or-1 sharpness of definition of bit number nineteen in a string of binary digits, or (2) the accordion property of a mathematical theorem, the length of which, that is, the number of supplementary lemmas contained in which, the analyst can stretch or shrink according to his convenience.

4. Survey one by one with an imaginative eye the powerful tools that mathematics—including mathematical logic—has won and now offers to deal with theorems on a wholesale rather than a retail level, and for each such technique work out the transcription into the world of bits. Give special attention to one and another deductive axiomatic system which is able to refer to itself (Smorynski 1985), one and another *self-referential deductive system*.

5. From the wheels-upon-wheels-upon-wheels evolution of computer programming dig out, systematize and display every feature that illuminates the level-upon-level-upon-level structure of physics.

6. Capitalize on the findings and outlooks of information theory (Chaitin 1987; Delahaye 1989; Young 1987; Traub, Wasilkowski, and Woznaikowski 1988), algorithmic entropy (Zurek 1989a), evolution of organisms (Eigen and Winkler 1975; Elsasser 1987;

Wheeler is ahead of his time in emphasizing the potential utility of information theory and the study of inference to understand the origin and structure of the universe. This is an expansive take on the physics of information.

Nicols and Prigogine 1989), and pattern recognition.[15] Search out every link each has with physics at the quantum level. Consider, for instance, the string of bits 1111111... and its representation as the sum of the two strings 1001110... and 0110001.... Explore and exploit the connection between this information-theoretic statement and the findings of theory and experiment on the correlation between the polarizations of the two photons emitted in the annihilation of singlet positronium (Wheeler 1946) and in like Einstein–Podolsky–Rosen experiments (Bohm 1950). Seek out, moreover, every realization in the realm of physics of the information-theoretic triangle inequality recently discovered by Zurek (1989b).

Finally: Deplore? No, celebrate the absence of a clean clear definition of the term "bit" as the elementary unit in the establishment of meaning. We reject "that view of science which used to say, 'Define your terms before you proceed.' The truly creative nature of any forward step in human knowledge," we know, "is such that theory, concept, law and method of measurement—forever inseparable—are born into the world in union (Taylor and Wheeler 1963)." If and when we learn how to combine bits in fantastically large numbers to obtain what we call existence, we will know better what we mean both by bit and by existence.

A single question animates this report: Can we ever expect to understand existence? Clues we have, and work to do, to make headway on that issue. Surely someday, we can believe, we will grasp the central idea of it all as so simple, so beautiful, so compelling that we will all say to each other, "Oh, how could it have been otherwise! How could we all have been so blind so long!"

Acknowledgments

For discussion, advice or judgment on one or another issue taken up in this review, I am indebted to Nandar Balazs, John D. Barrow, Charles H. Bennett, David Deutsch, Robert H. Dicke, Freeman Dyson and the late Richard P. Feynman as well as David Gross, James B. Hartle, John

---

[15]See Watanabe (1967), Tou and Gonzalez (1974), Haken (1979), Small and Garfield (1985), Agu (1988), Minsky and Papert (1988), Steen (1988), and Bennett, Hoffman, and Prakash (1989).

J. Hopfield, Paul C. Jeffries, Bernulf Kanitscheider, Arkady Kheyfets and Rolf W. Landauer; and to Warner A. Miller, John R. Pierce, Willard Van Orman Quine, Benjamin Schumacher and Frank J. Tipler as well as William G. Unruh, Morton White, Eugene P. Wigner, William K. Wootters, Hans Dieter Zeh and Wojciech H. Zurek. For assistance in preparation of this report I thank E. L. Bennett and NSF grant PHY 245-6243 to Princeton University. I give special thanks to the sponsors of the 28-31 August 1989 conference ISQM Tokyo '89 at which the then current version of the present analysis was reported.

This report evolved from presentations at Santa Fe Institute Conferences, 29 May–2 June and 4–8 June 1989 and at the 3rd International Symposium on Foundations of Quantum Mechanics in the Light of New Technology, Tokyo, 28–31 August 1989, under the title "Information, Physics, Quantum: The Search for Links"; and headed "Can We Ever Expect to Understand Existence?", as the Penrose Lecture at the 20–22 April 1989 annual meeting of Benjamin Franklin's "American Philosophical Society, held at Philadelphia for Promoting Useful Knowledge," and the Accademia Nazionale dei Lincei Conference on La Verità nella Scienza, Rome, 13 October 1989; submitted to proceedings of all four in fulfillment of obligation and in deep appreciation for hospitality. Preparation for publication assisted in part by NSF Grant PHY 245-6243 to Princeton University.

Discussion

A discussion followed:

N. G. van Kampen: Did you mean to say that the observer influences the observed object?

J. A. Wheeler: The observer does not influence the past. Instead, by his choice of question, he decides about what feature of the object he shall have the right to make a clear statement.

J.P. Vigier: Two problems.

1. The first is that the QSO raise lots of unsolved problems, i.e.— strange quantized $N_0/\log(1+z)$ relation—correlation with galaxies (Arp)—angular correlation with brightest nearby galaxies (Burbidge et al.)

2. The second is that the idea (Einstein et al.) of the reality of fields has led (assuming that "particles" are field singularities) to the only known justification of the geodesic law. To contest it is to make the meaning of dynamical behaviour purely observer-dependent, i.e., to kill the reality of the physical world.

J.A. Wheeler:

1. The book by Thorne and colleagues, "Black Holes: The Membrane Paradigm," describes how a supermassive black hole, endowed via accretion with great angular momentum inside and an accretion disk outside, produces counterdirected jets and radiation of great power. I know no other mechanism able to produce quasars.

2. No one has discovered a way to get a particle of wave length λ from point A through empty flat space to a point B at a great distance L without its undergoing on the way a transverse spread of the order $\sqrt{L\lambda}$. This spread imposes an inescapable limitation on the classical concept of "worldline."

REFERENCES

Agu, M. 1988. "Field Theory of Pattern Recognition." *Physical Review A* 37:4415–18.

Aharonov, Y., and D. Bohm. 1959. "Significance of Electromagnetic Potentials in the Quantum Theory." *Physical Review* 115:485–91.

Anandan, J. 1988. "Comment on Geometric Phase for Classical Field Theories." *Physical Review Letters* 60:2555.

Anandan, J., and Y. Aharonov. 1988. "Geometric Quantum Phase and Angles." Includes references to the literature of the subject. *Physical Review D* 38:1863–70.

Anderson, P. W. 1972. "More is Different." *Science* 177:393–96.

Atiyah, M. 1988. *Collected Papers, Vol. 5: Gauge Theories*. Oxford, UK: Clarendon Press.

Barrow, J. D., and F. J. Tipler. 1986. *The Anthropic Cosmological Principle*. Also the literature cited therein. New York, NY: Oxford University Press.

Beck, F. [pseudonym of the early nuclear-reaction-rate theorist Fritz Houtermans], and W. Godin. 1951. *Russian Purge and the Extraction of Confessions* [in German]. Translated by E. Mosbacher and D. Porter. London, UK: Hurst and Blackett.

Bekenstein, J. D. 1972. "Black Holes and the Second Law." *Lettere al Nuovo Cimento* 4:737–40.

———. 1973. "Generalized Second Law of Thermodynamics in Black-Hole Physics." *Physical Review D* 8:3292–300.

———. 1980. "Black-Hole Thermodynamics." *Physics Today* 33:24–31.

Bell, J. S. 1987. *Collected Papers in Quantum Mechanics*. Cambridge, UK: Cambridge University Press.

Bennett, B. M., D. D. Hoffman, and C. Prakash. 1989. *Observer Mechanics: A Formal Theory of Perception*. San Diego, CA: Academic Press.

Berger, J. O., and D. A. Berry. 1988. "Statistical Analysis and the Illusion of Objectivity." *American Scientist* 76:159–65.

Berkeley, G. 1734. *Treatise Concerning the Principles of Understanding*. 2nd ed. Cf. article on Berkeley by R. Adamson, Encyclopedia Britannica, Chicago 3 (1959), 438. Dublin, Ireland.

Bohm, D. 1950. "The Paradox of Einstein, Rosen, and Podolsky." Chap. 22, sect. 15–19 in *Quantum Theory*. Reprinted in Wheeler and Zurek (1983), 356–68. Englewood Cliffs, NJ: Prentice-Hall.

Bohr, N. 1928. "The Quantum Postulate and the Recent Development of Atomic Theory." *Nature* 121:580–90.

———. 1935. "Can Quantum-Mechanical Description of Physical Reality be Considered Complete?" Reprinted in Wheeler and Zurek (1983), 145–51, *Physical Review* 48:696–702.

Bohr, N., and L. Rosenfeld. 1933. "Zur Frage der Messbarkeit der elektromagnetischen Feldgrossen." Translated by A. Petersen. Reprinted in Wheeler and Zurek (1983), 479–534, *Mat.-fys Medd. Dan. Vid. Selsk.* 12 (8).

Brink, L., and M. Henneaux. 1988. *Principles of String Theory: Studies of the Centro de Estudios Cientificos de Santiago*. New York, NY: Plenum.

Burke, J. 1985. *The Day the Universe Changed*. Boston, MA: Little, Brown, & Co.

Calvin, W. H. 1990. *The Cerebral Symphony*. New York, NY: Bantam.

Cartan, E. 1925a. *La Geometrie des Espaces de Riemann, Memorial des Sciences Mathematiques*. Paris, France: Gauthier-Villars.

———. 1925b. *Legons sur la Geometric des Espaces de Riemann*. Paris, France: Gauthier-Villars.

Chaitin, G. J. 1987. *Algorithmic Information Theory*. Rev. ed. Cambridge, UK: Cambridge University Press.

Christodoulou, D. 1970. "Reversible and Irreversible Transformations in Black-Hole Physics." *Physical Review Letters* 25:1596–97.

Christodoulou, D., and R. Ruffini. 1971. "Reversible Transformations of a Charged Black Hole." *Physical Review D* 4:3552–55.

Collins, W. W. 1868. *The Moonstone*. London, UK: Tinsley Brothers.

d'Espagnat, B. 1989. *Reality and the Physicist: Knowledge, Duration, and the Quantum World.* Cambridge, UK: Cambridge University Press.

Darwin, C. W. 1859. *On the Origin of Species by Means of Natural Selection, or the Preservation of Favoured Races in the Struggle for Life.* London, UK: John Murray.

Delahaye, J.-P. 1989. "Chaitin's Equation: An Extension of Gödel's Theorem." *Notices of the American Mathematical Society* 36:984–87.

Denning, P. J. 1989. "Bayesian Learning." *American Scientist* 77:216–18.

Edelman, G. M. 1987. *Neural Darwinism.* New York, NY: Basic Books.

Ehresmann, C. 1965. *Categories et Structures.* Paris, France: Dunod.

Eigen, M., and R. Winkler. 1975. *Das Spiel: Naturgesetze sieuem den Zufall.* München, Germany: Piper.

Einstein, A. 1915. "Zur allgemeinen Relativitätstheorie." *Preuss. Akad. Wiss. Ber,* 799–801, 832–39, 844–47.

———. 1916. "Zur allgemeinen Relativitätstheorie." *Preuss. Akad. Wiss. Ber,* 688–96.

———. 1917. "Zur allgemeinen Relativitätstheorie." *Preuss. Akad. Wiss. Ber,* 142–52.

———. 1995. *The Collected Papers of Albert Einstein, Volume 5: The Swiss Years: Correspondence, 1902–1914.* Translated by A. Beck. Einstein, A., to J. J. Laub, 1908, undated, Einstein Archives. Princeton, NJ: Princeton University Press.

Elsasser, W. M. 1987. *Reflections on a Theory of Organisms.* Frelighsburg, Quebec, Canada: Orbis.

Feferman, Solomon. 1988. "Turing in the land of O(z)." https://api.semanticscholar.org/CorpusID:116348646.

Fisher, R. A. 1922. "On the Dominance Ratio." *Proceedings of the Royal Society of Edinburgh* 42:321–41.

———. 1956. *Statistical Methods and Statistical Inference.* 8–17. New York, NY: Hafner.

Føllesdal, D. 1975. "Meaning and Experience," edited by S. Guttenplan, 25–44. Oxford, UK: Clarendon Press.

Forsee, A. 1963. *Albert Einstein, Theoretical Physicist.* New York, NY: Macmillan.

Fuller, R. W. 1967. "Causal and Moral Law: Their Relationship as Examined in Terms of a Model of the Brain." *Monday Evening Papers* (Middletown, CT).

Fuller, R. W., and R. Putnam. 1966. "On the Origin of Order in Behavior." *General Systems* 12:111–21.

Green, M. B., J. H. Schwarz, and E. Witten. 1987. *Superstring Theory.* Cambridge, UK: Cambridge University Press.

Greenberger, D. M., ed. 1986. *New Techniques and Ideas in Quantum Measurement Theory.* Vol. 480. Annals of the New York Academy of Sciences. New York, NY: New York Academy of Sciences.

Gross, D. J. 1989. "On the Calculation of the Fine-Structure Constant." *Physics Today* 42 (12).

Haken, H., ed. 1979. *Pattern Formation by Dynamic Systems and Pattern Recognition.* Berlin, Germany: Springer.

———. 1988. *Information and Self-Organization: A Macroscopic Approach to Complex Systems.* Berlin, Germany: Springer.

Hartle, J. B. 1989. "Progress in Quantum Cosmology." Preprint from the Physics Department, University of California, Santa Barbara.

Hartle, J. B., and S. W. Hawking. 1983. "Wave Function of the Universe." *Physical Review D* 28:2960–75.

Hawking, S. W. 1975. "Particle Creation by Black Holes." *Communications in Mathematical Physics* 43:199–220.

———. 1976. "Black Holes and Thermodynamics." *Physical Review* 13:191–97.

———. 1982. "The Boundary Conditions of the Universe." In *Astrophysical Cosmology,* edited by H. A. Brück, G. V. Coyne, and M. S. Longair, 563–74. Vatican City: Pontificia Academia Scientiarum.

Heisenberg, W. 1927. "Über den anschaulichen Inhalt der quantentheoretischen Kinematik und Mechanik." English translation in Wheeler and Zurek (1983), 62–84, *Zeitschrift für Physik* 43:172–98.

Herken, R., ed. 1988. *The Universal Turing Machine: A Half-Century Survey.* New York, NY: Oxford University Press.

Hetherington, N. S. 1988. *Science and Objectivity: Episodes in the History of Astronomy.* Ames, IA: Iowa State University Press.

Hobson, J. A. 1989. *Sleep.* 86, 89, 175, 185, 186. Scientific American Library. New York, NY: Freeman.

Jastrow, R. 1989. *Journey to the Stars: Space Exploration—Tomorrow and Beyond.* New York, NY: Bantam.

Jaynes, E. T. 1986. "Bayesian Methods: General Background." In *Maximum Entropy and Bayesian Methods in Applied Statistics,* edited by J. H. Justice, 1–25. Cambridge, UK: Cambridge University Press.

Joos, E., and H. D. Zeh. 1985. "The Emergence of Classical Properties through Interaction with the Environment." *Zeitschrift für Physik* B59:223–43.

Kepler, J. 1619. *Harmonices Mundi.* 5 books.

———. 1621. *Utriusque Cosmo Maioris scilicet et Minoris Metaphysica, Physica atque technica Historia.* 1st ed. Oppenheim, Germany.

Kheyfets, A., and J. A. Wheeler. 1986. "Boundary of a Boundary Principle and Geometric Structure of Field Theories." *International Journal of Theoretical Physics* 25:573–80.

Kohonen, T. 1989. *Self-Organization and Associative Memory.* 3rd ed. New York, NY: Springer.

Langmuir, I. 1989. "Pathological Science." 1953 colloquium, transcribed and edited, *Physics Today* 42 (12): 36–48.

Larson, H. J. 1974. *Introduction to Probability Theory and Statistical Inference.* 2nd ed. New York, NY: Wiley.

Leibniz, G. W. 1951. "Animadversiones ad Joh. George Wachteri librum de recondita Hebraeorum philosophia." In *Leibniz Selections,* edited by P. P. Wiener, 488. New York, NY: Scribner.

Lohmer, D. 1989. *Phänomenologie der Maihematik: Elemenie einer Phänomenologischen Aufklärung der Mathematischen Erkenntnis nach Husserl.* Norwell, MA: Kluwer.

Mann, T. 1937. *Freud, Goethe, Wagner.* Translated by H. T. Lowe-Porter from *Freud und die Zukunft,* Vienna (1936). New York, NY.

Mather, J., E. S. Cheng, R. E. Eplee, Jr., R. B. Isaacman, S. S. Meyer, R. A. Shafer, R. Weiss, et al. 1990. "A Preliminary Measurement of the Cosmic Microwave Background Spectrum by the Cosmic Background Explorer (COBE) Satellite." *Astrophys. J. Lett.* 354:L37–L40.

Mead, C., and L. Conway. 1980. *Introducton to VLSI [very large-scale integrated circuit design] Systems.* Reading, MA: Addison-Wesley.

Mendel, G. 1866. "Versuche über Plflanzenhybriden." In *Verhandlungen des naturforschenden Vereines in Brünn, Bd. IV für das Jahr 1865,* 3–47.

Miller, W. A., and J. A. Wheeler. 1984. "Delayed-Choice Experiments and Bohr's Elementary Quantum Phenomenon." In *Proceedings of International Symposium on Foundations of Quantum Mechanics in the Light of New Technology, 1983,* edited by S. Kamefuchi et al., 140–51. Tokyo, Japan: Physical Society of Japan.

Minsky, M., and S. Papert. 1988. *Perceptrons: An Introduction to Computational Geometry.* 2nd ed. Cambridge, MA: MIT Press.

Misner, C. W., K. S. Thorne, and J. A. Wheeler. 1973. *Gravitation.* See the paragraph on participatory concept of the universe, 1217. New York, NY: Freeman.

Mittelstaedt, P., and E. W. Stachow, eds. 1985. *Recent Developments in Quantum Logic.* Zurich, Switzerland: Bibliographisches Institut.

Mott, N. F. 1929. "The Wave Mechanics of α-Ray Tracks." Reprinted in Wheeler and Zurek (1983), 129–34, *Proceedings of the Royal Society of London* A126:74–84.

Newman, J. R. 1956. *The World of Mathematics.* Leibniz, G. W., as cited. New York, NY: Simon and Schuster.

Newton, I. 1687. *Philosophiae Naturalis Principia Mathematica.* 1st ed. London, UK.

Nicols, G., and I. Prigogine. 1989. *Exploring Complexity: An Introduction.* New York, NY: Freeman.

O'Neill, G. K. 1989. *The High Frontier.* 4th ed. Princeton, NJ: Space Studies.

Parmenides of Elea [ca. 515–450 BCE]. 1959. "Poem "Nature," Part "Truth"." In *Encyclopedia Britannica,* vol. 17. As summarized by A. C. Lloyd in the article on Parmenides. Chicago, IL: William Benton.

Patton, C. M., and J. A. Wheeler. 1975. "Is Physics Legislated by Cosmogony?" In *Quantum Gravity,* edited by C. Isham, R. Penrose, and D. Sciama, 538–605. Reprinted in part in Duncan, R. and M. Weston-Smith, eds., *Encyclopaedia of Ignorance,* 19–35, Oxford, UK: Pergamon (1977). Oxford, UK: Clarendon Press.

Pauli, W. 1952. "Der Einfluss archetypischer Vorstellungen auf die Bildung natruwissenschaftlicher Theorien bei Kepler." In *Naturerklärung und Psyche,* 109–94. Reprinted in Kronig, R. and V. F. Weisskopf, eds., *Wolfgang Pauli: Collected Scientific Papers,* vol. 1, 1023, New York, NY: Interscience-Wiley (1964). Zurich, Switzerland: Rascher.

Peirce, C. S. 1940. *The Philosophy of Peirce: Selected Writings.* Edited by J. Buchler. Selected passages reprinted in Patton and Wheeler (1975), 593–95. London, UK: Routledge and Kegan Paul.

Penrose, R. 1969. "Gravitational Collapse: The Role of General Relativity." *La Rivista del Nuovo Cimento* I:252–76.

Pierce, J. R. 1961. *Symbols, Signals, and Noise: The Nature and Process of Communication.* New York, NY: Harper and Brothers.

Planck, M. 1900. "Zur Theorie des Gesetzes der Energieverteilung im Normalspektrum." *Verhandlungen der Deutschen Physikalischen Gesellschaft* 2:237–45.

Popper, K. 1962. *Conjectures and Refutations: The Growth of Scientific Knowledge.* New York, NY: Basic Books.

Pugh, G. E. 1976. *On the Origin of Human Values.* See the chapter "Human Values, Free Will, and the Conscious Mind," reprinted in *Zygon* 11 (1976): 2–24. New York, NY.

Quine, W. V. O. 1980. "On What There Is." In *From a Logical Point of View,* 2nd ed., 18. Cambridge, MA: Harvard University Press.

Roden, M. S. 1988. *Digital Communication Systems Design.* Englewood Cliffs, NJ: Prentice Hall.

Rosenkrantz, R. D., ed. 1989. *E. T. Jaynes: Papers on Probability, Statistics, and Statistical Physics.* Hingham, MA: Reidel-Kluwer.

Saxon, D. S. 1964. *Elementary Quantum Mechanics.* San Francisco, CA: Holden.

Schneck, P. B. 1987. *Supercomputer Architecture.* Norwell, MA: Kluwer.

Schrödinger, E. 1947. "The Foundation of the Theory of Probability—I, II." *Proceedings of the Royal Irish Academy* 51 A:51–66, 141–46.

Schwartz, M. 1987. *Telecommunication Networks: Protocols, Modeling, and Analysis.* Reading, MA: Addison-Wesley.

Shakespeare, W. *The Tempest.* Act IV, Scene I, lines 148 ff.

Sheehan, W. 1988. *Planets and Perception: Telescopic Views and Interpretations.* Tucson, AZ: University of Arizona Press.

Small, H., and E. Garfield. 1985. "The Geography of Science: Disciplinary and National Mappings." *Journal of Information Science* 11:147–59.

Smorynski, C. 1985. *Self-Reference and Model Logic.* Berlin, Germany: Springer.

Spanier, E. H. 1966. *Algebraic Topology.* New York, NY: McGraw-Hill.

Steen, L. A. 1988. "The Science of Patterns." *Science* 240:611–16.

Steenrod, N. E. 1962. *Cohomology Operations.* Princeton, NJ: Princeton University Press.

Streete, T. 1661. *Astronomia Carolina: A New Theorie of the Celestial Motions.* London.

Stueckelberg, E. C. G. 1952. "Theoreme H et unitarite de S." *Helvetica Physica Acta* 25:577–80.

———. 1960. "Quantum Theory in Real Hilbert Space." *Helvetica Physica Acta* 33:727–52.

Taylor, E. F., and J. A. Wheeler. 1963. *Spacetime Physics.* 102. San Francisco, CA: Freeman.

Tou, J., and R. C. Gonzalez. 1974. *Pattern Recognition Principles.* Reading, MA: Addison-Wesley.

Traub, J. F., G. W. Wasilkowski, and H. Woznaikowski. 1988. *Information-Based Complexity.* San Diego, CA: Academic Press.

Tukey, J. W. 1947. *Sequential Conversion of Continuous Data to Digital Data.* Bell Laboratories memorandum of Sept. 1, 1947. Marks the introduction of the term "bit." Reprinted in Ikopp, H. S., "Origin of the Term Bit," Annals Hist. Computing 6 (1984): 152–55.

Unruh, W. G., and W. H. Zurek. 1989. "Reduction of a Wave Packet in Quantum Brownian Motion." *Physical Review D* 40:1071–94.

Viertl, R., ed. 1987. *Probability and Bayesian Statistics*. Singapore: World Scientific.

Vilenkin, A. 1982. "Creation of Universes from Nothing." *Physics Letters B* 117:25–28.

von Neumann, J., and O. Morgenstern. 1944. *Theory of Games and Economic Behavior*. Princeton, NJ: Princeton University Press.

von Schelling, F. W. J. 1958-1959. *Schellings Werke, nach der Originalausgabe in neuer Anordnung herausgegben*. Edited by M. Schroter. München, Germany: Beck.

Wang, J. 1988. *Theory of Games*. New York, NY: Oxford University Press.

Watanabe, S., ed. 1967. *Methodologies of Pattern Recognition*. New York, NY: Academic Press.

Watson, J. D., and F. H. C. Crick. 1953. "Molecular Structure of Nucleic Acids: A Structure for Deoxyribose Nucleic Acid." *Nature* 171:737–38.

Weil, A. 1979. "De la Metaphysique aux mathematiques." Reprinted in A. Weil, *Ouevres Scientifiques: Collected Works, Vol. 2, 1951–64*, New York: Springer (1979), 408–12, *Sciences,* 52–56.

Weyl, H. 1928. *Gnippentheorie und Quantenmechanik*. Leipzig, Germany: Hirzel.

———. 1946. "Mathematics and Logic." A brief survey serving as a preface to a review of *The Philosophy of Bertrand Russell, American Mathematics Monthly* 53:2–13.

Wheeler, J. A. 1946. "Polyelectrons." *Annals of the NY Academy of Sciences* 46:219–38.

———. 1957a. "Assessment of Everett's 'Relative State' Formulation of Quantum Theory." *Reviews of Modern Physics* 29:463–65.

———. 1957b. "On the Nature of Quantum Geometrodynamics." *Annals of Physics* 2:604–14.

———. 1968. "Superspace and the Nature of Quantum Geometrodynamics." In *Battelle Rencontres: 1967 Lectures in Mathematics and Physics,* edited by C. M. DeWitt and J. A. Wheeler, 242–307. Reprinted as "Le Superespace et la Nature de la Géométrodynamique Quantique," in *Fluides et Champ Gravitationnel en Relativité Générale, No. 170, Colloques Internationaux*, Paris, France: Editions du Centre National de la Recherche Scientifique (1969). New York, NY: Benjamin.

———. 1971. "Transcending the Law of Conservation of Leptons." In *Atti del Convegno Intemazionale sul Tema: The Astrophysical Aspects of the Weak Interactions,* 133–64. (Cortona "II Palazzone," 10-12 Guigno 1970). Rome, Italy: Accademia Nationale dei Lincei.

———. 1975. "The Universe as Home for Man." In *The Nature of Scientific Discovery,* edited by O. Gingerich, 261–96. Preprinted in part in *Am. Scientist* 62 (1974): 683–91. Reprinted in part in T. P. Snow, *The Dynamic Universe,* St. Paul, MN: West (1983), 108–9. Washington, DC: Smithsonian Institution Press.

———. 1976. "Include the Observer in the Wave Function?" Reprinted in J. Leite Lopes and M. Paty, eds., *Quantum Mechanics, A Half Century Later,* Dordrecht, Netherlands: Reidel (1977), 1–18, *Fundamenta Scientiae: Seminaire sur les Fondements des Sciences (Strasbourg)* 25:9–35.

———. 1977. "Genesis and Observership." In *Foundational Problems in the Special Sciences,* edited by R. Butts and J. Hintikka, 1–33. Dordrecht, Netherlands: Reidel.

———. 1978. "The 'Past' and the 'Delayed-Choice' Double-Slit Experiment." In *Mathematical Foundations of Quantum Theory,* edited by A. R. Marlow, 9–48. Reprinted in part in Wheeler and Zurek (1983), 182–200. New York, NY: Academic Press.

———. 1979a. "Frontiers of Time." In *Problems in the Foundations of Physics, Proceedings of the International School of Physics "Enrico Fermi" (Course 72),* edited by N. Toraldo di Francia, 395–497. Reprinted in part in Wheeler and Zurek (1983), 200–208. Amsterdam, Netherlands: North Holland.

———. 1979b. "The Quantum and the Universe." In *Relativity, Quanta, and Cosmology in the Development of the Scientific Thought of Albert Einstein,* edited by M. Pantaleo and F. deFinis, II:807–25. New York, NY: Johnson Reprint Corp.

———. 1980a. "Beyond the Black Hole." In *Some Strangeness of the Proportion: A Centennial Symposium to Celebrate the Achievements of Albert Einstein,* edited by H. Woolf, 341–75. Reprinted in part in Wheeler and Zurek (1983), 208–10. Reading, MA: Addison-Wesley.

———. 1980b. "Delayed-Choice Experiments and the Bohr–Einstein Dialog." In *American Philosophical Society and the Royal Society: Papers Read at a Meeting, June 5, 1980,* 9–40. Reprinted in slightly abbreviated form and translated into German as "Die Experimente der verzögerten Entscheidung und der Dialog zwischen Bohr und Einstein," in B. Kanitschedier, ed., *Moderne Naturphilosophie,* Würzburg, Germany: Konigshausen and Neumann (1984), 203–22. Reprinted in A. N. Mitra, L. S. Kothari, V. Singh, and S. K. Trehan, eds., *Niels Bohr: A Profile,* New Delhi, India: Indian National Science Academy (1985), 139–68. Philadelphia, PA: American Philosophical Society.

———. 1980c. "Law without Law." In *Structure in Science and Art,* edited by P. Medawar and J. Shelley. New York, NY and Amsterdam, Netherlands: Elsevier North and Exerpta Medica.

———. 1980d. "Pregeometry: Motivations and Prospects." In *Quantum Theory and Gravitation, Proceedings of a Symposium Held at Loyola University, New Orleans, May 23–26, 1979,* edited by A. R. Marlow, 1–11. New York, NY: Academic Press.

———. 1980e. "The Elementary Quantum Act as Higgledy-Piggledy Building Mechanism." In *Quantum Theory and the Structures of Time and Space: Papers Presented at a Conference Held in Tutzing, July, 1980,* edited by L. Castell and C. F. von Weizsacker, 27–30. Munich, Germany: Carl Hanser.

———. 1981. "Not Consciousness but the Distinction between the Probe and the Probed as Central to the Elemental Quantum Act of Observation." In *The Role of Consciousness in the Physical World,* edited by R. G. Jahn, 87–111. Boulder, CO: Westview.

———. 1982a. "Blackholes and New Physics." *Discovery: Research and Scholarship at the University of Texas at Austin* 7 (2): 4–7.

———. 1982b. "Bohr, Einstein, and the Strange Lesson of the Quantum." In *Mind in Nature, Nobel Conference XVII, Gustavus Adolphus College, St. Peter, Minnesota,* edited by R. Q. Elvee, 1–30, 88, 112–13, 130–31, 148–49. New York, NY: Harper and Row.

———. 1982c. "Particles and Geometry." In *Unified Theories of Elementary Particles,* edited by P. Breitenlohner and H. P. Durr, 189–217. Berlin, Germany: Springer.

———. 1982d. *Physics and Austerity* [in Chinese]. Reprinted in part in *Krisis,* I. Marculescu, ed., vol. 1, no. 2, Lecture II, Paris, France: Klinckscieck (1983), 671–75. Anhui, China: Anhui Science and Technology Publications.

———. 1982e. "The Computer and the Universe." *International Journal of Theoretical Physics* 21:557–71.

Wheeler, J. A. 1983a. "Elementary Quantum Phenomenon as Building Unit." In *Quantum Optics, Experimental Gravitation, and Measurement Theory,* edited by P. Meystre and M. Scully, 141–43. London, UK: Plenum.

———. 1983b. "Jenseits aller Zeitlichkeit." In *Die Zeit, Schriften der Carl Friedrich von Siemens–Stiftung,* edited by A. Peisl and A. Mohler, 6:17–34. München, Germany: Oldenbourg.

———. 1983c. "On Recognizing Law without Law." *American Journal of Physics* 51:398–404.

———. 1984. "Quantum Gravity: The Question of Measurement." In *Quantum Theory of Gravity,* edited by S. M. Christensen, 224–233. Bristol, UK: Hilger.

———. 1985. "Bohr's 'Phenomenon' and 'Law without Law'." In *Chaotic Behavior in Quantum Systems,* edited by G. Casati, 363–378. New York, NY: Plenum.

———. 1986a. "'Physics as Meaning': Three Problems." In *Frontiers of Non-Equilibrium Statistical Physics,* edited by G. T. Moore and M. O. Scully, 25–32. New York, NY: Plenum.

———. 1986b. "Interview on the Role of the Observer in Quantum Mechanics." In *The Ghost in the Atom,* edited by P. C. W. Davies and J. R. Brown, 58–69. Cambridge, UK: Cambridge University Press.

———. 1987. "How Come the Quantum." In *New Techniques and Ideas in Quantum Measurement Theory,* edited by D. M. Greenberger, 480:304–316. Annals of the New York Academy of Sciences.

———. 1988a. "Hermann Weyl and the Unity of Knowledge." In *Exact Sciences and Their Philosophical Foundations,* edited by W. et al. Deppert, 469–503. Appeared in abbreviated form in *Am. Scientist* 74 (1986): 366–375. Frankfurt, Germany: Lang.

———. 1988b. "World as System Self-Synthesized by Quantum Networking." Reprinted in E. Agazzi, ed., *Probability in the Sciences,* Amsterdam, Netherlands: Kluwer (1988), 103–29, *IBM J. Res. & Dev.* 32:4–25.

———. 1990. *A Journey into Gravity and Spacetime.* Scientific American Library. New York, NY: Freeman.

Wheeler, J. A., and W. H. Zurek. 1983. *Quantum Theory and Measurement.* Princeton, NJ: Princeton University Press.

———. 1984. "Bits, Quanta, Meaning." In *Problems in Theoretical Physics,* edited by A. Giovannini, F. Mancini, and M. Marinaro, 121–41. Also in *Theoretical Physics Meeting: Atti del Convegno, Amalfi, 6–7 maggio 1983,* Naples, Italy: Edizioni Scientifiche Italiane (1984), 121–34. Also in A. Giovannini, F. Mancini, M. Marinaro, and A. Rimini, *Festschrift in Honour of Eduardo R. Caianiello*, Singapore: World Scientific (1989). Salerno, Italy: University of Salerno Press.

White, M. 1972. *Science and Sentiment in America: Philosophical Thought from Jonathan Edwards to John Dewey.* New York, NY: Oxford University Press.

Wootters, W. K. 1980. "The Acquisition of Information from Quantum Measurements." PhD diss., University of Texas at Austin.

———. 1981. "Statistical Distribution and Hilbert Space." *Physical Review* 23:357–62.

Wootters, W. K., and W. H. Zurek. 1979. "Complementarity in the Double-Slit Experiment: Quantum Nonseparability and a Quantitative Statement of Bohr's Principle." *Phys. Rev. D* 19:474–84.

———. 1982. "A Single Quantum Cannot Be Cloned." *Nature* 279:802–3.

———. 1983. "On Replicating Photons." *Nature* 304:188–89.

Yates, F. E., ed. 1987. *Self-Organizing Systems: The Emergence of Order.* New York, NY: Plenum.

Young, P. 1987. *The Nature of Information.* Westport, CT: Prager-Greenwood.

Zeh, H. D. 1970. "On the Interpretation of Measurement in Quantum Theory." *Foundations of Physics* I:69–76.

———. 1989. *The Physical Basis of the Direction of Time.* Berlin, Germany: Springer.

Zel'dovich, Y. B., and I. D. Novikov. 1971. *Relativistic Astrophysics, Vol. I: Stars and Relativity.* Chicago, IL: University of Chicago Press.

Zurek, W. H. 1981. "Pointer Basis of Quantum Apparatus: Into What Mixture Does the Wavepacket Collapse?" *Physical Review D* 24:1516–25.

———. 1982. "Environment-Induced Superselection Rules." *Physical Review D* 26:1862–80.

———. 1983. "Information Transfer in Quantum Measurements: Irreversibility and Amplification." In *Quantum Optics, Experimental Gravitation and Measurement Theory,* edited by P. Meystre and M. O. Scully, 87–116. NATO ASI Series. New York, NY: Plenum.

———. 1989a. "Algorithmic Randomness and Physical Entropy." *Physical Review A* 40:4731–51.

———. 1989b. "Thermodynamic Cost of Computation: Algorithmic Complexity and the Information Metric." *Nature* 34:119–24.

Zurek, W. H., and K. S. Thorne. 1985. "Statistical Mechanical Origin of the Entropy of a Rotating, Charged Black Hole." *Physical Review Letters* 20:2171–75.

ROBERT DE RUYTER VAN STEVENINCK
May 31, 1954
The Hague

DAVID WARLAND
April 8, 1963
Los Angeles, CA

FRED RIEKE
January 4, 1966

WILLIAM SAMUEL BIALEK
August 14, 1960
Los Angeles, CA

1960

1962: LISP 1.5 Programmer's Manual is published by MIT Press.

1970

1970: Gödel's ontological proof is circulated for the first time, as a result of Kurt Gödel's fear that he might be dying. He died in 1978.

1980

1989: Robert Tappan Morris becomes the first person indicted for violating the American Computer Fraud and Abuse Act as a result of his computer virus, the Morris worm.

BIALEK, RIEKE, DE RUYTER VAN STEVENINCK, AND WARLAND 1991
University of California, Berkeley, CA, USA
University of California, Berkeley, CA, USA
University of California, Berkeley, CA, USA
Rijksuniversiteit Groningen, Groningen, The Netherlands

1996

1997: Fred Gage and Peter Eriksson definitively demonstrate that neurogenesis occurs in the human hippocampus.

2000

2010

2019: SpiNNaker, a novel massively parallel supercomputer patterned after the human brain with the goal of assisting neuroscience research, officially begins operating.

2020

BIALEK, RIEKE, DE RUYTER VAN STEVENINCK & WARLAND

[71]

WHAT A NEURON SAYS AND HOW IT SAYS IT

Vijay Balasubramanian, University of Pennsylvania and Santa Fe Institute

Neurons are information-processing devices. Their inputs are environmental signals like light and pressure, or chemical ligands like volatile molecules or neurotransmitters released by pre-synaptic neurons in a circuit. These signals activate receptors embedded in the cell membrane, often on a neuron's dendritic arbor, triggering currents that change the membrane potential. In some neurons, a sufficient increase in the potential triggers a sharp voltage "spike" that propagates down the axon, leading to neurotransmitter release at the output synapses and transmission to the postsynaptic neurons. The net effect is that the neuron reconfigures and transmits some information in its inputs to the next stage.

In the mid-twentieth century, scientists realized that if we regard neurons in this way, communications theory should provide a framework for understanding the organization and function of neural circuits. In particular, the brain sciences quickly absorbed Claude Shannon's invention of information theory in 1948. This led to the seminal works of Fred Attneave in the mid-1950s and Horace Barlow in the early 1960s, which proposed applying ideas from information theory to interpreting sensory perception and its neural basis. For example, Barlow proposed that the lateral inhibition seen ubiquitously in early mammalian sensory circuits provided a mechanism for removing correlations in natural stimuli, thereby compressing them for maximal transmission through bottlenecks like the optic nerve.

The next important step took about twenty years: in 1981, Simon Laughlin applied the idea of information optimization quantitatively to a neuron in the fly visual system. He showed that the neuron had adapted its nonlinear transfer function from local contrast in an image to spike firing rates to maximize information transmitted from

W. Bialek, F. Rieke, R. de Ruyter van Steveninck, and D. Warland, "Reading a Neural Code," *Science* 252 (5014), 1854–1857 (1991).

Reprinted with permission from AAAS.

natural images. In the late 1980s, Joseph Atick and Norman Redlich returned to Barlow's proposal by showing quantitatively via the use of optimal coding theory that maximizing information transfer from natural scenes predicts the different structures of lateral inhibition in the retinas of terrestrial and aquatic vertebrates. Around the same time, William Bialek, with collaborators such as Anthony Zee, began to assess the quality of sensory coding by neurons in terms of information-transmission rates, along with stimulus reconstruction techniques pioneered in earlier decades in the fields of cybernetics, control theory, and communications theory (Bialek and Zee 1990).

Most of these pioneering ideas focused on the firing rate of neurons, that is, the number of spikes per second, as the variable carrying information. Against this context, the foundational paper "Reading a Neural Code" by Bialek, Fred Rieke, R. R. de Ruyter van Steveninck, and David Warland, started from the observation that no behaving animal has ready access to such firing rates. First of all, determining a rate requires observing multiple spikes over a period of time, and the required durations often exceed the behavioral latency of animals. Second, real neurons show variable responses to repetitions of the same inputs, so that determining the firing rate by definition requires averaging either over repeated instances of the same input (a procedure that is ethologically irrelevant), or over a population of neurons that redundantly carry the same information (which seems wasteful). Thus, Bialek *et al.* proposed that the neural code must carry information in the timing of individual spikes, and developed an approach to determine whether, and how much, information is carried in this way.

The key insight in this paper is that the fundamental object carrying the neural code is the probability distribution over the set of spike times t_i conditioned on the time-varying input to the neuron $s(t)$. The firing rate is a temporally coarse-grained average over this distribution. However, as the authors point out, the actual problem an animal faces is to make inferences about $s(t)$ that guide future actions based on a *single* draw from this distribution. The central result of the paper is to show that this is possible. In fact, the authors consider the most stringent possible inference problem—complete reconstruction, or decoding, of $s(t)$ from the spike times. What is more, they demonstrate that a simple linear filter of the spiking outputs of the wide-field, movement-sensitive H1 neuron of the blowfly suffices to recover the input motion field with high fidelity. In fact, the responses of this single neuron conveyed some 64 bits/second about motion, and permitted estimation of the amplitude of 20Hz jitter in the visual input with an accuracy of 0.1 degrees, which is an order of magnitude finer than the angular resolution of the photoreceptors!

The results in this paper were a revelation to the field. First, the authors showed definitively that the timing of spikes was informative despite the timing noise evident in neural activity. Second, they showed that simple methods like linear filtering, artfully applied, could immensely clarify the structure of the neural code. Third, they showed that it was possible to quantify the notion that the brain could reconstruct behaviorally relevant inputs from neural responses. And finally, they demonstrated the power of using information theory as a tool to assess the dynamics of information coding by neurons, while showing in particular that the information rates of individual spike trains could be measured and compared.

In this way, Bialek *et al.*'s paper transformed how neuroscientists thought about the neural code, and provided powerful methods for analyzing experiments from this new perspective. The authors later collected their ideas into a very influential book, *Spikes: Exploring the Neural Code*, published in 1999. Their insights opened the way for a flood of work analyzing information coding in many parts of the brains of species ranging from the worm *C. elegans* to mice and humans. These techniques have spread so widely that they have become part of the "standard" repertoire of the field. In neuroscience, for example, their insights have enabled scholars to address topics as diverse as accounting for how much the eye tells the brain, how large correlated networks represent information, and how the sensory repertoires of the early olfactory and visual systems adapt to the environments they sense. A beautiful synthesis of many of these developments appears in the book *Principles of Neural Design* by Sterling and Laughlin (2015).

In the three decades since their paper appeared, the ideas of Bialek *et al.* have spread across branches of biology, for example, enabling scientists to quantify how signal transduction networks in single cells encode information about environmental signals. Sociologically, this paper also had a remarkable effect. It helped to open a door into neuroscience for young scientists from fields like physics and computer science, including the present author, who had previously considered problems in biology to be too intractable to understand or address using quantitative methods.

REFERENCES

Atick, J. J., and A. N. Redlich. 1992. "What Does the Retina Know about Natural Scenes?" *Neural Computation* 4 (2): 196–210. https://doi.org/10.1162/neco.1992.4.2.196.

Attneave, F. 1959. *Applications of Information Theory to Psychology: A Summary of Basic Concepts, Methods, and Results.* New York, NY: Holt, Rinehart, / Winston.

Barlow, H. B. 1961. "Possible Principles Underlying the Transformation of Sensory Messages." In *Sensory Communication,* edited by W. A. Rosenblith, 217–233. Cambridge, MA: MIT Press. https://doi.org/10.7551/mitpress/9780262518420.003.0013.

Bialek, W., and A. Zee. 1990. "Coding and Computation with Neural Spike Trains." *Journal of Statistical Physics* 59:103–115. https://doi.org/10.1007/bf01015565.

Koch, K., J. McLean, R. Segev, M. A. Freed, M. J. Berry, V. Balasubramanian, and P. Sterling. 2006. "How Much the Eye Tells the Brain." *Current Biology* 16 (14): 1428–1434. https://doi.org/10.1016/j.cub.2006.05.056.

Laughlin, S. 1981. "A Simple Coding Procedure Enhances a Neuron's Information Capacity." *Zeitschrift für Naturforschung C* 36 (9–10): 910–912. https://doi.org/10.1515/znc-1981-9-1040.

Meister, M., and M. J. Berry. 1999. "The Neural Code of the Retina." *Neuron* 22 (3): 435–450. https://doi.org/10.1016/s0896-6273(00)80700-x.

Ratliff, C. P., B. G. Borghuis, Y. H. Kao, P. Sterling, and V. Balasubramanian. 2010. "Retina is Structured to Process an Excess of Darkness in Natural Scenes." *Proceedings of the National Academy of Sciences* 107 (40): 17368–17373. https://doi.org/10.1016/j.neuron.2014.08.040.

Rieke, F., D. Warland, R. de Ruyter van Steveninck, and W. Bialek. 1999. *Spikes: Exploring the Neural Code.* Cambridge, MA: MIT Press.

Schneidman, E., M. J. Berry, R. Segev, and W. Bialek. 2006. "Weak Pairwise Correlations Imply Strongly Correlated Network States in a Neural Population." *Nature* 440 (7087): 1007–1012. https://doi.org/10.1038/nature04701.

Shannon, C. E. 1948. "A Mathematical Theory of Communication." *The Bell System Technical Journal* 27:379–423, 623–656.

Sterling, P., and S. Laughlin. 2015. *Principles of Neural Design.* Cambridge, MA: MIT Press.

Teşileanu, T., S. Cocco, R. Monasson, and V. Balasubramanian. 2019. "Adaptation of Olfactory Receptor Abundances for Efficient Coding." *eLife* 8:e39279. https://doi.org/10.7554/eLife.39279.

READING A NEURAL CODE

William Bialek, University of California Berkeley
Fred Rieke, University of California Berkeley
Rob R. de Ruyter van Steveninck, Rijksuniversiteit Groningen
and David Warland, University of California Berkeley

Abstract

Traditional approaches to neural coding characterize the encoding of known stimuli in average neural responses. Organisms face nearly the opposite task—extracting information about an unknown time-dependent stimulus from short segments of a spike train. Here the neural code was characterized from the point of view of the organism, culminating in algorithms for real-time stimulus estimation based on a single example of the spike train. These methods were applied to an identified movement-sensitive neuron in the fly visual system. Such decoding experiments determined the effective noise level and fault tolerance of neural computation, and the structure of the decoding algorithms suggested a simple model for real-time analog signal processing with spiking neurons.

Prior to this work, most studies treated neurons as encoding information in their response rate, averaged over an ensemble of repeated experiments or neurons.

All of an organism's information about the sensory world comes from real-time observation of the activity of its own neurons. Incoming sensory information is represented in sequences of essentially identical action potentials, or "spikes." To understand real-time signal processing in biological systems, one must first understand this representation: Does a single neuron signal only discrete stimulus "features," or can the spike train represent a continuous, time-varying input? How much information is carried by the spike train? Is the reliability of the encoded signal limited by noise at the sensory input or by noise and inefficiencies in the subsequent layers of neural processing? Is the neural code robust to errors in spike timing? Clear experimental answers to these questions have been elusive (Perkel and Bullock 1968; de Ruyter van Steveninck and Bialek 1988). We present an approach to the characterization of the neural code that provides explicit and sometimes surprising answers to these questions.

The first recordings from single sensory neurons demonstrated that the intensity of a static stimulus can be coded in the firing

The authors raise a fundamental issue: animals must make decisions on the basis of individual firing events in unique spike trains.

rate of a sensory neuron (Adrian 1926). This concept of rate coding, extended to time-dependent stimuli, provides the framework for most studies of neural coding, leading to the definition of receptive fields, temporal filter characteristics, and so on. Beyond rate coding, a variety of different statistical measures have been proposed—interval distributions, correlation functions, and so forth .[1] As with the rate itself, these quantities can be seen as moments of the probability distribution $P[\{t_i\} \mid s(\tau)]$ that describes the likelihood of different spike trains $\{t_i\}$, given the stimulus $s(\tau)$ (Bialek 1990). These moments, however, are not properties of a single spike train; they are average properties of an ensemble of spike trains.[2] Organisms rarely have the opportunity to compute these averages: To say that information is coded in firing rates is of no use to the organism unless one can explain how the organism could estimate these firing rates from real-time observation of the spike trains of its own neurons.[3]

The simplest problem of real-time signal processing is decoding the spike train to estimate the signal waveform. If one chooses an inappropriate definition of the signal this reconstruction will fail; for example, we have studied auditory neurons that provide enough information for reconstruction of the envelope of the acoustic stimulus but not of the waveform itself (Rieke *et al.* 1992). One can define the signal encoded by a particular neuron to be that signal which is reconstructed most accurately from observation of the spike train. If

---

[1] Perkel and Bullock (1968). Recent work has focused attention on the time variation of firing rates in the auditory (for example, Miller, Barta, and Sachs 1987) and visual (for example, Richmond, Optican, and Spitzer 1990) systems and on the possible significance of temporal structure and correlated firing in the visual cortex (Gray and Singer 1989). All these observations, however, involve measurements of average properties of spike trains over repeated presentations of a stimulus. Although these experiments are suggestive of coding strategies that different systems might use, none directly answers the question of how the organism could extract this coded information from single spike trains in real time.

[2] The most general of such approaches are the white-noise methods, such as reverse correlation (Marmarelis and Marmarelis 1978). Although these methods do not rely on repeated presentation of identical stimuli, they nonetheless yield models that ideally predict the firing rate in response to arbitrarily chosen stimuli. Reverse correlation thus suffers the same limitations as other rate-based approaches.

[3] Other authors have realized the importance of approaching neural coding from the point of view of the organism. In early work, R. Fitzhugh (1958) discusses real-time decision making. As far as we know, P. I. M. Johannesma (1981) comes closest to our approach.

one can reconstruct analog signals, then one can begin to understand how spike trains could be manipulated in subsequent stages of neural circuitry to perform more complex processing of these signals. It may not be possible, however, to interpolate between the discrete spikes to estimate a continuous stimulus.

The decoding problem is completely specified by the probability $P[s(\tau) \mid \{t_i\}]$ of a particular stimulus waveform $s(\tau)$ conditional on the spike train $\{t_i\}$. From this distribution, one can estimate the stimulus, for example, by finding the function of time that maximizes $P[s(\tau) \mid \{t_i\}]$. Thus, one approach to the decoding problem is to design experiments that directly measure $P[s(\tau) \mid \{t_i\}]$ (de Ruyter van Steveninck and Bialek 1988). An alternative approach is to model the encoding process and analytically develop encoding process and analytically develop decoding algorithms within the context of the model; this approach (Bialek and Zee 1990) indicates that there is a broad regime in which linear filtering of the spike train results in an optimal estimate of the stimulus waveform.[4] Are such simple decoding algorithms applicable to a real neuron?

The authors quantify a notion of stimulus decoding as a means of assessing a neural code.

The problem of reading the neural code is essentially the problem of building a (generally nonlinear) filter that operates continuously on the spike train to produce a real-time estimate of the unknown stimulus waveform (Fig. 1). If the spikes arrive at times $\{t_i\}$, our estimate of the signal is

$$s_{\text{est}}(t) = \sum_i F_1(t - t_i) \\ + \sum_{i,j} F_2(t - t_i, t - t_j) + \ldots \quad (1)$$

Note that the times of the individual spikes matter for the decoding. This is a timing code.

To optimize the reconstructions one chooses the filters $\{F_n\}$ to minimize $\chi^2 = \int dt \mid s(t) - s_{\text{est}}(t) \mid^2$, where $s(t)$ is the true stimulus and

---

[4]Linear decoding might work for a trivial reason. Many sensory neurons have a regime in which the firing rate varies linearly with stimulus amplitude. This is an example of linear encoding and linear decoding might work as well in this regime. However, Bialek and Zee (1990) predict that linear decoding is possible beyond the regime of linear input–output relations. The cell we studied in this work has a linear input–output regime, but the stimuli we used drove the cell outside this limit into saturation. See, for example, Eckert (1980), who finds nonmonotonic rate-velocity profiles in the range of stimuli used here.

FOUNDATIONAL PAPERS IN COMPLEXITY SCIENCE

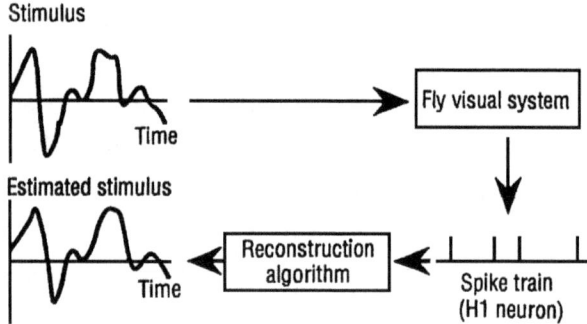

Figure 1. Schematic view of the decoding process. The "black box" filters the spike train input $\{t_i\}$ to produce an estimate $s_{\text{est}}(\tau)$ of the stimulus.

In practice one rarely has enough data to go beyond the linear order.

the integration is over the duration of the experiment.[5] The stimulus $s(t)$ is not restricted to simple sine waves or Gaussian noise, and thus one can study the coding of complex and naturalistic signals (Rieke *et al.* 1992; Warland *et al.* 1992). In this initial experiment, however, we used relatively simple stimuli.

We applied these ideas in experiments on a single, wide-field, movement-sensitive neuron (H1) in the visual system of the blowfly *Calliphora erythrocephela*. H1 encodes rigid horizontal movements over the entire visual field (Hausen 1984). Flies and other insects exhibit visually guided flight; during chasing behavior, course corrections can occur on time scales as short as 30 ms (Land and Collett 1974). The

---

[5]To ensure that the reconstruction process could be implemented in real time, we required that the filters be causal, for example, $F_1(\tau < 0) = 0$. We calculated the minimum χ^2 causal filters in two ways: (i) The best filter is first calculated without the causality constraint. An explicit formula can be written for this filter in terms of the spike trains and the stimulus

$$F_1(\tau) = \int \frac{d\omega}{2\pi} e^{-i\omega\tau} \frac{\left\langle \tilde{s}(\omega) \sum_j e^{-1\omega t_j} \right\rangle}{\left\langle \sum_{i,j} e^{i\omega(t_i - t_j)} \right\rangle} \quad (3)$$

where $\tilde{s}(\omega) = \langle d\tau \to \tilde{s}(\omega) = /d\tau \rangle$. The averages (inside triangular brackets) are over the ensemble of stimuli $s(\tau)$ used in the experiment. This filter can be shifted by a delay τ_{delay} and causality can be imposed by setting the shifted filter to zero at negative times. (ii) χ^2 can be minimized with respect to purely causal functions by expansion of the filters $\{F_n\}$ in a complete set of functions that vanish at negative times. In this method, a delay time must be explicitly introduced that measures the lag between the true stimulus and the reconstruction, so $\chi^2(\tau_{\text{delay}}) = \int dt |s(t - \tau_{\text{delay}}) - s_{\text{est}}(t)|^2$ is minimized. These two methods together ensure that the optimal causal filters are calculated.

maximum firing rate in H1 is 100 to 200 s^{-1}, so behavioral decisions are based on just a few spikes from this neuron. Furthermore, the horizontal motion detection system consists of only a handful of cells, so the fly has no opportunity to compute average responses (for example, firing rates).

In our experiments, the stimulus $s(\tau)$ was the angular velocity of a rigidly moving random pattern. We chose $s(\tau)$ from an ensemble that approximated Gaussian noise with standard deviation 132 deg s^{-1}; the spectrum of $s(\tau)$ was constant up to a cutoff frequency of 1 kHz. We recorded the spike arrival times $\{t_i\}$ extracellularly from the H1 neuron (de Ruyter van Steveninck 1986; de Ruyter van Steveninck and Bialek 1988). We began by trying to reconstruct $s(\tau)$ with just the linear term of the general expansion, Eq. 1 (Fig. 2). Reconstructions including higher order terms in Eq. 1 were not significantly different, as quantified below. The filters used in the reconstruction integrated over short time intervals, so the optimal estimate of angular velocity at each instant of time was controlled by just a handful of spikes, as expected from behavioral studies.

They look at a system where the speed of response, and small number of cells, mean that the neural code must involve individual spike times.

How good are the reconstructions? The reconstructions consist of a piece that is deterministically related to the stimulus and a random noise piece. We separated these by introducing a frequency-dependent gain $g(\omega)$ such that $\tilde{s}_{\rm est}(\omega) = g(\omega)[\tilde{s}(\omega) + \tilde{n}(\omega)]$, where $\tilde{n}(\omega)$ is the noise referred to the input. In a plot of $\tilde{s}(\omega)$ versus $\tilde{s}_{\rm est}(\omega)$, the gain is the slope of the best linear fit and $\tilde{n}(\omega)$ is the scatter about this line. The distribution of $\tilde{n}(\omega)$ is approximately Gaussian.

Plotting the spectral density of the angular displacement noise, we found (Fig. 3) a peak signal-to-noise ratio (SNR) of better than 5:1, and an SNR of greater than one across a bandwidth of roughly 25 Hz. Using Shannon's formula (Shannon 1948), shown in Eq. 2, we converted these spectra into an estimate of the average rate at which we gained information about the stimulus ($R_{\rm info}$) by virtue of observing the spike train; the result was 64 ± 1 bits per second (baud).

The authors helped to pioneer the use of information theory to quantify the quality of neural representations. They use this method to argue that the linear decoder is sufficient.

$$R_{\rm info} = \int \frac{d\omega}{2\pi} \log_2 \left[1 + \frac{\langle |\tilde{s}(\omega)|^2 \rangle}{\langle |\tilde{n}(\omega)|^2 \rangle} \right] \qquad (2)$$

FOUNDATIONAL PAPERS IN COMPLEXITY SCIENCE

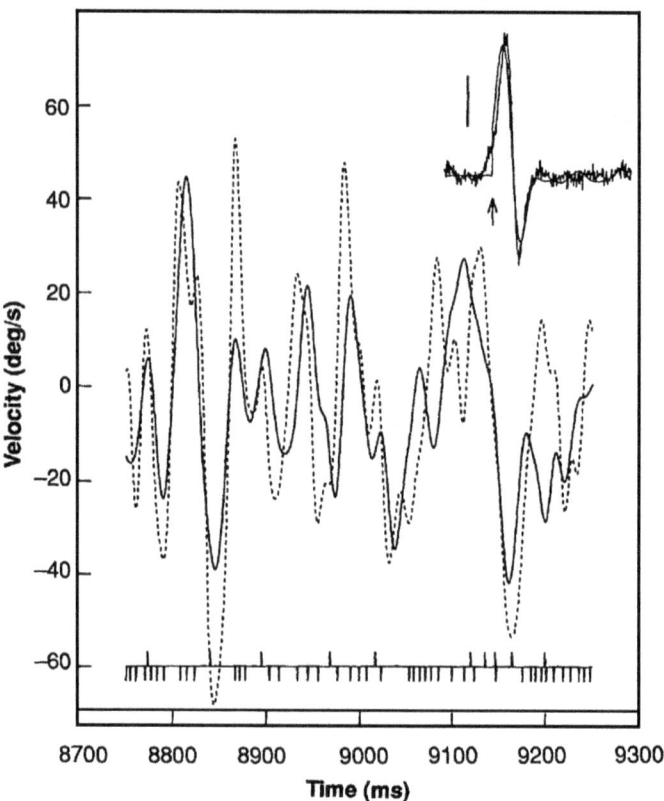

Figure 2. First-order reconstruction (solid line) using method 1 (see footnote 5). The stimulus is shown as a dotted line for comparison, and the spike train is shown at the bottom. This example is from a segment of the spike train that was not used in the filter calculations. Stimulus and reconstruction were smoothed with a 5-ms half-width Gaussian filter. (Inset) Filters calculated from methods 1 and 2 (footnote 5); the time scale is the same as the main figure, the scale bar = 10 deg s^{-1}, and the arrow marks $t = 0$. H1 has a highly asymmetric response profile, with a much larger dynamic range for movement in the excitatory direction. To compensate for this asymmetry, we recorded spike trains in response to $s(t)$ and to the inverted stimulus $-s(t)$; these are shown as positive and negative spikes in the spike train. These two spike trains approximate the trains that would be generated by H1 cells on opposite sides of the head during a rigid rotation of the fly (de Ruyter van Steveninck and Bialek 1988). Our reconstruction is then $s_{\text{est}}(t) = \sum_i \left[F_1 \left(t - t_i^+ \right) - F_1 \left(t - t_i^- \right) \right]$.

Bialek et al. (1991)

The second term in the expansion improved the information rate by less than 5%. We have not explored conditions that might maximize this information transmission.

The noise level achieved in the reconstructions is the noise against which an observer of the H1 spike trains (such as the fly) must discriminate to estimate horizontal motion. The absolute noise level of the reconstructions is very low. With a behaviorally relevant integration time of 30 ms, one could judge the amplitude of a 20-Hz dither to within ~ 0.1 deg, which should be compared to the photoreceptor spacing of $\phi_0 = 1.35°$. This angular resolution corresponds to the phenomenon of hyperacuity in human vision (Shannon 1948) and is in quantitative agreement with direct measures of discriminability for stepwise displacements in H1 (de Ruyter van Steveninck 1986). Defining an equivalent spectral density of noise in a spiking neuron allows one to exhibit hyperacuity in a real-time estimation task.

Information about movement across the visual field is carried in the spatiotemporal correlations of photoreceptor outputs, but these correlations are degraded by noise in the photoreceptors. How accurately can one estimate rigid motion if one optimally processes these noisy photoreceptor signals? In our stimulus ensemble, the angular displacements were small ($\delta\theta \ll \phi_0$) for frequencies above 10 Hz. In this limit, the optimal movement estimator involves multiplying the direct current voltage in one cell by the alternating current voltage in its nearest neighbor (Bialek 1990). This is essentially the "correlation" scheme for movement detection proposed by Reichardt (1961); in our case, this algorithm was not a minimal model but rather the optimal computational strategy. Analysis of the correlation scheme (Bialek 1990) led to the limiting angular displacement noise level shown in Fig. 3, where the displacement noise from the linear reconstruction approaches the limits imposed by the photoreceptor noise, at least at frequencies above 10 Hz, where our theory of the limiting noise level is valid. The fly visual system thus performs an optimal and nearly noiseless extraction of movement signals from the array of photoreceptor voltages.

Coding is often used to reduce the effects of noise on signal transmission. Does the neural code have any such noise immunity?

They argue that this fly motion-detection system is optimal. It achieves a performance bound set by noise in the photoreceptors.

FOUNDATIONAL PAPERS IN COMPLEXITY SCIENCE

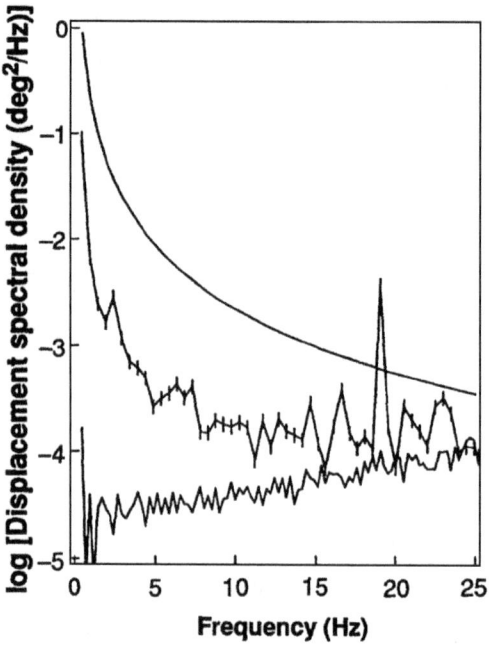

Figure 3. Stimulus level (smooth curve) and spectral density of displacement noise from the reconstruction (middle curve). The bottom curve is the limit to the resolution of small displacements (valid for frequencies > 10 Hz) set by noise in the photoreceptor array (Bialek 1990)

$$S_\theta^{\text{eff}}(\omega) = \frac{S_v(\omega)}{|\widetilde{T}(\omega)|^2} \frac{32\pi^{1/2}}{NS_c} \frac{(\Delta\phi)^5}{\phi_0^2} \\ \sinh\left[(1/2)(\phi_0/\Delta\phi)^2\right] \quad (4)$$

where S_c is the spectral density of the random contrast pattern used in the experiment, $\Delta\phi$ is the width of the photoreceptor aperture, ϕ_0 is the photoreceptor spacing along the direction of movement, $\widetilde{T}(\omega)$ is the frequency response of the photoreceptor (millivolts per unit contrast), and $S_V(\omega)$ is the spectral density of voltage noise in the photoreceptor. The limiting noise power spectrum varies as the inverse of the number N of photoreceptors. These quantities were measured in photoreceptor recordings under conditions identical to those used for the H1 experiments (de Ruyter van Steveninck 1986), so we can make meaningful comparisons of theory and experiment.

They use their information-theoretic measure of decoding quality to argue that the neural code is robust to errors of timing, deletion, and addition.

For several noise sources, such as timing errors, dropped spikes, and spontaneously generated spikes, we created an ensemble of spike trains which were randomly corrupted versions of the original data and then treated these as new data that required decoding. We were able to recover 95% of the original information R_{info} for the following noise levels: (i) spikes added to increase the firing rate by 20%, (ii) 5% of the spikes deleted, and (iii) Gaussian timing jitter with standard deviation of 2 ms introduced to each spike time. One objection to "spike timing" as a coding strategy is the need for precise measurement of spike arrival times. In the case of H1, this objection is irrelevant. The code is robust to errors of several milliseconds in spike timing and to other corruptions of the spike train.

Preliminary results from four studies suggest that linear decoding is not a special property of H1 under particular stimulus conditions but a more general property of sensory neurons.

1) In H1, we performed reconstruction experiments using stimuli with different spatial characteristics. These different stimulus ensembles provided different SNRs, and these differences were reflected in the reconstructions. Despite large changes in SNR, linear reconstruction continued to work and the reconstruction filters were essentially identical up to a constant scale factor.

2) In the mechanoreceptor cells of the cricket cercal system, the displacement waveform for motions of the filiform hairs was reconstructed from the spike trains of primary afferent neurons (Warland *et al.* 1992). For this system, the information transfer rates exceeded 300 baud ($\sim$ 3 bits per spike).

3) In simulations on realistic models (Koch and Segev 1989) for spike initiation, we reconstructed the waveform of injected currents by linearly filtering the spike train.

4) In vibratory receptors of the bullfrog sacculus, we reconstructed the waveform of groundborne vibration using a linear decoding algorithm, although in this case the reconstructions improved substantially (by $\sim$ 10 to 15%) with the addition of second-order terms (Rieke *et al.* 1992). We again measured information rates close to 3 bits per spike.

We have learned in later years that many sensory neurons, even in the sensory periphery, actually have substantial nonlinearities.

It is, of course, not known if organisms perform the sort of reconstructions demonstrated in Fig. 2. Because linear reconstruction is possible, however, analog processing of the encoded signals can be done in a simple way. It is not unusual for the postsynaptic voltage response to a single presynaptic spike to have the qualitative form of the optimal filter, with a relatively sharp positive peak followed by a slower negative tail. Thus simple synapses could serve as decoders. With this decoding done, cells could then perform analog computations using the nonlinearities contributed by voltage-gated channels along the dendrites and cell body, much as envisioned for non-spiking cells (Koch, Poggio, and Torre 1983). The results of such analog computations could then be encoded by the spike-generating region of the cell, and the process could then begin again. In this view of computation with spike trains, the combination of nonlinearities in spike generation and the filter characteristics of the synapse results in an essentially linear transmission of analog signals from presynaptic cell bodies to postsynaptic dendrites. The dramatic "all-or-none" nonlinearities of spike generation (the focus of so many models for neural computation) are then not as important as the more subtle analog dynamic properties of nonspiking regions of the cell.

This proposal of filtering and stimulus reconstruction by synapses, followed by analog, nonlinear computation in dendrites, provides a simple and powerful baseline description of at least some sensory neurons.

It is surprising that time-dependent signals can be recovered so simply from neural spike trains. Reconstruction of the stimulus waveform permitted quantification of the fault tolerance of the code and allowed us to show that the fly visual system approaches optimal real-time computation. These results demonstrate that the representation of time-dependent sensory data in the nervous system is simpler than might have been expected. Correspondingly simpler models of sensory signal processing may be appropriate.

Acknowledgments

We thank W. J. Bruno, M. Crair, W. Gerstner, L. Kruglyak, J. P. Miller, W. G. Owen, A. Zee, and G. Zweig for helpful discussions and D. A. Baylor, D. Glaser, and M. Meister for thoughtful comments on the manuscript. Preliminary work on simulations of model neurons was done with L. Kruglyak, the results on the cricket cercal system were obtained in collaboration with M. A. Landolfa and J. P. Miller, and the experiments on the bullfrog auditory system were done in collaboration with W. Yamada and E. R. Lewis. Work at Berkeley was supported in part by the NSF through a Presidential Young Investigator Award to W. B., supplemented by funds from Cray Research, Sun Microsystems, and the NEC Research Institute, and through a Graduate Fellowship to F. R. D. W. was supported in part by the Systems and Integrative Biology Training Program of the NIH. Initial work was supported by the Netherlands Organization for Pure Scientific Research (ZWO). Research also performed at NEC Research Institute by W. B. and F. R.

REFERENCES

Adrian, E. D. 1926. "The Impulses Produced by Sensory Nerve Endings. Part I." *The Journal of Physiology* 61 (1): 49–72.

Bialek, W. 1990. "Theoretical Physics Meets Experimental Neurobiology." In *1989 Lectures in Complex Systems,* edited by E. Jen, 2:513–595. SFI Studies in the Sciences of Complexity. Reading, MA: Addison-Wesley.

Bialek, W., and A. Zee. 1990. "Coding and Computation with Neural Spike Trains." *Journal of Statistical Physics* 59:103–115.

de Ruyter van Steveninck, R. 1986. *Real-Time Performance of a Movement-Sensitive Neuron in the Blowfly Visual System.* Groningen, Netherlands: Rijksuniversiteit Groningen.

de Ruyter van Steveninck, R., and W. Bialek. 1988. "Real-Time Performance of a Movement-Sensitive Neuron in the Blowfly Visual System: Coding and Information Transfer in Short Spike Sequences." *Proceedings of the Royal Society of London B* 234:379–414.

Eckert, H. 1980. "Functional Properties of the H1-Neurone in the Third Optic Ganglion of the Blowfly, Phaenicia." *Journal of Comparative Physiology* 135:29–39.

Fitzhugh, R. 1958. "A Statistical Analyzer for Optic Nerve Messages." *Journal of General Physiology* 41 (4): 675–692.

Gray, C. M., and W. Singer. 1989. "Stimulus-Specific Neuronal Oscillations in Orientation Columns of Cat Visual Cortex." *Proceedings of the National Academy of Sciences of the USA* 86 (5): 1698–1702.

Hausen, K. 1984. "The Lobula-Complex of the Fly: Structure, Function and Significance in Visual Behaviour." In *Photoreception and Vision in Invertebrates,* edited by M. Ali, 523–559. Boston, MA: Springer.

Johannesma, P. 1981. "Neural Representation of Sensory Stimuli and Sensory Interpretation of Neural Activity." In *Neural Communication and Control,* edited by G. Székely, E. Lábos, and S. Damjanovich, 103–125. Advances in Physiological Sciences. Budapest, Hungary: Akademiai Kiado.

Koch, C., T. Poggio, and V. Torre. 1983. "Nonlinear Interactions in a Dendritic Tree: Localization, Timing, and Role in Information Processing." *Proceedings of the National Academy of Sciences of the USA* 80 (9): 2799–2802.

Koch, C., and I. Segev. 1989. *Methods in Neuronal Modeling.* Cambridge, MA: MIT Press.

Land, M. F., and T. S. Collett. 1974. "Chasing Behaviour of Houseflies (*Fannia canicularis*)." *Journal of Comparative Physiology* 89:331–357.

Marmarelis, P., and V. Marmarelis. 1978. *Analysis of Physiological Systems: The White-Noise Approach.* New York, NY: Plenum.

Miller, M. I., P. E. Barta, and M. B. Sachs. 1987. "Strategies for the Representation of a Tone in Background Noise in the Temporal Aspects of the Discharge Patterns of Auditory-Nerve Fibers." *The Journal of the Acoustical Society of America* 81 (3): 665–679.

Perkel, D. H., and T. H. Bullock. 1968. "Neural Coding." *Neurosciences Research Program Bulletin* 6 (3): 221–348.

Reichardt, W. 1961. "Autocorrelation, a Principle for the Evaluation of Sensory Information by the Central Nervous System." In *Principles of Sensory Communication,* edited by W. Rosenblith, 303–317. Cambridge, MA: MIT Press.

Richmond, B. J., L. M. Optican, and H. Spitzer. 1990. "Temporal Encoding of Two-Dimensional Patterns by Single Units in Primate Primary Visual Cortex. I. Stimulus-Response Relations." *Journal of Neuropsychology* 64 (2): 351–369.

Rieke, F., W. Yamada, K. Moortgat, E. R. Lewis, and W. Bialek. 1992. "Real Time Coding of Complex Sounds in the Auditory Nerve." In *Auditory Physiology and Perception: Proceedings of the 9th International Symposium on Hearing Held in Carcens, France, on 9–14 June 1991,* edited by Y. Cazals, K. Horner, and L. Demany, 315–322. Oxford, UK: Pergamon Press.

Shannon, C. E. 1948. "A Mathematical Theory of Communication." *The Bell System Technical Journal* 27:379–423, 623–656.

Warland, D., M. Landolfa, J. P. Miller, and W. Bialek. 1992. "Reading Between the Spikes in the Cereal Filiform Hair Receptors of the Cricket." In *Analysis and Modeling of Neural Systems,* edited by F. H. Eeckman, 327–333. Boston, MA: Springer.

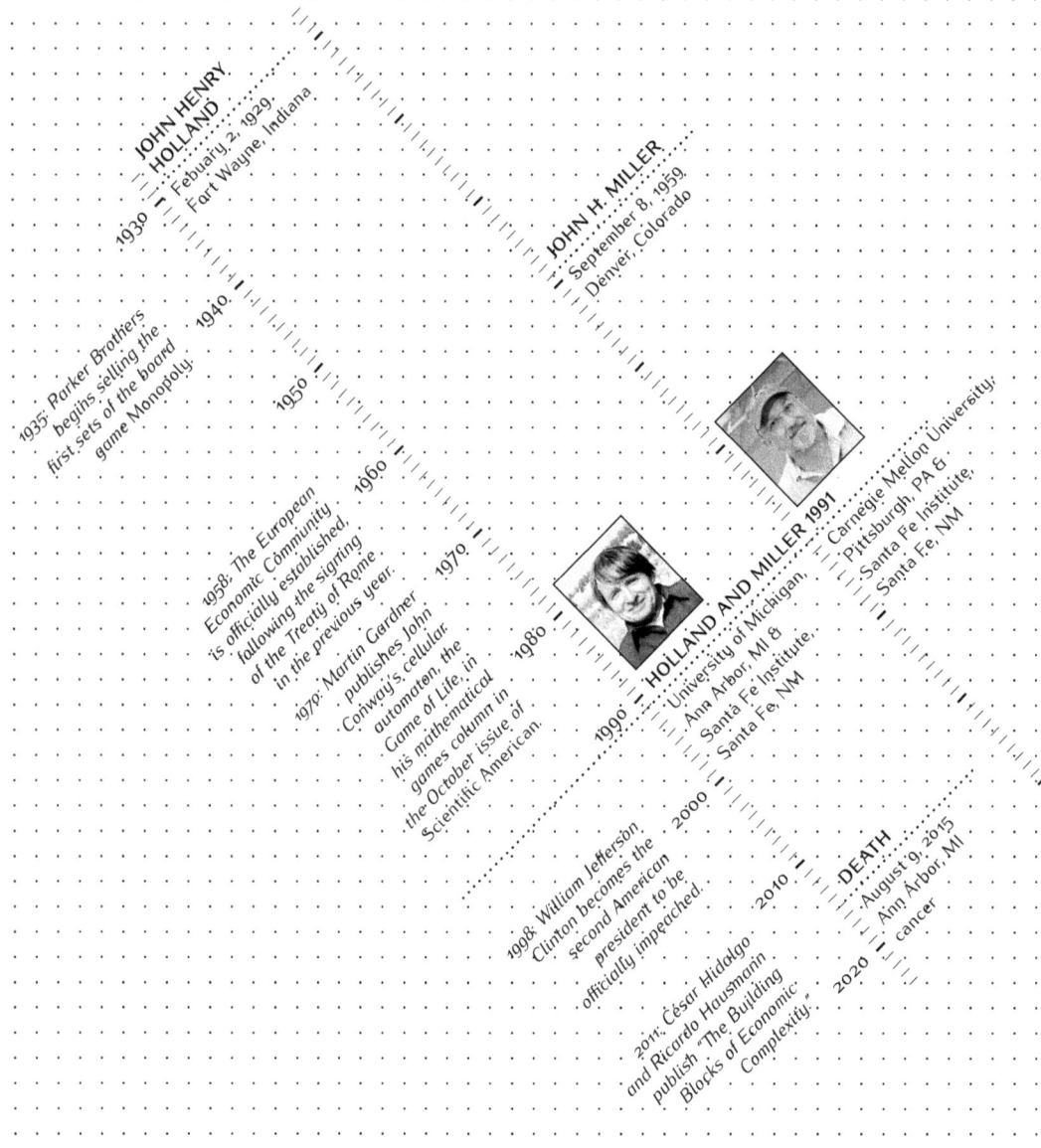

HOLLAND & MILLER

[72]

ECONOMICS FOR HUMANS

Richard Bookstaber

It seems self-evident that people are different from one another. We each have a different way of looking at the world, a different approach to making decisions. People also change. We are shaped by our individual experience and by changes in our environment, so how we make decisions today might not be the way we do so tomorrow. Such is the world in which we reside. It is also a world in which the artificial adaptive agents described by John Holland and John Miller reside. Their brief paper speaks to an approach to economic modeling based on simulations rather than mathematical solutions, an approach that can address problems that are beyond the reach of the tools common in economic theory.

Holland and Miller's models can be summarized as executing in three steps:

1. Each agent observes its environment and takes action based on its heuristic.

2. The actions of the agents change the environment.

3. Each agent observes the change in its environment, adjusts its heuristic accordingly, and takes action in the next time step.

Two key features of the agents are that they are adaptive and heterogeneous: adaptive in that they react to changes in the environment and do so based on heuristics that might themselves change based on the new environment; heterogeneous in that the various agents see different aspects of the environment and act based on different heuristics.

The heuristics can be simple rules of thumb or can be determined in a process that provides for "emergent, learned capabilities" as the agents interact with their environment (Gigerenzer, Reb, and Luan 2022).

J. H. Holland and J. H. Miller, "Artificial Adaptive Agents in Economic Theory," *American Economic Review* 81 (2), 365–370 (1991).

Copyright American Economic Association; reproduced with permission of the authors.

The genetic algorithm described in Holland and Miller is one example. Period by period, the agents meet pairwise to exchange information—or, more evocatively, to mate and share their genetic code—as they adapt step by step toward a mutual goal. The agent's heuristic is thus one of generating rules that are continually changing with the objective of improving its fitness.

Much like modern-day crowdsourcing, this adaptive agent-based algorithm extracts useful information from the entire agent population, rather than only finding and exploiting the best agents. From a practical standpoint, agent-based models share an important characteristic of physical systems, in that "aggregate behavior can be described without a detailed knowledge of the behavior of the individual agents." This means the important aggregate dynamics of the system do not require heuristics to be specified in great detail.[1]

Holland and Miller note a number of characteristics of these models that are essential in facing the real world but are left wanting by standard economic theory.

The first is addressing emergent phenomena. Emergence means the behavior of the overall system can be different from the actions of the individuals within the system. Different, and possibly complex and unexpected. "Possibilities are so rich that it is often difficult to predict on a priori grounds what behaviors and structures will emerge." Emergent phenomena are a natural result of the heterogeneity and interactions of the agents. But in economic theory emergence is constrained away due to the technical requirements of its mathematics and its use of a representative agent. In mathematical formulations, there is usually some rubber band keeping the train from running off the track, whereas in an agent-based model, the train might derail and roll down the hill as the agents interact, even though no agent is taking action with that as its objective. Because of this, emergence can lead to operating "far from a global optimum."

The second is dealing with problems that are computationally irreducible, meaning they cannot be solved mathematically but only through following a simulation period by period to the end result. Even almost trivial systems can be beset with computational irreducibility.

---

[1] Gualdi *et al.* (2015) discuss this point and show similar behavior in statistical physics.

As Holland and Miller note in footnote 1, simple models of cellular automata, such as John Conway's Game of Life, are in this category. The three-body problem, which tries to describe the path of three planets interacting through their gravitational pull, is another example (although it can be solved for a small set of special cases). Given that simple problems like these can be impervious to mathematical solutions, certainly the same is true for complex economic problems. We can only know where things are going by walking down the path step by step. Holland and Miller note that emergence and computational irreducibility pose limitations to developing a general mathematical theory of complex systems. And, indeed, thirty years later, such a theory remains elusive.[2]

Another important characteristic of these methods is to give transparency to complex dynamical systems. On an intuitive level, an agent-based model can be visualized as a network with brains in the nodes, and period by period, as the nodes interact, the topology of the network changes. The model's unfolding can be observed step by step, and because the models are expressed linguistically through a computer language, they are open to a narrative. As a practical matter, economic decisions are unlikely to be based on a model where the results cannot be explained and discussed in plain English, and Holland and Miller's models are suited to that.

Agent-based models have expanded from their early application in building artificial stock markets to have widespread application across the economic sphere.[3] They overcome some severe limitations of economic theory in meeting features of the real world. However, adaptive agent models have been slow to gain acceptance in the economic academic community. Holland and Miller write of looking forward

---

[2] Bookstaber (2017) presents four key weaknesses in the standard neoclassical approach—emergence and computational irreducibility discussed here, as well as non-ergodic processes and radical uncertainty—and ways agent-based models help to resolve those weaknesses.

[3] An early application of agent-based models in economics is the artificial stock market in Palmer et al. (1994). This model includes learning through a genetic algorithm in the agents' heuristics. Steinbacher et al. (2021) provide a broad survey on the use of agent-based models in economics. Testing applications along the lines of Holland and Miller's suggested approach "to check the various unfolding behaviors for plausibility" is discussed for financial markets by Lux (2009).

to a meeting of these methods with economic theory, but to date that remains elusive. Although this article appeared in the *American Economic Review*, it is perplexing and perhaps illustrative of the current state of economics that ongoing economic research using agent-based models has yet to make major headway in the mainstream economic literature and finds its most welcome home in scientific journals outside the field.

REFERENCES

Bookstaber, R. 2017. *The End of Theory.* Princeton, NJ: Princeton University Press.

Gigerenzer, G., J. Reb, and S. Luan. 2022. "Smart Heuristics for Individuals, Teams, and Organizations." *Annual Review of Organizational Psychology and Organizational Behavior* 9:171–198. https://doi.org/10.1146/annurev-orgpsych-012420-090506.

Gualdi, S., M. Tarzia, F Zamponi, and J. P. Bouchaud. 2015. "Tipping Points in Macroeconomic Agent-Based Models." *Journal of Economic Dynamics and Control,* 50:29–61. https://doi.org/10.1016/j.jedc.2014.08.003.

Lux, T. 2009. "Hens, T. and Schenk-Hoppé, K. R." In *Handbook of Financial Markets: Dynamics and Evolution,* 161–215. North-Holland. https://doi.org/10.1016/B978-012374258-2.50007-5.

Palmer, R., W. B. Arthur, J. H. Holland, B. LeBaron, and P. Tayler. 1994. "Artificial Economic Life: A Simple Model of a Stock Market." *Physica D* 75:264–274. https://doi.org/10.1016/0167-2789(94)90287-9.

Steinbacher, M., M. Raddant, F. Karimi, E. C. Camacho-Cuena, S. Alfarano, G. Iori, and T. Lux. 2021. "Advances in the Agent-Based Modeling of Economic and Social Behavior." *SN Business and Economics* 1:99. https://doi.org/10.1007/s43546-021-00103-3.

ARTIFICIAL ADAPTIVE AGENTS IN ECONOMIC THEORY

*John H. Holland, University of Michigan
and John H. Miller, Carnegie Mellon University*

Economic analysis has largely avoided questions about the way in which economic agents make choices when confronted by a perpetually novel and evolving world. As a result, there are outstanding questions of great interest to economics in areas ranging from technological innovation to strategic learning in games. This is so, despite the importance of the questions, because standard tools and formal models are ill-tuned for answering such questions. However, recent advances in computer-based modeling techniques, and in the subdiscipline of artificial intelligence called machine learning, offer new possibilities. Artificial adaptive agents (AAA) can be defined and can be tested in a wide variety of artificial worlds that evolve over extended periods of time. The resulting *complex adaptive systems* can be examined both computationally and analytically, offering new ways of experimenting with and theorizing about adaptive economic agents.

Many economic systems can be classified as complex adaptive systems. Such a system is *complex* in a special sense: *(i)* It consists of a network of interacting agents (processes, elements); *(ii)* it exhibits a dynamic, aggregate behavior that emerges from the individual activities of the agents; and *(iii)* its aggregate behavior can be described without a detailed knowledge of the behavior of the individual agents. An agent in such a system is *adaptive* if it satisfies an additional pair of criteria: the actions of the agent in its environment can be assigned a value (performance, utility, payoff, fitness, or the like); and the agent behaves so as to increase this value over time. A complex adaptive system, then, is a complex system containing adaptive agents, networked so that the environment of each adaptive agent includes other agents in the system.

A complex adaptive system can be viewed as a network, the agents being the nodes, the edges the interactions, with the topology of the network changing period by period.

Complex adaptive systems usually operate far from a global optimum or attractor. Such systems exhibit many levels of aggregation, organization, and interaction, each level having its own time scale and characteristic behavior. Any given level can usually be described in terms of local niches that can be exploited by particular adaptations. The niches are various, so it is rare that any given agent can exploit all of them, as rare as finding a universal competitor in a tropical forest. Moreover, niches are continually created by new adaptations. It is because of this ongoing evolution of the niches, and the perpetual novelty that results, that the system operates far from any global attractor. Improvements are always possible and, indeed, occur regularly. The ever-expanding range of technologies and products in an economy, or the ever-improving strategies in a game like chess, provide familiar examples. Adaptive systems may settle down temporarily at a local optimum, where performance is good in a comparative sense, but they are usually uninteresting if they remain at that optimum for an extended period.

A dynamic, adaptive system does not fit an objective of seeking a global optimum. By construction, the agents operate in niches that shift in nature and operate with limited knowledge of the environment. Furthermore, there are no conditions that act like a rubber band to pull the system toward a stable configuration, much less hold it there.

A theory of complex adaptive systems based on AAA makes possible the development of well-defined, yet flexible, models that exhibit emergent behavior. Such models can capture a wide range of economic phenomena precisely, even though the development of a general mathematical theory of complex adaptive systems is still in its early stages.[1] The AAA models complement current theoretical directions; they are not intended as a substitute. Many of the most interesting questions concern points of overlap between AAA models and classical theory. As a minimal requirement, wherever the new approach overlaps classical theory, it must include verified results of that theory in a way reminiscent of the way in which the formalism of general relativity includes the powerful results of classical physics.

I. Why Study Artificial Adaptive Agents?

The AAA models have several characteristics that are not available in traditional modeling techniques. Models based on pure linguistic descriptions, while infinitely flexible, often fail to be logically

---

[1] It is important in this research to determine just where the potential for general solutions exists. There are simple models of cellular automata, for example, wherein the solutions to particular questions are computationally irreducible—the shortest way to analyze the system is to run the complete computation.

consistent. Mathematical models lose flexibility, but gain a consistent structure and general solution techniques. The AAA models, specified in a computer language, retain much of the flexibility of pure linguistic models, while having precision and consistency enforced by the language. The resulting models are dynamic and are "executable" in the sense that the unfolding behavior of the model can be observed step by step. This makes it possible to check the plausibility of the behavior implied by the assumptions of the model. The precision of the definitions also opens AAA models to mathematical analysis. The ability to explore a wide range of phenomena involving learning and adaptation, linked with the rigor imposed by a computer language, provides a powerful modeling technique.[1]

Both the agents' heuristics and the evolution of the system are open to a narrative. At any period during a run, the model can be opened up to see what the agents are doing. This provides a mode for common-sense application and verification, and is in contrast to black-box machine-learning models as well as mathematical models that blur a narrative behind requisite abstraction.

The AAA models offer a way of approaching one of the major questions of present theory. Current theoretical constructs, based on optimization principles, often require technically demanding derivations. It is an obvious criticism of these constructs that real agents lack the behavioral sophistication necessary to derive the proposed solutions. This dilemma is resolved if it is postulated that adaptive mechanisms, driven by market forces, lead the agents to act as if they were optimizing (see, for example, Milton Friedman 1953). AAA explicitly model this link between adaptation and market forces, and can thus be used to analyze the conditions under which optimization behavior will (not) occur.

Insofar as human behavior is driven by adaption, an understanding of AAA may prove to be a useful benchmark for, and provide insights into, existing human experiments (see, for example, J. A. Andreoni and Miller 1990; Arthur 1990).[2] An experiment consisting of artificial agents allows the utility, risk aversion, information, knowledge, expectations, and learning of each subject to be carefully controlled. Moreover, at any point in the experiment, the knowledge and learning of the artificial agents can be "reset" to any desired previous state, and subtle variations of the environment can be analyzed. The strategy

The constraints imposed by building a mathematical model require sometimes-heroic assumptions that distance the model from reality. For example, assuming a representative agent, which imposes homogeneity in an obviously heterogeneous world or the need for regularity conditions to allow for a stable and unique solution. These constraints are loosened with an agent-based model.

---

[1] Programming even a simple market is instructive on the limitations of both the pure linguistic and mathematical approaches.

[2] Artificial agents could also be used as "subjects" in pilot studies to identify potentially interesting new human experiments.

(as well as the behavior) of the AAA can always be explicitly analyzed, something not usually possible with human subjects. Finally, the infinite patience and low motivational needs of AAA "subjects" implies that large-scale experiments can be conducted at a relatively low cost.

A major feature of AAA models is their ability to produce emergent behavior. A wide variety of behaviors can arise endogenously, even though these behaviors, as with any model, are constrained by the initial structure. The possibilities are so rich that it is often difficult to predict on a priori grounds what behaviors and structures will emerge. It thus becomes possible to explore realms that were unanticipated when the model was defined. Analysis of these emergent phenomena should offer both insights and suggestions for new theorems about the effects of adaptive agents in economic systems.

The key feature of an adaptive system is emergence: simple rules at the agent level can lead to complex and unanticipated results at the larger scale. An example from everyday experience is how the actions of individual drivers can lead to a traffic jam. The nature of emergence is unpredictable, and only becomes evident as the model is run.

The AAA models may also prove useful in studying economic systems that have either an absence or a plethora of theoretical solutions. Many important economic problems, such as double-auction strategies, multisectoral general equilibrium models, and the like, have no easily derived analytic solutions. Several AAA techniques were originally designed as optimization methods for environments that are nonlinear, noisy, discontinuous, or involve enormous search spaces. As a result, they offer useful numerical techniques for such problems in economics. At the opposite extreme are systems with multiple solutions. For example, in repeated games, the Folk theorem often admits a vast number of potential solutions. In these cases, the interaction of the adaptive systems with the economic environment may narrow the set of potential solutions. Different equilibria may have different degrees of *adaptive complexity*.

Beyond complementing current theoretical and empirical work, AAA offer the potential for unique extensions of current theory. The mechanisms generating the global behavior of a complex adaptive system can be directly observed when the computer is an integral part of the theory. For such theories, the computer plays a role similar to the role the microscope plays for biology: It opens up new classes of questions and phenomena for investigation. Problems that prove difficult for traditional mathematical approaches are often easily implemented as an AAA system. In that form, they can be dissected and modified with ease, providing new opportunities for theory generation and testing. More generally, the potential for the development of a general calculus of "adaptive mechanics" exists. A calculus of these systems would combine the advantages of analytic perspicacity with computer-driven hypothesis testing.

II. Some Current Artificial Adaptive Agent Techniques

A wide range of computer-based adaptive algorithms exist for exploring AAA systems, including classifier systems, genetic algorithms, neural networks, and reinforcement learning mechanisms. The multiplicity of techniques presents a problem for analysis. How sensitive are the results to a particular incarnation of the adaptive agent? This problem, of course, confronts any attempt to lessen the rationality postulates traditionally used in economic theory. Usually, there is only one way to be fully rational, but there are many ways to be less rational. It is important in building a theory based on AAA to construct agents that exhibit robust behavior across algorithmic choices. Current economic studies of adaptive agents rely on genetic algorithms (R. M. Axelrod 1987; Miller 1989; Andreoni and Miller 1990) and classifier systems (R. Marimon, McGratten, and Sargent 1990; Arthur 1990).

Genetic algorithms (GAs) were developed by Holland (1986) as a way of studying adaptation, optimization, and learning. They are modeled on the processes of evolutionary genetics. A basic GA manipulates a set of structures, called a *population*. Structures are usually coded as strings of characters drawn from some finite alphabet (often binary). For example, in a game context, a string might be interpreted either as a simple strategy (a rule table) or as a computer program for playing the game (a finite automaton). Depending upon the model, an agent may be represented by a single string, or it may consist of a set of strings corresponding to a range of potential behaviors. For example, a string that determines an oligopolist's production decision could either represent a single firm operating in a population of other firms, or it could represent one of many possible decision rules for a given firm. Whatever the interpretation, each string is assigned a measure of performance, called its *fitness*, based on the performance of the corresponding structure in its

As with agents in the real world, adaptive agents do not operate with full knowledge of the environment, and often apply rule-of-thumb heuristics. In a complex dynamical system, being robust is more important than being fine-tuned for the current period, because the current period will not reflect what might be soon to emerge. To quote the iconoclastic military strategist John Boyd, "If you can model it, you're wrong."

The genetic algorithm is a startling innovation. It applies the bedrock concept of evolution to adaptation and learning. It works robustly on a huge span of problems. And it illustrates how simple heuristics at the individual level can lead to unexpected and complex results at the aggregate level. Not only is the system dynamic, but the heuristics being applied within the system adapt and change.

This anticipates the concept of crowdsourcing.

environment. The GA manipulates this population in order to produce a new population that is better adapted to the environment.

In execution, a GA first makes copies of strings in the population in proportion to their observed performance, fitter strings being more likely to produce copies. As a result, fitter strings are more likely to contribute to the new population. After the copies are produced, they are modified by the application of genetic operators. The genetic operators provide for the introduction of new strings (structures) that still retain some of the characteristics of the fitter strings in the parent population.

The primary genetic operator for a GA is the *crossover* operator. The crossover operator is executed in three steps: 1) a pair of strings is chosen from the set of copies; 2) the strings are placed side by side and a point is randomly chosen somewhere along the length of the strings; 3) the segments to the left of the point are exchanged between the strings. For example, crossover of 111000 and 010101 after the second position produces the offspring strings 011000 and 110101. Crossover, working with reproduction according to performance, turns out to be a powerful way of biasing the system toward certain patterns, *building blocks*, that are consistently associated with above-average performance.

It can be proved (see Holland 1975) that GAs are a powerful technique for locating improvements in complicated high-dimensional spaces. They exploit the mutual information inherent in the population, rather than simply trying to exploit the best individual in the population. We can liken each of the potential building blocks to one arm of an n-armed bandit. Under this interpretation, each successive generation samples the building blocks in a way that closely corresponds to the optimal solution of an n-armed bandit problem. The GA learns by biasing the search toward combinations of above-average building blocks. Reproduction and crossover are very simple operations that impose low-information and processing requirements on the agents employing them.

A *classifier system* (CS) (Kauffman 1984) is an adaptive rule-based system that models its environment by activating appropriate clusters of rules. It uses a GA to revise its rules. Each rule is in condition/action form, and many rules can be active simultaneously. The action part of a rule specifies a message that is to be posted when the rule is activated. The condition part of a rule specifies messages that must be present for it to be activated. Thus, each rule is a simple message-processing device that emits a specific message when certain other messages are present. Overt actions affecting the environment are the result of messages directed to the system's output devices (effectors), while information from the environment is received via messages generated by its input devices (detectors). The overall system is computationally complete in the sense that any program written in a programming language, such as FORTRAN, can also be implemented by a CS.

A CS-rule does not automatically post its message when its condition part is satisfied. Rather, it enters a competition with other rules having satisfied conditions. The outcome of this competition is based on a quantity, called *strength*, assigned to each rule. A rule's strength measures its past usefulness, and it is modified over time by one of the system's learning algorithms (see below). There may be more than one winner of the competition at any given time—hence a cluster of rules can react to external situations. A CS operates on large numbers of rules, with a small number of simple, domain-independent mechanisms. It provides emergent, learned capabilities for reacting to its environment.

A CS adapts or learns through the application of two well-defined machine-learning algorithms. The first algorithm, called a *bucket-brigade algorithm*, adjusts rule strengths. Each rule is treated as an intermediate producer in a complex economy, buying input messages and selling output messages. When a satisfied rule R succeeds in the competition to post its own message, it pays the rule(s) that supplied the messages satisfying its condition part. This amount is subtracted from R's strength. On the next time-step, if other rules are satisfied by R's message, and win the competition in turn, then R receives the rules' payment. R's strength is increased accordingly. The net effect of the two transactions is R's profit (loss). Some rules also act directly on the environment in a way that produces direct payoff from the environment to the system. Their strength is increased in proportion to that payoff. A rule's strength will increase over time only if it earns a profit, on average, in these transactions. Generally this happens only if the rule directly produces payoff, or else belongs to one or more causal chains leading to payoff. Under appropriate conditions, the strengths assigned by the bucket-brigade algorithm do converge to a useful measure of the rule's contributions to system performance (Holland *et al.*).

In order to generate and test new approaches to the environment, the CS needs a second learning algorithm, a *rule discovery algorithm*. A GA can be used for this purpose, because the rules of a CS can be represented by strings in an appropriate alphabet, and a rule's strength amounts to a measure of its performance. The GA, by forming new rules in terms of tested, above-average building blocks, transfers experience from the past to new situations. Plausible new rules result—rules to be tested and retained or discarded on the basis of their ability to enhance the performance of the CS.

Under the combined effects of the bucket-brigade and genetic algorithms, rules become coupled in complex networks. Clusters and hierarchies of rules emerge. Over time, these substructures serve as building blocks for still more complex substructures. A CS agent can: 1) generate broad categories for describing its environment (so that experience can be brought to bear on novel situations); 2) progressively refine and elaborate the relation between categories (using experience to make distinctions and associations not previously possible); 3) use these categories to build internal models that supply the agent with expectations about the world; 4) treat all internal models as provisional (subject to confirmation or refutation as experience accumulates); and 5) generate new hypotheses that are plausible in terms of accumulated experience. Moreover, because of the bucket-brigade algorithm, these activities can proceed in an environment where payoff is intermittent or rare. Such capacities enable a CS agent that is not omniscient to act with increasing rationality.

III. Towards a Mathematics of Complex Adaptive Systems

A mathematical calculus appropriate to the study of complex adaptive systems must meet distinctive requirements. The usual mathematical tools, exploiting linearity, fixed points, and convergence, provide only an entering wedge. In addition we need a mathematics that works in close conjunction with computer modeling techniques—one that puts more emphasis on combinatorics and algorithms. We require techniques that emphasize the emergence of structure, particularly internal models, through the generation, combination, and interaction of building blocks. The present situation seems quite similar to that of evolutionary theory prior to the development of a mathematical theory of genetic selection (R. A. Fisher 1930).

Thirty years later, no such general mathematical theory has emerged in the field of economics. Because many complex dynamical systems generated by adaptive agents cannot be reduced to a set of equations, this might be unattainable.

Though there is nothing like an overall theory, there are some extant pieces of mathematics that are relevant. The schema theorems for genetic algorithms (Holland 1975) offer some insight into processes that discover and recombine building blocks. It appears that schema theorems are special cases of a much more general formulation of the effects of recombination in evolution. This formulation should bring some of the more sophisticated tools of mathematical genetics to bear on adaptive agent models. Mathematical work aimed at understanding the evolution of CS may also be useful. The progressive development of hierarchical organization can be treated as the addition of levels to a quasi homomorphism (Holland *et al.*).

Perpetual novelty can be modeled by a regular Markov process in which each of the states has a recurrence time that is large with respect to any feasible observation time. Equivalence classes can be imposed and used as the states of a *derived* Markov process (Holland *et al.* 1986). Work by Miller and Forrest (1989), based on S. A. Kauffman's (1984) studies of random graphs, provides additional insights into the emergent structures of CSs.

IV. Conclusions

The AAA research complements ongoing theoretical and empirical work, allowing exploration and analysis of previously inaccessible phenomena. What are the future prospects for this line of inquiry? Early work with AAA in economics has shown that they can acquire sophisticated behavioral patterns. Observation of the course of learning in these AAA has already increased our understanding of some economic issues. Even limited AAA open up new avenues for analyzing decentralized, adaptive, and emergent systems. Steady advances in computation and AAA modeling offer ever more powerful tools for programming artificial worlds. By executing these models on a computer we gain a double advantage: *(i)* An experimental format allowing

The use of reality checks includes comparing the results of the model with "stylized facts," relevant characteristics of the world that a model should meet. For example, two stylized facts for equity markets are that returns have kurtosis (fat tails) and return volatility is autocorrelated.

Holland and Miller repeatedly write of looking forward to a meeting of these methods with economic theory, but to date that remains elusive. Although this article appeared in the *American Economic Review*, ongoing economic research using agent-based models has yet to make major headway in the mainstream economic literature, and finds its most welcome home in scientific journals outside the field.

free exploration of system dynamics, with complete control of all conditions; and *(ii)* an opportunity to check the various unfolding behaviors for plausibility, a kind of "reality check." Whether or not agents in such worlds behave in an optimal manner, the very act of contemplating such systems will lead to important questions and answers.

REFERENCES

Andreoni, J. A., and J. H. Miller. 1990. *Auctions with Adaptive Artificially Intelligent Agents*. Technical report 90-01-004. Santa Fe Institute.

Arthur, W. B. 1990. *A Learning Algorithm that Replicates Human Learning*. Technical report 90--026. Santa Fe Institute.

Axelrod, R. 1987. "The Evolution of Strategies in the Iterated Prisoner's Dilemma." In *Genetic Algorithms and Simulated Annealing*, edited by L. D. Davis, 32–41. Los Altos, CA: Morgan-Kaufmann.

Fisher, R. A. 1930. *The Genetical Theory of Natural Selection*. Oxford, UK: Clarendon Press.

Friedman, M. 1953. *Essays in Positive Economics*. Chicago, IL: University of Chicago Press.

Holland, J. H. 1975. *Adaptation in Natural and Artificial Systems*. Ann Arbor, MI: University of Michigan Press.

———. 1986. "A Mathematical Framework for Studying Learning in Classifier Systems." In *Evolution, Games and Learning*, edited by J. D. Farmer, A. S. Lapedes, N. H. Packard, and B. Wendroff, 307–17. Amsterdam: North-Holland.

Holland, J. H., K. J. Holyoak, R. E Nisbett, and P. R. Thagard. 1986. *Induction: Processes of Inference, Learning, and Discovery*. Cambridge, MA: MIT Press.

Kauffman, S. A. 1984. "Emergent Properties in Random Complex Automata." *Physica D* 10:145–156.

Marimon, R., E. McGratten, and T. J. Sargent. 1990. "Money as a Medium of Exchange in an Economy with Artificially Intelligent Agents." *Journal of Economic Dynamics and Control* 14 (May): 329–73.

Miller, J. H. 1989. *The Coevolution of Automata in the Repeated Prisoner's Dilemma*. Technical report 89-003. Santa Fe Institute.

Miller, J. H., and S. Forrest. 1989. "The Dynamical Behavior of Classifier Systems." In *Proceedings of the Third International Conference on Genetic Algorithms*, edited by J. D. Schaffer, 304–10. San Mateo, CA: Morgan-Kaufmann.

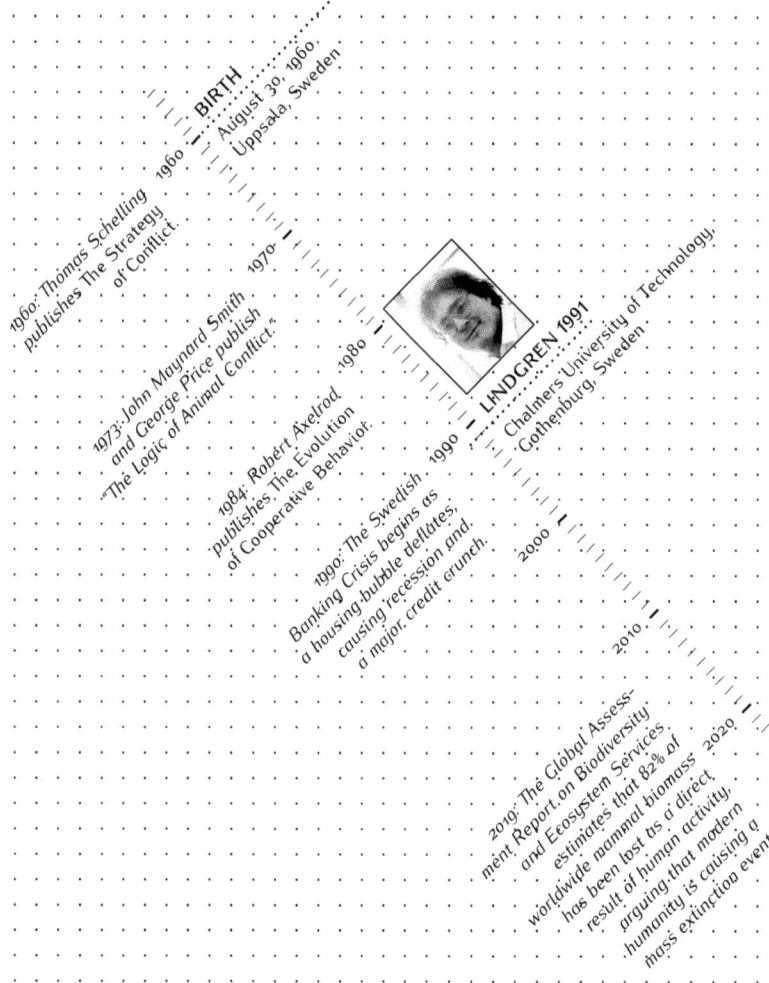

BIRTH
August 30, 1960.
Uppsala, Sweden.

1960: Thomas Schelling publishes The Strategy of Conflict.

1973: John Maynard Smith and George Price publish "The Logic of Animal Conflict."

1984: Robert Axelrod publishes The Evolution of Cooperative Behavior.

1990: The Swedish Banking Crisis begins as a housing bubble deflates, causing recession and a major credit crunch.

LINDGREN 1991
Chalmers University of Technology,
Gothenburg, Sweden.

2019: The Global Assessment Report on Biodiversity and Ecosystem Services, estimates that 82% of worldwide mammal biomass has been lost as a direct result of human activity, arguing that modern humanity is causing a mass extinction event.

KRISTIAN LINDGREN

[73]

SIMPLE DYNAMICS WITH RICH IMPLICATIONS

*W. Brian Arthur, SRI International
and Santa Fe Institute*

Every so often a paper comes along with a simple setup but a richness of implications. Kristian Lindgren's (1991) is such a paper. It appeared thirty years ago, and I still find it striking.

Lindgren constructed a computerized tournament where strategies compete in pairs to play an iterated prisoner's dilemma game. Players play one-against-one in a repeated sequence of rounds, and each time they have two options: "cooperate" or "defect." There's a tension here. Neither player knows what the other will choose. If you and your opponent choose to cooperate, both will do quite well. But if you choose to defect—take advantage—you can do a bit better. However, if your opponent defects as well, you will both be harmed. Cooperating, defecting, and retaliating are all possible; and you need to choose judiciously if you want to do well.

The number of players in the tournament is kept fixed, and each plays a given strategy in the repeated game, a set of fixed instructions for how to act given its own and its opponent strategy's immediate past actions. All strategies play all strategies in a given round or "generation." If a strategy performs well over its encounters, it gets to replicate; if it does poorly, it dies and is removed. Existing strategies can mutate their instructions with a small probability, thus creating new ones. Lindgren adds two important ingredients to the mix: strategies can make occasional mistakes, so there is some "noise," and they can occasionally "deepen"—that is, mutate by using deeper memory of their opponent's immediate past moves and their own. This allows them to "read" their opponents' moves better, anticipate them, and become smarter.

Evolution here arises in a natural way. We can think of a given strategy as a species, each with its own numbers of players. Species

K. Lindgren, "Evolutionary Phenomena in Simple Dynamics," *Artificial Life II*, eds. C. G. Langton, C. Taylor, J. D. Farmer, and S. Rasmussen, Santa Fe Institute Studies in the Sciences of Complexity, Redwood City, CA: Addison–Wesley, 365–370 (1991).

(strategies) occasionally mutate, so they compete for survival and co-evolve in an "ecology" of other species (strategies) competing for survival and occasionally mutating.

The model's dynamics are simple enough that Lindgren can write them as stochastic equations, but these give far from a full picture; we need computation to see how things might unfold over time.

§

In Lindgren's computerized tournament, 100 strategies play each other for a very large number of rounds in one generation, and he runs the system for 60,000 generations. In each run—I'll call it an experiment—the outcome differs. Typically at the start, simple strategies such as tit-for-tat dominate, but over time, more sophisticated ones appear that exploit them or in some cases have a better mechanism to deal with occasional mistakes. In some experiments, complicated strategies show up early on; in others, only later. Figure 1 shows a typical experiment. We see long epochs of dominance by one or a few strategies, with periods of turbulence in between. In this experiment there's a period of Lotka–Volterra oscillations between strategies (for example, in the initial phase involving the simplest ones).

What I find interesting is that if Lindgren had stopped the experiment toward the end of a given epoch, you might suppose the resulting pattern to be the final one, the final word, be it equilibrium or chaos. Yet beyond it, the system keeps evolving and changing. What we might accept as a status quo proves to be merely an introduction to the next epoch.

Other experimental runs of course randomly tell different stories. But in spite of variations, Lindgren's world shows consistent phenomena: emergence of mutual support among strategies, exploitation of strategies by other strategies, sudden large extinctions, periods of stasis followed by ones of tumultuous change. A distinct biological, evolutionary theme emerges. In fact the overall scene, if it resembles anything at all, reminds me of species competition in paleozoological times.

Lindgren's paper appeared in 1991; three years later he and Mats Nordahl (1994) generalized it to a spatial version.

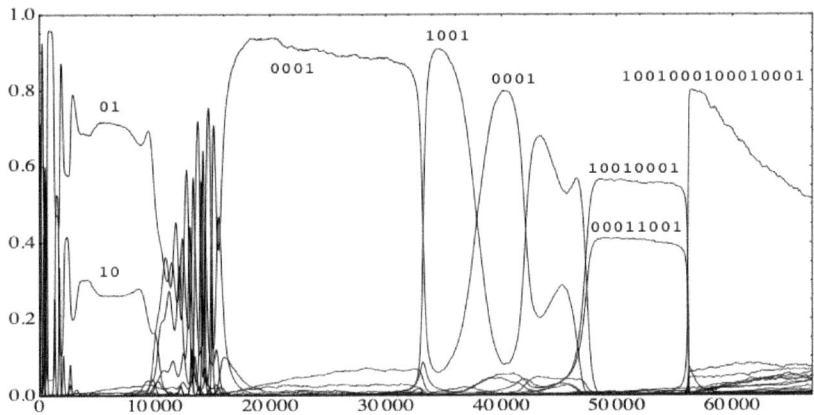

Figure 1. Prevalence of strategies in a computerized tournament of the prisoner's dilemma. The horizontal axis denotes time, the vertical axis numbers using a particular strategy. The length of labels indicate the memory depth of strategies, that is, how many previous moves in the game they take into account. (*Modified from fig. 2 in Lindgren 1991.*)

Lindgren's paper wasn't the first to explore prisoner's dilemma tournaments. Robert Axelrod had proposed such a tournament in 1980, where invited scientists sent in computer-coded strategies to compete, and a small but significant collection of studies had followed (Lindgren 1991).

What interests me is not where Lindgren's paper fits into this literature but what it has to say about economics and the doing of economics.

I would classify this paper as an early example of complexity economics (Arthur 1994, 2021), though the term wasn't used until 1999. Complexity economics relaxes the main assumptions of standard economics. It views the economy as not necessarily in equilibrium, its agents as not hyperrational, the problems they face as not necessarily well-defined, and the economy not as a perfectly purring machine but as an always changing ecology of actions, forecasts, and strategies. When we were developing this approach at the Santa Fe Institute in the late 1980s, Lindgren's paper hadn't been written, but I've found it useful ever since because it illustrates much of what complexity economics is about: agents updating their beliefs and actions in a setting they mutually create and that changes as they update. I would also say Lindgren's setup is an early agent-based model, though that term arrived only around the same time as Lindgren's paper.

What is striking about this paper is that from a very few assumptions it constructs a computational world, an ecology of strategies competing, deepening, and changing. The world created is an economic one: it is one of competition for scarce resources (here, membership in the set of strategies that survive), though there is no equilibrium reached, no rationality assumed, no "optimality" arrived at.

Lindgren's world displays a number of properties, all of which are now familiar in complexity economics but still remarkable. I'll draw attention to three.

1. It shows that significant structural change can arise endogenously. The model assumes no outside shocks, no novel technologies, no new territory discovered. Yet the "structure" or character of the economy constantly changes, often significantly. Change happens from within as exploration uncovers better strategies, and the system lurches between epochs of relative equilibrium and turbulence. It might seem that small changes in a strategy should cause small changes, but occasionally they cause large changes, new epochs in this strategy world. If it happened in real life, each epoch would have its own character, its own culture, its own explanation. History forms, slowly and endogenously, but not always smoothly. Nature does make leaps.

2. It shows "intelligence" emerging without any built-in assumption of rationality. The strategies automatically get deeper and smarter as the computation progresses. The method by which they get smarter—random trial and survival of successful ones—may appear clunky, but it is the backbone of methods used these days to teach computers to play Go at a grand master level. The kind of intelligence arrived at in Lindgren is neither perfectly rational nor necessarily optimal. It is merely pretty good. But it is not forced, and not assumed. It emerges naturally.

3. It shows that computation in economics can sometimes be regarded as "theory." Lindgren's study needs to use computation to track outcomes of his dynamical system. He calls these "simulations," so it would be easy to dismiss this as an exercise in "mere simulation." But his model is rigorously described

mathematically; it is a well-specified stochastic process with well-described realizations (particular series of events that result from a run of this stochastic process). As such, the process and results are just as mathematical—just as theoretical—as any stochastic process and its implications. Yet the implications are arrived at computationally. The boundary here between computation and mathematics becomes fuzzy. This isn't unusual in modern mathematics. Theoretical systems like the one that generates Mandelbrot's set are undoubtedly mathematical, yet their implications need to be explored computationally. In this sense I would say that Lindgren's system is theory, explored computationally.

Lindgren's paper is now over thirty years old, and I still find it fascinating. One reason I gave earlier: from simple assumptions come complicated outcomes and a plethora of lessons. The paper is an allegory, a demo, a parable worth pondering. The other reason is that the outcomes give us—or me at least—a feeling of economic "realism." What emerges is not a world of stasis and perfection but a world of exploratory trials, "discoveries" and setbacks, endogenous adjustment, openness, long-lasting epochs followed by sudden collapses.

Good theory in complexity, I often think, is like a Chekhov play. It sets up characters that interact within a context that contains some tension, some unresolved issue. The interest is to see what plays out and how things play out. We learn not by extrapolating the tendencies of the individual characters but by seeing them reacting and re-reacting to the situation their actions mutually create.

REFERENCES

Arthur, W. B. 1994. "Inductive Reasoning and Bounded Rationality: the El Farol Problem." *American Economic Review* 84:406–411. https://www.jstor.org/stable/2117868.

———. 2021. "Foundations of Complexity Economics." *Nature Reviews Physics* 3:136–45.

Lindgren, K. 1991. "Evolutionary Phenomena in Simple Dynamics." In *Artificial Life II,* edited by C. G. Langton, C. Taylor, J. D. Farmer, and S. Rasmussen, 295–312. Reading, MA: Addison-Wesley.

Lindgren, K., and M. G. Nordahl. 1994. "Evolutionary Dynamics of Spatial Games." *Physica D: Nonlinear Phenomena* 75 (1-3): 292–309. https://doi.org/10.1016/0167-2789(94)90289-5.

EVOLUTIONARY PHENOMENA IN SIMPLE DYNAMICS

Kristian Lindgren, Chalmers University of Technology

Abstract

We present a model of a population of individuals playing a variation of the iterated Prisoner's Dilemma in which noise may cause the players to make mistakes. Each individual acts according to a finite memory strategy encoded in its genome. All play against all, and those who perform well get more offspring in the next generation. Mutations enable the system to explore the strategy space, and selection favors the evolution of cooperative and unexploitable strategies. Several kinds of evolutionary phenomena, like periods of stasis, punctuated equilibria, large extinctions, coevolution of mutualism, and evolutionary stable strategies, are encountered in the simulations of this model.

Introduction

In the construction of simple models of abstract evolutionary systems, game theory provides a large number of concepts and examples of games that can be used to model the interaction between individuals in a population. Originally, game theory was developed by von Neumann and Morgenstern for the application to economic theory, (1944) but it has now spread to other disciplines as well. The work of Maynard-Smith and Price (1982, 1973) has lead to an increasing use of game theory in evolutionary ecology. In the social sciences game theoretical methods have been accepted for a long time. A renewed interest in the Prisoner's Dilemma followed the work of Axelrod and Hamilton (1984,1981) who performed a detailed analysis of the iterated version of that game, and this has lead to several game theoretic models based on the iterated Prisoner's Dilemma. In large computer networks the presence of interacting agents may lead to computational ecosystems, (Kephardt, Hogg, and Huberman 1989) which can be analyzed from a game-theoretical point of view.

For a population with a fixed number of species, natural selection drives the system towards a fixed point, limit cycle, or strange attractor, assuming an unchanged environment. This process can be modelled by population dynamics, where one usually uses the number of individuals for the different species as variables, so that the dimensionality of the system equals the number of species. Population dynamics models the reproduction, survival, and death of individuals. If the behavior of the individuals (or species) depends on a genetic description inherited by the offspring, the introduction of mutations in the replication process may totally change the dynamic behavior of the system. One way to characterize such a dynamical system is to interpret mutations leading to new species as creations of new variables and extinction of species as the disappearance of present variables. But in both cases these events are due to the (stochastic) dynamic system itself. If there is no limit on the length of the genetic description and the number of phenotypic characters this is coded into, the system may be considered a potentially infinite-dimensional dynamical system. Evolution can then be viewed as a transient phenomenon in a potentially infinite-dimensional dynamical system. (Farmer, Kauffman, and Packard 1986; Rössler 1971) If the transients continue for ever, we have *open-ended evolution*. Of course, we may still get the same behavior as in the mutation-free population dynamics. Therefore, one of the main problems in the construction of evolutionary models is how to model the interactions between species (and/or environment) so that the transients are infinite or at least long enough for evolutionary phenomena to appear. In this construction one is faced with the dilemma that one wants to achieve both high complexity, which is necessary for evolution to occur, and simplicity, which makes simulation possible for evolutionary time scales. Note that the dynamics used to model the behavior in prebiotic or chemical evolution is usually a form of population dynamics. Such systems have been analyzed by, e.g., Farmer, Kauffman, and Packard (1986), Schuster (1986) and Eigen, McCaskill, and Schuster (1988) in models for evolution of macromolecules.

We have constructed a model of a population of individuals playing the iterated Prisoner's Dilemma. The game is modified so that noise may disturb the actions performed by the players, which makes the problem of the optimal strategy more complicated. This increases the potential for having long transients showing evolutionary behavior. We construct a suitable coding for all deterministic strategies with finite memory, and let such a code serve as the genome for an individual playing the corresponding strategy. By adding mutations to the population dynamics we get a potentially infinite-dimensional dynamical system in which evolution is possible. The "artificial" selection in the model is determined by the result in the game—those individuals who get high scores also have higher fitness.

The idea of using the iterated Prisoner's Dilemma in evolutionary situations is not new, see, e.g., the studies by Axelrod (1987) and Miller (1989) and a variety of other kinds of evolutionary models can be found in Langton (1989). The novel approach in this study is the combination of noisy games, simple population dynamics, analytically solvable interactions, and the possibility of increase in genome length, and it appears that this leads to a richness in evolutionary behavior that has not been observed in such models before.

The Prisoner's Dilemma

The Prisoner's Dilemma is a two-person non-zerosum game, which has been used in both experimental and theoretical investigations of cooperative behavior. The game is based on the following situation. Two persons have been caught and are suspected of having committed a crime together. There is not enough evidence to sentence them, unless at least one of them confesses. So, if both stay quiet (cooperate, C) they will be released. If one confesses (defects, D) but the other does not, the one who confesses will be released and rewarded, while the other one will get a severe punishment. Finally, if both confess, they will be imprisoned but for a shorter period. It is assumed that they make their choice of action simultaneously without knowing the others decision.

This problem is formalized by assigning numerical values for each pair of choices. An example of such a payoff matrix for the players is shown in Table 1.

If the game is viewed as a single event, each player finds defection to be the optimal behavior, regardless of the opponents action. However, if there is a high probability that the two players will meet again in the same type of game, the question of the most optimal choice of action is more delicate. This kind of "iterated Prisoner's Dilemma" has been

Table 1. The payoff matrix we use in the Prisoner's Dilemma is the same as the one used by Axelrod (1984). The pair (s_1, s_2) denotes the scores to players 1 and 2, respectively.

		Player 2	
		Cooperate	Defect
Player 1	Cooperate	(3, 3)	(0, 5)
	Defect	(5, 0)	(1, 1)

extensively studied by Axelrod (1984). From the results of a computer tournament, he found that a simple strategy called Tit-for-Tat (TFT) showed the best performance in the iterated game. Tit-for-Tat starts with cooperation and then repeats the opponent's last action. Thus, two TFT players meeting each other in a series of games, share the highest possible total payoff and each gets an average score of 3.

In our model we shall let noise interfere with the actions of the players. With probability p the performed action is opposite to the intended one. (We shall assume that the average length T of the game is much longer than the average time between noise-modified actions, $T \gg 1/(2p)$.) For two players using the TFT strategy the result is that they will alternate between three modes of behavior. First they will play the ordinary TFT actions (C, C), but when an error occurs they will shift to alternating (C, D) and (D, C). The third possibility of behavior is sequences of (D, D). The average probability for the three modes are $1/4, 1/2$, and $1/4$, respectively, giving an average payoff of $9/4$. None of the strategies in Axelrod's tournament was able to deal with noise *and* resist exploitation, and TFT turned out to be the best one in that set of strategies (Axelrod 1984). A simple strategy that is more stable to noise is Tit-for-Two-Tats, which defects only if the opponent defects twice in a row, but this strategy is vulnerable to exploiting strategies, and in an evolutionary context it should perform worse. Another way to decrease the sensitivity to noise is to allow for the strategies to choose among different actions according to a certain probability (mixed strategies). This approach has been analyzed by Molander (1985), who found that a strategy which mixes TFT with ALLC (always cooperate) can reach an average score very close to 3. In our model we shall assume that the strategies are deterministic (pure strategies), and in the simulations we shall see that there are deterministic, noise-robust, unexploitable strategies that reach an average score of almost 3.

There's a whole prisoner's dilemma literature here worth looking at. See the paper's references.

Finite Memory and Infinite Games

GENETIC CODING OF STRATEGIES

In the model we allow for deterministic finite memory strategies. This means that a finite history determines the next intended action, although the performed action can be changed by the noise. An m-length history consists of a series of previous actions starting with the opponent's last action a_0, the individual's own last action a_1, the opponent's next to last action a_2, etc. By introducing a binary coding for the actions, 0 for defection and 1 for cooperation, we can label an m-length history by a binary number

$$h_m = (a_{m-1}, \ldots, a_1, a_0)_2 \, .$$

Since a deterministic strategy of memory m associates an action to each m-length history, it can be specified by a binary sequence

$$S = [A_0, A_1, \ldots, A_{n-1}] \, .$$

This sequence then serves as the genetic code for the strategy that chooses action A_k when history k turns up. The length n of the genome equals 2^m.

In the population dynamics we shall allow for three kinds of mutations: point mutations, gene duplications, and split mutations. The point mutation changes a symbol in the genome, e.g., [01] → [00], the gene duplication attaches a copy of the genome to itself, e.g., [01] → [0101], and the split mutation randomly removes the first or second half of the genome, e.g., [1001] → [01]. Note that gene duplication does not change the phenotype. The memory capacity is increased by one but the additional information is not used in the choice of action. For point mutations we have used the rate 2×10^{-5} per symbol and genome, and the other mutations occur with probability 10^{-5} per genome. Regarding a position in the genome as a locus and a symbol as an allele rather than a base pair, the point mutation rate we use has the order of magnitude that has been estimated for mutation rates at loci in living system (Futuyma 1986).

For strategies of memory one, the histories are labeled 0 and 1, corresponding to the opponent defecting and cooperating, respectively.

Lindgren creates a code that specifies strategy "genomes" succinctly. This makes mutation easy to implement.

The four memory 1 strategies are [00], [01], [10], and [11]. The strategy [00] always defects (ALLD), [01] cooperates only if history 1 turns up (i.e., the opponent cooperated), and we recognize it as Tit-for-Tat, [10] does the opposite and we denote it Anti-Tit-for-Tat (ATFT), and [11] always cooperates (ALLC). We use equal fractions of these strategies as the initial state in the simulations.

Solving the Game

If the length of the game is infinite, the stationary distribution over finite histories can be solved analytically. This solution is unique if noise disturbing the actions is present. Although the game is infinite, the strategies can only take into account a finite history when choosing an action, which means that the infinite game is a Markov process. The average payoff for two players meeting in this game can be derived from the probabilities p_{00}, p_{01}, p_{10}, and p_{11} for all possible pairs of action (11), (10), (01), and (00). These can be found if we solve the equation

$$H = MH, \tag{1}$$

where $H^T = (h_0, h_1, \ldots, h_{n-1})$ is the vector of probabilities for different histories $0, 1 \ldots, n-1$, and M is a transfer matrix. The elements of M are determined by the strategies involved in the game including the possibility of making mistakes. The minimal size n of the matrix is given by the memory sizes of the involved strategies and is 2^m if the largest memory is m (or 2^{m+1} if m is odd and both players have the same memory size). Then one gets p_{ij} by summing the appropriate components in H, and the average payoff is

$$s = 3p_{11} + 5p_{01} + p_{00}, \tag{2}$$

according to the payoff matrix in Table 1.

The average payoff between two strategies is solved analytically, assuming they play an unlimited (infinite) number of times. This reduces computation greatly.

Population Dynamics

We shall consider a system consisting of a population of N individuals interacting according to the iterated Prisoner's Dilemma with noise. Each individual acts according to a certain strategy encoded in its genome. We think of a population sharing the same niche, fighting or cooperating with each other, to get a part of the available resources for survival and reproduction. In *each* generation all individuals play the infinitely iterated Prisoner's Dilemma against all, and the score s_i for individual i is compared to the average score of the population, and those above average will get more offspring in the next generation. In the reproduction, mutations may occur leading to the appearance of new strategies.

We model this situation as follows. First, we identify the different genotypes present in the population, and let them meet in the game described above. Let g_{ij} be the score for the strategy of genotype i playing against the strategy of j, and let x_i be the fraction of the population occupied by genotype i. Then, the score s_i for an individual with genotype i is

$$s_i = \sum_j g_{ij} x_j, \qquad (3)$$

and the average score is

$$s = \sum_i s_i x_i. \qquad (4)$$

The fitness w_i of an individual is defined as the difference between its own score and the average score,

$$w_i = s_i - s. \qquad (5)$$

From one generation t to the next $t+1$, we assume that due to the result of the interactions, the fraction x_i of the population for genotype i changes according to

$$x_i(t+1) - x_i(t) = d w_i x_i(t), \qquad (6)$$

where d is a growth constant. This equation can also be written in the following form

$$x_i(t+1) - x_i(t) = d s_i x_i(t) \left(1 - \sum_j \frac{s_j x_j(t)}{s_i}\right), \qquad (7)$$

which is a logistic equation for a population of competing species (Futuyma 1986). The carrying capacity is normalized to 1, and the competition coefficients for species i are $s_j/s_i (j = 1, 2, \ldots)$. Note that this growth equation conserves the total population size. If x_j falls below $1/N$ for a certain genotype j, we set $x_j = 0$ and that species has

Lindgren (1991)

died out. When this happens, the fractions x_i have to be renormalized for the population size to be constant. When mutations are present there is an additional stochastic term m_i in the growth equation. If the mutation rates are small ($p_p + p_d + p_s \ll 1/N$), the additional term is well approximated by

$$m_i = \frac{1}{N} \sum_j \left(\mathcal{Q}_{ij} - \mathcal{Q}_{ji} \right), \tag{8}$$

where $\mathcal{Q}_{ij}$ is a stochastic variable taking the value 1 if a gene j mutates to the gene i, and 0 otherwise. The probability for $\mathcal{Q}_{ij}$ to be 1 is

$$P\left(\mathcal{Q}_{ij} = 1 \right) = N x_j q_{ij}, \tag{9}$$

where q_{ij} is the probability that genotype j mutates to i, obtained from the mutation rates and the genotypes i and j. (This mutation may be composed of one gene duplication and several point mutations, although this is less frequent.) Due to the term m_i, new genotypes may appear in the time evolution, and we get a model with a potentially infinite state space.

Simulation Results and Discussion

The system described above consists of a population of N individuals interacting according to the iterated Prisoner's Dilemma with a probability p for mistake (noise). Individuals who get high scores get more offspring in the next generation than those who get low scores. In this reproduction we allow for mutations to occur and new strategies to enter the game.

We model the dynamics of this system by Eqs. (6) – (9), and the parameters that enter are the growth rate d, the mutation rates p_p, p_d, and p_s, the population size N, and the error probability p. In the simulation example the parameter values are $N = 1000, p = 0.01, p_p = 2 \times 10^{-5}, p_d = p_s = 10^{-5}$, and $d = 0.1$, and we have also restricted the length of the genetic code to be at most 32, i.e., at most strategies of memory 5. For the first generation we have chosen equal fractions of the four strategies with memory one, i.e., $x_{00} = x_{01} = x_{10} = x_{11} = 1/4$.

Almost all simulations have in common that during the evolution the system passes a number of long-lived metastable states (periods

Lindgren describes the results of one run of his model in detail in this section. As with examining the results of a lab experiment, this is where we achieve insight.

Models like this must be tuned by careful selection of parameters if we want to produce interesting results.

of stasis) that appear in a certain order. These periods are usually interrupted by fast transitions to unstable dynamic behavior or to new periods of stasis. Below we shall discuss the evolutionary phenomena observed in a typical simulation of the model. In the four most common periods of stasis we find examples of coexistence between species, exploitation, spontaneously emerging mutualism (symbiosis), and unexploitable cooperation.

The Evolution of Strategies of Memory 1.

In Figure 1 the development of the population for the first 600 generations is shown. During the first 150 generations, the dynamics drives the system of the 4 strategies towards a population mainly consisting of TFT strategies. The All-D strategy [00] exploits the kind All-C strategy [11] and the ATFT strategy [10], and consequently [00] increases its fraction of the population. When the strategies [11] and [10] are extinct, the average score for All-D is close to 1 and the more cooperative Tit-for-Tat strategy takes over the population.

However, Tit-for-Tat only reaches an average score of 9/4 since noise interferes with the interaction. Then, through a point mutation [01] → [11], the All-C strategy enters the scene again. The mutant gets an average score of almost 3, and thus the fraction of [11] rapidly increases. Next, it is favorable for a mutant [11] → [10] to survive, since ATFT exploits ALLC and plays fairly well against TFT. Actually, ATFT gets the same score $s = 9/4$ as TFT when playing against ATFT or TFT. When the population of ATFT has grown large enough, mutations from [01] and [10] to [00] will survive, and the fraction of ALLD increases again. The system oscillates, driven by the relatively fast population dynamics in combination with the point mutations.

In Figure 2 the time scale is compressed by a factor of 50, and the evolution of the first 30000 generations is shown. The picture we get is a history with stable periods interrupted by fast transitions or unstable dynamics. The average score for the same simulation is drawn in Figure 3, which shows that there is no general tendency towards higher scores, although the simulation seems to end in a stable high score state. In the same figure the number of species per generation is depicted, showing

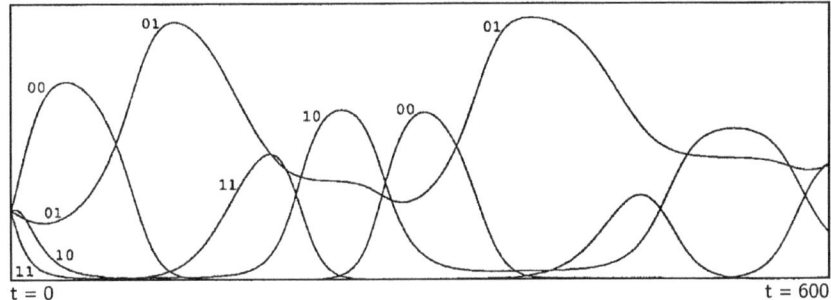

Figure 1. The evolution of a population of strategies starting with equal fractions of the memory one strategies [00], [01], [10], and [11] is shown for the first 600 generations. The fractions of different strategies are shown as functions of time (generation).

that the dimensionality of the system can increase and decrease in the evolution.

After some thousands of generations the oscillations observed in Figure 1 are damped out, and the system stabilizes with a mixture of TFT [01] and ATFT [10]. If only the four simplest strategies are taken into account, this situation is easily analyzed. Assume that the population is divided into two fractions, one consisting of TFT and one of ATFT, and denote the fraction of the first by x. Then, for a large population, if $x < 7/16$ a mutant [00] will start to replicate, and if $x > 3/4$ any mutation to [11] will survive and replicate. But, if $7/16 < x < 3/4$ there is a meta-stable state consisting of a mixture of TFT and ATFT. This state is long-lived because none of the one-step mutations [00], [11], [0101], and [1010] are able to disturb the system. Actually, a detailed analysis shows that the only strategy with memory 2 that can invade this population alone and survive is the strategy [1100] which alternates between C and D, regardless of the opponent's action. However, this is not the usual way the stasis collapses, since one gene duplication and two point mutations are needed to get [1100] from [01] or [10]. Usually, a number of strategies, all having small fractions of the population, have a combined effect and cause the destabilization.

The Evolution of Strategies of Memory 2.

The first stasis is usually followed by a period of unstable behavior, as is exemplified in Figure 2. When the system stabilizes the strategy $A = [1001]$ manages to dominate the population for some time. This

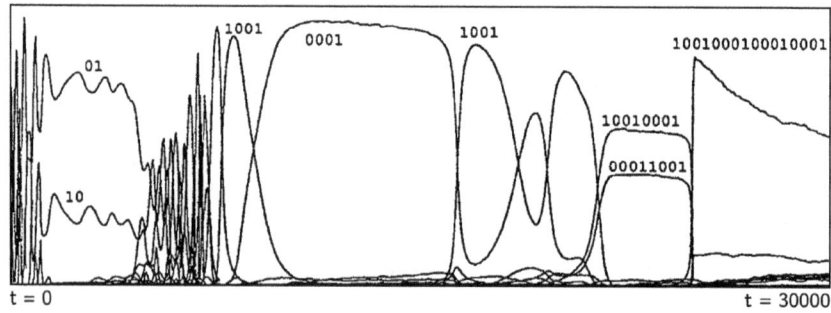

Figure 2. The simulation of Figure 1 is continued for 30000 generations, showing that four periods of stasis appear in the evolution. The oscillations observed in Figure 1 are damped and the system reaches a period of stasis with coexistence between [01] (TFT) and [10] (ATFT). This stasis is punctuated by a number of memory 2 strategies, and after a period of unstable behavior the system slowly stabilizes when the strategy [1001] increases in the population. This strategy cooperates if both players performed the same action last time. For two individuals using this strategy, an accidental defection by one of the players leads to both players defecting the next time, but in the round after that they return to cooperative behavior. Thus, the strategy [1001] is cooperative and stable against mistakes, but it can be exploited by uncooperative strategies. Actually, one of its mutants [0001] exploits the kindness of [1001], which results in a slow increase of [0001] in the population. This leads to a long-lived stasis dominated by the uncooperative behavior of [0001]. A slowly growing group of memory 3 strategies is then formed by mutations, and the presence of these species causes the fractions of the strategies [0001] and [1001] to oscillate. Two of the memory 3 strategies, $M_1 = [10010001]$ and $M_2 = [00011001]$, manage to take over the population, leading to a new period of stasis. Neither M_1 nor M_2 can handle mistakes when playing against individuals of their own kind, but if M_1 meets M_2 they are able to return to cooperative behavior after an accidental defection. This polymorphism is an example of mutualism which spontaneously emerges in this model. The stasis is destabilized by a group of mutants, and we get a fast transition to a population of memory 4 strategies which are both cooperative and unexploitable.

strategy chooses C when the last pair of actions (the own and the opponent's) was CC or DD, which means that two individuals, both playing this strategy, get scores close to 3 when playing against each other. A typical history including a misaction **D** looks as follows (CC, CD, DD, CC, CC, ...), showing that the strategy is not sensitive to the noise. On the other hand the strategy can be exploited by one of its mutants, B=[0001]. When the strategy A plays against B, there are two modes of behavior, exemplified by the following types of histories: (CC, CC, CC, ...) and (DD, CD, DD, CD, ...) where the second action in each pair is due to B. The second mode appears with frequency 0.80 and its average payoff is 3 for B and only 1/2 for A. Although the strategies A and B have totally different behavior (cooperative and uncooperative, respectively), the scores they receive are very close. This leads to a slow increase of B, while A decreases in the population, see Figure 2. Even a small group of mutants can then influence their scores so that the

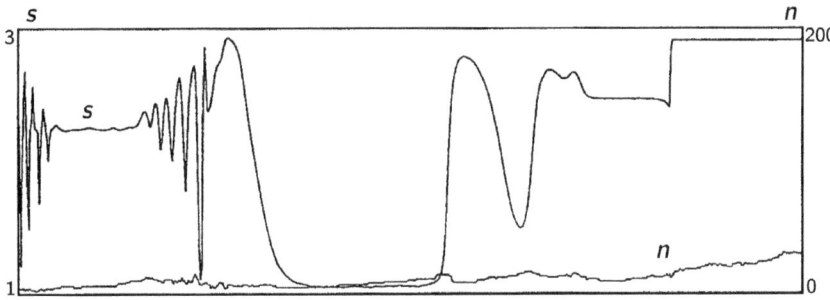

Figure 3. The average score s (continuous line) and the number of genotypes n (broken line) are shown for the simulation of Figure 2. When the exploiting memory 2 strategy dominates the scene, the average score drops close to 1. The last stasis, populated by the evolutionary stable memory 4 strategies, reaches a score of 2.91, close to the score of 3 achieved by the best strategies in a noise-free environment. Before the transitions and in the periods of unstable behavior, it appears that there are more mutants that survive and the number of genotypes increases, suggesting that most of the evolution takes place in these intervals.

dominant strategy scores less than the rival species, which explains the oscillatory pattern that follows.

The Evolution of Strategies of Memory 3.

During the time period dominated by the memory 2 strategies, a group of mutants containing memory 3 strategies is slowly growing. In Figure 2 we see two new strategies $M_1 = [10010001]$ and $M_2 = [00011001]$ spread in the population. A new stasis is reached between M_1 and M_2, and we shall analyze their behavior in more detail. The histories below exemplify how these strategies act when a single noise-induced **D**-action occurs.

$M_1 : M_1$	$M_2 : M_2$	$M_1 : M_2$	$M_2 : M_1$
C C	C C	C C	C C
C **D**	C **D**	C **D**	C **D**
D D	D D	D D	D D
C D	D C	C C	D D
D D	D D	C C	D C
C D	C D	C C	D D
D D	D D	C C	C C
C D	D C	C C	C C
.	.	.	.
.	.	.	.
.	.	.	.

Individuals playing against the same strategy type are not able to handle the noise, but when the strategies M_1 and M_2 play against each other they manage to return to a cooperative mode after a series of intermediate actions. The strategies respond to a disturbance D with a certain pattern of actions which fits to the opponent's actions. This leads to a payoff close to 3 when they meet, but the payoff when M_1 meets M_1 is $S_{1:1} = 2.17$, and this is even worse for M_2, $s_{2:2} = 1.95$, because M_2 also has a mode consisting of a series of defect actions. Obviously, this strategy mix is an example of mutualism. The success of one of them is dependent on the success of the other one, and in Figure 2 we see that they spread simultaneously in the population.

The Evolution of Strategies of Memory 4.

During the stasis of the two symbiotic strategies a group of mutants is formed and their fraction of the population is slowly increasing. The stasis ends with a fast transition to a new meta-stable state, consisting of two leading strategies and a growing group of mutants. All of these strategies have memory 4, i.e., they take into account the actions performed by both players the previous 2 time steps. There are several genotypes that can take the role of the leading one in this transition, because there is a class of genotypes coding into phenotypes or strategies that have practically the same behavior. All of them are cooperative, and if one player accidentally defects both players defect twice before returning to the cooperative mode again. This assures that the strategy cannot be exploited by evil strategies at the same time as the mistakes only marginally decrease the average payoff. In the schematic genome E= [1xx10xxx0xxxx001] the most frequently used positions are shown and each x corresponds to a history occurring with a probability of order p^2 or less. There are 512 strategies fitting this mask, which explains the formation of a large genetic variety in this population, although some of these may have imperfections that can be exploited by other strategies. A typical game involving an accidental defect action **D** is shown below.

In Figure 2 [1001000100010001] has taken the lead, but there are others present in the growing group of quasi-species. The fact that the fraction of the leading genotype decreases can be explained by the

```
E  :  E
C     C
C     D
D     D
D     D
C     C
C     C
   .
   .
   .
```

small difference between the leading strategy and many of the strategies among the mutants. It should also be noted that since the length of the genome doubles each time the memory capacity is increased by 1, the probability for point mutations also doubles.

An important stability criterion for a strategy in a population dynamics model is given by the concept of an *evolutionary stable strategy* (Maynard Smith 1982). Assume that all individuals in a large population play a certain strategy S. The strategy S is evolutionary stable if any sufficiently small invading group of strategies dies out. It has been shown that, in the iterated Prisoner's Dilemma without noise, the Tit-for-Tat strategy is not evolutionary stable, because there are other strategies playing on equal terms with TFT at the same time as they perform better against other strategies. It has been shown by Boyd and Lorberbaum (1987) that there is no pure strategy that is evolutionary stable in the iterated Prisoner's Dilemma. A generalization of their result shows that this also holds for any finite population mixture of pure strategies (Farrell and Ware 1989).

For the iterated Prisoner's Dilemma used in our model the presence of noise implies that every strategy can be regarded as a mixture of two opposite pure strategies, which allows for evolutionary stable strategies to exist (Boyd 1989). Actually, the leading strategy in Figure 2 is evolutionary stable. A strategy that is simpler to analyze is $E_0 =$ [1001000000000001], which defects whenever the behavior deviates from the pattern in the game example above. This implies that no strategy can exploit it, and no strategy can invade a population of these by trying to be more cooperative, because any such attempt would be favorable to E_0 and it would reduce the payoff for the intruder.

(Note that E_0 actually exploits the kind strategy [11].) However, even if the one-step mutants play slightly worse than the master species the mutation rate may be large enough for a net increase of these mutants, which leads to a growing group of quasi-species. In the simulations of our model we find that a large group of quasi-species is formed.

Pathways for Open-Ended Evolution?

The scenario described above, passing periods of stasis dominated by strategies of increasing memory and then getting stuck in the evolutionary stable stasis, occurs with a probability of about 0.9. There are, however, evolutionary pathways that avoid the evolutionary stable memory 4 strategies. In Figure 4 an example of such a simulation is shown, and instead of getting to the stasis of the symbiotic species (see Figure 4(a)), the system takes a new way in state space and in Figure 4(b) we find the population dominated by memory 4 strategies not present in the ordinary simulations. The bottom diagram of Figure 4(b) shows that the number of genotypes (most of these are also of different phenotype) may increase to more than 200. In the figure it is seen that the system undergoes a collapse in which most of the genotypes disappear in a few hundred generations. Similar extinctions occur also in Figure 4(c), but they do not involve that many genotypes. In all these events the average score drops fast, suggesting that the extinctions are due to a mutant that exploits the present strategies but is unable to establish a cooperative behavior with its own species.

Conclusions

The presence of mutations in the population dynamics leads to intrinsical changes of the dimensionality of the system. The dynamic behavior observed is highly complicated with extremely long transients. One important characteristic of the model is that the game-theoretic problem used is complicated enough for complex strategies to evolve, at the same time as we can solve the game analytically, letting us simulate the population dynamics over evolutionary time scales. If one instead uses the iterated Prisoner's Dilemma without noise the potentiality for evolutionary transients is essentially lost. Another important aspect is that we use an effective way to code the strategies in genomes, and that

Lindgren (1991)

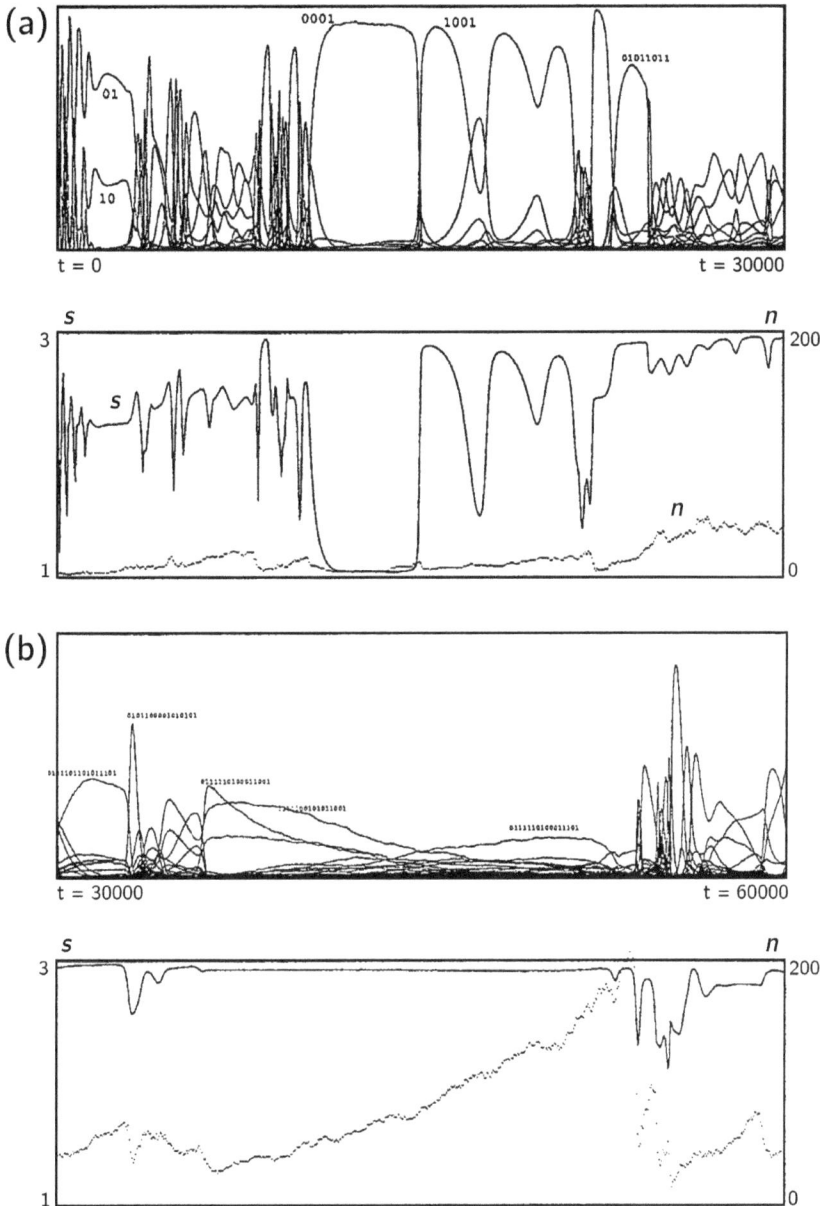

Figure 4. See caption next page.

the genome is easily modified by mutations. Having these aspects in mind it should be possible to model other situations as well, for example evolutionary models with more realistic assumptions, including, e.g., spatial dependence and sexual reproduction.

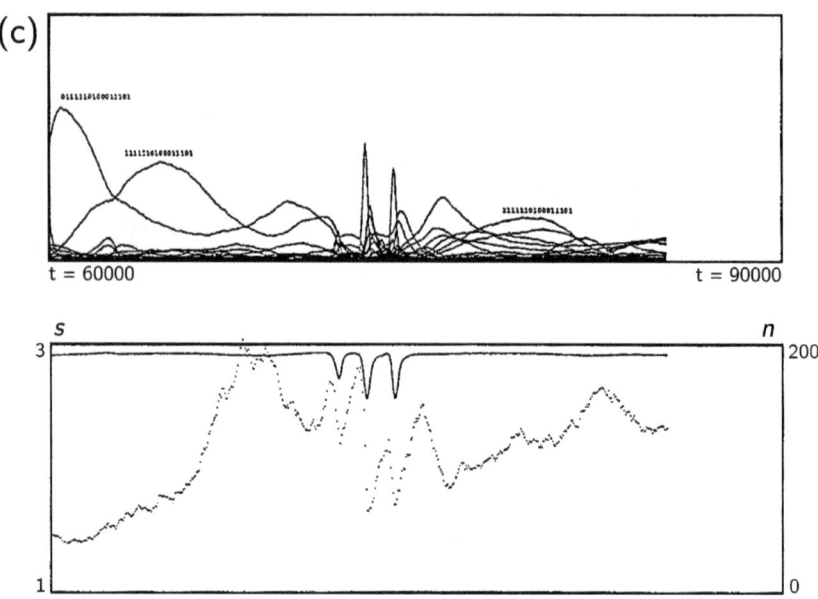

Figure 4. In (a) through (c) the evolution of a system avoiding the stable memory 4 stasis is shown for more than 80000 generations. The bottom graphs show the average score and the number of genotypes (cf. Figure 3). (a) In this simulation the system never reaches the symbiotic stasis but finds another way in state space leading to new strategies dominating the population. (b) Several new memory 4 strategies appear and dominate the population. The system reaches a dimensionality of more than 200, and after that a collapse occurs in which most of the genotypes disappears. At the same time, the average score drops, indicating that this large extinction is caused by a parasite mutant exploiting the present species. (c) Some new large extinctions occur, and a few of them are accompanied by a decrease in the average score.

Notice that the conclusions here are not quite game theoretic, not quite economic. The model shows long-term biological phenomena, issuing squarely from a nonbiological system.

From the game-theoretical point of view we have found that when the iterated Prisoner's Dilemma is modified by noise, there is an unexploitable strategy that is cooperative. The evolutionary simulation, which actually is a kind of genetic algorithm (Holland 1975) for finding good strategies for the noisy iterated Prisoner's Dilemma, indicates that the minimal memory for this kind of strategy is 4, i.e., the strategy should take into account the action of both players the previous 2 time steps. By answering a single defection by defecting twice the strategy is prevented from exploitation by intruders.

We have found periods of stasis punctuated by rapid transitions to new stasis or to periods of unstable dynamics. These rapid transitions are reminiscent of punctuated equilibria (Eldredge and Gould 1972), and it appears that the destabilization usually is due to a slowly growing group of mutants reaching a critical level. The coevolution of mutualism emerges spontaneously, and it serves as an example of a higher level of cooperation than the actions on the single round level provide. The appearance of an evolutionary stable strategy is interesting from the game-theoretic point of view, but in the construction of models possessing open-ended evolution one tries to eliminate such stabilizing phenomena. Therefore, from the evolutionary point of view, one should pay more attention to the less probable evolutionary pathways that avoid this evolutionary stable stasis. In particular, the large extinctions that appear in these simulations should be studied in more detail, since these collapses are triggered by the dynamical system itself and do not need external catastrophes for their explanation. The analysis of these results is in progress and shall be reported elsewhere. The major result of this model is that it establishes the fact that several evolutionary phenomena, like those described above, can emerge from very simple dynamics.

Acknowledgments

I thank Tomas Kåberger for inspiring me to start the construction of this model, Mats Nordahl and Doyne Farmer for stimulating discussions, Gunnar Eriksson and Karl-Erik Eriksson for collaboration in game theory that also provided inspiration for the present study, and Torbjörn Fagerström for useful biological remarks. Financial support from Erna and Victor Hasselblad Foundation is gratefully acknowledged.

REFERENCES

Axelrod, R. 1984. *The Evolution of Cooperation*. New York, NY: Basic Books.

———. 1987. *In Genetic Algorithm and Simulated Annealing*. Edited by D. Davies. 32–42. London, UK: Pitman.

Axelrod, R., and E. Dion. 1988. "The Further Evolution of Cooperation." *Science* 242:1385–1390.

Axelrod, R., and W. D. Hamilton. 1981. "The Evolution of Cooperation." *Science* 211:1390–1396.

Boyd, R. 1989. "Mistakes Allow Evolutionary Stability in the Repeated Prisoner's Dilemma Game." *Journal of Theoretical Biology* 136:47–56.

Boyd, R., and J.P. Lorberbaum. 1987. "No Pure Strategy is Evolutionarily Stable in the Iterated Prisoner's Dilemma Game." *Nature* (London, UK) 327:58–59.

Eigen, M., J. S. McCaskill, and P. Schuster. 1988. "Molecular Quasi-Species." *Journal of Physical Chemistry* 92:6881–6891.

Eldredge, N., and S. J. Gould. 1972. *Models in Paleobiology*. Edited by T. J. M. Schopf. 82–115. San Fransisco, CA: Freeman, Cooper and Company.

Farmer, J. D., S. A Kauffman, and N. H. Packard. 1986. "Autocatalytic Replication of Polymers." *Physica 22D,* 50–67.

Farrell, J., and R. Ware. 1989. "Evolutionary Stability in the Repeated Prisoner's Dilemma." *Theoretical Population Biology* 36:161–166.

Futuyma, D. J. 1986. *Evolutionary Biology*. 2nd ed. Sunderland, MA: Sinauer Associates.

Holland, J. H. 1975. *Adaptation in Natural and Artificial Systems*. Ann Arbor, MI: University of Michigan Press.

Kephardt, J. O., T. Hogg, and B. A. Huberman. 1989. "Dynamics of Computational Ecosystems." *Physical Review,* no. A40, 404–420.

Langton, C. G., ed. 1989. *Artificial Life*. Redwood City, CA: Addison-Wesley.

Maynard Smith, J. 1982. *Evolution and the Theory of Games*. Cambridge, UK: Cambridge University Press.

Maynard Smith, J., and G. R. Price. 1973. "The Logic of Animal Conflict." *Nature* (London) 246:15–18.

Miller, J. H. 1989. *The Coevolution of Automata in the Repeated Prisoner's Dilemma*. Technical report 89-003. Santa Fe Institute.

Molander, P. 1985. "The Optimal Level of Generosity in a Selfish Uncertain Environment." *Journal of Conflict Resolution* 29:611–618.

Rössler, O. E. 1971. "A System Theoretic Model of Biogenesis." *Zeitschrift für Naturforschung* 26b:741–746.

Schuster, P. 1986. "Dynamics of Molecular Evolution." *Physica D* 22:1–3.

von Neumann, J., and O. Morgenstern. 1944. *Theory of Games and Economic Behavior*. Princeton, NJ: Princeton University Press.

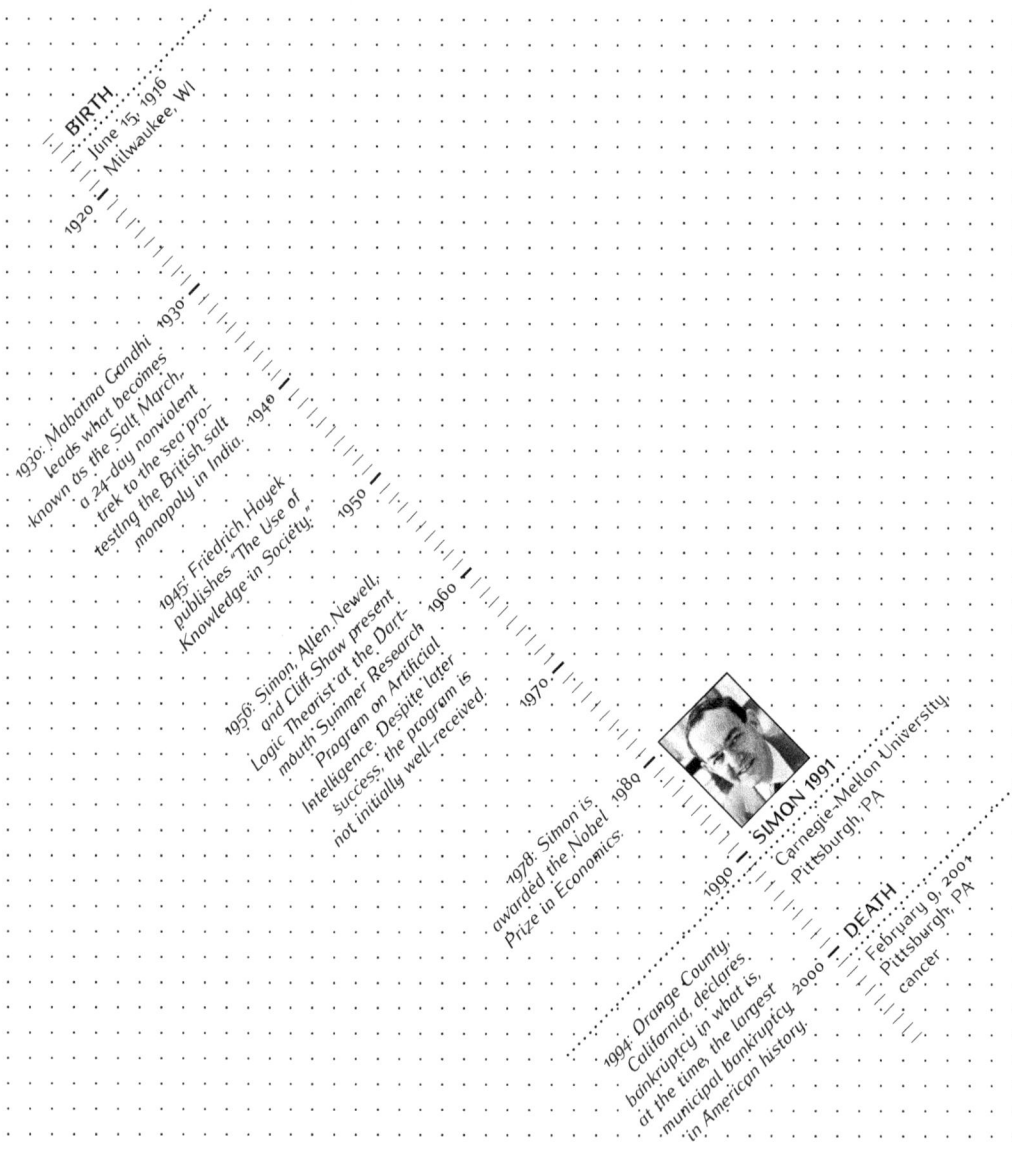

BIRTH
June 15, 1916
Milwaukee, WI

1930: Mahatma Gandhi leads what becomes known as the Salt March, a 24-day nonviolent trek to the sea protesting the British salt monopoly in India.

1945: Friedrich Hayek publishes "The Use of Knowledge in Society."

1956: Simon, Allen Newell, and Cliff Shaw present Logic Theorist at the Dartmouth Summer Research Program on Artificial Intelligence. Despite later success, the program is not initially well-received.

1978: Simon is awarded the Nobel Prize in Economics.

SIMON 1991
Carnegie-Mellon University
Pittsburgh, PA

1994: Orange County, California, declares bankruptcy in what is, at the time, the largest municipal bankruptcy in American history.

DEATH
February 9, 2001
Pittsburgh, PA
cancer

HERBERT ALEXANDER SIMON

[74]

WHAT SIMON SAYS: INTRODUCING ORGANIZATIONS AND MARKETS

Suresh Naidu, Columbia University and Santa Fe Institute

Herbert Simon's "Organizations and Markets" offers a sprinkling from the conceptual gusher coming from the pen of this legendary polymath author. It focuses on organizations and highlights his differences from the "optimization + efficiency" heuristics used by economists to account for organizational patterns. Simon instead lays out a world of networks, norms, and heuristics as the basis for economic organization, stressing the role of organizations in allocating power and authority, engendering loyalty and group identity, and managing expectations and facilitating coordination, none of which would be required in a world of perfect rational agents, costless economic contracting, and efficient selection of organizational forms. I think much of Simon's view summarized in the paper has been internalized by students of organizations, and so this paper is foundational.

Simon is clearly arguing with the dominant organizational theories in economics at the time he was writing, which invoked Panglossian arguments at every turn. These include the Alchian–Demsetz theory of efficient hierarchy, the Coase view on the efficient boundary of the firm, and the Chicago/Hayek optimism about the informational power and efficiency of market selection. Simon's view qualifies all these positions. Instead of firms being evolutionarily optimized organizations designed to monitor team production and minimize transaction costs, Simon suggests that the adaptive pressures function weakly at best. Instead of the price mechanism now being the primary source of information through a modern economy, Simon suggests "orders," both in the sense of inventories and in the sense of commands, as the actual flow of information in a day-to-day sense.

H. A. Simon, "Organizations and Markets," *Journal of Economic Perspectives* 5 (2), 25–44 (1991).

Copyright American Economic Association; reproduced with permission of the Journal of Economic Perspectives.

FOUNDATIONAL PAPERS IN COMPLEXITY SCIENCE

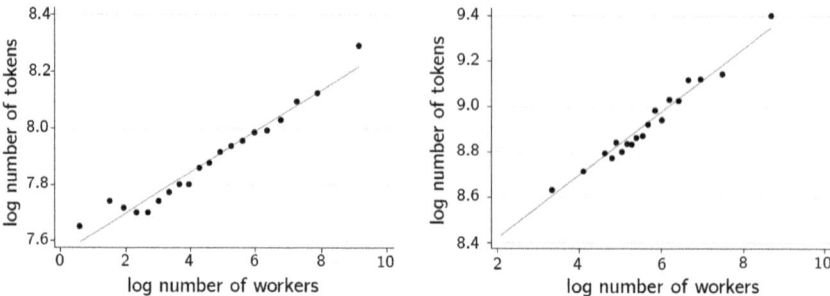

Figure 1. Workplace governance and scale. On the left is a binned scatterplot of ventiles for Brazilian collective-bargaining agreements, plotting the log number of workers covered by a collective-bargaining agreement on the log number of words in the collective-bargaining agreement. On the right is the same figure from Canada, a very different institutional environment.

The visual image Simon draws, of green organizations containing blue lines of authority connected by red threads of market interactions, is an extremely vivid model of the whole economy. It contains the seed of a massive empirical project, mapping social space more in terms of the relationships between people, with organizations and market relationships appearing as coarse-grained clusters of these relationships. But one thing that eluded Simon was how to measure the various edges between people. Modern administrative datasets can let us see who is in which organization and even which organizations are in market relationships with each other. Most administrative datasets miss the lines of authority within the firm—and how to distinguish them from lines of communication and information and capture the fact that some take orders from others on pain of termination or lack of promotion, which Simon thought was so essential. Exceptions exist. For example, the Enron email corpus, for example, has given us rich insight into the flows of information and authority within organizations, but much of the data necessary to operationalize Simon's model is proprietary and belongs to companies like Google.

Simon himself gave us a clue as to how to fix this for unionized workplaces, writing in 1952 that who has authority is generally spelled out in collective-bargaining agreements. In my own research I have tried to operationalize the allocation of power within organizations, using natural language processing (again, something Simon was an early pioneer in suggesting in "Verbal Reports as Data" with Ericsson) applied to union contracts and employee handbooks. In figure 1, I present some data from Brazilian and Canadian collective-bargaining agreements showing how the length of contracts scales with the number of workers covered by the contract: Larger bargaining units have longer texts delineating the rights and duties of workers and managers, rejecting both the additional differentiation in these organizations but also the larger variety of conflicts that need to be adjudicated. Given Simon's pioneering work on power laws in the distribution of firm sizes, a striking finding is the scaling relationship, where the log-log plot is virtually identical to a linear fit.

One limitation, reflecting, I think, a twentieth-century focus on markets and hierarchies as substitutes rather than complements, is that there is little discussion of how authority operates between organizations, for example, a vertical restraint in retailing. Further, the red lines in Simon's vision are only sometimes spot transactions; they are often complex legal relationships, with long documents and well-paid lawyers governing the terms between organizations. Simon mentions the role of dynamic inventories as equally, if not more, informative modes of communicating information between economic agents than prices. But he neglects the role of law and contract and authority even on the red threads between organizations. A red line between Apple and its suppliers in China is not simply a transfer of goods and services but also pages and pages of specifications, contractual conditions, with the threat of termination lying the background. Finally there are noninstitutional relationships: friendships and family networks that span and cross-cut organizations and are important sources of information, reputation, and community. There is little room in Simon's account for solidarity, reciprocity, and trust as sentiments that sustain organizational cultures as well as informational ties between organizations.

The second is the origins and historical path-dependence of organizations at all scales. Organizations will often reflect the circumstances of their founding well after the original founders or even business purpose has been transformed. This is of course obvious at the level of states, where an enormous literature has documented the long hand of history in influencing modern patterns of governance. But this path dependence is all the more remarkable because organizations are also malleable: They merge, they are conquered, and they dissolve. The ecology of organizations, capturing both the homeostatis as well as dynamic evolution of patterns of durable relationships, remains filled with models that are either too descriptive or too parsimonious, missing the complex sweet spot in-between.

REFERENCES

Weitzman, M. L. 1974. "Prices vs. Quantities." *The Review of Economic Studies* 41 (4): 477–491. https://doi.org/10.2307/2296698.

ORGANIZATIONS AND MARKETS

Herbert A. Simon, Carnegie Mellon University

In classical and neoclassical economic theory, markets are at the center of the stage. The actors in these markets are workers and consumers (sometimes combined into households), firms, owners of resources, governments, and perhaps others. The economic world of the neoclassical textbooks is a world of transactions, and these transactions typically involve an exchange of goods, services, and/or money that both parties to the transaction find advantageous to achieve these goals. Along with consumption, work and leisure are important components of the utility functions of households. Often, profit is assumed to be the sole objective of firms and their owners.

The description of the parties who participate in these transactions is minimal. However, as soon as firms are elaborated to become more than simple nodes in a network of transactions, to be producers—transformers of "factors" into products—difficult and important questions arise for the theory. A large part of the behavior of the system now takes place inside the skins of firms, and does not consist just of market exchanges. Counted by the head, most of the actors in a modern economy are employees, who either do not spend their days in trading, or if they do (for example, if they are salesmen or purchasing agents) are assumed to trade as agents of the firm rather than in their own interest, which might be quite different.

This raises the question of why there are firms at all. Why are not all the actors independent contractors? Why do most of them enter into employment contracts, selling their labor for a wage? What determines the make-or-buy decisions of firms, hence the boundaries between them and markets? When will two domains of activity lie within a single firm, and when will they be handled by separate contracting firms?

A second set of questions asks how the employees of firms are motivated to work for the maximization of the firm's profit. What's in it for them? How are their utility functions reconciled with those of the firm? In the employee's utility function, work is usually assumed to have negative utility and leisure (including loafing and working lackadaisically) to have positive utility. Why do employees often work hard?

The simple (neoclassical) answer to the motivational question derives from the employment contract, under which workers maximize their utility by accepting the authority of the firm; that is, by agreeing to accept orders from the profit maximizers in charge. But this answer leads to the new question of how the employment contract is enforced by the employer. In particular, how are employees induced to work more than minimally, and perhaps even with initiative and enthusiasm? Why should employees attempt to maximize the profits of their firms when making the decisions that are delegated to them?

This is where Simon lays out the theories of organizations and markets popular in economics in the 1970s and 1980s. Some of the theories focused on efficient hierarchies in organizations, along with efficient boundaries of the firm, because market pressures select for efficiency. Others focused on inefficiencies due to information or contracting frictions and rationalized organizations as second-best solutions to these problems.

These questions about the scope of activity and operation of firms have spawned a vigorous cottage industry, a branch of which is sometimes called "the new institutional economics," which tries to explain when activities will be carried out through the market and when they will be carried out within the skins of firms, and tries to explain also how it is possible for firms to operate efficiently. In the literature of the new institutional economics, two ideas that play a major role in the explanations are "transaction costs" and "opportunism" (for example, Williamson 1975, 1985). Sometimes the explanations are couched in terms of "information asymmetry" or "incomplete information" (Ross 1973; Stiglitz 1974). In other writings these topics are subsumed under agency theory, which treats the employment contract as an optimal contract between principal and agents, and studies how contractual arrangements can deal with shirking and other motivational problems.

The idea behind these ideas is that a proper explanation of an economic phenomenon will reduce it to maximizing behavior of parties who are engaged in contracting, given the circumstances that surround the transaction. The terms of the contract will be influenced by the access of the parties to information, by the costs of negotiating, and by the opportunities for cheating. Access to information, negotiation costs, and opportunities for cheating are most often treated as exogenous variables that do not themselves need to be explained. It has been observed that they even introduce a sort of bounded rationality into the behavior, with the exogeneity of the limits of rationality allowing the theory to remain within the magical domains of utility and profit maximization.

A fundamental feature of the new institutional economics is that it retains the centrality of markets and exchanges. All phenomena are to be explained by translating them into (or deriving them from) market transactions based upon negotiated contracts, for example, in which employers become "principals" and employees become "agents." Although the new institutional economics is wholly compatible with and conservative of neoclassical theory, it does greatly multiply the number of auxiliary exogenous assumptions that are needed for the theory to work. For example, to explain the presence or absence of certain kinds of insurance contracts, moral risk is invoked; the incompleteness of contracts is assumed to derive from the fact that information is incomplete or distributed asymmetrically between the parties to the contract. Since such constructs are typically introduced into the analysis in a casual way, with no empirical support except an appeal to introspection and common sense, mechanisms of these sorts have proliferated in the literature, giving it a very *ad hoc* flavor.

In general, the new institutional economics has not drawn heavily from the empirical work in organizations and decision-making for its auxiliary assumptions. (For introductions to that literature, see March and Simon 1958; Cyert and March 1963; Kornai 1971; Simon 1979). Nevertheless, it is appropriately subversive of neoclassical theory in that it suggests a whole agenda of microeconomic empirical work that must be performed to estimate the exogenous parameters and to test the theory empirically. Until that research has been carried out (and the existing literature on organizations and decision making taken into account), the new institutional economics and related approaches are acts of faith, or perhaps of piety.

Since Simon's writing, both economists and sociologists have conducted an enormous volume of empirical work on organizations. Some of this has fallen under "personnel economics" and "organizations." But it has mostly focused on issues of pay and compensation, not the empirical details of communication and authority inside organizations.

The Ubiquity of Organizations

A mythical visitor from Mars, not having been apprised of the centrality of markets and contracts, might find the new institutional economics rather astonishing. Suppose that it (the visitor—I'll avoid the question of its sex)

> This is a great visual metaphor of the economy as a network of people. If you cluster or coarse-grain the network, you get a network of organizations.
>
>
>
> Simon presents market transactions as uniform red lines. I think in a modern, complex economy with contracts, we should distinguish between "thin" red lines of spot transactions and "thick" red lines of ongoing economic contracts and repeat interactions.
>
> GDP and the national accounts can be viewed through Simon's metaphor: Households give money to firms as customers with thin red lines, who then send money to workers and owners/shareholders (with thick red lines) and other firms (both thick and thin). GDP is the flow from households to firms and back to households.
>
>
>
> The only edges in Simon's network are market transactions between organizations and authority within organizations. Nonmarket relationships of trust and information across organizations don't exist. Maybe he would just say they are contained in other, higher-level organizations.

approaches the Earth from space, equipped with a telescope that reveals social structures. The firms reveal themselves, say, as solid green areas with faint interior contours marking out divisions and departments. Market transactions show as red lines connecting firms, forming a network in the spaces between them. Within firms (and perhaps even between them) the approaching visitor also sees pale blue lines, the lines of authority connecting bosses with various levels of workers. As our visitor looked more carefully at the scene beneath, it might see one of the green masses divide, as a firm divested itself of one of its divisions. Or it might see one green object gobble up another. At this distance, the departing golden parachutes would probably not be visible.

No matter whether our visitor approached the United States or the Soviet Union, urban China or the European Community, the greater part of the space below it would be within the green areas, for almost all of the inhabitants would be employees, hence inside the firm boundaries. Organizations would be the dominant feature of the landscape. A message sent back home, describing the scene, would speak of "large green areas interconnected by red lines." It would not likely speak of "a network of red lines connecting green spots."

Of course, if the vehicle hovered over central Africa, or the more rural portions of China or India, the green areas would be much smaller, and there would be large spaces inhabited by the little black dots we know as families and villages. But the red lines would be fainter and sparser in this case, too, because the black dots would be close to self-sufficiency, and only partially immersed in markets. But let us, for the present, restrict our attention to the landscape of the developed economies.

When our visitor came to know that the green masses were organizations and the red lines connecting them were market transactions, it might be surprised to hear the structure called a market economy. "Wouldn't 'organizational economy' be the more appropriate term?" it might ask. The choice of name may matter a great deal. The name can affect the order in which we describe its institutions, and the order of description can affect the theory. In particular, it may strongly affect our choice of the variables that are important enough to be included in a first-order theory of the phenomena.

Simon (1991)

How does the economy look when it is viewed as an organizational economy, with market relations among organizations? I have already suggested some of the more prominent features.

First, most producers are employees of firms, not owners. Viewed from the vantage point of classical theory, they have no reason to maximize the profits of firms, except to the extent that they can be controlled by owners. Moreover, profit-making firms, nonprofit organizations, and bureaucratic organizations all have exactly the same problem of inducing their employees to work toward the organizational goals. There is no reason, a priori, why it should be easier (or harder) to produce this motivation in organizations aimed at maximizing profits than in organizations with different goals. If it is true in an organizational economy that organizations motivated by profits will be more efficient than other organizations, additional postulates will have to be introduced to account for it.

Where is the government (upward of 30% of GDP in advanced countries) in Simon's map? Where is regulation and regulatory relationships between organizations and the government? Or flows of tax income? The central bank? Seems strange to draw such a rich, granular picture of the economy and ignore the largest player in it.

Second, the system is nearly in neutral equilibrium between the use of market transactions and authority relations to handle any particular matter: that is to say, very small changes in the situation can tip the equilibrium one way or the other. It is hard to explain degrees of integration of economic activities. In many instances, transaction cost analysis is not applicable, and even where it is, there often remains considerable latitude for different degrees of integration. For example, why are auto dealerships not a part of auto manufacturing companies, rather than having contractual relations with them?[1] Why did General Motors manage its own tool design for many years, but recently decide to contract most of it out? Under constant returns to scale and reasonably competitive markets, which characterize many manufacturing situations, make-or-buy decisions become ambiguous. The possibility of using internal division-by-division balance sheets, and internal pricing in negotiation between components of an organization further blurs the boundary between organizations and markets.

The make-or-buy decision is a classic question in economics, and Simon here poses an answer not predicated on efficiency.

Without the introduction of very particular ad hoc assumptions, unbuttressed by empirical evidence, neoclassical theory provides no explanation for the repeated appearance of Pareto distributions of business

---

[1] Williamson's explanation—actually, Alfred P. Sloan's explanation (see Williamson 1985, p. 10)—that employees could not be supervised adequately in their offers for used cars, is not convincing. Dealerships are also organizations, and their salesmen are employees.

firm sizes in virtually all situations where size distributions have been studied (Ijiri and Simon 1977; Simon 1979). (In a Pareto distribution, the logarithm of the number of firms above any given size decreases linearly with the logarithm of the size.) These observed distributions are difficult to reconcile with any notions that have been proposed for optimal firm size, but are easily explained by simple, plausible probabilistic mechanisms that make no appeal to optimality.

The tension between randomness and optimality recurs in complex systems analysis. Benoit Mandelbrot and Simon had a lengthy and sometimes vicious debate over whether power laws were best explained as optimal. Mandelbrot argued Simon's stochastic models were circular and didn't constitute a valid explanation. Mandelbrot (and many economists) explained power laws as solutions to an optimal design problem, where either the market or evolution or human design exert optimizing pressure.

In sum, an organizational economy poses the questions of why the larger part of a modern economy's business is done by organizations, what role markets play in connecting these organizations with each other, and what role markets play in connecting organizations with consumers. Moreover, the boundary between markets and organizations varies greatly from one society to another and from one time to another. What mechanism maintains the highly fluid equilibrium between them? Until these questions are answered, it will be difficult to draw conclusions about the relative efficiencies of different forms of ownership and control of organizations, or the relative efficiency of markets versus central planning.

Motivation and Efficiency in Organizations

There are three different questions of social organization that are usually confounded, but which need to be considered separately. The first is the question of the relative efficiency of markets and organizations. The second is the question of the consequences of having a society's organizations owned by profit-making organizations, by nonprofit organizations, or by public organizations, respectively. The third is the question of the consequences of using central planning instead of markets to regulate relations among organizations. At present, our concern is only with the first question: what makes organizations work as well or badly as they do?

Simon wrote this article before the collapse of the Soviet Union, and so the question of central planning remained on his horizon. But today, the recent rise of AI and platforms makes questions of central planning relevant again, sometimes framed in more computational terms.

In particular, for whom is profit the motive? Adolf Berle and Gardiner Means posed the problem very sharply in their famous book, *The Modern Corporation and Private Property* (1933), by showing that even at the top executive levels of the modern corporation there is a great gap between ownership and control, and a correspondingly great opportunity for discrepancy between the goal of owners (profit) and the goals of managers (career status, wealth, a quiet life, and so on).

Demsetz and Lehn (1985) have contested the argument of Berle and Means on the ground that even large corporations show considerable concentration of ownership. Typically, a half dozen owners (or fewer) own 10 or 20 percent of the shares, enough to retain controlling power. Often, these owners are also the active top executives. But the objection does not hold water. If a company has an executive bonus plan, and if an executive's percentage share in bonus awards is greater than his or her percentage share of dividends, then it pays that executive to divert earnings from dividends to bonuses. Most companies have executive award systems that make this conflict of interest very real. Golden parachutes and leveraged buyouts are other significant examples of transactions where the interests of shareholders and executives may diverge strongly.[2]

If even top executives may be conflicted in their motives, the problem should be still greater for employees who are not owners at all, or only insignificantly. Principal-agent theory, on which the new institutional economists often rely, assumes that agents within firms will shirk unless their actions contribute directly to their own economic self-interest. It is only via monitoring combined with contracts that appeal to their self-seeking nature that such shirking may be mitigated. But the assumption that executives (and perhaps other employees) would choose to advance their own careers and wealth and consumption, rather than pursuing organizational goals like maximizing profit, is not prescribed by neoclassical theory, which leaves the specification of the utility function completely open.

Why not assume that maximizing the firm's profit is precisely what maximizes the utilities of executives and other workers? In a society of robots, an owner would not settle for less. But most of us would think this an unrealistic assumption to make for a human society. An organization theory with an unspecified utility function is not a theory at all. And one with an unrealistic utility function does not provide a basis for understanding real organizations. Instead, we should begin

Mainstream economics holds as a benchmark model that firms simply maximize profits, with little in the way of internal corporate governance built into the model. The problem of defining a firm's objective function is an old one (it motivated Ken Arrow's famous social choice theorem in his dissertation).

---

[2] Demsetz and Lehn (1985) cite evidence to show that corporations where ownership is widely distributed have, on average, profits as large as those with concentrated ownership. This fact does not undermine the argument of Berle and Means for conflict of economic interest; on the contrary, it raises the question—which I will undertake to answer below—of why executives with small stakes as shareholders do appear to work for company profits.

with empirically valid postulates about what motivates real people in real organizations. I shall argue that such postulates can be derived from four organizational phenomena whose roles are amply documented in the literature on organizations: authority, rewards, identification, and coordination.

Authority: The Employment Relation

The employment contract is an example of what is now sometimes called an "incomplete contract;" that is to say, some of its terms are unspecified. Employees agree to do, over the life of the contract, what they are ordered to do; but the orders will not be issued until some time after the contract is negotiated (Simon 1951; Williamson 1975).

The usual argument (within the neoclassical framework) for the existence of incomplete contracts is that in a world of uncertainty actions will have to be taken as the situation calls for them, without time for negotiation. The employee is rewarded, in the level of the wage, for willingness to bear the brunt of this uncertainty as to what actions will be chosen, and to do, when the time comes, whatever the employer thinks the situation calls for. This argument does not imply that uncertainty is replaced by complete certainty at the time of decision. On the contrary, taking decisions under conditions of uncertainty may be one of the important skills demanded of the decision maker. The essential point is that the uncertainty for the employer is decreased by delaying the commitment to specific actions from the time employment begins until the time when action is called for.

An employment contract contains all sorts of implicit (and explicit) limitations that set boundaries to the range of actions the employee will be directed to perform. These boundaries define the "zone of acceptance" within which an employee can be expected to obey orders. The zone of acceptance is also sometimes called a "zone of indifference," for the choice among alternative behaviors, while of major importance to the employer, may be of little or no concern to the employee. A secretary, for example, usually has little or no preference for typing a letter to one of the company's customers rather than another, and little interest in the content of the letter. Even a factory manager will accept, within wide limits, whatever mix of products the factory is ordered to produce in a given month.

The combination of uncertainty on the part of the employer (as to what will need to be done in the future) and broad acceptance of the employee (of what he or she will be ordered to do) makes the employment contract a very attractive bargain for both parties. The new institutional economics finds that employment achieves great savings in transaction costs—the costs of negotiating separate contracts for each action.

But this theory of the employment contract must be elaborated. Authority in organizations is not used exclusively, or even mainly, to command specific actions. Most often, the command takes the form of a result to be produced ("repair this hinge"), or a principle to be applied ("all purchases must be made through the purchasing department"), or goal constraints ("manufacture as cheaply as possible consistent with quality"). Only the end goal has been supplied by the command, not the method of reaching it. The mechanic must apply all kinds of knowledge and skill to repairing the hinge. The section chief must initiate purchases of supplies needed for the work of that section; however, the company's standard procedures must be taken as ground rules for the way the purchases are made. The factory manager must control manufacturing cost and quality.

Simon argues that authority here specifies not exact actions but instead broad collections of tasks that are consistent with the organization's goals.

Employees, especially but not exclusively at managerial and executive levels, are responsible not only for evaluating alternatives and choosing among them but also for recognizing the need for decisions, putting them on the agenda, and seeing to the generation of possible actions. Doing the job well is not mainly a matter of responding to commands, but is much more a matter of taking initiative to advance organizational objectives.

Commands do not usually specify concrete actions but, instead, define some of the premises that are to be used by employees in making the decisions for which they are responsible (Simon 1947). Hence, seeing that commands are obeyed is not simply a matter of observing behavior, but of affecting the thought processes and the decision premises of employees. Further, it is usually difficult or impossible to ascertain what these decision premises have been without reviewing the whole decision—thus causing an almost complete loss of the economy that was sought in delegating it in the first place.

The command an employer might like to issue is: "Always decide in such a way as to maximize company profit!" But that would simply reintroduce the question of how the extent of obedience to the command is to be observed without losing the benefit of delegation. Even if the employees were robots, whose loyalty could be guaranteed, the problem would not be solved. For giving each robot complete discretion would surrender large efficiencies usually attainable from specialization in decision-making work. We need to delegate within guidelines, which creates the problem of monitoring the observance of guidelines without recentralizing what has just been delegated.

If authority is used to transmit premises for making decisions rather than commands for specific behaviors, then many different experts can contribute their knowledge to a single decision. Information and policy rules can flow through the organization along many channels, serving as inputs—decision premises—for many organizational behaviors.

The accounting department gathers cost data, which it supplies to the head of the blast furnace department to help make operating decisions in that department. At the same time, the blast furnace manager is receiving instructions from metallurgical specialists on the technical aspects of the operation. The faint blue lines that our visitor from Mars saw within the green areas were not just streams of orders, but flows of all kinds of decision premises (constraints and information as well as orders) from one point in the organization to another.

This explication of the employment contract and authority takes us back to the question of motivation. For the organization to work well, it is not enough for employees to accept commands literally. In fact, obeying operating rules literally is a favorite method of work slowdown during labor-management disputes, as visitors to airports when controllers are unhappy can attest. What is required is that employees take initiative and apply all their skill and knowledge to advance the achievement of the organization's objectives.

I find it interesting that the example Simon gives here is of a unionized shop. Non-union workers (and even union workers not during a negotiation) would probably be fired for overly literal interpretations of orders. Because the authority of the employer also has some discretion.

We should not assume without evidence that organizations *do* work well. But "well" is a relative term. In most organizations, employees contribute much more to goal achievement than the minimum that could be extracted from them by supervisory enforcement of the (vague) terms of the employment contract. Why do employees not substitute leisure for work more consistently than they do? Why do they often work so vigorously for the welfare of the organization?

Rewards as Motivations

One obvious answer to the motivational question is that employees may be motivated to accept authority by giving them material rewards, promotion, and recognition for advancing the organization's goals as defined by management. Such rewards certainly provide motivation, but they only operate satisfactorily when certain conditions are met. The most important condition is that the employee's contribution to the organization's goals must be measurable with reasonable accuracy. For example, salesmen are frequently compensated (at least partly) on a commission basis. Blue-collar employees are sometimes compensated on a piecework basis, albeit in a continually decreasing number of situations. Executives, and sometimes others, receive bonuses that are supposed to be related to their contributions to profits.

But such reward systems are effective only to the extent that success can be attributed accurately to individual behaviors. If the indices used to measure outcomes are inappropriate, either because they do not measure the right variables, or because they do not properly identify individual contributions, then reward systems can be grossly inefficient or even counterproductive. Where output quantities are measured with inadequate attention to quality, response to rewards will cause quantities to grow at the cost of lowered quality. Where compliance with company policies that constrain action is not measured, constraints will be ignored and violated. Salesmen may misrepresent the product, workmen may ignore safety rules, managers may buck difficulties to other departments.

An important question is how the ability to attribute rewards to individuals has changed with the economy. Is it higher or lower now than in Simon's day? What has remote work done to it?

In general, the greater the interdependence among various members of the organization, the more difficult it is to measure their separate contributions to the achievement of the organizational goals. But of course, intense interdependence is precisely what makes it advantageous to organize people instead of depending wholly on market transactions. The measurement difficulties associated with tying rewards to contributions are not superficial, but arise from the very nature and rationale of organization.

Many large U.S. corporations attempted to respond to this problem in the years after World War II by slicing their organization into components that were relatively self-contained. Then, separate balance sheets could be maintained for each division, and these balance sheets could be used to evaluate results and to compute rewards.

An economist might say the difficulty in monitoring just means the costs inflicted on the worker in event of shirking must be correspondingly higher.

Of course, divisionalization can be successful only to the extent that the divisions are actually self-contained. If one division operates mainly as a supplier of parts to other divisions, then policies have to be laid down for setting the prices for items "sold" by the one division to the others, and for determining under what conditions a division may go outside the company to purchase items at a lower price. For these and similar reasons, divisionalization can only be carried a short distance down the structure of a typical corporation, and solves the problem of attributing outcomes to individuals only at the higher levels, if at all.

Although economic rewards play an important part in securing adherence to organizational goals and management authority, they are limited in their effectiveness.[3] Organizations would be far less effective systems than they actually are if such rewards were the only means, or even the principal means, of motivation available. In fact, observation of behavior in organizations reveals other powerful motivations that induce employees to accept organizational goals and authority as bases for their actions. We turn next to the most important of these mechanisms: organizational identification.

Loyalty: Identification with Organizational Goals

Pride in work and organizational loyalty are widespread phenomena in organizations (Simon 1947). These traits are more strongly evident among skilled and managerial employees than among employees engaged in very routine work. (The latter are also more easily supervised, and can sometimes be rewarded on a piecework basis.) In part, these attitudes can be attributed to the linkage between an organization's overall success and the personal careers and monetary rewards it can provide its employees. But this explanation ignores the problem of the commons—of benefits that are jointly gained and shared by all, non-contributors along with contributors—and the consequent possibilities for free-riding. The quality and success of an organization depends very little on the energy of any

---

[3] Everything said here about economic rewards applies equally to privately owned, nonprofit, and government-owned organizations. The opportunity for, and limits on, the use of rewards to motivate activities toward organizational goals are precisely the same in all three kinds of organizations. For sophisticated discussions of motivation and efficiency in profit-making and nonprofit organizations, see Weisbrod (1988, 1989).

single employee (except possibly an executive at or near the very top). Why will employees work hard if they can gain almost as much by loafing?

Of course free-riding can be observed in organizations. The elimination of free-riding is generally thought to be the principal reason for the success of the Chinese agricultural reforms after 1980, when responsibility and reward for agricultural production were transferred from the commune to the family. The question is not whether free riders exist—much less employees who exert something less than their maximum—but why there is anything *besides* free-riding. Why do many workers, perhaps most, exert more than minimally enforceable effort? Why do employees identify with organizational goals at all?

My sense is that organizational identification has declined since Simon's writing. In extreme form we have gig work, where there is very little identification with employers. I think the relative role of rewards vs. loyalty may have moved toward rewards since Simon's writing.

Contemporary evolutionary theory has cautioned us against postulating altruistic motives for people. In models of natural selection, nice guys generally aren't fit; they don't multiply as rapidly as their more selfish brethren. The argument from natural selection has often been used, explicitly or implicitly, to fill the utility function with selfish personal goals. But models of natural selection do not actually provide strong support for the idea that people will only pursue selfish personal economic goals. In fact, such models in no way foreclose the possibility (indeed, the probability) that people will be strongly motivated by organizational loyalty, even when they can expect no "selfish" rewards from it (Simon 1983, 1990).

First, it should be emphasized that what natural selection increases is fitness, the number of progeny of the successful competitor. But in modern society, the attainment of wealth or other selfish rewards is not directly connected to number of progeny. In fact, first-world societies generally display a negative correlation between income level and size of family. But let us waive this point, as distracting us from the main argument, and suppose that attainment of the goals usually described as selfish (especially personal economic goals) contributes to evolutionary fitness.

We come then to the second point: each human being depends for survival on the immediate and broader surrounding society. Human beings are not the independent windowless Leibnitzian monads sometimes conjured up by libertarian theory. Society is not imposed on humans; rather, it provides the matrix in which we survive and mature and act on the environment. Families and the rest of society provide nutrition, shelter, and safety during childhood and youth, and then the knowledge and skills for adult performance. Moreover, society can react to a person's activities at every stage of life, either facilitating them or severely impeding them. Society has enormous powers, enduring though a person's lifetime, to enhance or reduce evolutionary fitness.

What kinds of traits, in addition to personal strength and intelligence, would contribute to the fitness of this socially dependent creature? One such trait, or combination of traits, might be called docility. To be docile is to be tractable, manageable, and above all, teachable. Docile people tend to adapt their behavior to norms and pressures of the society. I am not satisfied that "docile" conveys my meaning precisely, but I know of no better word.

That fitness is derivable from being docile becomes evident when we consider the opposite of docility: intractability, unmanageability, unteachability, incorrigibility. The argument is not that people are totally docile, nor that they are totally selfish, but that fitness calls for a measured but substantial responsiveness to social influence. In some contexts, this responsiveness implies motivation to learn or imitate; in other contexts, willingness to obey or conform. From an evolutionary standpoint, having a considerable measure of docility is not altruism but enlightened selfishness.

To survive as a trait, docility must contribute on average to the fitness of the individual who possesses it. Yet it may still lead to self-injurious behavior in particular cases. Thus, docile individuals may do better at earning a living, but loyalty to the nation may lead them to sacrifice their lives in wartime. Once docility is present, society may exploit it by teaching values that are truly altruistic; that is, which contribute to the society's fitness, but not to the individual's. The only requirement is that *on balance* and on the average the docile individual must be fitter than the one who is not docile.[4]

---

[4]This is not the place to describe in detail how docility and altruism induced through the docility mechanisms can be incorporated in a formal model of evolution by natural selection. I will simply sketch the general idea. Let k be the average number of offspring of an individual in the absence of docility or altruistic behavior; $d > 0$ the gross increase in offspring due to docility; $c > 0$ the cost to a docile individual in offspring of the socially induced altruistic behavior; p the percentage of individuals in the population which are docile and hence altruistic; and b the number of offspring added to the

Of course, showing that a configuration of traits or genes would contribute to fitness, if they existed, does not prove they exist. But ample empirical evidence shows that most human beings are gifted with a considerable measure of docility. The purpose of the present argument is to show that this docility and the altruism it induces is wholly consistent with the premise of selection of the fittest. In fact, the theory of natural selection strongly predicts the appearance of docility and altruism in social animals.

Docility is used to inculcate individuals with organizational pride and loyalty. These motives are based upon a discrimination between a "we" and a "they." Identification with the "we," which may be a family, a company, a city, a nation, or the local baseball team, allows individuals to experience satisfactions (to gain utility) from successes of the unit thus selected. Thus, organizational identification becomes a motivation for employees to work actively for organizational goals. Of course, identification is not an exclusive source of motivation; it exists side by side with material rewards and enforcement mechanisms that are part of the employment contract. But a realistic picture of how organizations operate must include the importance of identification in the motivations of employees.

The strength of organizational identifications will depend upon the extent to which a society uses the docility mechanism to inculcate them, and this appears to vary considerably from one society to another. For instance, it would probably be agreed by ethnographers that in Chinese society greater pressure is exerted to induce identification with the family than with employing organizations, while the reverse is true of Japanese society. Such conjectures can be tested, for example, by examining practices of nepotism, and attitudes toward it, in the two societies.

The strength of the organizational loyalties of employees is not to be attributed only to motivation induced by docility. There is also an

---

population by an individual's altruistic behavior. Assume further, that the parentage of offspring contributed by altruism is distributed randomly through the population. Then it is easy to show that the difference between the net fitness of altruists and non-altruists (non-docile individuals), respectively, will equal $d - c$. Hence, provided that d is larger than c, altruists will be fitter than non-altruists. Moreover, a society will grow more rapidly the greater the fraction of altruists in it, the increase in average fitness being $(d - c + b)p$.

important cognitive component. The bounded rationality of humans does not allow us to grasp the complex situations that provide the environments for our actions in their entirety. The first step in rational action is to focus attention on specific (strategic) aspects of the total situation, and to form a model of the situation in terms of those aspects that lie in that focus of attention. Rational computation takes place in the context of this model, rather than in the response to the whole external reality.

One dimension of simplification is to focus on particular goals, and one form of focus is to attend to the goals of an organization or organization unit. Having defined that unit as the "we," actions are evaluated in terms of their contribution to the unit's objectives. The ubiquity of this narrowing of attention is easily demonstrated. As one example, Dearborn and Simon (1958) presented a group of business executives with a description of the current situation of a large company, and asked them to identify the most serious problem facing the company. In their own companies, some of the executives were responsible for manufacturing, others for sales, others for finance. In almost every case, the "most serious problem" identified by the executive lay in the domain of his or her own department—manufacturing problems for manufacturing executives, sales problems for sales executives, and so on.

Organizational identification (or "loyalty") is an important adhesive for Simon's organization. Simon sets it up as an answer to the question of why people follow orders. But while loyalty is important, most people follow orders because they will be punished if they do not. Loyalty is perhaps a social convention that evolves to rationalize the structure of authority.

It is a commonplace of organizational life that a person's organizational identification will shift with his or her position, although the motivational basis for the shift is perhaps more widely recognized than the cognitive basis. But a shift in organizational position exposes the employees to new "facts" and phenomena, to a new network of communications, and to new goals. A different model is inevitably formed of the decision-making situation, a model that emphasizes local components of the environment and local goals. Behavior is very much a function of position.

Because of cognitive limits, the precise form that goals take may depend on what can be measured in the situation. In business organizations, the accounting statements provide stylized measurements of profits, size, growth, market share, and so on. Even if these measurements are only rough approximations of the things they are supposed to be measuring, they are likely to replace the "real" unmeasured concepts in the decision-making process.

Willingness of employees at all levels to assume responsibility for producing results—not simply "following the rules"—is generally believed to be a major determinant of organizational success. This discussion implies that acceptance of responsibility will be affected both by the reward system and by the strengths of organizational identifications. Here again, large intercultural differences may exist. The recent establishment of a substantial number of international joint ventures, with managements and employees recruited from different cultures, provides an excellent research environment for studying these differences and their effects upon organizational efficiency.

Since the developments are quite new, little information is yet available about them. But one example where data are available is the joint venture between Toyota and General Motors in northern California (Krafcik and Womack 1987). Here Toyota took over a former General Motors plant, equipped it with standard state-of-the-art machinery, rehired employees mainly from the previous work force and accepted the same union. They have been able to produce automobiles with about 45 percent fewer labor hours than an entirely comparable GM plant that uses American managers and management methods, and about 30 percent fewer hours than a new GM plant having more modern "hitech" equipment, and only about 15 percent *more* labor hours than a comparable Toyota plant in Japan.

The causes for these enormous differences in efficiency have almost nothing to do with the classical physical production function. They also appear to have little to do with cultural differences at the blue-collar level.[5] They seem to have nothing to do, either, with material reward structures, which are not significantly different in the various plants. They must be attributed in large part to differences in management practices (for example, quality control practices, and

---

[5] These two statements should be qualified slightly. With regard to the first, components imported by the Toyota plant from Japan may be more uniform in quality than components purchased by the other GM plants. With respect to the second, applicants interviewed for employment in the Toyota plant were screened for problem solving attitudes and skills. Note that both of these factors, whether important or not, are matters of management practice. Finally, I would not wish to claim that the factors I have mentioned were the only ones affecting the comparison.

inventory policies), perhaps bolstered by differences in management attitudes and motivations.

Coordination

This examination of authority and organizational identification should help explain how organizations can be highly productive even though the relation between their goals and the material rewards received by employees, if it exists at all, is extremely indirect and tenuous. In particular, it helps explain why careful comparative studies have generally found it hard to identify systematic differences in productivity and efficiency between profit-making, nonprofit, and publicly-controlled organizations (Weisbrod 1988, 1989). Also, it explains why Demsetz and Lehn (1985) found no difference in profits between corporations that were managed or controlled by owners and those with diffuse stock ownership.

But to understand the relative advantages of organizations and markets, and the circumstances under which one would operate more effectively than the other, one further component must be added to our description of organizations. Organizations, through the authority mechanism, provide a means for *coordinating* the activities of groups of individuals in ways that are not always easily achieved by markets.

Coordination is a rather slovenly word, often abused in organizations. An experienced executive cringes when he or she learns that someone has been appointed to "coordinate" a set of activities, since calling for coordination without specifying just what it means is simply a lazy way of passing off problems to someone else. I will try to make the concept more precise by using it to designate a specific kind of activity.

The theory of games has sharply underscored that decisions are usually indeterminate when each party in a situation is uncertain about the actions of the others. This result is quite independent of whether their goals are complementary or competitive. One simple example of this indeterminacy is that it is rational for a motorist to drive on the *same* side of the road as other drivers headed the same direction, whichever that may be. There is no question of correct behavior in relation to the environment, but only of coordinating the behaviors of all the actors. Such rules of the road, or standardization, can greatly improve the performance of systems in those (ubiquitous) situations where the correctness of an action depends on what the other actors are doing.

A more complex example of coordination is provided by a university. Conceive of a university that consisted only of some rooms, some teachers, and some students. Students and teachers would "simply" negotiate to meet at certain times and places for their classes. The resulting chaos would probably be resolved by inventing the institutions of a registrar's office and a class schedule. While it would be extravagant to urge that class schedules provide the raison d'etre for education by universities, rather than by contractual tutoring arrangements negotiated through markets, nevertheless, the coordinating function of schedules is not trivial.

A major use of authority in organizations is to coordinate behavior by promulgating standards and rules of the road, thus allowing actors to form more stable expectations about the behavior of the environment (including the behavior of other actors). Since organizations provide a mechanism (authority) for establishing rules of the road, which markets do not, one might even expect organizations rather than markets to be the environments in which the behavior called "rational expectations" would be most often observed.[6]

Organizations have a function similar to that of conventions and law: They coordinate expectations about what will happen if a set of actions are taken, enabling individuals to form accurate predictions of each others' behavior.

In a book on central planning during World War II, Ely Devons (1950) raised the question of why prices are supplanted by government plans, expressed as quantity goals for production and allocation, as coordinating mechanisms during wartime. The usual argument for markets, as in the well-known 1945 paper of von Hayek, is that they simplify the decision process by reducing the need of each actor to know what the other actors are doing or what situations confront them. To the extent that markets and prices perform this simplifying function, we would expect them to replace centralized decisions when a situation becomes more complex—for example, during the rapid changes that take place in shifting from a peacetime to a wartime economy. Yet, as Devons points out, it is just at such times that central planning tends to increase. Is this irrationality, or are there valid reasons for the shift?

Martin Weitzmann (1974) provides a classic answer. The idea is that prices are good tools when there is high uncertainty in costs relative to demand. When you need tanks (or carbon emission targets) to be a certain number, and you don't care so much about the cost of getting that target, then quantities are a better tool.

---

[6]Of course, perfectly competitive markets do provide stable expectations of prices at least in equilibrium, and thereby permit Pareto optima to be achieved in principle. But prices are only one of many dimensions along which uncertainty of expectations may complicate rational decision making.

The answer is rather obvious. Prices perform their informational function when they are known or reasonably predictable. Uncertain prices produced by unpredictable shifts in a system reduce the ability of actors to respond rationally. This point is often made by economists in arguing the costs of unexpected inflation, but its implication for the choice between organizations and markets is less often noted. Nor is it often noted that many kinds of uncertainties other than price uncertainties may make coordination through organizational procedures advantageous.

The difficulty that economics has had in giving a good account of organizations and their predominance is traceable in no small part to the fascination of economists with systems in equilibrium. Analysis under assumptions of perfect knowledge and certain expectations has little relevance, surely, for such issues of economic organization as explaining how an economy is structured between organizations and markets. Prices provide only one of the mechanisms for coordination of behavior, either between organizations or within them. Coordination by adjustment of quantities is probably a far more important mechanism from a day-to-day standpoint, and in many circumstances will do a better job of allocation than coordination by prices. For example, inventory control systems record the amounts of inputs for the organization's activities, and place orders when quantities fall below specified levels. The orders, recorded by the control systems of suppliers, initiate the scheduling of new production and are used to adjust aggregate production levels as well.

From a conceptual standpoint, it is entirely feasible to construct economies in which prices are based on costs and final demands are limited wholly by budget constraints, with demand vectors that are otherwise insensitive to prices. Quantities of goods sold and inventories, not prices, provide the information for coordinating these systems. The Leontief input-output models, with exogenous vectors of final demands, are examples, and the Hawkins-Simon theorem (1949) states the conditions under which such systems have non-negative solutions. They possess the same information-conserving virtues as price-regulated systems (von Hayek 1945). Each actor need only know his or her own business.

> That changes in quantities (e.g., inventories) have as much information as changes in prices is a great observation. The Austrian response might be that firm information about costs is in monetary terms, and so changes in prices are "more useful" information for determining production decisions.

Many observers of business scheduling and pricing practices have claimed that (with the possible exception of the agricultural and mining sectors) models that use quantities as signals approximate first-world national economies more closely than do models in which prices are the principal mechanisms of coordination. I don't wish to argue that point here: but simply observe that quantity adjustments play a very large role in the real world in equilibrating the operations of different organizations and different parts of organizations.

The stylized market exchanges of neoclassical economic theory generally involve only prices and quantities, which is the foundation for their parsimony in information. But actual contracts negotiated between business firms—putting consumer products aside, for the moment—usually specify far more than prices and quantities. Contracts for construction of a building or of a product of engineering (like a generator or an airplane) specify in enormous detail the specifics of the product to be delivered. They require a massive exchange of information in both negotiation and execution. The red market traces that our Martian visitors observed from space are not narrow tracks along which only money and goods flow, but broad highways to accommodate a vast flow of detailed information as well.

These are the thin vs. thick red lines described earlier. I would add that there is a heavy dose of law in these contracts, and so the legal and court systems exert a shadow influence on all the lines in Simon's network.

Thus, the assertion that markets permit each firm to do its business with little knowledge of its partners is a fiction. In construction, in heavy industry, in manufacturing involving high technology, and in other areas, contracting partners carry on communication at a level comparable to the levels observed between departments of a firm. When products are manufactured to specifications, a great deal of information must flow among the various groups of people involved in the manufacture. But the widespread use of subcontracting in the automobile and construction industries, just to mention two, demonstrates that it is often quite feasible to transport this information across organizational boundaries, so that vertical integration is unnecessary. From this perspective, the distinction between market communications and internal communications, and the criteria for choosing between the two alternative arrangements, become correspondingly vague.

The choice between prices or quantities to coordinate the activity levels of different organizations or parts of organizations does not by itself dictate the respective roles of organizations and markets. Prices may be used to coordinate the activities of different parts of single organizations, provided that some way can be found to determine what the market prices should be, and quantity adjustments can be made between different organizations as well as within them.

There is one important difference in the operation of coordination mechanisms within and between organizations. Coordination between organizations depends almost wholly on economic motivations and rewards, and becomes seriously imperfect wherever major externalities are present that cannot be removed by enforceable contract arrangements. Within organizations, on the other hand, identification is a powerful force for combatting externalities produced by attachment to subgoals, by virtue of the loyalty it can produce to the goals of the whole system. A department will be less likely to skimp on quality to cut costs if its members identify with the final product. In particular, identification becomes an important means for removing or reducing those inefficiencies that are labeled by the terms "moral hazard" and "opportunism."

These observations nudge us toward the conclusion that organization size and degree of integration, and the boundaries between organizations and markets, are determined by rather subtle forces. The wide range of organizational arrangements observable in the world suggests that the equilibrium between these two alternatives may often be almost neutral, with the level highly contingent on a system's history. A traditional arrangement may be preserved until its inefficiencies become overwhelming—or even beyond. The same conclusion is suggested by the constant flux of mergers and spinoffs in the business world, many of these transformations being governed by considerations quite unrelated to productive or allocative efficiency, and many having consequences for efficiency that even those involved in them cannot evaluate.

Over a span of years, a large fraction of all economic activity has been gathered within the walls of large and steadily growing organizations. The green areas observed by our Martian have grown steadily. Ijiri and I have suggested that the growth of organizations may have only a little to do with efficiency (especially since, in most large-scale enterprises, economies and diseconomies of scale are quite small), but may be produced mainly by simple stochastic growth mechanisms (Ijiri and Simon 1977).

But if particular coordination mechanisms do not determine exactly where the boundaries between organizations and markets will lie, the existence and effectiveness of large organizations does depend on some adequate set of powerful coordinating mechanisms being available. These means of coordination in organizations, taken in combination with the motivational mechanisms discussed earlier, create possibilities for enhancing productivity and efficiency through the division of labor and specialization.

In general, as specialization of tasks proceeds, the interdependency of the specialized parts increases. Hence a structure with effective mechanisms for coordination can carry specialization further than a structure lacking these mechanisms. It has sometimes been argued that specialization of work in modern industry proceeded quite independently of the rise of the factory system. This may have been true of the early phases of the industrial revolution, but would be hard to sustain in relation to contemporary factories. With the combination of authority relations, their motivational foundations, a repertory of coordinative mechanisms, and the division of labor, we arrive at the large hierarchical organizations that are so characteristic of modern life.

Conclusions

The economies of modern industrialized society can more appropriately be labeled organizational economies than market economies. Thus, even market-driven capitalist economies need a theory of organizations as much as they need a theory of markets. The attempts of the new institutional economics to explain organizational behavior solely in terms of agency, asymmetric information, transaction costs, opportunism, and other concepts drawn from neoclassical economics ignore key organizational mechanisms like authority, identification, and coordination, and hence are seriously incomplete.

The theory presented here is simple and coherent, resting on only a few mechanisms that are causally linked. Better yet, it agrees with empirical observations of organizational phenomena. Large organizations, especially governmental ones, are often caricatured as "bureaucracies," but they are

Game theory and behavioral economics have filled out a lot of what Simon saw as missing from economics. Authority in the labor market stems from moral hazard, identification with an organization comes from identity and reciprocity, and coordination comes from shared expectations and information. I do wonder if Simon would agree that progress has been made.

often highly effective systems, despite the fact that the profit motive can penetrate these vast structures only by indirect means.

This theory of organizations calls for reexamining some of the classical questions of political economy. The primacy of profit as the enforcer of organizational efficiency is replaced by organizational goals, combined with organizational identifications and with material rewards and supervision, all of which motivate employees to work toward these goals. This framework makes it necessary to reopen the question of when profit-making, nonprofit, and governmental organizations should be expected to operate well, and when market competition is needed to discipline organizations to perform efficiently.

The reopening of these questions is important for both capitalist and socialist economies. On the one side, capitalist economies are actually mixed economies, faced with a multitude of problems of regulation and deregulation, of socialization and privatization. On the other side, many socialist economies have had mediocre success in maintaining the efficiency of their organizations, and are experimenting with the reintroduction of markets, often while trying to avoid extensive privatization. Good answers to the policy questions that face all industrialized societies depend on having empirically sound theories of the behavior of large organizations. Such theories cannot be developed from the armchair. They call for fact-gathering that will carry researchers deep into the green areas, the organizations, that dominate the terrain of our economic systems.

Simon here is asking economists to throw out profit maximization as both an objective of firms and a market pressure. Surely some form of this is right, but it is also the case that capital markets exert pressures on organizations to produce profits.

As I write this, the comparison in postsocialist trajectories between Russia and China looms charge due to current events. I wonder what a Simon-esque take on Chinese and Russian economic development would look like: Where does the Chinese Communist Party show up in Simon's map of the economy? Where do the Russian oligarchs?

REFERENCES

Berle, A. A., and G. C. Means. 1933. *The Modern Corporation and Private Property.* New York, NY: Macmillan.

Cyert, R. M., and J. G. March. 1963. *A Behavioral Theory of the Firm.* Englewood Cliffs, NJ: Prentice Hall.

Dearborn, D. C., and H. A. Simon. 1958. "Selective Perception: The Identifications of Executives." Reprinted in Simon, Administrative Behavior, chapter 15, *Sociometry* 21:140–144.

Demsetz, H., and K. Lehn. 1985. "The Structure of Corporate Ownership: Causes and Consequences." *Journal of Political Economy* 93:1155–1177.

Devons, E. 1950. *Planning in Practice.* Cambridge, UK: Cambridge University Press.

Hawkins, D., and H. A. Simon. 1949. "Note: Some Conditions of Macroeconomic Stability." *Econometrica* 17:245–248.

Ijiri, Y., and H. A. Simon. 1977. *Skew Distributions and the Sizes of Business Firms.* Amsterdam, Netherlands: North Holland.

Kornai, J. 1971. *Anti-Equilibrium.* Amsterdam, Netherlands: North Holland.

Krafcik, J., and J. P. Womack. 1987. *Comparative Manufacturing Practice: Imbalances and Implications.* Appendix A, working paper International Motor Vehicle Program, MIT, May 1987.

March, J. G., and H. A. Simon. 1958. *Organizations.* New York, NY: Wiley.

Ross, S. 1973. "The Economic Theory of Agency: The Principal's Problem." *American Economic Review* 63:134–139.

Simon, H. A. 1947. *Administrative Behavior.* 3rd (1976). New York, NY: Macmillan.

———. 1951. "A Formal Theory of the Employment Relationship." Reprinted in Simon, H. A., Models of Bounded Rationality, Vol. II, Chapter 5.2. Cambridge, UK: MIT Press, 1982, *Econometrica* 19:293–305.

———. 1979. "Rational Decision Making in Business Organizations." *American Economic Review* 69:493–513.

———. 1983. *Reason in Human Affairs.* Stanford, CA: Stanford University Press.

———. 1990. "A Mechanism for Social Selection and Successful Altruism." *Science* 250:1665–1668.

Stiglitz, J. E. 1974. "Incentives and Risk-Sharing in Sharecropping." *Review of Economic Studies* 41:219–255.

von Hayek, F. A. 1945. "The Use of Knowledge in Society." *American Economic Review* 35:519–530.

Weisbrod, B. A. 1988. *The Nonprofit Economy.* Cambridge, MA: Harvard University Press.

———. 1989. "Rewarding Performance that is Hard to Measure: The Private Nonprofit Sector." 244:541–546.

Williamson, O. E. 1975. *Markets and Hierarchies.* New York, NY: The Free Press.

———. 1985. *The Economic Institutions of Capitalism.* New York, NY: The Free Press.

Winter, S. 1964. "Economic 'Natural Selection' and the Theory of the Firm." *Yale Economic Essays* 4:225–272.

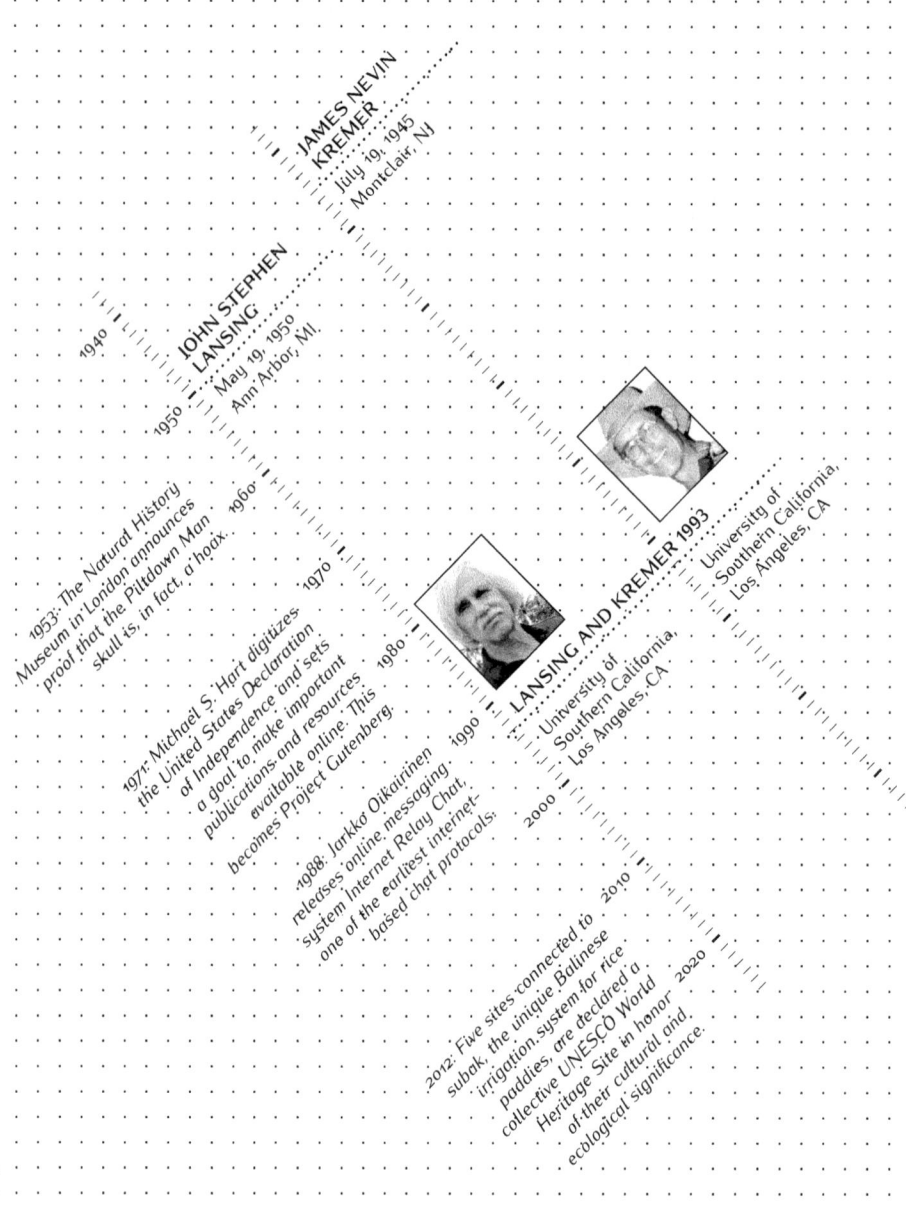

JAMES NEVIN KREMER
July 19, 1945
Montclair, NJ

JOHN STEPHEN LANSING
May 19, 1950
Ann Arbor, MI

1953: The Natural History Museum in London announces proof that the Piltdown Man skull is, in fact, a hoax.

1971: Michael S. Hart digitizes the United States Declaration of Independence and sets a goal to make important publications and resources available online. This becomes Project Gutenberg.

1988: Jarkko Oikarinen releases online messaging system Internet Relay Chat, one of the earliest internet-based chat protocols.

LANSING AND KREMER 1993
University of Southern California, Los Angeles, CA

University of Southern California, Los Angeles, CA

2012: Five sites connected to subak, the unique Balinese irrigation system for rice paddies, are declared a collective UNESCO World Heritage Site in honor of their cultural and ecological significance.

LANSING & KREMER

[75]

EVOLVING CENTRALIZED INSTITUTIONS FROM THE BOTTOM UP

*Mirta Galesic, Santa Fe Institute
and Complexity Science Hub Vienna*

The central question of this paper has never been more important: How do we organize and coordinate our collective action to increase joint benefits and reduce harms? This question is particularly critical now, when institutions we took for granted seem to be failing, from communication norms and epistemic frameworks to democratic elections and independent courts, even as our natural and social environments keep on rapidly and irreversibly changing. How do we develop new institutions to manage our collective response to catastrophic weather events, biodiversity loss, global pandemics, spread of misinformation, and extensive migration? Can we rely on bottom-up self-organization of autonomous individuals or do we need an externally imposed, top-down governance?

In this seminal 1993 paper, Stephen Lansing and James Kremer describe a wonderful example of the success of bottom-up organization. Groups of farmers growing rice on colorful terraces of Bali need to solve a tradeoff between water shortage and pest damage. If they all plant rice at the same time, there might not be enough water for all farmers, especially for those farther from the central water source. But if they stagger their planting schedule to avoid water shortages, another problem arises: pests. If newly planted fields are always available, pests can migrate from one field to another, causing a lot of damage. To avoid this, farmers should ideally all plant at the same time. Pests would do some damage, but after harvest there would be a long period when there would be nothing for them to eat and they would perish.

We learn from Lansing and Kremer that Balinese farmers minimize both water shortages and pest damage by an intricate system of coordinated crop schedules. The number of fields that are planted, harvested, flooded,

J. S. Lansing and J. N. Kremer, "Emergent Properties of Balinese Water Temple Networks: Coadaptation on a Rugged Fitness Landscape," *American Anthropologist* 95 (1), 97–114 (1993).

Used with permission of American Anthropologist; permission conveyed through Copyright Clearance Center, Inc.

or drained at the same time can self-organize to minimize both problems and enable good yields. This physical coordination of production between farmers is maintained by a symbolic structure of religious rituals performed at water temples placed throughout the water system, by regular meetings of the whole community, and by using elaborate calendrical systems. The rituals such as mixing of holy water from different parts of the water system serve as a collective memento of the interdependency between farmers, the meetings enable occasional adjustments of cropping schedules to reflect climate and pest trends, and the calendars are physical artifacts that allow for monitoring and execution of precisely coordinated farming activities.

In the Balinese example, the physical production and the social organization are intertwined in an elaborate sociophysical system that has developed through self-organization. The religious practices are not simply a commentary on the physical structures sensu Marx. They are the institutional structure that enables the physical production. They transcend other demarcation lines of the local communities, crossing the lines of former kingdoms, and operate outside governmental control.

Although self-organization is sometimes mistaken for a lack of governance, this work shows that self-organization can evolve centralized institutions, in this case the worship of the Goddess of Waters and the recognition of her human representative, high priest Jero Gde (as described in Lansing's 1991 book *Priests and Programmers*). The point is that centralized institutions needed to manage the commons can be evolved from the bottom up rather than being imposed to the system from the outside.

Lansing and Kremer were not the only ones at that time who noticed that commons can evolve jointly beneficial institutions from the bottom up. For example, Elinor Ostrom (1990) and her colleagues have described a number of case studies in which communities developed institutions that prevent overexploitation of resources. However, while self-organization is possible, no single solution is a panacea, and what works for one community might not work for others (Ostrom, Janssen, and Anderies 2007). Prior to Lansing and Kremer's paper, there were no computational models that would enable rigorous exploration of the conditions for self-organization.

And this is why Lansing and Kremer's paper is foundational: it not only offers an excellent anthropological analysis of Balinese water temples, but it also develops a quantitative, computational model that shows how such systems could have evolved and under what conditions. The model demonstrates that a simple social learning mechanism—copying the cropping schedules of successful neighbors—changes future payoffs of different options (cropping schedules) and generates a structure of farming associations (subaks) that is very similar to the one observed in reality.

This paper opened doors to a more rigorous quantitative exploration of complex, realistic social systems. Before it, human complex social systems were often described in qualitative terms, but nobody had tried to build them from the ground up, starting from a population of individuals following simple rules in realistic natural environments. Computational models existing at that time, such as genetic algorithms and cellular automata (Langton *et al.* 1991), were theoretically illuminating but very abstract and mostly not applicable to any specific human social system. Although there were a few applications in economics, major treatments of how agent-based models could be used to model human societies were yet to be published (Epstein and Axtell 1996; Kohler and Gumerman 2000; Miller and Page 2007).

Lansing and Kremer offer a blueprint for building computational models of real-world societies: a thorough qualitative understanding of the system combined with a simple model that captures main empirical trends. This quantitative portrait of a real-world situation not only enables understanding and exploration of the underlying mechanisms but also allows scientists from other disciplines to engage with the data and compare modeling approaches with different underlying assumptions. For example, the model presented in this paper has been revisited from the perspective of economics (Lansing and Miller 2005), ecology (Lansing *et al.* 2017), genetics (Lansing and Cox 2019), evolutionary biology (Lansing *et al.* 2021), and statistical physics (Gandica *et al.* 2021).

As with every good piece of science, this paper and the model raise many further questions. For example, when is such self-organization preferable, and when is it more efficient to use external interventions to manage the system? How much does this depend on clear, jointly agreed-upon payoffs such as crop yields in the Bali example vs. more vaguely defined and often conflicting payoffs in problems we are facing today, such as climate change and misinformation? What is the role of the time available to societies to solve a problem, and of costs and irreversibility of failure?

Lansing and Kremer's key insight became the point of departure for many subsequent studies, showing that human–environment interactions at larger spatial and temporal scales are an emergent property of coevolved social-ecological systems maintained by short-term benefits to individual agents (Goldstone and Janssen 2005; Bliege Bird 2015; Downey, Haas Jr., and Shennan 2016; Moritz *et al.* 2018). These models can have important real-world impacts. In 2012, thanks in large part to Lansing and Kremer's work, which pointed out the importance of the traditional water temple networks, the Balinese subak system was recognized as a UNESCO World Heritage Site.

REFERENCES

Bliege Bird, R. 2015. "Disturbance, Complexity, Scale: New Approaches to the Study of Human–Environment Interactions." *Annual Review of Anthropology* 44:241–257. https://doi.org/10.1146/annurev-anthro-102214-013946.

Downey, S. S., W. R. Haas Jr., and S. J. Shennan. 2016. "European Neolithic Societies Showed Early Warning Signals of Population Collapse." *Proceedings of the National Academy of Sciences* 113 (35): 9751–6. https://doi.org/10.1073/pnas.1602504113.

Epstein, J. M., and R. Axtell. 1996. *Growing Artificial Societies: Social Science from the Bottom Up*. Cambridge, MA: MIT Press and Brookings Institution Press.

Gandica, Y., J. S. Lansing, N. N. Chung, S. Thurner, and L. Y. Chew. 2021. "Bali's Ancient Rice Terraces: A Hamiltonian Approach." *Physical Review Letters* 127:168301.

Goldstone, R. L., and M. A. Janssen. 2005. "Computational Models of Collective Behavior." *Trends in Cognitive Sciences* 9:424–430.

Kohler, T. A., and G. G. Gumerman, eds. 2000. *Dynamics in Human and Primate Societies: Agent-Based Modeling of Social and Spatial Processes*. Oxford, UK: Oxford University Press.

Langton, C. G., C. Taylor, J. D. Farmer, and S. Rasmussen. 1991. *Artificial life II*. Redwood City, CA: Addison–Wesley Longman Publishing Co., Inc.

Lansing, J. S., N. N. Chung, L. Y. Chew, and G. S. Jacobs. 2021. "Averting Evolutionary Suicide from the Tragedy of the Commons." *International Journal of the Commons* 15.

Lansing, J. S., and M. P. Cox. 2019. *Islands of Order*. Princeton, NJ: Princeton University Press.

Lansing, J. S., and J. Miller. 2005. "Cooperation, Games, and Ecological Feedback: Some Insights From Bali." *Current Anthropology* 46:328–334.

Lansing, J. S., S. Thurner, N. N. Chung, A. Coudurier-Curveur, Ç. Karakaş, K. A. Fesenmyer, and L. Y. Chew. 2017. "Adaptive Self-Organization of Bali's Ancient Rice Terraces." *Proceedings of the National Academy of Sciences* 114:6504–9.

Miller, J. H., and S. E. Page. 2007. *Complex Adaptive Systems: An Introduction to Computational Models of Social Life*. Princeton, NJ: Princeton University Press.

Moritz, M., R. Behnke, C. M. Beitl, R. Bliege Bird, R. M. Chiaravalloti, J. K. Clark, S. A. Crabtree, *et al*. 2018. "Emergent Sustainability in Open Property Regimes." *Proceedings of the National Academy of Sciences* 115:12859. https://doi.org/10.1073/pnas.1812028115.

Ostrom, E. 1990. *Governing the Commons: The Evolution of Institutions for Collective Action*. Cambridge, UK: Cambridge University Press.

Ostrom, E., M. A. Janssen, and J. M. Anderies. 2007. "Going Beyond Panaceas." *Proceedings of the National Academy of Sciences* 104:15176–8.

EMERGENT PROPERTIES OF BALINESE WATER TEMPLE NETWORKS: COADAPTATION ON A RUGGED FITNESS LANDSCAPE

J. Stephen Lansing
and James N. Kremer

Abstract

For over a thousand years, generations of Balinese farmers have gradually transformed the landscape of their island, clearing forests, digging irrigation canals, and terracing hillsides to enable themselves and their descendants to grow irrigated rice. Paralleling the physical system of terraces and irrigation works, the Balinese have also constructed intricate networks of shrines and temples dedicated to agricultural deities. Ecological modeling shows that water temple networks can have macroscopic effects on the topography of the adaptive landscape, and may be representative of a class of *complex adaptive systems* that have evolved to manage agroecosystems.

Compared to many computational modeling papers today, this paper develops its arguments much more thoughtfully and extensively, situating the model deeply in the existing debates in anthropology and in social sciences more broadly.

In 1984, Eric Alden Smith published a devastating critique of the uses of systems ecology and simulation modeling in anthropology. While this article is in part a defense of these methods, we do not take issue with any of Smith's conclusions. Instead, we hope to demonstrate that systems models can serve a different heuristic purpose than the naive functionalist, energy-maximization or group-selection models skillfully demolished by Smith. In particular, we hope to show that simulation models are uniquely appropriate for addressing the issues of adaptation and determinism in the development of complex social systems like the water temples of Bali. But before we turn to the uses of simulation models, it may be useful to sketch out how our approach differs from those criticized by Smith.

Although simulation models have always been a rarity in anthropology, they continue to be used extensively in biology as a tool to investigate complex interactive processes. For example, we recently served on the doctoral committee of a graduate student who was interested in the growth of algae in Antarctic sea ice, a major source of fixed carbon in the Antarctic

Ocean. The student built a model to study the interactive effects of processes thought to influence the growth of the algae, such as temperature, nutrient flow, and available sunlight. The result was a system of differential equations that predicted, on purely theoretical grounds, variations in the growth of algae depending on the relationships among these causal factors. The model's predictions were then compared with observations, helping the student fine-tune his understanding of the mechanistic processes that drive the growth of the algae (Arrigo 1991).

However, an obvious problem in extending this kind of analysis from biology to anthropology is that natural ecosystems evolve through a process of "blind" natural selection, while the systems of most interest to anthropologists are by definition shaped by conscious human intentions. Life in the sea ice of Antarctica is thought to have evolved opportunistically through the random effects of natural selection. On the other hand, centuries-old Balinese rice terraces would cease to function in a matter of days if Balinese farmers stopped managing them. It is precisely the introduction of human agency into natural ecologies that blunts the tools designed for the study of mindless processes.

Marx drew a distinction between two models of "nature" that is relevant to the point we are trying to make. On the one hand, according to Marx, there is "external nature," or nature apart from society: remote islands, or distant galaxies. Of greater interest to the social theorist is "humanized nature": those portions of the natural world that have been shaped by human intention (Habermas 1971: 34). Significantly, this distinction does not depend on the presence or absence of people in an ecosystem, but rather on whether the natural system in question has been purposefully modified by human activities. For example, one of the unexpected results of the model of Antarctic sea ice was the realization that human society may soon have a significant negative impact on the Antarctic food chain. The reason is that global warming may shorten the growing season for algae by several weeks. From a methodological standpoint, there is no difficulty in modeling this effect in a computer simulation: it can be expressed in terms of a seasonal change in the rate of light absorption. But a new and different set of questions emerges when we turn from an ecosystem that has evolved through natural selection to one that has also been deliberately modified to suit human purposes. The question of

And yet, the concept of purpose might not have much explanatory power when explaining human societies either. The model described later seems to work even without assuming that agents have a conscious purpose to achieve good yields—at least not more than algae are assumed to strive for growth.

purpose is meaningless in the context of the growth of algae in Antarctic sea ice, but essential to understanding the growth of rice in an irrigated terrace. In the latter case we are confronted with Hegel's "active practical reality of consciousness" shaping a natural landscape over the course of many generations.

"It is as clear as noon-day," Marx wrote in 1844, "that man, by his industry, changes the forms of the materials furnished to him by Nature, in such a way as to make them useful to him" (Marx 1961 [1844]: 71). So much may be obvious, but Marx added an important corollary: in the process of reshaping nature, society gradually reshapes itself. Anthony Giddens neatly summarizes this point: "Marx emphasizes that social development must be examined in terms of an active interplay between human beings and their material environment" (Giddens 1981: 59). A similar insight led Fernand Braudel to formulate the concept of the "structures of the longue durée," although unlike Marx, Braudel emphasized the passivity of human societies held nearly immobilized by their environments:

> *For centuries, man has been a prisoner of climate, of vegetation, of the animal population, of a particular agriculture, of a whole slowly established balance from which he cannot escape without the risk of everything's being upset.... There is the same element of permanence or survival in the vast domain of cultural affairs.*
> (Braudel [1969]1980: 31)

Yet despite their differing perspectives on the active relationship of societies to the natural world, on a deeper level both Marx and Braudel agree that the intentionality of individual social actors, or even whole generations, is insufficient to explain the historical evolution of modes of production. Instead, both posit a historical evolution of consciousness as societies gradually reshape the natural world. Humanized nature is actively transformed by social action, and gradually acquires a purposive structure, which sharply distinguishes it from the universe of "external nature." This distinction is particularly relevant when we address the question of change. If we could rewind an imaginary videotape of the evolution of life, as Stephen Jay Gould observes in *Wonderful Life: The Burgess Shales and the Nature of History*, each sequence would show us not design not contingency. The paleontologist cannot predict which species will thrive, and which become extinct, because the process of change is always local, contingent, and unpredictable:

> *The divine tape player holds a million scenarios, each perfectly sensible. Little quirks at the outset, occurring for no particular reason, unleash cascades of consequences that make a particular future seem inevitable in retrospect. But the slightest early nudge contacts a different groove, and history veers into another plausible channel, diverging continually from its original pathway.* (Gould 1989: 320–321)

But consider a tape showing the historical evolution of irrigation systems, rice terraces, and water temples in Bali over the past millennium. We would witness the engineering of the landscape, as generations of Balinese farmers cleared forests, dug irrigation canals, and terraced hillsides to enable themselves and their descendants to grow irrigated rice. We might also see false starts, abandoned irrigation works, and conflicts between groups of farmers. But over time, the historical record would show us precisely what is missing from Gould's *pre*historic record: the traces of conscious design. We would see the results of a stochastic process in which the realization of each generation's plans changed the world for their descendants.

In this article we try to show how the techniques of ecological simulation modeling can help to illuminate this historical process. The major difference between the modeling process that we describe, and the sorts of models used by biologists, is that here our interest focuses on the effects of human agency in reshaping ecosystems. History only happened once, of course, but with simulation models we can "rewind the tapes" and investigate the consequences of changing social and ecological parameters. This approach can help avoid one of the most common pitfalls in materialist approaches to social theory: the assumption that whatever social institutions happen to exist at a particular place and time are the deterministic results of environmental circumstances. For while the evolution of productive systems like irrigation networks is undoubtedly shaped by material constraints, it does not follow that such constraints mandate a specific set of cultural and ecological responses.

The need for an approach that will enable us to analyze the effects of interactions between social and ecological variables over time is the reason for our disagreement with Eric Smith. Smith objects to simulation models on the grounds that they appear to be teleological, depicting ecosystems as self-regulating or functionally integrated (1984: 66–70). In preference to systems ecology, which tries to study whole ecosystems, Smith urges us to adopt the perspective of evolutionary ecology, which focuses on calculations of the "fitness value" or payoff of specific behavioral strategies to individual social actors. But successful applications of evolutionary ecology in anthropology have been confined to studies of hunting-and-gathering societies, where arguably we are concerned with "external" rather than "humanized" nature. In this article, our major goal is to comprehend the emergence of cooperative behavior among Balinese farmers. Here, the "fitness value" or payoff of different farming strategies changes as a result of complex interactions between irrigation networks and the domesticated ecology of the rice terraces. To foreshadow the most interesting results, a spontaneous process of self-organization occurred when we allowed water temples to react to changing environmental conditions over time in a simulation model. Artificial cooperative networks appeared that bore a very close resemblance to actual temple networks in the study area. As these networks formed, average harvest yields rose to a new plateau. Subsequently, irrigation systems organized into artificial temple networks were able to withstand ecological perturbations (such as pest outbreaks or drought) much better than in otherwise identical models that lacked temple networks. It appears, then, that networks of water temples may have a definite structure, which leads to higher sustained productivity than would be the case if they were randomly ordered. Further, these structures can emerge without conscious planning, through a stochastic process of coadaptation. Thus water temple networks may represent a hitherto unnoticed type of social organization: a self-organizing managerial system, shaped by a process of coadaptation on a *rugged fitness landscape* (Palmer 1991; Kauffman and Johnsen 1991).

We begin with a brief analysis of the ecological role of water temples and a description of a simulation model we developed to explore the ecological role of water temples along two rivers. Tests of the model's predictions against two years of historical data added empirical support to the theoretical argument in *Priests and Programmers* (Lansing 1991: 117–126) that water temples optimize rice harvests. In the second half of the article, we turn to the question of how the water temple networks manage to find optimal or near-optimal scales of coordination in water management. The entire ecological simulation model described in the first part of the article becomes a single time-step in a nonlinear model designed to explore the relative importance of trial and error versus conscious design in the evolution of temple networks. Finally, we return to the methodological issues raised by Smith.

The Ecological Role of Water Temples

In Bali, rice is grown in paddy fields fed by irrigation systems dependent on rainfall (there are no storage dams in the rivers). Rainfall varies by season and elevation and, in combination with groundwater inflow, determines river flow. Traditional Balinese irrigation systems begin with a weir in a river, which shunts all or part of the flow into a tunnel that emerges some distance downstream, at a lower elevation, where the water is routed through a system of canals and aqueducts to the summit of a terraced hillside. Thus, the flow of water into each farmer's fields depends on the seasonal flow of water in the rivers and streams, which in turn depends on rainfall and groundwater flow.

The role of water in the rice paddy ecosystem goes far beyond providing water to the roots of the rice plants. By controlling the flow of water into the terraced fields, farmers are able to create pulses in several important biogeochemical cycles. The cycle of wet and dry phases alters soil pH, induces a cycle of aerobic and anaerobic conditions in the soil that determines the activity of microorganisms, circulates mineral nutrients, fosters the growth of nitrogen-fixing cyanobacteria, excludes weeds, stabilizes soil temperature, and over the long term governs the formation of a plough pan that prevents nutrients from being leached into the subsoil. On a larger scale, the flooding and draining of blocks of terraces also has important effects on pest populations. If farmers with adjacent fields can synchronize their cropping patterns to create a uniform fallow period over a sufficiently large area, rice pests are temporarily deprived of their habitat and pest populations can be sharply reduced. How large an area must be fallow, and for how long, depends on the species characteristics of the rice pests. However, if too many farmers follow an identical cropping pattern in an effort to control pests, they will experience peak irrigation demand at the same time, and there might not be enough water for all, especially because the distance between weirs on Balinese rivers is usually only a few kilometers. Water sharing and pest control are thus opposing constraints, and the optimal scale for the coordination of cropping patterns depends on local conditions.

Paralleling the physical system of terraces and irrigation works, the Balinese have also constructed intricate networks of shrines and temples dedicated to agricultural deities and the Goddess of Waters. These temples play an instrumental role in the productive process by providing farmers with a structure to coordinate cropping patterns and the phases of agricultural labor. An analysis of the "ritual technology" that makes this possible is beyond the scope of this article (see Lansing 1987, 1991), but the ecological effects of temple networks can be clearly seen in the relationship between neighboring irrigation systems in the upper reaches of the Petanu River in southern Bali (Figure 1).

As the map indicates, the Bayad weir provides water for 100 hectares of rice terraces organized as a single *subak*, or farmer's association. A few kilometers downstream from the Bayad weir, the Manuaba weir provides water for 350 hectares or terraces, organized into ten subaks. The water temple hierarchy at Bayad consists of a weir-altar (*pura ulun empelan*) and a "Head of the Ricefields" temple *(pura ulun swi)* situated above the terraces. The larger Manuaba system also begins with a weir-altar, but includes two Ulun Swi temples, one for each major block of terraces. The congregations of both Ulun Swi temples also belong to a larger Masceti temple that is symbolically identified with the entire Manuaba irrigation system. Representatives of all ten subaks meet once a year at the Masceti temple to decide on a cropping pattern. Subsequently, the ritual calendar carried out at the Ulun Swi temples provides a template for the phases of agricultural labor.

For example, all ten subaks belonging to the Masceti temple planted IR 64, a high-yielding Green Revolution rice, in mid-September 1988 and harvested an average of 6.5 tons per hectare in mid-December. Subsequently, they planted kruing (another high-yielding rice) in early February and harvested 6 tons/ha in May. In June, they all planted vegetables and harvested approximately 2 tons/ha in August. During this time period, the flow of irrigation water into the Manuaba weir was as

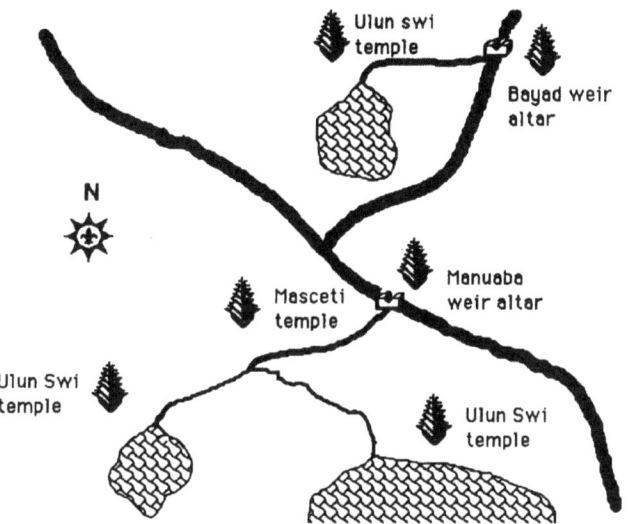

Figure 1. Bayad and Manuaba irrigation systems (from Lansing 1991: 60).

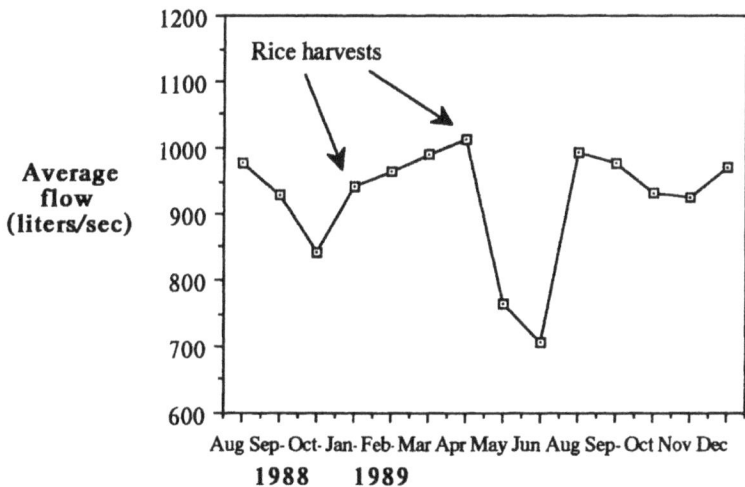

Figure 2. Flow at the Manuaba weir (1988–89).

shown in Figure 2. This cropping pattern synchronized harvests for all ten subaks, encompassing 350 hectares of rice terraces, thereby possibly helping to keep down pest populations. Pest infestations for this period were reported to be minimal: less than 1% damage to the crops, primarily from brown planthoppers. This compares to pest losses of up to 50% of the crop in the late 1970s, when each subak planted rice continuously and cropping patterns were very disorganized (Lansing 1991: 112–117). However, the average flow of approximately 3 liters per second per hectare is less than the recommended average flow of 5 liters/sec/ha, and suggests that the crops may have experienced some water stress. Certainly there was never any excess water. In that light, it is interesting to note that the Bayad subak upstream followed exactly the same cropping pattern as Manuaba, except that they began two weeks earlier. In general, irrigation demand is highest at the beginning of a new planting cycle, because the dry fields must become saturated. By starting two weeks after their upstream neighbors, the Manuaba subaks could help avoid water shortages at the time when irrigation demand peaks.

There would thus appear to be good ecological reasons for the Manuaba subaks to coordinate their cropping patterns with their upstream neighbors. The Bayad subak might also find that it is in their interest to coordinate their fallow periods with Manuaba, so as to keep down pest populations. As it happens, the Manuaba subaks regularly send a delegation to the Bayad weir to request holy water, and the interdependency of the two irrigation systems is given symbolic expression by ritual ties between the deities of the two weirs.

A Simulation Model of Two Rivers

In the case we have just considered, the water temples play their part by helping the subaks to balance two opposing constraints: water sharing and pest control. The effect of these constraints varies by location. In 1988, we built a simulation model to explore the effects of synchronized cropping patterns along two entire rivers, the Oos and the Petanu (Kremer and Lansing, 1992a, 1992b).

The model allowed us to simulate the effects of coordination by water temples under varying ecological conditions, and also to simulate other possible levels of coordination. At one extreme, all subaks follow

> The modeling was conducted in two phases. In the first phase, described in this section, the authors build a model of the current sociophysical system they are trying to explain, with basic relationships between the amount of water, pest levels, social organization, and yields.

exactly the same cropping pattern; at the other, each subak sets its own unique cropping pattern. The actual water temple scale of coordination lies between these extremes.

The watershed of the Oos and Petanu rivers includes approximately 6,136 hectares of irrigated rice terraces (Figure 3). Based on topographical maps, we divided the Oos-Petanu watershed into 12 subsections, specifying the catchment basins for each weir for which hydrological data was available. For each of the 172 subaks located in these basins, we specified the name, the area, the basin in which it resides, the weir from which it receives irrigation water, and the weir to which any excess is returned. We also defined the real spatial mosaic connecting these subaks. Given this geographical setting, the program simulates the rainfall, river flow, irrigation demand, rice growth stage, and pest levels for all watersheds and all subaks. At the appropriate times, the harvest is adjusted for cumulative water stress and pest damage, yields are tallied, and the next crop cycle is initiated (Kremer 1991).

In order to test the predictions of this model, in 1989 Lansing began to work with a team of Balinese students to gather real data on rainfall, irrigation flow, crop yields, water stress, and pest damage. As Figure 4 shows, there was a great deal of variation in actual harvest yields reported by the subaks (a point we return to later on). Subsequently, we loaded the model with rainfall records based on real monthly averages, assigned each subak a cropping pattern that approximated its real cropping pattern, and compared the results with the actual distribution of yields (see Figure 5). Considering the simplicity of the model, yields per hectare were also well correlated, with $\tau = 0.5$. To assess the possibility that model results were simply not very responsive to variations in cropping plans, we ran additional simulations in which we disrupted the local coordination implicit in the planting schedules followed by the subaks in 1989. When planting dates were randomized, but the actual crops planted remained the same, the correlation for the second crop in 1989 dropped from 0.50 to 0.01 (Kremer and Lansing, 1992b).

Comparison of the effects of different scales of coordination by the subaks for many simulation runs showed that the scale of coordination that most closely resembles the actual pattern of water temple control

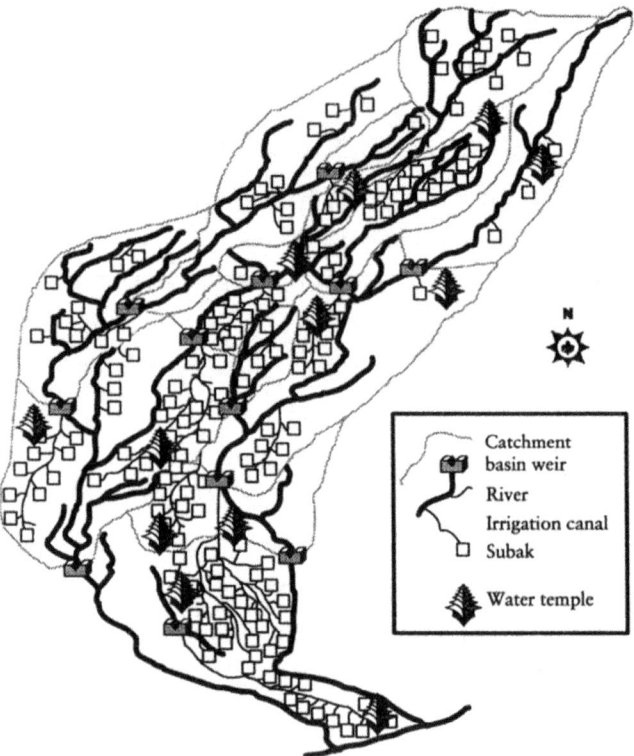

Figure 3. The Oos and Petanu rivers in south-central Bali (not to scale; from Lansing 1991: 119).

achieves the highest rice yields by optimizing the trade-off between water sharing and pest control, as shown in Figure 6.

On the far right in Figure 6, when the cropping pattern is set by each individual subak, there is high pest damage. On the far left, a single cropping pattern for the whole watershed reduces pest damage but maximizes water stress. The highest peak is achieved by the scale of coordination that most closely approximates the temple scale of coordination. These results suggest two initial conclusions: (1) most of the observed variation in harvest yields is explainable by reductions in yields caused by water stress or pest damage (Figure 5); and (2) the scale of social coordination in the management of irrigation has important effects on both of these variables (Figure 6; Kremer and Lansing 1992b).

Lansing and Kremer (1993)

Modeling Adaptation on a Rugged Fitness Landscape

These results encouraged one of us (Lansing) to shift his attention from the mechanisms at work at the level of individual water temples, to the possible existence of system-level properties of temple networks. Our approach to this question borrows from the theory of "fitness landscapes" in biology (Wright 1932; Palmer 1991; Kauffman and Johnsen 1991). As we have just seen, it is possible to calculate theoretical optima for rice production, and then compare these results to the actual system of temple coordination along each river. In biology, the idea of a fitness landscape is based on the notion that the fitness of an organism (species, population) depends not only on its own intrinsic characteristics ("genotype") but also on its interaction with a local environment. The term "landscape" comes from visualizing a geographical landscape of fitness "peaks," where each peak represents an adaptive solution to a problem of optimization. Figure 6 may therefore be viewed as an example of a fitness landscape, in which the highest fitness peak is achieved by the temple scale of coordination. Note, however, that in biological models optimization occurs through "blind" natural selection, whereas here we propose a different mechanism: the deliberate efforts of farmers to cooperate in setting cropping patterns so as to maximize their harvests.

In the second modeling phase, the authors investigate how this sociophysical system adapts over time and how it manages to find the appropriate level of coordination. Now it gets really interesting!

But it was not yet clear from our earlier analysis how the temple networks manage to find an optimal scale of coordination in the management of irrigation. Are differences in the structure of temple networks from one river to the next the result of deliberate planning by farmers, priests, or royal engineers? Or are they the product of trial-and-error adjustments by generations of farmers? Could the water temple network of River A do as well managing River B, or is each temple network a uniquely optimal solution to the specific ecological conditions of a particular region? And, finally, how do temple networks respond to changing conditions?

The concept of rugged fitness landscapes continues to inspire much of the work on modeling complex social systems today. It is very suitable for investigations of how people manage tradeoffs between exploration and exploitation of their environments.

A different type of simulation analysis provides a way to investigate such questions, which relate to the dynamic behavior of temple networks. Kremer's model (described above) became a single time-step in a program developed by Lansing to explore the process of adaptation on a changing fitness landscape. Imagine that the water temple system

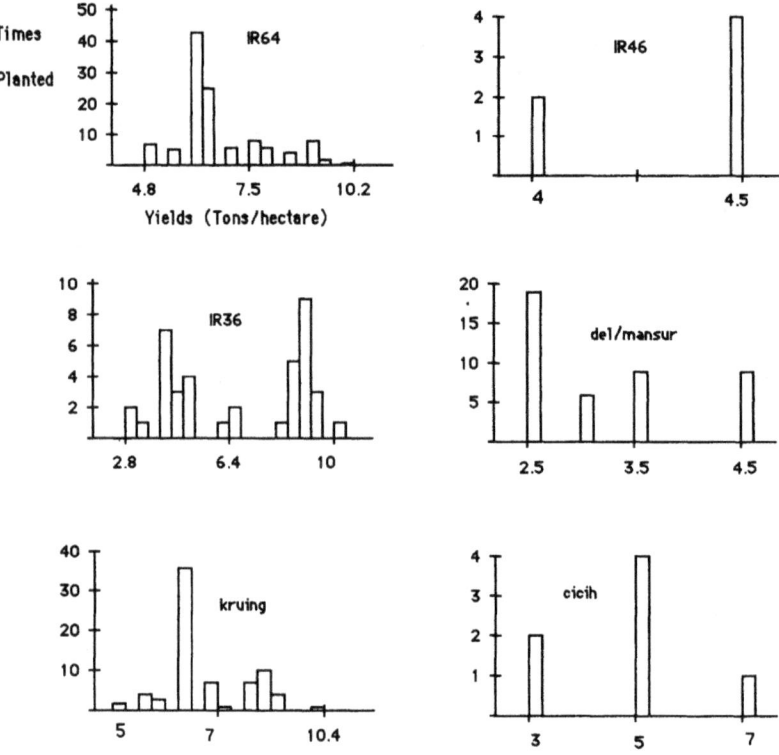

Figure 4. Distribution of rice harvest yields (tons/hectare) for six varieties planted in 1988-89.

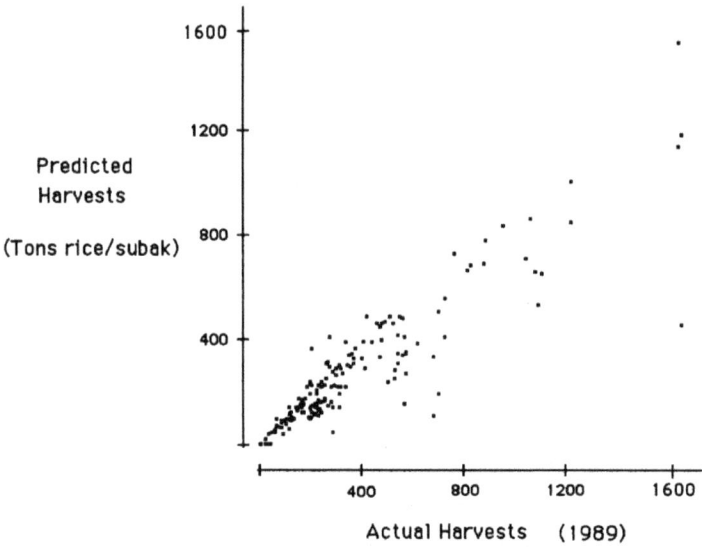

Figure 5. Comparison of model predictions with actual rice harvests by subak (1989). Pearson's product-moment correlation $r = .90$ (from Kremer and Lansing 1992b).

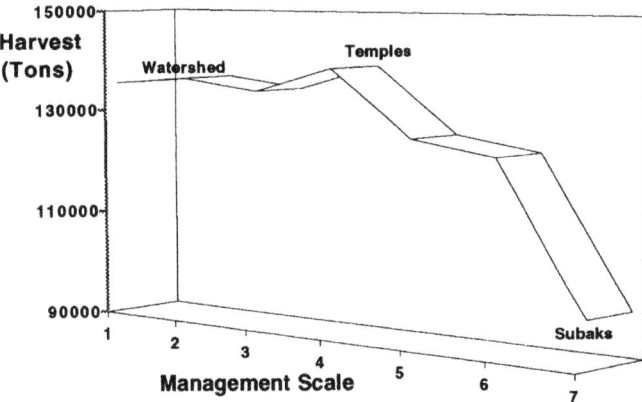

Figure 6. Effects of seven different scales of social coordination on rice yields, pest damage, and water stress.

does not exist, but that all the known ecological conditions remain unchanged along both rivers. As a new year begins, each of the 172 subaks in the model begins to plant rice or vegetables. At the end of the year, harvest yields are calculated for each subak. Subsequently, each subak checks to see whether any of its neighbors got higher yields. If so, the target subak copies the cropping pattern of its (best) neighbor. The model then simulates another year of growth, tabulates yields, and continues to run until each subak has reached its local optimum. What will happen?

Figure 7 shows the results of such a simulation for the traditional Balinese cropping pattern, *kerta masa*. This cropping pattern begins with a long-maturing rice variety (del or mansur), followed by a fallow period, and a faster-maturing second rice crop (cicih). In this simulation, all subaks follow this cropping pattern. However, starting dates (when to begin planting the first crop) are randomly assigned to each subak.

After the first run, the average yield (tons rice/hectare/year) for the subaks was slightly more than 5 tons. Each subak then compared its yield with those of its four closest neighbors. Eighty-six subaks discovered that one of their neighbors had a higher yield, and copied their neighbor's cropping pattern. The next year, average yields went up dramatically, and 94 subaks changed their cropping patterns. After eight years, average yields peaked, and all but 20 subaks stopped

The authors assume a beautifully simple rule for updating the cropping choices of each subak: copy the best neighbor. In combination with basic ecological regularities between the amount of water, pests, and cropping schedules, this simple decision-making rule produces patterns similar to those in the real world.

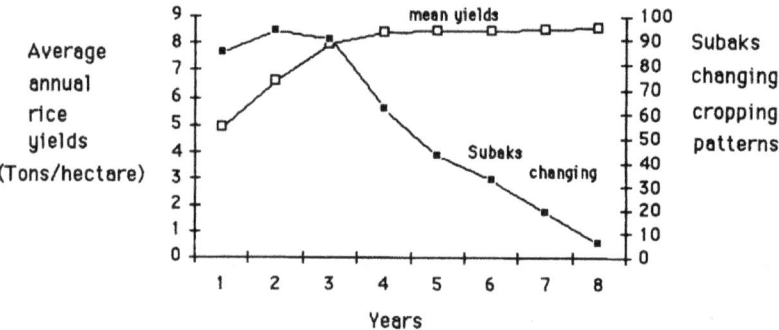

Figure 7. Increase in average rice yields as artificial temple networks appear.

changing their cropping patterns. One hundred and fifty-two subaks had reached a local optimum, in which their yields would not improve if they adopted a neighbor's cropping pattern. The remaining 20 subaks keep swapping cropping patterns with their neighbors indefinitely, as first one and then another obtains a slightly better yield.

Figure 8 maps the distribution of cropping patterns for subaks in the first run, and Figure 9 in the last run of the simulation; Figure 10 maps the pattern of coordinated cropping patterns achieved by the water temple network. The resemblance between the last run of the hill-climbing program (Figure 9) and the temple system (Figure 10) is evident. Competition to achieve maximal yields led to the formation of cooperative units with synchronized cropping patterns that bear a very close resemblance to actual water temple networks.

Should we therefore conclude that patches of coordination resembling water temples will spontaneously develop as subaks seek to optimize their harvest yields? It is worthwhile to note that the number of possible distributions of cropping patterns that could occur in this model is astronomically large, and the chances that a temple-like system of coordination would occur by chance are correspondingly small. But perhaps the temple networks are uniquely fitted to the traditional cropping pattern as a result of centuries of trial and error by the subaks? To test this possibility, many more simulations were conducted in order to vary not only the cropping patterns (crop varieties and start dates), but also the ecological parameters (rainfall, evapotranspiration, pest growth rates, dispersal rates, and damage coefficients). The same pattern occurred in all cases: after 8-35 years, a complex structure

In subsequent work (Lansing and Miller 2005; Lansing et al. 2017) Lansing and his colleagues built on this initial model to probe its dynamics. Counterintuitively, the threat of pests in the fields actually promotes cooperation because of the need to reduce their numbers by synchronizing harvests and temporarily removing their preferred habitat. This result—an adaptive process triggering a phase transition—has recently been generalized (Gandica et al. 2021). Interactions between farmers at the local level trigger a transition from local to global-scale connectivity, which maximizes rice harvests.

Figure 8. First run of a model of coadaptation. Each symbol indicates a different cropping pattern, randomly distributed among subaks in the Oos-Petanu watersheds. Average harvest was 4.9 tons rice/ha.

Figure 9. Last run of the model; average harvest rose to 8.57 tons/ha.

Figure 10. Distribution of synchronized cropping patterns in the traditional system of water temple networks.

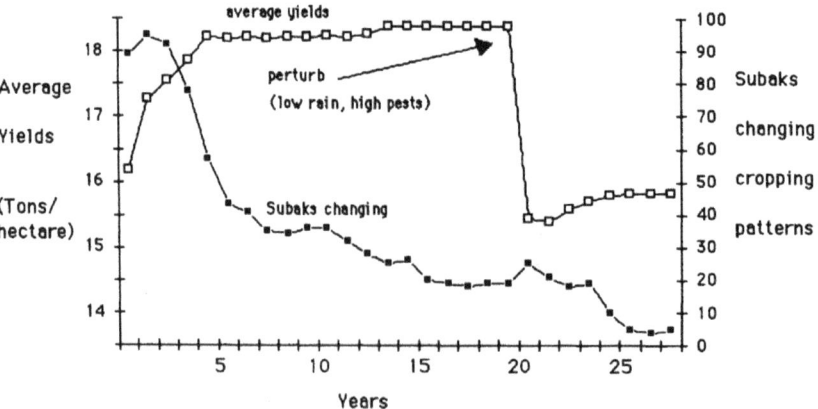

Figure 11. Effects of perturbation on yields.

of coordinated cropping patterns emerged, which bore a remarkable similarity to the actual pattern of water temple coordination along these rivers.

Finally, consider the outcome of an experiment in which the network is perturbed: Figure 11 shows the results of a simulation in which each subak grew two crops of high-yielding rice, as well as a vegetable crop. Using normal values for all ecological variables, average yields improved from 16 to 18.5 tons/hectare/year after 20 years, and stabilized with 20 subaks still changing their cropping patterns. In the 21st year a plague of pests was visited on the region by increasing the pest growth, dispersal, and damage rates. Simultaneously, rainfall was decreased to 80% of normal. Yields fell to 15.3 tons/hectare/year immediately but recovered to 15.8 within 7 years. Yet when the conditions of low rain and high pests occurred from the very beginning of an otherwise identical simulation, it took twice as long to reach the same optimal yield value.

Discussion

Before pursuing the implications of these results, it is worthwhile to linger for a moment over the mathematical logic that produced them. In the sequences of simulation runs described above, each subak alters ecological conditions for its neighbors when it varies its cropping pattern. A coadaptive process begins as each subak responds to these changing conditions by seeking to optimize rice yields. The actions of the subaks affect conditions for their neighbors, and the fitness landscape changes for most subaks with each run of the model. In other words, the actions of the subaks influence ecological variables such as irrigation flows, which in turn affect future decisions by other subaks. For example, Figure 12 shows differences in water shortages at one of the 12 major irrigation systems in the model (Klutug) for a typical simulation. Major water shortages in March and April, which led to reduced yields for some subaks in the early runs, have disappeared by the 54th run. A similar pattern always occurs at the other 12 weirs: water shortages are gradually reduced. Note that this coadaptive process does not find the ideal cropping pattern for all; rather, it finds locally optimal scales of coordination in the synchronization of cropping patterns.

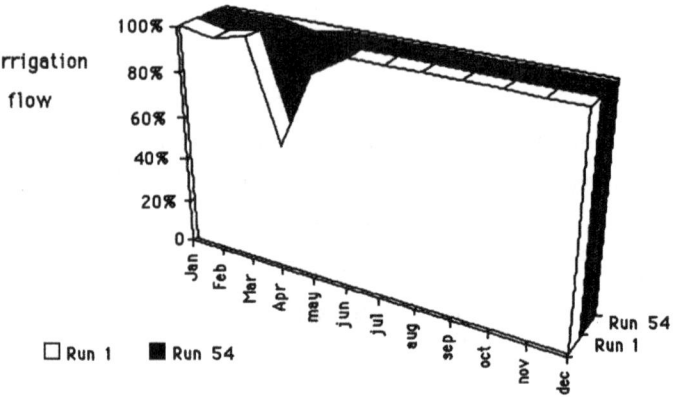

Figure 12. Water shortages at the Klutug dam.

Let us call these artificial water temples, to distinguish them from real water temples. The interesting result here is that the same phenomenon occurs every time, regardless of the initial distribution of cropping patterns, or ecological parameters such as flow rates or pest biology. Within 8-35 years, depending on ecological conditions, the subaks spontaneously self-organize into a network of artificial water temples where all subaks are at or near a local optimum.

Once this structure appears, the entire network displays an interesting emergent property: the ability to recover from external perturbations (such as low rainfall or high pest levels), as shown in Figure 11. Such disturbances initially generate a cascade of changes that propagate through one or more clusters of subaks, but soon lead to a new equilibrium. The ability of the system as a whole to react to changes is thus a property of the network itself, rather than the accidental consequence of the actions of individual subaks or temples. Put another way, the ability of each local group of farmers to reclimb the shifting peaks of the fitness landscape depends not only on their own initiative, but on the ability of the system as a whole to respond to changes.

And indeed, the system managed to recover from a devastating disturbance, the Green Revolution starting in the 1970s, which disrupted the traditional way of setting cropping schedules through water temples.

Recent work in the mathematical theory of optimization on rugged fitness landscapes sheds an interesting light on this process, suggesting that these results are not artifacts of our computer program, but predictable outcomes from a process of coadaptation on a rugged fitness landscape. Both of the properties we have noted—the maximization of sustained yields and the enhanced ability of the entire network to cope with perturbations—appear to occur for a wide range of coevolving complex systems. Artificial water temple networks fulfill the formal definition of a *complex adaptive system*:

> *(i) It consists of a network of interacting agents (processes, elements); (ii) it exhibits a dynamic, aggregate behavior that emerges from the individual activities of the agents; and (iii) its aggregate behavior can be described without a detailed knowledge of the behavior of the individual agents.*
>
> *An agent in such a system is adaptive if it satisfies an additional pair of criteria: the actions of the agent in its environment can be assigned a value (performance, payoff, fitness or the like); and the agent behaves so as to increase this value over time. A complex adaptive system, then, is a complex system containing adaptive agents, networked so that the environment of each adaptive agent includes others in the system.* (Holland and Miller 1991: 365)

If each agent (in this case, each subak) always acts independently, the system as a whole behaves chaotically. Alternatively, if each agent is linked to all the others, the system is stable but only massive perturbations can cause alterations in behavior. Metaphorically, the system is frozen into place. Complex behaviors occur when these frozen components begin to "melt," and different-sized islands of synchronized agents emerge. This transitional state between order and chaos is the narrow zone of complex or periodic behavior (Langton 1991; Weisbuch 1991). Recent work in the theory of complex systems suggests that coevolution on a rugged fitness landscape may drive many kinds of networks toward this zone or class of behaviors. As the biologist Stuart Kauffman recently observed, "as if by an invisible hand, coevolving complex entities may mutually attain the poised boundary between order and chaos. Here, mean sustained payoff, or fitness, or profit, is optimized" (Kauffman 1993).

By now we have come a long way from our original point of departure, and it may be helpful to say a few concluding words about the relationship between the models described here and evolutionary ecology. One major difference is the level of analysis: evolutionary

	CROPPING PATTERNS		
	TRADITIONAL	HIGH-YIELDING	LOW RAIN, HIGH PESTS
FIRST RUN	4.9	15.91	13.67
LAST RUN	8.57	18.08	17.66

Table 1. Increase in mean fitness ($\bar{f}$) after temple networks appear.

ecology focuses on determining the fitness value or payoff of specific strategies for individuals, whereas from a systems perspective we are interested in whether it is possible to predict changes in these values over time by analyzing the interaction of social and ecological variables. Our model tracks the behavior of 172 subaks along two rivers, each of which sets an annual cropping pattern (rice, vegetables, fallow interludes). In the model, for $i = 1$ to N cropping patterns: x_i = frequency of cropping pattern i; f_i = fitness payoff (yield, expressed as tons rice/hectare/year); and f_{opt} = a local optimum fitness value.

In the first year, we calculate the fitness payoff (yield) for each cropping pattern, f_i. In succeeding years, x_i varies as its fitness value changes with respect to the average fitness ($\bar{f}$). For each cropping pattern tested:

$$\dot{x}_i = f_i - \bar{f}$$

where

$$\bar{f} = \sum_{i=1}^{N} f_i / N$$

As time goes on, and patches of coordinated cropping patterns appear which improve the balance between water sharing and pest control, the mean fitness level of the entire system increases:

$$\bar{f} \to f_{\text{opt}}$$

A three-dimensional illustration of this process is shown below in Figure 13. Ultimately, regardless of the initial choice of cropping patterns, the mean value $\bar{f}$ increases to a local optimum after artificial water-temple networks come into existence. For example, compare the average yields for the initial and final runs with three different cropping patterns in Table 1.

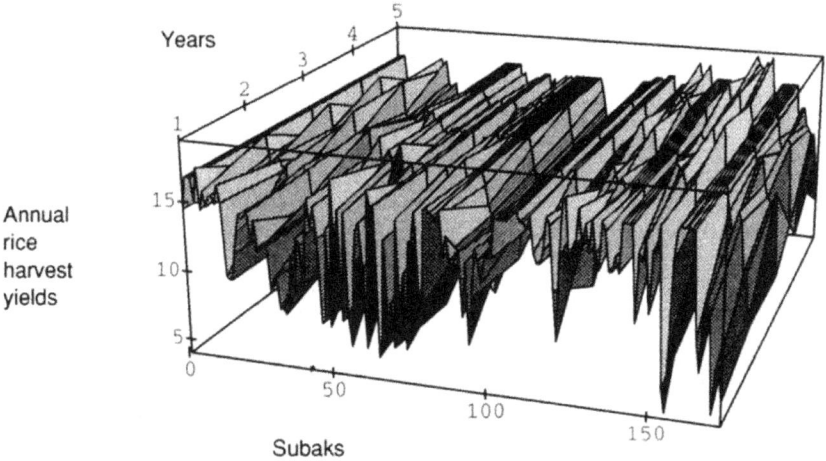

Figure 13. First five years of a hill-climbing simulation of the Oos-Petanu watersheds. Yields change every year for most subaks, in a process of coadaptation.

While evolutionary ecology also begins with the concept of fitness (f_i), it makes different assumptions about how these values are found. The key issue is the process of selection: as Eric Smith observes, "in evolutionary ecology adaptation via natural selection in a finite environment is the primary causal force" (1984: 67). Natural selection is thus obliged to try out most of the possible variants (Eigen 1992: 27). In our model, we could mimic this process by randomly changing cropping patterns for each subak every year and selecting good ones only as they appear, a procedure that would take a very long time, even on a computer. Instead, we try to model the effects of deliberate selection, in this case the annual meetings of the subaks, where the farmers discuss the outcomes of last year's cropping patterns in their vicinity and try to pick a good one for the next cycle. In mathematical terms, they are searching the peaks of their local fitness landscape. This process of selection is not blind, but on the contrary highly efficient.

A second key distinction is the change in the fitness landscape itself brought about by this search strategy. As we have seen, the emergence of artificial temple networks increases the height of the fitness peaks, an effect that occurs regardless of our initial assumptions about the physical and biological systems. Our model demonstrates that cultural systems like the water temple networks can have macroscopic effects on the topography of the adaptive landscape. But such effects are not

And yet, the model reproduces cropping schedules based on just a simple social learning rule coupled with a few ecological regularities between the water, pest, and cropping systems. It does not evolve intricate religious rituals that were observed in the real world. Do these rituals evolve in response to the natural forgetfulness and random decay inherent to human individual and collective memories? Would the models reproduce the rituals if such forgetfulness would be assumed?

apparent from within the horizons of evolutionary ecology, since they are properties of systems rather than individuals.

Conclusion

The analytic techniques described in this article have enabled us to shift the level of analysis progressively from the individual farmer and field to the subak and water temple, and ultimately to the historical development of temple networks. We have shown that the structure of water temple networks could have developed through a process of spontaneous self-organization, rather than deliberate planning by royal engineers or other planners. This idea could be subjected to a more rigorous test, since our methodology is capable of predicting the response of temple networks to historical changes. The resulting patterns of spatial organization could be compared with the actual sequence of development as revealed by archeological investigations.

We have also shown that the emergence of temple networks leads to higher average harvest yields, and improvement in sustainability (the ability to cope efficiently with ecological perturbations). Since these effects occur in our model as consequences of a process of coadaptation, they are probably not unique. Balinese water temples may be representative of a class of complex adaptive systems that have evolved to manage agroecosystems. As Roy Rappaport argued a generation ago, it is likely that ritual often plays an important role in such "traditional" systems of resource management (Rappaport 1971).

It also follows from our analysis that the wide range of harvest yields reported by the subaks for 1988–89 (Figure 4) is likely to be the signature of suboptimal conditions, in which the subaks are prevented from rediscovering an optimal distribution of cropping patterns. In other words, under present conditions new crops and cropping patterns are tried out each year, in an effectively random process. Some subaks do well, others poorly, and the network as a whole remains in a state of continuous perturbation, corresponding to the early runs in our simulation model. As we have argued elsewhere, these conditions are aftereffects of the Green Revolution in Bali: the replacement of native Balinese rice varieties with high-yielding imported varieties, coupled with new management plans based on the assumption that production will be optimized if each farm or subak is an autonomous productive unit (Lansing 1991). Our results suggest, to the contrary, that self-organizing temple networks are intrinsically capable of a better job of water management than either autonomous subaks or centralized hierarchical control (Figures 8—10). But as long as research on agroecosystems remains focused on the behavior of individuals, the productive role of "traditional" systems of resource management such as the water temples will remain invisible. Ironically, recent plans for the improvement of Balinese irrigation systems by international development agencies foresee an end to the productive role of water temples "as an almost inevitable result of technical progress" (Asian Development Bank 1988: 47).

Acknowledgments

The research described in this article was supported by grants from the National Science Foundation (BNS 8705400; BNS 9005919). The simulation model of the Oos and Petanu rivers was tested with data collected by a team of Balinese undergraduate students from Udayana University: I Gde Suarja, Dewa Gde Adi Parwata, Ni Made Sri Tutik Andayani, and I Made Cakranegara. Many friends and colleagues have contributed to this research, including Lene Crosby, Charles Taylor, Kevin Arrigo, Gary Seaman, Liane Gabora, John Miller, Chris Langton, Glynnis Collins, Walter Fontana, Mike Simmons, Tyde Richards, Thierry Bardini, David Jefferson, Stuart Plattner, Jill Schroeder, and David Rudner; and in Bali the Jero Gde Mekalihan, Wayan Pageh, Dr. Nyoman Sutawan, Ida Pedanda Sidemen, Dr. Gusti Ngurah Bagus, Guru Nengah Tekah, Ir. Cokorde Raka, Ir. Jelantik Sushila, and Dr. Andrew Toth.

REFERENCES

Arrigo, K. R. 1991. "A Simulated Antarctic Fast-Ice Ecosystem." PhD diss., University of Southern California.

Asian Development Bank. 1988. *1988 Project Performance Audit Report, Bali Irrigation Project in Indonesia.* Technical report PE-241 L 352INO. Manila, Philippines: Post Evaluation Office, Asian Development Bank.

Braudel, F. [1969]1980. *On History.* Translated by S. Matthews. Chicago, IL: University of Chicago Press.

Eigen, M., and R. Winkler-Oswatitsch. 1992. *Steps Towards Life: A Perspective on Evolution.* Translated by P. Wooly. Oxford, UK: Oxford University Press.

Giddens, A. 1981. *A Contemporary Critique of Historical Materialism.* Berkeley, CA: University of California Press.

Gould, S. J. 1989. *Wonderful Life: The Burgess Shale and the Nature of History.* New York, NY: Norton.

Habermas, J. 1971. *Knowledge and Human Interests.* Boston, MA: Beacon Press.

Holland, J. H., and J. H. Miller. 1991. "Artificial Adaptive Agents in Economic Theory." *American Economic Association* 81 (2).

Kauffman, S. A. 1993. *The Sciences of Complexity and the Origins of Order: Philosophy of Science.* (In press.)

Kauffman, S. A., and S. Johnsen. 1991. "Co-Evolution to the Edge of Chaos: Coupled Fitness Landscapes, Poised States, and Co-Evolutionary Avalanches." In *Artificial Life II,* edited by C. G. Langton, J. D. Farmer, S. Rasmussen, and C. Taylor, 325–370. Redwood City, CA: Addison-Wesley.

Kremer, J. N. 1991. "Technical Report on the Ecological Simulation Model." In *Priests and Programmers: Technologies of Power in the Engineered Landscape of Bali,* edited by J. S. Lansing, 153–158. Princeton, NJ: Princeton University Press.

Kremer, J. N., and J. S. Lansing. 1992b. "Landscape Coordination by Temples Optimizes Rice in Bali(Indonesia)." (Unpublished ms. in authors' possession.)

———. 1992a. "Modelling Water Temples and Rice Irrigation in Bali: A Lesson in Socioecological Communication." In *Maximum Power: The Application of the Ideas of Howard Odum to Ecology, Environment and Engineering,* edited by C. A. S. Hall. Boulder, CO: University of Colorado Press.

Langton, C. G. 1991. "Life at the Edge of Chaos." In *Artificial Life II,* edited by G. L. Christopher, T. Charles, J. D. Farmer, and S. Rasmussen, 41–92. Redwood City, CA: Addison-Wesley.

Lansing, J. S. 1987. "Balinese Water Temples and the Management of Irrigation." *American Anthropologist* 89:326–341.

———. 1991. *Priests and Programmers: Technologies of Power in the Engineered Landscape of Bali.* Princeton, NJ: Princeton University Press.

Marx, K. 1961 [1844]. *Economic and Philosophical Manuscripts of 1844.* Moscow, Russia: Foreign Languages Publishing House.

Palmer, R. 1991. "Optimization on Rugged Landscapes." In *Molecular Evolution on Rugged Fitness Landscapes,* 3–25. Redwood City, CA: Addison-Wesley.

Rappaport, R. A. 1971. "The Sacred in Human Evolution." *Annual Review of Ecology and Systematics* 2:23–44.

Smith, E. A. 1984. "Anthropology, Evolutionary Theory and the Explanatory Limits of the Ecosystem Concept." Edited by E. Moran. *The Ecosystem Concept in Anthropology* (Boulder, CO), 51–86. https://doi.org/10.4324/9780429310386-3.

Weisbuch, G. 1991. *Complex System Dynamics: An Introduction to Automata Networks.* Redwood City, CA: Addison-Wesley.

Wright, S. 1932. "The Roles of Mutation, Inbreeding, Crossbreeding and Selection in Evolution." *Proceedings of the XI International Congress of Genetics* 1:356–366.

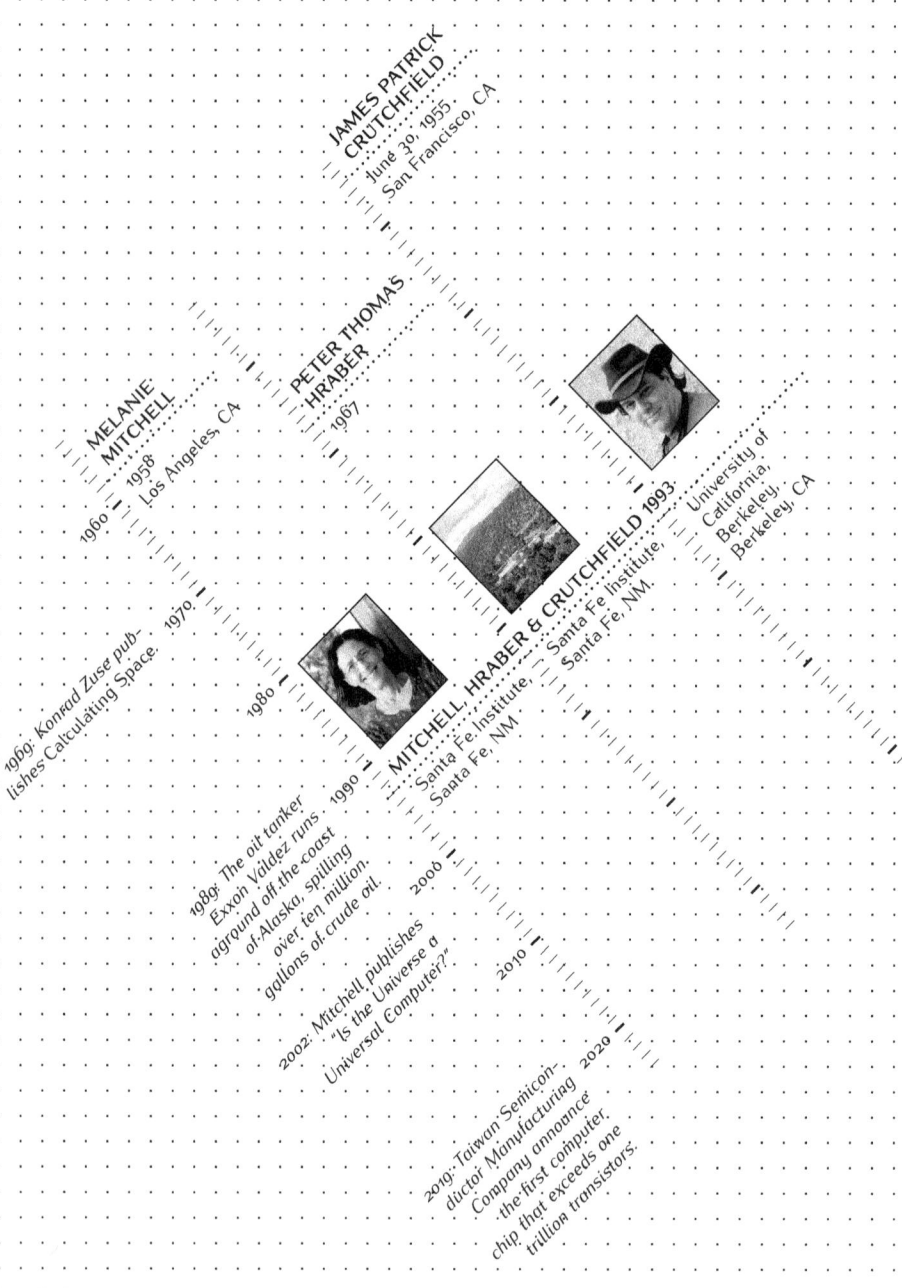

JAMES PATRICK CRUTCHFIELD
June 30, 1955
San Francisco, CA

PETER THOMAS HRABER
1967

MELANIE MITCHELL
1958
Los Angeles, CA

1960

1969: Konrad Zuse publishes Calculating Space.

1970

1980

MITCHELL, HRABER & CRUTCHFIELD 1993
Santa Fe Institute, Santa Fe, NM
Santa Fe Institute, Santa Fe, NM
University of California, Berkeley, Berkeley, CA

1989: The oil tanker Exxon Valdez runs aground off the coast of Alaska, spilling over ten million gallons of crude oil.

1990

2002: Mitchell publishes "Is the Universe a Universal Computer?"

2006

2010

2019: Taiwan Semiconductor Manufacturing Company announce the first computer chip that exceeds one trillion transistors.

2020

MITCHELL, HRABER & CRUTCHFIELD

[76]

LIFE AND COMPUTATION: FROM STATISTICAL PHYSICS TO EMERGENT PHYSICS

David H. Ackley, Living Computation Foundation and University of New Mexico

From one point of view, what we have here is a common science story, played out in three acts over an eight-year span: Christopher Langton (1986; see *Foundational Papers* vol. 3, ch. 61) made a conjecture about a general quantity he called lambda (λ). Then Norman Packard (1988) built an example that supported that hypothesis. But later, Melanie Mitchell, Peter Hraber, and James Crutchfield (1993) were unable to replicate Packard's detailed results, offered reasons to question some of its interpretations, and suggested useful alternatives.

In that telling, it's just science making progress and working as designed. It might sound bad for Packard, but it turns out his contribution was key. For me at the time, as a fresh computer-science PhD, it was exciting research to follow. Looking back now, it feels even a bit glorious.

Writ large, I see people daring to draw bold new connections between two vast conceptual domains: here's technology and computation, central to our species' success and modern society, and there's life in all its variety, central to how we see ourselves and the natural world. To frame a satisfying and effective unified understanding of life and computation would—and I believe will—be a tremendous human accomplishment. Of course, hypotheses will sometimes fail, and simple assumptions will need to be refined—or we're just not trying hard enough.

For me, this particular story is about a shift from theoretical physics to empirical computer science. The idea that persists throughout is that physics and life are fundamentally linked, somehow, via computation and programming. The change is that later work focuses less on global properties to be derived or formally proven and more on concrete

M. Mitchell, P. T. Hraber, and J. P. Crutchfield, "Revisiting the Edge of Chaos: Evolving Cellular Automata to Perform Computations," *Complex Systems* 7, 89–130 (1993).

Reprinted under CC BY 4.0.

examples to be implemented and studied. In this story, statistical physics as a formal framework gives way to "emergent physics" observed in empirical results... but let's start at the beginning.

In 1987, Chris Langton convened the first "Workshop on the Synthesis and Simulation of Living Systems" in Los Alamos, New Mexico (Langton 1989). His call to explore "life as it could be" drew participants from biology, computer science, anthropology, and much more. Though each discipline brought its own research styles, theoretical physics was a key perspective from the outset—as exemplified here precisely by Langton (1986).

At the time I was finally pounding out my dissertation (Ackley 1987), but I've always regretted passing up that first "artificial life" workshop. In 1990, though, I did attend the second workshop, "ALIFE 2" in Santa Fe, where turnout more than doubled while interdisciplinary participation remained extreme. There I first met Melanie Mitchell, who three years later would be the first author of "Revisiting the Edge of Chaos: Evolving Cellular Automata to Perform Computations."

That title introduces our three featured players: the "edge of chaos" alludes to Langton's statistical physics framework; its key traits are generality and a global perspective. Cellular automata (CA) are a kind of programmable machinery; they are easy to implement and can have a simple program format. Finally, "evolving" refers to the use of genetic algorithms (GA), which are a method of searching for improved solutions to some prespecified problem; GAs don't guarantee success, but they are easy to implement and quite general.

As the curtain rises on Act I, theoretical physics is famous for studying minimal simplified systems, such as a frictionless pendulum or a single hydrogen atom, where some overall set of states and behaviors can be accounted for exactly. Statistical physics, though, builds theoretical models of systems that have many independently moving parts, creating a staggering number of possible states for the system as a whole.

Rather than deal with each system state individually, statistical physics views them all together as an ensemble and tries to derive or estimate various ensemble properties using simplifying assumptions and sampling. For example, in Ludwig Boltzmann's [1896]1964 kinetic theory of gases, the ensemble was the possible states of a volume of gas molecules, and the ensemble statistics were global properties like temperature and pressure. Today, statistical physics applications range far beyond gas molecules to planetary orbits, neural and social networks, and complex systems broadly.

Langton's idea, building on work by Stephen Wolfram (1983, 1984) and others, was to explore an ensemble of computer programs, using cellular automata as the machines to be programmed. A CA consists of many simple computers—sometimes physical machines, but more commonly simulated—laid out in space, typically in a regular pattern like a row or a grid. Each "cell" is connected only to its neighbors, according to some chosen floor plan. Each cell gets just one or a few modifiable memory bits, and all the cells run the exact same program, usually in a flawless lockstep fashion. The job of the CA program is, over and over, to update the cell's memory based on the memories of the neighboring cells.

If a chosen CA is simple enough, its "program" can simply be a list of updated cell memory for each combination of input memories. In Langton's λ experiments, for example, each program was a list of 32,768 integers from 0 to 7. Handling individual programs is easy enough, but examining all $8^{32,768}$ of them one by one is utterly hopeless.

Instead, in the statistical physics style, Langton organized and randomly sampled that program ensemble according to a parameter he called λ, which specified what fraction of the 32,768 program outputs would be non-zero. Then, by executing the sampled CA programs on random inputs and observing the resulting behaviors, Langton developed his famous "life at the edge of chaos" hypothesis: that complex computations and artificial life would tend to be found near a critical λ value. That's the gun on the mantle in this play: "interesting" CA programs, viewed properly, might have one specific number in common. Curtain.

Act II takes place the following year. Packard (1988) loads the gun by making several changes and simplifications to Langton's approach. Crucially, he focuses on a single computational task—the "majority function"—as a case study. Second, rather than sampling the global ensemble across λ values, he uses a genetic algorithm to optimize CA programs for the majority function. In this framework, λ is not a choice to make but a result to observe—and most intriguingly, Packard's results appear to suggest the resulting optimized CA programs tend to have λ values close to Langton's "edge of chaos" prediction.

Alas, in Act III the gun misfires, as Mitchell, Hraber, and Crutchfield fail to replicate Packard's empirical results. Worse, their analysis makes clear that effective CA programs for the majority task will never have λ values close to the edge of chaos prediction. Of course, one data point does not invalidate a global statistic, but it does highlight a conceptual gap between studying collective program properties, on the one hand, and finding their relevance to any computation we might actually care about, on the other.

If we lower our sights, follow Packard, and accept results that are conditioned on a task we choose, Mitchell, Hraber, and Crutchfield showed how effective program evolution can be—and how worthy of close study, as evolution sometimes deploys quite subtle computational strategies. A wonderful subsequent paper (Hordijk, Crutchfield, and Mitchell 1996), for example, showed how a clever emergent "particle physics" can explain much of the behavior of CA programs evolved for multiple tasks. Final curtain.

Looking back, a significant question for me is: Where did Langton's statistical physics framework go astray? Yes, Langton's CA programs were quite simple—but we understand more clearly today that they can still be viciously complex beasts, largely due to the uncritically adopted assumption of "flawless lockstep operation." In such deterministic cellular automata, the CA program is hugely leveraged in both space and time: it is copied perfectly to every processing cell, which then executes it perfectly over and over. Whereas perturbing any single molecule in a volume of gas will hardly affect the global temperature and pressure, changing even one bit in a deterministic CA program can potentially alter that entire space-time computational volume.

There are computational frameworks that don't assume deterministic execution, though. In the Boltzmann machine (Ackley, Hinton, and Sejnowski 1985), for example, a statistical physics framework was key to its learning algorithm. Various forms of stochastic cellular automata have been studied, and some of them might be more amenable to the statistical physics style.

Still, much of modern computing requires deterministic execution, with its ability to compute sharply defined and extremely high-order functions quickly. Back in the 1990s, in part because of work like Mitchell, Hraber, and Crutchfield, there was a groundswell of artificial life researchers following the empirical computer science path via selected tasks and software simulations. That "soft alife" community has grown and persisted to this day.

REFERENCES

Ackley, D. H. 1987. *A Connectionist Machine for Genetic Hillclimbing*. Boston, MA: Kluwer Academic Publishers.

Ackley, D. H., G. E. Hinton, and T. J. Sejnowski. 1985. *A Learning Algorithm for Boltzmann Machines*. 9:147–169. 1.

Boltzmann, L. [1896]1964. *Lectures on Gas Theory*. Berkeley, CA: University of California Press.

Hordijk, W., J. P. Crutchfield, and M. Mitchell. 1996. *Embedded-Particle Computation in Evolved Cellular Automata*. Technical report 96-09-073. Santa Fe, NM: Santa Fe Institute.

Langton, C., ed. 1989. *Artificial Life: The Proceedings of an Interdisciplinary Workshop on the Synthesis and Simulation of Living Systems: Held September, 1987 in Los Alamos, New Mexico*. Vol. 6. Santa Fe Institute Studies in the Sciences of Complexity. Redwood City, CA: Addison-Wesley.

Langton, C. G. 1986. "Studying Artificial Life with Cellular Automata." A preliminary investigation of the potential of CA for supporting life, *Physica D* 22:120–149.

Mitchell, M., P. T. Hraber, and J. P. Crutchfield. 1993. "Revisiting the Edge of Chaos: Evolving Cellular Automata to Perform Computations." *Complex Systems* 7 (2).

Packard, N. H. 1988. "Adaptation Toward the Edge of Chaos." Edited by J. A. S. Kelso, A. J. Mandell, and M. F. Shlesinger, 293–301.

Wolfram, S. 1983. "Statistical Mechanics of Cellular Automata." *Reviews of Modern Physics* 55:601–644.

———. 1984. "Universality and Complexity in Cellular Automata." *Physica D* 10:1–35.

REVISITING THE EDGE OF CHAOS: EVOLVING CELLULAR AUTOMATA TO PERFORM COMPUTATIONS

Melanie Mitchell, Santa Fe Institute
Peter T. Hraber, Santa Fe Institute
and James P. Crutchfield, University of California, Berkeley

Abstract

We present results from an experiment similar to one performed by Packard (Packard 1988), in which a genetic algorithm is used to evolve cellular automata (CAs) to perform a particular computational task. Packard examined the frequency of evolved CA rules as a function of Langton's λ parameter (Langton 1990); he interpreted the results of his experiment as giving evidence for two hypotheses: (1) CA rules that are able to perform complex computations are most likely to be found near "critical" λ values (which have been claimed to correlate with a phase transition between ordered and chaotic behavioral regimes for CAs); (2) When CA rules are evolved to perform a complex computation, evolution will tend to select rules with λ values close to the critical values. Our experiment produced quite different results, and we suggest that the interpretation of the original results is not correct. We also review and discuss issues related to λ, dynamical-behavior classes, and computation in CAs. The primary constructive results of our study are the identification of the emergence and competition of computational strategies, and the analysis of the central role of symmetries in an evolutionary system. In particular, we demonstrate how symmetry breaking can impede evolution toward higher computational capability.

1. Introduction

The notion of "computation at the edge of chaos" has attracted considerable attention in the study of complex systems and artificial life (see, for example, Crutchfield and Young 1989, 1990; Kauffman and Johnsen 1992; Langton 1990; Packard 1988; Wolfram 1984). This notion is related to the broad question of the relationship between a computational system's ability for complex information processing and other measures of the system's behavior. In particular, does the ability to perform nontrivial computation require that a system's dynamical behavior be "near a transition to chaos"? Likewise, much attention

> The fundamental question is sounded immediately. The implication that "other measures" will be in the statistical physics style is adopted uncritically, but that is just following Langton's lead.

has been given to the notion of "the edge of chaos" in the context of evolution. In particular, it has been hypothesized that when biological systems must perform complex computation to survive, the process of evolution under natural selection tends to select such systems near a phase transition from ordered to chaotic behavior (Kauffman 1990; Kauffman and Johnsen 1992; Packard 1988).

In this paper, we reexamine one study that addressed these questions in the context of cellular automata (Packard 1988). The results of the original study were interpreted as evidence that an evolutionary process in which cellular-automata rules had been selected to perform a nontrivial computation preferentially selected rules near transitions to chaos. We show that this conclusion is neither supported by our experimental results nor consistent with basic mathematical properties of the computation being evolved. We also review and clarify, in the context of cellular automata, notions relating to such terms as "computation," "dynamical behavior," and "edge of chaos."

2. Cellular Automata and Dynamics

Cellular automata (CAs) are discrete, spatially extended dynamical systems that have been studied extensively as models of physical processes and as computational devices (Farmer, Toffoli, and Wolfram 1984; Gutowitz 1990; Preston and Duff 1984; Toffoli and Margolus 1987; Wolfram 1986). In their simplest form, CAs consist of spatial lattices of *cells*, each of which, at time t, can be in one of k states. We denote the lattice size or number of cells as N. A CA has a single fixed rule, which is used to update each cell; the rule maps from the states in a neighborhood of a cell—for example, the states of a cell and its nearest neighbors—to a single state, which is the update value for that cell. The lattice begins with an initial configuration of local states and, at each time step, the states of all cells in the lattice are synchronously updated. We use the term "state" to refer to the value of a single cell—for example, 0 or 1—and "configuration" to mean the pattern of states over the entire lattice.

The CAs that we discuss in this paper are all one-dimensional, with two possible states per cell (0 and 1). In a one-dimensional CA, the neighborhood of a cell includes the cell itself and some *radius r* of

Again, following Langton and the custom at the time, the assumption that CAs will provide deterministic execution is considered too obvious even to mention here.

FOUNDATIONAL PAPERS IN COMPLEXITY SCIENCE

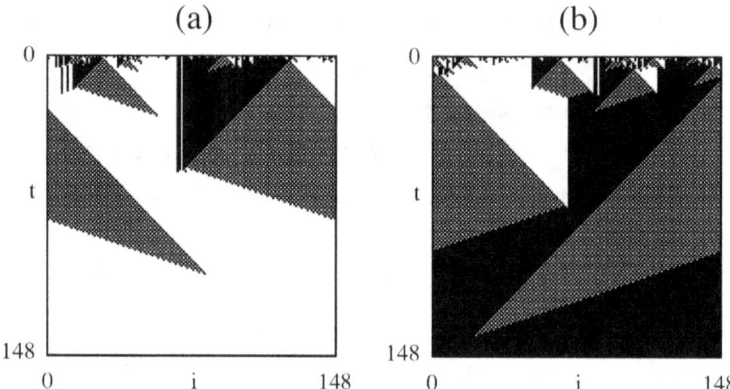

Figure 1. Two space-time diagrams for the binary-state Gacs–Kurdyumov–Levin CA. $N = 149$ sites are shown evolving over 149 time steps, with time increasing down the page, from each of two different initial configurations. In (a), the initial configuration has a density of 1s of approximately 0.48; in (b), a density of approximately 0.52. By the last time step the CA has converged to a fixed pattern of (a) all 0s and (b) all 1s. In this way the CA has classified the initial configurations according to their density.

neighbors on either side of the cell. All of the simulations will be of CAs with spatially periodic boundary conditions (in other words, the one-dimensional lattice is viewed as a circle, with the right neighbor of the rightmost cell being the leftmost cell, and vice versa).

The equations of motion for a CA are often expressed in the form of a *rule table*: a lookup table listing each of the neighborhood patterns, and the state to which the central cell in that neighborhood is mapped. For example, the following displays one possible rule table for a one-dimensional, two-state CA with radius $r = 1$. Each possible neighborhood η is given, along with the "output bit" $s = \phi(\eta)$ to which the central cell is updated.

η	000	001	010	011	100	101	110	111
s	0	0	0	1	0	1	1	1

In words, this rule says that for each neighborhood of three adjacent cells, the new state is decided by a majority vote among the three cells. To run the CA, this lookup table is applied to each neighborhood in the current lattice configuration, respecting the choice of boundary conditions, to produce the configuration at the next time step.

A method commonly used to examine the behavior of a two-state, one-dimensional CA is the display of its space-time diagram (a two-dimensional picture that vertically strings together the one-dimensional

CA lattice configurations at the successive time steps, with white squares corresponding to cells in state 0, and black squares corresponding to cells in state 1). Two such space-time diagrams, reproduced in Figure 1, show the actions of the Gacs–Kurdyumov–Levin (GKL) binary-state CA on two random initial configurations of different densities of 1s (Gonzaga de Sá and Maes 1992; Gacs, Kurdyumov, and Levin 1978). In both cases, the CA relaxes to a fixed pattern over time—in one case, all 0s, and in the other case, all 1s. These patterns are, in fact, fixed points of the GKL CA. That is, once they are reached, further applications of the CA do not change the pattern. We will discuss the GKL CA in further detail below.

CAs are of interest as models of physical processes because, like many physical systems, they consist of a large number of simple components (cells) that are modified only by local interactions, but which, acting together, can produce global complex behavior. Like the class of dissipative dynamical systems, the class of one-dimensional CAs exhibit the full spectrum of dynamical behavior: from fixed points, as seen in Figure 1, to limit cycles (periodic behavior), to unpredictable ("chaotic") behavior. Wolfram considered a coarse classification of CA behavior in terms of these categories; he proposed the following four classes with the intention of capturing all possible CA behavior (Wolfram 1984).

Class 1: Almost all initial configurations relax after a transient period to the same fixed configuration (for example, all 1s).

Class 2: Almost all initial configurations relax after a transient period, either to some fixed point or to some temporally periodic cycle of configurations, depending on the initial configuration.

Class 3: Almost all initial configurations relax after a transient period to chaotic behavior. (The term "chaotic" refers, in this paper, to apparently unpredictable space-time behavior.)

Class 4: Some initial configurations result in complex localized structures, sometimes long-lived.

> Over the years, Wolfram's classes have had a durable, if rough-and-ready, appeal, but as suggested here, they can give the impression that there's more structure within and relationships between the purported classes than can actually be found.

Wolfram does not state the requirements for membership in Class 4 any more precisely than this. Thus, unlike the categories derived from dynamical systems theory, Class 4 is not rigorously defined.

It should be pointed out that, on finite lattices, there is only a finite number (2^N) of possible configurations, so all rules ultimately lead to periodic behavior. Class 2 refers not to this type of periodic behavior, but to cycles with periods much shorter than 2^N.

3. Cellular Automata and Computation

CAs are of interest also as computational devices, both as theoretical tools and as practical, highly efficient parallel machines (Preston and Duff 1984; Rosenfeld 1983; Toffoli and Margolus 1987; Wolfram 1986).

"Computation" has several possible meanings in the context of CAs. The most common meaning is that a CA does some "useful" computational task. In that case, the rule is interpreted as the "program," the initial configuration is interpreted as the "input," and the CA runs for some specified number of time steps or until it reaches some "goal" pattern—possibly a fixed point pattern. The final pattern is interpreted as the "output." An example of this meaning is the use of CAs to perform image-processing tasks (Rosenfeld 1983).

> This paper, following Packard, reports results primarily in this first sense.

A second meaning of computation by CAs is that a CA, given particular initial configurations, is capable of universal computation. That is, given the right initial configuration, the CA can simulate a programmable computer—complete with logical gates, timing devices, and so on. Conway's Game of Life is such a CA; one construction for universal computation in the Game of Life is given in Berlekamp, Conway, and Guy (1982). Similar constructions have been made for one-dimensional CAs (Lindgren and Nordahl 1990). Wolfram speculated that all Class 4 rules have the capacity for universal computation (Wolfram 1984); however, given the informality of the definition of Class 4 (not to mention the difficulty of proving that a given rule is, or is not, capable of universal computation), this hypothesis is impossible to verify.

A third meaning of computation by CAs involves the behavior of a given CA on an ensemble of initial configurations, interpreted as a kind of "intrinsic" computation. Such computation is not interpreted as the performance of "useful" transformations of input to produce output; rather, it is measured in terms of generic, structural computational elements such as memory, information production, information transfer, logical operations, and so on. It is important to emphasize that the measurement of such intrinsic computational elements does not rely on a semantics of utility (as do the preceding computation types). That is, these elements can be detected and quantified without reference to any specific "useful" computation performed by the CA—such as enhancing edges in an image or computing the digits of π. This notion of intrinsic computation is central to the work of Crutchfield, Hanson, and Young (Crutchfield and Young 1989; Hanson and Crutchfield 1992).

This third sense is in the spirit of Langton's statistical mechanics formulation, but here the ensemble is formed not over possible programs but over possible inputs to a single program. Despite the disclaiming of utility here, though, subsequent work tended to focus on identifying computational mechanisms that did indeed improve a program's performance—and thus utility—on the given problem.

In general, CAs have the capacity for all kinds of both dynamical and computational behaviors. For this reason—in addition to the computational ease of simulating them—CAs have been considered a good class of models for use in the study of how dynamical behavior and computational ability are related. Similar questions have been addressed in the context of other dynamical systems, including continuous-state dynamical systems (such as iterated maps and differential equations) (Crutchfield and Young 1989, 1990), Boolean networks (Kauffman 1990), and recurrent neural networks (Pollack 1991). We confine our discussion to CAs.

With this background in mind, the broad questions presented in Section 1 can now be rephrased in the context of CAs, as follows.

- What properties must a CA possess to perform nontrivial computation?

- In particular, does a capacity for nontrivial computation (in any of the three senses previously described) require that a CA be in a region of rule space near a transition from ordered to chaotic behavior?

- When CA rules are evolved to perform nontrivial computation, will evolution tend to select rules near such a transition to chaos?

Using "require" here may be a bit stronger than Langton implied, but indeed this paper's results are a dagger straight into the heart of this question.

4. Structure of CA Rule Space

A number of studies conducted during the last decade have addressed our first question. We focus on Langton's empirical investigations of the second question in terms of the structure of the space of CA rules (Langton 1990). The relationship of the first two questions to the third—evolving CAs—is described subsequently.

One of the primary difficulties in understanding the structure of the space of CA rules (and its relation to computational capability) is its discrete nature. In contrast to the well-developed theory of bifurcations for continuous-state dynamical systems (Guckenheimer and Holmes 1983), there appears to be little or no geometry in CA space, and no notion of smoothly changing a CA to get another that is "nearby in behavior." In an attempt to emulate such a change, however, Langton defined a parameter, λ, that varies incrementally as single output bits are turned on or off in a given rule table. For a given CA rule table, λ is computed as follows. For a k-state CA, one state, q, is arbitrarily chosen to be "quiescent." In Langton (1990), all states obeyed a "strong quiescence" requirement: for any state $s \in \{0, \ldots, k-1\}$, the neighborhood consisting entirely of state s must map to s.) The λ of a given CA rule is the fraction of nonquiescent output states in the rule table. For a binary-state CA, if 0 is chosen to be the quiescent state, then λ is simply the fraction of output 1 bits in the rule table. Typically, there are many CA rules with a given λ value. For a binary CA, the number is strongly peaked at $\lambda = 1/2$, due to the combinatorial dependence on the radius r and the number of states k. It is also symmetric about $\lambda = 1/2$, due to the symmetry of exchanging 0s and 1s. Generally, as λ is increased from 0 to $1 - (1/k)$, the associated CAs shift from those having the most homogeneous rule tables to those having the most heterogeneous.

Again, this lack of smoothness can fundamentally be seen as an unrecognized consequence of the assumption of deterministic execution.

Langton performed a range of Monte Carlo samples of two-dimensional CAs, in an attempt to characterize their average behavior as a function of λ (Langton 1990). The notion of "average behavior" was intended to capture the most likely behavior observed with a randomly chosen initial configuration for CAs randomly selected in a fixed-λ subspace. His observation was that the average behavior of rules passed through the following regimes, as λ was incremented from 0 to $1 - (1/k)$:

$$\text{fixed point} \Rightarrow \text{periodic} \Rightarrow \text{``complex''} \Rightarrow \text{chaotic}.$$

That is, the average behavior at low λ was for a rule to relax to a fixed point after a relatively short transient phase (see Figure 16 in Langton 1990, for example). As λ was increased, rules tended to relax to periodic patterns, again after a relatively short transient phase. As λ reached a "critical value" λ_c, rules tended to have longer and longer transient phases. Additionally, the behavior in this regime exhibited long-lived, "complex" patterns—nonperiodic, but nonrandom. As λ was increased further, the average transient length decreased, and rules tended to relax to apparently random space-time patterns. The actual value of λ_c depended on r, k, and the actual path of the CA found as λ was incremented.

These four behavioral regimes roughly correspond to Wolfram's four classes. Langton's claim was that, as λ was increased from 0 to $1 - (1/k)$, the classes were passed through in the order $1, 2, 4, 3$. He noted that, as λ increases, "... one observes a *phase transition* between highly *ordered* and highly *disordered* dynamics, analogous to the phase transition between the *solid* and *fluid* states of matter." (Langton 1990, p. 13). According to Langton, as λ is increased from $1 - (1/k)$ to 1, the four regimes will occur in the reverse order, subject to some constraints for $k > 2$ (Langton 1990). For two-state CAs, there are two values of λ_c at which the complex regime will occur, since behavior is necessarily symmetric about $\lambda = 1/2$.

How is λ_c determined? Following standard practice, Langton used various statistics (such as single-site entropy, two-site mutual information, and transient length) to classify CA behavior. His additional step was to correlate behavior with λ via these statistics. Langton's Monte Carlo samples showed that there was some correlation between the statistics and λ. But the averaged statistics did not reveal a sharp transition in average behavior, a basic property of a phase transition in which macroscopic highly averaged quantities do make marked changes. We note that Wootters and Langton (1990) gave evidence that the transition region narrows in the limit of an increasing number of states. Their main result indicates that there is a sharp transition in single-site entropy at $\lambda_c \approx 0.27$, in one class of two-dimensional infinite-state stochastic CAs.

The existence of a critical λ, and the dependence of the critical region's width on r and k, are less clear for finite-state CAs. Nonetheless, Packard empirically determined rough values of λ_c for $r = 3, k = 2$ CAs by looking at the *difference-pattern spreading rate*, γ, as a function of

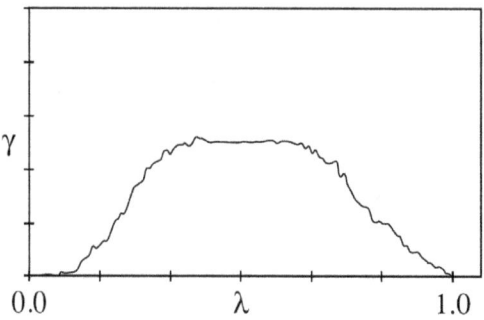

Figure 2. A graph of the average difference-pattern spreading rate, γ, of a large number of randomly chosen $r = 3$, $k = 2$ CAs, as a function of λ. (Adapted from [24] (Packard 1988), with the author's permission. No vertical scale was provided in the original.)

λ (Packard 1988). The spreading rate is a measure of unpredictability in spatiotemporal patterns, and is thus one possible measure of chaotic behavior (Packard 1984; Wolfram 1984). It is analogous (but not identical) to the Lyapunov exponent for continuous-state dynamical systems. In the case of CAs, it indicates the average propagation speed of information through space-time, though not the rate of production of local information.

At each λ, a large number of rules was sampled, and γ was estimated for each CA. The average for γ over the selected CA was taken as the average spreading rate at the given λ (the results are reproduced in Figure 2). At low and high λ, γ vanishes; at intermediate λ, it is maximal; and in the "critical" λ regions—centered about $\lambda \approx 0.23$ and $\lambda \approx 0.83$—it rises or falls gradually. (Li, Packard, and Langton 1990, see Appendix B, define λ_c as the onset of nonzero γ, and use mean-field theory to estimate λ_c in terms of r for two-state CAs. The value from their formula, setting $r = 3$, is $\lambda_c = 0.146$, which roughly matches the value for the onset of nonzero γ seen in Figure 2.)

Though not shown in Figure 2, the variance of γ is high for most values of λ. The same is true for single-site entropy and two-site mutual information as a function of λ (Langton 1990). In other words, the behavior of any *particular* rule at a given value of λ might be very different from the *average* behavior at that value. Thus, interpretation of these averages is somewhat problematic. The preceding account of the behavioral structure of CA rule space (as parameterized by λ) is based on statistics taken from Langton's and Packard's Monte Carlo simulations.

And that is fundamentally the death knell for expecting easy insights from statistical properties of an ensemble of all possible programs, at least with deterministic computational architectures.

(Various problems in correlating λ with behavior will be discussed in Section 8; a detailed analysis of some of these problems can be found in Crutchfield and Hanson (1993).) Other investigations of the structure of CA rule space are reported in Li and Packard (1990) and Li, Packard, and Langton (1990).

It is claimed in Langton (1990) that λ accurately predicts dynamical behavior only when the space of rules is large enough. Apparently, λ is not intended to be a good behavioral predictor for the space of elementary ($r = 1, k = 2$) CA rules (and possibly not for $r = 3, k = 2$ rules either).

5. CA Rule Space and Computation

Langton (1990) hypothesized that a CA's computational capability is related to its average dynamical behavior, which λ is claimed to predict. In particular, he hypothesized that CAs capable of performing nontrivial computation—including universal computation—are most likely to be found in the vicinity of "phase transitions" between order and chaos; that is, near λ_c values. This hypothesis relies on a basic observation of computation theory, that any form of computation requires memory (information storage) and communication (information transmission), and interaction between stored and transmitted information. In addition, however, universal computation requires memory and communication over arbitrary distances in time and space. Thus, complex computation requires significantly long transients and space-time correlation lengths; in the case of universal computation, arbitrarily long transients and correlations are required. Langton claimed that these phenomena are most likely to be seen near λ_c values—near "phase transitions" between order and chaos. This intuition was behind Langton's notion of "computation at the edge of chaos" for CA. (This should be contrasted with the analysis in Crutchfield and Young (1989, 1990) of computation at the onset of chaos and, in particular, with the discussion, also in Crutchfield and Young (1989, 1990), of the structure of CA space.)

Here "requirements" yields to "likelihoods," sliding from deterministic prescriptions to statistical descriptions.

Switching to universality, the second sense above, is risky. In the end, finite systems cannot be universal, strictly speaking, and all real systems are finite at any given moment. The actual issue is about which particular approach to universality is implemented, and what properties that implementation displays as it is scaled up.

6. Evolving CA

The empirical studies that we have just described addressed only the relationship between λ and the dynamical behavior of CA, as revealed by several statistics. They did not correlate λ, or behavior, with an

And is to be commended for putting the vague "edge of chaos" notions on a firm enough footing to be tested and ultimately refuted.

independent measure of computation. Packard (1988) addressed this issue by using a genetic algorithm (GA) (Goldberg 1989; Holland 1975) to evolve CA rules to perform a particular computation. His experiment was meant to test two hypotheses: (1) CA rules able to perform complex computations are most likely to be found near λ_c values; and (2) when CA rules are evolved to perform a complex computation, evolution will tend to select rules near λ_c values.

6.1 THE COMPUTATIONAL TASK, AND AN EXAMPLE CA

Packard's experiment consisted of evolving two-state ($s \in \{0,1\}$) one-dimensional CAs with $r = 3$. The computational task for the CA was to decide whether or not the initial configuration consisted of more than half 1s. If so, the desired behavior for the CA was to relax, after some number of time steps, to a fixed-point pattern of all 1s. If the initial configuration consisted of less than half 1s, the desired behavior for the CA was to relax, after some number of time steps, to a fixed-point pattern of all 0s. If the initial configuration contained exactly half 1s, then the desired behavior was undefined. (This situation can be avoided in practice by requiring that the CA lattice be of odd length.) Thus, the desired CA had only two invariant patterns, all 1s or all 0s. In the following, we denote the density of 1s in a lattice configuration by ρ, the density of 1s in the configuration at time t by $\rho(t)$, and the threshold density for classification by ρ_c.

Does the $\rho_c = 1/2$ classification task qualify as a nontrivial computation for a small-radius ($r \ll N$) CA? Though the term "nontrivial" was not rigorously defined in Langton (1990) or Packard (1988), one possible definition might be any computation for which the memory requirement increases with N (that is, any computation which corresponds to the recognition of a nonregular language), and in which information must be transmitted over significant space-time distances (on the order of N). Under this definition, the $\rho_c = 1/2$ classification task can be thought of as a nontrivial computation for a small-radius CA. The effective minimum amount of memory required is proportional to $\log(N)$, because the equivalent of a counter register is required to track the excess of 1s in a serial scan of the initial pattern. And because the 1s can be distributed throughout the lattice, information transfer over long space-time distances must occur. This is supported in a CA by the nonlocal interactions among many different neighborhoods after some period of time.

Packard cited a $k = 2, r = 3$ rule constructed by Gonzaga de Sá and Maes (1992) and Gacs, Kurdyumov, and Levin (1978), which purportedly performs this task. The Gacs–Kurdyumov–Levin (GKL) CA is defined by the following rule.

If $s_i(t) = 0$, then $s_i(t+1) = $ majority $[s_i(t), s_{i-1}(t), s_{i-3}(t)]$
If $s_i(t) = 1$, then $s_i(t+1) = $ majority $[s_i(t), s_{i+1}(t), s_{i+3}(t)]$

where $s_i(t)$ is the state of site i at time t. In words, this rule says that for each neighborhood of seven adjacent cells, if the state of the central cell is 0, then its new state is decided by a majority vote among itself, its left neighbor, and the cell three sites to the left. Likewise, if the state of the central cell is 1, then its new state is decided by a majority vote among itself, its right neighbor, and the cell three sites to the right.

Figure 1 gives space-time diagrams for the action of the GKL rule on two initial configurations: with $\rho < \rho_c$, and with $\rho > \rho_c$. Although the CA eventually converges to a fixed point, it can be seen that there is a transient phase during which a spatial and temporal transfer of information about local neighborhoods takes place; this local information interacts with other local information to produce the desired final state. Stated crudely, the GKL CA successively classifies "local" densities, with the locality range increasing with time. In regions where there is ambiguity, a "signal" is propagated. This is seen as either a checkerboard pattern propagated in both spatial directions or a vertical white-to-black boundary. These signals indicate that the classification is to be made at a larger scale. Note that both signals locally have $\rho = \rho_c$; as a result, the signal patterns can propagate, since the density of patterns with $\rho = \rho_c$ is not increased or decreased under the rule. In a simple sense, this is the CA's "strategy" for performing the computational task.

It has been claimed that the GKL CA performs the $\rho_c = 1/2$ task (Li 1992); in fact, this is true only to an approximation. The GKL rule was not developed for the purpose of performing any particular computational task, but as part of studies of reliable computation and phase transitions in one spatial dimension. (The goal in the computation studies, for example, was to find a CA whose behavior is robust to small errors in the rule's

Here we see inklings of the "emergent physics" approach: certain patterns of activations are redescribed as "signals" within an implicit computational framework.

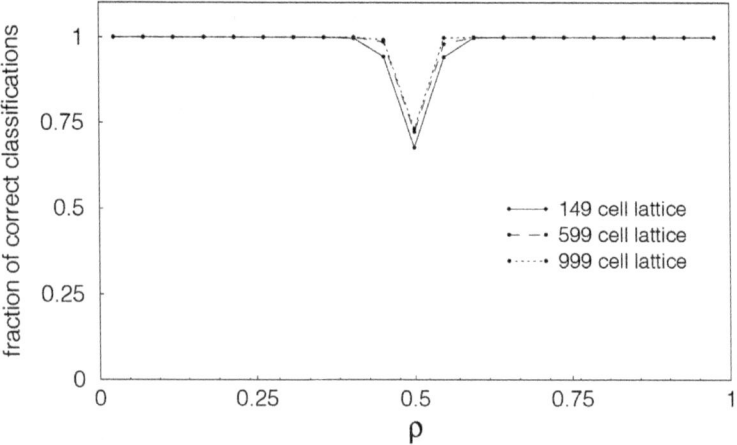

Figure 3. Experimental performance of the GKL rule as a function of $\rho(0)$ for the $\rho_c = 0.5$ task. Performance plots are given for three lattice sizes: $N = 149$ (the size of the lattice used in the GA runs), 599, and 999.

update of the configuration.) It has been proved that the GKL rule has only two attracting patterns, all 1s or all 0s (Gonzaga de Sá and Maes 1992). Attracting patterns in this context are those invariant patterns which return to the same pattern when slightly perturbed. It turns out that the basins of attraction for the all-1 and all-0 patterns are not precisely the initial configurations with $\rho > 0.5$ or $\rho < 0.5$, respectively. On finite lattices the GKL rule does classify most initial configurations according to this criterion, but on a significant number the "incorrect" attractor is reached. (The terms "attractor" and "basin of attraction" are used here in the sense of Gonzaga de Sá and Maes (1992) and Hanson and Crutchfield (1992). This differs substantially from the notion used, for example, in Wuensche and Lesser (1992), where "attractor" refers to any invariant or time-periodic pattern, and "basin of attraction" refers to that set of finite lattice configurations relaxing to an attractor.)

One set of experimental measures of the GKL CA's classification performance is displayed in Figure 3. To make this plot, we ran the GKL CA on 500 randomly generated initial configurations close to each of 21 densities $\rho \in [0.0, 1.0]$. The fraction of correct classifications was then plotted at each ρ. The rule was run either until a fixed point was reached or for a maximum number of time steps equal to $10 \times N$. This was done for CA with three different lattice sizes: $N \in \{149, 599, 999\}$. Note that approximately 30% of the initial configurations with $\rho \approx \rho_c$

were misclassified. All the incorrect classifications are made for initial configurations with $\rho \approx \rho_c$. In fact, the worst performances occur at $\rho = \rho_c$. The error region narrows with increasing lattice size.

The GKL rule table has $\lambda = 1/2$, not $\lambda = \lambda_c$. Given that it appears to perform a computational task of some complexity, at a minimum it is a deviation from the "edge of chaos" hypothesis for CA computation. The GKL rule's $\lambda = 1/2$ puts it right at the center of the "chaotic" region in Figure 2. This may seem puzzling, because the GKL rule clearly does not produce chaotic behavior during either its transient or asymptotic epochs—far from it, in fact. However, the λ parameter was intended to correlate with "average" behavior of CA rules at a given λ value. Recall that γ in Figure 2 represents an *average* over a large number of randomly chosen CA rules and, though not shown in that plot, the variance in γ for most λ values is high. Thus, as previously mentioned, the behavior of any *particular* rule at its λ value might be very different from the *average* behavior at that value.

More to the point, we *expect* a λ value close to $1/2$ for a rule that performs well on the $\rho_c = 1/2$ task. This is the case primarily because the task is symmetric with respect to the exchange of 1s and 0s. Suppose, for example, that a rule that carries out the $\rho_c = 1/2$ task has $\lambda < 1/2$. This implies that there are more neighborhoods in the rule table that map to output bit 0 than to output bit 1. This, in turn, means that there will be *some* initial configurations with $\rho > \rho_c$ on which the action of the rule will *decrease* the number of 1s, which is the opposite of the desired action. However, if the rule acts to *decrease* the number of 1s on an initial configuration with $\rho > \rho_c$, it risks producing an intermediate configuration with $\rho < \rho_c$, which then would lead (under the original assumption that the rule carries out the task correctly) to a fixed point of all 0s, misclassifying the initial configuration. A similar argument holds in the other direction if the rule's λ value is greater than $1/2$. This informal argument shows that a rule with $\lambda \neq 1/2$ will misclassify certain initial configurations. Generally, the further away from $\lambda = 1/2$ that the rule is, the greater the number of such initial configurations. Such rules may perform fairly well, classifying most initial configurations correctly. However, we expect any rule that performs reasonably well on the this task—in the sense of being close to the GKL CA's average performance shown in Figure 3—to have a λ value close to $1/2$.

This analysis points to a problem with using the $\rho_c = 1/2$ task as an evolutionary goal for the study of the relationship between computation and λ. As shown in Figure 2, for an $r = 3, k = 2$ CA the λ_c values occur at roughly 0.23 and 0.83; one hypothesis that was to be tested by Packard's original experiment is that the GA will tend to select rules close to these λ_c values. But for the ρ-classification tasks, the range of λ values required for good performance is simply a function of the task and, specifically, of ρ_c. For example, the underlying 0–1 exchange symmetry of the $\rho_c = 1/2$ task implies that if a CA exists that can perform the task at an acceptable level, then it has $\lambda \approx 1/2$. Though this does not directly invalidate the adaptation hypothesis or claims about λ's correlation with *average* behavior, it presents problems for the use of ρ-classification tasks to gain evidence about a generic relation between λ and computational capability.

6.2 THE ORIGINAL EXPERIMENT

Packard (1988) used a GA to evolve CA rules to perform the $\rho_c = 1/2$ task. His GA began with a randomly generated initial population of CA rules. Each rule was represented as a bit string containing the output bits of the rule table. That is, the bit at position 0 (i.e., the leftmost position) in the string is the state to which the neighborhood 0000000 is mapped, the bit at position 1 in the string is the state to which the neighborhood 0000001 is mapped, and so on. The initial population was randomly generated, but it was constrained to be uniformly distributed across λ values between 0.0 and 1.0.

A given CA rule in the population was evaluated for ability to perform the classification task by choosing an initial configuration at random, running the CA on that initial configuration for some specified number of time steps, and measuring the fraction of cells in the lattice that had the correct state at the final time step. (For initial configurations with $\rho > \rho_c$, the correct final state for each cell is 1, and for initial configurations with $\rho < \rho_c$, the correct final state for each cell is 0.) For example, if the CA were run on an initial configuration with $\rho > \rho_c$ and at the final time step the lattice contained 90%1s, the CA's score on that initial configuration would be 0.9. The fitness of a rule was simply the rule's average score over a set of initial configurations. For each rule in the population, Packard generated a set of initial configurations that were uniformly distributed across ρ values from 0 to 1.

Actually, a slight variation on this method was used in Packard (1988). Instead of measuring the fraction of correct states in the final lattice, the GA measured the fraction of correct states over configurations from a small number n of final time steps (Packard). This prevented the GA from evolving rules that were temporally periodic; for example, those with patterns that alternated between all 0s and all 1s. Such rules obtained higher than average fitness at early generations by often landing at the "correct" phase of the oscillation for a given initial configuration. On the next time step the classification would have been incorrect. In our experiments we used a slightly different method to address this problem, which is explained in subsection 7.1.

Packard's GA worked as follows. At each generation

1. The fitness of each rule in the population was calculated,
2. The population was ranked by fitness,
3. Some fraction of the lowest fitness rules were removed,
4. The removed rules were replaced by new rules formed by crossover and mutation from the remaining rules.

Crossover between two strings involves randomly selecting a position in the strings and exchanging parts of the strings before and after that position. Mutation involves flipping one or more bits in a string, with some low probability. A diversity-enforcement scheme was also used to prevent the population from converging too early and losing diversity (Packard). If a rule was formed that was too close in Hamming distance (i.e., the number of matching bits) to existing rules in the population, its fitness was decreased.

The results from Packard's experiment are displayed in Figure 4. The two histograms display the observed frequency of rules in the GA population as a function of λ, with rules merged from a number of different runs. The top graph gives this data for the initial generation; the rules are uniformly distributed over λ values. The middle graph gives the same data for the final generation—in this case, after the GA has run for 100 generations. The rules now cluster around the two λ_c regions, as can be seen by comparison with the difference-pattern spreading rate plot, reprinted at the bottom of the figure. Note that each individual run produced rules at one or the other peak in the middle graph, so when the runs were merged together, both peaks appear (Packard). Packard interpreted these results as evidence for the hypothesis that, when an ability for complex computation is required, evolution tends to select rules near the transition to chaos. Like Langton, he argues that this result intuitively makes sense, because "rules near the transition to chaos have the capability to selectively communicate information with complex structures in space-time, thus enabling computation" (Packard 1988).

7. New Experiments

As the first step in a study of the reliability of these general conclusions, we carried out a set of experiments similar to those that we have just described. We were unable to obtain precise details of some of the original experiment's parameters, such as the exact population size for the GA, the mutation rate, and so on. As a result, we used what we felt were reasonable values for these parameters. We carried out a number of parameter sensitivity tests which indicated that varying the parameters within small bounds did not change our qualitative results.

To this day, across the sciences, replication studies remain hard to do, although the increasing availability of code repositories has helped a little. In this case, though, it seems unlikely that exact parameters were critical.

7.1 DETAILS OF OUR EXPERIMENTS

In our experiments, as in the original, the CA rules in the population all had $r = 3$ and $k = 2$. Thus, the bit strings that represented the rules were of length $2^{2r+1} = 128$. The size of this search space is huge—the number of possible CA rules is 2^{128}. The tests for each CA rule were carried out on lattices of length $N = 149$ with periodic boundary conditions. The population size was 100, which was roughly the population size used in the original experiment (Packard). The initial population was generated at random, but constrained to be uniformly distributed among different λ values. A rule's fitness was estimated by running the rule on 300 randomly generated initial configurations that were uniformly distributed over $\rho \in [0.0, 1.0]$. Exactly half the initial configurations had $\rho < \rho_c$, and the other half had $\rho > \rho_c$.

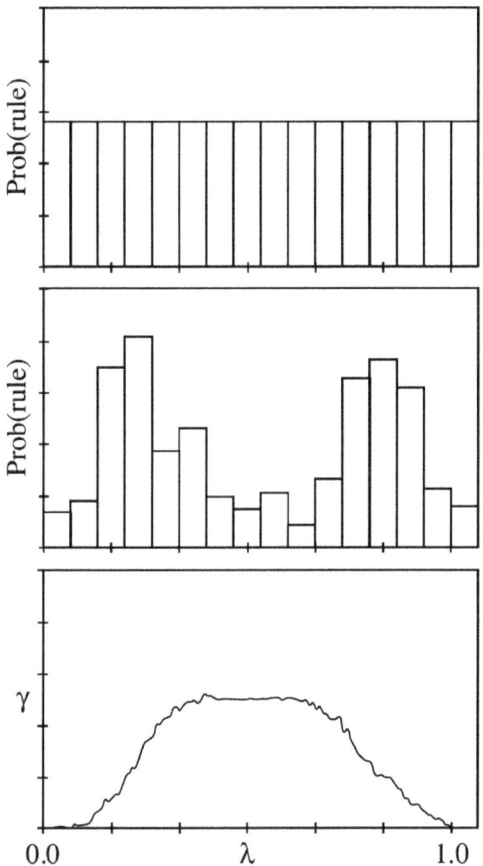

Figure 4. Results from Packard's original experiment on GA evolution of CA for the $\rho_c = 1/2$ classification task. The top two figures are populations of CA at generations 0 and 100, respectively, versus λ. The bottom figure is Figure 2, reproduced here for reference. (Adapted from (Packard 1988), with the author's permission.)

Exact symmetry in the initial configurations at each generation was necessary to avoid early biases in the λ of selected rules. If 49%, say, of the initial configurations had $\rho < \rho_c$, and 51% had $\rho > \rho_c$, rules with λ close to 1 would obtain slightly higher fitness than rules with λ close to 0, because rules with λ close to 1 would map most initial configurations to all 1s. A rule with, say, $\lambda \approx 1$ would in this case classify 51% of the initial configurations correctly whereas a rule with $\lambda \approx 0$ would classify only 49% correctly. But such slight differences in fitness have a large effect in the initial generation, when all rules have fitness close to 0.5, because the GA selects the 50 *best* rules, even if they are only very slightly better than the 50 *worst* rules. This biases the representative rules in the early population. And this bias can persist well into the later generations.

We allowed each rule to run for a maximum number M of iterations, where a new M was selected for each rule from a Poisson distribution with mean 320. This is the measured maximum amount of time for the GKL CA to reach an invariant pattern over a large number of initial configurations on lattice size 149.[1] A rule's fitness was its average score—the fraction of cell states correct at the last iteration—over the 300 initial configurations. We termed this fitness function *proportional fitness*, to contrast with a second fitness function—*performance fitness*—which we describe subsequently. A new set of 300 initial configurations was generated every generation; at each generation, all the rules in the population were tested on this set. Notice that this fitness function is stochastic—the fitness of a given rule may vary a small amount from generation to generation depending on the particular set of 300 initial configurations used in testing it.

Our GA was similar to Packard's: the fraction of new strings each new generation—the "generation gap"—was 0.5. In other words, once the population was ordered according to fitness, the top half of the population—the set of "elite" strings—was copied without modification into the next generation. To GA practitioners more familiar with nonoverlapping generations, this may sound like a small generation gap. However, testing a rule on 300 "training cases" does not necessarily provide a very reliable gauge of what the fitness would be over a larger set of training cases; our selected gap was a good way of making a "first cut," and allowing rules that survived to be tested over more initial configurations. As a new set of initial configurations was produced every generation, rules that were copied without modification were always retested on this new set. If a rule performed well and thus survived over a large number of generations,

---

[1] It may not be necessary to allow the maximum number of iterations to vary. However, in some early tests with smaller sets of fixed initial configurations, we found the same problem that Packard reported (Packard): if M was fixed, then period-2 rules evolved that alternated between all 0s and all 1s. These rules adapted to the small set of initial configurations and the fixed M by landing at the "correct" pattern for a given initial configuration at time step M, only to move to the opposite pattern and wrong classification at time step $M + 1$. These rules performed very poorly when tested on a different set of initial configurations—evidence for "overfitting."

then it was likely to be a genuinely better rule than those that were not selected, since it had been tested with a large set of initial configurations. An alternative method would have been to test every rule in every generation on a much larger set of initial configurations but, given the amount of computer time involved, that method seemed unnecessarily wasteful. Too much effort, for example, would have gone into testing very weak rules, which could safely be weeded out early using our method.

The remaining half of the population for each new generation was created by crossover and mutation from the previous generation's population. (This method of producing non-elite strings differs from that in Packard (1988), where the non-elite strings were formed from crossover and mutation among the elite strings only, rather than from the entire population. We observed no statistically significant differences in our tests using the latter mechanism, other than a modest difference in time scale.) Twenty-five pairs of parent rules were chosen at random with replacement from the entire previous population. For each pair, a single crossover point was selected at random, and two offspring were created by exchanging the subparts of each parent before and after the crossover point. The two offspring then underwent mutation, which consisted of flipping a randomly chosen bit in the string. The number of mutations for a given string was chosen from a Poisson distribution with a mean of 3.8 (this is equivalent to a per-bit mutation rate of 0.03). Again, to GA practitioners this may seem to be a high mutation rate, but one must take into account that half the population was copied without modification at every generation.

7.2 RESULTS OF THE PROPORTIONAL-FITNESS EXPERIMENT

We performed 30 runs of the GA with the parameters described above, each with a different random-number seed. On each run the GA was iterated for 100 generations. We found that running the GA longer (up to 300 generations) did not result in improved fitness. The results of this set of runs are displayed in Figure 5. Figure 5(a) is a histogram of the frequency of rules in the initial populations as a function of λ, merging the rules from all 30 initial populations; thus, the total number of rules represented in this histogram is 3000. The λ bins in this histogram are

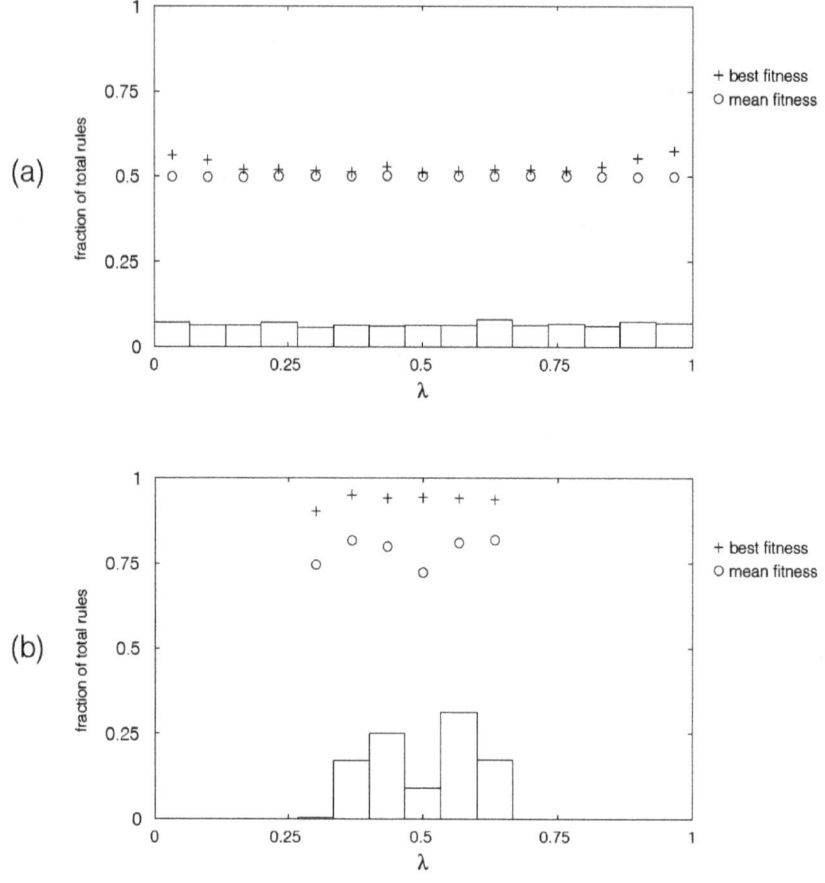

Figure 5. Results from our experiment with proportional fitness. Histogram (a) plots the frequencies of rules merged from the initial generations of 30 runs as a function of λ. Following Packard (1988), the x-axis is divided into 15 bins of length 0.0667 each. The rules with $\lambda = 1.0$ are included in the rightmost bin. Histogram (b) plots the frequencies of rules merged from the final generations (generation 100) of these 30 runs. In each histogram the best and mean fitnesses are plotted for each bin. (The y-axis interval for fitnesses is also $[0, 1]$).

those that were used by Packard, each of width 0.0667. Packard's highest bin contained only rules with $\lambda = 1$ (that is, rules that consist of all 1s). We have merged this bin with the immediately lower bin.

The initial populations consisted of randomly generated rules uniformly spread over the λ values between 0.0 and 1.0. The mean and best fitness values for each bin are also plotted. These are all near 0.5, which is to be expected for a set of randomly generated rules under this fitness function. The best fitnesses are slightly higher in the very low and very high λ bins, because rules with output bits that are almost all 0s (or 1s) correctly classify all low density (or all high density) initial

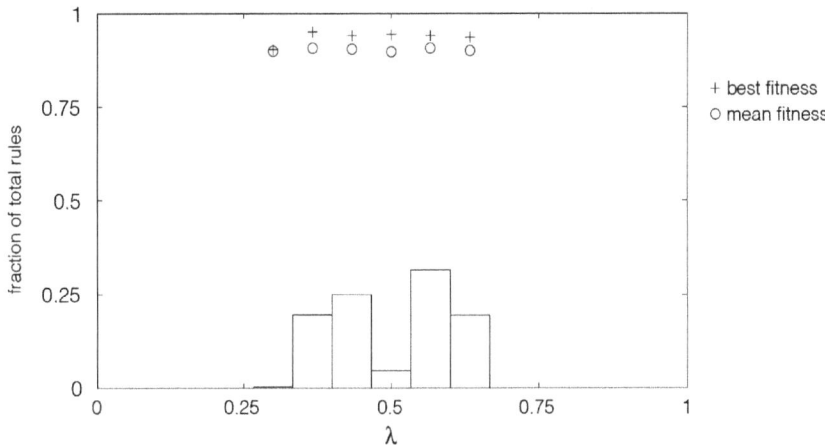

Figure 6. Histogram including only the elite rules from the final generations of the 30 runs with the proportional-fitness function.

configurations. In addition, such CAs obtain small partial credit on some high density (low density) initial configurations and, thus, have fitnesses sightly higher than 0.5.

Figure 5(b) shows the histogram for the final generation (100), merging rules from the final generations of all 30 runs. Again, the mean and best fitness values for each bin are plotted. In the final generation the mean fitnesses are all near 0.8. The exceptions are the central bin (with a mean fitness of 0.72) and the leftmost bin (with a mean fitness of 0.75). The leftmost bin contains only five rules—each at $\lambda \approx 0.33$, right next to the bin's upper λ limit. The standard deviations of mean fitness for each bin (not shown in the figure) are all approximately 0.15—except for the leftmost bin, which has a standard deviation of 0.20. The best fitnesses for each bin are all between 0.93 and 0.95—except for the leftmost bin, which has a best fitness of 0.90. Under this fitness function the GKL rule has fitness ≈ 0.98; the GA never found a rule with fitness above 0.95.

The fitness function is stochastic: a given rule might be assigned a different fitness each time the fitness function is evaluated. The standard deviation for a given rule under the present fitness scheme is approximately 0.015. This indicates that the differences among the best fitnesses plotted in the histogram are not significant, except for that in the leftmost bin.

The lower mean fitness in the central bin is due to the fact that the rules in that bin predominantly come from non-elite rules generated by crossover and mutation in the final generation. This is a combinatorial effect: the density of CA rules as a function of λ is very highly peaked about $\lambda = 1/2$. (We will return to this "combinatorial drift" effect.) Many of the rules in the middle bin have not yet undergone selection and, thus, tend to have lower fitnesses than rules that have been selected in the elite. This effect disappears in Figure 6, which includes only the elite rules at generation 100 for the 30 runs: the difference in mean fitness disappears and the height of the central bin is decreased by half.

The results presented in Figure 5(b) are strikingly different from the results of Packard's experiment. In the final generation histogram in Figure 4, most of the rules clustered around $\lambda \approx 0.23$ or $\lambda \approx 0.83$. In Figure 5(b), however, there are no rules in these λ_c regions. Rather, the rules cluster much more closely around $\lambda = 1/2$—with a ratio of variances of 4 between the two distributions. Recall that this clustering is what we would expect from the basic 0–1 exchange symmetry of the $\rho_c = 1/2$ task.

One rough similarity between the results of the two experiments is the presence of two peaks centered around a dip at $\lambda \approx 0.5$—a phenomenon which we will explain shortly, and which is a key to understanding the GA's behavior on this problem. Nonetheless, there are significant differences, even within this similarity. As was already noted in Packard's experiment, the peaks are in bins centered about $\lambda \approx 0.23$ and $\lambda \approx 0.83$, but in Figure 5(b), the peaks are very close to $\lambda = 1/2$, being centered in the neighboring bins—those with $\lambda \approx 0.43$ and $\lambda \approx 0.57$. Thus, the ratio of original to current peak spread is roughly a factor of 4. Additionally, in the final-generation histogram of Figure 4, the two highest bin populations are roughly five times as high as the central bin, whereas in Figure 5(b) the two highest bins are roughly three times as high as the central bin. Finally, the final-generation histogram in Figure 4 shows the presence of rules in every bin; in Figure 5(b), there are rules in six of the central bins only.

As in Packard's experiment, we found that on any given run the population was clustered about one or the other peak but not both. (Thus, in the histograms that merge all runs, two peaks appear.) This is illustrated in Figure 7, which displays histograms from the final generations of two individual runs. In one of these runs, the population clustered to the left of the central bin and in the other run it clustered to the right of the center. The fact that different runs resulted in different clustering locations was our reason for performing many runs and merging the results, rather than performing a single run with a much larger population—the latter method might have yielded only one peak. In other words, independent of the population size, a given run will be driven by (and the population organized around) the fit individuals that appear earliest. Thus, examining an ensemble of individual runs reveals additional details of the evolutionary dynamics.

The asymmetry in the heights of the two peaks in Figure 5(b) results from a small statistical asymmetry in the results of the 30 runs. In 14 runs, the rules clustered at the lower λ bin, and in 16 runs, the rules clustered at the higher λ bin. This difference is not significant, but it explains the small asymmetry in the peak heights.

We extended 16 of the 30 runs to 300 generations, and found that the basic shape of the histogram does not change significantly (just as the fitnesses do not increase).

7.3 EFFECTS OF DRIFT

The results of our experiments suggest that an evolutionary process modeled by a genetic algorithm tends to select rules with $\lambda \approx 1/2$, for the $\rho_c = 1/2$ task. This is what we had expected, given our prior theoretical discussion concerning this task and its symmetries. We postpone until the next section a discussion of the curious feature near $\lambda = 1/2$ (the dip surrounded by two peaks). In this section, we focus on the larger-scale clustering in that λ region.

To understand that clustering we need to understand the degree to which the selection of rules close to $\lambda = 1/2$ is due to an intrinsic selection pressure, and the degree to which it is due to "drift." By drift we refer to the force that derives from the combinatorial aspects of CA space as explored by random selection ("genetic drift"), combined with the effects of crossover and mutation. The intrinsic effect of random selection with crossover and mutation is to move the population, irrespective of any selection pressure, to $\lambda = 1/2$. This is illustrated in Figure 8 by a histogram mosaic. These histograms show the frequencies

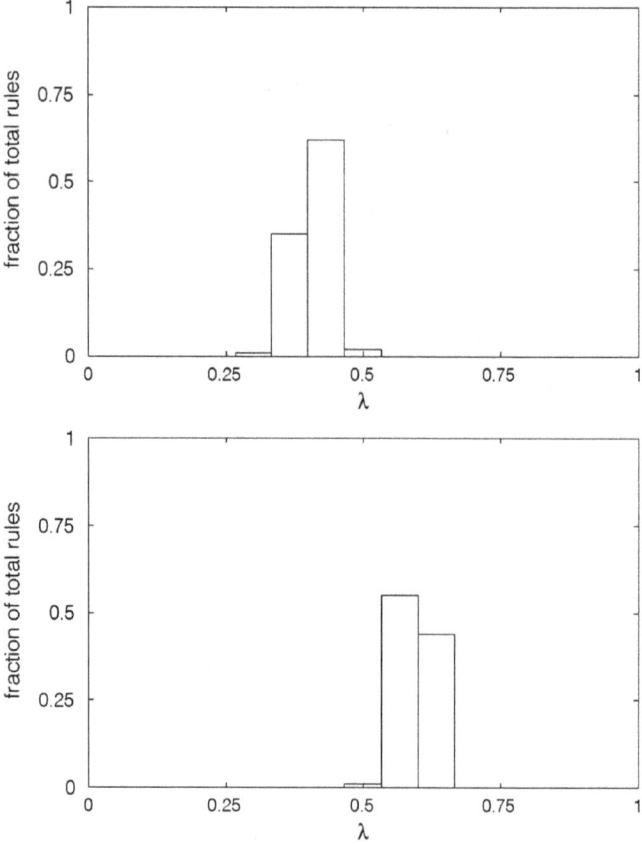

Figure 7. Histograms from the final generations of two individual runs of the GA employing proportional fitness. Each run had a population of 100 rules. The final distribution of rules in each of the 30 runs resembled one of these two histograms.

of the rules in the population as a function of λ, for every 5 generations. Rules were merged from 30 runs in which selection according to fitness was turned off—that is, the fitnesses of the rules in the population were never calculated, and at each generation the selection of the elite group of strings was performed at random. Otherwise, the runs remained the same as previously. Because there is no fitness-based selection, drift is the only force at work. Under the effects of random selection, crossover, and mutation, by the tenth generation the population has largely drifted to the region of $\lambda = 1/2$, and this clustering becomes increasingly pronounced as the run continues.

This drift to $\lambda = 1/2$ is related to the combinatorics of the space of bit strings. For binary CA rules with neighborhood size $n(= 2r + 1)$,

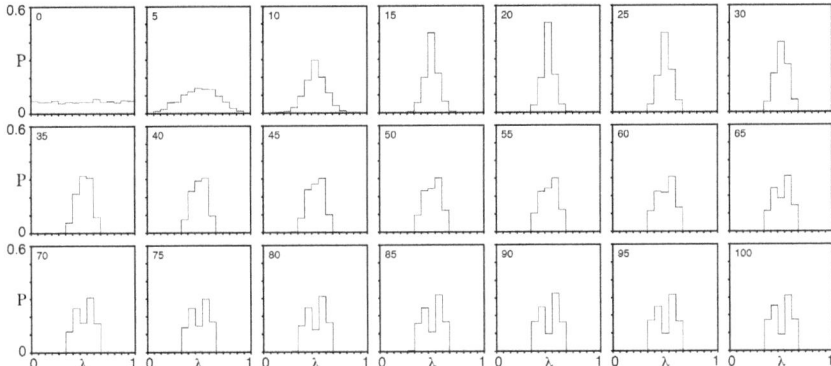

Figure 8. *No-selection mosaic*. Rule-frequency-versus-λ histograms, given every five generations, for populations evolved under the genetic algorithm with no selection (that is, the fitness function was not calculated and the selection of elite rules was performed at random). Each histogram was merged from 30 runs; each run had a population of 100 rules. The generation number is given in the upper left corner of each histogram.

this space consists of all 2^{2^n} binary strings of length 2^n. Denoting the subspace of CAs with a fixed λ and n as $CA(\lambda, n)$, we point out that the size of the subspace is binomially distributed with respect to λ, as follows.

$$|CA(\lambda, n)| = \binom{2^n}{\lambda 2^n}$$

The distribution is symmetric in λ and tightly peaked about $\lambda = 1/2$, with variance $\propto 2^{-n}$. Thus, the vast majority of rules are found at $\lambda = 1/2$. The steepness of the binomial distribution near its maximum gives an indication of the magnitude of the drift "force." Note that the last histogram in Figure 8 gives the GA's rough approximation of this distribution.

Drift is thus a powerful force moving the population to cluster around $\lambda = 1/2$. For comparison, Figure 9 gives the rule-frequency-versus-λ histograms for the merged populations from 30 runs of our proportional-fitness experiment, for every five generations. (A similar mosaic plotting only the elite strings at each generation looks qualitatively similar.) The last histogram in this figure is the same as the one that was displayed in Figure 5(b).

The histograms in Figure 9 look roughly similar to those in Figure 8, up to generation 35. The primary difference in generations 0–30 is that Figure 9 indicates a more rapid peaking about $\lambda = 1/2$. The increased speed of movement to the center is presumably due to the additional

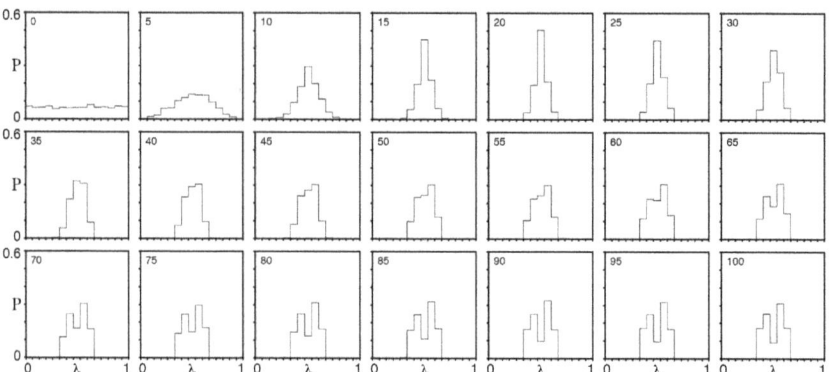

Figure 9. *Proportional-fitness mosaic.* Rule-frequency-versus-λ histograms, given every five generations, merged from the 30 GA runs with proportional fitness. Each run had a population of 100 rules.

evolutionary pressure of proportional fitness. At generation 35, a new feature appears. The peak in the center has begun to shrink significantly and the two surrounding bins are beginning to rival it in magnitude. By generation 40 the right-of-center bin has exceeded the central bin, and by generation 65 the histogram has developed two peaks surrounding a dip in the center. The dip becomes increasingly pronounced as the run continues, but stabilizes by generation 85 or so.

The differences between Figures 8 and 9, over all 100 generations, show that the population's structure in each generation in Figure 9 is not solely due to drift. Indeed, after generation 35, the distinctive features of the population indicate new, qualitatively different, and unique properties due to the selection mechanism. The two peaks represent a symmetry breaking in the evolutionary process—the rules in each individual run initially are clustered around $\lambda = 1/2$, but move to one side or the other of the central bin by about generation 35. We discuss the causes of this symmetry breaking in the next subsection.

7.4 EVOLUTIONARY MECHANISMS: SYMMETRY BREAKING AND THE DIP AT $\lambda = 1/2$

At this point we move away from questions related to Packard's experiment, and concentrate on the mechanisms involved in producing our results. Two major questions must be answered: Why are there significantly fewer rules in the central bin than in the two surrounding

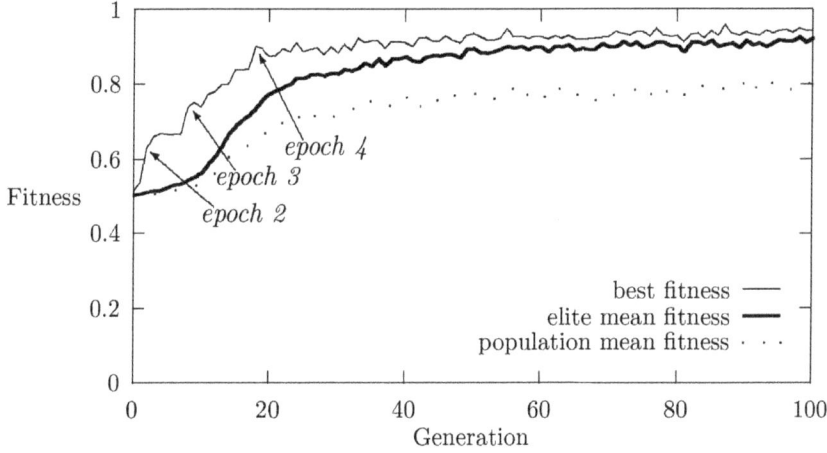

Figure 10. Best fitness, elite mean fitness, and population mean fitness, versus generation for one typical run. The beginnings of epochs 2–4 are pointed out on the best-fitness plot. Epoch 1 begins at generation 0.

bins, in the final generation? What causes the symmetry breaking that begins near generation 35 (as seen in Figure 9)?

The answers (in the briefest terms) obtained by detailed analysis of the 30 GA runs, are as follows. The course of CA evolution under our GA falls roughly into four "strategy" epochs. Each epoch is associated with an innovation discovered by the GA for solving the problem. Though the absolute time at which these innovations appear in each run varies somewhat, each run essentially passes through the four epochs in succession. The epochs are shown in Figure 10 (which plots the best fitness, the mean fitness of the elite strings, and the mean fitness of the population, versus generation, for one typical run of the GA). The beginnings of epochs 2 through 4 are indicated on the best-fitness plot. Epoch 1 begins at generation 0.

Epoch 1: Randomly Generated Rules

The first epoch starts at generation 0, when the best fitness in the initial generation is approximately 0.5 and the λ values are uniformly distributed between 0.0 and 1.0. No rule is much fitter than any other rule, though rules with very low and very high λ tend to have slightly higher fitness, as shown in Figure 5(a). The strategy in this epoch—if it can be called a strategy at all—derives from only the most

Drilling this deeply into understanding the behavior of a genetic algorithm was commendable and somewhat unusual (then and now), and in the spirit of the emergent physics style.

The scare-quotes around "strategy" announce our shifting fully into the emergent physics framework, with its "as-if-intentional" descriptive stance.

elementary aspect of the task. Rules either specialize for $\rho > \rho_c$ configurations by mapping high-density neighborhoods in the CA rule table to 1, or specialize for $\rho < \rho_c$ configurations by mapping low-density neighborhoods to 0.

Epoch 2: Discovery of the Two Halves of the Rule Table

The second epoch begins when a rule is discovered for which most neighborhood patterns in the rule table that have $\rho < \rho_c$ map to 0, and most neighborhood patterns in the rule table that have $\rho > \rho_c$ map to 1. Under the coding scheme we have used, this is roughly correlated with the left and right halves of the rule table: neighborhoods 0000000 to 0111111, and 1000000 to 1111111, respectively. Such a strategy is (presumably) easy for the GA to discover, due to the tendency of single-point crossover to preserve contiguous sections of the rule table. It differs from the accidental strategy of epoch 1 in that there is an organization to the rule table: output bits are roughly associated with densities of neighborhood patterns. It is the first significant attempt at distinguishing initial configurations with more 1s than 0s, and vice versa. Under our fitness function, the fitness of such rules is, approximately, between 0.6 and 0.7, which is significantly higher than the fitness of the initial random rules. This innovation typically occurs between generations 1 and 10; in the run displayed in Figure 10 it occurred in generation 2, and can be seen as the steep rise in the best-fitness plot at that generation. All such rules tend to have λ close to 1/2. There are many possible variations on these rules, with similar fitness, so such rules—all close to $\lambda = 1/2$—begin to dominate in the population. This fact, along with the natural tendency for the population to drift toward $\lambda = 1/2$, is the cause of the clustering around $\lambda = 1/2$ seen by generation 10 in Figure 9. For the next several generations the population tends to explore small variations on this broad strategy. This can be seen in Figure 10 as the leveling off in the best-fitness plot between generations 2 and 10.

Epoch 3: Growing Blocks of 1s or 0s

The next epoch begins when the GA discovers either of two new strategies. The first strategy is to increase the size of a sufficiently large

Mitchell, Hraber, and Crutchfield (1993)

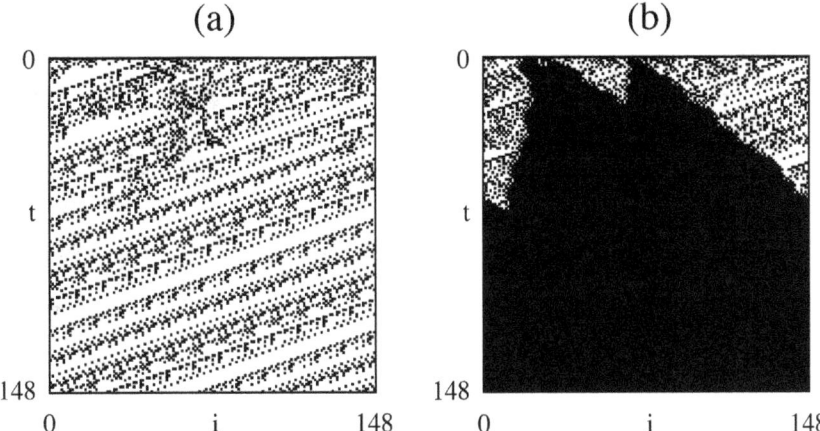

Figure 11. Space-time diagrams of one epoch-3 rule with $\lambda \approx 0.41$ that increases sufficiently large blocks of adjacent or nearly adjacent 1s. Both diagrams have $N = 149$ and are iterated for 149 time steps (the time displayed here is shorter than the actual time allotted under the GA). In (a), $\rho(0) \approx 0.40$ and $\rho(148) \approx 0.17$. In (b), $\rho(0) \approx 0.54$ and $\rho(148) = 1.0$. Thus, in (a) the classification is incorrect, but partial credit is given; in (b) it is correct.

block of adjacent or nearly adjacent 1s; the second strategy is to increase the size of a sufficiently large block of adjacent or nearly adjacent 0s.

Examples of these two strategies are illustrated in Figures 11 and 12. These figures give space-time diagrams from two rules that marked the beginning of this epoch in two different runs of the GA. Figure 11 illustrates the action of a rule discovered at generation 9 of one run. This rule has $\lambda \approx 0.41$, which means that the rule maps most neighborhoods to 0. Its strategy is to map initial configurations to mostly 0s—the configurations it produces have $\rho < \rho_c$, unless the initial configuration contains a sufficiently large block of 1s, in which case it increases the size of that block. Figure 11(a) shows how the rule evolves an initial configuration with $\rho < \rho_c$ to a final lattice with mostly 0s. This produces a fairly good score. Figure 11(b) shows how the rule evolves an initial configuration with $\rho > \rho_c$. The initial configuration contains a few sufficiently large blocks of adjacent or nearly adjacent 1s, and the size of these blocks is quickly increased to yield a final lattice with all 1s for a perfect score. The fitness of this rule at generation 9 was ≈ 0.80.

Figure 12 illustrates the action of a second rule, discovered at generation 20 in another run. This rule has $\lambda \approx 0.58$, which means that the rule maps most neighborhoods to 1. Its strategy is the inverse of the previous rule. It maps initial configurations to mostly 1s unless the

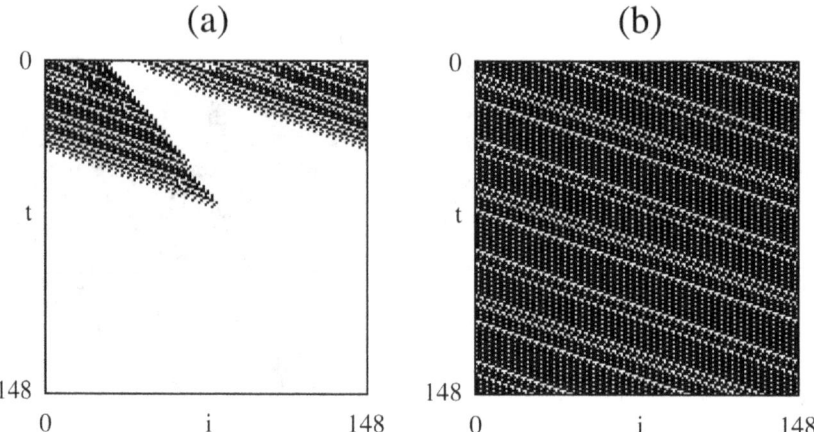

Figure 12. Space-time diagrams of one epoch-3 rule with $\lambda \approx 0.58$ that increases sufficiently large blocks of adjacent or nearly adjacent 0s. In (a), the initial configuration with $\rho \approx 0.42$ maps to a correct classification pattern of all 0s. In (b), the initial configuration with $\rho \approx 0.56$ is not correctly classified ($\rho(148) \approx 0.75$), but partial credit is given.

initial configuration contains a sufficiently large block of 0s, in which case it increases the size of that block. Figure 12(a) illustrates this for an initial configuration with $\rho < \rho_c$; here a sufficiently large block of 0s appears in the initial configuration and is increased in size, yielding a perfect score. Figure 12(b) shows the action of the same rule on an initial configuration with $\rho > \rho_c$. Most neighborhoods are mapped to 1 so the final configuration contains mostly 1s, yielding a fairly high score. The fitness of this rule at generation 20 was ≈ 0.87.

The general idea behind these two strategies is to rely on statistical fluctuations in the initial configurations. An initial configuration with $\rho > \rho_c$ is likely to contain a sufficiently large block of adjacent or nearly adjacent 1s. A rule like the one illustrated in Figure 11 then increases this region's size to yield the correct classification. This holds similarly for rules like the one illustrated in Figure 12, with respect to blocks of 0s in initial configurations with $\rho < \rho_c$. In short, these strategies are assuming that the presence of a sufficiently large block of 1s or 0s is a good predictor of $\rho(0)$.

Similar strategies were discovered in every run; they typically emerged by generation 20. Any single strategy increased blocks of 0s or blocks of 1s, but not both. These strategies result in a significant jump in fitness: typical fitnesses for the first instances of such strategies range from 0.75 to 0.85. This jump in fitness can be seen in the run of Figure 10 at approximately generation 10, and is marked as the beginning of epoch 3. This is the first epoch in which a substantial increase in fitness is associated with a symmetry breaking in the population. The symmetry breaking involves deciding whether to increase blocks of 1s or blocks of 0s. The GKL rule is perfectly symmetric with respect to the increase of blocks of 1s and 0s. The GA on the other hand tends to discover one or the other strategy, and the one that is discovered first tends to take over the population, moving the population λ's to one or the other side of 1/2. The causes of the symmetry breaking are explained following the description of epoch 4.

Typically, the first instances of epoch-3 strategies have a number of problems. As shown in Figures 11 and 12, the rules often rely on partial credit to achieve fairly high fitness on structurally incorrect classification. They do not get perfect scores on many initial configurations, and they often make mistakes in classification. Three common types of classification errors are illustrated in Figure 13. Figure 13(a) illustrates a rule increasing a too-small block of 1s and thus misclassifying an initial configuration with $\rho < \rho_c$. Figure 13(b) illustrates a rule that does not increase blocks of 1s fast enough on an initial configuration with $\rho > \rho_c$, leaving many incorrect bits in the final lattice. Figure 13(c) illustrates the *creation* of a block of 1s that did not appear in an initial configuration with $\rho < \rho_c$, ultimately leading to a misclassification. The rules that produced these diagrams come from epoch 3 in various GA runs.

The increase in fitness seen in Figure 10 between generations 10 and 20 or so is due to further refinements of the basic strategies, which correct these problems to some extent.

Epoch 4: Reaching and Staying at a Maximal Fitness

In most runs, the best fitness is at its maximum value of 0.90 to 0.95 by generation 40 or so. In Figure 10 this occurs at approximately generation 20, and is marked as the beginning of epoch 4. The best fitness does not increase significantly after this; the GA simply finds a number of variations of the best strategies, which all have roughly the

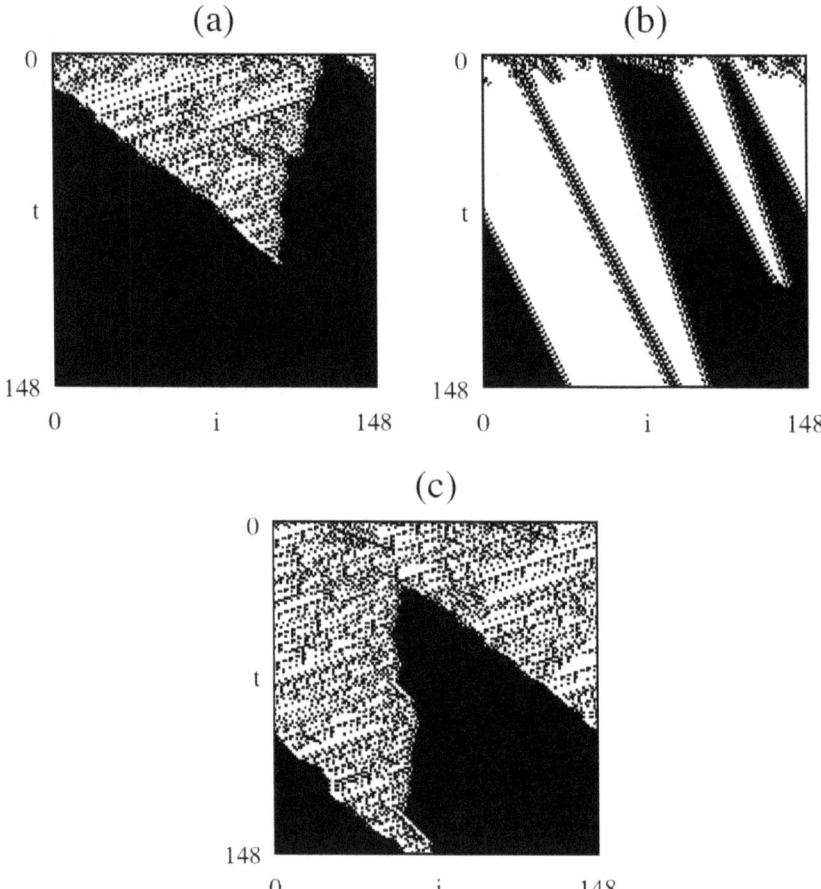

Figure 13. Space-time diagrams illustrating three types of classification errors committed by epoch-3 rules: (a) growing a block of 1s in a sea of $\rho < \rho_c$, (b) growing blocks of 1s too slowly for an initial configuration with $\rho > \rho_c$ (the correct fixed point of all 1s does not occur until iteration 480), and (c) generating a block of 1s from a sea of $\rho < \rho_c$ and growing it so that $\rho > \rho_c$ (the incorrect fixed point of all 1s occurs at iteration 180). The initial configuration densities are (a) $\rho(0) \approx 0.39$, (b) $\rho(0) \approx 0.59$, and (c) $\rho(0) \approx 0.45$.

Mitchell, Hraber, and Crutchfield (1993)

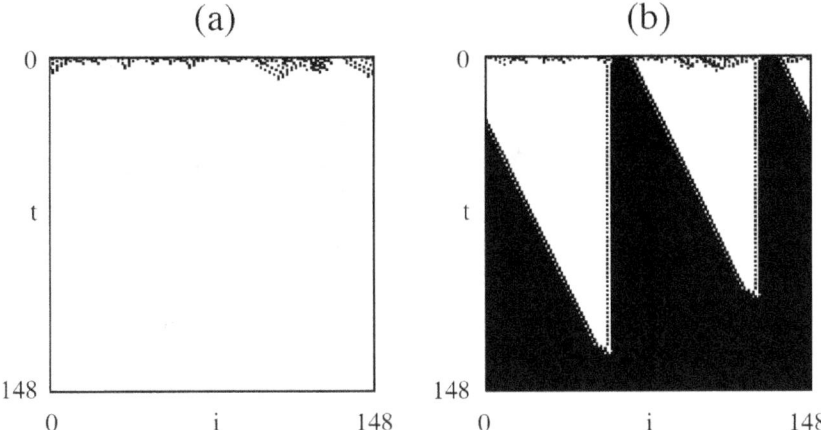

Figure 14. Space-time diagrams of one epoch-4 rule with $\lambda \approx 0.38$ that increases sufficiently large blocks of adjacent or nearly adjacent 1s. In (a), $\rho(0) \approx 0.44$; in (b), $\rho(0) \approx 0.52$. Both initial configurations are correctly classified.

same fitness. When we extended 16 of the 30 runs to 300 generations, we did not see any appreciable increase in the best fitness.

The actions of the best rules from generation 100 of two separate runs are shown in Figures 14 and 15. The space-time diagrams on the left in each figure are for initial configurations with $\rho < \rho_c$, and the diagrams on the right are for initial configurations with $\rho > \rho_c$. The rule illustrated in Figure 14 has $\lambda \approx 0.38$; its strategy is to map initial configurations to 0s unless there is a sufficiently large block of adjacent or nearly adjacent 1s, which if present is increased. The rule shown in Figure 15 has $\lambda = 0.59$ and has the opposite strategy. Each of these rules has fitness ≈ 0.93. They are better tuned versions of the rules in Figures 11 and 12.

Symmetry Breaking in Epoch 3

Notice that the λ values of the rules that have been described are in the bins centered around 0.43 and 0.57 rather than $1/2$. In fact, it seems to be much easier for the GA to discover versions of the successful strategies close to $\lambda = 0.43$ and $\lambda = 0.57$ than to discover them close to $\lambda = 1/2$, though some instances of the latter rules were found. Why is this the case? One reason is that rules with high or low λ work well by *specializing*. The rules with low λ map most neighborhoods to 0s and then increase sufficiently large blocks of 1s when they appear. Rules

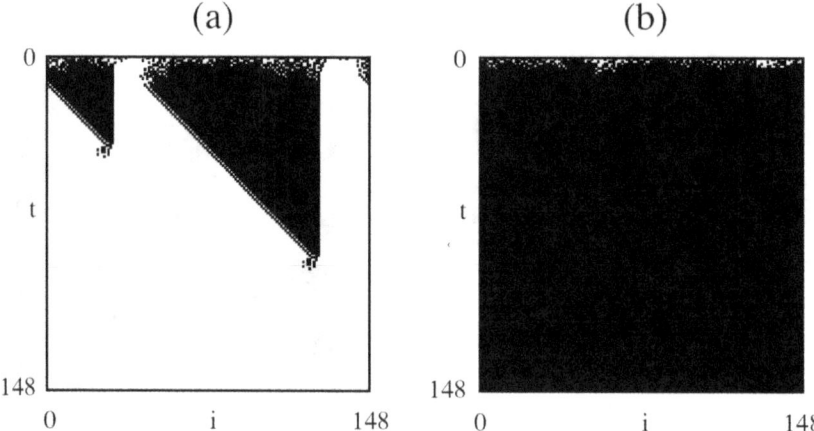

Figure 15. Space-time diagrams of one epoch-4 rule with $\lambda \approx 0.59$ that increases sufficiently large blocks of adjacent or nearly adjacent 0s. In (a), $\rho(0) \approx 0.40$; in (b), $\rho(0) \approx 0.56$. Both initial configurations are correctly classified.

with high λ specialize in the opposite direction. A rule at $\lambda = 1/2$ cannot easily specialize in this way. Another reason is that a successful rule that grows sufficiently large blocks of (say) 1s must avoid *creating* a sufficiently large block of 1s from an initial configuration with less than half 1s. Doing so will lead it to increase the block of 1s and produce an incorrect answer, as was seen in Figure 13(c). An easy way for a rule to avoid creating a sufficiently large block of 1s is to have a low λ. This ensures that low-density initial configurations will quickly map to all 0s, as was seen in Figure 14(a). Likewise, if a rule increases sufficiently large blocks of 0s, it is safer for the rule to have a high λ value so it will avoid creating sufficiently large blocks of 0s where none existed. A rule close to $\lambda = 1/2$ will not have this safety margin, and may be more likely to inadvertently create a block of 0s or 1s that will lead it to a wrong answer. A final feature that contributes to the difficulty of finding good rules with $\lambda = 1/2$ is the combinatorially large number of rules there. In effect, the search space is much larger, which makes the global search more difficult. Locally, about a given adequate rule at $\lambda = 1/2$, there are many rules close in Hamming distance, and thus reachable via mutation, that are not markedly better than the given rule.

Once the more successful versions of the epoch-3 strategies are discovered in epoch 4, their variants spread in the population, and the most successful rules have λ on the low or high side of $\lambda = 1/2$.

This explains the shift from the clustering around $\lambda = 1/2$, as seen in generations 10–30 in Figure 9, to a two-peaked distribution that becomes clear around generation 65. The rules in each run cluster around one or the other peak, specializing in one or the other way. We believe this type of symmetry breaking may be a key mechanism that determines much of the population dynamics and the GA's success—or lack thereof—in optimization.

How does the preceding analysis of the symmetry breaking jibe with the argument, given earlier, that the best rules for the $\rho_c = 1/2$ task must be close to $\lambda = 1/2$? None of the rules found by the GA had a fitness as high as 0.98—the fitness of the GKL rule, whose λ is exactly 1/2. That is, the evolved rules make significantly more classification errors than the GKL rule and, as will be seen below, the measured fitness of the best evolved rules is much worse on larger lattice sizes, whereas the GKL rule's fitness increases with increasing lattice sizes. To obtain the fitness of the GKL rule a number of careful balances in the rule table must be achieved. This is evidently very hard for the GA to do, especially in light of the symmetries in the task and their suboptimal breaking by the GA.

7.5 PERFORMANCE OF THE EVOLVED RULES

Recall that the proportional fitness of a rule is the fraction of correct cell states at the final time step, averaged over 300 initial configurations. This calculation of fitness gives a rule partial credit for getting some final cell states correct. However, the actual task is to relax to either all 1s or all 0s, depending on the initial configuration. In order to measure how well the evolved rules actually perform the task, we define the *performance* of a rule to be the fraction of times the rule correctly classifies initial configurations, averaged over a large number of initial configurations. In this case, credit is given only if the initial configuration relaxes to exactly the correct fixed point after some number of time steps. We measured the performance of each of the elite rules in the final generations of the 30 runs, by testing each rule on 300 randomly generated initial configurations that were uniformly distributed in the interval $0 \leq \rho \leq 1$, and letting the rule iterate on each initial condition for 1000 time steps. Figure 16 displays the mean performance (diamonds) and best performance (squares) in each λ bin. This figure shows that while the mean performances in each bin are much lower than the mean fitnesses for the elite rules shown in Figure 6, the best performance in each bin is roughly the same as the best fitness in that bin. (In some cases the best performance in a bin is slightly higher than the best fitness shown in Figure 6. This is because different sets of 300 initial conditions were used to calculate fitness and performance. This difference can produce small variations in the fitness or performance values.) The best performance we measured was ≈ 0.95. Under this measure the performance of the GKL rule is ≈ 0.98. Thus the GA never discovered a rule that performed as well as the GKL rule, even up to 300 generations. In addition, when we measure the performance of the fittest evolved rules on larger lattice sizes, their performances decrease significantly, while that of the GKL rule remains roughly the same.

7.6 USING PERFORMANCE AS THE FITNESS CRITERION

Can the GA evolve better-performing rules on this task? To find out, we conducted an additional experiment in which performance (as defined in the previous section) was the fitness criterion. As in the previous experiments, at each generation each rule was tested on 300 initial configurations that were uniformly distributed over density values. However, for this experiment, a rule's fitness was defined as the fraction of initial configurations that were correctly classified. An initial configuration was considered to be incorrectly classified if any bits in the final lattice were incorrect. Aside from this modified fitness function, the GA remained the same as in the proportional-fitness experiments. We performed 30 runs of the GA for 100 generations each. The results are given in Figure 17, which gives a histogram plotting the frequencies of the elite rules from generation 100 of all 30 runs, as a function of λ. The shape of the histogram again has two peaks centered around a dip at $\lambda = 1/2$. This shape results from the same symmetry-breaking effect that occurred in the proportional-fitness case; these runs evolved essentially the same strategies as the epoch-3 strategies described previously. The best performances found were ≈ 0.95; these are comparable to the best performances in the proportional-fitness case.

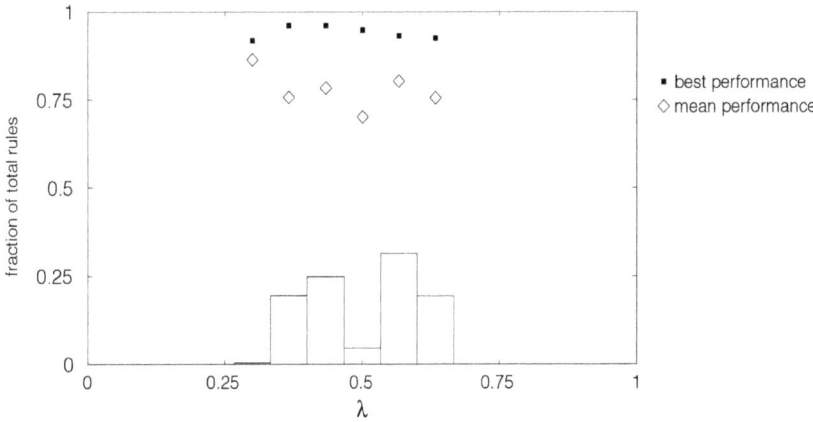

Figure 16. Performances of the final generation elite rules (merged from the 30 runs using the proportional fitness function). The mean and best performances in each bin are plotted on the same histogram as that in Figure 6

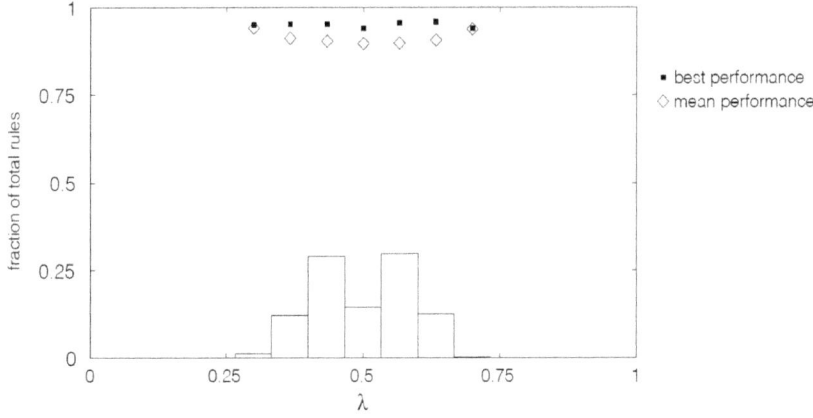

Figure 17. Results from our experiment with performance as the fitness criterion. The histogram plots the frequencies of elite rules merged from the final generations (generation 100) of 30 runs in which the performance-fitness function was used.

The performance of one of the best rules evolved with performance fitness is plotted as a function of $\rho(0)$ in Figure 18, for lattice sizes of 149 (the lattice size used for testing the rules in the GA runs), 599, and 999. This rule has $\lambda \approx 0.54$, and its strategy is similar to that shown in Figure 15: it increases sufficiently large blocks of adjacent or nearly adjacent 0s. We used the same procedure to make these plots as described for Figure 3. The performance, according to this measure, is significantly worse than that of the GKL rule (see Figure 3), especially on larger lattice sizes. The worst performances for the larger lattice sizes are centered slightly above $\rho = 1/2$. On such initial configurations the CA should relax to a fixed point of all 1s, but more detailed inspection of these results revealed that on almost every initial configuration with ρ slightly above $1/2$, the CA relaxed to a fixed point of all 0s. This is a result of the rule's strategy of increasing "sufficiently large" blocks of 0s: the appropriate size to increase was evolved for a lattice with $N = 149$. With larger lattices, the probability of such blocks in initial configurations with $\rho > 1/2$ increases, and the closer the ρ of such initial conditions to $1/2$, the more likely such blocks are to occur. In the CA we tested with $N = 599$ and $N = 999$, such blocks occurred in most initial configurations with ρ slightly above $1/2$, and these initial conditions were always classified incorrectly. This shows that keeping the lattice size fixed during GA evolution can lead to overfitting for the particular lattice size. We plan to experiment with lattice-size variation during evolution in an attempt to prevent such overfitting.

7.7 ADDING A DIVERSITY-ENFORCEMENT MECHANISM

Our description of the four epochs in the GA's search explains the results of our experiment, but it does not explain the difference between our results and those of Packard's experiment reported in Packard (1988). One difference between our GA and the original was the inclusion in the original of a diversity-enforcement scheme that penalized newly formed rules that were too similar in Hamming distance to existing rules in the population. To test the effect of such a scheme on our results, we included a similar scheme in one set of experiments. In our scheme, every time a new string is created through crossover and mutation, the average Hamming distance between the new string and the elite strings—the 50 strings that are copied

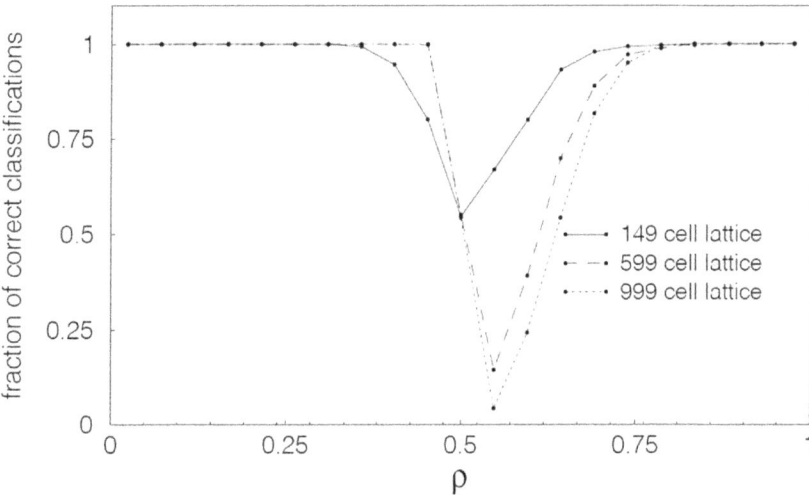

Figure 18. Performance of one of the best rules evolved using performance fitness, plotted as a function of $\rho(0)$. Performance plots are given for three lattice sizes: 149 (the size of the lattice used in the GA runs), 599, and 999. This rule has $\lambda \approx 0.54$.

unchanged—is measured. If this average distance is less than 30% of the string length (here 38 bits), then the new string is not allowed in the new population. New strings continue to be created through crossover and mutation until 50 new strings have met this diversity criterion. We note that many other diversity-enforcement schemes have been developed in the GA literature; one example is "crowding" (Goldberg 1989).

The results of this experiment are given in Figure 19. The histogram in that figure represents the merged rules from the entire population at generation 100 of 20 runs of the GA, using the proportional-fitness function and our diversity-enforcement scheme. The histogram in this figure is very similar to that in Figure 5(b). The only major difference is the significantly lower mean fitness in the middle and leftmost bins, which results from the increased requirement for diversity in the final non-elite population. We conclude that the use of a similar diversity-enforcement scheme was not the factor responsible for the difference between the results in (Packard 1988) and our results.

7.8 DIFFERENCES BETWEEN OUR RESULTS AND THE ORIGINAL EXPERIMENT

As shown in Figure 5(b), our results are strikingly different from those reported in Packard (1988). These experimental results, along

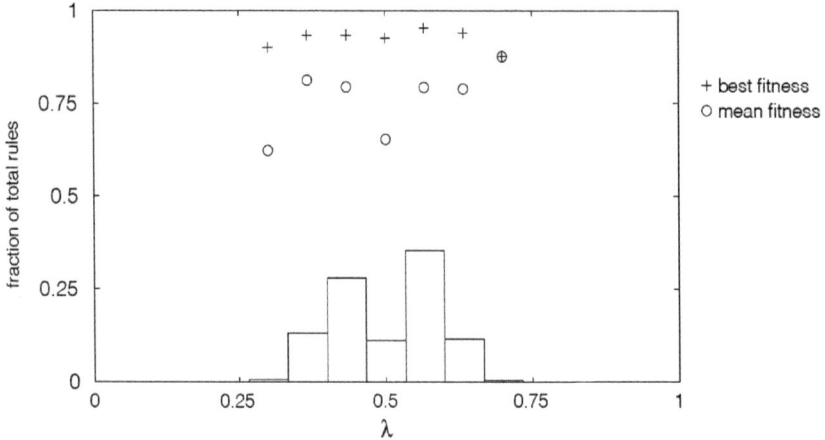

Figure 19. Results from our experiment in which a diversity-enforcement mechanism was added to the GA. The histogram plots the frequencies of rules merged from the entire population at generation 100 of 20 runs with the diversity-enforcement scheme.

with the theoretical argument that the most successful rules for this task should have λ close to $1/2$, lead us to conclude that Packard's interpretation of his results (as giving evidence for the two hypotheses concerning evolution, computation, and λ) is not correct. However, we do not know what accounts for the differences between our results and those obtained in the original experiment. We speculate that the differences are due to additional mechanisms in the GA used in Packard's experiment, which were not reported in Packard (1988). For example, the original experiment included a number of additional sources of randomness, such as the regular injection of new random rules at various λ values and a much higher mutation rate than that in our experiment (Packard). These sources of randomness may have slowed the GA's search for high-fitness rules, and prevented it from converging on rules close to $\lambda = 1/2$. Our experimental results and theoretical analysis give strong reason to believe that the clustering close to λ_c seen in Figure 4 is an artifact of mechanisms in the particular GA that was used, rather than a result of any computational advantage conferred by the λ_c regions.

Although the results were very different, there is one qualitative similarity: the rule-frequency-versus-λ histograms in both cases contained two peaks separated by a dip in the center. As we have noted, the two peaks in our histogram were closer to $\lambda = 1/2$ by a factor of 4, but it is possible that Packard's original results were due to a mechanism similar to either the epoch-1 sensitivity to initial configuration and population asymmetry about $\lambda = 1/2$, or the symmetry breaking we observed in epoch 3. Perhaps these were combined with additional forces, such as additional sources of randomness, that kept rules far away from $\lambda = 1/2$. Unfortunately, the best and mean fitnesses for the λ bins were not reported in Packard (1988). As a consequence we do not know whether the peaks in the original histogram contained high-fitness rules, or even if they contained rules that were more fit than rules in other bins. Our results, and the basic symmetry in the problem, suggest that they did not.

8. General Discussion

8.1 WHAT WE HAVE SHOWN

The results reported in this paper have demonstrated that Packard's results are not reproduced by our experiments. We conclude that the original experiment does not give firm evidence for the hypotheses it was meant to test: first, that rules capable of performing complex computation are most likely to be found close to λ_c values; second, that when CA rules are evolved by a GA to perform a nontrivial computation, evolution will tend to select rules close to λ_c values.

As we argued theoretically, and as our experimental results suggest, the most successful rules for performing a given ρ-classification task will be close to a particular value of λ that depends on the particular ρ_c of the task. Thus, for this class of computational tasks, the λ_c values associated with an "edge of chaos" are not correlated with the ability of rules to perform the task.

The results that we have presented do not disprove the hypothesis that computational capability can be correlated with phase transitions in CA rule space. Individual CAs have been known for some time to exhibit phase transitions with the requisite divergence of correlation length required for infinite memory capacity (Creutz 1986). Indeed, a correlation between computational capability and phase transitions has been noted for other dynamical systems. In the context of continuous-state dynamical systems, it has been shown that there is a direct relationship between the intrinsic computational capability of a process and the degree of randomness of that process at the phase transition from order to chaos. Computational capability was quantified in terms of the statistical complexity, a measure of the amount of memory of a process, and via the detection of an embedded computational mechanism equivalent to a stack automaton (Crutchfield and Young 1989, 1990). More generally, the computational capacity of evolving systems may very well require dynamical properties characteristic of phase transitions, if they are to increase their complexity. We have shown only that the published experimental support cited for hypotheses relating λ_c and computational capability in CA was not reproduced.

In the remainder of this section, we step back from particular experiments and discuss in more general terms the ideas that motivated these studies.

8.2 λ, DYNAMICAL BEHAVIOR, AND COMPUTATION

As noted in section 4, Langton presented evidence that, given certain caveats regarding the radius r and number of states k, there is some correlation between λ and the behavior of an "average" CA on an "average" initial configuration (Langton 1990). Behavior was characterized in terms of such quantities as single-site entropy, two-site mutual information, difference-pattern spreading rate, and average transient length. The correlation is quite strong for very low and very high λ values, which predict fixed-point or short-period behavior. However, for intermediate λ values, there is a large degree of variation in behavior. Moreover, there is no precise correlation between these λ values and the location of a behavioral "phase transition," other than that described by Wootters and Langton (1990) in the limit of infinite k.

These remarks, and the experimental results in Langton (1990), are concerned with the relationship between λ and the dynamical behavior of CAs—they do not directly address the relationship between λ and computational capability of CAs. The basic hypothesis was that λ correlates with computational capability, in the sense that rules capable of complex (and in particular, universal) computation must be, or at least are most likely to be, found near λ_c values. As far as CAs are concerned, the hypothesis was based on the intuition that complex computation cannot be supported in the short-period or chaotic regimes because phenomena such as long transients and long space-time correlation, necessary to support complex computation, apparently occur in "complex" (nonperiodic, nonchaotic) regimes only. Thus far, there has been no experimental evidence correlating λ with an independent measure of computation. Packard's experiment was intended to address this issue, as it involved an independent measure of computation—performance on a particular complex computational task—but, as we have shown, it did not provide evidence for the hypothesis linking λ_c values with computational ability.

One problem is that these hypotheses have not been rigorously formulated. If the hypotheses put forth in Langton (1990) and Packard (1988) are interpreted to mean that *any* rule performing complex computation (as exemplified by the $\rho = 1/2$ task) must be close to λ_c, then we have shown them to be false with our argument that correct performance on the $\rho = 1/2$ task requires $\lambda = 1/2$. If the hypotheses are concerned instead with generic, statistical properties of CA rule space—the "average" behavior of an "average" CA at a given λ—then the notion of "average behavior" must be better defined. Additionally, more appropriate measures of dynamical behavior and computational capability must be formulated, and the notion of the "edge of chaos" must also be well defined.

The argument that complex computation cannot occur in chaotic regimes may seem intuitively correct, but there is a theoretical framework and strong experimental evidence to the contrary. Hanson and Crutchfield (Crutchfield and Hanson 1993; Hanson and Crutchfield 1992) have developed a method for filtering out chaotic "domains" in the space-time diagram of a CA, sometimes revealing "particles" that have the nonperiodic, nonchaotic properties of structures in Wolfram's Class 4 CA. In other words, with the application of the appropriate filter, complex structures can be uncovered in a space-time diagram that, to the human eye (and to the statistics used in Langton 1990; Packard 1988) appears to be completely random. As an extreme example, it is conceivable that such filters could be applied to a seemingly chaotic CA and reveal that the CA is actually implementing a universal computer (with glider guns implementing AND, OR, and NOT gates, and so on). Hanson and Crutchfield's results strikingly illustrate the fact that apparent complexity of behavior—and apparent computational capability—can depend on the implicit "filter" imposed by one's chosen statistics.

Again, the emergent physics stance is made strikingly explicit here: complex structures, some with clearly computational uses, that appear as mere randomness to statistical physics.

8.3 WHAT KIND OF COMPUTATION IN CA DO WE CARE ABOUT?

In the previous section, the phrases "complex computation" and "computational capability" were used somewhat loosely. As was discussed in section 3, there are at least three different interpretations of the notion of computation in CAs. The notion of a CA being able to perform a "complex computation" such as the $\rho_c = 1/2$ task, where the CA performs the same computation on all initial configurations, is very different from the notion of a CA being capable of simulating, under some special set of initial configurations, a universal computer. Langton's speculations regarding the relationship between dynamical behavior and computational capability seem to be more concerned with the latter than the former, though they imply that the capability to sustain long transients, long correlation lengths, and so on, is necessary for both notions of computation.

If "computationally capable" is taken to mean "capable, under some initial configurations, of universal computation," then one might ask why this is a particularly important property of CAs on which to focus attention. In Langton (1990), CAs were used as a vehicle to study the relationship between phase transitions and computation, with an emphasis on universal computation. But for those wishing to use CAs as scientific models or practical computational tools, a focus on the capacity for universal computation may be misguided. If a CA is being used as a model of a natural process (e.g., turbulence), then it is of limited interest to know whether or not the CA is, in principle, capable of universal computation (especially if universal computation will arise only under some specially engineered initial configurations that the natural process is extremely unlikely ever to encounter). To understand emergent computation in natural phenomena as modeled by CAs, one should try to understand what computation the CA does "intrinsically" (Crutchfield and Hanson 1993; Hanson and Crutchfield 1992), rather than what it is capable of doing "in principle" (and only under some very special initial configurations). Thus, understanding the conditions under which a capacity for universal computation is possible will not be of much value in understanding the natural systems modeled by CAs.

Indeed, it could be argued that, to this day, focusing on "computability"—what can in principle be computed deterministically—has misled many researchers and science authors, old and new. Thirty years later, this whole section remains a potent call to look beyond universality and focus on specific problems and the computational complexity of their solutions.

This general point is neither new nor deep. Analogous arguments have been put forward in the context of neural networks, for example. While many constructions of universal computation in neural networks have been made (e.g., Siegelman and Sontag 1991), some psychologists (e.g., Rumelhart and McClelland 1986) have argued that this has little to do with understanding how brains or minds work in the natural world.

Similarly, if one wishes to use a CA as a parallel computer for solving a real problem—such as face recognition—it would be very inefficient, if not practically impossible, to solve the problem by (say) programming Conway's Game of Life CA to be a universal computer that simulates the action of the desired face recognizer. Thus, understanding the conditions under which universal computation is possible in CAs is not of much practical value either.

In addition, it is not clear that anything like a drive toward universal computational capabilities is an important force in the evolution of biological organisms. It seems likely that substantially less computationally-capable properties play a more frequent and robust role. Thus, asking under what conditions evolution will create entities (including CAs) that are capable of universal computation may not be of great importance in understanding natural evolutionary mechanisms.

In short, it is mathematically important to know that some CAs are, in principle, capable of universal computation. But we argue that this is by no means the most scientifically interesting property of CAs. More to the point, this property does not help scientists much in understanding the emergence of complexity in nature, or in harnessing the computational capabilities of CAs to solve real problems.

9. Conclusion

The main purpose of this study was to examine and clarify the evidence for various hypotheses related to evolution, dynamics, and the computational capability of cellular automata. As a result of our study we have identified a number of evolutionary mechanisms, such as the role of combinatorial drift, and the role of symmetry. We have also found that the breaking of the goal task's symmetries in the early generations can be an impediment to further optimization of individuals in the population. Symmetry breaking results in a kind of suboptimal speciation in a population that is stable (or at least metastable) over long times. The symmetry-breaking effects we have described may be similar to symmetry-breaking phenomena that emerge in biological evolution, such as brain hemispheric dominance and handedness, or the breaking of the spherical symmetry of a blastula which results in bilateral symmetry. It is our goal to develop a more rigorous framework for understanding these mechanisms in the context of evolving CAs. We believe that a deep understanding of these mechanisms in this relatively simple context can yield insights for understanding evolutionary processes in general, and for successfully applying evolutionary computation methods to complex problems.

Though our experiments did not reproduce the results reported in Packard (1988), we believe that Packard's original strategy of using GAs to evolve computation in CAs is an important idea. In addition to its potential for the study of various theoretical issues, it has a practical potential that could be significant. As previously mentioned, CAs are increasingly being studied as a class of efficient parallel computers; the main bottleneck in applying CAs more widely to parallel computation is *programming*—in general, it is very difficult to program CAs to perform complex tasks. Our results suggest that the GA has promise as a method for accomplishing such programming automatically. In order to test further the GA's effectiveness when compared with other search methods, we performed an additional experiment, comparing the performance of our GA on the $\rho_c = 1/2$ task with the performance of a simple steepest-ascent hill-climbing method. We found that the GA significantly outperformed hill climbing, reaching much higher fitnesses for an equivalent number of fitness evaluations. This gives some evidence for the relative effectiveness of GAs when compared with simple gradient ascent methods for programming CAs. Koza (1993) has also evolved CA rules using a very different type of representation scheme; the relationship between representation and GA success on such tasks is a topic of substantial practical interest.

Cramming in more results right down to the wire!

Acknowledgments

This research was supported by the Santa Fe Institute, under the Adaptive Computation, Core Research, and External Faculty Programs, and by the University of California, Berkeley, under contract AFOSR 91-0293. Thanks to Doyne Farmer, Jim Hanson, Erica Jen, Chris Langton, Wentian Li, Cris Moore, and Norman Packard for many helpful discussions and suggestions concerning this project. Thanks also to Emily Dickinson and Terry Jones for technical advice.

This paper has held up well! I enjoyed this opportunity to look back it.

I apologize that my comments have dwelt so on deterministic execution, even though this paper and the field today usually accept that implicitly. I accepted it too for most of my career; the reformed sinner often prays the loudest.

REFERENCES

Berlekamp, E., J. H. Conway, and R. Guy. 1982. *Winning Ways for Your Mathematical Plays.* New York, NY: Academic Press.

Creutz, M. 1986. "Deterministic Ising Dynamics." *Annals of Physics* 167:62–72.

Crutchfield, J. P., and J. E. Hanson. 1993. *Turbulent Pattern Bases for Cellular Automata.* Technical report. Technical Report 93-03-010; Physica D, in press. Santa Fe Institute.

Crutchfield, J. P., and K. Young. 1989. "Inferring Statistical Complexity." *Physical Review Letters* 63:105–108.

———. 1990. "Computation at the Onset of Chaos." In *Complexity, Entropy, and the Physics of Information,* edited by W. H. Zurek. Redwood City, CA: Addison-Wesley.

Farmer, J. D., T. Toffoli, and S. Wolfram, eds. 1984. *Cellular Automata: Proceedings of an Interdisciplinary Workshop.* Amsterdam, Netherlands: North Holland.

Gacs, P., G. L. Kurdyumov, and L. A. Levin. 1978. "One-Dimensional Uniform Arrays that Wash Out Finite Islands." *Problemy Peredachi Informatsii* 14:92–98.

Goldberg, D. E. 1989. *Genetic Algorithms in Search, Optimization, and Machine Learning.* Reading, MA: Addison-Wesley.

Gonzaga de Sá, P., and C. Maes. 1992. "The Gacs–Kurdyumov–Levin Automaton Revisited." *Journal of Statistical Physics* 67 (3/4): 507–522.

Guckenheimer, J., and P. Holmes. 1983. *Nonlinear Oscillations, Dynamical Systems, and Bifurcations of Vector Fields.* New York, NY: Springer-Verlag.

Gutowitz, H. A., ed. 1990. *Cellular Automata.* Cambridge, MA: MIT Press.

Hanson, J. E., and J. P. Crutchfield. 1992. "The Attractor-Basin Portrait of a Cellular Automaton." *Journal of Statistical Physics* 66 (5/6): 1415–1462.

Holland, J. H. 1975. *Adaptation in Natural and Artificial Systems.* Ann Arbor, MI: University of Michigan Press.

Kauffman, S. A. 1990. "Requirements for Evolvability in Complex Systems: Orderly Dynamics and Frozen Components." *Physica D* 42:135–152.

Kauffman, S. A., and S. Johnsen. 1992. "Co-Evolution to the Edge of Chaos: Coupled Fitness Landscapes, Poised States, and Co-Evolutionary Avalanches." In *Artificial Life II,* edited by C. G. Langton, J. D. Farmer, S. Rasmussen, and C. Taylor, 325–370. Redwood City, CA: Addison-Wesley.

Koza, J. R. 1993. *Genetic Programming: On the Programming of Computers by Means of Natural Selection.* Cambridge, MA: MIT Press.

Langton, C. G. 1990. "Computation at the Edge of Chaos: Phase Transitions and Emergent Computation." *Physica D* 42:12–37.

Li, W. 1992. "Non-Local Cellular Automata." In *1991 Lectures in Complex Systems,* edited by L. Nadel and D. Stein. Redwood City, CA: Addison-Wesley.

Li, W., and N. H. Packard. 1990. "The Structure of the Elementary Cellular Automata Rule Space." *Complex Systems* 4:281–297.

Li, W., N. H. Packard, and C. G. Langton. 1990. "Transition Phenomena in Cellular Automata Rule Space." *Physica D* 45:77–94.

Lindgren, K., and M. G. Nordahl. 1990. "Universal Computation in a Simple One-Dimensional Cellular Automaton." *Complex Systems* 4:299–318.

Packard, N. H. 1984. "Complexity of Growing Patterns in Cellular Automata." In *Dynamical Behavior of Automata: Theory and Applications,* edited by J. Demongeot, E. Golès, and M. Tchuente. New York, NY: Academic Press.

———. 1988. "Adaptation Toward the Edge of Chaos." In *Dynamic Patterns in Complex Systems,* edited by J. A. S. Kelso, A. J. Mandell, and M. F. Shlesinger, 293–301. Singapore: World Scientific.

———. Personal communication.

Pollack, J. B. 1991. "The Induction of Dynamical Recognizers." *Machine Learning* 7:227–252.

Preston, K., and M. Duff. 1984. *Modern Cellular Automata.* New York, NY: Plenum.

Rosenfeld, A. 1983. "Parallel Image Processing Using Cellular Arrays." *Computer* 16:14–20.

Rumelhart, D. E., and J. L. McClelland. 1986. "PDP Models and General Issues in Cognitive Science." In *Parallel Distributed Processing, Vol. 1,* edited by D. E. Rumelhart and J. L. McClelland. Cambridge, MA: MIT Press.

Siegelman, H., and E. D. Sontag. 1991. *Neural Networks are Universal Computing Devices.* Technical report SYCON-91-08. Rutgers Center for Systems and Control, Rutgers University, New Brunswick, NJ, 08903.

Toffoli, T., and N. Margolus. 1987. *Cellular Automata Machines: A New Environment for Modeling.* Cambridge, MA: MIT Press.

Wolfram, S. 1984. "Universality and Complexity in Cellular Automata." *Physica D* 10:1–35.

———, ed. 1986. *Theory and Applications of Cellular Automata.* Singapore: World Scientific.

Wootters, W. K., and C. G. Langton. 1990. "Is There a Sharp Phase Transition for Deterministic Cellular Automata?" *Physica D* 45:95–104.

Wuensche, A., and M. Lesser. 1992. *The Global Dynamics of Cellular Automata*. Santa Fe Institute Studies in the Sciences of Complexity: Reference Volumes. Reading, PA: Addison-Wesley.

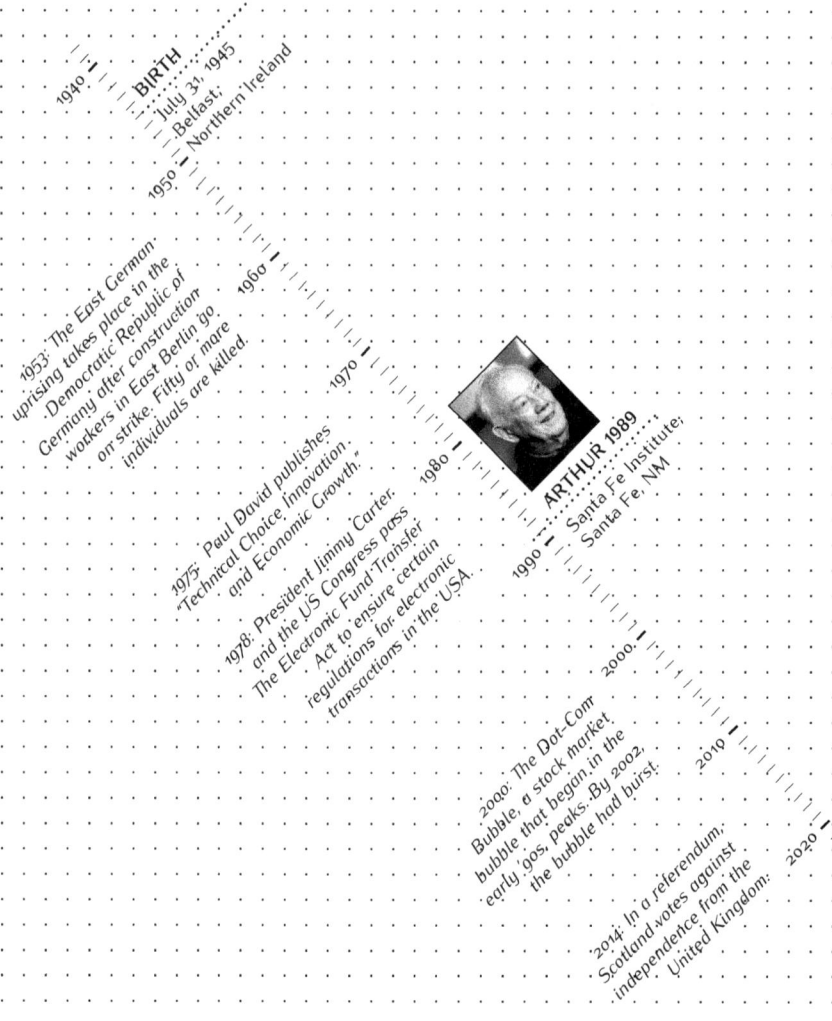

BIRTH — July 31, 1945, Belfast, Northern Ireland

1953: The East German uprising takes place in the Democratic Republic of Germany after construction workers in East Berlin go on strike. Fifty or more individuals are killed.

1975: Paul David publishes "Technical Choice, Innovation, and Economic Growth."

1978: President Jimmy Carter and the US Congress pass The Electronic Fund Transfer Act to ensure certain regulations for electronic transactions in the USA.

ARTHUR 1989 — Santa Fe Institute, Santa Fe, NM

2000: The Dot-Com Bubble, a stock market bubble that began in the early '90s, peaks. By 2002, the bubble had burst.

2014: In a referendum, Scotland votes against independence from the United Kingdom.

WILLIAM BRIAN ARTHUR

[77]

THE COSTS OF MISCOORDINATION

Willemien Kets, Utrecht University and Santa Fe Institute

Nobody goes there anymore. It's too crowded. —Yogi Berra

With some papers, it is clear that they will become an instant classic. W. Brian Arthur's 1994 "Inductive Reasoning and Bounded Rationality (The El Farol Problem)" is such a paper. Like many classics, it starts with a question anyone can understand: Suppose going to a bar is only pleasant if not too many people show up. How do you decide whether to go? This "bar problem" was inspired by the popular El Farol bar in Santa Fe.[1] But, of course, this toy problem is an exemplar of a much more general class of problems: How do people divide a scarce resource among themselves? Yet game theory—the discipline that studies interactive decision-making—is ill equipped to address it. Classical game theory posits that players will coordinate on one of the equilibria of the game. The argument, roughly speaking, is this: If the system is not in equilibrium, then at least one of the players could gain by changing their action. For example, if few people are going to the bar one night, then at least one player would be better off if they attended rather than staying home. And even if people face this problem repeatedly (say, every Thursday night), it is far from clear the system would converge to an equilibrium. Of course, people can look at past attendance to predict how many people will go. But if everyone uses the same forecasting model, then predictions are bound to be invalidated: If all believe few will go, all will go; and if all believe most will go, nobody will go. So, if some stationary state is to emerge, it will involve persistent heterogeneity.

W. B. Arthur, "Inductive Reasoning and Bounded Rationality," *American Economic Review Papers and Proceedings* 84, 406–411 (1994).

Reprinted by permission of the author.

---

[1] At the time, the Santa Fe Institute—where Arthur was the first resident fellow—was just around the corner from El Farol.

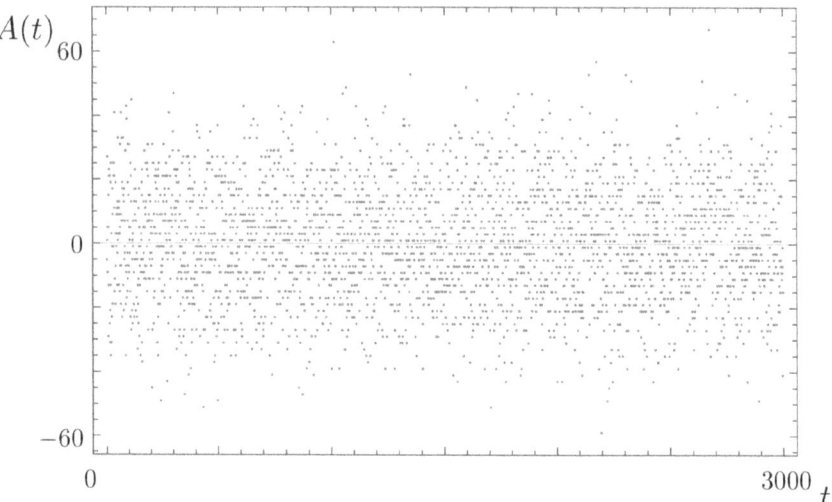

Figure 1.

What type of behavior can we expect in such an environment? Arthur's key insight is that the problem players face in such a setting is not unlike the one we face when playing chess. Like the bar problem, chess has an equilibrium (Zermelo 1913) and is therefore considered "solved" by classical game theory. But of course no one, not even grand masters or the most advanced chess computer, plays chess by engaging in a deductive equilibrium analysis. Rather, players use induction: They try to detect patterns in their opponent's play (e.g., try to make out which opening their opponent is using), construct mental representations or models based on these, use these models to carry out localized deductions, and strengthen or replace these models as they receive feedback on their models' performance. In the bar problem, patterns are past attendance levels, and mental models take the form of simple hypotheses like "If the bar was crowded last week, it'll likely be quiet this week." To capture this, Arthur developed a simple dynamic model: There are N players, and each player is randomly assigned k possible strategies, that is, functions that map the past m weeks' attendance levels into an action: Go to the bar or stay home. Players keep track of how successful each strategy is—does it prescribe staying home when the bar is crowded but not otherwise?—and use the one with the current best performance to decide whether to attend tonight.

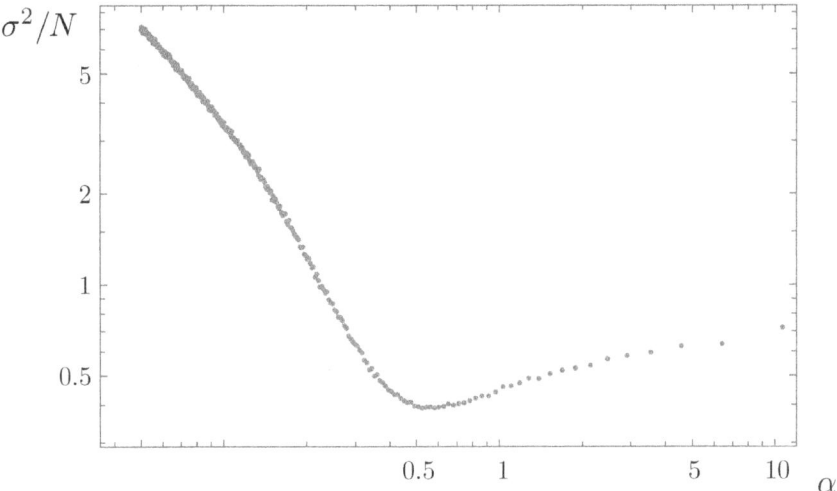

Figure 2.

This simple model defines an adaptive complex system: How well each strategy does depends on how well it is adapted to the environment created by the other strategies. The resulting ecology of predictors displays rich behavior. For example, while the average attendance level is consistent with game-theoretic predictions, deviations from equilibrium can be large (fig. 1). This is consistent with experimental evidence but cannot be explained by other models of adaptive play (Kets 2011). These fluctuations are more than a curiosity: When there are big swings in attendance, the bar is either overcrowded or nearly empty, and payoffs are low.

An important question, therefore, is how the magnitude of these fluctuations—measured by the volatility—depends on model parameters. Subsequent research has shown that the key parameter is $\alpha := 2^m/N$ (Challet, Marsili, and Zhang 2004). For large α, the volatility approaches that for the "binomial" case where each player makes their decision by flipping a coin; for small α, volatility is much higher than in the binomial case; and for intermediate α, volatility is minimized (fig. 2).

What drives this striking relation between the volatility and α? The parameter α measures the number of maximally differentiated strategies per player (i.e., strategies that prescribe different actions after every possible history). When α is small, there is little room for differentiation: A large share of players use the same strategy in any given period, so choices are highly correlated. This translates into high volatility and low payoffs. When α is large, on the other hand, there are so few strategies per player that players are acting almost independently. The case of intermediate α ($\alpha \simeq \alpha c$) is the sweet spot: Players are able to anticorrelate their actions by self-organizing into groups that always do the opposite. If one group goes, the other stays home, and vice versa. This minimizes volatility and maximizes efficiency.

While the system itself is easy to describe (and simulate), analyzing the system in a mathematically rigorous way turned out to be challenging. A major leap forward was the insight that the bar problem has much in common with disordered systems in condensed matter physics, with its heterogeneity, quenched disorder, and frustration. This made it possible to use statistical mechanics techniques to analyze the problem (Coolen 2005) and led to novel insights. For example, the system undergoes a phase transition at α_c, from an informationally efficient regime where past history contains no information that players can exploit ($\alpha < \alpha c$) to an informationally inefficient regime full of arbitrage opportunities (fig. ??). Not only is this conceptually interesting, it also has practical implications: To the extent that speculators in financial markets face similar incentives to differentiate their actions as our patrons (e.g., sell when everyone else is buying), this result suggests that, depending on the strategic complexity of the market, agents entering the market can exploit the predictability in the time series, thus eliminating any arbitrage opportunities (cf. Arthur *et al.* 1997).

At a conceptual level, Arthur's paper is one of the few in economics to highlight the costs of miscoordination: Even if behavior is consistent with equilibrium at the macro level (i.e., average attendance is close to equilibrium), nonequilibrium behavior at the micro-level can give rise to large deviations and thus to inefficiency. But while miscoordination is pervasive, it is often difficult to explain with standard models of adaptive play. An emerging insight is that the key to stabilizing miscoordination in dynamic models of behavior is to allow precisely for the type of correlation induced by players sharing mental models as in Arthur's conceptualization of inductive reasoning (Kets, Kager, and Sandroni 2022).

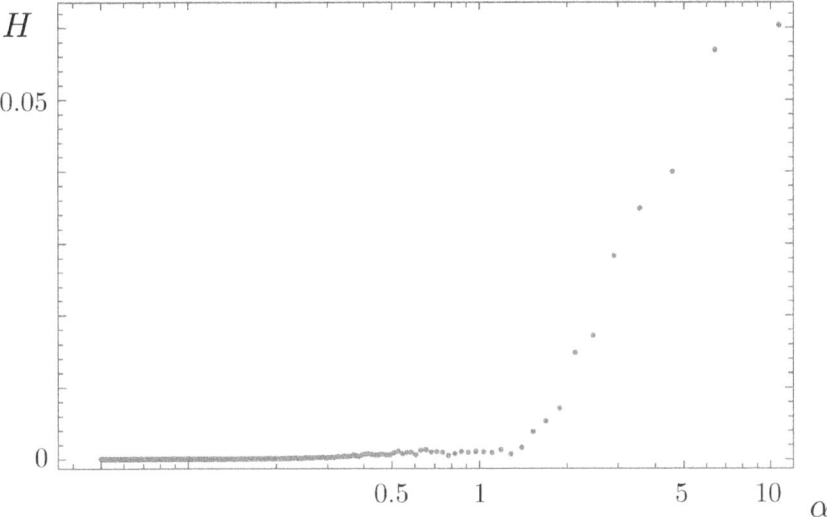

Figure 3.

This only scratches the surface of the profound impact Arthur's simple question—to go or not to go?—has had on game theory, economics, and other fields. For example, with its simple model and rich patterns of behavior, it was a major impetus for the then-nascent literature on agent-based computational economics (Miller and Page 2009). One thing the paper seems to have had no impact on, though: While the Santa Fe Institute is always crowded, this does not seem to deter anyone from visiting.

REFERENCES

Arthur, W. B. 1989. "Competing Technologies, Increasing Returns, and Lock-in by Historical Events." *Economic Journal* 99 (394): 116–131. https://doi.org/10.2307/2234208.

Arthur, W. B., J. H. Holland, B. LeBaron, R. Palmer, and P. Tayler. 1997. "Asset Pricing under Endogenous Expectations in an Artificial Stock Market." In *The Economy as an Evolving Complex System II*, edited by W. B. Arthur, S. Durlauf, and D. Lane. Boston, MA: Addison-Wesley.

Challet, D., M. Marsili, and Y. C. Zhang. 2004. *Minority Games: Interacting Agents in Financial Markets*. Oxford, UK: Oxford University Press.

Coolen, A. C. 2005. *The Mathematical Theory of Minority Games: Statistical Mechanics of Interacting Agents*. Oxford, UK: Oxford University Press.

Dafoe, A., E. Hughes, Y. Bachrach, T. Collins, K. R. McKee, J. Z. Leibo, K. Larson, and T. Graepel. 2020. *Open Problems in Cooperative AI.* ArXiv preprint arXiv:2012.08630. https://doi.org/10.48550/arXiv.2012.08630.

Kets, W. 2011. "Learning with Fixed Rules: The Minority Game." *Journal of Economic Surveys* 26:865–878. https://doi.org/10.1111/j.1467-6419.2011.00686.x.

Kets, W., W. Kager, and A. Sandroni. 2022. "The Value of a Coordination Game." *Journal of Economic Theory* 201:105419. https://doi.org/10.1016/j.jet.2022.105419.

Miller, J. H., and S. E. Page. 2009. *Complex Adaptive Systems.* Princeton, NJ: Princeton University Press.

Neumann, Von, and Morgenstern. 1944. *Theory of Games and Economic Behavior.* Princeton, NJ: Princeton University Press.

Sandholm, W. H. 2010. *Population Games and Evolutionary Dynamics.* Cambridge, MA: MIT Press.

Zermelo, Ernst. 1913. "Über eine Anwendung der Mengenlehre auf die Theorie des Schachspiels." In *Proceedings of the Fifth Congress of Mathematicians,* 501–504. Cambridge University Press. Cambridge.

INDUCTIVE REASONING AND BOUNDED RATIONALITY

W. Brian Arthur

The type of rationality assumed in economics—perfect, logical, deductive rationality—is extremely useful in generating solutions to theoretical problems. But it demands much of human behavior, much more in fact than it can usually deliver. If one were to imagine the vast collection of decision problems economic agents might conceivably deal with as a sea or an ocean, with the easier problems on top and more complicated ones at increasing depth, then deductive rationality would describe human behavior accurately only within a few feet of the surface. For example, the game tic-tac-toe is simple, and one can readily find a perfectly rational, minimax solution to it; but rational "solutions" are not found at the depth of checkers; and certainly not at the still modest depths of chess and go.*†

There are two reasons for perfect or deductive rationality to break down under complication. The obvious one is that beyond a certain level of complexity human logical capacity ceases to cope—human rationality is bounded. The other is that in interactive situations of complication, agents cannot rely upon the other agents they are dealing with to behave under perfect rationality, and so they are forced to guess their behavior. This lands them in a world of subjective beliefs, and subjective beliefs about subjective beliefs. Objective, well-defined, shared assumptions then cease to apply. In turn, rational, deductive reasoning (deriving

---

Discussants: W. Brian Arthur, Stanford University and Santa Fe Institute; Paul Krugman, Massachusetts Institute of Technology; Michael Kremer, Massachusetts Institute of Technology.

†Santa Fe Institute, 1660 Old Pecos Trail, Santa Fe, NM 87501, and Stanford University. I thank particularly John Holland, whose work inspired many of the ideas here. I also thank Kenneth Arrow, David Lane, David Rumelhart, Roger Shepard, Glen Swindle, Nick Vriend, and colleagues at Santa Fe and Stanford for discussions. A lengthier version is given in Arthur (1992). For parallel work on bounded rationality and induction, but applied to macroeconomics, see Thomas J. Sargent (1994).

For much of the 1980s, game theorists attempted to show that equilibrium could be deduced from the specification of the game and the hypothesis that players are rational (and this is common knowledge). By the late '80s, the limitations of such a purely rationality-based approach were becoming increasingly clear, opening the door to alternative approaches, such as evolutionary game theory.

The difficulties in chess (or Go) are arguably of a somewhat different nature than the challenges in the El Farol bar problem: Given enough computational capability, computers can become perfect chess players, yet the bar problem would likely still elude them. That is, while the complexity in chess is structural, the complexity of the bar problem is strategic and cannot be overcome using computational power alone. Strategic complexity is now an active research area in AI (see Dafoe et al. 2020).

a conclusion by perfect logical processes from well-defined premises) itself cannot apply. The problem becomes ill-defined.

This important idea goes back to the seminal work of Von Neumann and Morgenstern (1944).

This is exacerbated in the El Farol bar problem: Even if the equilibria would be easy to find, there are so many of them (with none being an obvious focal point) that players may not be able to coordinate on the same equilibrium, thus invalidating the assumption of equilibrium behavior. As Arthur shows, the bar problem indeed gives rise to substantial miscoordination.

Economists, of course, are well aware of this. The question is not whether perfect rationality works, but rather what to put in its place. How does one model bounded rationality in economics? Many ideas have been suggested in the small but growing literature on bounded rationality; but there is not yet much convergence among them. In the behavioral sciences this is not the case. Modern psychologists are in reasonable agreement that in situations that are complicated or ill-defined, humans use characteristic and predictable methods of reasoning. These methods are not deductive, but *inductive*.

I. Thinking Inductively

How *do* humans reason in situations that are complicated or ill-defined? Modern psychology tells us that as humans we are only moderately good at deductive logic, and we make only moderate use of it. But we *are* superb at seeing or recognizing or matching patterns—behaviors that confer obvious evolutionary benefits. In problems of complication then, we look for patterns; and we simplify the problem by using these to construct temporary internal models or hypotheses or *schemata* to work with.[1] We carry out localized deductions based on our current hypotheses and act on them. As feedback from the environment comes in, we may strengthen or weaken our beliefs in our current hypotheses, discarding some when they cease to perform, and replacing them as needed with new ones. In other words, when we cannot fully reason or lack full definition of the problem, we use simple models to fill the gaps in our understanding. Such behavior is inductive.

One can see inductive behavior at work in chess playing. Players typically study the current configuration of the board and recall their opponent's play in past games to discern patterns (Adriann De Groot 1965). They use these to form hypotheses or internal models about

---

[1] For accounts in the psychological literature, see R. Schank and R. P. Abelson (1977), David Rumelhart (1980), Gordon H. Bower and Ernest R. Hilgard (1981), and John H. Holland et al. (1986). Of course, not all decision problems work this way. Most mundane actions like walking or driving are subconsciously directed, and for these pattern-cognition maps directly in action. In this case, connectionist models work better.

each other's intended strategies, maybe even holding several in their minds at one time: "He's using a Caro–Kann defense." "This looks a bit like the 1936 Botvinnik–Vidmar game." "He is trying to build up his mid-board pawn formation." They make local deductions based on these, analyzing the possible implications of moves several moves deep. And as play unfolds they hold onto hypotheses or mental models that prove plausible or toss them aside if not, generating new ones to put in their place. In other words, they use a sequence of pattern recognition, hypotheses formation, deduction using currently held hypotheses, and replacement of hypotheses as needed.

This type of behavior may not be familiar in economics; but one can recognize its advantages. It enables us to deal with complication: we construct plausible, simpler models that we *can* cope with. It enables us to deal with ill-definedness: where we have insufficient definition, our working models fill the gap. It is not antithetical to "reason," or to science for that matter. In fact, it is the way science itself operates and progresses.

Modeling Induction.—If humans indeed reason in this way, how can one model this? In a typical problem that plays out over time, one might set up a collection of agents, probably heterogeneous, and assume they can form mental models, or hypotheses, or subjective beliefs. These beliefs might come in the form of simple mathematical expressions that can be used to describe or predict some variable or action; or of complicated expectational models of the type common in economics; or of statistical hypotheses; or of condition/prediction rules ("If situation Q is observed, predict outcome or action D"). These will normally be subjective, that is, they will differ among the agents. An agent may hold one in mind at a time, or several simultaneously.

Each agent will normally keep track of the performance of a private collection of such belief-models. When it comes time to make choices, he acts upon his currently most credible (or possibly most profitable) one. The others he keeps at the back of his mind, so to speak. Alternatively, he may act upon a combination of several. (However, humans tend to hold in mind many hypotheses and act on the most plausible one, Feldman 1962). Once actions are taken, the aggregative picture is updated, and agents update the track record of all their hypotheses.

This is a system in which learning takes place. Agents "learn" which of their hypotheses work, and from time to time they may discard poorly performing hypotheses and generate new "ideas" to put in their place. Agents linger with their currently most believable hypothesis or belief model but drop it when it no longer functions well, in favor of a better one. This causes a built-in hysteresis. A belief model is clung to not because it is "correct"—there is no way to know this—but rather because it has worked in the past and must cumulate a record of failure before it is worth discarding. In general, there may be a constant slow turnover of hypotheses acted upon. One could speak of this as a system of *temporarily fulfilled expectations*—beliefs or models or hypotheses that are temporarily fulfilled (though not perfectly), which give way to different beliefs or hypotheses when they cease to be fulfilled.

If the reader finds this system unfamiliar, he or she might think of it as generalizing the standard economic learning framework which typically has agents sharing one expectational model with unknown parameters, acting upon the parameters' currently most plausible values. Here, by contrast, agents differ, and each uses several subjective models instead of a continuum of one commonly held model. This is a richer world, and one might ask whether, in a particular context, it converges to some standard equilibrium of beliefs; or whether it remains open-ended, always leading to new hypotheses, new ideas.

One might think that this metaphor is not entirely apt, since hypotheses may not be subject to mutation and selection in the same way species are. Subsequent research, however, has shown that there is a mathematical equivalence between evolutionary models and models of learning in games (see Sandholm 2010).

It is also a world that is evolutionary, or more accurately, coevolutionary. Just as species, to survive and reproduce, must prove themselves by competing and being adapted within an environment created by other species, in this world hypotheses, to be accurate and therefore acted upon, must prove themselves by competing and being adapted within an environment created by other agent's hypotheses. The set of ideas or hypotheses that are acted upon at any stage therefore coevolves.[3]

A key question remains. Where do the hypotheses or mental models come from? How are they generated? Behaviorally, this is a deep question in psychology, having to do with cognition, object

---

[3] A similar statement holds for strategies in evolutionary game theory; but there, instead of a large number of private, subjective expectational models, a small number of strategies compete.

representation, and pattern recognition. I will not go into it here. However, there are some simple and practical options for modeling. Sometimes one might endow agents with *focal* models: patterns or hypotheses that are obvious, simple, and easily dealt with mentally. One might generate a "bank" of these and distribute them among the agents. Other times, given a suitable model-space one might allow the genetic algorithm or some similar intelligent search device to generate ever "smarter" models. One might also allow agents the possibility of "picking up" mental models from one another (in the process psychologists call *transfer*). Whatever option is taken, it is important to be clear that the framework described above is independent of the specific hypotheses or beliefs used, just as the consumer-theory framework is independent of the particular products chosen among. Of course, to use the framework in a particular problem, some system of generating beliefs must be adopted.

II. The Bar Problem

Consider now a problem I will construct to illustrate inductive reasoning and how it might be modeled. N people decide independently each week whether to go to a bar that offers entertainment on a certain night. For concreteness, let us set N at 100. Space is limited, and the evening is enjoyable if things are not too crowded—specifically, if fewer than 60 percent of the possible 100 are present. There is no sure way to tell the numbers coming in advance; therefore a person or agent *goes* (deems it worth going) if he expects fewer than 60 to show up or *stays home* if he expects more than 60 to go. Choices are unaffected by previous visits; there is no collusion or prior communication among the agents; and the only information available is the numbers who came in past weeks. (The problem was inspired by the bar El Farol in Santa Fe which offers Irish music on Thursday nights; but the reader may recognize it as applying to noontime lunch-room crowding, and to other commons or coordination problems with limits to desired coordination.) Of interest is the dynamics of the numbers attending from week to week.

Notice two interesting features of this problem. First, if there were an obvious model that all agents could use to forecast attendance and

This is the hallmark of strategic complexity: With so many possible outcomes, how can agents know which one will be selected? This very general problem also arises in games where players aim to coordinate, rather than differentiate, their actions.

This is a key distinction between the bar problem and coordination games (and more generally games with strategic complementarities). "Anti-coordination" games like the bar problem are difficult to analyze using standard evolutionary game theory models. There is little work on the meta problem of which reasoning models adopt in different strategic environments.

base their decisions on, then a deductive solution would be possible. But this is not the case here. Given the numbers attending in the recent past, a large number of expectational models might be reasonable and defensible. Thus, not knowing which model other agents might choose, a reference agent cannot choose his in a well-defined way. There is no deductively rational solution—no "correct" expectational model. From the agents' viewpoint, the problem is ill-defined, and they are propelled into a world of induction. Second, and diabolically, any commonalty of expectations gets broken up: if all believe *few* will go, all will go. But this would invalidate that belief. Similarly, if all believe *most* will go, *nobody* will go, invalidating that belief.[4] Expectations will be forced to differ.

At this stage, I invite the reader to pause and ponder how attendance might behave dynamically over time. Will it converge, and if so to what? Will it become chaotic? How might predictions be arrived at?

A. A DYNAMIC MODEL

To answer the above questions, I shall construct a model along the lines of the framework sketched above. Assume the 100 agents can individually form several predictors or hypotheses, in the form of functions that map the past d weeks' attendance figures into next week's. For example, recent attendance might be:

$$\ldots, 44, 78, 56, 15, 23, 67, 84,$$
$$34, 45, 76, 40, 56, 22, 35.$$

Particular hypotheses or predictors might be: *predict next week's number to be*

- the same as last week's [35]
- a mirror image around 50 of last week's [65]
- a (rounded) average of the last four weeks [49]
- the trend in last 8 weeks, bounded by 0, 100 [29]
- the same as 2 weeks ago (2-period cycle detector) [22]

---

[4] This is reminiscent of Yogi Berra's famous comment, "Oh, that place. It's so crowded nobody goes there anymore."

- the same as 5 weeks ago (5-period cycle detector) [76]

- etc.

Assume that each agent possesses and keeps track of a individualized set of *k* such focal predictors. He decides to go or stay according to the currently most accurate predictor in his set. (I will call this his *active* predictor.) Once decisions are made, each agent learns the new attendance figure and updates the accuracies of his monitored predictors.

Notice that in this bar problem, the set of hypotheses currently most credible and acted upon by the agents (the set of active hypotheses) determines the attendance. But the attendance history determines the set of active hypotheses. To use John Holland's term, one can think of these active hypotheses as forming an *ecology*. Of interest is how this ecology evolves over time.

B. COMPUTER EXPERIMENTS

For most sets of hypotheses, analytically this appears to be a difficult question. So in what follows I will proceed by computer experiments. In the experiments, to generate hypotheses, I first create an "alphabet soup" of predictors, in the form of several dozen focal predictors replicated many times. I then randomly ladle out *k* (6 or 12 or 23, say) of these to each of 100 agents. Each agent then possesses *k* predictors or hypotheses or "ideas" he can draw upon. We need not worry that useless predictors will muddy behavior. If predictors do not "work" they will not be used; if they do work they will come to the fore. Given starting conditions and the fixed set of predictors available to each agent, in this problem the future accuracies of all predictors are predetermined. The dynamics here are deterministic.

The results of the experiments are interesting (Fig. 1). Where cycle-detector predictors are present, cycles are quickly "arbitraged" away so there are no persistent cycles. (If several people expect many to go because many went three weeks ago, they will stay home.) More interestingly, mean attendance converges always to 60. In fact the predictors self-organize into an equilibrium pattern or "ecology" in which, of the active predictors (those most accurate and therefore acted upon), on average 40 percent are forecasting above 60, 60 percent below

Arthur (1994)

This is a key difference from other adaptive dynamics in game theory, which typically assume that agents' choice is over *actions* (go/stay home), not predictors (which effectively map histories into actions).

There is a slight bias in the update rule: Players do not take into account the effect of their action on the aggregate outcome. That is, when updating the accuracy of a predictor, players consider only whether it would have predicted the actual outcome correctly, not taking into account that using the predictor might have changed the outcome. This is crucial for behavior (see Kets 2011).

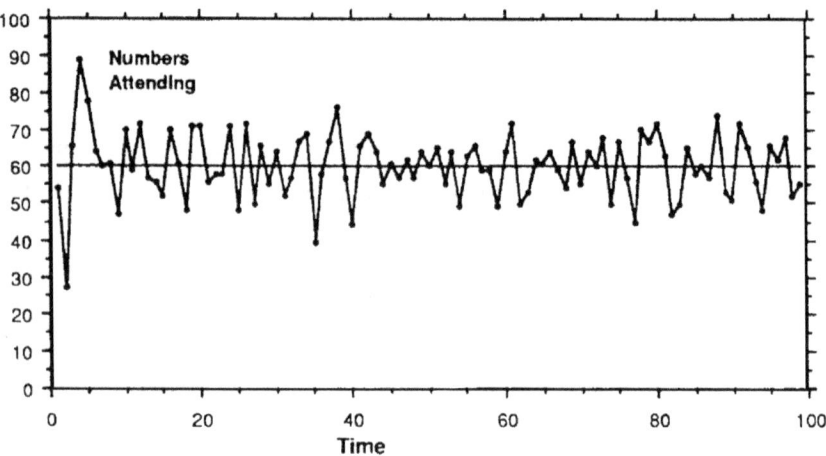

Figure 1. Bar attendance in the first 100 weeks

60. This emergent ecology is almost organic in nature. For, while the population of active predictors splits into this 60/40 average ratio, it keeps changing in membership forever. This is something like a forest whose contours do not change, but whose individual trees do. These results appear throughout the experiments and are robust to changes in types of predictors created and in numbers assigned.

How do the predictors self-organize so that 60 emerges as average attendance and forecasts split into a 60/40 ratio? One explanation might be that 60 is a natural "attractor" in this bar problem; in fact, if one views it as a pure game of predicting, a mixed strategy of forecasting above 60 with probability 0.4 and below it with probability 0.6 is a Nash equilibrium. Still, this does not explain how the agents approximate any such outcome, given their realistic, subjective reasoning. To get some understanding of how this happens, suppose that 70 percent of their predictors forecasted above 60 for a longish time. Then on average only 30 people would show up; but this would validate predictors that forecasted close to 30 and invalidate the above-60 predictors, restoring the "ecological" balance among predictions, so to speak. Eventually the 40–60-percent combination would assert itself. (Making this argument mathematically exact appears to be nontrivial.) It is important to be clear that one does not need any 40–60 forecasting balance in the predictors that are set up. Many could have a tendency to predict high, but aggregate behavior calls the equilibrium predicting ratio to the fore.

Indeed, progress on this important issue was only possible by making the connection with disordered systems in condensed-matter physics (Coolen 2005).

Of course, the result would fail if all predictors could only predict below 60; then all 100 agents would always show up. Predictors need to "cover" the available prediction space to some modest degree. The reader might ponder what would happen if all agents shared the same set of predictors.

It might be objected that I lumbered the agents in these experiments with fixed sets of clunky predictive models. If they could form more open-ended, intelligent predictions, different behavior might emerge. One could certainly test this using a more sophisticated procedure, say, genetic programming (John Koza 1992). This continually generates new hypotheses, new predictive expressions, that adapt "intelligently" and often become more complicated as time progresses. However, I would be surprised if this changes the above results in any qualitative way.

The bar problem introduced here can be generalized in a number of ways (see Grannan and Swindle 1994). I encourage the reader to experiment.

III. Conclusion

The inductive-reasoning system I have described above consists of a multitude of "elements" in the form of belief-models or hypotheses that adapt to the aggregate environment they jointly create. Thus it qualifies as an *adaptive complex* system. After some initial learning time, the hypotheses or mental models in use are mutually coadapted. Thus one can think of a *consistent* set of mental models as a set of hypotheses that work well with each other under some criterion—that have a high degree of mutual adaptedness. Sometimes there is a unique such set, it corresponds to a standard rational expectations equilibrium, and beliefs gravitate into it. More often there is a high, possibly very high, multiplicity of such sets. In this case one might expect inductive-reasoning systems in the economy—whether in stock-market speculating, in negotiating, in poker games, in oligopoly pricing, or in positioning products in the market—to cycle through or temporarily lock into psychological patterns that may be nonrecurrent, path-dependent, and increasingly complicated. The possibilities are rich.

Economists have long been uneasy with the assumption of perfect, deductive rationality in decision contexts that are complicated and

So, there is a kind of equilibrium at a higher level: an equilibrium not so much in actions but in mental models. This generates much richer behavior than the usual kind, including persistent volatility.

potentially ill-defined. The level at which humans can apply perfect rationality is surprisingly modest. Yet it has not been clear how to deal with imperfect or bounded rationality. From the reasoning given above, I believe that as humans in these contexts we use *inductive* reasoning: we induce a variety of working hypotheses, act upon the most credible, and replace hypotheses with new ones if they cease to work. Such reasoning can be modeled in a variety of ways. Usually this leads to a rich psychological world in which agents' ideas or mental models compete for survival against other agents' ideas or mental models—a world that is both evolutionary and complex.

REFERENCES

Arthur, W. B. 1992. *On Learning and Adaptation in the Economy.* Technical report 92-07-038. Santa Fe, NM: Santa Fe Institute.

Bower, G. H., and E. R. Hilgard. 1981. *Theories of Learning.* Englewood Cliffs, NJ: Prentice-Hall.

De Groot, A. 1965. *Thought and Choice in Chess (Psychological Studies).* The Hague, Netherlands: De Gruyter, M.

Feldman, J. 1962. "Computer Simulation of Cognitive Processes." In *Computer Applications in the Behavioral Sciences,* edited by H. Borko, 336–359. Englewood Cliffs, NJ: Prentice- Hall.

Grannan, E. R., and G. H. Swindle. 1994. *Contrarians and Volatility Clustering.* Technical report 94-03-010. Santa Fe, NM: Santa Fe Institute.

Holland, J. H., K. J. Holyoak, R. E. Nisbett, and P. R. Thagard. 1986. *Induction.* Appears in the Computational Models of Cognition and Perception series. Cambridge, MA: MIT Press.

Koza, J. 1992. *Genetic Programming.* Cambridge, MA: MIT Press.

Rumelhart, D. 1980. "Schemata: The Building Blocks of Cognition." In *Theoretical Issues in Reading Comprehension,* edited by R. Spiro, Bruce B., and W. Brewer, 35–58. Hillside, NJ: Erlbaum.

Sargent, T. J. 1994. *Bounded Rationality in Macroeconomics.* Oxford, UK: Oxford University Press.

Schank, R., and R. P. Abelson. 1977. *Scripts, Plans, Goals, and Understanding: An Inquiry Into Human Knowledge Structures.* Hillside, NJ: Erlbaum.

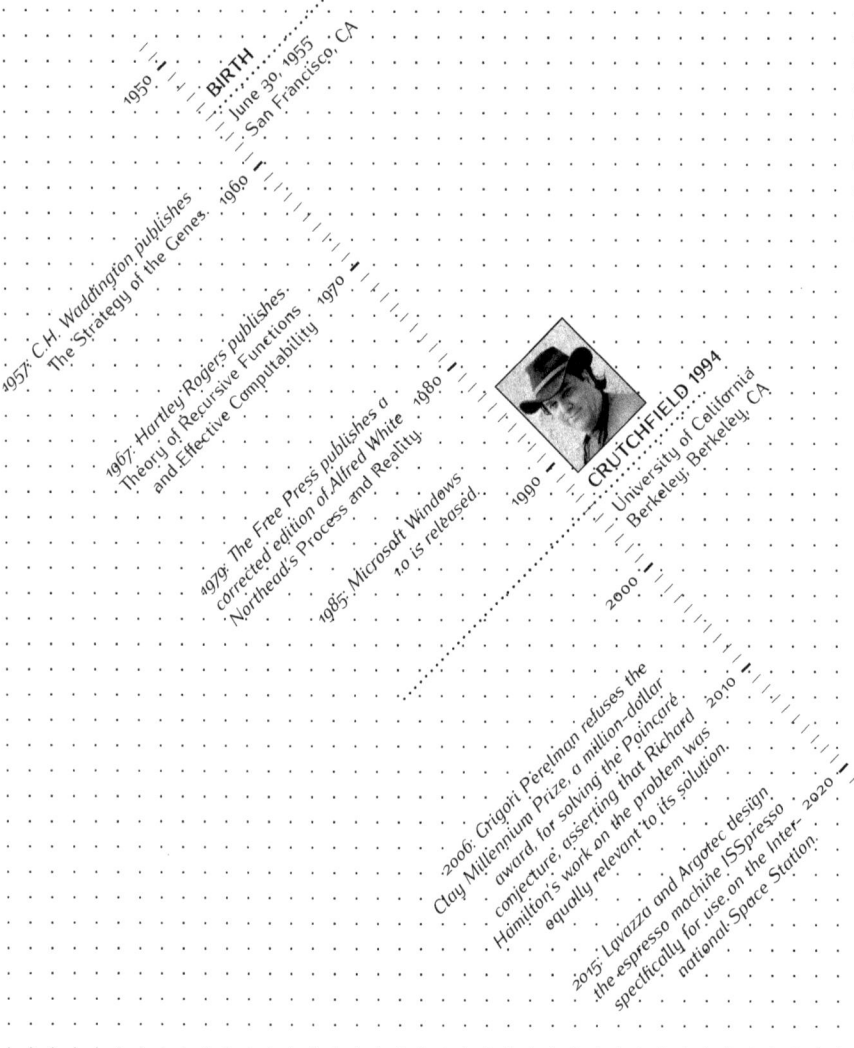

BIRTH
June 30, 1955
San Francisco, CA

1957: C.H. Waddington publishes
The Strategy of the Genes.

1967: Hartley Rogers publishes
Theory of Recursive Functions
and Effective Computability.

1979: The Free Press publishes a
corrected edition of Alfred White
Northead's Process and Reality.

1985: Microsoft Windows
1.0 is released.

CRUTCHFIELD 1994
University of California
Berkeley, CA

2006: Grigori Perelman refuses the
Clay Millennium Prize, a million-dollar
award, for solving the Poincaré
conjecture, asserting that Richard
Hamilton's work on the problem was
equally relevant to its solution.

2015: Lavazza and Argotec design
the espresso machine ISSpresso
specifically for use on the Inter-
national Space Station.

JAMES PATRICK CRUTCHFIELD

[78]

NATURAL INFORMATION PROCESSING

Peter M. A. Sloot, University of Amsterdam
Rick Quax, University of Amsterdam, and
Mile Gu, Nanyang Technological University

Measuring the emergent complexity of a complex system has itself become a complex process—and is still ongoing. Over the past few decades, an ever-expanding realm of researchers from various disciplines have come up with a wide variety of different metrics, starting from different viewpoints and answering different questions that can often somehow be related to each other. A root cause of this expansion is the difficulty of pinning down the exact problem. Or as Seth Lloyd (2001) aptly put it: "A historical analog to the problem of measuring complexity is the problem of describing electromagnetism before Maxwell's equations."

Initially, many researchers were in pursuit of *the* complexity measure: one formula or algorithm that quantifies the amount of complexity in any given program or pattern. The sheer variety of measures that resulted has shifted the focus to look for *a* complexity measure: a choice that depends on the context, the research question, and the assumptions one is willing to make.

In this light, James Crutchfield's 1994 paper can be seen as a novel approach in the statistical description of complexity measures, but with a key distinction that has crucial consequences.

A Key Distinction

What sets Crutchfield's framework apart from related work at the time is what he coins the "intrinsic" emergence of complexity. A complexity measure could be said to be "extrinsic": for instance, the category "difficulty of creation" revolves, roughly speaking, around the amount of effort (primarily energy and/or heat) that an external observer (typically living in a vacuum or in an otherwise regular and hypothetical

J. P. Crutchfield, "The Calculi of Emergence: Computation, Dynamics, and Induction," *Physica D* 75: 1–3, 11–54 (1994).

Reprinted with permission from Elsevier.

universe) must expend to create an instance of the pattern or process under scrutiny. As a typical example from the "difficulty of description" category, in which this paper fits, an external observer could calculate the formula of stochastic complexity by Jorma Rissanen (1987) of a given process. To do that, the observer is thought to either be able to observe the process's mechanics directly and perfectly or to be able to observe all relevant data (values of all variables over space and time) produced by the process. Such external observers are unburdened by limits of computational power, observation, and, if needed, actuation. In addition, whether a process is called "complex" depends crucially on the prior knowledge and interests of this hypothetical observer, which is invariably some reflection of the authors who devised it.

Instead, in Crutchfield's work, the observer becomes an intrinsic part of the computational process. Think of an agent, with certain (finite) computational capabilities, that must observe another process through some potentially limited sensors and may even be part of the process that it observes. This agent is tasked with describing the observed process as efficiently as possible, that is, by constructing an internal model that uses a minimal number of states to predict the observed future of the process. The overarching idea is that the extent to which the process is deemed complex is defined by the agent's autonomous deliberation. More specifically, it is defined through the evolution of the size of the minimal model that the agent can formulate, during an algorithm of "innovation." That is, a sudden decrease in size of the agent's internal model could be an innovation such as the realization that there is an emergent pattern that can be exploited. Comparable, for instance, to the description of a critical process by its critical exponent instead of its mechanistic details.

A Crucial Consequence of the Key Distinction

The crucial consequence of imbuing a participant agent with a computational process and analyzing its resulting predictive power leads naturally to analyzing so-called *information processing* (James, Ellison, and Crutchfield 2011). That is, one can analyze how predictive power (i.e., information!) can arise from, and percolate through, the states and state transitions of a stochastic model. How predictable is the next state given the current state? How much additional predictive power is gained synergistically by considering a sequence of states versus each state separately (Quax, Har-Shemesh, and Sloot 2017)? How much information, stored in the model at one time point, is destroyed in the next time point (replaced by uncorrelated noise) on average? Which states contain more information about the future than others? And so on.

These are important questions on their own with deep consequences. The language of information theory lends itself perfectly for characterizing a process, or indeed its emergent properties, in a model-agnostic way. To illustrate this and its importance, consider two hypothetical models: one is formulated as a collection of neurons connected by synapses and its dynamics (rules of change over time) in terms of ion gates and potentials; another is a collection of agents in a social environment and its dynamics in terms of involvement in certain discussion topics and posting messages to friends. It is entirely possible that, for a particular choice of parameter values, interaction structure, and so on, the information processing implemented by both models is the same—even though each model is mechanistically completely different. The importance of this observation cannot be overstated. Any emergent property that is detectable using a combination of information-processing features would then become free from mechanistic details. Analyzing a model in terms of neurons and synapses for signs of, say, deterministic chaos is far from trivial and involves analyzing the governing dynamics equations in addition to some manual searching for degrees of freedom, at which temporal scales a phenomenon such as period doubling occurs. In contrast, the information-processing route, though still an early work in progress today and by no means implemented in practice already, has the promise to look something like this: (1) write down the mechanistic model; (2) convert the model to its informational processing structure, which acts as an equivalence class over mechanistic models; (3) run a pattern-matching algorithm to discover whether the information structure harbors the (emergent) property of interest somewhere.

Some Critical Notes

Taking Crutchfield's article literally, the agents are not so finite after all. An agent is tasked with constructing and comparing exact joint probability distributions of possibly infinite futures and pasts. For this it would need an infinite memory in general, as well as an infinite amount of time for calculations. This essential computing mechanism that the agents need to perform to construct and innovate the internal model of the process is not discussed at all. It is also left unspecified how an agent should perform innovations in a practical sense.

As a second point of debate, it may still be worth discussing whether the external observer is indeed fully eliminated. Recall that the notion of an emergent complex property depends on an agent innovating (reducing the size of) its internal model of the process at hand. But who is to observe such an event? As it is written now, the agent simply keeps attempting to innovate but does not have an additional cognitive process that oversees the innovations and does not know how many innovations have occurred previously, how these were achieved, or even whether a particular innovation is to be considered "large and unexpected" versus "typical and nothing special." Moreover, some of the steps of the agent depend on varying the epsilon parameter (the measurement of the noise of the agent's sensors) and studying the resulting model size. This would in practice be very difficult for an agent to do itself and would in the very least require a significant cognitive process to be embedded, not to mention a powerful sensory apparatus with a tunable noise level.

Placing into Context

Regarding the first point—that is, the practical issue of inferring the causal states from finite data and using finite resource—is a special case (Ray 2004) of a much older concept: hidden Markov models (HMM). That is, the causal states are required to be equivalence classes of the measured microstates, whereas in the general HMM this mapping may be any conditional probability distribution. Crutchfield does mention HMMs but only in the example section, whereas to us it appears Crutchfield's framework can piggyback on this domain to bring it closer to practice (e.g., Brodu 2011). Meanwhile, there are now a crowing collection of works illustrating that HMMs and causal states can naturally emerge from trained recurrent neural networks or transformer models. Such connection provides a means towards explainable AI, while simultaneously offering news tools for discovery of causal states (Shai *et al.* 2024; Casey 1996).

A similar argument can be made for the second point, that is, formulating practical algorithms for learning new internal representations by an agent. Predating Crutchfield's article is a famous discussion of how to best do this, commonly referred to as "reactive versus deliberative." Crutchfield's sketch of an algorithm leans toward the deliberative side, which involves having agents that maintain and update an internal representation and implement some reasoning for achieving some goal(s) (Arkin 1998). Although some elements of the proposed algorithm could well be new, such as inference based on the model size as function of the sensor noise levels, the research field of agent deliberation is far too large and rich to not be leveraged on for the innovation part of the framework.

Section B: Contemporary Impact

Crutchfield's work discussed multiple ideas that significantly impacted our views on intelligent agents and natural information processing. Since the paper's release, these fields have undergone significant development themselves. Agents play an ever-greater role in artificial intelligence, while advancements in quantum information science and information thermodynamics have greatly influenced the science of natural information processing. As such, there is significant value in revisiting some of Crutchfield's ideas in the context of these contemporary developments.

Of Agents and Games

Crutchfield's initial discourse was very much motivated by the perspective of agents surviving and thriving in complex environments. Yet its focus was primarily on how these agents could form the most efficiently represent relevant information about their environment. While this is undoubtedly a critical aspect for the survival of such agents, how the agents would leverage such knowledge is equally important. A predator, for example, would greatly benefit from having an accurate model of how its prey moves. However, such a model would be useless without taking appropriate action based on such knowledge.

A complete model of input-output response for agents has since emerged in the context of reinforcement learning. In this context, agents continually interact with a complex environment and receive feedback through observable input stimuli and possible rewards. Thus, the goal of the agent is more clearly specified—and success in achieving this goal will generally entail the agent building a sufficiently accurate model of its environment. While traditional reinforcement learning algorithms typically assume that the agent has essentially unlimited memory (as they are capable of storing continuous parameters), there has been recent interest in what agents with finite memory can do (Koul, Greydanus, and Fern 2018). Meanwhile, Crutchfield has provided means to generalize ϵ-machines to devices that also received inputs, termed ϵ-transducers (Barnett and Crutchfield 2015). Recent work has begun exploring the connections between these two distinct fields. Such developments have the potential to both better understand the complexity of optimizing certain reward functions and introduce methods to automate the learning of efficient agents (Zhang *et al.* 2021).

The Thermodynamic Perspective

While agents are often associated with computational science, its first scientific treatment arguably came from the field of thermodynamics. In 1867, James Clerk Maxwell proposed the idea of a hypothetical entity that can observe and manipulate physical systems at the scale of individual molecules (now known as Maxwell's demon). Maxwell argued that such an entity could partition a gas chamber into two halves and sort the particles within such that hotter particles gravitate toward only one half. The resulting partitions would then exhibit a temperature differential, leading to the ability to create free energy in apparent violation of the thermodynamic second law.

Charles Bennett's (1982) resolution of this paradox indicated a close relationship between information and free energy. The demon can do extra work from its environment by collecting information through observation and taking appropriate action based on these observations. Maxwell's demon is essentially an agent. Like the agents discussed by Crutchfield, such demons must also necessarily have memory—which is used to predict which of its future actions can best extract environmental free energy.

While the original Maxwell's demon thought experiment involved a simple system, subsequent works explored demons interacting with systems of ever-greater complexity. For example, in information ratchets (Boyd, Mandal, and Crutchfield 2016), the out-of-equilibrium resources are not simply temperature gradients but rather correlations structured by time-series data. We can then consider the demon as an agent that sees one element of this time series at a time. To leverage free energy in future correlations, it must retain information about what it has seen and make appropriate predictions. These ideas have resulted in thermodynamic principles of pattern manipulation and requisite variety (Boyd, Mandal, and Crutchfield 2016, 2017; Garner *et al.* 2017).

This thermodynamic perspective immediately bestows Crutchfield's framework with physical consequences. Whereas in the current work, Crutchfield described memory as an abstract resource that is finite, the thermodynamic connection implies that this memory can also have an energetic cost. Thus, if we envision agents as biological organisms that wish to leverage free energy within their environments, innovation may become an energetic necessity.

Enter Quantum Agents

Crutchfield's work assumed that all agents were classical in that their means of representing relevant past information and methods of the process this information believed that all information is classical. This was natural. At the time of publication, quantum information science was very much a nascent field. Yet the rapid development of quantum information in the subsequent quarter-century makes it worthwhile to revisit Crutchfield's innovative ideas from a quantum perspective.

In particular, all information—in principle—is quantum. Every bit of information represents some physical system with two distinguishable states, which we label 0 and 1. Quantum theory then tells us that this system can—in principle—also be configured in some quantum superposition of both, much as Schrödinger's cat is simultaneously dead and alive. For large macroscopic systems, such states are very fragile but not forbidden by the laws of physics. When we aim for continued innovation, it then makes sense that agents should also be able to leverage such properties for more concise representations of their environment.

In this context, one of the most relevant discoveries is that quantum systems—storing past information within such superposition states—can offer even more concise descriptions of the relevant past (Gu *et al.* 2012). Furthermore, the difference is that the amount of memory required for classical and quantum agents can scale without bound. When it comes to approximating observable statistics to some accuracy ϵ, there are many scenarios of the memory cost of ϵ—machines scale inversely with ϵ and diverge to infinity as ϵ tends to 0. In contrast, recent work demonstrates the existence of quantum agents that may represent the same processes using memory whose information content remains bounded by a finite number that is ϵ-independent (Elliott *et al.* 2020; Garner *et al.* 2017). Such quantum agents can have thermodynamic benefits (Loomis and Crutchfield 2020), while scaling advantaging also persist in complex decision-making (Elliott *et al.* 2020). This raises an exciting possibility: Could the need to innovate ultimately motivate agents to reason quantum mechanically?

Future Possibilities

We see from these contemporary developments that while Crutchfield's work left many open questions, the ideas it raised—agents, innovation, prediction—remain relevant today. Notable recent developments include: (1) results showing standard quantifiers of model efficacy in machine learning coincide with measures of thermodynamic efficiency should agents use these models to harvest energy (Boyd, Crutchfield, and Gu 2022); (2) quantum transducers can execute particular complex adaptive strategies to arbitrary precision with finite memory—while memory requirements of their classical counterparts scale toward infinity; and (3) analysis showing that the memory advantages of quantum models can also translate to thermodynamic advantages (Loomis and Crutchfield 2020). These results illustrate the central role agency plays in interfacing modern ideas from information thermodynamics, machine learning, and quantum information science, and presents a compelling demonstration of the diverse disciplines that have benefited from Crutchfield's initial ideas.

REFERENCES

Arkin, R. C. 1998. *Behavior-Based Robotics*. Cambridge, MA: MIT Press.

Barnett, N., and J. Crutchfield. 2015. "Computational Mechanics of Input–Output Processes: Structured Transformations and the ϵ-Transducer." *Journal of Statistical Physics* 161:404–451. https://doi.org/10.48550/arXiv.1412.2690.

Bennett, C. 1982. "The Thermodynamics of Computation—A Review." *International Journal of Theoretical Physics* 21:905–40. https://doi.org/10.1007/BF02084158.

Boyd, A., D. Mandal, and J. P. Crutchfield. 2016. "Identifying Functional Thermodynamics in Autonomous Maxwellian Ratchets." *New Journal of Physics* 18 (2): 023049. https://doi.org/10.1088/1367-2630/18/2/023049.

———. 2017. "Leveraging Environmental Correlations: The Thermodynamics of Requisite Variety." *Journal of Statistical Physics* 167:1555–85. https://doi.org/10.1007/s10955-017-1776-0.

Boyd, A. B., J. P. Crutchfield, and M. Gu. 2022. "Thermodynamic Machine Learning through Maximum Work Production." *New Journal of Physics* 24. https://doi.org/10.1088/1367-2630/ac4309.

Brodu, N. 2011. "Reconstruction of Epsilon-Machines in Predictive Frameworks and Decisional States." *Advances in Complex Systems* 14 (05): 761–94. https://doi.org/10.1142/S0219525911003347.

Casey, M. 1996. "The Dynamics of Discrete-Time Computation, with Application to Recurrent Neural Networks and Finite State Machine Extraction." *Neural Computation* 8 (6): 1135–1178. https://doi.org/10.1162/neco.1996.8.6.1135.

Crutchfield, J. P. 1994. "The Calculi of Emergence: Computation, Dynamics, and Induction." *Physica D: Nonlinear Phenomena* 75 (1-3): 11–54. https://doi.org/10.1016/0167-2789(94)90273-9.

Elliott, T., C. Yang, F. C. Binder, A. J. P. Garner, J. Thompson, and M. Gu. 2020. "Extreme Dimensionality Reduction with Quantum Modeling." *Physical Review Letters* 125:260501. https://doi.org/10.1103/PhysRevLett.125.260501.

Garner, A., Q. Liu, J. Thompson, and V. Vedral. 2017. "Provably Unbounded Memory Advantage in Stochastic Simulation Using Quantum Mechanics." *New Journal of Physics* 19:103009. https://doi.org/10.1088/1367-2630/aa82df.

Gu, M., K. Wiesner, E. Rieper, and V. Vedral. 2012. "Quantum Mechanics Can Reduce the Complexity of Classical Models." *Nature Communications* 3 (1): 762. https://doi.org/10.1038/ncomms1761.

James, R. G., C. J. Ellison, and J. P. Crutchfield. 2011. "Anatomy of a Bit: Information in a Time Series Observation." *Chaos: An Interdisciplinary Journal of Nonlinear Science* 21 (3). https://doi.org/10.48550/arXiv.1105.2988.

Jurgens, A. M., and J. P. Crutchfield. 2021. "Divergent Predictive States: The Statistical Complexity Dimension of Stationary, Ergodic Hidden Markov Processes." *Chaos* 31 (8): 083114. https://doi.org/10.1063/5.0050460.

Koul, A., S. Greydanus, and A. Fern. 2018. *Learning Finite State Representations of Recurrent Policy Networks*. https://doi.org/10.48550/arXiv.1811.12530. arXiv: 1811.12530 [cs.LG]. https://arxiv.org/abs/1811.12530.

Lloyd, S. 2001. "Measures of Complexity: A Non-Exhaustive List." *IEEE Control Systems Magazine* 21 (4): 7–8. https://doi.org/10.1109/MCS.2001.939938.

Loomis, S. P., and J. P. Crutchfield. 2020. "Thermal Efficiency of Quantum Memory Compression." *Physical Review Letters* 125 (2): 020601. https://doi.org/10.1103/PhysRevLett.125.020601.

Quax, R., O. Har-Shemesh, and P. M. A. Sloot. 2017. "Quantifying Synergistic Information Using Intermediate Stochastic Variables." *Entropy* 19 (2): 85. https://doi.org/10.3390/e19020085.

Ray, A. 2004. "Symbolic Dynamic of Complex Systems for Anomaly Detection." *Signal Processing* 84 (7): 1115–30. https://doi.org/10.1016/j.sigpro.2004.03.011.

Rissanen, J. 1987. "Stochastic Complexity." *Journal of Royal Statistical Society* 49 (3). https://doi.org/10.1111/j.2517-6161.1987.tb01694.x.

Shai, A. S., S. E. Marzen, L. Teixeira, A. G. Oldenziel, and P. M. Riechers. 2024. "Transformers Represent Belief State Geometry in their Residual Stream." *arXiv,* https://doi.org/10.48550/arXiv.2405.15943.

Yang, Y., G. Chiribella, and M. Hayashi. 2018. "Quantum Stopwatch: How to Store Time in a Quantum Memory." *Proceedings of the Royal Society A* 474 (2213): 20170773. https://doi.org/10.1098/rspa.2017.0773.

Zhang, A., Z. C. Lipton, L. Pineda, K. Azizzadenesheli, A. Anandkumar, L. Itti, J. Pineau, and T. Furlanello. 2021. *Learning Causal State Representations of Partially Observable Environments.* https://doi.org/10.48550/arXiv.1906.10437. arXiv: 1906.10437 [cs.LG]. https://arxiv.org/abs/1906.10437.

THE CALCULI OF EMERGENCE: COMPUTATION, DYNAMICS, AND INDUCTION

James P. Crutchfield, University of California, Berkeley

Abstract

Defining structure and detecting the emergence of complexity in nature are inherently subjective, though essential, scientific activities. Despite the difficulties, these problems can be analyzed in terms of how model-building observers infer from measurements the computational capabilities embedded in nonlinear processes. An observer's notion of what is ordered, what is random, and what is complex in its environment depends directly on its computational resources: the amount of raw measurement data, of memory, and of time available for estimation and inference. The discovery of structure in an environment depends more critically and subtly though on how those resources are organized. The descriptive power of the observer's chosen (or implicit) computational model class, for example, can be an overwhelming determinant in finding regularity in data.

This paper presents an overview of an inductive framework—hierarchical ϵ-machine reconstruction—in which the emergence of complexity is associated with the innovation of new computational model classes. Complexity metrics for detecting structure and quantifying emergence, along with an analysis of the constraints on the dynamics of innovation, are outlined. Illustrative examples are drawn from the onset of unpredictability in nonlinear systems, finitary nondeterministic processes, and cellular automata pattern recognition. They demonstrate how finite inference resources drive the innovation of new structures and so lead to the emergence of complexity.

Order is not sufficient. What is required, is something much more complex. It is order entering upon novelty; so that the massiveness of order does not degenerate into mere repetition; and so that the novelty is always reflected upon a background of system.

—A.N. Whitehead on "Ideal Opposites" in *Process and Reality*.

1. Introduction

How can complexity emerge from a structureless universe? Or, for that matter, how can it emerge from a completely ordered universe? The following proposes a synthesis of tools from dynamical systems, computation, and inductive inference to analyze these questions.

The central puzzle addressed is how we as scientists—or, for that matter, how adaptive agents evolving in populations—ever "discover" anything new in our worlds, when it appears that all we can describe is expressed in the language of our current understanding. This dilemma is analyzed in terms of an open-ended modeling scheme, called hierarchical ϵ-machine reconstruction, that incorporates at its base inductive inference and quantitative measures of computational capability and structure. The key step in the emergence of complexity is the "innovation" of new model classes from old. This occurs when resource limits can no longer support the large models—often patchworks of special cases—forced by a lower-level model class. Along the way, complexity metrics for detecting structure and quantifying emergence, together with an analysis of the constraints on the dynamics of innovation, are outlined.

The presentation is broken into four sections. Section 2 is introductory and attempts to define the problems of discovery and emergence. It delineates several classes of emergent phenomena in terms of observers and their internal models. It argues that computation theory is central to a proper accounting of information processing in nonlinear systems and in how observers detect structure. Section 2 is intended to be self-contained in the sense that the basic ideas of the entire presentation are outlined. Section 3 reviews computation theory—formal languages, automata, and computational hierarchies—and a method to infer computational structure in nonlinear processes. Section 4, the longest, builds on that background to show formally, and by analyzing examples, how innovation and the emergence of complexity occur in hierarchical processes. Section 5 is a summary and a look forward.

2. Innovation, Induction, and Emergence

2.1. EMERGENT?

Some of the most engaging and perplexing natural phenomena are those in which highly structured collective behavior emerges over time from the interaction of simple subsystems. Flocks of birds flying in lockstep formation and schools of fish swimming in coherent array abruptly turn together with no leader guiding the group (Reynolds 1987). Ants form complex societies whose survival derives from specialized laborers, unguided by a central director (Holldobler and Wilson 1990). Optimal pricing of goods in an economy appears to arise from agents obeying the local rules of commerce (Fama 1991). Even in less manifestly complicated systems emergent global information processing plays a key role. The human perception of color in a small region of a scene, for example, can depend on the color composition of the entire scene, not just on the spectral response of spatially-localized retinal detectors (Land 1964; Wandell 1993). Similarly, the perception of shape can be enhanced by global topological properties, such as whether or not curves are opened or closed (Kovacs and Julesz 1993).

How does global coordination emerge in these processes? Are common mechanisms guiding the emergence across these diverse phenomena? What languages do contemporary science and mathematics provide to unambiguously describe the different kinds of organization that emerge in such systems?

Emergence is generally understood to be a process that leads to the appearance of structure not directly described by the defining constraints and instantaneous forces that control a system. Over time "something new" appears at scales not directly specified by the equations of motion. An emergent feature also cannot be explicitly represented in the initial and boundary conditions. In short, a feature emerges when the underlying system puts some effort into its creation.

These observations form an intuitive definition of emergence. For it to be useful, however, one must specify what the "something" is and how it is "new". Otherwise, the notion has little or no content, since almost any time-dependent system would exhibit emergent features.

Defining and quantifying "emergence" and "complexity" has been a popular research topic for a few decades. The various metrics devised so far differ along many dimensions, such as: What is considered "structure" (or "pattern")? How different is a larger pattern compared to smaller patterns?

2.1.1. Pattern!

One recent and initially baffling example of emergence is deterministic chaos. In this, deterministic equations of motion lead over time to apparently unpredictable behavior. When confronted with chaos, one question immediately demands an answer: Where in the determinism did the randomness come from? The answer is that the effective dynamic, which maps from initial conditions to states at a later time, becomes so complicated that an observer can neither measure the system accurately enough nor compute with sufficient power to predict the future behavior when given an initial condition. The emergence of disorder here is the product of both the complicated behavior of nonlinear dynamical systems and the limitations of the observer (Crutchfield *et al.* 1986).

Consider instead an example in which order arises from disorder. In a self-avoiding random walk in two-dimensions the step-by-step behavior of a particle is specified directly in stochastic equations of motion: at each time it moves one step in a random direction, except the one it just came from. The result, after some period of time, is a path tracing out a self-similar set of positions in the plane. A "fractal" structure emerges from the largely disordered step-by-step motion.

This is the paper's central question: Can we formulate emergence properties in an objective way?

Deterministic chaos and the self-avoiding random walk are two examples of the emergence of "pattern". The new feature in the first case is unpredictability; in the second, self-similarity. The "newness" in each case is only heightened by the fact that the emergent feature stands in direct opposition to the systems' defining character: complete determinism underlies chaos and near-complete stochasticity, the orderliness of self-similarity. But for whom has the emergence occurred? More particularly, to whom are the emergent features "new"? The state of a chaotic dynamical system always moves to a unique next state under the application of a deterministic function. Surely, the system state doesn't know its behavior is unpredictable. For the random walk, "fractalness" is not in the "eye" of the particle performing the local steps of the random walk, by definition. The newness in both cases is in the eye of an observer: the observer whose predictions fail or the analyst who notes that the feature of statistical self-similarity captures a commonality across length scales.

Such comments are rather straightforward, even trivial from one point of view, in these now familiar cases. But there are many other phenomena that span a spectrum of novelty from "obvious" to "purposeful" for which the distinctions are less clear. The emergence of pattern is the primary theme, for example, in a wide range of phenomena that have come to be labeled "pattern formation". These include, to mention only a few, the convective rolls of Bénard and Couette fluid flows, the more complicated flow structures observed in weak turbulence (Swinney and Gollub 1981), the spiral waves and Turing patterns produced in oscillating chemical reactions (Turing 1952; Winfree 1980; Ouyang and Swinney 1991), the statistical order parameters describing phase transitions, the divergent correlations and long-lived fluctuations in critical phenomena (Stanley 1971; Bak and Chen 1990; Binney *et al.* 1992), and the forms appearing in biological morphogenesis (Turing 1952; Thompson 1917; Meinhardt 1982).

Although the behavior in these systems is readily described as "coherent", "self-organizing", and "emergent", the patterns which appear are detected by the observers and analysts themselves. The role of outside perception is evidenced by historical denials of patterns in the Belousov–Zhabotinsky reaction, of coherent structures in highly turbulent fluid flows and of the energy recurrence in anharmonic oscillator chains reported by Fermi, Pasta, and Ulam. Those experiments didn't suddenly start behaving differently once these key structures were appreciated by scientists. It is the observer or analyst who lends the teleological "self" to processes which otherwise simply "organize" according to the underlying dynamical constraints. Indeed, the detected patterns are often *assumed* implicitly by analysts via the statistics they select to confirm the patterns' existence in experimental data. The obvious consequence is that "structure" goes unseen due to an observer's biases. In some fortunate cases, such as convection rolls, spiral waves, or solitons, the functional representations of "patterns" are shown to be consistent with mathematical models of the phenomena. But these models themselves rest on a host of theoretical assumptions. It is rarely, if ever, the case that the appropriate notion of pattern is extracted from the phenomenon itself using minimally-biased discovery procedures. Briefly stated, in the realm of pattern formation "patterns" are guessed and then verified.

This implies once more the subjective aspect of identifying emergent properties. Crutchfield's goal is to avoid that and come up with a "machine" that can do that in an objective, unique, and cost-minimal way.

2.1.2. Intrinsic Emergence

For these reasons, pattern formation is insufficient to capture the essential aspect of the emergence of coordinated behavior and global information processing in, for example, flocking birds, schooling fish, ant colonies, financial markets, and in color and shape perception. At some basic level, though, pattern formation must play a role. The problem is that the "newness" in the emergence of pattern is always referred outside the system to some observer that anticipates the structures via a fixed palette of possible regularities. By way of analogy with a communication channel, the observer is a receiver that already has the codebook in hand. Any signal sent down the channel that is not already decodable using it is essentially noise, a pattern unrecognized by the observer.

When a new state of matter emerges from a phase transition, for example, initially no one knows the governing "order parameter". This is a recurrent conundrum in condensed matter physics, since the order parameter is the foundation for analysis and, even, further experimentation. After an indeterminant amount of creative thought and mathematical invention, one is sometimes found and then verified as appropriately capturing measurable statistics. The physicists' codebook is extended in just this way.

In the emergence of coordinated behavior, though, there is a closure in which the patterns that emerge are important *within* the system. That is, those patterns take on their "newness" with respect to other structures in the underlying system. Since there is no external referent for novelty or pattern, we can refer to this process as "intrinsic" emergence. Competitive agents in an efficient capital market control their individual production-investment and stock-ownership strategies based on the optimal pricing that has emerged from their collective behavior. It is essential to the agents' resource allocation decisions that, through the market's collective behavior, prices emerge that are accurate signals "fully reflecting" all available information (Fama 1991).

What is distinctive about intrinsic emergence is that the patterns formed confer additional functionality which supports global information processing, such as the setting of optimal prices. Recently, examples of this sort have fallen under the rubric of "emergent computation" (Forrest 1990). The approach here differs in that it is based on explicit methods of detecting computation embedded in nonlinear process. More to the point, the hypothesis in the following is that during intrinsic emergence there is an increase in intrinsic computational capability, which can be capitalized on and so lends additional functionality.

In summary, three notions will be distinguished:

1. The intuitive definition of emergence: "something new appears";

2. Pattern formation: an observer identifies "organization" in a dynamical system; and

3. Intrinsic emergence: the system itself capitalizes on patterns that appear.

2.2. EVOLUTIONARY PROCESSES

One arena that frames the question of intrinsic emergence in familiar terms is biological evolution, which presumes to explain the appearance of highly organized systems from a disorganized primordial soup. Unfortunately, biological evolution is a somewhat slippery and difficult topic; not the least reason for which is the less-than-predictive role played by evolutionary theory in explaining the present diversity of life forms. Due to this, it is much easier to think about a restricted world whose structure and inhabitants are well-defined. Though vastly simplified, this world is used to frame all of the later discussion, since it forces one to be clear about the nature of observers.

The prototype universe I have in mind consists of an environment and a set of adaptive observers or "agents". (See Fig. 1.) An agent is a stochastic dynamical system that attempts to build and maintain a maximally-predictive internal model of its environment. The environment for each agent is the collection of other agents. At any given time an agent's sensorium is a projection of the current environmental state. That is, the environmental state is hidden from the agent by its sensory apparatus. Over time the sensory apparatus produces a series of measurements which guide the agent's use of its available resources—the "substrates" of Fig. 1—in the construction of an internal model. Based on the regularities captured by its internal model, the agent then takes actions via effectors that ultimately change the environmental state. The "better" its internal model, the more regularity in the environment the agent can take advantage of. Presumably,

Here the description of the system is largely identical to how the field of robotics describes their systems, which is in rapid development around the same time as this article.

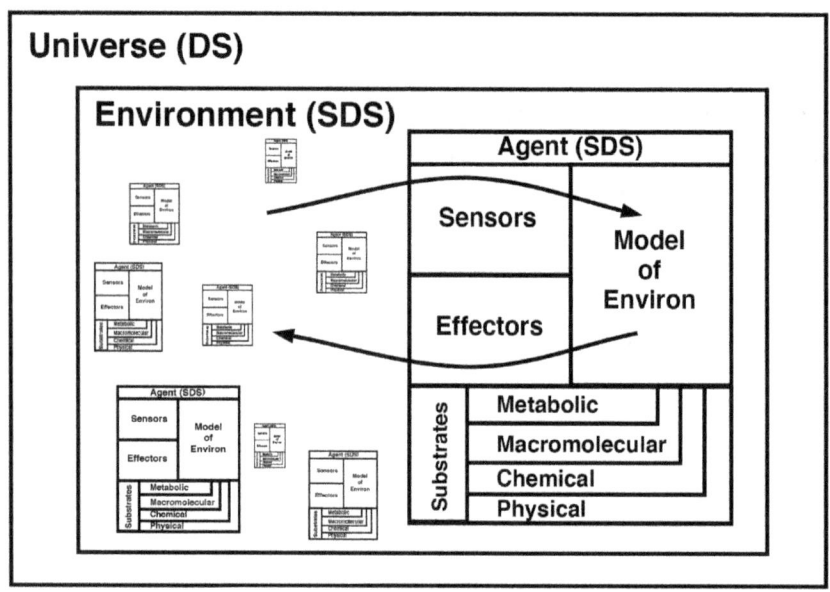

Figure 1. Agent-centric view of the environment: The universe can be considered a deterministic dynamical system (DS). The environment, as seen by any one agent, is a stochastic dynamical systems (SDS) consisting of all other agents. Its apparent stochasticity results from several effects—some intrinsic and some due to an agent's limited computational resources. Each agent is itself a stochastic dynamical system, since it may sample, or be plagued by, the uncontrollable randomness in its substrates and in environmental stimuli. The substrates represent the available resources that support and limit information processing, model building, and decision making. The arrows indicate the flow of information into and out of the agent.

that advantage increases the agent's survivability. If the available inference resources are limited, then the internal model may fail to capture useful environmental states.

The basic problem facing an agent is the prediction of future sensory input based on modeling the hidden environmental states and on selecting possible actions. The problem facing the designer of such a prototype universe is how to know if the agents have adapted and how they did so. This requires a quantitative theory of how agents process information and build models.

2.3. WHAT'S IN A MODEL?

In moving from the initial intuitive definition of emergence to the more concrete notion of pattern formation and ending with intrinsic emergence, it became clear that the essential novelty involved had to be referred to some evaluating entity. The relationship between novelty and its evaluation can be made explicit by thinking always of some observer that builds a model of a process from a series of measurements. At the level of the intuitive definition of emergence, the observer is that which recognizes the "something" and evaluates its "newness". In pattern formation, the observer is the scientist that uses prior concepts—e.g. "spiral" or "vortex"—to detect structure in experimental data and so to verify or falsify their applicability to the phenomenon at hand. Of the three, this case is probably the most familiarly appreciated in terms of an "observer" and its internal "model" of a phenomenon. Intrinsic emergence is more subtle. The closure of "newness" evaluation pushes the observer inside the system, just as the adaptive agents are inside the prototype universe. This requires in turn that intrinsic emergence be defined in terms of the "models" embedded in the observer. The observer in this view is a subprocess of the entire system. In particular, the observer subprocess is one that has the requisite information processing capability with which to take advantage of the emergent patterns.

This refers back to notions from cybernetics, like Ashby's law of requisite variety, which here would mean that the observer needs to have enough information processing capabilities to characterize the system it studies.

"Model" is being used here in a sense that is somewhat more generous than found in daily scientific practice. There it often refers to an explicit representation—an analog—of a system under study. Here models will be seen in addition as existing implicitly in the dynamics and behavior of a process. Rather than being able to point to (say) an agent's model of its environment, the designer of the prototype universe may have to excavate the "model". To do this one might infer than an agent's responses are in co-relation with its environment, that an agent has memory of the past, that the agent can make decisions, and so on. Thus, "model" here is more "behavioral" than "cognitive".

2.4. THE MODELING DILEMMA

The utility of this view of intrinsic emergence depends on answering a basic question: How does an observer understand the structure of natural processes? This includes both the scientist studying nature and an organism trying to predict aspects of its environment in order to survive. The answer requires stepping back to the level of pattern formation.

A key modeling dichotomy that runs throughout all of science is that between order and randomness. Imagine a scientist in the laboratory confronted after days of hard work with the results of a recent experiment—summarized prosaically as a simple numerical recording of instrument responses. The question arises: What fraction of the particular numerical value of each datum confirms or denies the hypothesis being tested and how much is essentially irrelevant information, just "noise" or "error"?

A fundamental point is that *any* act of modeling makes a distinction between data that is accounted for—the ordered part—and data that is not described—the apparently random part. This distinction might be a null one: for example, for either completely predictable or ideally random (unstructured) sources the data is explained by one descriptive extreme or the other. Nature is seldom so simple. It appears that natural processes are an amalgam of randomness and order. It is the organization of the interplay between order and randomness that makes nature "complex". A complex process then differs from a "complicated" process, a large system consisting of very many components, subsystems, degrees of freedom, and so on. A complicated system—such as an ideal gas—needn't be complex, in the sense used here. The ideal gas has no structure. Its microscopic dynamics are accounted for by randomness.

Experimental data are often described by a whole range of candidate models that are statistically and structurally consistent with the given data set. One important variation over this range of possible "explanations" is where each candidate draws the randomness-order distinction. That is, the models vary in the regularity captured and in the apparent error each induces.

It turns out that a balance between order and randomness can be reached and used to define a "best" model for a given data set. The balance is given by minimizing the model's size while minimizing the amount of apparent randomness. The first part is a version of Occam's dictum: causes should not be multiplied beyond necessity. The second part is a basic tenet of science: obtain the best prediction of nature. Neither component of this balance can be minimized alone, otherwise absurd "best" models would be selected. Minimizing the model size alone leads to huge error, since the smallest (null) model captures no regularities; minimizing the error alone produces a huge model, which is simply the data itself and manifestly not a useful encapsulation of what happened in the laboratory. So both model size and the induced error must be minimized together in selecting a "best" model. Typically, the sum of the model size and the error is minimized (Kemeny 1953;

Wallace and Boulton 1968; Rissanen 1978; Crutchfield and McNamara 1987; Rissanen 1989).

From the viewpoint of scientific methodology the key element missing in this story of what to do with data is how to measure structure or regularity. Just how structure is measured determines where the order-randomness dichotomy is drawn. This particular problem can be solved in principle: we take the size of the candidate model as the measure of structure. Then the size of the "best" model is a measure of the data's intrinsic structure. If we believe the data is a faithful representation of the raw behavior of the underlying process, this then translates into a measure of structure in the natural phenomenon originally studied.

Not surprisingly, this does not really solve the problem of quantifying structure. In fact, it simply elevates it to a higher level of abstraction. Measuring structure as the length of the description of the "best" model assumes one has chosen a language in which to describe models. The catch is that this representation choice builds in its own biases. In a given language some regularities can be compactly described, in others the same regularities can be quite baroquely expressed. Change the language and the same regularities could require more or less description. And so, lacking prior God-given knowledge of the appropriate language for nature, a measure of structure in terms of the description length would seem to be arbitrary.

And so we are left with a deep puzzle, one that precedes measuring structure: How is structure discovered in the first place? If the scientist knows beforehand the appropriate representation for an experiment's possible behaviors, then the amount of that kind of structure can be extracted from the data as outlined above. In this case, the prior knowledge about the structure is verified by the data if a compact, predictive model results. But what if it is not verified? What if the hypothesized structure is simply not appropriate? The "best" model could be huge or, worse, appear upon closer and closer analysis to diverge in size. The latter situation is clearly not tolerable. At the very least, an infinite model is impractical to manipulate. These situations indicate that the behavior is so new as to not fit (finitely) into current understanding. Then what do we do?

This is the problem of "innovation". How can an observer ever break out of inadequate model classes and discover appropriate ones? How can incorrect assumptions be changed? How is anything new ever discovered, if it must always be expressed in the current language?

If the problem of innovation can be solved, then, as the preceding development indicated, there is a framework which specifies how to be quantitative in detecting and measuring structure.

2.5. A COMPUTATIONAL VIEW OF NATURE

Contemporary physics does not have the tools to address the problems of innovation, the discovery of patterns, or even the practice of modeling itself, since there are no physical principles that define and dictate how to measure natural structure. It is no surprise, though, that physics does have the tools for detecting and measuring complete order—equilibria and fixed point or periodic behavior—and ideal randomness—via temperature and thermodynamic entropy or, in dynamical contexts, via the Shannon entropy rate and Kolmogorov complexity. What is still needed, though, is a definition of structure and way to detect and to measure it. This would then allow us to analyze, model, and predict complex systems at the emergent scales.

One recent approach is to adapt and extend ideas from the theory of discrete computation, which has developed measures of information-processing structure, to inferring complexity in dynamical systems (Crutchfield and Young 1989). Computation theory defines the notion of a "machine"—a device for encoding the structures in discrete processes. It has been argued that, due to the inherent limitations of scientific instruments, all an observer can know of a process in nature is a discrete-time, discrete-space series of measurements. Fortunately, this is precisely the kind of thing—strings of discrete symbols, a "formal" language—that computation theory analyzes for structure.

How does this apply to nature? Given a discrete series of measurements from a process, a machine can be constructed that is the best description or predictor of this discrete time series. The structure of this machine can be said to be the best approximation to the original process's information-processing structure, using the model size and apparent error minimization method discussed above. Once we have

> The ultimate machine model of a structure is the structure itself. So we aim for a machine that becomes indistinguishable from the thing it represents.
>
>

reconstructed the machine, we can say that we understand the structure of the process.

But what kind of structure is it? Has machine reconstruction discovered patterns in the data? Computation theory answers such questions in terms of the different classes of machines it distinguishes. There are machine classes with finite memory, those with infinite one-way stack memory, those with first-in first-out queue memory, those with counter registers, and those with infinite random access memory, among others. When applied to the study of nature, these machine classes reveal important distinctions among natural processes. In particular, the computationally distinct classes correspond to different types of pattern or regularity.

Given this framework, one talks about the structure of the original process in terms of the complexity of the reconstructed machine. This is a more useful notion of complexity than measures of randomness, such as the Kolmogorov complexity, since it indicates the degree to which information is processed in the system, which accords more closely to our intuitions about what complexity should mean. Perhaps more importantly, the reconstructed machine describes *how* the information is processed. That is, the architecture of the machines themselves represents the organization of the information processing, that is, the intrinsic computation. The reconstructed machine is a model of the mechanisms by which the natural process manipulates information.

2.6. COMPUTATIONAL MECHANICS: BEYOND STATISTICS, TOWARD STRUCTURE

That completes the general discussion of the problem of emergence and the motivations behind a computational approach to it. A number of concrete steps remain to implement and test the utility of this proposal. In particular, a key step concerns how a machine can be reconstructed from a series of discrete measurements of a process. Such a reconstruction is a way that an observer can model its environment. In the context of biological evolution, for example, it is clear that to survive agents must detect regularities in their environment. The degree to which an agent can model its environment in this way depends on its own computational resources and on what machine class or language it implicitly is restricted to or explicitly chooses when making a model. The second key step concerns how an agent can jump out of its original assumptions about the model class and, by induction, can leap to a new model class which is a much better way of understanding its environment. This is a formalization of what is colloquially called "innovation".

The overall goal, then, concerns how to detect structures in the environment—how to form an "internal model"—and also how to come up with true innovations to that internal model. There are applications of this approach to time series analysis and other areas, but the main goal is not engineering but scientific: to understand how structure in nature can be detected and measured and, for that matter, discovered in the first place as wholly new innovations in one's assumed representation.

What is new in this approach? Computation theorists generally have not applied the existing structure metrics to natural processes. They have mostly limited their research to analyzing scaling properties of computational problems; in particular, to how difficulty scales in certain information processing tasks. A second aspect computation theory has dealt with little, if at all, is measuring structure in stochastic processes. Stochastic processes, though, are seen throughout nature and must be addressed at the most basic level of a theory of modeling nature. The domain of computation theory—pure discreteness, uncorrupted by noise—is thus only a partial solution. Indeed, the order-randomness dichotomy indicates that the interpretation of any experimental data has an intrinsic probabilistic component which is induced by the observer's choice of representation. As a consequence probabilistic computation must be included in any structural description of nature. A third aspect computation theory has considered very little is measuring structure in processes that are extended in space. A fourth aspect it has not dealt with traditionally is measuring structure in continuous-state processes. If computation theory is to form the foundation of a physics of structure, it must be extended in at least these three ways. These extensions have engaged a number of workers in dynamical systems recently, but there is much still to do (Crutchfield and Young 1989; Wolfram 1984; Blum, Shub, and Smale 1989; Nordahl 1989; Hanson and Crutchfield 1992; Crutchfield 1992c; Moore 1993).

2.7. AGENDA

The remainder of the discussion focuses on temporal information processing and the first two extensions—probabilistic and spatial computation—assuming that the observer is looking at a series of measurements of a continuous-state system whose states an instrument has discretized. The phrase "calculi of emergence" in the title emphasizes the tools required to address the problems which intrinsic emergence raises. The tools are (i) dynamical systems theory with its emphasis on the role of time and on the geometric structures underlying the increase in complexity during a system's time evolution, (ii) the notions of mechanism and structure inherent in computation theory, and (iii) inductive inference as a statistical framework in which to detect and innovate new representations. The proposed synthesis of these tools develops as follows.

First, Section 3 defines a complexity metric that is a measure of structure in the way discussed above. This is called "statistical complexity", and it measures the structure of the minimal machine reconstructed from observations of a given process in terms of the machine's size. Second, Section 3 describes an algorithm—ϵ-machine reconstruction—for reconstructing the machine, given an assumed model class. Third, Section 4 presents an algorithm for innovation—called hierarchical ϵ-machine reconstruction—in which an agent can inductively jump to a new model class by detecting regularities in a *series* of increasingly accurate models. Fourth, the remainder of Section 4 analyzes several examples in which these general ideas are put into practice to determine the intrinsic computation in continuous-state dynamical systems, recurrent hidden Markov models, and cellular automata. Finally, Section 5 concludes with a summary of the implications of this approach for detecting and understanding the emergence of structure in evolving populations of adaptive agents.

In a way, statistical complexity is a measure of how much information of the past is needed to have a complete capability to replicate the statistics of the future.

3. Mechanism and Computation

Probably the most highly developed appreciation of hierarchical structure is found in the theory of discrete computation, which includes automata theory and the theory of formal languages (Hopcroft and Ullman 1979; Lewis and Papadimitriou 1981; Brookshear 1989). The many diverse types of discrete computation, and the mechanisms that implement them, will be taken in the following as a framework whose spirit is to be emulated and extended. The main objects of attention in discrete computation are strings, or words, $\omega = s_0 s_1 s_2 \cdots s_{L-1}, s_i \in \mathscr{A} = \{0, 1, \ldots, k-1\}$. Sets of words are called formal languages; for example, $\mathscr{L} = \{\omega_0, \omega_1, \ldots, \omega_m\}$. One of the main questions in computation theory is how difficult it is to "recognize" a language—that is, to classify any given string as to whether or not it is a member of the set. "Difficulty" is made concrete by associating with a language different types of machines, or automata, that can perform the classification task. The automata themselves are distinguished by how they utilize various resources, such as memory or logic operations or even the available time, to complete the classification task. The amount and type of these resources determine the "complexity" of a language and form the basis of a computational hierarchy—a road map that delineates successively more "powerful" recognition mechanisms. Particular discrete computation problems often reduce to analyzing the descriptive capability of an automaton, or of a class of like-structured automata, in terms of the languages it can recognize. This duality, between languages as sets and automata as functions which recognize sets, runs throughout computation theory.

Although discrete computation theory provides a suggestive framework for investigating hierarchical structure in nature, a number of its basic elements are antithetical to scientific practice. Typically, the languages are countable and consist of arbitrary length, but finite words. This restriction clashes with basic notions from ergodic theory, such as stationarity, and from physics, such as the concept of a process that has been running for a long time, that is, a system in equilibrium. Fortunately, many of these deficiencies can be removed, with the result that the concepts of complexity and structure in computation theory can be usefully carried over to the empirical sciences to describe how a process's behavioral complexity is related to the structure of its underlying mechanism. This type of description will be one of the main points of review in the following. Examples later on will show explicitly how nonlinear dynamical systems have various computational elements embedded in them.

But what does it mean for a physical device to perform a computation? How do its dynamics and the underlying device physics support information processing? Answers to these questions need to distinguish two notions of computation. The first, and probably more familiar, is the notion of "useful" computation. The input to a computation is given by the device's initial physical configuration. Performing the computation corresponds to the temporal sequence of changes in the device's internal state. The result of the computation is read off finally in the state to which the device relaxed. Ultimately, the devices with computational utility are those we have constructed to implement input-output mappings of interest to us. In this type of computation an outside observer must interpret the end product as useful: it involves a semantics of utility. One of the more interesting facets of useful computation is that there are universal computers that can emulate any discrete computational process. Thus, in principle, only one type of device needs to be constructed to perform any discrete computation.

In contrast, the second notion—"intrinsic" computation—focuses on how structures in a device's state space support and constrain information processing. It addresses the question of how computational elements are embedded in a process. It does not ask if the information produced is useful. In this it divorces the semantics of utility from computation. Instead, the analysis of a device's intrinsic computation attempts to detect and quantify basic information processing elements—such as memory, information transmission and creation, and logical operations (Crutchfield 1992b).

3.1. ROAD MAPS TO INNOVATION

With this general picture of computation the notion of a computational hierarchy can be introduced. Fig. 2 graphically illustrates a hierarchy of discrete-state devices in terms of their computational capability. Each circle there denotes a class of languages. The abbreviations inside indicate the class's name and also, in some cases, the name of the grammar and/or automaton type. Moving from the bottom to the top one finds successively more powerful grammars and automata and harder-to-recognize languages. The interrelationships between the classes is denoted with a line: if class M_i is below and connected to M_j, then M_j recognizes all of the languages that M_i does and more. The hierarchy itself is only a partial ordering of descriptive capability. Some classes are not strictly comparable. The solid lines indicate inclusion: a language lower in the diagram can be recognized by devices at higher levels, but there are languages at higher levels not recognizable at lower levels. The least powerful models, at the hierarchy's bottom, are those with finite memory—the finite automata (DFA/NFA). At the top are the universal Turing machines (UTM) which have infinite random-access tape memories. In between, roughly speaking, there are two broad classes of language: context-free languages that can be recognized by machines whose infinite memories are organized in a stack, and context-sensitive languages recognized by machines whose memory accesses are limited by a linear function of the initial input's length. What is remarkable about this hierarchy is the wealth of intervening model classes and the accumulated understanding of their relative language classification powers. Fig. 2 includes more detail than is necessary for the following discussion, but it does demonstrate some of the diversity of computational mechanisms that have been studied (Hopcroft and Ullman 1979).

Fig. 2 includes the formal grammar models of Chomsky and others, the associated finite and stack automata, and the arbitrary-access tape machines of Turing. Hierarchical structure should not be thought of as being limited to just these, however. Even staying within the domain of discrete symbol manipulation, there are the (Lindenmayer) parallel-rewrite (Rozenberg and Salomaa 1986) and queue-based (Allevi, Cherubini, and Crespi-Reghizzi 1988; Cherubini et al. 1991) computational models. There are also the arithmetic and analytic hierarchies of recursive function theory (Rogers 1967). The list of discrete computation hierarchies seems large because it is and needs to be to capture the distinct types of symbolic information processing mechanisms.

Although the discrete computation hierarchy of Fig. 2 can be used to describe information processing in some dynamical systems, it is far from adequate and requires significant extensions. Several sections in Section 4 discuss three different extensions that are more appropriate to computation in dynamical systems. The first is a new hierarchy for stochastic finitary processes. The second is a new hierarchy for discrete spatial systems. And the third is the ϵ-machine hierarchy of causal inference. A fourth and equally important hierarchy, which will not

Despite the use of the phrase "causal inference" and calling the macrostates S "causal states," this article is actually not concerned with causal inference (or even causal discovery), a field popularized by, e.g., Judea Pearl, who had already published early work on the topic. Causal inference is concerned with finding a model that is not only predictive but also causal: predicting the counterfactual outcome of a hypothetical intervention. Crutchfield here is only concerned with prediction.

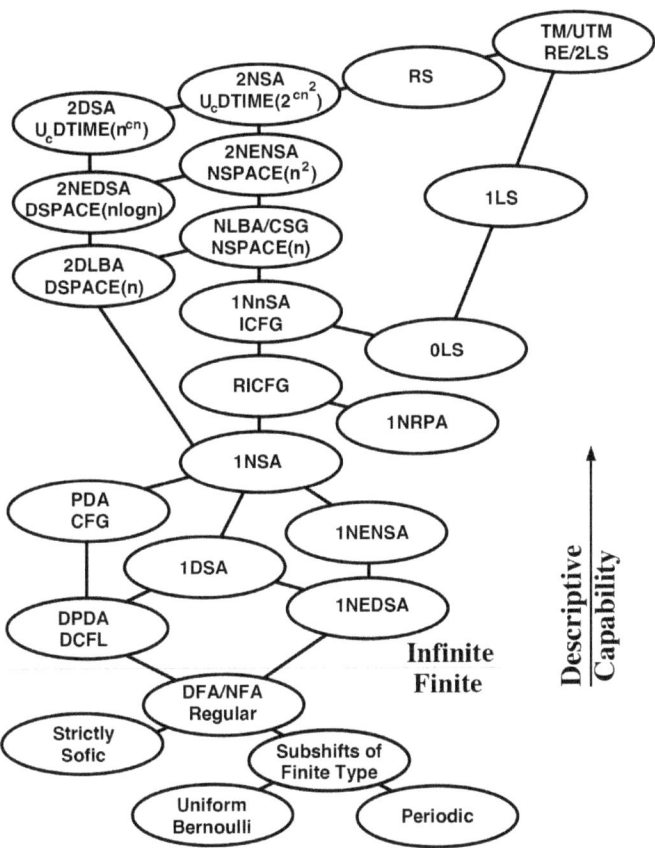

Figure 2. The discrete computation hierarchy. *Adjective legend*: 1 = one way input tape, 2 = two way input tape, D = deterministic, N = nondeterministic, I = indexed, RI = restricted I, n = nested, NE = nonerasing, CF = context free, CS = context sensitive, R = recursive, RE = R enumerable, and U = universal. *Object legend*: G = grammar, A = automata, FA = finite A, PDA = pushdown A, SA = stack A, LBA = linear bounded A, RPA = Reading PDA, TM = Turing machine, LS = Lindenmayer system, 0L = CFLS, 1L = CSLS, and RS = R set. (After Hopcroft and Ullman 1979; Aho 1968, 1969; Weiss 1973; Walker 1974; Lindenmayer and Prusinkiewicz 1989.)

be discussed in the following, classifies different types of continuous computation (Blum, Shub, and Smale 1989; Moore 1993). The benefit of pursuing these extensions is found in what their global organization of classes indicates about how different representations or modeling assumptions affect an observer's ability to build models. What a natural scientist takes from the earliest hierarchy—the Chomsky portion of Fig. 2—is the spirit in which it was constructed and not so much its details. On the one hand, there is much in the Chomsky hierarchy that is deeply inappropriate to general scientific modeling. The spatial and stochastic hierarchies introduced later give an idea of those directions in which one can go to invent computational hierarchies that explicitly address model classes which are germane to the sciences. On the other hand, there is a good deal still to be gleaned from the Chomsky hierarchy. The recent proposal to use context-free grammars to describe nonlocal nucleotide correlations associated with protein folding is one example of this (Searls 1992).

3.2. COMPLEXITY ≠ RANDOMNESS

The main goal here is to detect and measure structure in nature. A computational road map only gives a qualitative view of computational capability and so, within the reconstruction framework, a qualitative view of various types of possible natural structure. But empirical science requires quantitative methods. How can one begin to be quantitative about computation and therefore structure?

Generally, the metrics for computational capability are given in terms of "complexity". The complexity $C(x)$ of an object x is taken to be the size of its minimal representation $M_{\min}(x \mid \mathscr{V})$ when expressed in a chosen vocabulary $\mathscr{V}$: $C(x) = \|M_{\min}(x \mid \mathscr{V})\|$. x can be thought of as a series of measurements of the environment. That is, the agent views the environment as a process which has generated a data stream x. Its success in modeling the environment is determined in large part by the apparent complexity $C(x)$. But different vocabularies, such as one based on using finite automata versus one based on pushdown stack automata, typically assign different complexities to the same object. This is just the modeling dilemma discussed in Section 2.

Probably the earliest attempt at quantifying information processing is due to Shannon and then later to Chaitin, Kolmogorov, and Solomonoff. This led to what can be called a "deterministic" complexity, where "deterministic" means that no outside, e.g. stochastic, information source is used in describing an object. The next subsection reviews this notion; the subsequent one introduces a relatively new type called "statistical complexity" and compares the two.

3.2.1. Deterministic Complexity

In the mid-1960s it was noted that if the vocabulary was taken to be programs for universal Turing machines, then a certain generality obtained to the notion of complexity. The Kolmogorov–Chaitin complexity $K(x)$ of an object x is the number of bits in the smallest program that outputs x when run on a universal deterministic Turing machine (UTM) (Chaitin 1966; Kolmogorov 1965; Solomonoff 1964). The main deficiency that results from the choice of a universal machine is that $K(x)$ is not computable in general. Fortunately, there are a number of process classes for which some aspects of the deterministic complexity are well understood. If the object in question is a string s^L of L discrete symbols produced by an information source, such as a Markov chain, with Shannon entropy rate h_μ, (Shannon and Weaver 1948) then the growth rate of the Kolmogorov–Chaitin complexity is

$$\frac{K\left(s^L\right)}{L} \xrightarrow[L\to\infty]{} h_\mu. \tag{1}$$

The growth rate h_μ is independent of the particular choice of universal machine. In the modeling framework it can be interpreted as the error rate at which an agent predicts successive symbols in s^L.

Not surprisingly, for chaotic dynamical systems with continuous state variables and for the physical systems they describe, we have

$$K\left(s^L_\epsilon\right) \underset{\substack{L\to\infty \\ \epsilon\to 0}}{\propto} h_\mu L, \tag{2}$$

where the continuous variables are coarse-grained at resolution ϵ into discrete "measurement" symbols $s_\epsilon \in \{0, 1, 2, \ldots, \epsilon^{-d} - 1\}$ and d is the state space dimension [49]. Thus, there are aspects of deterministic complexity that relate directly to physical processes. This line of investigation has led to a deeper (algorithmic) understanding of randomness in physical systems. In short, $K(x)$ is a measure of randomness of the object x and, by implication, of randomness in the process which produced it (Ford 1983).

The Shannon entropy is a measure on random variables that roughly translates to how many bits it takes to communicate the value of said random variable. For example, a random variable X that takes on values a, b, c, d with equal probability has a Shannon entropy of 2, as it can be represented in 2 bits ($a = 00$, $b = 01$, $c = 10$, $d = 11$). The entropy rate of a sequence of random variables, then, represents the additional memory cost to communicate each additional variable in the chain.

3.2.2. Statistical Complexity

Roughly speaking, the Kolmogorov–Chaitin complexity $K(x)$ requires accounting for all of the bits, including the random ones, in the object x. The main consequence is that $K(x)$, considered as a number, is dominated by the production of randomness and so obscures important kinds of structure in x and in the underlying process. In contrast, the statistical complexity $C_\mu(x)$ discounts the computational effort the UTM expends in simulating random bits in x. One of the defining properties of statistical complexity is that an ideal random object x has $C_\mu(x) = 0$. Also, like $K(x)$, for simple periodic processes, such as $x = 0000000\cdots 0$, $C_\mu(x) = 0$. Thus, the statistical complexity is low for both (simple) periodic and ideal random processes. If s^L denotes the first L symbols of x, then the relationship between the complexities is simply

$$K\left(s^L\right) \approx C_\mu\left(s^L\right) + h_\mu L. \tag{3}$$

This approximation ignores important issues of how averaging should be performed; but, as stated, it gives the essential idea.

One interpretation of the statistical complexity is that it is the minimum amount of historical information required to make optimal forecasts of bits in x at the error rate h_μ. Thus, C_μ is not a measure of randomness. It is a measure of structure above and beyond that describable as ideal randomness. In this, it is complementary to the Kolmogorov–Chaitin complexity and to Shannon's entropy rate.

Various complexity metrics have been introduced in order to capture the properties of statistical complexity. The "logical depth" of x, one of the first proposals, is the run time of the UTM that uses the minimal representation $M_{\min}(x)$ (Bennett 1988). Introduced as a practical alternative to the uncomputable logical depth, the "excess entropy" measures how an agent learns to predict successive bits of x (Crutchfield and Packard 1983). It describes how estimates of the Shannon entropy rate converge to the true value of h_μ. The excess entropy has been recoined twice, first as the "stored information" and then as the "effective measure complexity" (Shaw 1984; Grassberger 1986). Statistical complexity itself was introduced in Crutchfield and Young (1989). Since it makes an explicit connection with computation and with inductive inference, C_μ will be the primary tool used here for quantifying structure.

 For this reason, Kolmogorov–Chaitin complexity is now often referred to as algorithmic information. It is more a measure of information content contained in data than the complexity of its internal structure.

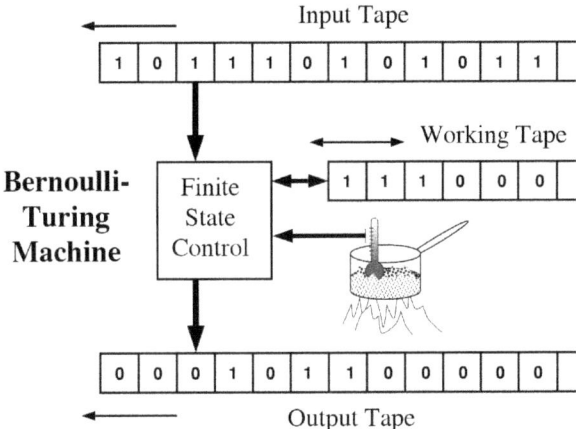

Figure 3. The Bernoulli–Turing Machine (BTM) is a deterministic Turing machine augmented by contact to an information source—a heat bath denoted as a boiling water pot. Like a Turing machine, it is a transducer that maps input tapes $(0+1)^*$ to output tapes $(0+1)^*$. The input (output) tape cells are read (written) sequentially and once only. Any intermediate processing and storage is provided by the working tape which allows bidirectional access to its contents. The BTM defines the most general model of discrete stochastic sequential computation.

3.2.3. Complexity Metrics

These two extremes of complexity metric bring us back to the question—What needs to be modified in computation theory to make it useful as a theory of structures found in nature? That is, how can it be applied to, say, physical and biological phenomena? As already noted, there are several explicit differences between the needs of the empirical sciences and formal definitions of discrete computation theory. In addition to the technical issues of finite length words and the like, there are three crucial extensions to computation theory: the inclusion of probability, inductive inference, and spatial extent. Each of these extensions has received some attention in theoretical computer science, coding theory, and mathematical statistics (Rissanen 1989; Angluin and Smith 1983). Each plays a prominent role in one of the examples to come later.

More immediately the extension to probabilistic computation gives a unified comparison of the deterministic and statistical complexities and so indicates a partial answer to these questions. Recall that the vocabulary underlying K consists of minimal programs that run on a deterministic UTM. We can think of C_μ similarly in terms of a Turing machine that can guess. Fig. 3 shows a probabilistic generalization—the Bernoulli–Turing machine (BTM)—to the basic Turing machine model of the discrete computation hierarchy (Crutchfield and Young 1990). The equivalent of the road map shown in Fig. 2 is a "stochastic" computation hierarchy, which will be the subject of a later section.

With the Bernoulli–Turing machine in mind, the deterministic and statistical complexities can be formally contrasted. For the Kolmogorov–Chaitin complexity we have

$$K(x) = \|M_{\min}(x \mid \text{UTM})\| \qquad (4)$$

and for the statistical complexity we have

$$C_\mu(x) = \|M_{\min}(x \mid \text{BTM})\|. \qquad (5)$$

Not to be confused with stochastic complexity introduced by Rissanen (1987).

The difference between the two over processes that range from simple periodic to ideal random is illustrated in Fig. 4. As shown in Fig. 4a, the deterministic complexity is a monotonically increasing function of the degree of ideal randomness in a process. It is governed by a process's Shannon entropy rate h_μ. The statistical complexity, in contrast, is zero at both extremes and maximized in the middle. (See Fig. 4b.) The "complex" processes at intermediate degrees of randomness are combinations of ordered and stochastic computational elements. The larger the number of such irreducible components composing a process, the more "complex" the process. The interdependence of randomness as measured by the Shannon entropy rate and the statistical complexity is a surprisingly universal phenomenon. A later section analyzes two families of dynamical systems using the complexity–entropy diagram of Fig. 4b to describe their information processing capabilities.

This is a beautiful and central observation in this paper and strong advocacy for computing theory to understand complex systems.

It is notable, in this context, that current physical theory does not provide a measure of structure like statistical complexity. Instead one finds metrics for disorder, such as temperature and thermodynamic entropy. In a sense, physics has incorporated elements from the Kolmogorov–Chaitin framework, but does not include the elements of computation theory or of statistical complexity. There are, though, some rough physical measures of structure. These are seen in the use of group theory in crystallography and quantum mechanics. Group theoretic properties, though, only concern periodic, reversible processes or operations. Unlike ergodic theory and dynamical systems theory, contemporary physical theory is mute when it comes to quantitatively distinguishing, for example, the various *kinds* of

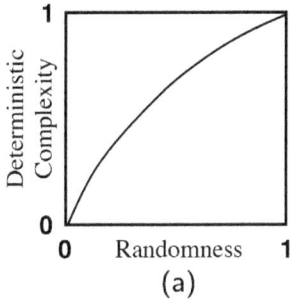
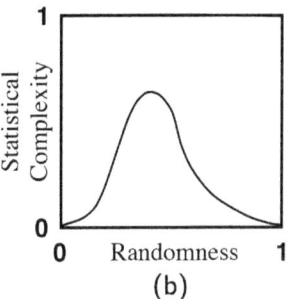

Figure 4. (a) Deterministic complexity—relative to (say) a deterministic universal Turing machine—is a measure of the degree of unpredictability of an information source. It indicates the degree of randomness which can be measured with the Shannon entropy rate h_μ. (b) Statistical complexity is based on the notion that randomness is statistically simple: an ideal random process has zero statistical complexity. At the other end of the spectrum, simple periodic processes have low statistical complexity. Complex processes arise between these extremes and are an amalgam of predictable and stochastic mechanisms. (After Crutchfield and Young 1990.)

chaotic and stochastic systems. This is what the statistical complexity is intended to provide.

The statistical complexity is a relative, not an absolute, measure of structure. It is relative to a source of ideal randomness—relative to a Random Oracle, in the parlance of computational complexity theory. A scientist might object to the use of statistical complexity, therefore, by arguing that it is important in a physical setting to account for all of the mechanisms involved in producing information. This is a fair enough comment. It acknowledges the study of randomness and it is compatible with the original spirit of Kolmogorov's program to investigate the algorithmic basis of probability. Deterministic chaos, though, has shown us that there are many sources of effective randomness in nature. One can simply use a chaotic system or appeal to the "heat bath" as an effective Random Oracle. In physics and most empirical sciences explicit accounting for random bits is neither necessary nor desirable. Ultimately, there is no contradiction between the deterministic and statistical views. Within each one simply is interested in answers to different questions.

The explication of the discrete computation hierarchy of Fig. 2 and the two notions of deterministic and statistical complexity begins to suggest how different types of structure can be investigated. In addition to the probabilistic extension to computation theory that shed some light on the distinction between $K(x)$ and $C_\mu(x)$, another important generalization is to spatially extended systems—those that

With the rapid development of quantum computing in the last two decades, the availability of true randomness in nature is now quite generally accepted. In particular, it is possible to engineer quantum bits, whose measurements are verifiable random using Bell tests (e.g., Pironio et al. 2010) Such a system would make an excellent randomness oracle.

generate "patterns"—will be the subject of later discussion. But before considering this or any other extension, the intervening sections review how complexity and randomness can be inferred from a measurement time series by an observer. The result of this will be the inductive hierarchy of ϵ-machines, which will capture the intrinsic computational structure in a process. This inductive hierarchy stands in contrast to the engineering-oriented hierarchy of Fig. 2.

3.3. ϵ-MACHINE RECONSTRUCTION

How can an agent detect structure—in particular, computation—in its measurements of the environment? To answer this, let us continue with the restriction to discrete-valued time series; that is, the agent reads off a series of discrete measurements from its sensory apparatus. If one is interested in describing continuum-state systems then this move should be seen as purely pragmatic: an instrument will have some finite accuracy, generically denoted ϵ, and individual measurements, denoted s, will range over an alphabet $\mathscr{A} = \{0, 1, 2, \ldots, \lceil \epsilon^{-1} \rceil - 1\}$. It is understood that the measurements $s \in \mathscr{A}$ are only indirect indicators of the hidden environmental states.

The notion of causal states is closely related with "belief states" in reinforcement learning. That is, they represent an agent's belief of the environment based on their past measurement outcomes. Some recent discussion of such connections can be found in Zhang *et al.* (2021).

The goal for the agent is to detect the "hidden" states $\mathbf{S} = \{S_0, S_1, \ldots, S_{V-1}\}$ in its sensory data stream that can help it predict the environment. The states so detected will be called "causal" states. For discrete time series a causal state is defined to be the set of subsequences that renders the future conditionally independent of the past. Thus, the agent identifies a state at different times in a data stream as being in identical conditions of knowledge about the future (Crutchfield and Young 1989). (See Fig. 5 for a schematic illustration that ignores probabilistic aspects.)

The notion of causal state can be defined as follows. Consider two parts of a data stream $\mathbf{s} = \cdots s_{-2} s_{-1} s_0 s_1 s_2 \cdots$. The one-sided forward sequence $\mathbf{s}_t^{\rightarrow} = s_t s_{t+1} s_{t+2} s_{t+3} \cdots$ and one-sided reverse sequence $\mathbf{s}_t^{\leftarrow} = \cdots s_{t-3} s_{t-2} s_{t-1} s_t$ are obtained from s by splitting it at time t into the forward- and reverse-time semi-infinite subsequences. They represent the information about the future and past, respectively. Consider the joint distribution of possible forward sequences $\{\mathbf{s}^{\rightarrow}\}$ and reverse

Crutchfield (1994)

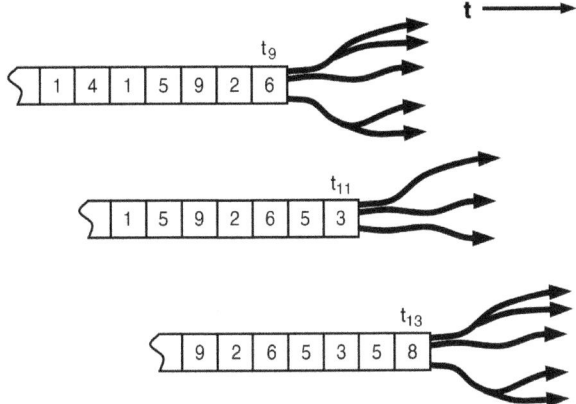

Figure 5. Within a single data stream, morph-equivalence induces conditionally-independent states. When the templates of future possibilities—that is, the allowed future subsequences and their past-conditioned probabilities—have the same structure, then the process is in the same causal state. At t_9 and at t_{13}, the process is in the same causal state since the future morphs have the same shape; at t_{11} it is in a different causal state. The figure only illustrates the non-probabilistic aspects of morph-equivalence. (After Crutchfield 1992a.)

sequences $\{s^\leftarrow\}$ over all times t:

$$\Pr(s) = \Pr(s^\rightarrow, s^\leftarrow) = \Pr(s^\rightarrow \mid s^\leftarrow) \Pr(s^\leftarrow). \quad (6)$$

The conditional distribution $\Pr(s^\rightarrow \mid \omega)$ is to be understood as a function over all possible forward sequences $\{s^\rightarrow\}$ that can follow the particular sequence ω wherever ω occurs in s.

Then the same causal state $S \in \mathcal{S}$ is associated with all those times $t, t' \in \{t_{i_1}, t_{i_2}, t_{i_3} \cdots : i_k \in \mathbb{Z}\}$ such that past-conditioned future distributions are the same. That is,

$$t \sim t' \text{ if and only if } \Pr(s^\rightarrow \mid s_t^\leftarrow) = \Pr(s^\rightarrow \mid s_{t'}^\leftarrow). \quad (7)$$

Here "$\sim$" denotes the equivalence relation induced by equivalent future morphs. If the process generating the data stream is ergodic, then there are several comments that serve to clarify how this relation defines causal states. First, the particular sequences $s_t^\leftarrow$ and $s_{t'}^\leftarrow$ are typically distinct. If $t \sim t'$, Eq. (7) means that upon having seen different histories one can be, nonetheless, in the same state of knowledge or ignorance about what will happen in the future. Second, $s_t^\leftarrow$ and $s_{t'}^\leftarrow$, when considered as particular symbol sequences, can each occur in s many times other than t and t', respectively. Finally, the conditional distributions $\Pr(s^\rightarrow \mid s_t^\leftarrow)$ and

There are some practical issues here. Even though the agent is said to have limited computational capabilities (hence the innovations later on), here an agent would require infinite memory or infinite time, and infinite data from specifically an ergodic process, to be able to decide this (eq. 7).

2483

$\Pr\left(\mathbf{s}^{\rightarrow} \mid \mathbf{s}_{t'}^{\leftarrow}\right)$ typically are functions over a nonempty range of "follower" sequences $\mathbf{s}^{\rightarrow}$.

This gives a formal definition to the set **S** of causal states as equivalence classes of future predictability: $\sim$ is the underlying equivalence relation that partitions temporal shifts of the data stream into equivalence classes. In the following the states will be taken simply as the labels for those classes. This does more than simplify the discussion. As integers ranging over $\{0, 1, 2, \ldots, \|\mathbf{S}\| - 1\}$, the states convey all of the information required to render the future conditionally independent of the past. For a given state S the set of future sequences $\{\mathbf{s}_s^{\rightarrow} : S \in \mathbf{S}\}$ that can be observed from it is called its "future morph". (Recall Fig. 5.) The set of sequences that lead to S is called its "past morph".

Note that a state and its morphs are the contexts in which an individual measurement takes on semantic content. Each measurement is anticipated or "understood" by the agent vis à vis the agent's internal model and, in particular, the structure of the states. This type of measurement semantics is discussed elsewhere (Crutchfield 1992b).

Once the causal states are found, the temporal evolution of the process—its symbolic dynamic—is given by a mapping $T : \mathbf{S} \to \mathbf{S}$ from states to states; that is, $S_{t+1} = TS_t$. The pair $\mathrm{M} = (\mathbf{S}, T)$ is referred to as an ϵ-machine; where ϵ simply reminds us that what we have reconstructed (i) is an approximation of the process's computational structure and (ii) depends on the measuring instrument's characteristics, such as its resolution. The procedure that begins with a data stream and estimates the number of states and their transition structure and probabilities is referred to as ϵ-machine reconstruction (Crutchfield and Young 1989).

What do these reconstructed machines represent? First, by the definition of future-equivalent states, the machines give the minimal information dependency between the morphs. It is in this respect that they represent the causal structure of the morphs considered as events. The machines capture the information flow within the given data stream. If state B follows state A then, as far as the observer is concerned A is a cause of B and B is one effect of A. Second, ϵ-machine reconstruction produces minimal models up to the given prediction

Since their inception, ϵ-machines have been used in multiple contexts independent of measurement resolution and approximation. For example, digital data stream analysis is naturally discrete, as are finite-dimensional quantum systems.

error level. The effective error level is determined by the available inference resources. Third, minimality guarantees that there are no other events (morphs) that intervene, at the given error level, to render A and B independent. In this case, we say that information flows from A to B. The amount of information that flows is the negative logarithm of the connecting transition probability: $-\log_2 p_{A \to B}$. Finally, time is the natural ordering captured by ϵ-machines.

3.4. MEASURING PREDICTABILITY AND STRUCTURE

With the modeling methodology laid out, several statistics can be defined that capture how information is generated and processed by the environment as seen by an agent. A useful coordinate-independent measure of information production has already been introduced—the Shannon entropy rate h_μ (Shannon and Weaver 1948). If the agent knows the distribution $\Pr(\omega)$ over infinite measurement sequences ω, then the entropy rate is defined as

$$h_\mu = \lim_{L \to \infty} \frac{H\left(\Pr\left(s^L\right)\right)}{L}, \tag{8}$$

in which $\Pr\left(s^L\right)$ is the marginal distribution, obtained from $\Pr(\omega)$, over the set of length L sequences s^L and H is the average of the self-information, $-\log_2 \Pr\left(s^L\right)$, over $\Pr\left(s^L\right)$. In simple terms, h_μ measures the rate at which the environment appears to produce information. Its units are bits per symbol. The higher the entropy rate, the more information produced, and the more unpredictable the environment appears to be.

Typically, the agent does not know $\Pr(\omega)$ and so the definition in Eq. (8) is not directly applicable. Assuming that the agent has observed a "typical" data stream **s** and that the process is ergodic, the entropy becomes

$$h_\mu = H\left(\Pr\left(s_{t+1} \mid \overleftarrow{s_t}\right)\right), \tag{9}$$

where $\Pr\left(s_{t+1} \mid \overleftarrow{s_t}\right)$ is the conditional distribution of the next symbol s_{t+1} given the semi-infinite past $\overleftarrow{s_t}$ and H averages the conditional distribution over $\Pr\left(\overleftarrow{s}\right)$. Using the agent's current set **S** of inferred

ϵ-machines are indeed the minimal machines when we assume that their state-space is classical. However, recent research indicates that there can be even simpler machines if quantum information processing is allowed (Gu et al. 2012).

Eq. 13: The statistical complexity can be readily generalized to quantum systems using the Von Neumann entropy. This generalization has allowed research in quantum models of stochastic processes and quantum statistical complexity (Gu et al. 2012).

In more modern contexts, $\|\mathbf{S}\|$ can also be referred to as the minimal memory dimension of the agent. Here the memory dimension of a system is the maximum number of mutually distinguishable states it can have. While these two definitions are synonymous, the latter can be more readily generalized to cases where S is stored in quantum memory. In such cases, it is possible to have K different memory states stored in a memory with dimension less than K. For example, the states $|0>, |1>, |+>$ can all be stored within a single qubit—a system with memory dimension 2.

As ϵ goes to 0, the statistical complexity tends to trend toward infinity.

causal states and finding the one to which $s_t^{\leftarrow}$ leads, the agent can estimate the entropy in a much simpler way using

$$h_\mu = H(\Pr(s \mid S)) \qquad (10)$$

in which $\Pr(s \mid S)$ is the conditional distribution of the next symbol s given the current state $S \in \mathbf{S}$.

Thinking about quantifying unpredictability in this way suggests there are other, perhaps more immediate, measures of the environment's structure. The topological complexity C_0 of a process is simply given in terms of the minimal number of causal states in the agent's model

$$C_0 = \log_2 \|\mathbf{S}\|. \qquad (11)$$

It is an upper bound on the amount of information needed to specify which state the environment is in. There is also a probabilistic version of the "counting" topological complexity. It is formulated as follows. The $\|\mathbf{S}\| \times \|\mathbf{S}\|$ transition probability matrix T determines the asymptotic causal state probabilities as its left eigenvector

$$p_s T = p_s, \qquad (12)$$

in which p_s is the causal states' asymptotic probability distribution: $\Sigma_{s \in \mathbf{s}} p_s = 1$. From this we have an informational quantity for the machine's size

$$C_\mu = H(p_\mathbf{s}). \qquad (13)$$

This is the statistical complexity. If, as provided by machine reconstruction, the machine is minimal, then C_μ is the amount of memory (in bits) required for the agent to predict the environment at the given level "ϵ" of accuracy (Crutchfield and Young 1989).

Let's step back a bit. This section reviewed how an agent can build a model from a time series of measurements of its environment. If one considers model building to be a dynamic process, then during model construction and refinement there are two quantities, entropy rate and statistical complexity, that allow one to monitor the effectiveness and size, respectively, of the agent's model. Since the absolute difference between the environment's actual entropy rate and that of the agent's internal model determines the agent's rate of incorrect predictions, the closer the model's entropy is to that of the environment, the higher the agent's chance for survival. This survivability comes at a cost determined by the resources the agent must devote to making the predictions. This, in turn, is measured as the model's statistical complexity.

4. Toward a Mathematical Theory of Innovation

4.1. RECONSTRUCTING LANGUAGE HIERARCHIES

Complexity, entropy, and ϵ-machine reconstruction itself concern incremental adaptation for an agent: the agent's "development" or its "interim" evolution when survival is viewed as an optimization and the environmental statistics are quasi-stationary. In contrast, innovation is associated with a change in model class. One would expect this change to correspond to an increase in computational sophistication of the model class, but it need not be. Roughly, innovation is the computational equivalent of speciation—recall that the partial ordering of a computational hierarchy indicates that there is no single way "up" in general. In concrete terms, innovation is the improvement in an agent's notion of environmental (causal) state. However it is instantiated in physical and biological processes, innovation seems to be an active process given the demonstrated robustness and creativity of life in the face of adversity. Innovation, in the narrow sense used here, should be distinguished from the passive, random forces of evolutionary change implied by mutation and recombination.

The computational picture of innovation, shown schematically in Table 1, leads to an enlarged view of the evolutionary dynamic. This can be described from the agent's view in terms of hierarchical ϵ-machine reconstruction as follows (Hanson and Crutchfield 1992; Crutchfield 1991):

1. Start at the lowest level of the computational hierarchy by building stochastic finite automata via ϵ-machine reconstruction. There are, in fact, transitions over three levels implicit in the previous introduction of ϵ-machine reconstruction; these are shown explicitly as levels 0 through 2 in Table 1 These go from the data stream (Level 0) to trees (Level 1) and then to stochastic finite automata (Level 2).

Level	Model Class	Machine	Model Size, if class is appropriate	Equivalence Relation
...	...	...	...	
3	String Production		$\mathcal{O}(\|\mathbf{V}\| + \|\mathbf{E}\| + \|\mathbf{P}\|)$	Finitary-Recursive Conditional Independence
2	Finite Automata		$\mathcal{O}(\|\mathbf{V}\| + \|\mathbf{E}\|)$	Conditional Independence
1	Tree		$\mathcal{O}\left(\|\mathcal{A}\|^D\right)$	Block Independence
0	Data Stream		m	Measurement

Table 1. A causal time-series modeling hierarchy. Each level is defined in terms of its model class. The models themselves consist of states (circles or squares) and transitions (labeled arrows). Each model has a unique start state denoted by an inscribed circle. The data stream itself is the lowest level. From it a tree of depth D is constructed by grouping sequential measurements into recurring subsequences. The next level models, finite automata (FA) with states V and transitions E, are reconstructed from the tree by grouping tree nodes. The last level shown, string production machines (PM), are built by grouping FA states and inferring production rules P that manipulate strings in register A.

2. At any given level, if the approximations continue increasing in size as more data and resources are used in improving the model's accuracy, then "innovate" a new class when the current representation hits the limits of the agent's computational resources.

The innovation step is the evolutionary dynamic that moves from less to more capable model classes by looking for similarities between state-groups within the lower level models. This is how the agent's notion of causal state changes: from states to state-groups. The effective dynamic is one of increasing abstraction. The process is open-ended, though a possible first four levels are shown in Table 1.

Consider a data stream **s** of m measurements. If the source is periodic, then Level 0, the data itself, gives a representation that depends on m. In the limit $m \to \infty$, Level 0 produces an infinite representation. Level 0, of course, is the most accurate model of the data, though it is largely unhelpful and barely worth the label "model". In contrast, a depth D tree will give a finite representation, though, of a data stream with period $\leq D$, even if the data stream is infinite in length. This tree has paths of length D given by the source's period. Each of these paths corresponds to a distinct phase of the repeating pattern in **s**.

If **s** is nonperiodic, then the tree model class will no longer be finite and independent of m. Indeed, if the source has positive entropy ($h_\mu > 0$) then the tree's size will grow exponentially, $\approx \|\mathscr{A}\|^{Dh_\mu}$, as D is increased to account for subsequences in **s** of increasing length D.

If the source has, roughly speaking, correlations that decay fast enough over time, then the next level of (stochastic) finite automata, will give a finite representation. The number of states $\|\mathbf{V}\|$ indicates the amount of memory in the source and so the typical time over which correlations can exist between the measurements in **s**. But it could very well be, and examples will show this shortly, that Level 2 does not give a finite representation. Then yet another level (Level 3) will be required.

The next subsection gives a more precise statement of this picture. And later subsections will go through several examples in detail to illustrate the dynamic of increasing abstraction. But briefly the idea is to move up the hierarchy in search of a representation that gives a finite model of the environment with optimal prediction of the environment's behavior.

4.2. AT EACH LEVEL IN A HIERARCHY

To be more precise about the innovation step, let's review the common aspects across the levels in the hierarchy of Table 1 and, for that matter, in the computational hierarchy of Fig. 2. At each level in a hierarchy there are a number of elements that can be identified, such as the following:

1. *Symmetries* reflecting the agent's assumptions about the environment's structure. These determine the semantic content of the model

class $\mathcal{M}$, which is defined by equivalence relations $\{\sim\}$ corresponding to each symmetry.

2. *Models* M, in some class $\mathcal{M}$, consisting of states and transitions observed via measurements.

3. *Languages* being the ensembles of finitely representable behaviors.

4. *Reconstruction* being the procedure for producing estimated models. Formally, reconstruction of model M $\in \mathcal{M}$ is denoted as M = **s**/ $\sim$. That is, reconstruction factors out a symmetry from a data stream **s**.

5. *Complexity* of a process being the size of the minimal reconstructed model M with respect to the given class $\mathcal{M}$: $C(\mathbf{s} \mid \mathcal{M}) = \|M_{min}\|$.

6. *Predictability* being estimated with reference to the distinguishable states as in Eq. (10).

It is crucial that reconstructed models M $\in \mathcal{M}$ be minimal. This is so that M contains no more structure than and no additional properties beyond those in the environment. The simplest example of this is to note that there are many multiple-state representations of an ideal random binary string. But if the size of representation is to have any meaning, such as the amount of memory, only the single state process can be allowed as the model from which complexity is computed.

4.3. THE ϵ-MACHINE HIERARCHY

At this level of analysis—namely, discussing the structure of a hierarchy of model classes—the relativity of information, entropy, and complexity becomes clear. They all depend on the agent's assumed representation. Indeed, the representation's properties determine what their values can mean to the agent.

ϵ-machine reconstruction was introduced above as a way for the agent to detect causal states. Although causal states as formulated here can be related to notions of state employed in other fields, it should be clear now that there is an inductive hierarchy delineated by different notions of state. Once this is appreciated, the full definition of an ϵ-machine can be given. An ϵ-machine is that

minimal model at the
least computationally powerful level yielding a
finite description.

The definition builds in an adaptive notion that the agent initially might not have the correct model class. How does it find a better representation? Moving up the inductive hierarchy can be associated with the innovation of new notions of causal state and so new representations of the environment's behavior. In formal terms, and ϵ-machine is reconstructed at some level in the computational hierarchy when hierarchical reconstruction—considered as an operator on representations—falls onto a fixed point. One can envision a procedure, analogous to the schematic view in Table 1, that implements this incremental movement up the hierarchy as follows:

1. At the lowest level, the data stream is its own, rather degenerate and uninformative, model: $\mathbf{M}_0 = \mathbf{s}$. Initially set the hierarchy level indicator to one step higher: $l = 1$.

2. Reconstruct the level l model M_l from the lower level model by factoring out the regularities—equivalence classes—in the state transition structure of the lower level model M_{l-1} : $M_l = M_{l-1}/\sim$, where $\sim$ denotes the equivalence relation defining the level l causal-state equivalence classes. Literally, one looks for regularities in groups of states in M_{l-1}. The groups revealing regularity in M_{l-1} become the causal states of M_l; the transitions between the M_{l-1}-state groups become the transitions in M_l.

3. Test the parsimony of the l-level class's descriptive capability by estimating successively more accurate models. As before, the degree of approximation is generally denoted ϵ, with $\epsilon \to 0$ being the limit of increasingly accurate models.

4. If the model complexity diverges, $\|M_l\| \xrightarrow[\epsilon \to 0]{} \infty$, then set $l \leftarrow l+1$ and go back to 2 and move up another level.

5. If $\|M_l\| \underset{\epsilon \to 0}{<} \infty$, then the procedure has found the first level that is the least computationally powerful and that gives a finite description. An ϵ-machine has been reconstructed. Quit.

The essential idea in moving up the hierarchy is that the symmetries assumed by the agent are broken by the data when reconstruction leads to an infinite model at some level of representation. The process of going from step 4 back to step 2—i.e. of jumping up the hierarchy to a new model class—is what has been referred to as "innovation". The key step in innovating a new model class is the discovery of new equivalence relations. A large part of this, though, is simply a reapplication of ϵ-machine reconstruction: discovering new structure is done by grouping lower-level states into equivalence classes of the same future morph. These equivalence classes then become the notion of causal state at the new higher level. A series of increasingly-accurate lower level models are, in this sense, a data stream—$M_{l-1}(\epsilon), M_{l-1}(\epsilon/2), M_{l-1}(\epsilon/4), M_{l-1}(\epsilon/8), \ldots$—for reconstruction at the next higher level M_l. A section to follow shortly will show that, for example, at the onset of chaos hierarchical ϵ-machine reconstruction goes across four levels—data, trees, finite automata, and stack automata—before finding a finite representation. The details in Table 1 were selected in anticipation of those results.

There is an additional element beyond the grouping of states according to their transition (morph) structure, though. This will be seen shortly in the section on hidden Markov models as the innovation of a resettable counter register (Crutchfield and Upper, n.d.), at the onset of chaos as the innovation of string productions (Crutchfield and Young 1990), and in discrete spatial processes as the innovation of regular domains, domain walls, and particles (Crutchfield and Hanson 1993b). It is also seen in the innovation of local state machines to break away from cellular automata look-up table representations; an example of this can be found elsewhere (Crutchfield 1992c). In each case it is quite straightforward to find the additional structural element riding on top of the higher-level causal states. But since, as far as is known, no one has delineated an exhaustive and ordered spectrum of basic computational elements, innovation must contain a component, albeit small, of undetermined discovery.

The meta-reconstruction algorithm results in a hierarchy of computation classes—the ϵ-machine hierarchy. Unlike the generative hierarchy of Chomsky (Hopcroft and Ullman 1979), this is a causal hierarchy for inductive inference. It takes into account the possibility, for example, that causal recognition might be distinct from the complexity of the generating process.

4.4. THE THRESHOLD OF INNOVATION

When should innovation occur? A basic premise here is that an agent can only call upon finite resources. The answer then is straightforward. Innovation should occur as the agent's modeling capacity, denoted $\|\mathbf{A}\|$, is approached by the complexity of the agent's internal model M. That is, the threshold of innovation is reached when $\|\mathbf{M}\| \approx \|\mathbf{A}\|$. To be more explicit about what is happening, one can use a diagnostic for innovating a new model class. Let $C_\mu^l(\epsilon)$ denote the complexity of one model $M_l(\epsilon)$ in the increasing-accuracy series. Then the innovation rate $\mathscr{T}_l$ at the given level is defined

$$\mathscr{T}_l = \lim_{\epsilon \to 0} -\frac{2^{C_\mu^l(\epsilon)}}{\log_2 \epsilon}. \tag{14}$$

The innovation rate monitors the increase in model size. If $\mathscr{T}_l > 0$ the model size at level l diverges and the agent will have to innovate a new model class at the first accuracy threshold ϵ' where $C_\mu^l(\epsilon') > \|\mathbf{A}\|$. Failure to do so is tantamount to precluding the use of an enhanced notion of environmental state to represent new forms of regularity. The ultimate result of failing to innovate is that some deterministic aspect of the environment will appear forever random. The consequence may be, nonetheless, a perfectly appropriate balance of evolutionary forces; there is a reason why houseflies and humans coexist in the same environment.

It turns out that $\mathscr{T}_l$ has a simpler interpretation. First, note that from Eq. (14) it can be rewritten

$$\mathscr{T}_l = \lim_{\epsilon \to 0} \left(2^{C_\mu^l(\epsilon/2)} - 2^{C_\mu^l(\epsilon)} \right). \tag{15}$$

Expanding this, one finds

$$\mathscr{T}_l = \lim_{\epsilon \to 0} \sum_{\substack{v \in \mathbf{V}(\epsilon/2) \\ v' \in \mathbf{V}(\epsilon)}} p_v \log_2 \frac{p_v}{p_{v'}} \tag{16}$$

where $\mathbf{V}(\epsilon)$ is the sets of states in $M_l(\epsilon)$. Thus, $\mathscr{T}$ is the information gain in going from one model to a more accurate one. Under ϵ-machine reconstruction the states $v' \in \mathbf{V}(\epsilon/2)$ of the more accurate model come from the "splitting" of states $v' \in \mathbf{V}(\epsilon)$ in the less accurate model.

One might be tempted to define a single number χ for hierarchical complexity, such as

$$\chi = l \cdot \hat{\mathcal{T}}_l \in [0, \infty), \tag{17}$$

where l is the (integer) level above the raw data stream at which an ϵ-machine is reconstructed and $\hat{\mathcal{T}}_l = 1 - 2^{-C^l} \in [0, 1]$ is the fractional complexity at that level. Although in some circumstances this could be useful, it is ultimately doomed, since there is no linear order of computational capability. The hierarchies are only partial orderings.

Casting innovation in this formal light emphasizes one important consequence: When confronted with hierarchical processes, finite computational resources fuel the drive toward higher complexity—toward agents with internal models of increasing computational power.

4.5. EXAMPLES OF HIERARCHICAL LEARNING

The preceding sections laid out an abstract framework for computation, dynamics, and innovation. The intention was to show how the different calculi of emergence are related and how together they address the problem of inadequate representations both qualitatively and quantitatively. The discussion was couched in terms of an agent that learns models of an environment via a data stream of sensory measurements.

The following subsections take a more concrete approach and demonstrate how several of these general ideas are put into practice. In a sense, the following examples put us in the position of the agents above. The examples analyze the intrinsic computation in a wide range of processes: continuous-state dynamical systems, hidden Markov models, and cellular automata. The intention here is not only to be explicit, but to also broaden the notion of computation that has been used up to this point.

4.5.1. The Cost of Chaos

The following three subsections review how intrinsic discrete computation is embedded in two well-known continuous-state dynamical systems. The connection between discrete computation and the continuous states is made via symbolic dynamics. In this approach a continuous-state orbit is observed through an instrument that produces very coarse, in fact binary, measurements. To detect the intrinsic computation the resulting binary data stream is fed into ϵ-machine reconstruction to produce a minimal computational model. The resulting ϵ-machine describes the intrinsic computational capability of the observed process—dynamical system plus instrument. Due to the choice of a particular type of instrument, the ϵ-machine also describes the computational capability of the hidden dynamical system.

Intrinsic Computation in the Period-Doubling Cascade

The first dynamical system to be analyzed for computational structure is the logistic map and, in particular, its period-doubling route to chaos. The data stream used for reconstructing models is derived from a trajectory of the logistic map when it is started with an initial condition on its attractor. This makes the observed process stationary. The trajectory is generated by iterating the map

$$x_{n+1} = f(x_n) \tag{18}$$

with the logistic function $f(x) = rx(1-x)$, with nonlinearity parameter $r \in [0, 4]$ and initial condition $x_0 \in [0, 1]$. Note that the map's maximum occurs at $x_c = \frac{1}{2}$. The orbit $\mathbf{x} = x_0 x_1 x_2 x_3 \cdots$ is converted to a discrete sequence by observing it via the binary partition

$$\mathscr{P} = \{x_n \in [0, x_c) \Rightarrow s = 0, x_n \in [x_c, 1] \Rightarrow s = 1\}. \tag{19}$$

This partition is "generating" which means that sufficiently long binary sequences come from arbitrarily small intervals of initial conditions. Due to this, the information processing in the logistic map can be studied using the "coarse" measuring instrument $\mathscr{P}$.

Many investigations of the logistic map concentrate on how its time-asymptotic behavior, its attractor, changes with the nonlinearity parameter r. Here, however, the interest is in how its various information processing capabilities are related to one another. The two basic measures of this that can be directly taken from the reconstructed ϵ-machines were introduced above. The first was the statistical complexity C_μ, which is the size of the reconstructed ϵ- machine or, equivalently, the effective amount of memory in the logistic map. The second measure of information processing

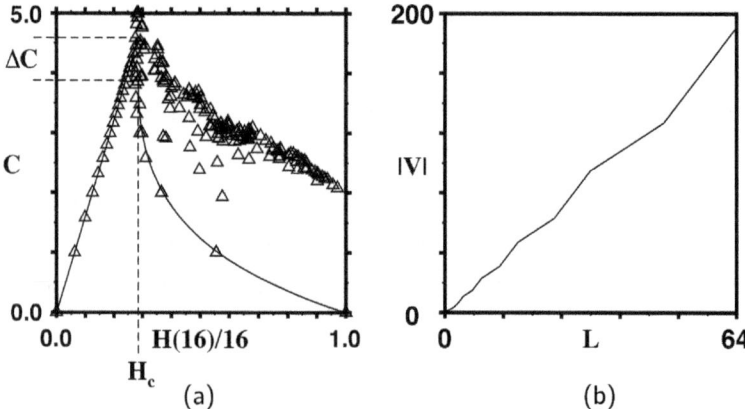

Figure 6. (a) Statistical complexity C_μ versus specific entropy $H(L)/L$ for the period-doubling route to chaos. Triangles denote estimated $(C_\mu, H(L)/L)$ at 193 values of the logistic map nonlinearity parameter. ϵ-machines were reconstructed using a subsequence length of $L = 16$. The heavy solid lines overlaying some of this empirical data are the analytical curves derived for C_0 versus $H_0(L)/L$. (After Crutchfield and Young 1989.) (b) At one of the critical parameter values of the period-doubling cascade in the logistic map the number $\|V\|$ of inferred states grows without bound. Here $r = r_c \approx 3.5699456718695445\ldots$ and the sequence length ranges up to $L = 64$ where $\|V\| = 196$ states are found. It can be shown, and can be inferred from the figure, that the per symbol density of states $\|V(L)\|/L$ does not have a limiting value as $L \to \infty$. (After Crutchfield and Young 1990.)

is the entropy rate h_μ, which is the rate in bits per time step at which information is produced. The net result of using just the complexity and entropy rate is that the original equations of motion and the nonlinearity parameter are simply forgotten. All that is of interest is how the complexity C_μ of the data stream depends on the rate h_μ of information production.

The complexity–entropy plot of Fig. 6a summarizes this relationship by showing the results of reconstructing ϵ-machines from data streams produced at different parameter values. For each data set produced, an ϵ-machine is reconstructed and its statistical complexity C_μ and entropy rate h_μ are estimated. In order to show the full range of behavior, from periodic to chaotic, the latter is estimated as $h_\mu(L) = H(L)/L$ where $H(L)$ is the Shannon information of length L sequences. Fig. 6a is simply a scatter plot of the estimated complexity–entropy pairs, in emulation of Fig. 4b.

There are a number of important features exhibited by the complexity–entropy diagram. (Details are given in Crutchfield and Young 1989, 1990.) The first is that the extreme values of entropy lead to zero complexity. That is, the simplest periodic process at $H(L)/L = 0$ and the most random one at $H(L)/L = 1$ are statistically simple. They both have zero complexity since they are described by ϵ-machines with a single state. Between the extremes the processes are noticeably more complex with an apparent peak about a critical entropy value denoted H_c. Below this entropy, it turns out, all of the data streams come from parameters at which the logistic map is periodic—including parameters within the "periodic windows" found in the map's chaotic regime. The data sets with $H(L)/L > H_c$ are produced at chaotic parameter values.

A theory was developed in Crutchfield and Young (1990) to explain the emergence of high computational capability between the ordered and disordered regimes. For processes with $H(L)/L < H_c$ the entropy and complexity are equivalent,

$$C_\mu = H. \tag{20}$$

This is shown as a solid straight line on the left portion of Fig. 6a. For processes with $H(L)/L > H_c$ the dependence of complexity on entropy is more interesting. In fact, the solution is given in terms of the dependence of the entropy on the topological complexity. The result, a lower bound, is that

$$H(L) = C_0 + \log_2\left(2^{L2-C_0} - 2^{-1}\right) \tag{21}$$

The curved solid line in Fig. 6a shows the relevant portion of Eq. (21).

Comparing the periodic and chaotic analyses—i.e. Eqs. (20) and (21)—provides a detailed picture of the complexity—entropy phase transition. The critical entropy H_c at each sequence length L is given

$$H_c(L) = C'(L) + \log_2\left(by - 2^{-1}\right), \tag{22}$$

where $C'(L) = \log_2 L - \log_2 \log_2 y$ is the complexity on the high entropy side at H_c, $y \approx 2.155535$ is the solution of $y \log_e y - y + 2^{-1} = 0$, and $1 \leq b \leq 3 \times 2^{-3/2} \approx 1.06066$ is a constant. From Eq. (20) it follows immediately that the complexity C'' on the low-entropy side of the transition is itself $H_c \times L$. The difference is a finite constant—the latent complexity of the transition $\Delta C = C'' - C' \approx 0.7272976$ bits. The latent complexity is independent of the sequence length.

The analysis of the interdependence of complexity and entropy is nonasymptotic in the sense that it applies at each sequence length L. If, as done for Fig. 6a, this length is fixed at $L = 16$, the preceding results predict the transition's location. The critical entropy there, for example, is $H_c \approx 0.286205$. But for any L the overall behavior is universal. All behaviors with specific entropy densities $H(L)/L < H_c$ are periodic. All behaviors with higher entropy densities are chaotic. The functional forms in Eqs. (20) and (21) are general lower bounds. The statistical complexity is maximized at the border between the predictable and unpredictable "thermodynamic phases". It is important to emphasize that the complexity—entropy diagram makes no explicit reference to the system's nonlinearity parameter. The diagram was defined this way in order to show those properties which depend only on the intrinsic information production and intrinsic computational structure.

This property—that statistical complexity first rises and then falls with entropy—makes it a particularly compelling measure of complexity. Sean Carroll gave an interesting talk about this in the opening of the FQXi conference Setting Time Aright in 2011.

Up to this point the overall interplay between complexity and entropy for the period-doubling cascade has been reviewed. But what happens at the phase transition; i.e. at the critical entropy density H_c? One parameter value, out of the many possible, corresponding to $H(L)/L = H_c$ is the first period-doubling onset of chaos at $r = r_c \approx 3.5699456718695445\ldots$. Fig. 7a shows the 47 state ϵ-machine reconstructed with window size $L = 16$ at this parameter setting. An improved approximation can be attempted by increasing the window length L to take into account structure in longer subsequences. Fig. 6b shows the result of doing just this: at the onset of period-doubling chaos the number $\|\mathbf{V}\|$ of states for the reconstructed ϵ-machines grows without bound.

The consequence is that the data stream produced at the onset of chaos leads to an infinite machine. This is consonant with the view introduced by Feigenbaum that this onset of chaos can be viewed as a phase transition at which the correlation length diverges (Feigenbaum 1983). The computational analog of the latter is that the process intrinsically has an infinite memory capacity. But there is more that the computational analysis yields. As will now be shown, for example, the infinite memory is organized in a particular way such that the logistic map is not a universal Turing machine, but instead is equivalent to a less powerful stack automaton.

The "explicit state" representation of Fig. 7a does not directly indicate what type of information processing is occurring at the phase transition. Nor does the unbounded growth of machine size shown in Fig. 6b give much help. A simple transformation of the 47 state machine in 7a goes some distance in uncovering what is happening. Replacing the unbranched "chains" in the machine with the corresponding sequences produces the "dedecorated" critical machine of Fig. 7b. In this representation is it

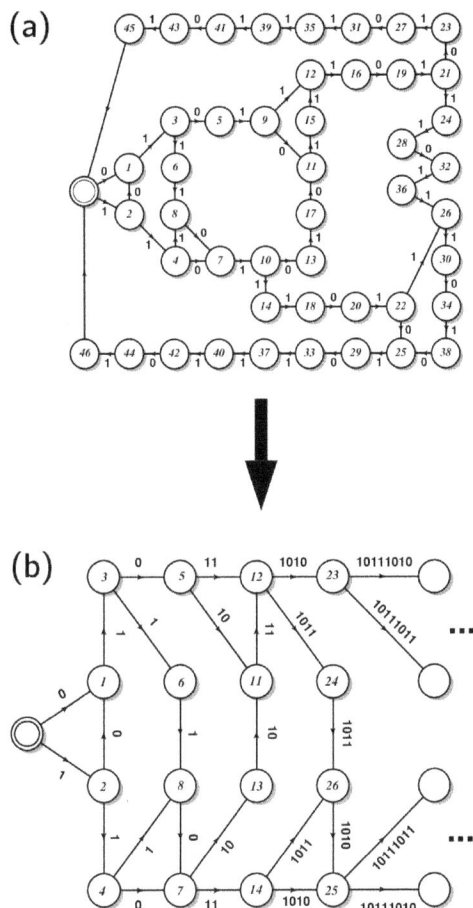

Figure 7. (a) Approximation of the critical ϵ-machine at the period-doubling onset of chaos. (After Crutchfield and Young 1989.) (b) The dedecorated version of the machine in (a). Here the deterministic state chains have been replaced by their equivalent strings. (After Crutchfield and Young 1990.)

evident that the branching states are quite regularly organized. Beyond the discovery of this higher-order regularity, there is an additional element that consists of manipulating the intervening strings between the branching states.

By following in detail the increasing-accuracy modeling experiment shown in Fig. 6b, one can ask *how* the machines in a series of successively-improved models grow in size. The result, as disclosed by the dedecorated machine, is that only the branching states and "string productions" are needed to describe the regularity in the growth of the machines. This in turn leads to the innovation, shown in Fig. 8a, of a finite machine with two kinds of states (the new type is denoted with squares) and two registers **A** and **B** that hold binary strings. Simple inspection of the dedecorated machine shows that the string manipulations can be described by appending a copy of **A**'s contents onto **B** and replacing the contents of **A** with two copies of **B**'s contents. These string productions are denoted **A** → **BB** and **B** → **BA**. At the outset, register **A** contains "0" and **B** contains "1".

One problem with the string production machine of Fig. 8a is that the length of strings in the registers grows exponentially fast, which contrasts sharply with the sequential production of symbols by the logistic map. Fig. 8b gives an alternative, but equivalent, serial machine that produces a single symbol at a time. It is called a one-way nondeterministic nested stack automaton and was denoted 1NnSA in Fig. 2. The memory in this machine is organized not as string registers, but as a pushdown stack. The latter is a type of memory whose only accessible element is on the top. In fact, the automaton shown has a slightly more sophisticated stack that allows the finite control to begin a new "nested" stack within the existing one. The only restriction is that the automaton cannot move on to higher levels in the outer stack(s) until it is finished with its most recently created stack.

The net effect of these constructions is that a finite representation has been discovered from an infinite one. One of the main benefits of this, aside from producing a manageable description and the attendant analytical results it facilitates, is that the type of information processing in the critical "state" of the logistic map has been made transparent.

Intrinsic Computation in Frequency-Locking Route to Chaos

The second route to chaos of interest, which also has received extensive study, is that through quasiperiodicity. In the simplest terms, this route to chaos and the models that exhibit it describe the coupling of two oscillators whose periods are incommensurate—the ratio of periods is not rational. The ratio of the number of periods of one oscillator to the other in order to complete a full cycle for both is called the winding number $\hat{\omega}$. This is a key parameter that controls the entire system's behavior: when $\hat{\omega}$ is rational the two oscillators are phase-locked. Quasiperiodic behavior is common in nature and underlies such disparate phenomena as cardiac arrhythmia,

Crutchfield (1994)

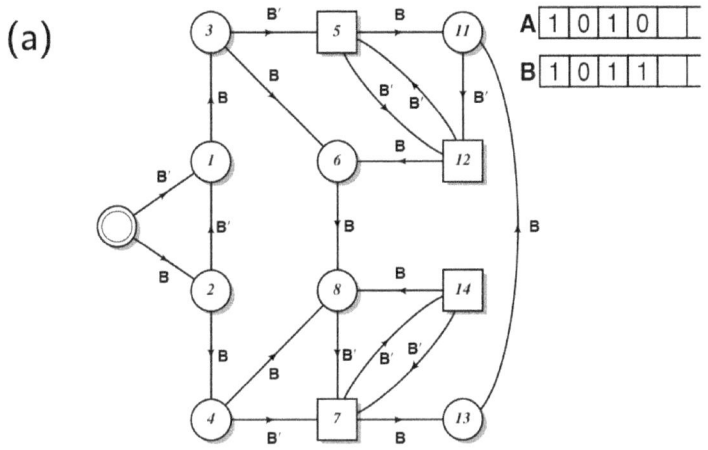

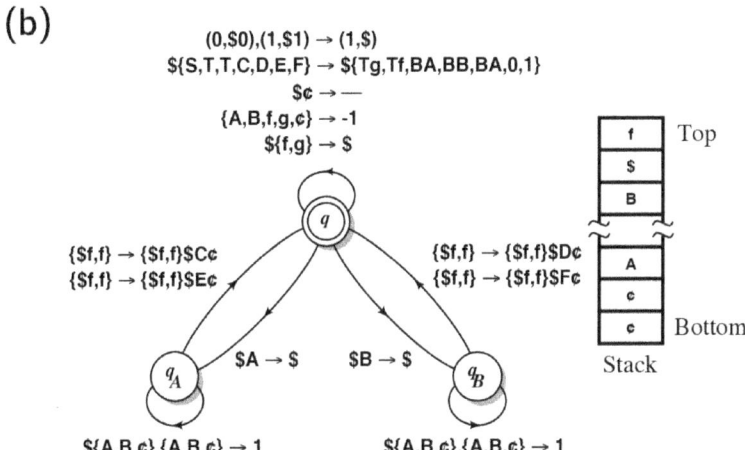

Figure 8. (a) The finite version of Fig. 7b's infinite critical ϵ-machine. This is a string production machine that, when making a transition from the square states, updates two string registers with the productions **A** → **BB** and **B** → **BA**. **B'** is the contents of **B** with the last bit flipped. (b) Another finite representation of the period-doubling critical ϵ-machine—a one-way nondeterministic nested stack automaton (1NnSA in Fig. 2)—that produces symbols one at a time. (After Crutchfield and Young 1990.)

the stability of the solar system, and the puzzling synchronization of two mechanical clocks located in close proximity.

The simplest model of two "competing" oscillators is the discrete-time circle map

$$\phi_{n+1} = f(\phi_n) \bmod 1, \quad \text{where}$$
$$f(\phi) = \omega + \phi + \frac{k}{2\pi}\sin(2\pi\phi). \tag{23}$$

The map's name derives from the fact that the mod 1 operation keeps the state ϕ_n on the unit circle. One thinks of ϕ_n then as a phase—or, more properly, the relative phase of the two original oscillators. There are two control parameters, ω and k. The former directly sets the phase advance and the latter the degree of nonlinearity, which can be roughly interpreted as the coupling strength between the two oscillators.

As a function of the nonlinearity parameter the behavior makes a transition to chaos. Like the logistic map, there is a signature to the path by which chaotic behavior is approached from periodic behavior. Furthermore, the circle map's signature has the basic character of a phase transition (Kadanoff, Feigenbaum, and Shenker 1982).

The following will investigate one arc through (ω, k)-space that exhibits just such a phase transition to chaos. This is a path that includes the golden mean circle map—so-called since its winding number is the golden mean $\hat{\omega} = \frac{1}{2}(1 + \sqrt{5})$. The easiest way to implement this is to set $\omega = \hat{\omega}$. Varying $k \in [0, 6]$ then gives a wide sample of behavior types on the quasiperiodic route to chaos. $k = 1$ is the threshold of nonlinear behavior, since the map for larger values becomes many-to-one; $k > 1$ is also a necessary, but not sufficient condition for deterministic chaos.

The measuring instrument uses three types of partition depending on the parameter range: $k = 0$, $k \in (0, 1]$, and $k > 1$. Generally, the instrument is a binary partition that labels $\phi_n \in (\phi', \phi'']$ with $s = 0$ and $\phi_n = (\phi'', \phi']$ with $s = 1$. For $k = 0$, $\phi' = \frac{1}{2}$ and $\phi'' = 0$; for $k \in (0, 1]$, $\phi' = 1$ and $\phi'' = f^{-1}\left(\frac{1}{2}\right)$; and, for $k > 1$, ϕ' is the larger and ϕ'' the smaller value of $(2\pi)^{-1}\cos^{-1}\left(-k^{-1}\right)$ on the interval. By iterating the map many times on an initial condition a time series $\phi = \phi_0\phi_1\phi_2\cdots$ is produced. When observed with an instrument the time series is converted to a binary string $\mathbf{s} = s_0 s_1 s_2 \cdots$ of coarse measurements $s_i \in \{0, 1\}$.

Fig. 9a shows the complexities and entropies estimated for the quasiperiodic route to chaos at several hundred settings along the chosen

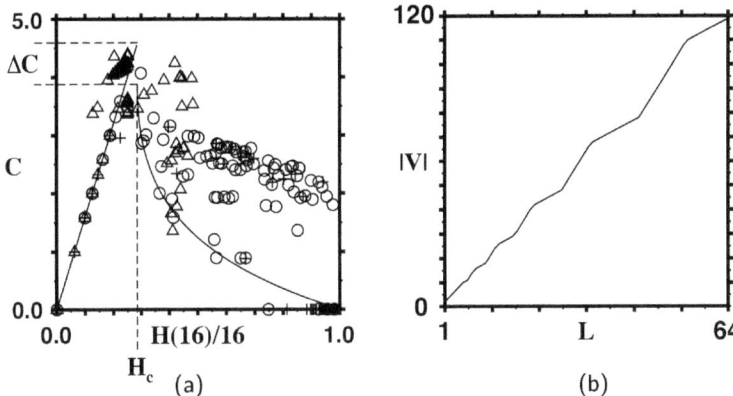

Figure 9. (a) Statistical complexity C_μ versus specific entropy $H(L)/L$ for the quasiperiodic route to chaos. Tokens denote estimated $(C_\mu, H/L)$ at 303 values of the circle map with $\omega = \frac{1}{2}(1+\sqrt{5})$ and nonlinearity parameter k in three different ranges: 101 values for $k \in [0, 2]$ (triangles), 101 values for $k \in [3.5, 3.9]$ (circles), and 101 values for $k \in [4.5, 6]$ (crosses). These are ranges in which the behavior is more than simple periodic. ϵ-machine reconstruction used a tree depth of $D = 32$ and a morph depth of $L = 16$ for the first range and $(D, L) = (16, 8)$ for the second two ranges, which typically have higher entropy rates. The entropy density was estimated with a subsequence length of $L = 16$. Refer to Fig. 6a for details of the annotations. (b) At the golden mean critical winding number (with $k = 1$) in the quasiperiodic route to chaos the number $\|\mathbf{V}\|$ of inferred states grows without bound. Here the sequence length ranges up to $L = 64$ where $\|\mathbf{V}\| = 119$ states are found.

parameter arc. As with period-doubling, the quasiperiodic behavior with entropies $H(L)/L < H_c$ are periodic. All those with higher entropies are unpredictable. The statistical complexity is maximized at the border between the ordered and chaotic "thermodynamic phases". The lower bounds, Eqs. (20) and (21), are shown again as solid lines for both phases. The circle map clearly obeys them, as did the logistic map, though the scatter differs. For example, there is a cluster of points just below H_c at high complexity. These are all due to the "irrational" quasiperiodic behavior that is predictable. The complexity derives from the fact that the map essentially "reads out" the digits of their irrational winding number. This leads to data streams that require large ϵ-machines to model. There is also some scatter at high entropy and low complexity. This is due to highly intermittent behavior that results in all subsequences being observed, but with an underlying probability distribution that is far from uniform. The result is that ϵ-machine reconstruction approximates the behavior as a biased coin-zero complexity, since it has a single state, and entropy less than unity.

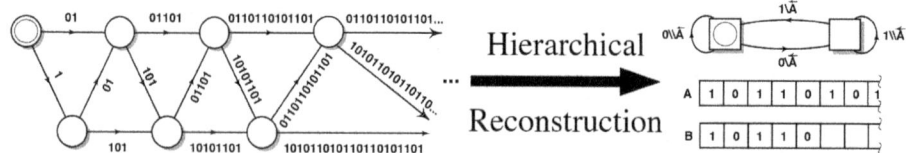

Figure 10. (a) A portion of the infinite critical machine for the quasiperiodic route to chaos at the golden mean winding number. Note that the dedecorated machine is shown—that is, the intervening states along deterministic chains have been suppressed. (b) The Fibonacci machine: the finite representation of the infinite machine in (a).

What happens at the quasiperiodic onset at $k = 1$? The metric entropy is zero here, since the number of length L subwords increases strictly linearly: $N(L) = L + 1$. The single symbol entropy is high, $H(1) \approx 0.959419$ bits, since the frequency of isolated zeros is $\lim_{L \to \infty} (F_{L-2}/F_L) = \hat{\omega}^{-2} \approx 0.381966$, where F_L is the Lth Fibonacci number.

ϵ-machine reconstruction applied to this "critical" data stream does not lead to a finite state machine. In fact, just as for the logistic map at the onset of chaos, the machine size keeps diverging. (See Fig. 9b.) A finite approximation to the presumably infinite "critical" machine is shown in Fig. 10a.

Notably, the intrinsic computation in quasiperiodicity can be finitely represented at a next higher level. When the average winding number is the golden mean, one finds the "Fibonacci" machine shown in Fig. 10b. There is a two state finite control automaton shown at the top portion of Fig. 10b that determines copying operations on two registers, **A** and **B**, holding binary strings. The finite control is started in the leftmost, double-circle state, **A** begins with "1", and **B** with "0". The finite control machine's edge are labelled with the actions to be taken on each state-to-state transition. The first symbol on each edge label is a zero or one read from the input data stream that is to be recognized. The symbol read determines the edge taken when in a given state. The backward slash indicates that a string production is performed on registers **A** and **B**. This consists of copying the previous contents of **A** to **B** and appending the previous contents of **B** to **A**. The string productions are denoted **A** → **AB** and **B** → **A**. They are applied simultaneously. If there are two backward slashes, then two "Fibonacci" productions are performed. The input string must match the contents of register **A**, when register **A** is read in reverse. The latter is denoted by the

t	A	B	‖A‖
1	1	0	1
2	10	1	2
3	101	10	3
4	10110	101	5
5	10110101	10110	8

Table 2. Contents of the Fibonacci machine registers **A** and **B** as a function of machine transitions. The registers contain binary strings and are modified by string concatenation: **A** → **AB** and **B** → **A**. That is, the previous contents of **A** are moved to **B** and the previous contents of **B** are appended to **A**.

left-going arrow above **A** in the edge label. Table 2 shows the temporal development of the contents of **A** and **B**.

The basic computation step describing the quasiperiodic critical dynamics employs a pair of string productions. The computational class here is quite similar to that for period-doubling behavior—that is, nested stack automata. It is at this higher level that a finite description of the golden mean critical behavior is found. This is demonstrated, as for period-doubling, by noting that the productions are context-free Lindenmayer productions and that these can be mapped first to an indexed context-free grammar and then to nested stack automaton (Hopcroft and Ullman 1979). Thus, rather than Fig. 10b the Fibonacci machine can be represented with a stack automaton analogous to that shown in Fig. 8b for the period-doubling onset of chaos.

The required length of the Fibonacci machine registers grows as a function of the number of applications of the productions at an exponential rate which is the golden mean, since the string length grows like the Fibonacci numbers—an observation directly following from the productions. Thus, with very few transitions in the machine input strings of substantial length can be recognized.

Another interpretation of the recognition performed by the Fibonacci machine in Fig. 10b is that if phase locks to the quasiperiodic data stream. That is, the Fibonacci machine can jump in at any point in the "critical" string, not necessarily some special starting time, and, from that symbol on, determines if the subword it is reading is in the language of all Fibonacci subwords.

Temporal Computation in Deterministic Chaos

This investigation of the computational structure of two well-known routes to chaos shows that away from the onset of chaos there are (at least) finite memory processes. Finite memory processes are all that is found below the onset—that is, with periodic processes. Above the onset the situation is much more interesting. There is a universal lower bound that the primary band-merging sequence obeys. But above this there can be more complex and highly unpredictable processes. These examples make it clear how to construct processes in this region of the complexity–entropy plane. Take a nonminimal representation of the all-sequences process, (say) one with 16 states. Add transition probabilities randomly to the outgoing edges, observing the need to have them sum to unity for each state. Typically, this machine will be minimal. And if the range of probabilities is restricted to be near $1/2$, then the entropy will be high and by construction the process has a statistical complexity of about 4 bits. Now an entire family of high complexity, moderate entropy machines can be constructed by applying the period-doubling operator to the high entropy machine just created. This results in processes of lower and lower entropy and higher and higher complexity. These move down to the onset of chaos. Finally, note that the analysis of this family's complexity versus entropy dependence is not so different from that for the lower bound.

The preceding subsections also showed that to get a simple model that captures the system's *true* computational capability, as determined by observations, it is sometimes necessary to jump up to a more powerful computational class. At both onsets of chaos the computational analysis identified structures that were higher than finite memory devices. The onset of chaos led to infinite memory and, just as importantly, to memory that is organized in a particular way to facilitate some types of computation and to proscribe others. The logistic and circle maps at their respective onsets of chaos are far less than Turing machines, especially ones that are universal. At the onset the information processing embedded in them jumps from the finitary level to the level of stack automata. One practical consequence of failing to change to a more powerful representation for these critical systems is that an observer will conclude that they are more random, less predictable, and less complex, than they actually are. More generally, appreciating how infinite complexity can arise at the onset of chaos leads one to expect that highly nonlinear systems can perform significant amounts of and particular forms of information processing.

4.5.2. The Cost of Indeterminism

This subsection explores the possible ill-effects of measurement distortion: the apparent complexity can diverge if the "wrong" instrumentation is used. (This subsection follows Crutchfield 1994.) Along the way a new class of processes will be considered—the stochastic nondeterministic finite automata, often called hidden Markov models. One of the main conclusions will be that an agent's sensory apparatus can render a simple environment apparently very complex. Thus, in an evolutionary setting the effects described here indicate that there should be a strong selection pressure on the quality of measurements produced by an agent's sensory apparatus.

The Simplest Example

Returning to the logistic map, let's fix its parameter to $r = 4$—where its attractor fills the interval and has the maximal entropy rate of $h_\mu = 1$. The probability density function for the invariant measure over "internal" real-valued states $x \in [0, 1]$ is

$$\Pr(x) = \frac{1}{\pi\sqrt{x - x^2}}. \tag{24}$$

Then, we associate a state **A** with the event $x_n \in [0, x_c)$ and a state **B** with the event $x_n \in [x_c, 1]$; recalling that $x_c = \frac{1}{2}$ is the map's maximum. Finally, we use a sliding-block code on the resulting **A**–**B** stream that outputs $s = 1$ when the length 2 subsequences **AA**, **AB**, or **BB** occur, and $s = 0$ when **BA** occurs. The 0–1 data stream that results is produced by the machine shown in Fig. 11—a stochastic nondeterministic finite automaton (SNFA).

That Fig. 11 gives the correct model of this source is seen by first noting that the intermediate states **A** and **B** have the asymptotic probabilities

$$\Pr(\mathbf{A}) = \int_0^{x_c} \mathrm{d}x \, \Pr(x) = \tfrac{1}{2}, \tag{25}$$

and, by symmetry, $\Pr(\mathbf{B}) = \tfrac{1}{2}$. The two inverse iterates of x_c, $x_\pm = x_c \pm 1/2\sqrt{2}$, delimit the interval segments corresponding to the occurrence

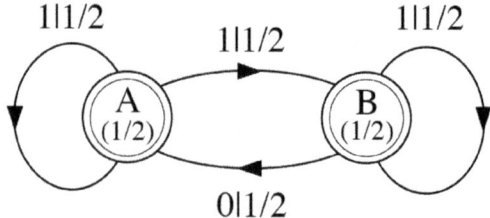

Figure 11. The source is a stochastic nondeterministic finite automaton—a class sometimes referred to as hidden Markov models. The hidden process consists of two states {**A**, **B**} and uniform branching between them—denoted by the fractions p on the edge labels $s \mid p$. The observer does not have access to the internal state sequences, but instead views the process through the symbols s on the edge labels $s \mid p$. The inscribed circle in each state indicates that both states are start states. The fractions in parentheses give their asymptotic probabilities, which also will be taken as their initial probabilities.

of **A**–**B** pairs. These then give the four state transition probabilities, such as

$$\Pr(\mathbf{A} \to \mathbf{A}) = \frac{\Pr(\mathbf{AA})}{\Pr(\mathbf{A})} = 2 \int_0^{x_-} dx \, \Pr(x). \qquad (26)$$

It turns out they are all equal to $\frac{1}{2}$.

With the use of the pairwise **A**–**B** coding this construction might seem somewhat contrived. But it can be reinterpreted without recourse to an intermediate code. It turns out that the 0–1 data stream comes directly from the binary partition.

$$\mathscr{P} = \{x_n \in [0, x_+) \Rightarrow s = 1, x_n \in [x_+, 1] \Rightarrow s = 0\}. \qquad (27)$$

This is a partition that is not much more complicated that the original. The main difference is that the "decision point", originally at x_c, has been moved over to x_+.

The result is that the environment seen by the agent is described by the two-state stochastic process shown in Fig. 11. There are two internal states {**A**, **B**}. Transitions between them are indicated with labeled, directed edges. The labels $s \mid p$ give the probability p of taking the transition. When the transition is taken the agent receives the measurement symbol $s \in \{0, 1\}$. In effect, the agent views the internal state dynamics through the instrument defined by the particular association of the measurement symbols and the transitions. The agent assumes no knowledge of the start state and so the environment could have started in either **A** or **B** with equal likelihood.

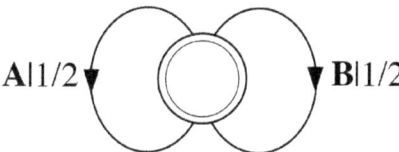

Figure 12. The minimal machine for Fig. 11's internal state process. It has a single state and equal branching probabilities. The topological and statistical complexities are zero and the topological and metric entropies are 1 bit per state symbol—a highly unpredictable, but low complexity process. That this is the correct minimal description of the internal state process follows directly from using machine reconstruction, assuming direct access to the internal state sequences **ABABBA**.... All state sequences are allowed and those of equal length have the same probability.

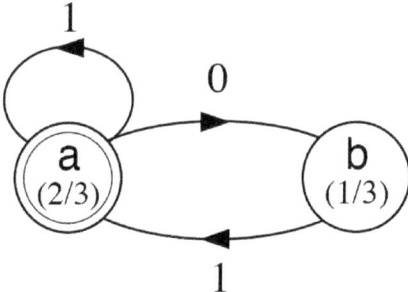

Figure 13. The process's topological structure is given by a deterministic finite automaton—the golden mean machine. The only rule defining the sequences is "no consecutive 0s". The number of sequences of length L is given by the Fibonacci number F_{L+2}; the growth rate or topological entropy h, by the golden mean $\phi = \frac{1}{2}(1+\sqrt{5}) : h = \log_2 \phi$. The numbers in parentheses give the states' asymptotic probabilities.

Fig. 12 shows the minimal machine for the environment's internal state dynamics. It is the single state Bernoulli process B $\left(\frac{1}{2}, \frac{1}{2}\right)$— a fair coin. From Eqs. (10) and (12) it is evident that the metric entropy is $h_\mu = 1$ bit per symbol, as is the topological entropy h. From Eqs. (11), (12), and (13) both the topological and statistical complexities are zero. It is a very random, but simple process.

The goal, of course, is for the agent to learn the causal structure of this simple process from the $\{0, 1\}$ data stream. It has no knowledge of Fig. 11, for example. The overall inference procedure is best illustrated in two steps. The first is learning a model of the "topological" process that produces the set of sequences in the data stream, ignoring the probabilities with which they occur. The second step is to learn a model that gives the sequences' probabilities.

The first step is relatively straightforward and can be explained briefly

FOUNDATIONAL PAPERS IN COMPLEXITY SCIENCE

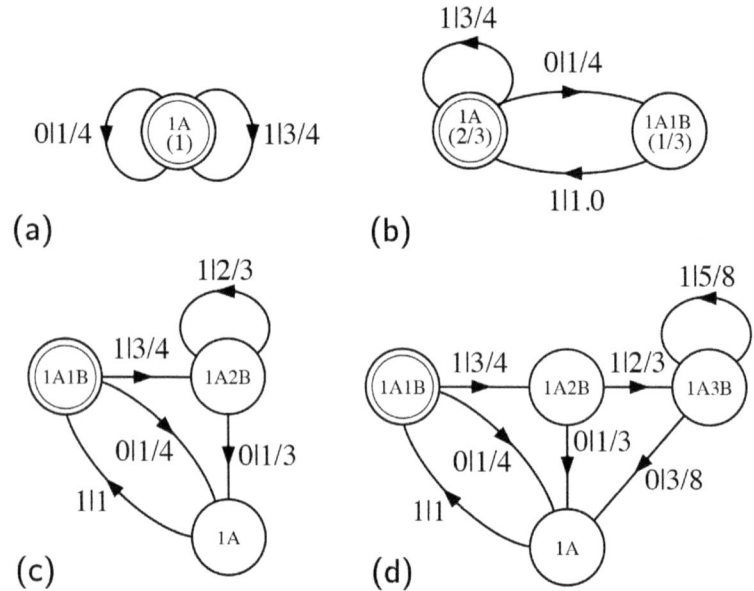

Figure 14. (a)-(d) The zeroth- through third-order causal approximations to the process of Fig. 11

in words. Inspection of the stochastic automaton's output symbols in Fig. 11 shows that if $s = 0$ is observed, then $s = 1$ must follow. Further reflection shows that this is the only restriction: consecutive 0s are not produced. All other binary sequences occur.

The automaton, again "topological", that captures this property is shown in Fig. 13. This automaton is also what machine reconstruction generates. There are several things to notice. First, the state **a** has a circle inscribed in it. This denotes that **a** is the start state; and it happens to be the unique start state. The reconstructed ϵ-machine has removed the first element of noncausality in the original process: ignorance of the start state. Second, the automaton is deterministic—a term used here as it is in formal language theory and which does *not* refer to probabilistic elements. Determinism means that from each state a symbol selects a unique successor state.

Note that the original process (Fig. 11) with its measurement labeling is not deterministic. If the process happens to be in state **A** and the observer then sees $s = 1$, then at the next time step the internal process can be in either state **A** or **B**. This ambiguity grows as one looks at longer

☞ The use of the word "determinism" here can be quite confusing, exactly as the manuscript mentioned. In contemporary texts, the property that each symbol selects a unique successor state is often referred to as being "unifilar."

and longer sequences. Generally, indeterminism leads to a many-to-one association between internal state sequences and measurement sequences. In this example, the observation of 0110 could have been produced from either the internal state sequence **BAABA** or **BABBA**.

The consequences of indeterminism, though, become apparent in the second inference step: learning the observed sequences' probabilities. To implement this, a series of increasingly-accurate machines approximating the process of Fig. 11 is reconstructed; these are shown in Fig. 14. Each gives a systematically better estimate of the original process's sequence distribution. The machine resulting from full reconstruction is shown in Fig. 15. It has an infinite number of causal states. All of their transitions are deterministic. Note that the infinite machine preserves the original process's reset property: when $s = 0$ is observed the machine moves to a unique state and from this state $s = 1$ must be seen. But what happened, in comparison to the finite machine of Fig. 13, to produce the infinite machine in Fig. 15? The indeterminism mentioned above for state **A** has lead to a causal representation that keeps track of the number of consecutive 1s since the last $s = 0$. For example, if 01 has been observed, then $\Pr(s = 0) = \frac{1}{4}$ and $\Pr(s = 1) = \frac{3}{4}$. But if 011 has been observed, $\Pr(s = 0) = \frac{1}{3}$ and $\Pr(s = 1) = \frac{2}{3}$. In this way the causal representation accounts for the agent's uncertainty in each internal states' contribution to producing the next symbol. The result is that as more consecutive 1s are seen the relative probability of seeing $s = 0$ or $s = 1$ continues to change—and eventually converges to a fair coin. This is reflected in the change in transition probabilities down the machine's backbone. Causal machine reconstruction shows exactly what accounting is required in order to correctly predict the transition probabilities. But it gives more than just optimal prediction. It provides an estimate of the process's complexity and a complete representation of the distribution $\Pr(\omega)$ over infinite sequences.

Interestingly, even if the agent has knowledge of Fig. 11, the infinite causal machine of Fig. 15 represents in a graphical way the requirements for achieving optimal predictability of the original process. There is no shortcut to computing, for example, the original process's entropy rate and complexities, since the machine in Fig. 15, though infinite, is minimal. That is, there is no smaller (causal) machine that correctly gives $\Pr(\omega)$. From the topological machine it follows that the topological entropy is

From an entropic perspective, indeterminism allows a model's internal states to contain oracular information—information about the future that is not contained in the past.

From the modern perspective of machine learning, this sequence of infinite states can be thought of as belief states of the agent. This encapsulates the various states of knowledge the agent can have about an environment that is in one of the two hidden states. Another interesting property is that if these these belief states were mapped into the quantum regime, the resulting agent can have finite memory (Elliott *et al.* 2020).

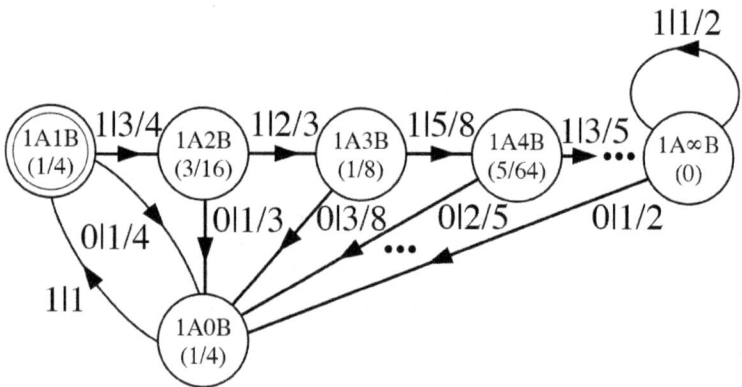

Figure 15. The infinite causal representation of the nondeterministic process of Fig. 11. The labels in the states indicate the relative weights of the original internal states $\{\mathbf{A}, \mathbf{B}\}$. The number in parentheses are the asymptotic state probabilities: $\Pr(v = 1\mathbf{A}i\mathbf{B}) = (i+1)2^{-i-2}$.

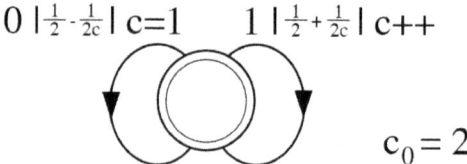

Figure 16. At a higher computational level a single state machine, augmented by a counter register, finitely describes the process of Figs. 11 and 15.

 The definition of causal is not explicitly stated here but in later texts was clarified to mean that the machine contains zero oracular information. Formally, the mutual information between past and future condition on the state of machine is zero.

$h = \log_2 \phi \approx 0.694242$ and from Eqs. (10) and (12) that the metric entropy is $h_\mu \approx 0.677867$ bits per symbol. Recall that the original process's topological and statistical complexities were zero. From Eqs. (11), (12), and (13) the causal machine's topological complexity is infinite, $C_0 = \log_2 \|\mathbf{S}\|$, and its statistical complexity is $C_\mu \approx 2.71147$ bits. These are rather large changes in appearance due to the instrumentation.

In this example, the agent can be considered to have simply selected the wrong instrument. The penalty is infinite complexity. Thus, the logistic map can appear to have an infinite number of causal states and so infinite topological complexity. In contrast to the preceding sections, which illustrated infinite intrinsic complexity, this example illustrates measurement-induced complexity.

Stochastic Counter Automata

The apparent infinite complexity of the deterministic denumerable-state machine of Fig. 15 gives way to a finite representation once the regularity of the change in transition probabilities is discovered. The resulting model—in the class of stochastic counter automata for this one example—is shown in Fig. 16. The structural innovation is a counter, denoted c, that begins with the value 2. c can be either incremented by one count or reset to 1. When $s = 0$ is observed, the counter is reset to 1. As long as $s = 1$ is observed, the counter is incremented. The nondeterminism of the original process is simulated in this deterministic representation using the counter to modify the transition probabilities: it keeps track of the number of consecutive 1s. The transition probabilities are calculated using the value stored in the counter: $\Pr(s = 0 \mid c) = \frac{1}{2} - 1/2c$ and $\Pr(s = 1 \mid c) = \frac{1}{2} + 1/2c$. The finite control portion of the machine is simply a single state machine, and so its complexity is zero. But the required register length grows like $\log_2 L$. The cost of nondeterminism in this example is this increment-reset counter.

Such processes are now known as renewal processes, which can interpolate between exact period clocks (when the probability of counter reset is a direct delta function) and memory Poissonian processes (when the probably of counter reset is independent of counter value). In fact, the ratio of the average duration between counter resets and the standard error of a counter reset has been used, as when quantifying the performance of such processes as clocks. See for example Yang, Chiribella, and Hayashi (2018).

Recurrent Hidden Markov Models

This example is just one from a rich class of processes called—depending on the field—recurrent hidden Markov models, stochastic nondeterministic finite automata, or functions of Markov chains. The difficulty of finding the entropy rate for these processes was first noted in the late 1950s (Blackwell and Koopmans 1957). In fact, many questions about this class are very difficult. It is only recently, for example, that a procedure for determining the equivalence of two such processes has been given (Ito, Amari, and Kobayashi 1992). The awareness that this problem area bears on the complexity of observation and the result that finite complexity processes can appear infinitely complex is also recent (Crutchfield and Upper, n.d.).

The new notion of state here that needs to be innovated involves the continuity of real variables. The causal states are no longer discrete. More precisely, an ϵ-machine state for a hidden Markov model is a *distribution* over the hidden states. Since this distribution is a vector of real numbers, the causal states are continuous. In addition, this vector is normalized since the components are state probabilities. The result is that the ϵ-machine

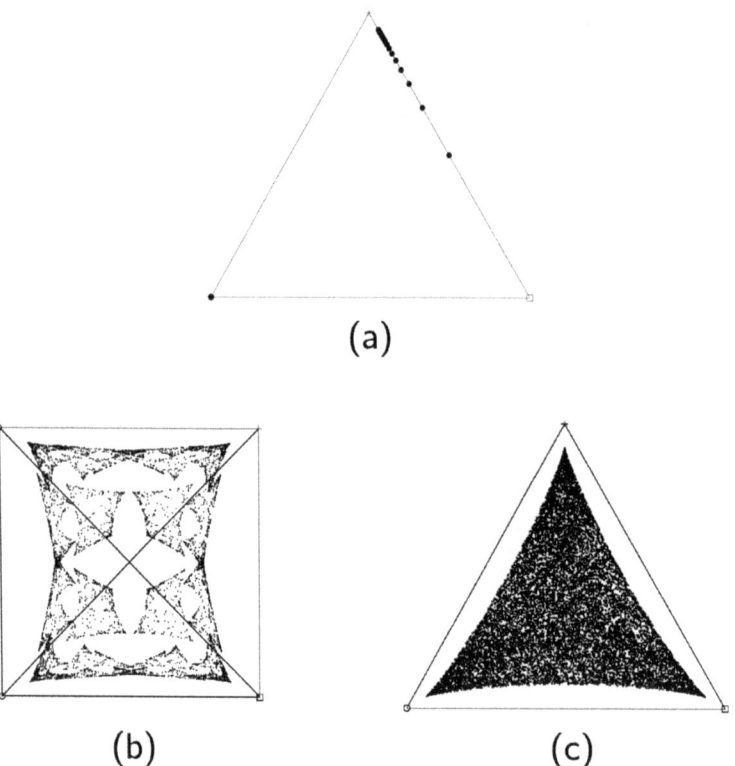

Figure 17. Stocastic deterministic automata (SDA): (a) Denumerable SDA: A denumerable ϵ-machine for the simple nondeterministic source of Fig. 11. It is shown here in the (two-dimensional) 3-simplex defining its possible deterministic states (indicated with enlarged dots). Since the state probability decays exponentially, the simulation only shows a very truncated part of the infinite chain of states that, in principle, head off toward the upper vertex. Those dots correspond to the 1s backbone of Fig. 15. The state on the lower lefthand vertex corresponds to the "reset" state 1 A0 B in that figure. (b) Fractal SDA: A nondenumerable fractal ϵ-machine shown in the 4-simplex defining the possible deterministic states. (c) Continuum SDA: A nondenumerable continuum ϵ-machine shown in the 3-simplex defining the possible deterministic states.

states for these processes can be graphically represented as a point in a simplex. The simplex vertices themselves correspond to δ-distributions concentrated on one of the hidden Markov model states; the dimension of the simplex is one less than the number of hidden states. In the simplex representation the ϵ-machine models are stochastic deterministic automata (SDA). Only a subset of the simplex is recurrent. This subset corresponds to an attracting set under the dynamics of the agent's step-by-step prediction of the hidden state distribution based on the current data stream. Fig. 17 shows the causal state simplices for three example processes.

The infinite, but countable state machine of Fig. 15 is shown in this simplex representation in Fig. 17a as a discrete point set in the 3-simplex. Each point, somewhat enlarged there, corresponds to one of the states in Fig. 15. Two more examples of ϵ-machines are given in Fig. 17b and 17c. They suggest some of the richness of stochastic finite processes. In Fig. 17b, for example, the ϵ-machine has a partial continuum of states; the causal states lie on a "fractal". Fig. 17c shows an ϵ-machine whose causal states limit on a full continuum of states.

The different classes shown in Fig. 17 are distinguished by a new complexity measure of the ϵ-machines's state structure—the ϵ-machine dimension $d_{\epsilon M}$. $d_{\epsilon M}$ is the information dimension of the state distribution on the simplex. In the case of the countable state ϵ-machine of Fig. 17a C_μ is finite due to the strong localization of the state distribution over the earlier states. But, since the states are a discrete point set, $d_{\epsilon M} = 0$. For the fractal and continuum SDAs the statistical complexity diverges, but $d_{\epsilon M} > 0$. $d_{\epsilon M}$ is noninteger in the first case and $d_{\epsilon M} = 2$ in the second. Further results will be presented elsewhere (Crutchfield and Upper, n.d.).

Recent results describing such statistical complexity dimension can be found in Jurgens and Crutchfield (2021).

The Finitary Stochastic Hierarchy

The preceding examples of stochastic finitary processes can be summarized using the computational model hierarchy of Fig. 18. This hierarchy borrows the finite memory machines of Fig. 2 and indicates their stochastic generalizations. As before each ellipse denotes a model class. As one moves up the diagram classes become more powerful in the sense that they can finitely describe a wider range of processes than lower classes. A class below and connected to a given one can finitely describe only a subset of the processes finitely described by the higher one. Again, this hierarchy is only a partial ordering of descriptive capability. It should be emphasized that "descriptive capability" above the Measure—Support line refers to finitely representing a *distribution* over the sequences; not just the distribution's supporting formal language.

In formal language theory it is well-known that deterministic finite automata (DFA) are as powerful as nondeterministic finite automata (NFA) (Hopcroft and Ullman 1979). This is shown in the hierarchy as both classes being connected at the same height. But the equivalence is just topological; that is, it concerns only the descriptive capability

FOUNDATIONAL PAPERS IN COMPLEXITY SCIENCE

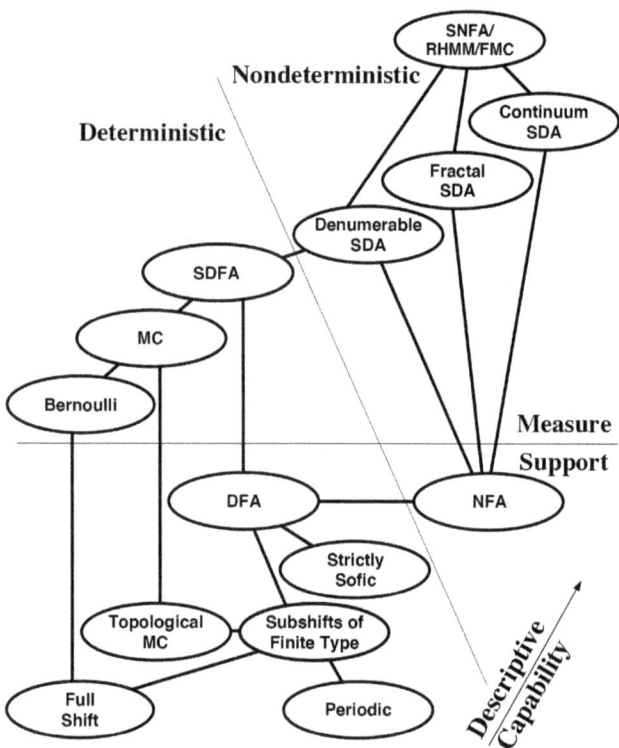

Figure 18. The computational hierarchy for finite-memory nonstochastic (below the Measure–Support line) and stochastic discrete processes (above that line). The nonstochastic classes come from Fig. 2, below the Finite-Infinite memory line. Here "Support" refers to the sets of sequences, i.e. formal languages, which the "topological" machines describe; "Measure" refers to sequence probabilities, i.e. what the "stochastic" machines describe. The abbreviations are: A is automaton, F is finite, D is deterministic, N is nondeterministic, S is stochastic, MC is Markov chain, HMM is hidden Markov model, RHMM is recurrent HMM, and FMC is function of MC.

of each class for sets of observed sequences. If one augments these two classes, though, to account for probabilistic structure over the sequences, the equivalence is broken in a dramatic way—as the above example for the "mismeasured" logistic map demonstrated. This is shown in Fig. 18 above the Measure–Support line. The class of SNFA is higher than that of the stochastic deterministic finite automata (SDFA). Crudely speaking, if a DFA has transition probabilities added to its edges, one obtains the single class of SDFA. But if transition probabilities are added to an NFA, then the class is qualitatively more powerful and, as it turns out, splits into three distinct classes. Each of these classes is more powerful than the SDFA class. The new causal classes—called stochastic deterministic automata (SDA)—are distinguished by having a countable infinity, a fractional continuum, or a full continuum of causal states: the examples of Fig. 17.

Initially, the original logistic map process as represented in Fig. 11 was undistinguished as an SNFA. Via the analysis outlined above its causal representation showed that it is equivalent to a denumerable stochastic deterministic automaton (DSDA). Generally, in terms of descriptive power DSDA $\subset$ SNFA as Fig. 18 emphasizes. Recall that we interpret the SNFA as the environment, which is a Markov chain (MC), viewed through the agent's sensory apparatus. So the computational class interpretation of the complexity divergence is that MC $\to$ DSDA under measurement distortion. That is, MC and DSDA are qualitatively different classes and, in particular, MC $\subset$ DSDA, as shown in the hierarchy of Fig. 18. The representational divergence that separates them is characteristic of the transition from a lower to a higher class.

4.5.3. The Costs of Spatial Coherence and Distortion

As an example of hierarchical structure and innovation for spatial processes this section reviews an analysis of two cellular automata (CA). CA are arguably the simplest dynamical systems with which to study pattern formation and spatio-temporal dynamics. They are discrete in time, in space, and in local state value. The two examples considered have two local states at each site of a one-dimensional spatial lattice. The local state at a site is updated according to a rule that looks only at itself and its two nearest neighbors.

Fig. 19a shows the temporal development of an arbitrary initial pattern under the action of elementary CA 18 (ECA 18). This rule maps all of the neighborhood patterns to 0; except 001 and 100, which are mapped to 1. A space-time diagram for elementary CA 54 (ECA 54) is shown in Fig. 20a.

ϵ-machine reconstruction applied to the patterns reveals much of the internal structure of these systems' state spaces and the intrinsic computation in their temporal development of spatial patterns (Crutchfield and Hanson 1993b). An important component of CA patterns are domains. CA domains are dynamically homogeneous regions in space-time defined in terms of the same set of pattern regularities, such as "every other site value is a 0". From knowledge of a

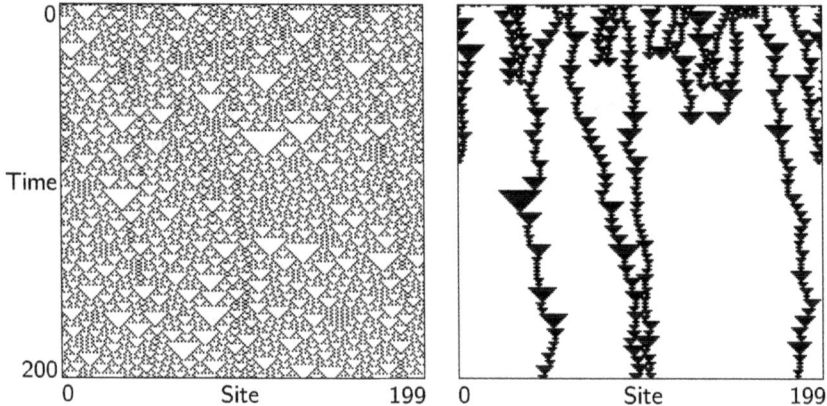

Figure 19. (a) Elementary cellular automaton 18 evolving over 200 time steps from an initial arbitrary pattern on a lattice of 200 sites. (b) The filtered version of the same space-time diagram that reveals the diffusive-annihilating dislocations obscured in the original. (After Crutchfield and Hanson 1993a.)

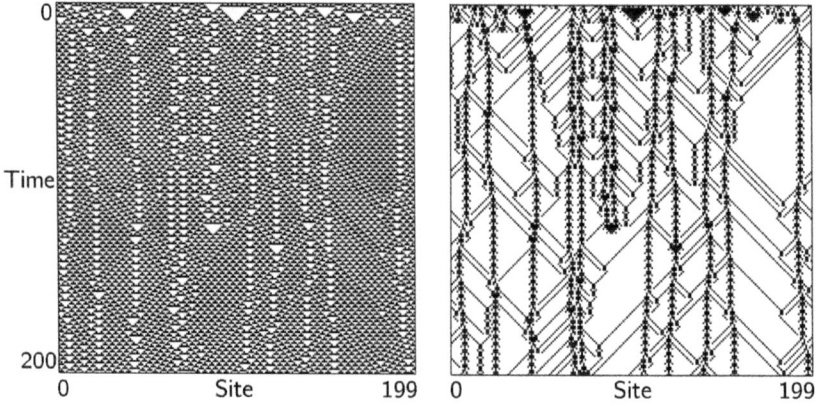

Figure 20. (a) Elementary cellular automaton 54 evolving over 200 time steps from an initial arbitrary pattern on a lattice of 200 sites. (b) The filtered version of the same space-time diagram that reveals a multiplicity of different particle types and interactions. (From Hanson 1993. Reprinted with permission of the author. Cf. Boccara, Nasser, and Roger 1991.)

CA's domains nonlinear filters can be constructed to show the motion and interaction of domains, walls between domains, and particles. The results of this filtering process—shown in Fig. 19b and 20b— is a higher level representation of the original space-time diagrams in Fig. 19a and 20a. This new level filters out chaotic (ECA 18) and periodic (ECA 54) backgrounds to highlight the various propagating space-time structures—dislocations, particles, and their interactions that are hidden in the unfiltered space-time diagrams.

In turns out that ECA 18 is described by a single type of chaotic domain—"every other local state is 0, otherwise they are random"—and a single type of domain wall—a dislocation that performs a diffusive annihilating random walk (Crutchfield and Hanson 1993a). At the level of particles one discovers an irreducibly stochastic description—the particle motion has vanishing statistical complexity. Because the description of ECA 18 is finite at this level, one can stop looking further for intrinsic computational structure. The correct ϵ-machine level, that providing a finite description, has been found by filtering the chaotic domains.

In ECA 54 there is a single periodic domain with a diversity of particles that move and interact in complicated—aperiodic, but nonstochastic—ways. ECA 54 is more complex than ECA 18, in the sense that even at the level of particles ECA 54's description may not be finite. Thus for ECA 54, one is tempted to look for a higher level representation by performing ϵ-machine reconstruction at this "particle" level to find further regularity or stochasticity and ultimately to obtain a finite description.

These hierarchical space-time structures concern different levels of information processing within a single CA. In ergodic theory terms, the unfiltered spatial patterns are nonstationary—both spatially and temporally. With the innovation of domains and particles, the new level of representation allows for stationary elements, such as the domains, to be separated out from elements which are nonstationary. For ECA 18, discovering the dislocations led to a stochastic model based on a diffusive annihilating random walk that could be analyzed in some detail and which explained the nonstationary aspects of the dislocation motion. For ECA 54 the description at this level is still not completely understood.

Intrinsic computation in spatio-temporal processes raises a number of interesting problems. Aside from the above, there is an analog for spatio-temporal processes of the apparent complexity explosion in the stochastic nondeterministic processes. This led to the introduction of a new hierarchy for spatial automata, in which CA are the least capable, that accounts for spatial measurement distortion (Crutchfield 1992c). This spatial discrete computation hierarchy is expressed in terms of automata rather than grammars (Crutchfield 1994).

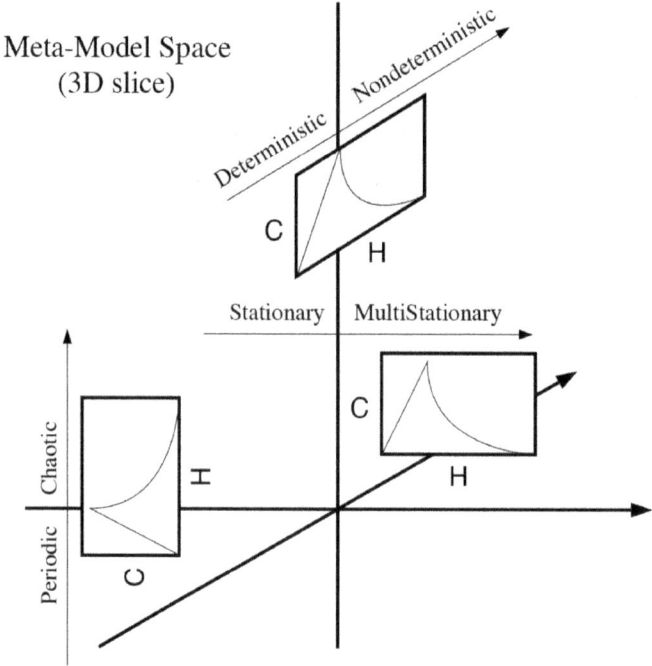

Figure 21. A schematic summary of the three examples of hierarchical learning in metamodel space. Innovation across transitions from periodic to chaotic, from stochastic deterministic to stochastic nondeterministic, and from spatial stationary to spatial multistationary processes were illustrated. The finite-to-infinite memory coordinate from Fig. 2 is not shown. The periodic to chaotic and deterministic to nondeterministic transitions were associated with the innovation of infinite models from finite ones. The complexity (C) versus entropy (H) diagrams figuratively indicate the growth of computational resources that occurs when crossing the innovation boundaries.

4.5.4. What to Glean from the Examples

Understanding the hierarchical structure embedded in the preceding examples required crossing a model class boundary: from determinism to indeterminism, from finitary support to finitary measure, from predictability to chaos, from undifferentiated patterns to domains and particles, and from observed states to hidden internal states. Each of these transitions is a phase transition in the sense that there is a divergence in a macroscopic observable—the representation's complexity—and in that the finitely-describable behaviors on either side of the divergence are qualitatively distinct. Fig. 21 attempts to summarize in a very schematic way their relationship.

The point that the examples serve to make is that innovation of new computational structures is required at each step. Just what needs to be innovated can be determined in very large measure by the state-grouping step in hierarchical ϵ-machine reconstruction. As was stated before, beyond the state-grouping it seems that there will always be some undetermined aspect during the innovation step required of the agent. The undetermined aspect has been isolated by the algorithmic portion of hierarchical ϵ-machine reconstruction that groups lower level states into a higher level state. To the extent it is undetermined, though, this step can be the locus of a now highly-constrained search over a space of new mechanisms. In any evolutionary process, such a space would be largely circumscribed by the agents's internal structure and the search, by how the latter can be modified. In this way, hierarchical ϵ-machine reconstruction reduces the search space for innovation by a substantial fraction and indicates where random search could be most informative and effective.

5. Observations and Hypotheses

5.1. COMPLEXITY AS THE INTERPLAY OF ORDER AND CHAOS

Neither order nor randomness are sufficient in and of themselves for the emergence of complexity. Nor is their naive juxtaposition. As Alfred North Whitehead and many before and after him knew, order and randomness are mere components in an ongoing process. That process *is* the subtlety in the problem of emergence. For Heraclitus, the persistence of change was the by-product of an "attunement of opposite tensions" (Russell 1945). Natural language, to take one ever-present example, has come to be complex because utility required it to be highly informative *and* comprehensibility demanded a high degree of structure. For the process that is human language, emergence is a simple question: How do higher semantic layers appear out of this tension?

As far as the architecture of information processing is concerned, these questions for natural language have direct analogues in adaptation, evolution, and even in the development of scientific theories (Kuhn 1962) [70]. While conflating these fields suggests a possible general theory of the emergence of complexity, at the same time it reveals an ignorance of fundamental distinctions. Due to the sheer difficulty of the nonlinear dynamics involved and to the breadth of techniques that must be brought to bear, one must be particularly focused on concrete results and established theory in discussing "emergence".

In spite of receiving much popular attention, the proposals for "computation at the edge of chaos" (Langton 1990), "adaptation toward the edge of chaos" (Packard 1988), and "life at the edge of chaos" (Kauffman 1993) are recent examples of the problems engendered by this sort of blurring of basic distinctions. Despite the preceding sections' review of the complexity versus entropy spectrum and their analyses of processes in which higher levels of computation arise at the onset of chaos, there is absolutely no general need for high computational capability to be near an "edge of chaos". The infinite-memory counter register functionality embedded in the highly chaotic hidden Markov model of Fig. 11 is a clear demonstration of this simple fact. More to the point, that stability and order are necessary for information storage, on the one hand, and that instability is necessary for the production of information and its communication, on the other hand, are basic requirements for any nontrivial information processing. The trade-off between these requirements is much of what computation theory is about. Moreover, natural systems are not constrained in some general way to move toward an "edge of chaos". For example, the very informativeness ($h_\mu > 0$) of natural language means, according to Shannon, that natural language is very far and must stay away from any "edge of chaos" ($h_\mu = 0$).

A critique of the first two "edge of chaos" proposals, which concern cellular automata, can be found elsewhere (Mitchell, Hraber, and Crutchfield 1993). Aside from the technical issues discussed there, all three proposals exhibit a fatal conceptual difficulty. To the extent that they presume to address the emergence of complexity, with reference neither to intrinsic computation nor to the innovation of new information-processing architectures, they can provide no grounding for the key concepts—"order", "chaos", "complexity", "computation"—upon which their arguments rely.

The preceding development of statistical complexity, hierarchical ϵ-machine reconstruction, and various extensions to discrete computation theory is an alternative. The presentation gave a broad overview of how three tool sets—the calculi of emergence: computation, dynamics, and innovation—interrelate in three different problem areas: various routes to chaos, measurement distortion and nondeterminism, and cellular automaton pattern formation. It suggested a way to frame the question of how structure appears out of the interplay of order and randomness. It demonstrated methods for detecting and metrics for quantifying that emergence. It showed how the information processing structure of nonlinear processes can be analyzed in terms of computational models and in terms of the effective information processing.

5.2. EVOLUTIONARY MECHANICS

The arguments to this point can be recapitulated by an operational definition of emergence. A process undergoes emergence if at some time the architecture of information processing has changed in such a way that a distinct and more powerful level of intrinsic computation has appeared that was not present in earlier conditions.

It seems, upon reflection, that our intuitive notion of emergence is not captured by the "intuitive definition" given in Section 2. Nor is it captured by the somewhat refined notion of pattern formation. "Emergence" is a groundless concept unless it is defined within the context of processes themselves; the only well-defined (non-subjective) notion of emergence would seem to be intrinsic emergence. Why? Simply because emergence defined without this closure leads to an infinite regress of observers detecting patterns of observers detecting patterns .... This is not a satisfactory definition, since it is not finite. The regress must be folded into the system, it must be immanent in the dynamics. When this happens complexity and structure are no longer referred outside. No longer relative and arbitrary, they take on internal meaning and functionality.

Where in science might a theory of intrinsic emergence find application? Are there scientific problems that at least would be clarified by the computational view of nature outlined here?

In several ways the contemporary debate on the dominant mechanisms operating in biological evolution seems ripe. Is anything ever new in the biological realm? The empirical evidence is interpreted as a resounding "yes". It is often heard that organisms today are more complex than in earlier epochs. But how did this emergence of complexity occur? Taking a long view, at present there appear to be three schools of thought on what the guiding mechanisms are in Darwinian evolution that produce the present diversity of biological structure and that are largely responsible for the alteration of those structures.

The Selectionists hold that structure in the biological world is due primarily to the fitness-based selection of individuals in populations whose diversity is maintained by genetic variation (Maynard Smith 1989). The second, anarchistic camp consists of the Historicists who hold fast to the Darwinian mechanisms of selection and variation, but emphasize the accidental determinants of biological form (Monod 1971; Gould 1989). What distinguishes this position from the Selectionists is the claim that major changes in structure can be and have been nonadaptive. Lastly, there are the Structuralists whose goal is to elucidate the "principles of organization" that guide the appearance of biological structure. They contend that energetic, mechanical, biomolecular, and morphogenetic constraints limit the infinite range of possible biological form (Thompson 1917; Kauffman 1993; Waddington 1957; Goodwin 1992; Fontana and Buss 1994b, 1994a). The constraints result in a relatively small set of structural attractors. Darwinian evolution serves, at best, to fill the waiting attractors or not depending on historical happenstance.

What is one to think of these conflicting theories of the emergence of biological structure? The overwhelming impression this debate leaves is that there is a crying need for a theory of biological structure and a qualitative dynamical theory of its emergence (Goodwin and Sanders 1992). In short, the tensions between the positions are those (i) between the order induced by survival dynamics and the novelty of individual function and (ii) between the disorder of genetic variation and the order of developmental processes. Is it just an historical coincidence that the structuralist—selectionist dichotomy appears analogous to that between order and randomness in the realm of modeling? The main problem, at least to an outsider, does not reduce to showing that one or the other view is correct. Each employs compelling arguments and often empirical data as a starting point. Rather, the task facing us reduces to developing a synthetic theory that balances the tensions between the viewpoints. Ironically, evolutionary processes themselves seem to do just this sort of balancing, dynamically.

The computational mechanics of nonlinear processes can be construed as a theory of structure. Pattern and structure are articulated in terms of various types of machine classes. The overall mandate is to provide both a qualitative and a quantitative analysis of natural information-processing architectures. If computational mechanics is a theory of structure, then innovation via hierarchical ϵ-machine reconstruction is a computation-theoretic approach to the transformation of structure. It suggests one mechanism with which to study what drives and what constrains the appearance of novelty. The next step, of course, would be to fold hierarchical ϵ-machine reconstruction into the system itself, resulting in a dynamics of innovation, the study of which might be called "evolutionary mechanics".

The prototype universe of Fig. 1 is the scaffolding for studying an abstract "evolutionary mechanics", which is distinguished from chemical, prebiotic, and biological evolution. That is, all of the "substrates" indicated in Fig. 1 are thrown out, leaving only those defined in terms of information processing. This more or less follows the spirit of computational evolution (Fogel, Owens, and Walsh 1966; Holland 1992). The intention is to expunge as much disciplinary semantic content as possible so that if novel structure emerges, it does so neither by overt design nor interpretation but (i) via the dynamics of interaction and induction and (ii) according to the basic constraints of information processing. Additionally, at the level of this view mechanisms for genotypic and phenotypic change are not delineated. In direct terms evolutionary mechanics concerns the change in the information-processing architecture of interacting adaptive agents. The basic components that guide and constrain this change are the following:

1. *Modeling*: Driven by the need for an encapsulation of environmental experience, an agent's internal model captures its knowledge, however limited, of the environment's current state and persistent structure.

2. *Computation*: Drive by the need to process sensory information and produce actions, computation is the adaptive agent's main activity. The agent's computational resources delimit its inferential, predictive, and semantic capabilities. They place an upper bound on the maximal level of computational sophistication. Indirectly, they define the language with which the agent expresses its understanding of the environment's structure. Directly, they limit the amount of

history the agent can store in its representation of the environment's current state.

3. *Innovation*: Driven by limited observational, computational, and control resources, innovation leads to new model classes that use the available resources more efficiently or more parsimoniously.

Subsequent research has shown that the memory requirements of quantum models can stay finite even when classical counterparts scale toward infinity (Garner et al. 2017). As such, one may argue that innovation would lead naturally to quantum models.

The computational mechanics approach to emergence attempted to address each component in turn, though it did not do so in the fully dynamic setting suggested by the prototype universe. But evolutionary mechanics is concerned specifically with this dynamical problem and, as such, it leads to a much wider range of questions.

At the very least, the environment appears to each agent as a hierarchical process. There are subprocesses at several different scales, such as those at the smallest scale where the agents are individual stochastic dynamical systems and those at larger scales at which coordinated global behavior may emerge. Each agent is faced with trying to detect as much structure as possible within this type of environment. How can an evolutionary system adapt when confronted with this kind of hierarchical process?

Fig. 22 gives a very schematic summary of one of the consequences following from this framework in terms of successive changes in information-processing architecture. The underlying representation of the entire prototype universe, of course, is the realization space of nonlinear processes—the orbit space of dynamical systems. The figure shows instead a temporal analog of the discrete computation hierarchy of Fig. 2 in which the trade-offs within each computational level are seen through the complexity–entropy plane. The overall evolutionary dynamic derives from agent interaction and survival. Species correspond in the realization space to metastable invariant languages: they are temporary (hyperbolic) fixed points of the evolutionary dynamic. Each species is dual to some subset of environmental regularities—a niche—whose defining symmetries some subpopulation of agents has been able to discover and incorporate into their internal models through innovation. Innovation then manifests itself as a kind of speciation. The invariant sets bifurcate into computationally distinct agents with new, possibly-improved capabilities. The statistical mechanical picture of this speciation is that of a phase transition between distinct computational "thermodynamic" phases. Through all of this the

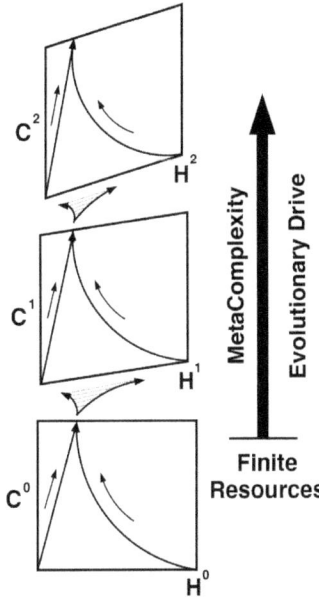

Figure 22. Schematic diagram of an evolutionary hierarchy in terms of the changes in information-processing architecture. An open-ended sequence of successively more sophisticated computational classes are shown. The evolutionary drive up the hierarchy derives from the finiteness of resources to which agents have access. The complexity–entropy diagrams are slightly rotated about the vertical to emphasize the difference in meaning at each level via a different orientation. (Cf. Table 1.)

macro-evolutionary observables—entropy rate h_μ, statistical complexity C_μ, and innovation $\mathcal{T}$—monitor changes in the agents' and the universe's information-processing architecture.

Evolutionary mechanics seems to suggest that the emergence of natural complexity is a manifestly open-ended process. One force of evolution appears in this as a movement up the inductive hierarchy through successive innovations. This force of evolution itself is driven at each stage by the limited availability of resources.

> *Some new principle of refreshment is required. The art of progress is to preserve order amid change, and to preserve change amid order. Life refuses to be embalmed alive.*
>
> A. N. Whitehead in *Process and Reality*.

Acknowledgments

Many thanks are due to Lisa Borland, Don Glaser, Jim Hanson, Blake LeBaron, Dan McShea, Melanie Mitchell, Dan Upper, and Karl Young for helpful discussions and comments on early drafts. This work was partially funded under AFOSR 91-0293 and ONR N00014-92-4024. For the opportunity to present this work thanks also are due to the organizers of the Oji International Seminar on *Complex Systems—from Complex Dynamics to Artificial Reality*, held 5–9 April 1993 Namazu, Japan; to the organizers of the Santa Fe Institute *Integrative Themes of Complex Adaptive Systems* Meeting held on 8 July 1992 in Santa Fe, New Mexico; and to the organizers of the *Artificial Life 2* conference, held during 5–9 February 1990.

REFERENCES

Aho, A. V. 1968. "Indexed Grammars—An Extension of Context-Free Grammars." *Journal of the Association for Computing Machinery* 15:647.

———. 1969. "Nested Stack Automata." *Journal of the Association for Computing Machinery* 16:383.

Allevi, E., A. Cherubini, and S. Crespi-Reghizzi. 1988. "Breadth-First Context-Free Grammars and Queue Automata." *Lecture Notes in Computational Science* 324:162.

Angluin, D., and C. H. Smith. 1983. "Inductive Inference: Theory and Methods." *ACM Computing Surveys* 15 (3): 237–269.

Bak, P., and K. Chen. 1990. "Self-Organized Criticality." *Physica A* 163:403–409.

Bennett, C. H. 1988. "Dissipation, Information, Computational Complexity, and the Definition of Organization." In *Emerging Syntheses in Science,* edited by D. Pines, 215–234. Redwood City, CA: Addison-Wesley.

Binney, J. J., N. J. Dowrick, A. J. Fisher, and M. E. J. Newman. 1992. *The Theory of Critical Phenomena.* Oxford, UK: Oxford University Press.

Blackwell, D., and L. Koopmans. 1957. "On the Identifiability Problem for Functions of Markov Chains." *Annals of Mathematical Statistics* 28:1011.

Blum, L., M. Shub, and S. Smale. 1989. "On a Theory of Computation Over the Real Numbers." *Bulletin of the American Mathematical Society* 21 (1).

Boccara, N., J. Nasser, and M. Roger. 1991. "Particle-Like Structures and Their Interactions in Spatio-Temporal Patterns Generated by One-Dimensional Deterministic Cellular Automaton Rules." *Physical Review A* 44:866–875.

Brookshear, J. G. 1989. *Theory of Computation: Formal Languages, Automata, and Complexity.* Redwood City, CA: Benjamin/Cummings.

Brudno, A. A. 1983. "Entropy and the Complexity of the Trajectories of a Dynamical System." *Transactions for the Moscow Mathematical Society* 44:127.

Chaitin, G. J. 1966. "On the Length of Programs for Computing Finite Binary Sequences." *Journal of the Association for Computing Machinery* 13:145.

Cherubini, A., C. Citrini, Crespi-Reghizzi S., and D. Mandrioli. 1991. "Quasi-Real-Time FIFO Automata, Breadth-First Grammars, and Their Relations." *Theoretical Computer Science* 85:171–203.

Crutchfield, J. P. 1991. "Reconstructing Language Hierarchies." In *Information Dynamics,* edited by H.A. Atmanspracher and H. Scheingraber, 45. New York, NY.

———. 1992a. "Knowledge and Meaning... Chaos and Complexity." In *Modeling Complex Phenomena,* edited by L. Lam and V. Naroditsky, 66–101. Berlin, Germany: Springer.

———. 1992b. "Semantics and Thermodynamics." In *Nonlinear Modeling and Forecasting,* edited by M. Casdagli and S. Eubank, XII:317–359. Santa Fe Institute Studies in the Sciences of Complexity. Reading, CA: Addison-Wesley.

———. 1992c. "Unreconstructible at Any Radius." *Physics Letters A* 171:52–60.

———. 1994. "Observing Complexity and the Complexity of Observation." In *Inside Versus Outside,* edited by H. Atmanspacher and G. J. Dalenoort, 235–272. Berlin: Springer.

Crutchfield, J. P., and J. E. Hanson. 1993a. "Attractor Vicinity Decay for a Cellular Automaton." *Chaos* 3:215–224.

———. 1993b. "Turbulent Pattern Bases for Cellular Automata." *Physica D* 69:279–301.

Crutchfield, J. P., and B. S. McNamara. 1987. "Equations of Motion From a Data Series." *Complex Systems* 1:417–452.

Crutchfield, J. P., and N. H. Packard. 1983. "Symbolic Dynamics of Noisy Chaos." *Physica D* 7:201–223.

Crutchfield, J. P., N. H. Packard, J. D. Farmer, and R. S. Shaw. 1986. "Chaos." *Scientific American* 255:46–57.

Crutchfield, J. P., and D. R. Upper. n.d. In preparation.

Crutchfield, J. P., and K. Young. 1989. "Inferring Statistical Complexity." *Physical Review Letters* 63:105–108.

———. 1990. "Computation at the Onset of Chaos." In *Complexity, Entropy, and the Physics of Information,* edited by W. H. Zurek, VIII:223–269. SFI Studies in the Science of Complexity. Reading, MA: Addison-Wesley.

Fama, E. F. 1991. "Efficient Capital Markets II." *The Journal of Finance* 46 (5): 1575–1617.

Feigenbaum, M. J. 1983. "Universal Behavior in Nonlinear Systems." *Physica D* 7:16.

Fogel, L. J., A. J. Owens, and M. J. Walsh, eds. 1966. *Artificial Intelligence Through Simulated Evolution.* New York, NY: Wiley.

Fontana, W., and L. Buss. 1994a. "The Arrival of the Fittest: Toward a Theory of Biological Organization." *The Bulletin of Mathematical Biology* 56:1–64.

———. 1994b. "What Would be Conserved if the Tape were Played Twice?" *Proceedings of the National Academy of Sciences* 91:757–761.

Ford, J. 1983. "How Random is a Coin Toss?" *Physics Today* 36 (4).

Forrest, S. 1990. "Emergent Computation: Self-organizing, Collective, and Cooperative Behavior in Natural and Artificial Computing Networks: Introduction to the Proceedings of the Ninth Annual CNLS Conference." *Physica D* 42:1–11.

Goodwin, B. 1992. "Evolution and the Generative Order." In *Theoretical Biology: Epigenetic and Evolutionary Order from Complex Systems,* edited by B. Goodwin and P. Sanders, 89–100. Baltimore, Maryland: Johns Hopkins University Press.

Goodwin, B., and P. Sanders, eds. 1992. *Theoretical Biology: Epigenetic and Evolutionary Order from Complex Systems.* Baltimore, Maryland: Johns Hopkins University Press.

Gould, S. J. 1989. *Wonderful Life: The Burgess Shale and the Nature of History.* New York, NY: Norton.

Grassberger, P. 1986. "Toward a Quantitative Theory of Self-Generated Complexity." *International Journal of Theoretical Physics* 25:907–938.

Hanson, J. E. 1993. "Computational Mechanics of Cellular Automata." PhD diss., University of California, Berkeley.

Hanson, J. E., and J. P. Crutchfield. 1992. "The Attractor-Basin Portrait of a Cellular Automaton." *Journal of Statistical Physics* 66:1415.

Holland, J. H. 1992. *Adaptation in Natural and Artificial Systems.* Cambridge, MA: MIT Press.

Holldobler, B., and E. O. Wilson. 1990. *The Ants.* Cambridge, MA: Belknap Press, Harvard University Press.

Hopcroft, J. E., and J. D. Ullman. 1979. *Introduction to Automata Theory, Languages, and Computation.* Reading, MA: Addison-Wesley.

Ito, H., S. Amari, and K. Kobayashi. 1992. "Identifiability of Hidden Markov Information Sources and Their Minimum Degrees of Freedom." *IEEE Transactions on Information Theory* 38:324–333.

Kadanoff, L. P., M. J. Feigenbaum, and S. J. Shenker. 1982. "Quasiperiodicity in Dissipative Systems: A Renormalization Group Analysis." *Physica D* 5:370.

Kauffman, S. A. 1993. *Origins of Order: Self-Organization and Selection in Evolution.* New York, NY: Oxford University Press.

Kemeny, J. G. 1953. "The Use of Simplicity in Induction." *The Philosophical Review* 62:391.

Kolmogorov, A. N. 1965. "Three Approaches to the Concept of the Amount of Information." *Problemy Peredachi Informatsii* 1 (1): 3–11.

Kovacs, I., and B. Julesz. 1993. "A Closed Curve is Much More Than an Incomplete One-Effect of Closure in Figure Ground Segmentation." *Proceedings of the National Academy of Sciences* 90:7495–7497.

Kuhn, T. S. 1962. *The Structure of Scientific Revolutions.* Chicago, IL: University of Chicago Press.

Land, E. 1964. "The Retinex." *American Scientist* 52:247–264.

Langton, C. G. 1990. "Computation at the Edge of Chaos: Phase Transitions and Emergent Computation." In *Emergent Computation,* edited by S. Forrest, 12. North-Holland, Amsterdam.

Lewis, H. R., and C. H. Papadimitriou. 1981. *Elements of the Theory of Computation.* Englewood Cliffs, NJ: Prentice-Hall.

Lindenmayer, A., and P. Prusinkiewicz. 1989. "Developmental Models of Multicellular Organisms: A Computer Graphics Perspective." In *Artificial Life,* edited by C. G. Langton, 221. Redwood City, CA: Addison-Wesley.

Maynard Smith, J. 1989. "Evolutionary Genetics." (Oxford, UK).

Meinhardt, H. 1982. *Models of Biological Pattern Formation.* London, UK: Academic Press.

Mitchell, M., P. Hraber, and J. P. Crutchfield. 1993. "Revisiting the Edge of Chaos: Evolving Cellular Automata to Perform Computations." *Complex Systems* 7:89–130.

Monod, J. 1971. *Chance and Necessity: An Essay on the Natural Philosophy of Modern Biology.* New York, NY: Vintage Books.

Moore, C. 1993. *Real-Valued, Continuous-Time Computers: A Model of Analog Computation, Part I.* Technical report 93-04-018. Santa Fe, NM: Santa Fe Institute.

Nordahl, M. G. 1989. "Formal Languages and Finite Cellular Automata." *Complex Systems* 3:63.

Ouyang, Q., and H. L. Swinney. 1991. "Transition from a Uniform State to Hexagonal and Striped Turing Patterns." *Nature* 352:610–612.

Packard, N. H. 1988. "Adaptation Toward the Edge of Chaos." In *Dynamic Patterns in Complex Systems,* edited by J. A. S. Kelso, A. J. Mandell, and M. F. Shlesinger, 293–301. Singapore: World Scientific.

Reynolds, C. W. 1987. "Flocks, Herds, and Schools: A Distributed Behavioral Model." *Computer Graphics* 21:25–34.

Rissanen, J. 1978. "Modeling by Shortest Data Description." *Automatica* 14:462.

———. 1989. "Stochastic Complexity in Statistical Inquiry." *World Scientific Series in Computer Science* 15.

Rogers, H. 1967. *Theory of Recursive Functions and Effective Computability.* New York, NY: McGraw-Hill.

Rozenberg, G., and A. Salomaa, eds. 1986. *The Book of L.* Berlin, Germany: Springer.

Russell, B. 1945. *A History of Western Philosophy.* New York, NY: Simon and Schuster.

Searls, D. B. 1992. "The Linguistics of DNA." *American Scientist* 80 (6): 579–591.

Shannon, C. E., and W. Weaver. 1948. "The Mathematical Theory of Communication." (Champaign-Urbana, IL).

Shaw, R. S. 1984. *The Dripping Faucet as a Model Chaotic System.* Santa Cruz, CA: Aerial Press.

Solomonoff, R. J. 1964. "A Formal Theory of Inductive Inference." *Information Control* 7:224.

Stanley, H. E. 1971. *Introduction to Phase Transitions and Critical Phenomena.* Oxford, UK: Oxford University Press.

Swinney, H. L., and J. P. Gollub, eds. 1981. *Hydrodynamic Instabilities and the Transition to Turbulence.* Berlin: Springer.

Thompson, D. W. 1917. *On Growth and Form.* Cambridge, UK: Cambridge University Press.

Turing, A. M. 1952. "The Chemical Basis of Morphogenesis." *Philosophical Transactions of the Royal Society B* 237:5.

Waddington, C. H. 1957. *The Strategy of the Genes.* Allen and Unwin.

Walker, A. 1974. "Adult Languages of L Systems and the Chomsky Hierarchy." In *L Systems,* edited by G. Rozenberg and A. Salomaa, 15:201–215.

Wallace, C. S., and D. M. Boulton. 1968. "An Information Measure for Classification." *The Computer Journal* 11:185.

Wandell, B. A. 1993. "Color Appearance: The Effects of Illumination and Spatial Patterns." *Proceedings of the National Academy of Sciences* 10:2458–2470.

Weiss, B. 1973. "Subshifts of Finite Type and Sofic Systems." *Monatshefte für Mathematik* 77:462.

Whitehead, A. N. 1978. *Process and Reality.* New York, NY: The Free Press.

Winfree, A. T. 1980. *The Geometry of Biological Time.* Berlin: Springer.

Wolfram, S. 1984. "Computation Theory of Cellular Automata." *Communications in Mathematical Physics* 96:15.

ALAN STUART PERELSON
April 11, 1947
Brooklyn, NY

RAJESH CHERUKURI

LAWRENCE ALLEN

STEPHANIE FORREST
1958

1950.
1960.

FORREST, PERELSON, ALLEN, AND CHERUKURI 1994
University of New Mexico, Albuquerque, NM

University of New Mexico, Albuquerque, NM

Santa Fe Institute, Santa Fe, NM

University of New Mexico, Albuquerque, NM

1962: Jacques Miller demonstrates that T cells, which develop from the thymus, play a crucial role within the immune system; mice that received thymectomies were able to tolerate foreign skin grafts.

1970.
1980.

1989: Charles Janeway proposes the Infectious Non-Self Model during the Cold Spring Harbor Symposium on Quantitative Biology.

1990.
2000.

2005: Researchers from the US Department of Agriculture announce the successful genetic sequencing of the formerly eradicated "Spanish flu" strain of the H1N1 virus.

2009: In the United States, the final switchover from analog television to digital television occurs in the month of June.

2010.
2020.

ALLEN & CHERUKURI PHOTOS COURTESY WIKIMEDIA COMMONS

FORREST, PERELSON, ALLEN & CHERUKURI

[79]

LEARNING FROM THE IMMUNE SYSTEM

Anil Somayaji, Carleton University

Immune systems are remarkable examples of complex adaptive systems. They are decentralized, massively distributed, and able to adaptively respond to novel circumstances. While a better understanding of immune systems has many potential benefits for human health, such understanding could also help with the creation of artificial systems with similar capabilities. "Self–Nonself Discrimination in a Computer" was the first attempt to take a specific model of the human immune system and apply it to a computer system. This paper presents a technique for detecting modified files based on a theoretical model of immune-system pattern recognition. The presented negative selection algorithm inspired significant follow-on work, as did the general idea of using the immune system as a template for novel computer defenses. It is foundational both because of the work it inspired and because it shows how much work remains to be done at the intersection of computer science and immunology.

This paper is part of a long-term collaboration between the co-authors Stephanie Forrest (a computer scientist) and Alan Perelson (an immunologist) that continues to this day. The bulk of this work has focused on modeling the human immune system, particularly on the functioning of immune-system memory. Unlike many other approaches that use differential equations to model immune-system dynamics, their work explores the dynamics of bitstrings (strings of 1s and 0s) and bitstring matching. Bitstrings are used to represent antigen/antibody interactions, with closer matches between them corresponding to stronger recognition of potential infectious agents. Bitstring matching had been previously used by Pentti Kanerva to model the associative memory of the human brain; the immune system can also be seen as a kind of associative memory, and so its dynamics can also be modeled using similar techniques (Smith, Forrest, and Perelson 1999).

S. Forrest, A. S. Perelson, L. Allen, and R. Cherukuri, "Self–Nonself Discrimination in a Computer," *Proceedings of 1994 IEEE Computer Society Symposium on Research in Security and Privacy*, 202–212 (1994).

Reprinted with permission from IEEE.

> **HASH FUNCTIONS**
> A standard approach to detecting changed data, particularly data stored in a file, is to calculate and compare hashes of their contents. A hash function takes an arbitrary-sized input and produces a fixed-sized (bounded) output. So if we are wondering if the contents of file A and file B are identical, we can calculate the hashes of A and B separately, HA and HB, and then if HA = HB, we can say that the contents of A and B are identical.
>
> (Note that the paper refers to checksums and message-digest algorithms—these are all types of hash functions.)

Rather than apply computer science concepts to understand the immune system, this paper explores the opposite question: namely, whether immunological concepts can be used in a computer science context effectively. Biology had already supplied metaphors for malicious code at this time. The Morris worm had spread throughout the Internet in 1988 and computer viruses (propagated using removable media) were commonplace on PCs. By 1994 the antivirus industry had already been established; the techniques antivirus engines used, however, bore little resemblance to how the human body defends against biological viruses. A natural question was whether it was possible to build a "computer immune system," a computer defense system that drew inspiration from how the human immune system worked. Contemporaneously Kephart (1994) and others were proposing to build computer immune systems; while such work used immune-system terminology, their architectures and algorithms were not based in then-current understandings of how the human immune system functioned.

Rather than a proposal for a computer immune system, this paper presented something very specific yet extremely versatile—an algorithm for detecting changes in data. Hash functions are the conventional approach to detecting changed files (see sidebar). We can use a hash function to detect malicious files either by maintaining a list of known malicious file hashes or a list of known good file hashes. While both of these strategies are in current use, both have significant limitations. With the first, we need to compare every local file to the set of all known malicious files. Lists of known malware are enormous and grow larger every day. With the known good file case, cryptographic hash functions impose very rigid conditions: even a one-bit change will cause the match to fail. The negative-selection algorithm, the key contribution of this paper, allows changes to be detected in a set of known good files without the brittleness of per-file cryptographic hashes while allowing for previously unseen malicious files to be identified. The ways it does so is directly inspired by antibody–antigen recognition in the immune system.

The bulk of the paper describes the algorithm in detail, so we won't repeat that description here. What we should point out, however, are its key properties, as outlined in the introduction: distributed detection, a tunable quality/cost tradeoff, detector/protected data symmetry, and expensive setup but inexpensive detection. Although other systems individually have aspects of each of these properties, this particular combination was novel because they all derive from how changes are detected. Rather than creating a fixed-length hash, a set of approximate patterns is generated for the information to be protected such that none of the patterns matches it. Later, if any of these patterns generates a match, we instantly know that something in the data has changed.

These "negative" detection patterns can be distributed because they can each operate independently—if any match, a change has occurred. They are tunable because the specificity and number of the patterns can be adjusted at initialization time. They are mutually protective because there is a symmetry between the size of data being checked and the size of the pattern: they are essentially the same size. If corruption happens in a detector or in the data, the amount of perturbed data is roughly the same. And the set of patterns is expensive to generate because all data must be checked against each pattern to ensure there are no matches; in contrast, finding a change later only requires a completely local comparison between a detector and a given portion of the protected data.

This paper was influential in three key ways. The first was its demonstration that computational models of the immune system can be used to create novel algorithms. Other immune system models such as idiotypic network theory, clonal selection of B-cells, and the danger model of the immune system have each served as templates, inspiring algorithms applied to optimization problems, computer defenses, and other applications. These, along with applications of the negative selection algorithm, form the field of computer immune systems (Dasgupta 1999).

The second key influence was its presentation of the negative selection algorithm. In the computer immune system literature, negative selection is largely used as a distributed change detection mechanism. It has also gained attention as an algorithm and approach in its own right, minus its biological inspiration. Negative selection is a remarkable way of representing data, in that you only know what is

---

HASH FUNCTIONS (cont.)
While we could also determine whether two files are identical by comparing them bit by bit, hash-based comparisons show their advantages when we want to do one-to-many or many-to-many comparisons. Hashes also allow large files to be compared on different systems without requiring large file transfers: hashes are computed on each system separately and then the hashes are exchanged. Hashed-based file comparisons do have one key limitation, namely that they are probabilistic. Because the output of hash functions is bounded but the input is not, there are always multiple inputs that can produce the same output. Longer hashes can make collisions arbitrarily rare, however, and the use of cryptographic hashes, such as SHA-256, can give us strong guarantees that document hashes are functionally unique.

being represented by looking at the whole set; each pattern, individually, tells you relatively little. Any entry in a database of phone numbers tells you somebody's phone number. If the same set of phone numbers was used to generate a set of negative detectors, though, no individual detector would tell you about a phone number; instead, it would only tell you a given pattern is not a phone number. The field of negative databases (Esponda *et al.* 2007) explores the privacy preserving possibilities of such negative representations of data, building upon the negative selection algorithm.

The third key contribution of this paper was the idea of self versus nonself as applied to computer systems, particularly computer security. Self–nonself discrimination (Perelson and Oster 1979) is the idea that the immune system recognizes the difference between self (the body) and nonself (viruses, bacteria, parasites) and defends the body against nonself. Further, it posits that the adaptive immune system models self much more than nonself—detectors (antibodies) are based on samples of self (bits of the body) rather than nonself (bits of virus, bacteria, and parasites). In 1996, Forrest and collaborators published "A Sense of Self for Unix Processes," a paper that took self–nonself discrimination and applied it to running programs rather than files on disk. This paper (of which I am a co-author) introduced the idea of program-level anomaly detection.

Programs on Unix systems make a sequential series of system calls in order to access files, communicate on the network, and to request access to local resources. This paper showed that anomalous program behavior could be detected by observing changes in short sequences of system calls, much as the human immune system detects unusual cellular activity through the presence of anomalous peptides. This idea came to be known as sequence-based anomaly detection, and its influence was significant enough that it received a "Test of Time" award at the 2020 IEEE Symposium on Security and Privacy. Note, however, that this paper did not make use of negative selection; the inspiration is instead in addressing a problem analogous to that faced by the immune system.

While this foundational paper has influenced many researchers, immune-inspired algorithms are not widely used in practice. This is in contrast with other biologically inspired algorithms, such as genetic algorithms and neural networks, that have seen significant adoption and considerable success. Adoption of any idea or technology, however, depends on many factors. Immune-inspired algorithms tend to work well in highly distributed, decentralized, autonomous systems. To date, even our most distributed systems are centralized in many ways, and autonomy is something that is more promised than actually delivered. As we increasingly attempt to build such systems, it is likely that others will continue to find inspiration in this work and the numerous works that it inspired.

REFERENCES

D'haeseleer, P., S. Forrest, and P. Helman. 1996. "An Immunological Approach to Change Detection: Algorithms, Analysis, and Implications." In *Proceedings 1996 IEEE Symposium on Security and Privacy*, 110–119. https://doi.org/10.1109/SECPRI.1996.502674.

Dasgupta, D., ed. 1999. *Artificial Immune Systems and Their Applications*. Berlin, Germany: Springer.

Esponda, F., E. S. Ackley, P. Helman, H. Jia, and S. Forrest. 2007. "Protecting Data Privacy Through Hard-to-Reverse Negative Databases." *International Journal of Information Security* 6:403–15. https://doi.org/10.1007/s10207-007-0030-1.

Forrest, S., S. Hofmeyr, and A. Somayaji. 2008. "The Evolution of System-Call Monitoring." In *2008 Annual Computer Security Applications Conference (ACSAC)*, 418–430. https://doi.org/10.1109/ACSAC.2008.54.

Forrest, S., S. A. Hofmeyr, A. Somayaji, and T. A. Longstaff. 1996. "A Sense of Self for Unix Processes." In *Proceedings 1996 IEEE Symposium on Security and Privacy*, 120–128. https://doi.org/10.1109/SECPRI.1996.502675.

Forrest, S., A. Somayaji, and D.H. Ackley. 1997. "Building Diverse Computer Systems." In *Proceedings. The Sixth Workshop on Hot Topics in Operating Systems (Cat. No.97TB100133)*, 67–72. https://doi.org/10.1109/HOTOS.1997.595185.

Kephart, Jeffrey O. 1994. "A Biologically Inspired Immune System for Computers." In *Artificial Life IV: Proceedings of the Fourth International Workshop on the Synthesis and Simulation of Living Systems*. MIT Press. https://doi.org/10.7551/mitpress/1428.003.0017.

Luo, W., Y. Hu, H. Jiang, and J. Wang. 2019. "Authentication by Encrypted Negative Password." *IEEE Transactions on Information Forensics and Security* 14 (1): 114–128. https://doi.org/10.1109/TIFS.2018.2844854.

Perelson, A. S., and G. F. Oster. 1979. "Theoretical Studies of Clonal Selection: Minimal Antibody Repertoire Size and Reliability of Self-non-self Discrimination." *Journal of Theoretical Biology* 4 (81): 645–670. https://doi.org/10.1016/0022-5193(79)90275-3.

Smith, D. J., S. Forrest, and A. S. Perelson. 1999. "Immunological Memory is Associative." In *Artificial Immune Systems and Their Applications*, edited by D. Dasgupta, 105–114. Springer.

SELF-NONSELF DISCRIMINATION IN A COMPUTER

Stephanie Forrest, University of New Mexico,
Alan S. Perelson,
Lawrence Allen, University of New Mexico, and
Rajesh Cherukuri, University of New Mexico

Abstract

The problem of protecting computer systems can be viewed generally as the problem of learning to distinguish *self* from *other*. We describe a method for change detection which is based on the generation of T cells in the immune system. Mathematical analysis reveals computational costs of the system, and preliminary experiments illustrate how the method might be applied to the problem of computer viruses.

1 Introduction

The problem of ensuring the security of computer systems includes such activities as detecting unauthorized use of computer facilities, guaranteeing the integrity of data files, and preventing the spread of computer viruses. In this paper, we view these protection problems as instances of the more general problem of distinguishing *self* (legitimate users, corrupted data, etc.) from *other* (unauthorized users, viruses, etc.). We introduce a change-detection algorithm that is based on the way that natural immune systems distinguish self from other. Mathematical analysis of the expected behavior of the algorithm allows us to predict the conditions under which it is likely to perform reasonably. Based on this analysis, we also report preliminary results illustrating the feasibility of the approach on the problem of detecting computer viruses (demonstrating that the algorithm can be practically applied remains an open problem), and finally, we suggest that the general principles can be readily applied to other computer security problems.

Self vs. nonself is directly taken from theoretical immunology and is here applied to computer security for the first time.

Practical application of the negative selection algorithm still remains an open problem.

Forrest et al. (1994)

Current commercial virus detectors are based on three distinct technologies: activity monitors, signature scanners, and file authentication programs. The system that we describe is essentially a file authentication method, or change detector. Although our initial testing has been in a virus detection setting, the algorithm may be more applicable to other change-detection problems. There are several significant differences between the algorithm described here and more conventional approaches to change detection, such as checksums and message-digest algorithms: (1) the checking activity can be distributed over many sites with each site having a unique signature, (2) the quality of the check can be traded off against the cost of performing check, (3) protection is symmetric in the sense that the change detector and protected data set are mutually protective, and (4) the algorithm for generating the change detectors is computationally expensive, although checking is cheap, so it would be difficult to modify a protected file and then alter the detectors in such a way that the modification could not be detected. As with other authentication methods, our method relies on the guarantee that the data to be protected are uncorrupted at the time that the detectors are generated.

These four points really capture what is different about the negative selection algorithm. Standard approaches to problems in computer science and computer security, however, don't really need these characteristics. Instead, these are requirements that are essential in natural adaptive immune systems.

There are several change-detection tools available which employ a variety of change-detection methods and signature functions, e.g., Tripwire (Kim and Spafford 1993). Tools such as Tripwire devote considerable attention to the important problems of administration, portability, and reporting. Our work is properly viewed as an algorithm, comparable in nature to a signature function, which might be incorporated into a tool like Tripwire. As we mentioned above, there are several features which distinguish our algorithm from conventional signature methods, in particular, our "signatures" are expensive to generate (although cheap to check, especially if the checking activity is distributed across multiple sites) and multiple signatures exist for each data set. These distinguishing features have advantages and disadvantages, and it remains to be seen what setting is most appropriate for an algorithm with these features.

Modern anti-malware systems still check for corrupted files using cryptographic hashes. The modern term is "file-integrity monitoring."

Our approach relies on three important principles:

- Each copy of the detection algorithm is unique. Most protection schemes need to protect multiple sites (e.g., multiple copies of software, multiple computers on a network, etc.). In these

Slow to generate hashes are now standard for password authentication in order to slow offline dictionary attacks. Luo (2019) proposed that negative databases be used to make password storage more resistant to offline dictionary attacks.

environments, we believe that any single protection scheme is unlikely to be effective, because once a way is found to avoid detection at one site, then all sites are vulnerable. Our idea is to provide each protected location with a unique set of detectors. This implies that even if one site is compromised, other sites will remain protected.

- Detection is probabilistic. One consequence of using different sets of detectors to protect each entity is that probabilistic detection methods are feasible. This is because an intrusion at one site is unlikely to be successful at multiple sites. By using probabilistic methods our system can achieve high system-wide reliability at relatively low cost (time and space). The price, of course, is a somewhat higher chance of intrusion at any one site.

Probabilistic methods are a mainstay of large-scale distributed detection systems today. Specifically, algorithms such as Bloom filters allow for fast and efficient probabilistic string matching against databases of known patterns. Probabilistic methods were not so common in 1994 as Internet-based systems were not large enough to benefit from them.

- A robust system should detect (probabilistically) any foreign activity rather than looking for specific known patterns of intrusion. Most virus detection programs work by scanning for unique patterns (e.g., digital signatures) that are known at the time the detection software is distributed. This leaves systems vulnerable to attack by novel means. Like other change detectors, our algorithm learns what self is and notices (probabilistically) any deviation from self.

Despite decades of effort, current computer defenses are still vulnerable to novel attacks because they still characterize malicious activity rather than "self."

1.1 SYSTEM OVERVIEW

The algorithm has two phases:

1. Generate a set of detectors. Each detector is a string that does not match any of the protected data (see below for a careful definition of "match"). This "censoring" phase is illustrated in Figure 1.

2. Monitor the protected data by comparing them with the detectors. If a detector is ever activated, a change is known to have occurred (shown in Figure 2).

This might seem to be an unpromising approach. If we view the set of data being protected as a set of strings over a finite alphabet, and a change to that data as any string not in the original set, then we are proposing to generate detectors for (almost) all strings not in the

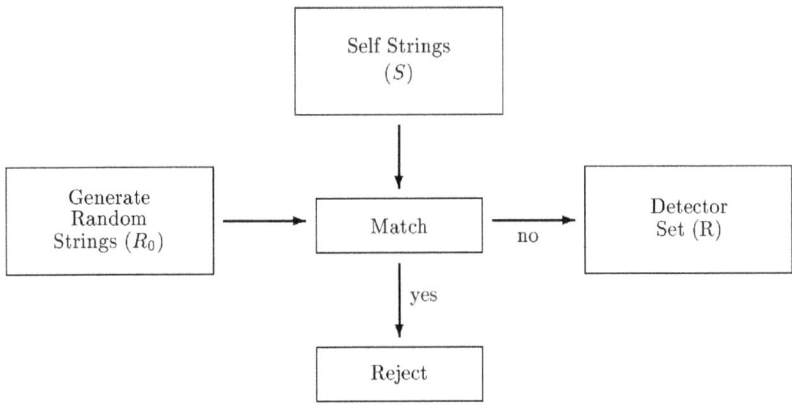

Figure 1. Generation of Valid Detector Set (Censoring).

original data set. Surprisingly, this algorithm turns out to be feasible mathematically—that is, a fairly small set of detector strings has a very high probability of noticing a random change to the original data. Further, the number of detectors can remain constant as the size of the protected data grows. Figures 1 and 2 illustrate how the algorithm works. Each copy of the detection system generates its own unique valid set of detectors once, and then runs the monitoring program regularly (for example, as a background process) to check for changes.

Before describing the procedure in detail, we need to describe what it is that we are trying to detect. We reduce the detection problem to the problem of detecting whether or not a string has been changed, where a change could be a modification to an existing string or a new string being added to self. The algorithm will fail to notice deletions. The string could be a string of bits (and hence, anything that can be represented in a digital computer), a string of assembler instructions, a string of data, etc. However, as will become apparent later, the method appears to be most relevant for strings that do not change over time, that is, the protected strings need to be fairly stable. We define *self* to be the string to be protected, and *other* to be any other string. Note that it will sometimes be convenient to view self as an unordered collection of substrings, and other times as one long string that is the concatenation of the substrings. We use the term "collection" instead of "set" because we do not remove or check for duplicates. Duplicates appear with extremely

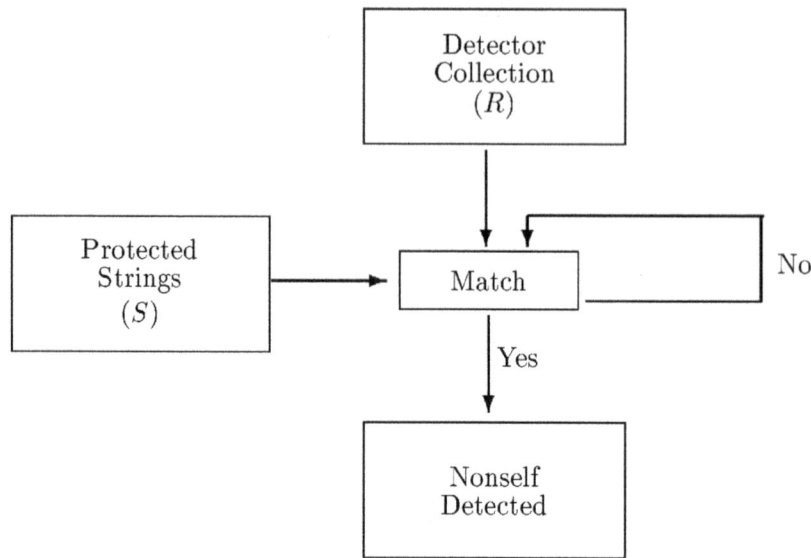

Figure 2. Monitor Protected Strings for Changes.

low frequency, but technically, the collections both of protected strings and detectors are "multisets."

To generate valid detectors, we first split (logically) the self string into equal-size segments. Originally, we chose to split the strings to facilitate our mathematical analysis of the system, which allows us to predict the probability of detection. However, it has turned out to have other advantages, such as making it much easier to detect certain kinds of computer viruses, and suggesting extensions to the system. As an example, we might break the following 32-bit string into eight substrings, each of length four:

0010 1000 1001 0000 0100 0010 1001 0011

This produces the collection S of self (sub)strings to be protected (S contains all of the substrings). The second step is to generate random strings (call this collection R_0), and then match the strings of R_0 against the strings in S. Strings from R_0 that match self (see Section 1.2) are eliminated. Strings that do not match any of the strings in S become members of the detector collection (R), also called the *repertoire*. This procedure is called *censoring*. Continuing the example, suppose R_0 contains the following four random strings: 0111, 1000, 0101, 1001.

Then, R will consist of two strings, 0111 and 0101, the strings 1000 and 1001 being eliminated because they each match a string in S.[1] The censoring procedure is illustrated in Figure 1.

Once a collection R of detector strings has been produced, the state of *self* can be monitored by continually matching strings in S against strings in R. This is achieved by choosing one string from S and one string from R and testing to see if they match. In our implementation, the pairings are made deterministically—each string is chosen for matching in a fixed order. The detectors are checked in the order they were produced. For the self strings, the order is determined, for example, by the order of instructions in the program. Alternatively, the procedure could be randomized. If ever a match is found, then it is concluded that S has changed.

In the example, suppose that one bit of the last self string (0011) is changed to produce 0111. Then, at some point in the monitoring process, it would be noticed that the "self" string (0111) matches one of the detector strings (the string 0111), and a change would be reported.

1.2 MATCHING

A perfect *match* between two strings of equal length means that at each location in the string, the symbols are identical. The example in Section 1 shows perfect matching between strings defined over the alphabet $\{0, 1\}$. Since perfect matching is extremely rare between strings of any reasonable length, a partial matching rule is needed. We relax the matching requirement by using a matching rule that looks for r contiguous matches between symbols in corresponding positions. Thus, for any two strings x and y, we say that *match (x, y)* is true if x and y agree (match) at at least r contiguous locations. See Figure 3 for an example.

The matching rule can be applied to strings defined over any alphabet of symbols. In the most general case, the strings will be over the alphabet $\{0, 1\}$, representing any bit pattern that can be stored in a computer. At a higher level, strings might be defined over a particular

---

[1] In practice, the procedure is to generate random strings sequentially, and to continue generating them until R has a sufficient number of elements. R_0 is useful conceptually for predicting how many strings must be generated to produce a R of a certain size.

X	ABADCBAB
Y	CAG<u>DCB</u>BA

Figure 3. Example Matching Rule. The two strings, x and y defined over the four-letter alphabet $\{A, B, C, D\}$ match at three contiguous locations (underlined). Thus, *match (x, y)* is false for $r = 4$ or greater, since x and y agree at 3 contiguous locations. *match (x, y)* is true for $r = 3$ or less.

machine instruction set. Figure 4 shows an example of censoring with $r = 2$.

It is useful to know the probability P_M that two random strings match at at least r contiguous locations. If:

m = the number of alphabet symbols,

l = the number of symbols in a string (length of the string), and

r = the number of contiguous matches required for a match,

then (Percus, Percus, and Perelson 1993a, 1993b),

$$P_M \approx m^{-r}[(l-r)(m-1)/m + 1].$$

The approximation is only good if $m^{-r} \ll 1$, so we use the exact formula for the cases in which the approximation fails (Uspensky 1937). Table 1 illustrates the effect of varying r and l on P_M for different values of m. The first row shows the configuration we have used in most of our experiments. Setting $r = 8$ corresponds to a one-byte change. The first four rows of the table show the linear increase in P_M as the length of the string (l) increases. Rows one and five show the exponential decrease in P_M as r increases. Finally, the last eight rows show the dramatic effect on P_M of increasing the alphabet size.

2 Probability of Detection

Since detection is probabilistic, we need to make accurate estimates of these probabilities for different configurations of the change-detection system. This section describes how we make our predictions. The following analysis is taken from (De Boer and Perelson 1993).

Suppose that we have some string that we want to protect. As we mentioned before, this string could be an application program, some data, or any other element of a computer system that is stored in memory.

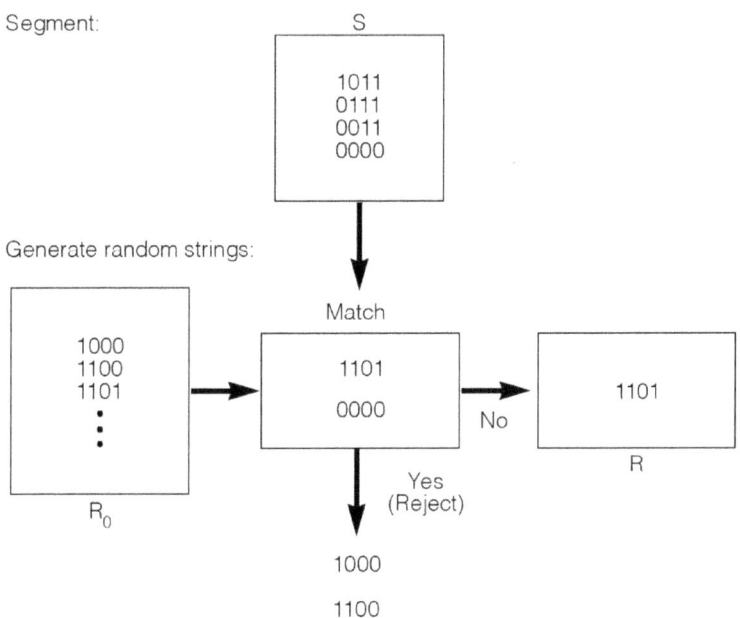

Figure 4. Generating the repertoire. The string to be protected is logically segmented into four equal-length "self" strings (stored in S). To generate the repertoire, random strings are produced in the box labeled R_0 and matched against each of the self strings. The first two strings, 1000 and 1100, are eliminated because they both match self string 0000 at at least two contiguous positions. The string 1101 fails to match any string in self at at least two contiguous positions, so it is accepted into the repertoire (box labeled "R").

m	r	l	P_M
2	8	32	0.0502023
2	8	64	0.108697
2	8	128	0.2151
2	8	256	0.391316
2	16	32	0.000137329
2	16	64	0.000381437
2	16	128	0.000869474
2	16	256	0.00184483
128	8	32	$3.33067 * 10^{-16}$
128	8	64	$7.77156 * 10^{-16}$
128	8	128	$1.66533 * 10^{-15}$
128	8	256	$3.44169 * 10^{-15}$
128	16	32	~ 0.0
128	16	64	~ 0.0
128	16	128	~ 0.0
128	16	256	~ 0.0

Table 1. Example values of P_M for varying values of m (alphabet size), r (number of contiguous matches required for a *match*), and l (string length).

Using the algorithm described in Section 1.1, we would like to estimate the number and size of detector strings that will be required to ensure that an arbitrary change to the protected string is detected with some fixed probability.

We make the following definitions and calculations:

N_{R_0} = The number of initial detector strings (before censoring).

N_R = The number of detector strings after censoring (size of the repertoire).

N_S = The number of self strings.

P_M = The probability of a match between 2 random strings.

f = The probability of a random string not matching any of the N_S self strings.

  = $(1 - P_M)^{N_S}$.

P_f = The probability that N_R detectors fail to detect an intrusion.

If P_M is small and N_S is large, then

$$f \approx e^{-P_M N_S}$$

and,

$$N_R = N_{R_0} \times f \tag{1}$$

$$P_f = (1 - P_M)^{N_R}. \tag{2}$$

$$\tag{3}$$

If P_M is small and N_R is large, then

$$P_f \approx e^{-P_M N_R}. \tag{4}$$

Thus,

$$N_R = N_{R_0} \times f = \frac{-\ln P_f}{P_M}.$$

Solving ?? and ?? for N_{R_0}, we get the following:

$$N_{R_0} = \frac{-\ln P_f}{P_M \times (1 - P_M)^{N_S}}. \tag{5}$$

This formula allows us to predict the number of initial strings (N_{R_0}) that will be required to detect a random change, as a function of the probability of detection $(1 - P_f)$, the number of self strings being protected (N_S), and the matching rule P_M. N_{R_0} is minimized by choosing a matching rule such that

$$P_M \approx \frac{1}{N_S}.$$

The foregoing analysis allows us to estimate the computational costs of the algorithm in the following way. First, we observe that the method relies on two basic operations: (1) generating a random string of fixed length and (2) comparing two strings to see if they meet the matching criterion (more than r contiguous matches). We assume that these operations take constant time. Then, the time complexity of Phase I will be proportional both to the number of strings in R_0 (i.e., N_{R_0}) and the number of strings in S (i.e., N_S). Equation 5 estimates N_{R_0} based on the size of the protected data set (N_S), the reliability of detection that is required (P_f), and the particular matching rule that is being used (P_M). The cost of complete checking in Phase II will be proportional to the number of strings in R (i.e., N_R) and the number of strings in S (i.e., N_S).

Based on the above analysis, we can make several observations about the algorithm:

1. It is tunable: we can choose a desired probability of detection (P_f), and then estimate the number of detector strings required as a

function of the size of N_S (the strings to be protected) by using Equations 2 and 80.

Since an increased probability of detection results in increased computational expense (due to the increased size of R_0 and R), one can choose a desired probability of detection by determining (a) how fatal a single intrusion would be, and (b) how much redundancy exists in the system (see Item 5 below).

2. N_R is independent of N_S for fixed P_M and P_f (Equation 2). That is, the size of the detector set does not necessarily grow with the number of strings being protected. This implies that it is possible to protect very large data sets efficiently.

3. If N_R, P_f, and P_M are fixed, then N_{R_0} grows exponentially with N_S. This exponential factor is unfortunate in one respect, but it does imply that once a set of detectors has been produced (say, using a supercomputer) that it would be virtually impossible for a malicious agent to change self and then change the detector set so that the change was unnoticed. N_{R_0} can be controlled by choosing $P_M = \frac{1}{N_s}$. Figure 5 illustrates the dramatic effect of minimizing N_{R_0} for each different value of N_S. We are currently studying several promising methods for reducing the size of N_{R_0} which do not rely on changing the matching rule for each new N_S.

Faster (linear time) detector generation methods were later described in D'haeseleer, Forrest, and Helman in 1996.

4. The probability of detection increases exponentially with the number of independent detection algorithms.

If N_t = the number of copies of the algorithm, then

$$P_{\text{system fails to detect}} = (P_f)^{N_t}.$$

We still do not have symmetric detection mechanisms, i.e., defenses that use their own protection mechanisms to protect themselves effectively, at least beyond simple methods such as file integrity protection. This is an ongoing problem as anti-malware and other computer defenses have themselves been targeted by attackers on numerous occasions.

This feature is the primary advantage of the algorithm. For example, Table 2 shows that with one copy of the detection algorithm $(N_t = 1)$, 46 detectors can protect a data set (of any size) with 90.6% reliability. With only ten different sites $(N_t = 10)$, the same system-level reliability can be obtained with less than four detectors per site.

5. Detection is symmetric. Changes to the detector set are detected by the same matching process that notices changes to self. This

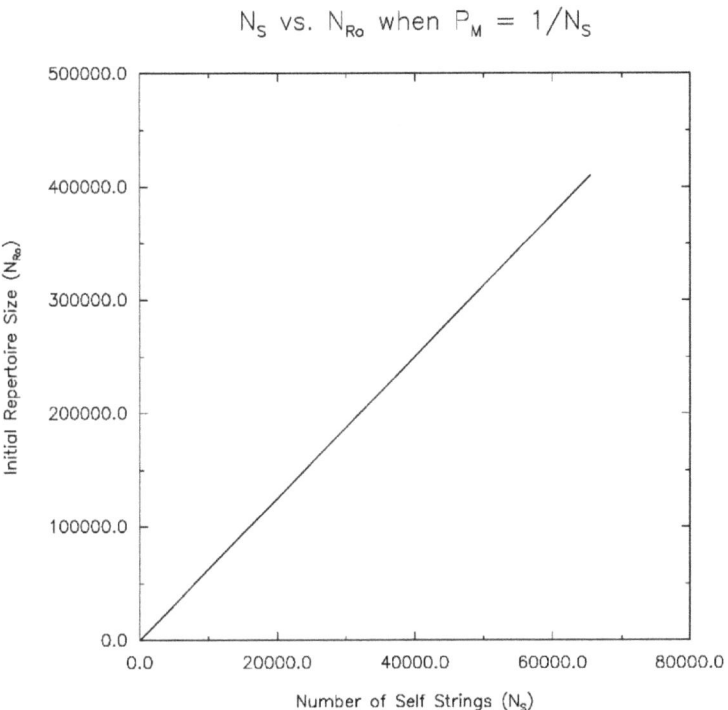

Figure 5. N_S vs. N_{R_0} when $P_M = \frac{1}{N_s}$. X-axis shows the number of self strings (N_S) and the Y-axis predicts the initial repertoire size (N_{R_0}) using the formula $N_{R_0} = \frac{-\ln P_f}{P_M \times (1-P_M)^{N_S}}$.

implies that when a change is detected there is no *a priori* way to decide if the change was to self or to the detectors. The advantage is that self confers the same protection to the detector set that the detector set provides to self.

3 Experiments

Based on the above analysis, it is possible to design a wide variety of detection systems, each with different properties. In this section, we report some preliminary results based on our investigations of different parameter settings. We report three classes of experiments: experiments using random binary strings, experiments on SPARC instructions generated by compiling C programs, and in the DOS environment, experiments on COM files infected with actual computer viruses.

The first set of experiments show some of the implications of Equation 5 and confirms the estimates provided by the theoretical analysis. The remaining two sets of experiments illustrate how the method might be applied to the problem of detecting computer viruses.

Table 2 compares the theoretical and experimental probabilities (P_f) that a fixed number (N_R) of detectors will fail to detect a random change to self. It also compares the theoretical and experimental values for the initial repertoire size (N_{R_0}), thus providing an estimate of how costly Phase I of the algorithm is. The repertoire size (N_R) is set to 46 (i.e., 46 detectors, each consisting of 32 bits) and the target failure rate (P_f) is set to 0.1. The experimental procedure was as follows:

1. Fix P_f to 0.1.

2. Compute P_M using $m = 2, l = 32, r = 8$. Setting $r = 8$ corresponds to a matching rule that notices 1-byte changes.

3. Compute N_R based on P_f and P_M (using Equation 2) and round to next largest integer.

4. Repeat the following 1000 times:

    a) Generate N_S random binary strings ($l = 32$).

    b) Determine N_{R_0} experimentally by generating random strings until N_R valid detectors are found.

    c) Test the detectors:

        i. Replace a string in self with one random string.

        ii. Compare the detector strings with the modified self strings.

        iii. If any of the detectors matches the new string (using the partial matching rule described earlier), report a modification.

5. Compare the mean P_f obtained over the 1000 trials with the P_f of (1).

Using this procedure, we obtained close agreement between the theoretical predictions and the observed results, as shown in Table 2. This experiment establishes a worst-case baseline for the algorithm. It is "worst case" in the sense that there is only one set of detectors (no advantage from distributing the detection task), the self strings are generated randomly, and changes to self consist of replacing a single string (changing at most 32 bits). For example, 128 self strings can be protected by a repertoire consisting of 46 detectors. These detectors detect one random change to self 84.3% of the time. Additionally, we can see that the exponential cost of generating the detector set is already significant (34, 915), even for 128 self strings (a modest amount of data to protect). However, if the detection task were distributed over 100 sites, then each site would need to generate only one valid detector, would use an initial repertoire of 269 and would achieve at least a 98% detection rate.

These and other similar experiments indicate that there is good agreement between experimental and predicted values for P_f. The desired P_f can be achieved either by fixing N_{R_0} or N_R.

We also conducted several tests using C programs compiled for a SPARC processor. These experiments are interesting because they illustrate the effect of using a larger alphabet. The tests differed in the method used to generate the infected program. In all of the tests a C source code file was first compiled. The resulting object file was then disassembled into SPARC instructions which were each mapped to a single ASCII character. This produced one long string in which each symbol represented a single op-code. The string was then split into substrings, each 32 ASCII symbols long, representing the collection S of self strings. Thus, each string was defined over an alphabet of size 104 ($m = 104$) and of length 32 ($l = 32$).[2]

Next, we constructed the collections R_0 (implicitly) and R (explicitly), defined over the same alphabet and also 32 symbols long. The detectors were selected by randomly generating strings, and comparing them to the program strings. If more than the specified number (r) of contiguous characters in the same positions matched those in a program string, the detector was rejected. The generate and match procedure continued until the specified number of detectors was generated.

Note that the detectors were generated to detect sequences of instructions rather than the raw bytes of program files. Thus the ability of this scheme to account for small variations in programs is only partially attributable to the negative selection algorithm; the rest is due to the representation used. As is known in the fields of machine learning and genetic algorithms, representation can have a huge impact on performance.

---

[2] The full SPARC instruction set is larger than 104, but we collapsed some variants into one symbol, e.g., different versions of certain floating point operations were treated uniformly.

N_S	EXPERIMENTAL N_{R_0}	THEORY N_{R_0}	EXPERIMENTAL P_f
8	69(6.06)	69	0.085(0.009)
16	105(11.99)	105	0.104(0.010)
24	156(20.09)	158	0.110(0.011)
32	240(32.49)	239	0.099(0.009)
40	360(53.49)	361	0.107(0.010)
48	549(82.98)	545	0.133(0.011)
56	829(133.06)	823	0.109(0.010)
64	1253(218.77)	1243	0.124(0.010)
72	1876(318.55)	1876	0.112(0.010)
80	2872(495.61)	2833	0.116(0.010)
88	4327(781.78)	4277	0.130(0.011)
96	6618(1343.56)	6458	0.130(0.011)
104	9903(2082.86)	9750	0.135(0.011)
112	15074(3140.29)	14722	0.124(0.010)
120	22878(5283.80)	22228	0.154(0.011)
128	34915(8513.26)	33561	0.157(0.012)

Table 2. Theoretical and Experimental P_f with fixed N_R. Numbers reported are the mean of 1000 trials. Numbers in parentheses are standard deviations. N_R is set to 46 and the theoretical P_f is 0.094 for all entries. The theoretical N_{R_0} is calculated using the formula $N_{R_0} = \frac{N_R}{(1-P_M)^{N_S}}$. Theoretical P_f is calculated using the formula $P_f = (1 - P_M)^{N_R}$ with $P_M = 0.0502$. It differs from 0.1 because N_R has been rounded in Step 3 of the algorithm.

To test for a modified file, a new file was constructed and the detectors were then compared to it. If any of the detectors matched the strings in the new file by more than the specified allowed maximum number of matches, a modification was reported. We used several methods to modify the program file: changing the source code and recompiling (say, to add a loop), changing a single character in the protected file (the minimal change possible), and changing 24 characters at the end of the code segment of the source file. These methods are labeled "Loops," "Single Mutation," and "Data Segment" respectively in Table 3. The last method is intermediate in difficulty and is probably the most realistic from a virus-detection viewpoint. It was the one used to generate the example shown in Table 5.

The first series of tests involved inserting a short loop into the source code of a program and recompiling the code (called Loop in the table). Although the Data Segment change inserts only 24 additional instructions into the compiled file, the insertion shifts the order of the instructions from the point of insertion forward. This has the effect of altering all of the

METHOD OF INFECTION	NUM. OF DETECTORS	P_F	$(P_F)^5$
LOOP	2	0.26	0.001
SINGLE MUTATION	50	0.62	0.092
SINGLE MUTATION	100	0.24	7.96e − 4
DATA SEGMENT	8	0.18	1.89e − 4
DATA SEGMENT	16	0.06	7.776e − 7
DATA SEGMENT	32	0.00	0.00

Table 3. System performance on high-cardinality alphabet (experimental results). Each experiment was run with 37 self strings, each of length 32 characters, and with $r = 1$. Each reported number is based on multiple repetitions of the detection system using a different R_0, but leaving the modification constant. The Loop and Single Mutation experiments were repeated 50 times, Data Segment experiments were repeated 100 times.

r	N_R	N_{Ro}	P_f
9	2	1435(1150.53)	0.270(0.044)
9	5	3229(1104.72)	0.111(0.074)
9	8	5910(1864.25)	0.010(0.010)
9	10	7274(2580.88)	0.000(0.000)
10	5	182(71.67)	0.150(0.036)
10	8	315(126.09)	0.040(0.020)
10	10	382(111.66)	0.020(0.014)
10	15	598(161.29)	0.020(0.014)
10	25	996(211.11)	0.000(0.000)
13	25	54 (7.18)	0.140(0.035)
13	50	86 (8.36)	0.110(0.031)
13	100	170 (12.35)	0.010(0.010)
13	125	205 (0.00)	0.000(0.000)

Table 4. Probability of failing to detect modification to more.com when infected by timid virus. $N_S = 655$. String Length $= 32$. r is match threshold. N_R is the number of detectors. N_{Ro} is the Initial repertoire size. N_M is the number of non-self strings. $N_M \approx 76$ P_f is the observed probability of failing to detect the virus.

character strings following the new code. For this type of change, tests run with only two detectors constructed with $r = 1$ (one matching symbol), could detect a change 74% of the time. We conclude that detecting viruses of this form is easy for our algorithm.

It may seem surprising that the algorithm works with $r = 1$. In fact, there are two constraints on the matching process: (1) the detector must have the correct character (out of a possible 104), and (2) the character must be in the correct position. It turns out that these constraints are restrictive enough that we do not need to require any contiguous matching.

The second series of tests created an infected file by changing only one character in the original file (labeled "Single Mutation" in the table). This represents the opposite extreme from the first case. In these tests, we performed two types of experiments, one with 100 detectors and one with 50. Each detector was 32 characters long. Again, r was set to one. In the test runs with 50 detectors, the modification was detected 38% of the time. With 100 detectors the modification was detected 76% of the time. Table 3 shows results from all three types of experiments. The rightmost column, labeled $(P_F)^5$, shows the dramatic improvement that is obtained if only five copies of the detection algorithm are present (i.e., five independent sites).

A final set of experiments tested a simple file-infector virus. As an example, consider the TIMID virus, described in (Ludwig 1991). This virus modifies the first five bytes of a COM file and appends 300 bytes of code to the end of the file. Viruses such as these turn out to be extremely easy for our algorithm to detect, for the same reason as the data segment change in the SPARC test. The testing method was as follows:

1. Generate detectors for a standard .com file, supplied with DOS 5.0.

2. Copy the virus into a directory containing the .com file.

3. Execute the virus, causing the original .com file to be infected.

4. Test modified .com file with detectors to see if a modification is detected.

Tests were conducted on three different files, more.com, loadfix.com (not shown), and edit.com (not shown). Table 4 shows the results for more.com, a file with 655 binary self strings, each of length 32. Values shown are the average of 100 trials (numbers in parentheses are the standard deviations). Considering the first four lines of the table, 73% reliability can be attained with only two detectors, and essentially 100% reliability is attained with ten detectors.

Finally, in order to compare the results from each of the preceding experiments, Table 5 displays typical results from each of the preceding experiments.

The most notable observation about the data in this table is that the algorithm performs much better in practice than in theory. The most obvious explanation for this discrepancy is that real programs are

PARAMS	BINARY STRINGS	SPARC ($m = 104$)	COM ($m = 2$)	TIMID ($m = 2$)
l	32	32	32	32
r	8	1	8	9
N_S	128	37	128	655
P_f	~ 0.0	~ 0.0	~ 0.0	~ 0.0
N_{R_0}	24081	68	1861	6576
N_R	46	8	25	10

Table 5. Typical experimental results on randomly generated binary strings (column 1), strings of SPARC instructions generated from a C program (column 2), .com files (column 3), and a file infected by the Timid virus (column 4). In each case, m, l, r, and N_S are predetermined. In the binary string and .com cases, P_f was fixed and N_{R_0} and N_R were determined experimentally. In the SPARC and Timid cases, N_R was fixed and P_f observed experimentally.

not collections of completely random strings. For example, our analysis assumed that each symbol of the alphabet occurs with equal probability, and this is certainly not the case with the SPARC instruction set. It is fairly straightforward to modify the analysis to account for different occurrence frequencies of symbols. However, there are other ways in which actual programs might deviate from random strings. For example, certain sequences of symbols may occur with some regularity, perhaps related to the particular compiler that was used. We have not yet investigated what these patterns are or how to extend the theory to account for them. A second way in which our experiments deviate from the analysis is in the method used to generate a modified string. The theory assumes that only one random string is added to self, possibly replacing an existing string, but the viruses and modifications we reported (except for single mutation in the SPARC case) all involve larger changes to self.

4 Discussion

The algorithm we have just presented takes its inspiration from the generation of T cells in the immune system. The immune system is capable of recognizing virtually any foreign cell or molecule. To do this, it must distinguish the body's own cells and molecules which are created and circulated internally (estimated to consist of on the order of 10^5 different proteins) from those that are foreign. T cells have receptors on their surface that can detect foreign proteins (called antigens). These receptors are made by a pseudo-random genetic process, and it is highly likely that some receptors will detect self molecules. T cells undergo a censoring process in the thymus, called negative selection, in which T cells that recognize self proteins are destroyed and not allowed to leave the thymus.[3] T cells that do not bind self peptides leave the thymus, and provide the basis for our immune protection against foreign antigens. Our artificial immune system works on similar principles, generating detectors randomly, and eliminating (censoring) the ones that detect self. We refer to the detectors as *antibodies*, even though our model was inspired more by the deletion of self reactive T cells than by the deletion of antibodies.

Note here how the presented algorithm was directly inspired by specific models of the human immune system.

The algorithm presented here is related to earlier immune-system models based on a universe in which antigens (foreign material) and antibodies (the cells that perform the recognition) are represented by binary strings (Farmer, Packard, and Perelson 1986; Stadnyk 1992; Forrest and Perelson 1991; Forrest *et al.* 1993). The complex chemistry of antibody/antigen recognition is highly simplified in these binary immune systems, being modeled as string matching. These binary models have been used to study several different aspects of the immune system, including its ability to detect common patterns in noisy environments (Forrest *et al.* 1993), its ability to discover and maintain coverage of diverse pattern classes (Smith, Forrest, and Perelson 1993), and its ability to learn effectively, even when not all antibodies are expressed and not all antigens are presented (Hightower, Forrest, and Perelson 1993). In the current algorithm, we logically split the self string into equal-size segments to generate valid antibodies (detectors), providing a collection of strings analogous to internal cells and molecules in the body.

The distributed nature of the algorithm was also inspired by the immune system in which each individual generates its own unique set of protective antibodies. This analogy is reflected in the change-detection algorithm because each copy of the detection system generates its own unique valid set of detectors.

5 Conclusions and Future Directions

We have described a general method for distinguishing self from other in the context of computational systems, and we have illustrated

---

[3] Just as our algorithm splits up a self string into smaller substrings, proteins are broken up into smaller subunits, called peptides, before recognition by T cells.

its feasibility as a change-detection method on the problem of computer virus detection. The major limitation appears to be the computational difficulty of generating the initial repertoire. Although this is potentially an advantage in that it protects the antibodies from being modified to conform to a modified form of self, we are currently investigating several possible ways to reduce this complexity. Our current investigations in this direction are based on ideas from immunology, although we believe that it may also be possible to apply more conventional algorithms, especially to take account of regularities in self. It should be noted, however, that any nonrandom method of generating detectors is likely to produce some regularities in the detector set. These regularities might be exploited by a malicious agent, thus compromising the security of the system.

One way to defeat our algorithm would be to design a virus that was composed from a subset of self (presumably in a different order). That is, if one could design a virus that used the same logical segments of the program as our method uses for its checking, we would be unable to detect it. Although we believe that this would be very difficult to do in practice, a slight modification to our method protects against this vulnerability. By simply choosing a different segment length (l) for each site, the number of common substrings available for the virus quickly diminishes. Interestingly, natural immune systems use a similar strategy. Proteins are broken up into a large pseudorandom collection of peptides. Major histocompatibility complex (MHC) molecules bind a subset of these peptides and present them to T cells for recognition. The genes that code for the MHC molecules are highly polymorphic and thus each individual may present a different set of peptides for recognition.

Here the authors describe what later came to be known as "mimicry attacks." Mimicry attacks were developed to counter sequence-based anomaly detection. For a history of this, see Forrest (2008).

The details of our partial matching rule and the segmentation of self into equal-size segments were arbitrary decisions. We designed our system with these features in order to simplify the mathematical analysis of its behavior. Although a matching rule based on contiguous matching regions makes sense immunologically, there may well be more appropriate rules for a computational environment. An important area of future research is to investigate other matching rules and to revisit the decision to partition self into equal-size segments. To date, we have only studied how the method can be applied to computer virus detection. However, we suspect that it is also applicable to a wide variety of network and operating system problems, and this is an area which we are currently investigating.

The work that followed this was "A Sense of Self for Unix Processes," in 1996.

Finally, our approach unifies a wide variety of computer and data security problems by treating them as the problem of distinguishing self from other. Negative selection is only one of many mechanisms that the immune system has evolved to distinguish self from other. We are interested in discovering other information-processing methods used by the immune system that can be translated into useful algorithms. ☙

Many researchers heeded this call, leading to the field of computer immune systems.

6 Acknowledgments

The authors gratefully acknowledge the Santa Fe Institute for encouraging and supporting the interdisciplinary research that produced the results reported here. Support was provided to Forrest by the National Science Foundation (grant IRI-9157644). David Mathews helped prepare the figures and John McHugh made many helpful suggestions about the manuscript.

REFERENCES

De Boer, R. J., and A. S. Perelson. 1993. "How Diverse Should the Immune System Be?" *Proceedings of the Royal Society of London B* 252.

Farmer, J. D., N. H. Packard, and A. S. Perelson. 1986. "The Immune System, Adaptation, and Machine Learning." In *Evolution Games and Learning*, vol. 22. Reprinted from Physica, 22D, 187-204. Amsterdam, Netherlands: North-Holland.

Forrest, S., B. Javornik, R. Smith, and A. S. Perelson. 1993. "Using Genetic Algorithms to Explore Pattern Recognition in the Immune System." *Evolutionary Computation* 1 (3): 191–211.

Forrest, S., and A. S. Perelson. 1991. "Genetic Algorithms and the Immune System." In *Parallel Problem Solving from Nature*. Berlin, Germany: Springer-Verlag.

Hightower, R., S. Forrest, and A. S. Perelson. 1993. *The Evolution of Secondary Organization in Immune System Gene Libraries*. Technical report 93-01-002. Santa Fe, NM: Santa Fe Institute.

Kim, G. H., and E. H. Spafford. 1993. *The Design and Implementation of Tripwire: A File System Integrity Checker*. Technical report. Lafayette, IN: Purdue University.

Ludwig, M. 1991. *The Little Black Book of Computer Viruses*. Show Low, AZ: American Eagle Publishing, Inc.

Percus, J. K., O. E. Percus, and A. S. Perelson. 1993a. "Predicting the Size of the Antibody Combining Region from Consideration of Efficient Self/Non-Self Discrimination." *Proceedings of the National Academy of Sciences* 90:1691–1695.

———. 1993b. "Probability of Self-nonself Discrimination." In *Theoretical and Experimental Insights into Immunology*. New York, NY: Springer-Verlag.

Smith, R., S. Forrest, and A. S. Perelson. 1993. "Searching for Diverse Cooperative Populations with Genetic Algorithms." *Evolutionary Computation* 1 (2).

Stadnyk, I. 1992. "Schema Recombination in Pattern Recognition Problems." In *Proceedings of the Second International Conference on Genetic Algorithms*, 27–35.

Uspensky, J. V. 1937. *Introduction to Mathematical Probability*. New York, NY: McGraw-Hill Book Co.

PETER K. SCHUSTER
1940 — March 7, 1941
Vienna, Austria

WALTER FONTANA
November 3, 1960
Meran/Merano, Italy

PETER FLORIAN STADLER
December 24, 1965
Vienna, Austria

IVO LUDWIG HOFACKER
November 23, 1964
Evanston, IL

1943: In the occupied Netherlands, Willem Johan Kolff creates the first artificial kidney and engineers the process of hemodialysis.

1950

1962: Walter Gilbert coins the term "RNA World."

1960

1970

1979: Schuster and Manfred Eigen publish "The Hypercycle: A Principle of Natural Self-Organization."

1980

1985: In the UK, Colin Pitchfork becomes the first person in the world to be convicted of murder on the basis of DNA evidence.

1990

SCHUSTER, FONTANA, STADLER, AND HOFACKER 1994
Institute of Theoretical Chemistry, University of Vienna, Vienna, Austria & Santa Fe Institute, Santa Fe, NM

Institute of Molecular Biotechnology, Jena University, Jena, Germany, & Institute of Theoretical Chemistry, University of Vienna, Vienna, Austria & Santa Fe Institute, Santa Fe, NM

2000

2010

Institute of Theoretical Chemistry, University of Vienna, Vienna, Austria

2015: Several lab companies begin producing $1,000 genomes for individuals, fulfilling a 2001 goal for affordable genetic research. As of the publication of this volume, the cost to sequence an individual's own genome can cost as little as $199 USD.

2020

HOFACKER PHOTO COURTESY WIKIMEDIA COMMONS

SCHUSTER, FONTANA, STADLER & HOFACKER

[80]

THE ARCHITECTURE OF GENOTYPE–PHENOTYPE MAPS

Evandro Ferrada, Universidad de Valparaíso

In this 1994 paper, Peter Schuster, Walter Fontana, Peter F. Stadler, and Ivo L. Hofacker reviewed a series of findings arising from a computational model of ribonucleic acid (RNA), published during the 1980s and early '90s. Insights from this model not only supported previous theoretical ideas on the early evolution of life, but also gave rise to a computational framework for the study of the molecular principles of the evolutionary process. Such framework was groundbreaking, combining notions of information and coding theory with ideas from biochemistry, genetics, structural biology, and bioinformatics.

The choice of RNA as a computational model was not serendipitous. Peter Schuster had previously collaborated with Manfred Eigen on the theory of quasi-species and the hypercycle (Eigen and Schuster 1977; Eigen, McCaskill, and Schuster 1988). A central theme of their contributions was the chemistry of prebiotic systems. By the early '70s molecular biology had established the universal role of RNA in the flow of genetic information, as well as uncovered molecular mechanisms for the replication of RNA viruses (Spiegelman 1971). These findings made RNA an interesting model for studying molecular evolution.

A second reason to choose RNA as a model was its relative simplicity compared to other biomolecules. For one, RNA folding is energetically dominated by the formation of secondary structure elements composed of hydrogen bonds between complimentary pairs of nucleotides. Thus, despite having a complex 3D structure, RNA folding is largely encapsulated by the planar pattern, or topology of these structural elements. In addition, functionally diverse RNA molecules are relatively short, often composed of 10 to 100 of monomers. In contrast, other biomolecules such as proteins, fold mainly driven by an entropic thermodynamic component, and reach average lengths of 200 to 300 monomers.

P. Schuster, W. Fontana, P. F. Stadler, and I. L. Hofacker, "From Sequences to Shapes and Back: A Case Study in RNA Secondary Structures," *Proceedings of the Royal Society B: Biological Sciences* 255 (1344), 279–284 (1994).

Used with permission of The Royal Society (UK); permission conveyed through Copyright Clearance Center, Inc.

A third reason that favored RNA was the availability of experimental information on the binding energy between short nucleotide fragments that compose the RNA secondary structure. This information was used to parameterize the computational model.

Lastly, and as outlined in this paper, RNA is the simplest example of a genotype–phenotype (GP) map. The nucleotide sequence is the genotype, while the phenotype can be construed as the arrangement or topology of RNA secondary structure elements, or conformation. Although simple in essence, it was thought (and later confirmed) that lessons learned from the RNA model might apply more generally to other GP maps, such as the sequence-structure map of proteins, or the GP map of gene regulatory networks or even man-made systems.

In retrospect, the model revealed that the evolutionary dynamics of populations of RNA molecules in sequence space was mainly determined by the overall statistical properties, or architecture, of the sequence-structure map. The main contribution of Schuster *et al.* was to identify and outline for the first time, and for a broader audience, the main features of such architecture. First, they showed that the same RNA conformation could be realized by vastly different sequences. In most cases, such sequences could reach each other by consecutive neutral mutations, that is, mutations that have no effect on the sequences' shared conformation. Schuster *et al.* used the term *neutral network* to refer to the collection of such sequences.

The fact that a large proportion of these neutral networks could traverse (i.e., percolate) through the entire space of sequences, was later found to be a common feature of GP maps. In 2001, Reidys and Stadler provided an analytical result for this phenomenon, by showing that the ability of a neutral network to percolate through sequence space depends on the degree of phenotypic neutrality of genotypes (i.e., the average fraction of neutral variants) and on the dimensionality of sequence space, defined by the alphabet size and the length of the sequences.

The existence of neutral networks revealed yet a more fundamental feature of the RNA GP map, namely a *many-to-one* relation between the sets of genotypes to phenotypes. In other words, each conformation can be realized by multiple sequences. This feature of the GP map is a prerequisite for the existence of neutral networks, and the basis of the resilience of phenotypes to mutations, or mutational robustness.

In the case of proteins, the many-to-one feature was first described in 1986 for structures studied through X-ray crystallography (Chothia and Lesk 1986), and had a profound impact on structural biology leading to the first reliable method for protein structure prediction, called comparative modeling (Šali and Blundell 1993). With the help of an exact computational model, Schuster *et al.* went further and characterized this relation for RNA exhaustively, showing that a rank distribution of the frequency of genotypes per phenotype follows a generalized Zipf's law.

Two years later, using a similar computational model, this distribution was also characterized for proteins, which demonstrated that the high frequency of certain protein folds in nature was not an artifact of biased sampling, as it was believed, but the result of the thermodynamics of folding, intrinsic to the architecture of the sequence-structure map of macromolecules (Li *et al.* 1996).

A third outstanding result of the model, so called *shape space covering*, answered the following question: how many mutations would take a sequence to reach the set of most common conformations in phenotype space? The answer turned out to be a deceptively small fraction of 15% of the total sequence length. In other words, only $\sim$ 15 mutations on a sequence of 100 nucleotides would suffice to reach virtually all other possible conformations.

Taken together, these three features revealed by the RNA model provided support to an early theoretical exercise by John Maynard Smith (1962) on the ability of proteins to evolve new functions. Maynard Smith identified neutrality as a fundamental condition for functional innovation. Independent support for this hypothesis had also been put forward using a lattice protein model (Lipman and Wilbur 1991).

Subsequent refinements and explorations of the RNA model inspired several analytical, numerical and experimental results, most significantly those exploring the punctuated nature of the evolutionary process as a phenomenon intrinsic to development (Fontana and Schuster 1998), the population genetic conditions for the evolution of mutational robustness (van Nimwegen, Crutchfield, and Huynen 1999), plastogenetic congruence (Ancel and Fontana 2000), the *in vitro* evolution of RNA and proteins (Schultes and Bartel 2000; Amitai, Gupta, and Tawfik 2007), and the interplay between robustness and evolvability (Wagner 2005).

REFERENCES

Amitai, G., R. D. Gupta, and D. S. Tawfik. 2007. "Latent Evolutionary Potentials under the Neutral Mutational Drift of an Enzyme." *HFSP Journal* 1 (1): 67–78. https://doi.org/10.2976/1.2739115/10.2976/1.

Ancel, L. W., and W. Fontana. 2000. "Plasticity, Evolvability, and Modularity in RNA." *Journal of Experimental Zoology* 288 (3): 242–283. https://doi.org/10.1002/1097-010X(20001015)288:3<242::AID-JEZ5>3.0.CO;2-O.

Chothia, C., and A. M. Lesk. 1986. "The Relation between the Divergence of Sequence and Structure in Proteins." *The EMBO Journal* 5 (4): 823–826. https://doi.org/10.1002/j.1460-2075.1986.tb04288.x.

Eigen, M., J. McCaskill, and P. Schuster. 1988. "Molecular Quasi-Species." *The Journal of Physical Chemistry* 92 (24): 6881–6891.

Eigen, M., and P. Schuster. 1977. "A Principle of Natural Self-Organization: Part A: Emergence of the Hypercycle." *Naturwissenschaften* 64:541–565. https://doi.org/10.1007/BF00450633.

Fontana, W., and P. Schuster. 1998. "Continuity in Evolution: On the Nature of Transitions." *Science* 280 (5368): 1451–1455. https://doi.org/10.1126/science.280.5368.1451.

Li, H., R. Helling, C. Tang, and N. Wingreen. 1996. "Emergence of Preferred Structures in a Simple Model of Protein Folding." *Science* 273 (5275): 666–669. https://doi.org/10.1126/science.273.5275.66.

Lipman, D. J., and W. J. Wilbur. 1991. "Modelling Neutral and Selective Evolution of Protein Folding." *Proceedings of the Royal Society of London B: Biological Sciences* 245 (1312): 7–11. https://doi.org/10.1098/rspb.1991.0081.

Maynard Smith, J. 1962. "The Limitations of Molecular Evolution." In *The Scientist Speculates: An Anthology of Partly Baked Ideas,* edited by I. J. Good, 252–256. New York, NY: Basic Books.

Reidys, C. M., and P. F. Stadler. 2001. "Neutrality in Fitness Landscapes." *Applied Mathematics and Computation* 117 (2–3): 321–350. https://doi.org/10.1016/S0096-3003(99)00166-6.

Šali, A., and T. L. Blundell. 1993. "Comparative Protein Modelling by Satisfaction of Spatial Restraints." *Journal of Molecular Biology* 234 (3): 779–815. https://doi.org/10.1006/jmbi.1993.1626.

Schultes, E. A., and D. P. Bartel. 2000. "One Sequence, Two Ribozymes: Implications for the Emergence of New Ribozyme Folds." *Science* 289 (5478): 448–452. https://doi.org/10.1126/science.289.5478.448.

Schuster, P., W. Fontana, P. F. Stadler, and I. L. Hofacker. 1994. "From Sequences to Shape and Back: A Case Study in RNA Secondary Structures." *Proceedings of the Royal Society of London B: Biological Sciences* 255 (1344): 279–284. https://doi.org/10.1098/rspb.1994.0040.

Spiegelman, S. 1971. "An Approach to the Experimental Analysis of Precellular Evolution." *Quarterly Reviews of Biophysics* 4 (2–3): 213–253. https://doi.org/10.1017/S0033583500000639.

van Nimwegen, E., J. P. Crutchfield, and M. Huynen. 1999. "Neutral Evolution of Mutational Robustness." *Proceedings of the National Academy of Sciences* 96 (17): 9716–9720. https://doi.org/10.1073/pnas.96.17.9716.

Wagner, A. 2005. *Robustness and Evolvability in Living Systems.* Princeton, NJ: Princeton University Press.

FROM SEQUENCES TO SHAPES AND BACK: A CASE STUDY IN RNA SECONDARY STRUCTURES

Peter Schuster, Institute of Molecular Biotechnology;
Walter Fontana, Los Alamos National Laboratory;
Peter F. Stadler, University of Vienna; and
Ivo L. Hofacker, University of Vienna

Abstract

RNA folding is viewed here as a map assigning secondary structures to sequences. At fixed chain length the number of sequences far exceeds the number of structures. Frequencies of structures are highly non-uniform and follow a generalized form of Zipf's law: we find relatively few common and many rare ones. By using an algorithm for inverse folding, we show that sequences sharing the same structure are distributed randomly over sequence space. All common structures can be accessed from an arbitrary sequence by a number of mutations much smaller than the chain length. The sequence space is percolated by extensive neutral networks connecting nearest neighbours folding into identical structures. Implications for evolutionary adaptation and for applied molecular evolution are evident: finding a particular structure by mutation and selection is much simpler than expected and, even if catalytic activity should turn out to be sparse in the space of RNA structures, it can hardly be missed by evolutionary processes.

1. Introduction

Folding sequences into structures is a central problem in biopolymer research. Both robustness and accessibility of structures, as functions of mutational change in the underlying sequence, are crucial to both natural and applied molecular evolution. Test-tube evolution experiments are based on properties of RNA molecules: as sequences they are genotypes, and as spatial structures they are phenotypes (Spiegelman 1971; Biebricher 1983). Our concern is the mapping from RNA sequences into structures being the simplest, and the only tractable, example of a genotype–phenotype mapping.

An RNA sequence is a point in the space of all 4^n sequences with fixed length n. This space has a natural metric induced by point

The advent of artificial intelligence methods in recent years has renewed interest in the folding problem.

This argument had important implications for both the use of simple computational models and for the generalization of the ideas in this work to other genotype–phenotype maps developed since.

mutations interconverting sequences known as the Hamming distance (Hamming 1950, 1986). The folding process considered here maps an RNA sequence into a secondary structure (figure 1a) minimizing free energy. A secondary structure is tantamount to a list of Watson–Crick type and GU base pairs, and can be represented as a tree graph (figure 1b). This emphasizes the combinatorial nature of secondary structures and allows for a canonical distance measure between structures (Tai 1979). Assuming elementary edit operations with pre-defined costs, such as deletion, insertion and relabelling of nodes, the distance between two trees is given by the smallest sum of the edit costs along any path that converts one tree into the other (Sankoff and Kruskal 1983).

An approximate upper bound on the number of minimum free-energy structures (of fixed chain length n) can be obtained along the lines devised by Stein and Waterman (1978). Counting only those planar secondary structures that contain hairpin loops of size three or more (steric constraint), and that contain no isolated base pairs (stacks of two or more pairs are essentially the only stabilizing elements), one finds:

$$S_n = 1.4848 \times n^{\frac{-3}{2}} (1.8488)^n,$$

which is consistently smaller than the number of sequences.

The consequences of this argument are the reason for the impact of this work: a reinterpretation of the folding problem in the context of the mapping between genotypes and phenotypes.

Folding can thus be viewed as a map between two metric spaces of combinatorial complexity, a sequence space and a shape space. (The notion of shape space was originally used in theoretical immunology in a similar context by Perelson and Oster (1979).) 'Shape' refers to a discretized (and hence coarse-grained) structure representation, such as the secondary structures or the tree graphs used here. The notion of secondary structure is but one among a spectrum of possible levels of resolution that can be used to define shape. It discards atomic coordinates, as well as the relative spatial orientation of the structural elements, taking into account only their number, size and relative connectedness. Nevertheless, secondary structure is a major component of whatever turns out to be an adequate shape definition for RNA: it covers the dominant part of the three-dimensional folding energies, very often it can be used successfully in the interpretation of function and reactivity, and it is frequently conserved in evolution (Sankoff, Morin, and Cedergren 1978; Konings and Hogeweg 1989; Le and Zuker 1990), sometimes together with a few tertiary interactions (Cech 1988).

The authors explain why their computational model of RNA secondary structure captures the essential aspects of RNA folding, a property that could not be generalized to other biomolecules.

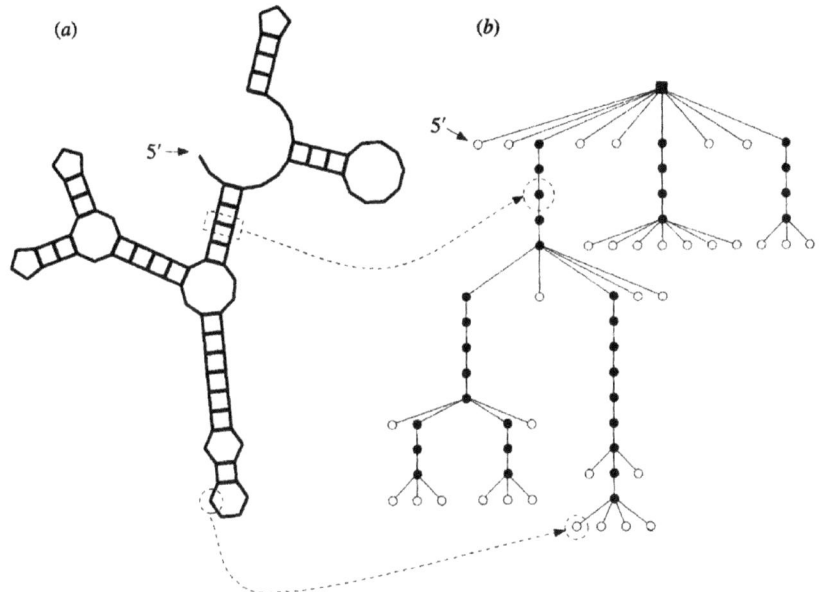

Figure 1. (*a*) A secondary structure on a sequence is any pattern of base pairs such that no bases inside a loop pair to bases outside it. Such a structure can be uniquely decomposed into structural elements that are: (i) base pair stacks; (ii) loops differing in size (number of unpaired bases) and branching degree, i.e. hairpin loops (degree one), internal loops (degree two or more); and (iii) bases which are not part of a stack or a loop, termed external (freely rotating joints and unpaired ends). Each stack or loop element contributes additively to the overall free energy of the structure according to empirically determined parameters that depend on the nucleotide sequence. A minimum free-energy structure is constructed according to an algorithm proposed by Zuker and Stiegler (1981) and Zuker and Sankoff (1984). (*b*) A secondary structure graph (*a*) is equivalent to an ordered rooted tree. An internal node (black) of the tree corresponds to a base pair (two nucleotides), a leaf node (white) corresponds to one unpaired nucleotide, and the root node (black square) is a virtual parent to the external elements. Contiguous base pair stacks translate into 'ropes' of internal nodes, and loops appear as bushes of leaves. Recursively traversing a tree by first visiting the root then visiting its subtrees in left to right order, finally visiting the root again, assigns numbers to the nodes in correspondence to the 5'–3' positions along the sequence. (Internal nodes are assigned two numbers reflecting the paired positions.)

This suggests that many of the relevant intermolecular interactions that collectively set a natural scale for shape are indeed strongly influenced by the secondary structure. The observation, then, is that—at least in the present case—the shape space is considerably smaller than the sequence space. (We remark that this is also true for protein models on lattices.)

2. Frequencies of Shapes and Inverse Folding

Frequencies of occurrence for individual shapes in sequence space were obtained from large samples derived by folding random sequences of fixed chain length. Ranking according to decreasing frequencies

This turned out to be one of the general properties of the sequence-structure map of biomolecules. In the case of proteins, empirical evidence was found in Chothia and Lesk (1986), and formed the basis of structure prediction methods.

Although this distribution could be explored through the RNA model, the equivalent distribution obtained from natural samples remains as an open question.

yields a distribution which obeys a generalized Zipf's law (figure 2). We are thus dealing with relatively few common shapes and many rare ones. How are the sequences which fold into the same shape distributed in sequence space? This distribution is evaluated with a heuristic inverse folding procedure, aimed at devising sequences that fold into an arbitrary pre-defined target shape (Hofacker *et al.* 1993). The obvious first step is to construct a compatible test sequence with nucleotide assignments such that the target shape is indeed a possible secondary structure, although typically not a minimum free-energy one. We choose at random among the many compatible sequences. The next step is to decompose the minimum energy structure on the chosen test sequence into substructures, and to mutate by trial and error the corresponding subsequences. When the individual substructures are as in the target, the entire sequence is reassembled. The procedure stops if the reassembled sequence folds into the target shape. This happens in about 50% of the cases. Several sequences that fold into the same structure are sampled by starting the procedure with different compatible sequences. The average number of mutations that converted a random compatible sequence of chain length $n = 100$ into one with the desired target shape was 7.2.

This prediction remains without experimental demonstration.

The resulting ensemble of compatible sequences that fold into a pre-defined target has been analysed for the target being the secondary structure of t-RNA^{Phe} and for three randomly constructed examples. In each case about 1000 sequences were derived by the inverse folding algorithm. The distribution of pairwise distances is not distinguishable from the one expected for random compatible sequences. The properties of the sequence sample as seen by statistical geometry (Eigen, Winkler-Oswatitsch, and Dress 1988) and split-decomposition (Bandelt and Dress 1992) yield the same result: sequences folding into the same structure are randomly distributed in the space of compatible sequences.

3. Structure Density Surfaces

Generalizing the previous question we ask how the possible shapes are distributed over the possible sequences. One insight is provided by considering the probability density (Fontana, Konings, *et al.* 1993;

Schuster et al. (1994)

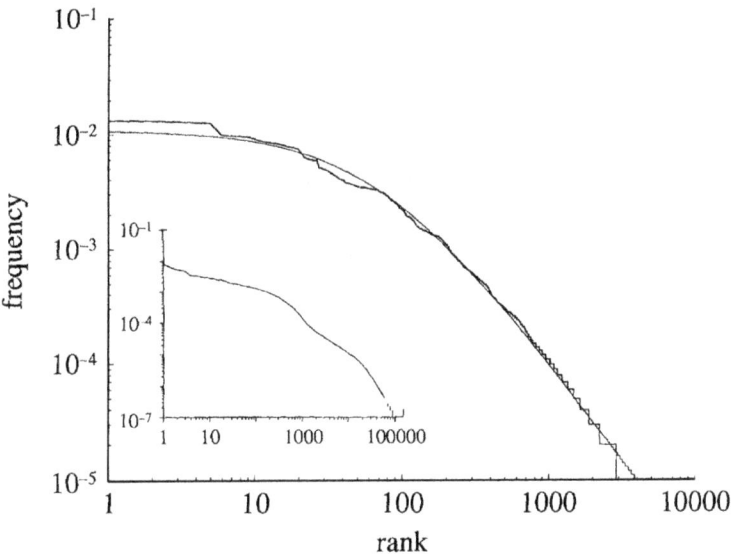

Figure 2. The frequency distribution of RNA secondary structures. Shapes are ranked by their frequencies. The particular example shown here deals with the loop structures (Shapiro and Zhang 1990) of 10^5 RNA molecules of chain length 100 which are derived from secondary structures by further coarse graining that eliminates all details concerning stack lengths and loop sizes. The diagram covers 97% of the total frequency. The frequencies follow a generalized form of Zipf's law: $f(x) = a(b+x)^{-c}$, with x being the rank of a shape and $f(x)$ its frequency. Parameter values of the best fit (thin curve) are $a = 1.25, b = 71.2$ and $c = 1.73$. The frequency distribution of full secondary structures is essentially the same as shown in the insert for chain length 30. Computation of the distribution for longer chains is hardly possible as the number of structures exceeds by far the available capacities (there are about 7×10^{23} full secondary structures of chain length $n = 100$).

Fontana, Stadler, *et al.* 1993) $P(t \mid h)$ of two structures being at (tree) distance t, given that the underlying sequences are at (Hamming) distance h. This structure density surface (SDS) shows how the distribution of structure differences changes as the sequences become more and more uncorrelated with increasing Hamming distance from the reference (figure 3 presents the SDS for sequences of chain length $n = 100$). Three observations are immediate: (i) although for very small Hamming distances ($h = 1, 2, 3$) the most probable structures are identical or very similar, there is none the less some probability that even a single mutation substantially alters the structure; (ii) beyond distance $h = 3$, identical or even closely related structures are extremely unlikely; and (iii) in the range $15 < h < 20$, the density becomes independent of h, thus approaching essentially what is expected for a sample of randomly drawn sequences ($h \approx 75$).

Although probably similar, this distribution remains uncharacterized for 3D RNA structures.

Today, we believe that this is the main distinction between the sequence-structure map of RNA and proteins. On average, samples of known protein structures tend to preserve their structure up to sequence distances of 30–40%.

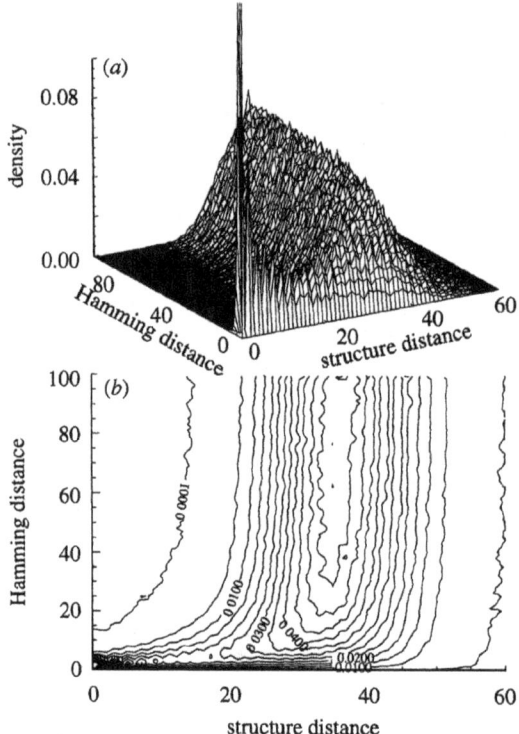

Figure 3. (*a*) The structure density surface (SDS) for RNA sequences of length $n = 100$ (upper). This surface was obtained as follows: (i) choose a random reference sequence and compute its structure; (ii) sample randomly ten different sequences in each distance class (Hamming distance 1–100) from the reference sequence, and bin the distances between their structures and the reference structure. This procedure was repeated for 1000 random reference sequences. Convergence is remarkably fast; no substantial changes were observed when doubling the number of reference sequences. This procedure conditions the density surface to sequences with base composition peaked at uniformity, and does not, therefore, yield information about strongly biased compositions. (*b*) Contour plot of the SDS.

The latter suggests that the structures of a reference sequence and its mutants at distances between 15 and 20 or larger are effectively uncorrelated. This suggests that memory of the reference structure is sufficiently lost to allow the mutants at that distance to acquire any frequent minimum energy structure, at least in its essential features. From the SDS the complete structure autocorrelation function can be recovered (Fontana, Konings, *et al.* 1993). This function is to a reasonable approximation a single decaying exponential with a characteristic length, $l = 7.6$ in the present case (chain length $n = 100$). From figure 3 it is seen that this corresponds essentially to the distance at which the dominant peak resulting from identical or very similar structures has disappeared.

Schuster et al. (1994)

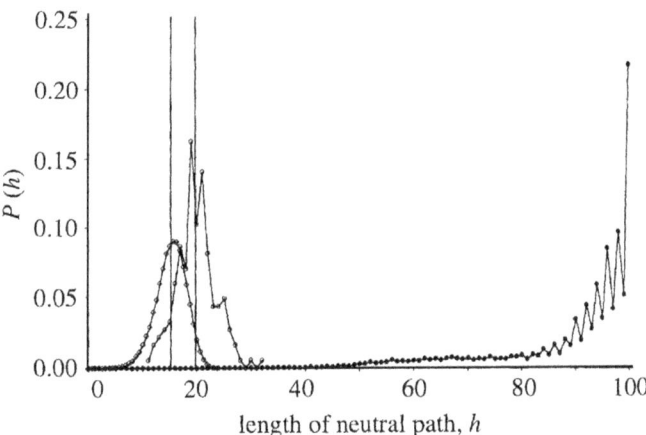

Figure 4. Neutral paths. A neutral path is defined by a series of nearest neighbour sequences that fold into identical structures. Two classes of nearest neighbours are admitted: neighbours of Hamming distance 1, which are obtained by single base exchanges in unpaired stretches of the structure, and neighbours of Hamming distance 2, resulting from base pair exchanges in stacks. Two probability densities of Hamming distances are shown that were obtained by searching for neutral paths in sequence space: (i) an upper bound for the closest approach of trial and target sequences (open circles) obtained as endpoints of neutral paths approaching the target from a random trial sequence (185 targets and 100 trials for each were used); (ii) a lower bound for the closest approach of trial and target sequences (open diamonds) derived from secondary structure statistics (Fontana *et al.* 1993a; see this paper, §4); and (iii) longest distances between the reference and the endpoints of monotonously diverging neutral paths (filled circles) (500 reference sequences were used).

4. Shape Space Covering

Combination of the previous results (showing the existence of relatively few common shapes which are minimum free-energy structures for sequences randomly distributed in sequence space) with the information from the SDS (showing the existence of a transition from local to global features) provides strong evidence for the existence of a neighbourhood (a high-dimensional ball) around every random sequence that contains sequences whose structures include almost all common shapes.

To verify the prediction of a characteristic neighbourhood covering almost all common shapes we did a computer experiment. A target sequence is chosen at random. A second random sequence serves as an initial trial sequence, and its structure as a reference structure. Next we search for a nearest neighbour of the trial sequence that folds into the reference structure but lies closer to the target. If such a sequence is found, it is accepted as the new trial sequence, and the procedure is repeated until no further approach to the target is possible. The final Hamming distance

Shape space covering is now believed to be a general property of genotype–phenotype maps.

to the target is an upper bound for the minimum distance between two sequences folding into the reference structure and the structure of the target, respectively. The probability density of this upper bound to the closest approach distances determined for RNA molecules of chain length 100 is shown in figure 4 (open circles). It yields a mean value of 19.8. (It is remarkable that this value coincides with the critical Hamming distance at which we observe the change from local to global features in the SDS; analogous investigations on RNA molecules with only GC or only AU base pairs have shown that the precise agreement is not generally valid.)

We can also compute a lower bound for the mean value of the closest approach distance. The probability that two arbitrarily chosen bases of an RNA sequence can form a base pair is given by the number of pairings divided by the number of possible combinations of two bases: $6/16 = 3/8$ (as we have six classes of base pairs: AU, GC, GU and inversions). The mean number of bases that have to be changed in a random sequence, to form a sequence which is compatible with the target structure (representing the lower bound), is obtained from the probability not to form a base pair by multiplication with the mean number of base pairs: $(1-\frac{3}{8}) \times \overline{n}_{Bp} = \frac{5}{8}\overline{n}_{Bp}$. For RNA molecules of chain length 100, the mean number of base pairs is 24.34 (Fontana, Konings, *et al.* 1993), and we obtain a mean Hamming distance of 15.2 for the lower bound. From the probability density of the number of base pairs, we derive a distribution of the lower bound also shown in figure 4 (open diamonds). The characteristic neighbourhood has a radius of $15 < h_c < 20$.

5. Neutral Paths Through Sequence Space

The structure of the RNA shape space over the sequence space is complemented by a second computer experiment. We search for neutral paths with monotonously increasing distance from a reference sequence. A neutral path ends when no sequence that forms the same structure is found among the nearest neighbours. The probability density of the lengths of these paths is shown in figure 4 (filled circles). The vast extension of the network of neutral paths came as a surprise: 21.7% of all paths percolate the entire sequence space and end in a sequence which has not a single base in common with the reference. (The existence of extensive neutral networks

The existence of neutral paths was yet another general property of genotype–phenotype maps. Reydis and Stadler (2001) provided an analytical result showing that the percolation of a neutral network is a function of both average neutrality and the dimension of sequence space.

Schuster et al. (1994)

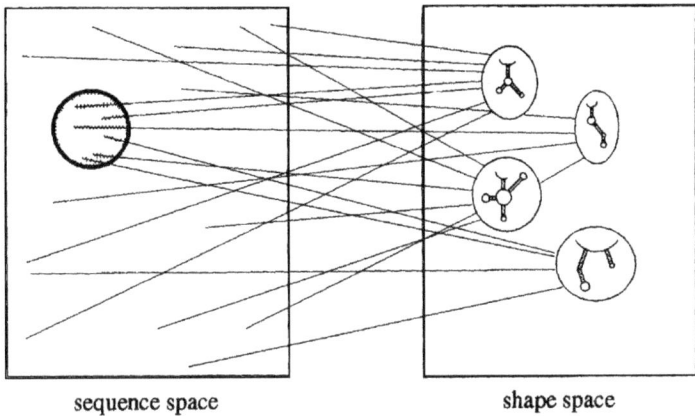

Figure 5. A sketch of the mapping from sequences into RNA secondary structures as derived here. Any random sequence is surrounded by a ball in sequence space which contains sequences folding into (almost) all common structures. The radius of this ball is much smaller than the dimension of sequence space.

meets a claim raised by Maynard Smith (1970) for protein spaces that are suitable for efficient evolution.)

6. Discussion

The existence of a ball with characteristic radius around any random sequence within which almost all common shapes are found (figure 5) is a robust phenomenon of the mapping from sequences into RNA secondary structures. It depends on the ratio of sequences to structures. Changes in the base-pairing alphabet, in particular the consideration of pure GC or pure AU sequences, may cause minor alterations that can be interpreted by much smaller values of this ratio as well as by differences in the topology of sequence space. Alphabet dependencies will be published elsewhere. The major features of the shape space structure depend on the generic properties of RNA folding, in particular on the non-local nature of base pairing, but they are insensitive to the empirical energy parameter sets used in folding algorithms, as well as to the distance measure between structures used in the SDS (essentially the same SDSs were obtained with a now superseded parameter set (Fontana *et al.* 1991), and also with a different structure distance measure (Hogeweg and Hesper 1984; Huynen, Konings, and Hogeweg 1993)).

An important distinction between Maynard Smith's claim and the conclusions of the RNA model is that Maynard Smith argued about function instead of structure. Experimental evidence was put forward by Schultes and Bartel (2000).

There are caveats to our approach.

This is an important drawback of computational models of biopolymers, which neglect the often rich dynamics of folding kinetics. However, over the years this has proven less so for RNA, for which relatively short sequences are of great biological relevance.

1. We use a thermodynamic criterion for RNA structure formation, not one that mimics the kinetics of folding. This does not constitute a problem for short sequences up to a few hundred nucleotides. Moreover, the number of possible kinetic structures is not entirely different from the number of thermodynamic structures, the principles of base pairing are the same, and thus the generic features of mappings of sequences into kinetic or thermodynamic structures will be essentially the same, too.

2. We consider only a single minimum free-energy structure for each sequence. Our approach can be carried over to ensembles comprising optimal and suboptimal foldings, as represented by partition functions (McCaskill 1990; Bonhoeffer *et al.* 1993). 'Shape', then, becomes a matrix of temperature-dependent base-pairing probabilities, and the concept of distance is changed accordingly. All qualitative features of the SDS remain essentially unchanged, and numerical corrections are in the range of 10% (as, for example, in the case of correlation lengths (Bonhoeffer *et al.* 1993)).

In the early 2000s, phenotypic plasticity became the subject of work using the RNA model (Ancel and Fontana 2000).

3. We do not consider three-dimensional structure. Nevertheless, the secondary structure defines an informative scale of resolution. In addition, it constitutes an approximation to a coarse-grained spatial structure (current algorithms for the modelling of RNA three- dimensional structures start from secondary structures, and introduce a few tertiary interactions (Major *et al.* 1991; Major, Gautheret, and Cedergren 1993)).

The consequences of our results for natural and artificial selection are immediate. We predict that there is no need to search systematically huge portions of the sequence space. In the particular example of RNA molecules of chain length 100, the characteristic ball contains some 10^{27} sequences, which is only a fraction of 10^{-33} of the entire sequence space. Almost all structures are within reach of a few mutations from a compatible sequence (average: 7.2), and even in reasonable proximity of any non-compatible random sequence ($\approx$ 18). The conclusion is thus that optimization of structures by evolutionary trial and error strategies is much simpler than is often assumed. It provides further support to the idea of widespread applicability of molecular evolution (Eigen 1971; Eigen and Schuster 1979; Eigen, McCaskill, and Schuster, 1988b, 1989). The existence of networks of neutral paths percolating the entire sequence space has strong implications for (molecular) evolution in nature, as well as in the laboratory. Populations replicating with sufficiently high error rates will readily spread along these networks and can reach more distant regions in sequence space.

If one were to design the ultimate evolvable molecule that carries information and is engaged in functional interactions, it would ideally require two features: (i) capability of drifting across sequence space without the necessity of changing shape; and (ii) proximity to any common shape everywhere. These are precisely the features that statistically characterize the mapping from RNA sequence to secondary structure.

As predicted by the authors, several of the properties first identified in the RNA model have been shown to apply to other genotype–phenotype maps (Wagner 2005).

Work by van Nimwegen and others (1999) presented analytic conditions for the evolution of mutational robustness in neutral networks.

Acknowledgments

The work reported here was supported financially by the Austrian Fonds zur Forderung der wissenschaftlichen Forschung (Projects S 5305-PHY, P 9942-PHY, and P 8526-MOB), the European Economic Community (Research Study Contract PSS*03960) and a National Science Foundation (Washington) core grant to the Santa Fe Institute (PHY-9021437). We thank Professor Leo Buss and Professor Andreas Dress for comments on the manuscript, and Dipl.- Chem. Erich Bornberg-Bauer for supplying the data for the insert of figure 2.

REFERENCES

Bandelt, H. J., and A. W. M. Dress. 1992. "A Canonical Decomposition Theory for Metrics on a Finite Set." *Advances in Mathematics* 92:47–105.

Biebricher, C. K. 1983. "Darwinian Selection of Self-Replicating RNA Molecules." *Evolutionary Biology* 16:1–52.

Bonhoeffer, S., J. S. McCaskill, P. F. Stadler, and P. Schuster. 1993. "RNA Multi-Structure Landscapes. A Study Based on Temperature Dependent Partition Functions." *European Biophysics Journal* 22:13–24.

Cech, T. R. 1988. "Conserved Sequences and Structures of Group I Introns: Building an Active Site for RNA Catalysis—A Review." *Gene* 73:259–271.

Eigen, M. 1971. "Selforganization of Matter and the Evolution of Biological Macromolecules." *Naturwissenschaften* 58:465–523.

Eigen, M., J. S. McCaskill, and P. Schuster. 1989. "The Molecular Quasi-Species." *Advances in Chemical Physics* 75:149–263.

———. 1988b. "Molecular Quasi-Species." *Journal of Physical Chemistry* 92:6881–6891.

Eigen, M., and P. Schuster. 1979. *The Hypercycle: A Principle of Natural Self-Organization.* Berlin, Germany: Springer-Verlag.

Eigen, M., R. Winkler-Oswatitsch, and A. Dress. 1988. "A Statistical Geometry in Sequence Space: A Method of Quantitative Comparative Sequence Analysis." *Proceedings of the National Academy of Sciences* 85:5913–5917.

Fontana, W., T. Griesmacher, W. Schnabl, P. F. Stadler, and P. Schuster. 1991. "Statistics of Landscapes Based on Free Energies, Replication and Degradation Rate Constants of RNA secondary Structures." *Chemical Monthly* 122:795–819.

Fontana, W., D. A. M. Konings, P. F. Stadler, and P. Schuster. 1993. "Statistics of RNA Secondary Structures." *Biopolymers* 33:1389–1404.

Fontana, W., P. F. Stadler, E. G. Bornberg-Bauer, T. Griesmacher, I. L. Hofacker, M. Tacker, P. Tarazona, E. D. Weinberger, and P. Schuster. 1993. "RNA Folding and Combinatory Landscapes." *Physical Review E* 47:2083–2099.

Hamming, R. W. 1950. "Error Detecting and Error Correcting Codes." *Bell Systems Technical Journal* 29:147–160.

———. 1986. "Coding and Information Theory." (Englewood Cliffs, NJ).

Hofacker, I. L., W. Fontana, P. F. Stadler, L. S. Bonhoeffer, M. Tacker, and P. Schuster. 1993. "Fast Folding and Comparison of RNA Secondary Structures." *Chemical Monthly* 125:167–188.

Hogeweg, P., and B. Hesper. 1984. "Energy Directed Folding of RNA Sequences." *Nucleic Acids Research* 12:67–74.

Huynen, M. A., D. A. M. Konings, and P. Hogeweg. 1993. "Multiple Coding and the Evolutionary Properties of RNA Secondary Structure." *Journal of Theoretical Biology* 165:251–267.

Konings, D. A. M., and P. Hogeweg. 1989. "Pattern Analysis of RNA Secondary Structure. Similarity and Consensus of Minimal-Energy Folding." *Journal of Molecular Biology* 207:597–614.

Le, S. Y., and M. Zuker. 1990. "Common Structures of the 5′ Non-Coding RNA in Enteroviruses and Rhinoviruses. Thermodynamical Stability and Statistical Significance." *Journal of Molecular Biology* 216:729–741.

Major, F., D. Gautheret, and R. Cedergren. 1993. "Reproducing the Three-Dimensional Structure of a tRNA Molecule from Structural Constraints." *Proceedings of the National Academy of Sciences* 90:9408–9412.

Major, F., M. Turcotte, D. Gautheret, G. Lapalme, E. Fillion, and R. Cedergren. 1991. "The Combination of Symbolic and Numerical Computation for Three-Dimensional Modeling of RNA." *Science* 253:1255–1260.

Maynard Smith, J. 1970. "Natural Selection and the Concept of a Protein Space." *Nature* (London, UK) 225:563–564.

McCaskill, J. S. 1990. "The Equilibrium Partition Function and Base Pair Binding Probabilities for RNA Secondary Structures." *Biopolymers* 29:1105–1119.

Perelson, A. S., and G. F. Oster. 1979. "Journal of Theoretical Biology." 81:645–670.

Sankoff, D., and J. B. Kruskal. 1983. *Time Warps, String Edits and Macro-Molecules. The Theory and Practice of Sequence Comparisons.* London, UK: Addison-Wesley.

Sankoff, D., A. M. Morin, and R. J. Cedergren. 1978. "The Evolution of 5S RNA Secondary Structures." *Canadian Journal of Biochemistry* 56:440–443.

Shapiro, B. A., and K. Zhang. 1990. "Comparing Multiple RNA Secondary Structures Using Tree Comparisons." *Bioinformatics* 6:309–318.

Spiegelman, S. 1971. "An Approach to the Experimental Analysis of Precellular Evolution." *Quarterly Review of Biophysics* 4:213–253.

Stein, P. R., and M. S. Waterman. 1978. "On Some New Sequences Generalizing the Catalan and Motzkin Numbers." *Discrete Mathematics* 26:261–272.

Tai, K. 1979. "The Tree-to-Tree Correction Problem." *Journal of the Association for Computing Machinery* 26:422–433.

Zuker, M., and D. Sankoff. 1984. "RNA Secondary Structures And Their Prediction." *Bulletin of Mathematical Biology* 46:591–621.

Zuker, M., and P. Stiegler. 1981. "Optimal Computer Folding of Large RNA Sequences Using Thermodynamic and Auxiliary Information." *Nucleic Acids Research* 9:133–148.

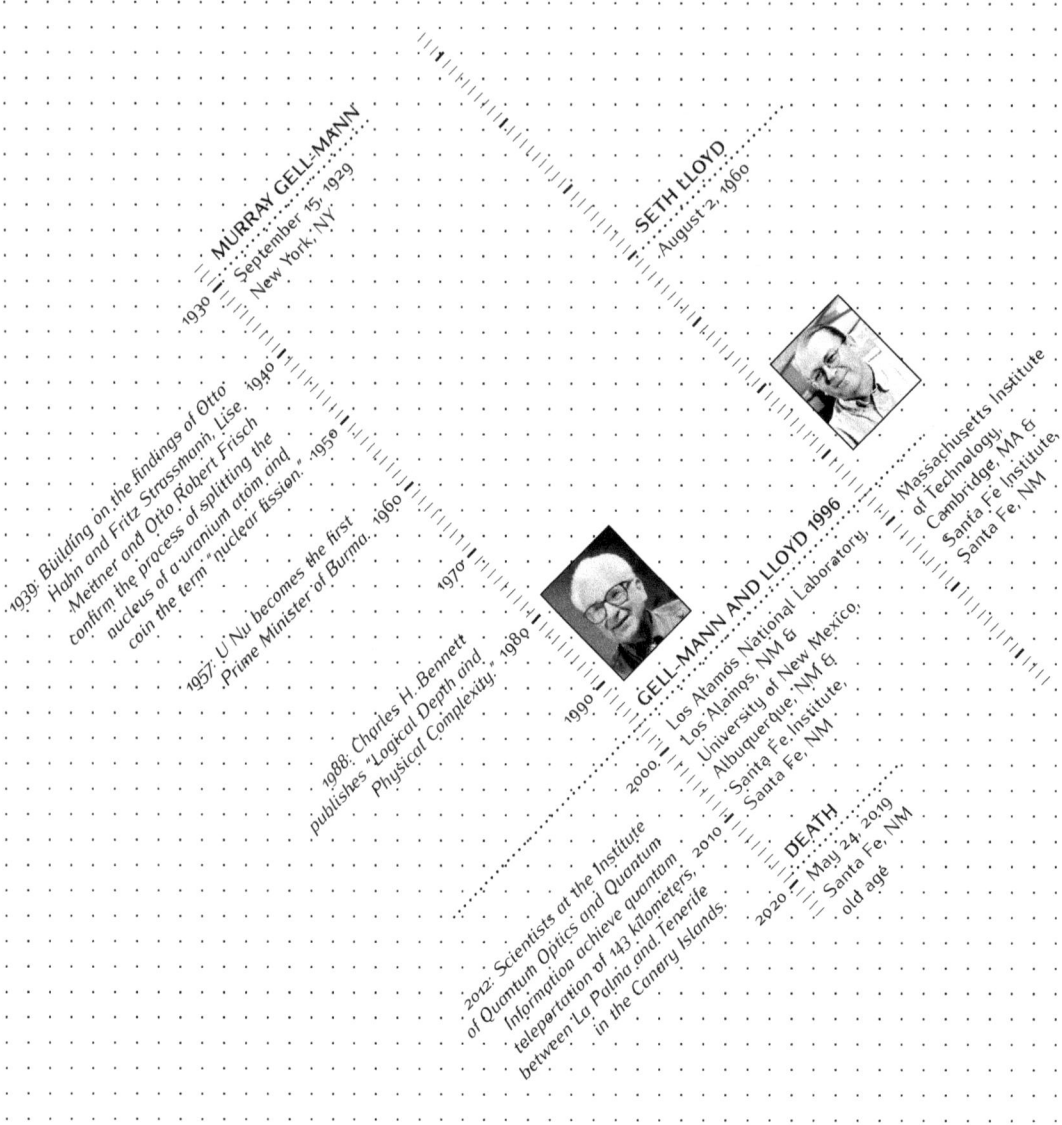

MURRAY GELL-MANN
1930: September 15, 1929
New York, NY

SETH LLOYD
August 2, 1960

1939: Building on the findings of Otto Hahn and Fritz Strassmann, Lise Meitner and Otto Robert Frisch confirm the process of splitting the nucleus of a uranium atom, and coin the term "nuclear fission." 1950

1957: U Nu becomes the first Prime Minister of Burma. 1960

1988: Charles H. Bennett publishes "Logical Depth and Physical Complexity." 1980

GELL-MANN AND LLOYD 1996

Los Alamos National Laboratory, Los Alamos, NM &
University of New Mexico, Albuquerque, NM &
Santa Fe Institute, Santa Fe, NM

Massachusetts Institute of Technology, Cambridge, MA &
Santa Fe Institute, Santa Fe, NM

2012: Scientists at the Institute of Quantum Optics and Quantum Information achieve quantum teleportation of 143 kilometers, between La Palma and Tenerife in the Canary Islands.

DEATH
May 24 2019
Santa Fe, NM
old age

GELL-MANN AND LLOYD

[81]

CHARACTERIZATION OF COMPLEX SYSTEMS AND THE EVOLUTION OF SCIENTIFIC THEORIES

Miguel Fuentes, Argentine Society of Philosophical Analysis and Santa Fe Institute

One goal of scientific work is to understand the essential character of natural phenomena, or at least those processes and principles that lend to these phenomena, a maximum of illumination. There is at present some consensus that life should be understood in relation to the flow of information. As discussed in this foundational 1996 paper by Murray Gell-Mann and Seth Lloyd, consideration of information content supports any simplification required for an understanding of nature and provides a rigorous basis for discussing complexity.

In their collaborative work, Gell-Mann and Lloyd discuss how information metrics can be used to both characterize complex systems and provide a calculational framework for inference. This contribution is made in the context of a long history where "from time to time, an author presents a supposedly new measure of complexity" that might inform a principled taxonomy of pattern and process (Gell-Mann and Lloyd 2010). Their paper supports concepts previously explored by Gell-Mann, including effective complexity, an essential measure of regularity. It also generates solid bridges spanning information theory, the philosophy of science, and a variety of interdisciplinary discussions relating to knowledge acquisition and scientific advancement.

Gell-Mann and Lloyd build their case carefully by inviting readers into their ideas with a minimum of technicalities, before introducing formal concepts associated with generalizations of information measures. They discuss challenges associated with defining probability, coarse-graining, and entropy, and explore how ambiguities percolate into informational measures, in an effort to keep their own formalism for (1) effective complexity and (2) total information clear from these hazards. These two

M. Gell-Mann and S. Lloyd, "Information Measures, Effective Complexity, and Total Information," *Complexity* 2 (1), 44–52 (1996).

Used with permission of The Royal Society (UK); permission conveyed through Copyright Clearance Center, Inc.

measures seek to account for the regularities that any observer encounters in encoding or describing a phenomenon (up to a certain limit) and the irregularities that any useful measure should exclude. They thereby achieve a factorization of complexity into two important epistemic features of any observation, by a human or otherwise, that must influence the calculation of compressed degrees of regularity.

Through a series of papers seeking to understand transformations occurring in all thermodynamic systems, Rudolf Clausius provided the basis for understanding the evolution of a key physical quantity, entropy. In his pioneering work from the 1860s, Clausius laid the foundations for developing what he would call entropy, conceived as growing inefficiencies leading to irreversibility in any work cycle. Entropy plays an essential role in Gell-Mann and Lloyd's proposal, which they connect explicitly with Claude Shannon's "A Mathematical Theory of Communication" (1948), which dealt with extensions of Clausius associated with technological problems of communication in telegraphy. Shannon discussed ideas such as signal-to-noise ratio and the statistical structure of an optimally encoded (fault-tolerant) message, as communicative analogs to thermodynamics.

Gell-Mann and Lloyd make the algorithmic information content (AIC) of a process their central measure. The AIC metric allows for an approach to the information content of phenomena—the length of a compact description of an algorithm generating any observable—that is able to bypass the need for statistical sets or frequentist definitions of information favored in Shannon's framework. Put somewhat differently, the AIC is the length of the shortest computer program that produces a target (finite) binary string on a universal Turing machine.

Some characteristics of a string or a message are truly random (that is, accidents or noise lacking functionally relevant information). These might be said to lie outside the regularities of a phenomenon we would want to describe using a scientific theory. And it is necessary to modify AIC to escape from this noise. The formalism of effective complexity (EC) thereby neglects some data that AIC must include. Effective complexity is restricted to the length of the compressed description of regularities.

The distinction between AIC and EC is critical. Pathological cases where all information is random (coin-flipping experiments, for example) are not complex in terms of EC since regularities are absent. However, practical applications of this distinction to the scientific process is immensely difficult; any differences between true regularities and presumed random content (or inessential content) requires judgments to made based on prior empirical values. Gell-Mann and Lloyd discuss how one might discover methods to distinguish between true regularities and incidental aspects of observation, but note that ultimately somewhat subjective judgment is inescapable.

The information in any message or object consists of two parts: the first are those characteristics taken as regular and predictable, and effective complexity captures this. The second part describes the irregular characteristics, which are captured through measures of entropy or the Shannon information. From the epistemological point of view, there is a tension between the "load" on a model (the size of algorithm used to describe a phenomenon) and the Shannon information (the random characteristics). The sum of the effective complexity and entropy is called total information—all the information required to describe a message (or phenomenon). The importance of having this form of description consists in the fact that it is very close to the task of any rigorous process of knowledge acquisition, providing a way to evaluate or compare alternative descriptions of a single phenomenon: the two terms in the formalism provide a basis for evaluating the quality of a theory versus its free-parameters. In many cases, as theories improve, there is a reduction in the description length of both terms, the description of regularities (the EC) and the entropy. This is especially true during times of critical theoretical change described by Thomas Kuhn as paradigm shifts (Kuhn 1962).

"Information Measures, Effective Complexity, and Total Information" introduces Murray Gell-Mann and Seth Lloyd's preferred means of describing a complex entity: separating regularity from randomness. It paved the way for subsequent work on "frozen accidents" seeking to make sense of the neutral and selective elements of any adaptive system, with extensions to linguistics and the evolution of scientific theories more generally. (Fuentes 2020, 2022).

REFERENCES

Fuentes, M. 2020. *Scientific Dynamics and Complexity Measures (Spanish Edition)*. Buenos Aires, Argentina: Ed. Sadaf.

———. 2022. *The Evolution of Scientific Theories (Spanish Edition)*. Valparaíso, Chile: Ediciones Universitarias de Valparaíso.

Gell-Mann, M., and S. Lloyd. 2010. "Effective Complexity." In *Murray Gell-Mann: Selected Papers*, 39–402. Singapore: World Scientific.

Kuhn, T. S. 1962. *The Structure of Scientific Revolutions*. Chicago, IL: University of Chicago Press.

Shannon, C. E. 1948. "A Mathematical Theory of Communication." *The Bell System Technical Journal* 27 (3): 379–423. https://doi.org/10.1002/j.1538-7305.1948.tb01338.x.

INFORMATION MEASURES, EFFECTIVE COMPLEXITY, AND TOTAL INFORMATION

Murray Gell-Mann, Santa Fe Institute, and
Seth Lloyd, Massachusetts Institute of Technology

Abstract

This article defines the concept of an information measure and shows how common information measures such as entropy, Shannon information, and algorithmic information content can be combined to solve problems of characterization, inference, and learning for complex systems. Particularly useful quantities are the effective complexity, which is roughly the length of a compact description of the identified regularities of an entity, and total information, which is effective complexity plus an entropy term that measures the information required to describe the random aspects of the entity. Mathematical definitions are given for both quantities and some applications are discussed. In particular, it is pointed out that if one compares different sets of identified regularities of an entity, the 'best' set minimizes the total information, and then, subject to that constraint and to constraints on computation time, minimizes the effective complexity; the resulting effective complexity is then in many respects independent of the observer. ©1996 John Wiley & Sons, Inc.

Introduction

The mathematical formulae used to measure information were developed in the latter half of the nineteenth century in order to measure another quantity: entropy. The thermodynamic definition of entropy put forward by Clausius in 1857 possessed no obvious relation to information (Clausius 1857). But by the turn of the century, natural scientists had realized that entropy also had a statistical interpretation: it effectively measured uncertainty about the actual microscopic state of matter (Maxwell 1860a, 1860b; Boltzmann 1866, 1871, 1872; Gibbs 1902; Ehrenfest and Ehrenfest 1959). Entropy measures ignorance.

Insofar as ignorance and information are opposites, what measures one measures the other as well. In the 1940s, Claude E. Shannon discussed a kind of information relevant to the capacity of a telegraph line to transmit messages. The mathematical formula that Shannon used to define information in terms of probabilities was the same, up to a multiplicative constant, as the formula that Boltzmann had used fifty years before to define entropy.

An older, more intuitive measure of information is the smallest number of petroglyphs, pictographs, kanji, or letters needed to convey a message. The development of computers over the last fifty years allows this measure to be made mathematically precise for a message expressed as a string of bits: the algorithmic information content of a such message is equal to the length of the most concise program that instructs a given universal computer how to produce that message and then halt.

This article provides a formal procedure for addressing the following complementary questions: What does one do with information? What does one do with ignorance? Starting from the historical examples of entropy and coarse—graining in statistical mechanics, the paper proceeds to develop the concept of an information measure, a more general definition of information than either Shannon information or algorithmic information content. Information measures are useful tools for dealing with complex systems: they can be used to measure both the amount of information needed to describe regular, rule—governed behavior, and the amount of information needed to describe irregular, apparently random behavior. The amount of information needed to describe the set of identified regularities of an entity is that entity's *effective complexity* (Gell-Mann 1994, 1995). An information or entropy term describing the random component can be added to the effective complexity to yield what we call the *total information*. Some examples from statistical mechanics, quantum mechanics, and dynamical systems theory are mentioned to demonstrate the key role that effective complexity and total information play in the characterization and control of complex systems.

Probabilities and Information

Originally developed for use in gambling, probability theory is a mathematical method for dealing with chance events. The third edition of the Oxford English Dictionary (1938) defines probability to be 'the likelihood of an event, measured by the ratio of the favourable cases to the whole number of cases possible, as "from a bag containing three red balls and seven white the probability of a red ball's being drawn first is 3/10."'

In this article, Gell-Mann and Lloyd define concepts relevant to complexity sciences. The so-called *effective complexity*, discussed by GM in his book *The Quark and the Jaguar*, and total information are of particular interest. It is essential to mention that these metrics' conceptual and quantitative structures have a substantial connection with two other earlier foundational works, one by Shannon (1948; see *Foundational Papers* vol. 1, ch. 9), and the other by Kolmogorov (1965; vol. 2, ch. 31).

From this point of view, a probability is not a frequency of occurrence, but a likelihood—a theoretical expectation of a frequency of occurrence, based on knowledge of physical law and on the historical events that set the stage for the event in question. In the example given, the bases for assigning the likelihood 3/10 to drawing a red ball are implicit, i.e., red and white balls have the same weight and texture, the bag is well-shaken, etc. These assumptions could just as well be made explicit.

An apparently more concrete point of view is that probabilities represent the actual frequencies of events. This concreteness is only apparent: such a purely frequentist definition of probability only makes formal sense for events that occur under exactly the same circumstances an arbitrarily large number of times. Many events to which probabilities are applied are not indefinitely repeatable; to the extent that they are, the likelihood definition of probability and the frequency definition tend to coincide as the number of trials becomes large. More precisely, to define probabilities as frequencies requires N independent, identically distributed (IID) trials, where N can be taken to be arbitrarily large. However, as $N \to \infty$, the fraction of sequences tends to zero for which the frequency of each outcome does not converge to the likelihood. Accordingly, in this article probability will be interpreted as likelihood.

Probability is a tool for dealing with uncertainty. Consider a set R of alternatives r, exactly one of which can occur. The probabilities p_r over such a set of exhaustive, mutually exclusive alternatives must obey $0 \leq p_r \leq 1$, and $\sum_{r \in R} p_r = 1$. Such a set of alternatives together with their probabilities will be called an ensemble, in analogy with the notion of an ensemble in statistical mechanics. Complete certainty corresponds to one of the alternatives having probability 1 and the remainder having probability 0. Complete uncertainty corresponds to all alternatives having the same probability.

Shannon gave an elegant answer to the question of how to measure uncertainty as a function of probabilities (Shannon and Weaver 1948). He demanded three requirements of such a measure, $I(\{p_r\})$: 1) I should be a continuous function of the p's; 2) for the case that the probabilities for n alternatives are equal (i.e., $p_r = 1/n$ for all r), I should be a monotonically increasing function of n; 3) when the alternatives make up a composite set, $U \times V$, with probabilities p_{uv}, then the uncertainty

The authors interpret probability as likelihood, directly connecting to Shannon's work and metrics for measuring either information or ignorance. An important point to consider is the extension of Shannon's results by relaxing one of the requirements used in his 1948 work. Relaxing the additive condition can generalize the functional form originally found by Shannon, which has usually been discussed for physical systems from the point of view of generalized entropies.

about $U \times V$ should be equal to the uncertainty about U plus the residual uncertainty about V: $I(\{p_{vu}\}) = I(\{p_u\}) + \sum_u p_u I(\{p_{v|u}\})$, where $p_u = \sum_v p_{uv}$ is the probability of the alternative u on its own, and $p_{v|u} = p_{uv}/p_u$ is the probability of v given u. 1) simply requires that nearby probabilities give nearby uncertainties; 2) requires that knowing nothing about $n+1$ alternatives corresponds to greater uncertainty than knowing nothing about n alternatives; 3) demands that ignorance be additive. 3) is the strongest constraint: requiring that uncertainty add up in this fashion completely determines the functional form of I. Shannon showed that the only I that obeys these three requirements is $I = -K \sum_r p_r \log p_r$, where K is an arbitrary positive constant. Choosing K sets the units by which uncertainty is measured: when $K = 1$ and the logarithm is $\log_2$, uncertainty is measured in *bits*.

As just defined, I measures ignorance or uncertainty. However, I also measures information. But ignorance and information are opposites: how can they be measured by the same quantity? Flip a fair coin. Before you know whether it has come up heads or tails, you assign to each alternative the probability $p_h = p_t = 1/2$. $I = 1$: you have one bit of uncertainty about the outcome. Now look at the coin. It is either heads ($p_h = 1, p_t = 0$) or tails ($p_h = 0, p_t = 1$). In either case, $I = 0$: you are certain of the result. By looking at the coin, you have lost a bit of ignorance and gained a bit of information. Similarly, the amount of information conveyed by a telegram can be measured by the recipient's uncertainty of the telegram's content prior to reading it.

Statistical Mechanics and Coarse Graining

In the latter half of the 19th century, Maxwell, Boltzmann, and Gibbs realized that thermodynamics could be expressed in statistical terms (Maxwell 1860a, 1860b; Boltzmann 1866, 1871, 1872; Gibbs 1902). Consider a container of gas. In general, information about the macroscopic features of the container—e.g., its volume, its mass, its temperature, the approximate number of molecules that it contains—is readily available, while information about the positions and velocities of the individual molecules in the gas is hard to get. That is, the *macroscopic state* of the gas can be certain or nearly so, while the *microscopic state* is typically almost completely uncertain.

How uncertain? Shannon's quantity can be used to measure how much information is missing about the microscopic state (or *microstate*) of the gas. Suppose that the volume, energy, and number v of molecules in the gas are known. These quantities may be taken to define a macroscopic state (or *macrostate*) α of the gas comprising a large number N_α of microstates denoted by α, j, where the index j runs from 1 to N_α. (The number N_α is determined by quantum mechanics: its value is roughly the total volume of phase-space accessible to the gas, divided by h^{3N}. In fact, N_α is just what Boltzmann called the number of 'elementary complexions' of the gas.) To measure uncertainty about the microstates, they must be assigned probabilities. The fundamental probabilities are given by quantum mechanics, the exact dynamics of the elementary particles, and the initial conditions for the universe. Of course, these probabilities are typically not available, and instead we make use of *coarse-grained* probabilities. Ideally, coarse graining is a process that assigns to microstates probabilities obtained from the fundamental ones by a kind of averaging that reflects our ignorance (Ehrenfest and Ehrenfest 1959; Tolman 1938). When the macrostate α is known, the simplest method of coarse graining is to assign equal probabilities $1/N_\alpha$ to all microstates in α. When α is specified, the amount of uncertainty about the actual microstate of the gas is then $\tilde{I} = \log N_\alpha$. For a mole of hydrogen confined to a liter at room temperature, $\bar{I}$ is about 10^{28} bits. In fact, the usual thermodynamic entropy of the gas is just $S = (k_B \ln 2)\,\bar{I}$, where k_B is Boltzmann's constant. For convenience, let us put $k_B \ln 2 = 1$ so that entropy and information may be treated on the same footing. Then entropy simply measures ignorance.

Although Gell-Mann and Lloyd discuss the coarse-grained case in the context of probability in statistical mechanics, it is worth mentioning the importance and high impact of this approach in the general study of complex systems.

Shannon information can be used to measure uncertainty about macrostates as well. If the macrostate α has probability p_α and all the microstates corresponding to α are assigned equal probability, as before, then the coarse-grained probability of the jth microstate corresponding to α is $\tilde{p}_{\alpha j} = p_\alpha / N_\alpha$, and the information over all microstates is

$$\tilde{I} = -\sum_{\alpha j} \tilde{p}_{\alpha j} \log \tilde{p}_{\alpha j} = -\sum_\alpha p_\alpha \log p_\alpha + -\sum_\alpha p_\alpha \log N_\alpha, \quad (1)$$

where the first term in the equation measures uncertainty about the macrostate, while the second measures the average uncertainty of the microstate when the macrostate is specified. Once again, the thermodynamic entropy of the gas is $S = \tilde{I}$, since $k_B \ln 2$ has been set equal to 1. The thermodynamic entropy limits the ability of the gas to do work and contains terms that measure both macroscopic and microscopic uncertainty. To identify the macroscopic state of a system generally requires only a few hundred bits, while to identify the microstate of the gas above takes 10^{28} bits. As a result, in most chemical and physical applications, the second term is the one emphasized. Systems that get and use information, such as Maxwell's famous 'demon,' require a careful accounting of both macroscopic and microscopic uncertainty. Both terms must be included in the thermodynamic entropy to preserve the validity of the second law of thermodynamics.

As noted above, the process by which effective probabilities are assigned to microstates is called coarse graining. That process replaces the actual probabilities p_r of the microstates of a system with coarse-grained probabilities $\tilde{p}_r$ that encapsulate what is known about the probabilities of the microstates and nothing more. A set of states together with their probabilities, whether actual or coarse-grained, is called an ensemble. In the example above, if all that is known about the macrostate is that the system's energy lies between E and $E + \delta E$, then, *faute de mieux*, all microstates with energy between E and $E + \delta E$ are assigned equal probabilities. The resulting ensemble of microstates and probabilities is called the microcanonical ensemble.

For a fixed macrostate, assigning equal probabilities to all microstates within it maximizes the entropy of the ensemble of microstates. Jaynes has developed a maximum entropy principle for assigning probabilities given incomplete prior knowledge (Jaynes 1982): Assign probabilities to alternatives r by maximizing $\tilde{I} = -\sum_r \tilde{p}_r \log \tilde{p}_r$, subject to all constraints given by prior knowledge of the alternatives.

Colloquially, Jaynes's principle reads, 'Don't claim you know any more than you know.' If it is known only that the energy of a system lies between E and $E + \delta E$, then Jaynes's principle gives the microcanonical ensemble: all states in this range are assigned equal probabilities. Now suppose instead that all that is known of a system is the expectation value for its energy, $\sum_r p_r E_r = E$. Jaynes's prescription is to maximize $-\sum_r \tilde{p}_r \log \tilde{p}_r - \beta \sum_r \tilde{p}_r E_r$ with respect both to the $\tilde{p}_r$ and the Lagrange multiplier β. The result is $\tilde{p}_r = e^{-\beta E_r}/Z(\beta)$, where $Z(\beta) = \sum_r e^{-\beta E_r}$, and β is chosen to that $\sum_r \tilde{p}_r E_r = E$: this is the familiar canonical ensemble of Gibbs. Further constraints can be added. For example, constraining the expectation value of the number of particles in the system results in the so-called grand canonical ensemble.

The general requirements for coarse graining are as follows. If the actual probability for the microstate r is p_r, then a coarse graining $p_r \to \tilde{p}_r$ of the p_r should have the following three properties, all natural for an averaging process: 1) the $\tilde{p}_r$ are probabilities; 2) $\tilde{\tilde{p}}_r = \tilde{p}_r$; and 3) $-\sum_r \tilde{p}_r \log \tilde{p}_r = -\sum_r p_r \log \tilde{p}_r$. The Ehrenfests (Ehrenfest and Ehrenfest 1959) and Tolman (Tolman 1938) noted that the averaging process of the microcanonical ensemble has these properties. It can be seen that Jaynes's maximum entropy method automatically respects the first two of these requirements and respects the third when the constraints are linear in the p's, as in the canonical and grand canonical ensemble above. All the usual statistical mechanical ensembles are the result of valid coarse grainings. Statistical mechanics works by assuming that one knows only what one knows.

The third property of coarse-grained probabilities, together with the theorem that $-\sum_r p_r \log p_r \leq -\sum_r p_r \log q_r$, where p_r, q_r are probabilities, implies that $-\sum_r p_r \log p_r \leq -\sum_r \tilde{p}_r \log \tilde{p}_r$. That is, whatever the value of the 'fine-grained entropy' $-\sum_r p_r \log p_r$, the coarse-grained entropy is at least as great. In statistical mechanics, the underlying Hamiltonian evolution of any closed system keeps fine-grained entropy constant over time. Thus the last inequality implies that when fine- and coarse-grained entropy are initially equal, the coarse-grained entropy cannot decrease, and will in general increase for some period of time. Boltzmann noted that this result is responsible for the second law of thermodynamics: the universe was once in a simple state, far from equilibrium, in which both fine- and coarse-grained entropy were both small and comparable in size. The universe is still young and the coarse-grained entropy continues to increase.

Information Measures

As the examples of the previous section show, statistical measures of information such as Shannon's proved very useful, even before they were thought to measure information. Are there other measures of information that are also useful? Shannon's theorem implies that $-\sum p \log p$ is the unique additive measure of information for statistical ensembles. But in many cases, such as in ordinary language, it is not clear how to define an appropriate ensemble to define Shannon information. This section generalizes the definition of information by identifying the features of Shannon information that are desirable to preserve in nonstatistical measures of information. A quantity that possesses these desirable features will be called an *information measure*.

Recall that I is a function of probabilities p_r over a set of alternatives $r \in R$. For compactness, we write $I(R)$ rather than $I(\{p_r\})$, to indicate that I is a property of the ensemble $\{r \in R, p_r\}$. Similarly, we write $I(A, B)$ for the *joint* information $-\sum p_{ab} \log p_{ab}$ of the ensemble $\{(ab \in A \times B, p_{ab})\}$. I obeys the following properties:

i. $I(A) \geq 0$. (Nonnegativity)

ii. $I(A, B) = I(B, A)$. (Symmetry)

iii. $I(A, B) \geq I(A)$. (Accumulation)

iv. $I(A) + I(B) \geq I(A, B)$. (Convexity)

As a result of these properties, information possesses associated nonnegative quantities: *conditional information*, $I(A \mid B) \equiv I(A, B) - I(B)$, measures the amount of information needed to describe A given that B is known; *mutual information*, $I(A : B) \equiv I(A) + I(B) - I(A, B) = I(A) - I(A \mid B) = I(B) - I(B \mid A)$, measures how much knowing B tells us about A, and *vice versa*; and *information distance*, $\Delta(A, B) \equiv 2I(A, B) - I(A) - I(B) = I(A, B) - I(A : B) = I(A \mid B) + I(B \mid A)$, measures how much of the information in A and B is *not* held in common. Properties (i–iv) can be used to show that Δ obeys the triangle inequality (Shannon 1993).

Any function, such as Shannon information, having properties (i–iv) is called an *information measure*. Shannon information is a function of an ensemble, but an *information measure* can be a function of a set without probabilities or of a single entity such as a string of symbols. All information measures possess analogues of conditional and mutual information and of information distance. One example of an information measure is the usual concept of measure on sets: if A and B are measurable sets with a measure μ, an information measure can be defined by $m(A) = \mu(A), m(A,B) = \mu(A \cup B)$. The analogue of mutual information $m(A : B) = \mu(A) + \mu(B) - \mu(A \cup B) = \mu(A \cap B)$ measures the intersection of the sets; conditional information $m(A \mid B) = \mu(A) - \mu(A \cap B)$ measures the part of A that does not lie in B; the analogue of information distance, $\Delta_m(A,B) = 2\mu(A \cup B) - \mu(A) - \mu(B) = \mu(A \cup B) - \mu(A \cap B)$, measures the part of $A \cup B$ that does not lie in the intersection. (Use of Venn diagrams permits the easy verification of the triangle inequality for Δ_m.)

A particularly useful *approximate* information measure is algorithmic information content (AIC) [1]. The algorithmic information content $K_U(s)$ of a string of symbols s with respect to a universal computer U is the length of the shortest program that instructs U to produce the output s and then halt. U could be a universal Turing machine or any general purpose digital computer with the ability to generate additional memory as needed. A Macintosh or PC could do as long as it has an unlimited supply of floppy disks. So would a human being who can always buy another legal pad. The program could be written in the language recognized by the Turing machine, or else in Fortran, English, or Hopi. For simplicity, we shall assume that the language used is one recognized by a universal Turing machine. Algorithmic information content clearly obeys exactly the first conditions above, and it obeys approximately the remaining three conditions: $\mid K_U(a,b) - K_U(b,a) \mid$ is no greater than the length of a program that instructs U to interchange a and b; $K_u(a) - K_U(a,b)$ is no greater than the length of a program that instructs U to erase b; and finally, $K_U(a,b) - K_U(a) - K_U(b)$ is no greater than the length of a program that instructs U to concatenate a and b. U may always be chosen so that these instructions take only a few bits (two, in fact). With respect to such a U, algorithmic information content is an information measure to a good approximation.

For an exact information measure, the conditional information $I(A \mid B)$ is exactly equal to $I(A,B) - I(B)$ and may be so defined. For AIC, they are not exactly equal. If $K_U(a \mid b)$ is defined as the conditional AIC, that is, as the length of the shortest program that produces a when given b as input,

Despite not being an information measure, AIC becomes a key concept in this work. It involves a point of view on complexity metrics that goes beyond statistical ideas and approaches the problem of algorithms and languages used to characterize an entity. This perspective is fundamental to connecting complexity measures with scientific endeavors and understanding how theories applied to the same phenomenon can be compared.

---

[1] Reviews of algorithmic information theory can be found in Cover and Thomas (1991), and in Li and Vitany (1993).

it is only approximately equal to $K_U(a,b) - K_U(b)$. In fact, the difference $K_U(a \mid b) + K_U(b) - K_U(a,b)$ is greater than or equal to -2 and less than or equal to $\log K_U(a)$.

Algorithmic information content makes formally precise the intuitive notion of information as the length of a compact description. AIC requires no probabilities over an ensemble of messages to define information content. Rather, it is a property of each individual message. If the set of messages does make up an ensemble, however, the average conditional algorithmic information content of a message in the ensemble is closely related to the Shannon information over the ensemble, via the following inequality: consider the ensemble E of messages $\{(r \in R, p_r)\}$. Since the shortest program for a message r is a (prefix-free) *code* for r, we have

$$-\sum_r p_r \log p_r \leq \sum_r p_r K_U(r \mid E) \leq -\sum_r p_r \log p_r + c_u(E), \quad (2)$$

where $K_U(r \mid E)$ is the length of the shortest program for U that specifies the individual message r, given a description of the ensemble E. Here $c_U(E)$ is the length of a program that instructs U to create a code for the members of R minimizing the expected value $\sum_r p_r \ell_r$ of the code word lengths. Schack (1997) has shown that for any ensemble $E = \{(r \in R, p_r)\}$ and any U there exists a slightly modified universal Turing machine U' such that

$$-\sum_r p_r \log p_r \leq \sum_r p_r K_{U'}(r \mid E) \leq -\sum_r p_r \log p_r + 1. \quad (3)$$

That is, one can modify the universal computer in such a way that the average conditional algorithmic information content over the ensemble is essentially equal to the information over the ensemble: $-\sum_r p_r \log p_r \approx \sum_r p_r K_{U'}(r \mid E)$. In other words, once a description of an ensemble has been given, the residual algorithmic information content of a member of the ensemble is essentially the length of the Shannon-Fano or Huffman code assigned to it by U'.

An analogous formula holds for the case where the probabilities p_r are replaced by coarse-grained probabilities $\tilde{p}_r$ and the ensemble E is correspondingly replaced by $\tilde{E} \equiv \{(r \in R, \tilde{p}_r)\}$:

$$-\sum_r p_r \log p_r \approx \sum_r p_r K_{U'}(r \mid E). \quad (4)$$

Information measures, exact and approximate, allow one to retain the virtues of a description in terms of information without being locked into using the Shannon information over an ensemble. For example, they are essential for specifying the amount of information conveyed by a single string or message. The idea of using algorithmic information content as a substitute for part of the entropy or Shannon information in the context of statistical mechanics is due to Zurek (1989b, 1989a), who noted that such a substitution was necessary to preserve the second law of thermodynamics from the point of view of a being, such as Maxwell's famous demon, that gets information about a system such as a gas and uses that information to perform useful work. Without some such modification, entropy would decrease in between the acquisition and erasure of information. A contribution by one of us (Lloyd 1991) to the 1989 Festschrift of the other (M.G.-M.) pointed out that the proper formulation of the preservation of the second law of thermodynamics is to preprogram the demon with the algorithmic information required to describe the ensemble of measurement results, and to have the demon assign Shannon-Fano or Huffman codes to the results. In a series of elegant articles detailing the connection between algorithmic information content and chaos, both classical and quantum, Caves (1990, 1994) showed the usefulness of expressing the overall AIC of a system and its measured state as the sum of the AIC of a description of the system together with the ensemble of states with their probabilities (which Caves called the 'algorithmic background information') plus the conditional AIC of the measured state. References Lloyd (1991) and Caves (1990, 1994) thus explain which piece of the AIC should be traded off for entropy in the Maxwell demon story: it is the average conditional AIC, which closely approximates the entropy it replaces, as in Eq. (4).

The idea of substitution is closely related to the Church–Turing thesis, as stated by Wojciech Zurek: "what is human-computable is also machine-computable." Concerning Maxwell's demon, Zurek showed that the second law of thermodynamics applies as long as information processing is subject to laws of the type of Turing's universal machines.

Effective Complexity

Information measures provide a precise method of dealing with trade-offs between knowledge and ignorance, and they supply a useful definition of complexity. People accumulate knowledge in the form of books, data, and computer programs, and deal with what they don't know using probabilities. We propose the following method for dealing quantitatively with the trade-off between knowledge and ignorance: measure knowledge using algorithmic information content of an ensemble and measure ignorance using Shannon information (or the average conditional AIC that is approximately equal to it).

How does the method work in practice? Suppose we want to discuss the effective complexity of a given entity, i.e., the length of a concise description of its perceived regularities. A prescription has been given by one of us (Gell-Mann 1995) that makes this idea more precise. Recognizing certain regularities of an entity can be regarded as equivalent to embedding it in a set of similar entities and specifying a probability for each member of the set. The entities differ in various respects but share the regularities.

For example, suppose a sequence of coin tosses is interpreted as representing a succession of independent events in each of which the probability of heads is 0.6 and that of tails is 0.4. Then the set in which the sequence is embedded consists of all possible sequences of the same length, where the probability of each one is 0.6 to the number of heads times 0.4 to the number of tails.

To take a very different example, suppose the regularities of a forest are described in terms of a particular number of trees belonging to each of its tree species, with a particular distribution of spacings and of ages and sizes and growth habits, ignoring for simplicity all the other organisms and all the other properties of the trees. That amounts to embedding the forest in a set of similar forests with the same number of trees of each species, but in various arrangements. For instance, if in the original forest a young Engelmann Spruce is interchanged with a mature Noble Fir, the resulting forest is also a member of the set. The perceived regularities assign a probability to each of the forests belonging to the set.

Thus the original entity e is replaced by an ensemble E that corresponds to certain regularities of e. Now a description of either e or E can be expressed as a bit string, although the resulting string is determined not only by intrinsic properties of e or E but also by a number of extrinsic factors that depend on who or what is doing the describing:

1. the level of detail of the description;

2. the knowledge and understanding of the world that is assumed before the description begins;

3. the language employed;

This point is one of the most relevant for effective complexity. As I have emphasized, the subtle step taken by Gell-Mann (1995a) opens a wide area of discussion that allows us to connect these concepts from information theory with epistemological aspects and scientific activity. In particular, the "arational" characteristic of Kuhn's ideas, criticized as irrationals but not accepted by him, finds a very appropriate response using Gell-Mann's ideas.

4. the coding convention for converting the language into bits.

Allowing for such sources of arbitrariness, one can utilize the resulting bit string to assign an AIC to the entity by equating it to the AIC of the string:

$$K_U(e) = K_U(s_e). \quad (5)$$

Some have seen this relation as a way of defining the complexity of the entity. However, a string of given length has maximal AIC when it is a 'random string'—one with no regularities—while calling something complex does not usually imply that it is random. Rather, the meaning of the term 'complexity' that corresponds most closely to its use in ordinary conversation and in scientific discourse corresponds to effective complexity, the length of a concise description not of the entity but of its identified regularities (Gell-Mann 1994, 1995). Thus we should equate the effective complexity to the AIC not of the entity but of the ensemble in which it is embedded as a typical member:

$$\mathcal{E} = K_{U'}(\tilde{E}). \quad (6)$$

By 'typical' we mean that $-\log \tilde{p}$ is no larger than the entropy or Shannon information of the ensemble. If $-\log \tilde{p}$ were greater than that, the entity would be a comparatively improbable member of the ensemble, which would not then serve very well to describe the entity's regularities. The algorithmic information content of an ensemble is the length of the shortest program required to specify the members of the ensemble together with their probabilities: by defining effective complexity $\mathcal{E}$ in this fashion we restrict our attention to ensembles whose membership and probabilities are computable.

As an example, suppose the entity is a dynamical system evolving in time and that, concerned with regularities in its behavior, we sample a set of variables Q for at times $t_1, t_2, \ldots$, getting the results $q_1, q_2, \ldots$. A *model* m for the system is an algorithm that assigns probabilities $p(q_1 q_2 \ldots \mid m)$ to the possible results of the sampling. Models must be intrinsically probabilistic in order to specify both predictions and noise-related deviations from those predictions. Specifying the ensemble $\{q_1 q_2 \ldots, p(q_1 q_2 \ldots \mid m)\}$ of possible results of the sampling is equivalent to embedding the entity consisting of the sequence of results in an ensemble of such sequences with the assigned probabilities. Hence the effective complexity of the behavior of the system as described by the model m is just the AIC of the model, $K_{U'}(m)$.

A model is a *schema* (Gell-Mann 1994) in the form of an algorithm for specifying an ensemble of possible data. It encapsulates a set of regularities in the data observed, and its AIC, the amount of information necessary to specify those regularities, is the effective complexity of the observed data as described by the model. But how can one compare the effectiveness of different models for the same observed data? More generally, how can one compare the regularities assigned to the same entity by different observers? Those questions bring us to the concept of *total information* or *augmented entropy*.

Total Information

Effective complexity measures knowledge, in the sense that it quantifies the extent to which an entity is taken to be regular, nonrandom, and hence predictable. The remaining features of the entity are taken to be irregular and probabilistic. Whereas the effective complexity is the AIC of the ensemble in which the entity is embedded, the information required to specify the residual probabilistic features of the entity is given by the entropy or Shannon information over the ensemble. Since that information is approximately equal to the ensemble average of the conditional AIC of the members, as in Eq. (4), that average could be substituted for the entropy. (Equation (4) holds just as well for an ensemble of abstract entities as it does for an ensemble of messages once the identification Eq. (5) is made.)

The sum Σ of the effective complexity $\mathcal{E}$ and the entropy (or Shannon information) S is the total information required to describe both the rule-based and the probabilistic features of the entity as perceived. If the entity e is embedded in a coarse-grained ensemble $\tilde{E} = \{(r, \tilde{p})\}$, then the total information or augmented entropy is

$$\Sigma = \mathcal{E} + s = K_{U'}(\tilde{E}) - \sum_r \tilde{p}_r \log \tilde{p}_r \approx K_{U'}(\tilde{E}) + \sum_r \tilde{p}_r K_{U'}(r \mid \tilde{E}), \tag{7}$$

which gives

$$\Sigma = \sum_r \tilde{p}_r \left[K_{U'}(\tilde{E}) + K_{U'}(r \mid \tilde{E}) \right]. \tag{8}$$

Making use of the discussion preceding the paragraph that contains Eq. (2), we see that Σ is roughly approximated by $\sum_r \tilde{p}_r K_{U'}(\tilde{E}, r)$.

Like algorithmic information content itself, Σ is an approximate information measure. It was introduced by Gell-Mann and Hartle (1994) to measure departures from classicality in ensembles of alternative decoherent coarse-grained quantum-mechanical histories of the universe. The smaller the total information in such an ensemble of histories (or "realm"), the closer it is to being quasiclassical, provided the ensemble satisfies certain conditions that exclude pathological cases. Shannon information applied to an ensemble of alternative histories was discussed earlier by Lloyd and Pagels (1988), who used this quantity as a measure of the resources required to pick a single history out of the ensemble.

The total information needed to describe the behavior of a system in time is a useful quantity in problems of characterization and control of complex systems (Lloyd and Slotine 1996). Any process that decreases the system's total information increases one's ability to describe and control the system.

We have the following results concerning the process of minimizing the total information. First, note

Theorem 1. *Σ achieves an approximate absolute minimum when $S = 0$, $K_{U'}(\tilde{E}) \approx K_{U'}(e)$.*

That is, the total information is within a few bits of a minimum for the ensemble $\tilde{E} = \{(e, \tilde{p}(e) = 1)\}$: one way of minimizing the total information is to replace it by algorithmic information alone. Of course this may not be the only ensemble $\tilde{E}$ that minimizes the total information. And in many cases it is not the most interesting one.

The proof of Theorem 1 is as follows: First, note that for any ensemble $\tilde{E}$ of which e is a typical member, we have $K_{U'}(e) \lesssim K_{U'}(e, \tilde{E}) \approx K_{U'}(\tilde{E}) + K_{U'}(e \mid \tilde{E})$, since K is an approximate information measure. But $K_{U'}(\tilde{E}) + K_{U'}(e \mid \tilde{E}) \lesssim K_{U'}(\tilde{E}) - \log \tilde{p}(e)$, since U' was explicitly constructed to allow efficient coding. The typicality of e in $\tilde{E}$ was defined above to mean that $-\log \tilde{p}(e) \lesssim S$, where $S = -\sum_r \tilde{p}_r \log \tilde{p}_r$ as above.

Combining these approximate inequalities gives:

$$\begin{aligned}K_{U'}(e) &\lesssim K_{U'}(e, \tilde{E}) \\ &\lesssim K_{U'}(\tilde{E}) - \log \tilde{p}(e) \\ &\lesssim K_{U'}(\tilde{E}) - \Sigma_r \tilde{p}_r \log \tilde{p}_r \\ &= \Sigma\end{aligned} \qquad (9)$$

There may in fact be many possible choices of ensemble $\tilde{E}$ containing e as a typical member that also approximately minimize the total information. From the proof of Theorem 1 we can infer the corollary that the set of ensembles that minimize the total information must obey the following approximate equalities:

$$(\text{1.i}) \; K_{U'}(e) \approx K_{U'}(e, \tilde{E}),$$

i.e., an ensemble minimizing the total information does not specify additional regularities that e lacks.

$$(\text{1.ii}) \; K_{U'}(e \mid \tilde{E}) \approx -\log \tilde{p}(e).$$

Since the construction of U' implies that $K_{U'}(e \mid \tilde{E}) < -\log \tilde{p}(e) - 2$, equality (1.ii) means that $\tilde{E}$ captures virtually all of the regularities of e: the only way for $K_{U'}(e \mid \tilde{E})$ to be much less than $-\log \tilde{p}(e)$ is for e to have regularities that are not specified by $\tilde{E}$.

$$(\text{1.iii}) \; -\log \tilde{p}(e) \approx -\sum_{r \in \bar{E}} \tilde{p}_r \log \tilde{p}_r,$$

i.e., not only is e a typical member of the ensemble, so that $-\log \tilde{p}(e)$ is not much greater than S, it is also in some sense a representative member, in that $-\log \tilde{p}(e)$ is not much less than S. Equality (1.iii) eliminates from consideration ensembles in which one, typical member has a large probability, and many other atypical members share the remaining small probability.

We can now answer the question of how different perceived sets of regularities of an entity can be compared. It is the smallness of the total information that gives the first indication of how well the regularities have been identified. The smaller the effective complexity term, the simpler the description of the regularities. The smaller the entropy term, the less spread there is among the possible entities possessing those regularities. In many

> In comparing competing theories, the discussion in this section is highly relevant. Historically, there has been a tension between observed regularities, their description, and the entropy associated with choosing a particular theory that accounts for the phenomenon. Two cases that may be valuable for discussing the epistemological scope of Gell-Mann and Lloyd's approach are the formulation of Maxwell's equations circa 1860, which unify a series of existing prior knowledge, and the paradigm shifts in cosmological models.

cases, reducing the total information involves a trade-off between the two terms—the description must be made more complex in order to reduce the entropy and thus specify more closely the character of the entity (a number of examples of this trade-off are given in reference 20). Such trade-offs occur frequently in connection with scientific theories. Sometimes, however, it is possible to reduce both terms at once, i.e., a simpler schema is also a more powerful one. That situation is encountered from time to time in the scientific enterprise, for example in formulating theories in fundamental physics. The minimization of total information is closely related to the minimum description length procedure for the analysis of dynamical systems (Rissanen 1989), but differs from it by its explicit use of the trade-off between algorithmic and probabilistic methods in Theorem 1 and its corollaries.

The smallness of the total information is thus the first criterion for comparing sets of perceived regularities of an entity e. In general, there are many different choices of $\tilde{E}$ that minimize Σ. To pick out a particular choice from this set requires further criteria. One possible criterion is to choose ensembles that minimize not only total information Σ but effective complexity as well, thus maximizing S at the expense of $\mathcal{E}$. Indeed, by looking at ensembles of strings with the same AIC as e, it is possible to put all but approximately $K_{U'}(e)$ of the total information into S, leaving only about $\log K_{U'}(e)$ for the effective complexity. That would make all entities effectively simple. If, however, e in fact possesses a wide variety of regularities, putting most of Σ into S results in unwieldy ensembles $\tilde{E}$ that typically require vast amounts of computer time to produce the elements of that ensemble from its shortest description. A more practical criterion for identifying the 'best' ensembles is to ask not only for economy of description, but for ease of expression as well, by placing a cut-off on the amount of computer time required to generate a typical member from the shortest description of the ensemble together with the Shannon-Fano code for the member. The cut-off should be at least as great as (and not much greater than) the amount of time required to compute e from its minimal program. Such a cut-off insures that the process of generating strings with the same regularities as e is not much more difficult than generating e itself, and has the effect of including hard-to-compute information in the description of the ensemble and easy-to-compute information in the codewords for the members of the ensemble. Minimizing total information and effective complexity for such ensembles is similar to including 'deep,' information (which must be used early in the computation) in $\mathcal{E}$, and 'shallow,' information (which can be used later) in S[2]. (An alternative cut-off procedure is to include deep bits in $\mathcal{E}$ and shallow bits in S explicitly.) Such a cut-off, along with minimization of $\sum$ and $\mathcal{E}$, results in an effective complexity that is much less subjective than the corresponding quantity for a particular set of regularities that an observer happens to perceive.

The following examples give some idea of how the criteria work in practice. First take the case of a long sequence of zeroes and ones representing the results of ℓ successive coin tosses, with 1 for heads and 0 for tails. If the sequence is random in the technical sense (with maximum AIC for its length ℓ), where ℓ is random as well, then its AIC, which we have called $K_{U'}(e)$, is about equal to $\ell + \log \ell$, or more precisely, to $\ell + \log \ell + \log \log \ell + \ldots$. (The logarithmic terms arise from the need to render self-delimiting the minimal program required to print out a random ℓ-bit number.) For simplicity, we shall ignore the terms beyond $\ell + \log \ell$. That is then the minimum value of Σ, and for the ensemble consisting only of e with probability 1, we have $S = 0$ and $\mathcal{E} = \Sigma \sim \ell + \log \ell$, where the symbol $\sim$ indicates equality to within $\mathcal{O}(\log \log \ell)$. But, for this random sequence, interpreting it as arising from independent tosses of a fair coin is not such a bad description. The ensemble is then the set of all sequences of length ℓ, with equal probabilities. The effective complexity, the AIC of that ensemble, is small, on the order of $\log \ell$, and the Shannon information is maximal, equal to ℓ. Thus, without changing Σ significantly, we can reduce $\mathcal{E}$ to near $\log \ell$, replacing almost all of it by S. Only a brief computation is required to produce a member of the ensemble. The random sequence then has very low effective complexity.

Now consider the case where the sequence happens, by a very rare accident, to consist entirely of ones. The AIC of the sequence is small, since the sequence can be produced by a program of length $\sim \log \ell$ that says 'print ℓ one's.' By Theorem 1, the minimal Σ is also small. Here utilizing the ensemble consisting of all sequences of length ℓ is a very poor choice of regularities, since it gives a large value for the total information ($\Sigma \sim \ell + \log \ell$), while the single member ensemble gives $S = 0$ and $\mathcal{E} =$

---

[2] Compare C. H. Bennett's notion of 'logical depth,' as described in Bennett (1988).

$\Sigma \sim \log \ell$, minimizing both total information and effective complexity with a trivial computation time.

A closely related example is that of a string of zeroes and ones representing π in binary notation out to a billion bits. An observer unaware of the connection with π might identify the string as a sequence of bits resulting from random tosses of a fair coin, with a low effective complexity and an entropy equal to a billion. An observer recognizing the sequence would assign a slightly higher complexity corresponding to a description of π, but with vanishing entropy, so that Σ and $\mathcal{E}$ are equal and small. The computation time is long, but no greater than the cut-off. Clearly, that observer has done a better job of identifying regularities. (This example shows that effective complexity is related to the algorithmic measures sophistication (Koppel 1988) and Kolmogorov sufficient statistic (Cover, Gacs, and Grey 1989), from both of which it differs by the explicit introduction of ensembles and probabilities.)

In all of these examples, the effective complexity turns out to be rather small once the regularities that minimize the total information are identified. In other cases, however, the optimal regularities correspond to an ensemble that requires a very long description, so that the effective complexity is high. The entity or bit string is then genuinely complex. We cannot give a simple explicit prescription for constructing a bit string with high effective complexity, because if we succeeded it would *ipso facto* not have high effective complexity. However, we can give a general description of what such a string might look like.

Consider the set of computable real numbers $\{\zeta\}$, where each ζ is generated by a program that, when given m as input, prints out the first m bits of ζ in an amount of time bounded by a computable function of m. Examples of computable real numbers include $\pi, e, \sqrt{2}, .11111\ldots$, etc. We obtain an ensemble V_m by taking v computable numbers, truncated to m bits, and assigning all of them the same probability $1/v$. Now imagine a string of length $\ell = km$ constructed by concatenating k numbers selected from V_m at random; since repetition is allowed, we can have $k = \alpha v$ where $\alpha > 1$. The algorithmic information content of such a string is $\sim K_{U'}(V_m) + \log k + k \log v$, which in turn provides the minimum value of the total information. $K_{U'}(V_m)$ comes out approximately $\gamma v \log v + \log m$, where γ is a constant and $\gamma \log v$ represents the average AIC of the computable numbers in V_m. We can now minimize the effective complexity while keeping the total information constant by taking the ensemble that describes the selected subset of programs and the method of concatenating their outputs, yielding $\mathcal{E} - K_{U'}(V_m) + \log k$, with $S = k \log v$: S is equal to the information required to describe the random order of the concatenation. Any attempt to reduce the effective complexity in this example further by subsuming in S the randomness inherent in generating the original ensemble V_m from the ensemble of computable real numbers in ensembles that violate the cut-off: brief programs for generating specific subsets of computable real numbers with the same average AIC have astronomically long running times.

We have $\mathcal{E} \sim \gamma v \log v + \log \ell$, $S \sim \alpha v \log v$, and $\Sigma_{\min} = \mathcal{E} + S \sim (\alpha + \gamma) v \log v + \log \ell$. The last quantity must of course be less than $\ell + \log \ell$, so that $(\alpha + \gamma) v \log v < \ell$. The effective complexity cannot grow more rapidly with ℓ that $c\ell + \log \ell$, where $c \leq 1$. Such a dependence is achieved for this example if $(\alpha + \gamma) v \log v = \beta \ell$ with $\beta < 1$: then $\mathcal{E} \sim (\alpha + \gamma)^{-1} \gamma \beta \ell + \log \ell$, $S \sim (\alpha + \gamma)^{-1} \alpha \beta \ell$, and $\Sigma_{\min} \sim \beta \ell + \log \ell$.

The effectively complex string of the previous example can be considered to be a mathematical analog of a sequence of DNA composed of a large number of subsequences each of which encodes some functional feature of an organism. A related example of an effectively complex system is a hierarchy in which random information at one level of the hierarchy is used to prescribe regular or rule-based behavior at the next level of the hierarchy. An example of such a hierarchy is the time evolution of the universe itself, in which the outcomes of random quantum fluctuations at one time become 'frozen accidents' that govern the dynamics by which quantum fluctuations form at later times (Gell-Mann 1994, 1995; Gell-Mann and Hartle 1994). The universe has large effective complexity.

Conclusion

This article has provided an introduction to information measures, a more general definition of information than either Shannon information or algorithmic information content. Information measures allow the identification of effective complexity, a measure of the amount of information required to describe a system's regularities or rule-governed behavior. Effective complexity corresponds naturally to human intuitions of complexity: wholly random processes and deterministic but easily specified processes are effectively simple, while processes such as evolution that combine deterministic behavior with a large number of historical accidents are effectively complex. Total information, which is the sum of effective complexity and Shannon information, is a quantity with applications to many fields. The smaller it is, the more a learning system has learned, the closer a set of alternative coarse-grained quantum-mechanical histories is to quasiclassicality, and the more a complex system can be characterized and controlled.

The 'best' regularities, which are independent of the cleverness of the observer identifying them, are encountered when, for the least value of the total information, the effective complexity is also minimized, or equivalently, the entropy or Shannon information is maximized, subject to constraints on computation time.

Acknowledgements

The authors would like to acknowledge the great value of conversations with C. H. Bennett.

The work of M. Gell-Mann was supported in part by grants to the Santa Fe Institute from Jeffrey Epstein, David Schiff, and Gideon Gartner and by Grant # N00014-95-1-1000 from the Office of Naval Research.

The work of Seth Lloyd is supported in part by Finmeccanica and by Grant # N00014-95-1-0975 from the Office of Naval Research.

REFERENCES

Bennett, C. H. 1988. *Dissipation, Information, Computational Complexity, and the Definition of Organization,* edited by D. Pines, 215–234. Redwood City, CA: Addison-Wesley.

Boltzmann, L. 1866. "Über die Mechanische Bedeutung des Zweiten Hauptsatzes der Wärmetheorie." *Wiener Berichte* 53:195–220.

———. 1871. "Über das Wärmegleichgewicht zwischen mehratomigen Gasmolekülen." *Wiener Berichte* 63:397–418.

———. 1872. "Weitere Studien über das Wärmegleichgewicht unter Gasmolekülen." *Wiener Berichte* 66:275–370.

Caves, C. M. 1990. "Entropy and Information: How much Information is Needed to Assign a Probability?" In *Complexity, Entropy, the Physics of Information,* edited by W. H. Zurek, 91–115. Redwood City, CA: Addison-Wesley.

———. 1994. "Information, Entropy, and Chaos." In *Physical Origins of Time Asymmetry,* edited by J. J. Halliwell, J. Pérez-Mercader, and W. H. Zurek, 47–89. Cambridge, UK: Cambridge University Press.

Clausius, R. 1857. "On the Nature of the Motion Which We Call Heat." *Philosophical Magazine* 14:108. https://doi.org/10.1142/9781848161337_0011.

Cover, T. M., P. Gacs, and R. M. Grey. 1989. "Kolmogorov's Contributions to Information Theory and Algorithmic Complexity." *Annals of Probability* 17 (3): 840–865.

Cover, T. M., and J. A. Thomas. 1991. "Elements of Information Theory." New York: Wiley.

Ehrenfest, P., and T. Ehrenfest. 1959. *The Conceptual Foundations of the Statistical Approach in Mechanics.* Ithaca, NY: Cornell University Press.

Gell-Mann, M. 1994. *The Quark and the Jaguar.* New York, NY: W. H. Freeman.

———. 1995. "What is Complexity?" *Complexity* 1:16–19.

Gell-Mann, M., and J. B. Hartle. 1994. "Strong Decoherence." In *The Quantum Classical Correspondence: Proceedings of the 4th Drexel Symposium on Quantum Non-integrability,* edited by B.-L. Hu and D.-H. Feng.

Gibbs, J. W. 1902. *Elementary Principles in Statistical Mechanics.* New Haven, CT: Yale University Press.

Jaynes, E. T. 1982. *Papers on Probability, Statistics and Statistical Physics.* Edited by R. D. Rosenkrantz. Dordrecht, Holland: Reidel.

Koppel, M. 1988. *Structure,* edited by R. Herken, 435–452. Oxford, UK: Oxford University Press.

Li, M., and P. M. B. Vitany. 1993. "An Introduction to Kolmogorov Complexity and its Applications." New York: Springer-Verlag.

Lloyd, S. 1991. "Uncomputability, Intractability, and the Efficiency of Heat Engines." In *Elementary Particles and the Universe: Essays in Honor of Murray Gell-Mann,* edited by J. H. Schwarz. Cambridge, UK: Cambridge University Press.

Lloyd, S., and H. Pagels. 1988. "Complexity as Thermodynamic Depth." *Annals of Physics* 188:186–213.

Lloyd, S., and J.-J. Slotine. 1996. "Algorithmic Lyapunov Functions for Stable Adaptation and Control." *International Journal of Adaptive Control and Signal Processing.*

Maxwell, J. C. 1860a. "Illustrations of the Dynamical Theory of Gases." *Philosophical Magazine* 19:19–32. https://doi.org/10.1142/9781848161337_0011.

———. 1860b. "Illustrations of the Dynamical Theory of Gases." *Philosophical Magazine* 20:21–37.

Rissanen, J. 1989. *Stochastic Complexity in Statistical Inquiry.* Singapore: World Scientific.

Schack, R. 1997. "Algorithmic Information and Simplicity in Statistical Physics." *International Journal of Theoretical Physics* 36:209–226.

Shannon, C. E. 1993. "The Lattice Theory of Information." In *Claude Elwood Shannon, Collected Papers.* Lecture delivered at the Symposium on Information Theory, held in the Lecture Theatre of the Royal Society Burlington House, Ministry of Supply, London, September 1950. New York, NY: IEEE Press.

Shannon, C. E., and W. Weaver. 1948. *The Mathematical Theory of Communication.* Urbana, IL: University of Illinois Press.

Tolman, R. C. 1938. *The Principals of Statistical Mechanics.* London, UK: Oxford University Press.

Zurek, W. H. 1989a. "Algorithmic Randomness and Physical Entropy." *Physical Review A* 40:4731–4751.

———. 1989b. "Thermodynamic Cost of Computation, Algorithmic Complexity and the Information Metric." *Nature* 341:119.

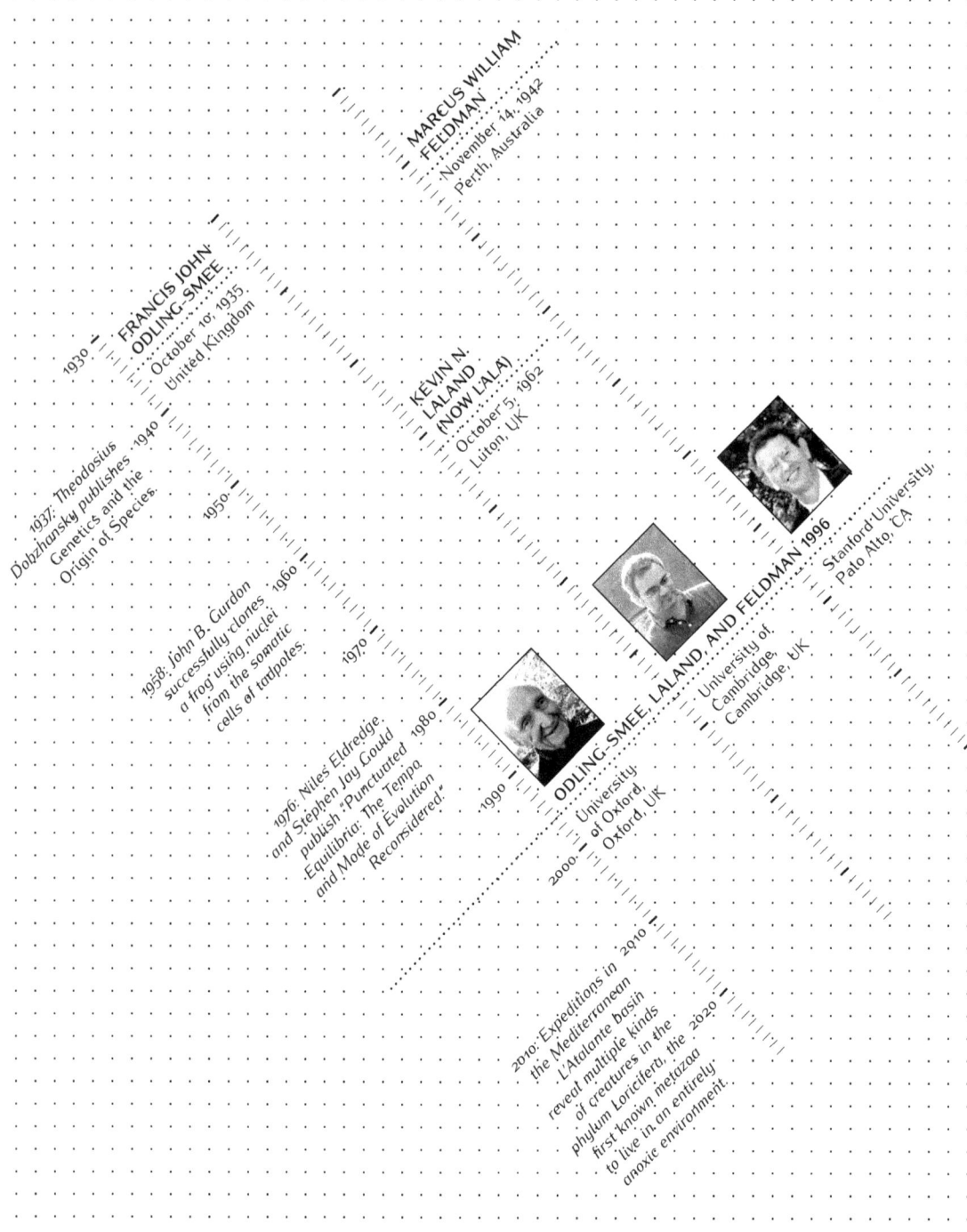

ODLING-SMEE, LALAND & FELDMAN

[82]

THE EVOLUTION OF SELECTION

David C. Krakauer, Santa Fe Institute

Abstract

The theory of niche construction emerged out of a growing realization that natural selection is itself selected for. The standard model of evolution separates the timescales of adaptation and selection in order that lineages evolve and optimize in stable selective conditions. Niche construction observes that organismal adaptation can directly modify selective feedback. This implies that the trajectory of evolution should not always be thought about in terms of simple optimization, and that it will respond to the strategies of diverse populations of organisms, tend to pursue elevated levels of exploration, and perhaps result in increasing degrees of complexity.

F. J. Odling-Smee, K. N. Laland, and M. W. Feldman, "Niche Construction," *The American Naturalist* 147 (4), 641–648 (1996).

Used with permission of the University of Chicago Press; permission conveyed through Copyright Clearance Center, Inc.

The prevailing understanding of natural selection originates on July 1, 1858, when Charles Lyell presented the work of Charles Darwin and Alfred Russell Wallace on natural selection to the Linnean Society of London. Shortly thereafter, on August 20, their presentation was published as a joint paper to the *Journal of the Proceedings of the Linnean Society* as "On the Tendency of Species to Form Varieties; and on the Perpetuation of Varieties and Species by Natural Selection" (Darwin and Wallace 1858).

The idea of natural selection was sufficiently novel to Victorian science that Darwin and Wallace felt it necessary to introduce a mechanical metaphor to elucidate its action: "The action of this principle is exactly like that of the centrifugal governor of the steam engine, which checks and corrects any irregularities almost before they become evident; and in like manner no unbalanced deficiency in the animal kingdom can ever reach any conspicuous magnitude, because it would make itself felt at the very first step, by rendering existence difficult and extinction almost sure soon to follow."

The centrifugal governor was invented by Christian Huygens to regulate grain size in windmills and subsequently adapted to steam engines by James Watt in 1788. The governor regulates the load on an engine by slowly throttling the flow of steam supplying the engine as power output increases. In this way it maintains a constant speed across variations in load through negative feedback. Crucially, the governor can be adjusted so as to establish an equilibrium speed and in this way the system is doubly adaptive: endogenously by means of centrifugal regulation, and exogenously by a human operator.

When Darwin published *On the Origin of Species* in 1859, all references to the governor were dropped and replaced with another metaphor—at once more grandiose, perhaps more unifying, but less accurate—Newton's law of gravity: "It has been said that I speak of natural selection as an active power or Deity; but who objects to an author speaking of the attraction of gravity as ruling the movements of the planets?" And the celebrated sentence, "There is grandeur in this view of life, with its several powers, having been originally breathed by the Creator into a few forms or into one; and that, whilst this planet has gone circling on according to the fixed law of gravity, from so simple a beginning endless forms most beautiful and most wonderful have been, and are being evolved" (Darwin 1859).

Governors and gravity have in common a separation of time scales, with slow processes—the fixed point of the governor and the value of the gravitational constant—regulating fast processes—the motion of the valve stem and valve or motion of the stars and planets.

Whereas the governor is a modifiable physical system that regulates a heat engine, gravitation is a universal and invariant interaction among all massive objects. By moving from governors to gravity, Darwin was on the one hand helping to explain how simple laws might govern complex realities, but, at the same time, effectively decoupling the action of natural selection from the behavior of organisms.

This is a rather strange turn of events in light of Darwin's earliest work on coral reefs. In his monograph of 1842, *The Structure and Distribution of Coral Reefs,* Darwin proposed that the form of every variety of reef could be explained through "reefs being formed of different species of corals, adapted to live at different depths," when combined with natural geological processes of subsistence. In this way, coral polyps are the architects of their own habitations, and thereby influence directly their selective environment: "The corals, also, on the margin of Keeling Island occurred in zones: thus the Porites and Millepora complanata grow to a large size, only where they are washed by a heavy sea, and are killed by a short exposure to the air; whereas, three species of Nullipora also live amidst the breakers, but are able to survive uncovered for a part of each tide" (Darwin 1842).

And Darwin revisited this idea in his final book, *On the Formation of Vegetable Molds by the Action of Earth Worms*, in which the worms themselves generate through their castings, "A layer, though a thin one, of fine earth, which probably long retains some moisture, is in all cases, as I believe, necessary for their existence; and the mere compression of the soil appears to be in some degree favorable to them" (Darwin 1881).

Nevertheless, these early ideas on what we know describe as niche construction, or the extended phenotype, or ecosystem engineering, did not inform the "modern synthesis" of evolution in the first decennaries of the twentieth century. The integration of simple Mendelian inheritance and natural selection through classical equations of motion in population genetics cleaved rather closely to the classical mechanics of gravity, in which selection pressures are assumed constant. Just like the gravitational constant in the Newtonian theory. In an interesting later development in physics, Einstein's general theory of relativity, a niche construction–like dynamic was discovered, that John Wheeler characterized as "Space-time tells matter how to move; matter tells space-time how to curve" (Misner, Thorne, and Wheeler 1973). But this insight was of another time and apparently very far from biological processes.

Across the second half of the twentieth century, a number of influential evolutionary books and papers were written that pursued the organism-built environment. Starting with the highly influential works of Karl von von Frisch (1974) on *Animal Architecture*, Richard Dawkins's (1982) prescient ideas on *The Extended Phenotype*, Richard Lewontin's (1978) introduction of the concept of niche construction itself, and the work of Jones and colleagues on ecosystem engineering, in which emphasis is placed on the collective contribution of organisms to macroecological patterns, both adaptive and maladative (Jones, Lawton, and Shachak 1994). Much of this work was empirical and conceptual. What was required was an effort to formalize these ideas mathematically and move them from a rather exotic periphery to the generative center of the field.

John Odling-Smee, Kevin N. Laland, and Marcus W. Feldman (SLF) describe niche construction in their as original paper as a "neglected process in evolution" because the environment and the organism have been treated as time-separable dynamical systems and their reciprocal interactions neglected. When traits of a phenotype capable of modifying the environment of an organism are considered and time scales overlap, then selection pressures can be seen to be, in part, encoded in the organism genome.

Among other things, this introduces a circular causality into evolutionary dynamics with implications for the evolution of development, sociality, and long-term macro-evolutionary patterns. For SLF, niche construction is not just a small missing part of the evolutionary puzzle, it is a full half of the puzzle. To the extent that natural selection allows us to make sense of the teleological or functional property of organisms, then niche construction is required to explain where selection comes from in the first place.

REFERENCES

Darwin, C. 1842. *The Structure and Distribution of Coral Reefs: Being the First Part of the Geology of the Voyage of the Beagle, Under the Command of Capt. Fitzroy, R.N. during the years 1832 to 1836.* London, UK: Smith, Elder and Co.

———. 1859. *On the Origin of Species.* London, UK: John Murray.

———. 1881. *The Formation of Vegetable Mould Through the Action of Worms, with Observations on Their Habits.* London, UK: John Murray.

Darwin, C., and A. Wallace. 1858. "On the Tendency of Species to Form Varieties; and on the Perpetuation of Varieties and Species by Natural Means of Selection." *Journal of the Proceedings of the Linnean Society of London. Zoology* 3 (9): 45–62. https://doi.org/10.1111/j.1096-3642.1858.tb02500.x.

Dawkins, R. 1982. *The Extended Phenotype: The Gene as the Unit of Selection.* Oxford, UK: Oxford University Press.

Jones, C. G., J. H. Lawton, and M. Shachak. 1994. "Organisms as Ecosystem Engineers." In *Ecosystem Management,* 130–147. Washington, DC: US Goverment Publishing Office.

Lewontin, R. C. 1978. "Adaptation." *Scientific American* 239 (3): 212–230. https://doi.org/10.1038/scientificamerican0978-212.

Misner, C. W., K. S. Thorne, and J. A. Wheeler. 1973. *Gravitation.* San Francisco, CA: W. H. Freeman and Company.

von Frisch, Karl. 1974. *Animal Architecture.* New York, NY: Harcourt Brace Jovanovich.

NICHE CONSTRUCTION

F. John Odling-Smee, Kevin N. Laland and Marcus W. Feldman

Niche construction relates not only to creating order but producing disorder. Arguably entropy production will dominate, and most NC is likely to generate much environmental damage indirectly.

Organisms, through their metabolism, their activities, and their choices, define, partly create, and partly destroy their own niches. We refer to these phenomena as "niche construction." Here we argue that niche construction regularly modifies both biotic and abiotic sources of natural selection in environments and, in so doing, generates a form of feedback in evolution that is not yet fully appreciated by contemporary evolutionary theory (Lewontin 1978; 1983; Odling-Smee 1988, in press).

Adaptation is generally thought of as a process by which natural selection shapes organisms to fit preestablished environmental "templates." Environments pose "problems," and those organisms best equipped to deal with the problems leave the most offspring. Despite the recognition that forces independent of organisms often change the worlds to which populations adapt (van Valen 1973), the changes that organisms bring about in their own worlds are rarely considered in evolutionary analyses. Yet, to varying degrees, organisms choose their own habitats, mates, and resources and construct important components of their local environments such as nests, holes, burrows, paths, webs, dams, and chemical environments. Many organisms also choose, protect, and provision "nursery" environments for their offspring. Organisms not only adapt to environments, but in part also construct them (Lewontin 1983). Hence, many of the sources of natural selection to which organisms are exposed exist partly as a consequence of the niche-constructing activities of past and present generations of organisms.

There are numerous cases of organisms modifying their own selective environments in nontrivial ways, by changing their surroundings or by constructing artifacts (von Frisch 1975; Hansell 1984). One early example was described by Darwin (1881). Earthworms, through their burrowing activities, their dragging organic material into the soil, their mixing it with inorganic material, and their casting, which serves as a basis for microbial activity, change both the structure and chemistry of soils (Lee 1985). As a result of the accumulated effects of past generations of earthworm niche construction, present generations inhabit radically altered environments and are exposed to changing sets of selection pressures.

There is also considerable evidence of evolutionary responses to self-induced selection pressures. For instance, social bees, wasps, ants, and termites construct elaborate nests that then mediate selection for many nest regulatory, maintenance, and defense behaviors (Rothenbuhler 1964; Spradbery 1973; von Frisch 1975; Mathews and Mathews 1978; Hansell 1984). Many species of fish, amphibians, reptiles, birds, and mammals construct nests and burrows, which then influence further selection. For example, comparative evidence suggests that the complex burrow systems excavated by moles, rats, badgers, and marmots, with their underground passages, interconnected chambers, and multiple entrances, have served as the source of selection for defense, maintenance, and regulation behaviors and components of mating rituals (von Frisch 1975; Hansell 1984).

Plants, too, change the chemical nature, pattern of nutrient cycling, temperature, humidity, and fertility of the soils in which they live (Ricklefs 1990). They may even affect local climates, the amount of precipitation, and the water cycle (Shukla, Nobra, and Sellars 1990). Many plants also change both their own and other species' local environments via allelopathy (Rice 1984), while pine and chaparral tree species facilitate forest fires by accumulating oils or litter (Mount 1964; Allen and Starr 1982). These species have evolved a resistance to fire and, in some pine species that require a fire before their seeds will germinate, a dependency on it (Allen and Starr 1982).

This relates to one of the most challenging aspects of niche construction: to demonstrate empirically direct and greatest adaptive benefit to constructors.

The Evolutionary Consequences of Niche Construction

These examples, and others (Lewontin 1982, 1983), suggest that niche construction may be a general phenomenon, that it is not restricted to a few isolated species or taxa, and that feedback from phenotypically modified sources of selection in environments has evolutionary as well as ecological consequences. Although several topics in contemporary population biology are already concerned with the evolutionary consequences of the changes that organisms bring about in their own environments (e.g., habitat, frequency- and density-dependent selection), so far these analyses have only focused on genetic loci that influence the production of the niche-constructing phenotype itself. What is missing is any exploration of the feedback effects on other genetic loci. A more general body of theory is required.

In order to encourage the development of this theory, in what follows we discuss some evolutionary consequences of niche construction and detail why it is likely to be of significance to the biological sciences.

Extragenetic Inheritance

Currently, evolutionary theory rests heavily on the assumption that only genes are transmitted from generation to generation, the principal exception being cultural inheritance (Feldman and Cavalli-Sforza 1976; Boyd and Richerson 1985). However, ancestral niche-constructing organisms effectively transmit legacies of modified natural selection pressures in their environments to their descendants. This extragenetic inheritance has previously been called an "exploitive system" (Waddington 1959), an "ontogenetic inheritance" (West, King, and Arberg 1988), and an "ecological inheritance" (Odling-Smee 1988). We will introduce it in stages.

If, in each generation, each organism only modifies its environment temporarily or inconsistently, then there will be no cumulative or consistent modification of any source of natural selection in its population's environment. If, however, in each generation, each organism repeatedly changes its own ontogenetic environment in the same way, because each individual inherits the same genes causing it to do so, then ancestral organisms can modify a source of natural selection for their descendants by repetitive niche construction. The environmental consequences of such niche construction may be transitory and may still be restricted to single generations only, but the same induced environmental change is reimposed sufficiently often, for sufficient generations, to serve as a significant source of selection.

For example, individual web spiders repeatedly make webs in their environments, generation after generation, because they repeatedly inherit genes instructing them to do so. Subsequently, the consistent presence of a web in each spider's environment may, over many generations, feed back to become the source of a new selection pressure for a further phenotypic change in the spiders, such as the building by Cyclosa of dummy spiders in their webs to divert the attention of avian predators (Edmunds 1974). Yet, this kind of feedback does not introduce an extragenetic inheritance in evolution, because no consequence of niche construction is transmitted through an external environment from one generation to the next.

In more complex cases, inherited genes instruct organisms to modify repeatedly the ontogenetic environments of their offspring as well as, or instead of, their own. Here the consequences of niche construction are effectively "transmitted" from one generation to the next via an external environment, in the form of a parentally modified source of natural selection for their offspring. This transmittal is sufficient to establish an extragenetic inheritance system in evolution. Offspring now receive a dual inheritance from their parents, genes relative to their selective environments, and at least some parentally modified sources of selection in their environments relative to their genes.

The cuckoo is an example. Cuckoo parents repeatedly select host nests for their offspring, generation after generation, thereby bequeathing modified selection pressures as well as genes to their offspring. These modified selection pressures have then apparently selected for changed adaptations in the offspring, such as a short incubation period or the behavioral ejection by newly hatched cuckoo chicks of host eggs from the parasitized nests. Also, cuckoos that are raised in the nests of a particular host species may preferentially parasitize that host species in whose nests they were originally raised themselves when they mature, possibly learning the host characteristics through imprinting (Krebs and Davies 1993). This kind of extragenetic inheritance is currently modeled as a non-Mendelian maternal inheritance (Cowley and Atchley 1992; Schluter and Gustafsson 1993). Maternal inheritance can generate some counterintuitive results, including temporarily reversed evolutionary responses to selection, and time lags that may result in populations continuing to evolve after selection has ceased by an "evolutionary momentum" (Feldman and Cavalli-Sforza 1976; Kirkpatrick and Lande 1989).

With niche construction, the easy distinctions between inheritance, selection, and adaptation can dissolve. According to the dominant timescales, environmental inheritance can be highly correlated with selection, and perhaps even indistinguishable from it.

Maternal inheritance is, however, a special case, and the effects of niche construction generalize to multiple generations and to multiple ancestors, not just mothers. For example, suppose a niche-constructing behavior is influenced by genetic variation at one set of loci, that it spreads through a population over many generations, and that it progressively changes the frequency of some resource in the environment as it does so. Suppose further that the frequency of the resource, now a part of an extragenetic inheritance, feeds back to the population to influence selection at a second set of loci. In these circumstances, a time lag should develop between the change in frequency of alleles at the niche construction loci and the response to a frequency-dependent modified selection pressure at the recipient loci, the only difference being that here the time lag is likely to take many more generations to build up.

Darwin's earthworms are an example. Suppose a first genetic locus influences some niche-constructing behavior, such as soil processing or burrow lining, which subsequently affects the amount of topsoil or the soil nutrients in the earthworms' environment. Another locus expresses a phenotype that is affected by soil conditions, such as the structure of the epidermis or the amount of mucus secreted. In this case, ancestral niche construction by many generations of earthworms, due to the first locus, will eventually feed back to the second locus and change its selection, but only after many generations of niche construction. Here again, the effect of the time lag should be to create an evolutionary "momentum," such that if the selection pressures at the first locus are relaxed or reversed, the response at the second locus should continue in the original direction for a number of additional generations. Moreover, assuming many generations are required to modify natural selection on a population, it might not be able to evolve fast enough to prevent the genetic variation on which its eventual response relies from being prematurely lost. This possibility also means that once a population reaches a stable equilibrium, it may take a greater period of time, or stronger selection, for the population to move away from it.

Odling-Smee, Laland, and Feldman (1996)

Indirect Gene Interactions

Niche construction provides a way in which the differential phenotypic expression of genotypes at one locus can be influenced by the genotype at another locus, indirectly via the external environment. For instance, the pink coloration characteristic of flamingo species is extracted from the carotenoid pigmentation of the crustacea they digest (Fox 1979). Here the genes influencing flamingo prey choice interact with those underlying pigment extraction and utilization, via the food resources in their environment. In several respects, the genetic basis of this feedback is reminiscent of epistasis. In contrast to conventional epistasis, however, niche construction can generate interactions between genes in different populations, even different species. For example, the genes that underlie those activities of earthworms that modify soil structure, thereby enhancing plant yields (Lee 1985), have influenced the expression of genes in the plant population affecting growth. Thus, the niche-constructing outputs of individuals not only change selective environments, which feed back to alter the fitnesses of alleles at other loci, but may also influence the phenotypic expression of those alleles in ontogenetic environments (West, King, and Arberg 1988). The effect of these interactions, which influence both the nature of the variants subjected to selection and the pattern of selection acting on those variants, is to introduce the kind of feedback loops between populations and their environments that Robertson (1991) suggests may make a considerable difference in evolution.

Synthesizing Evolutionary Biology and Ecology

Because niche construction may apply to interactions between genes in the same population, or in different populations, it provides a mechanism for driving populations to coevolve in ecosystems. Populations may affect each other not only directly in the ways that are already modeled by coevolutionary theory, as, for instance, in host-parasite coevolution (Futuyma and Slatkin 1983), but also indirectly via their impact on some intermediate abiotic component in a shared ecosystem, as in competition for a chemical or water resource. For example, if niche construction resulting from a gene in a plant population causes the soil chemistry to change in such a way that the

The possible feedback from niche construction to multiple loci within an individual, as well as across multiple species, creates nearly ideal conditions for group selection. By enforcing correlations across taxa, niche construction becomes a driver of cooperative effects.

selection of a gene in a second population, of plants or microorganisms, is changed, then the first population's niche construction will drive the evolution of the second population simply by changing the physical state of this abiotic ecosystem component. Since the dynamics of the intermediate abiotic component may be qualitatively quite different from either the frequency changes in the genes that underlie the niche construction or the number of niche-constructing organisms in the first population, this indirect feedback between species may generate some interesting—and as yet unexplored—behavior in coevolutionary systems.

> It is an open question how best to partition selection into direct niche-construction effects, historical constructive remnants, and construction-independent processes. These are likely have to changed considerably over the course of evolution, starting with the origin of life. In this way niche construction creates a non-uniformitarian evolution.

Accounting for the evolution and prevalence of mutualistic interactions is a stubborn problem for theoretical population biologists, who usually assume that there is some cost to the donor (Roughgarden 1975; Mesterton-Gibbons and Dugatkin 1992). Recent years, however, have seen a change in thinking about mutualism, with increasing emphasis placed on the fact that many mutualisms involve the transfer of incidental by-products, at no cost to the donor (West-Eberhard 1975; Brown 1983; Janzen 1985; Connor 1986; Mesterton-Gibbons and Dugatkin 1992). For example, seed predators often benefit the host plant through the dispersal of its seeds at no cost to themselves (Janzen 1985). These by-products, which are clearly a component of a population's niche construction, often serve as the source of selection for interspecific investment in their production (Connor 1986). For instance, fruit represents investment in seed dispersers (Thompson 1982). A thorough understanding of these coevolutionary dynamics may involve formal recognition that the niche-constructing activities of organisms can change selection pressures and thereby initiate mutualistic interactions.

A Second Role for Phenotypes in Evolution

When phenotypes niche construct, they can no longer be thought of as simply "vehicles for genes," since they are now also responsible for modifying some of the sources of selection in their environments that may subsequently feed back to select their genes. Moreover, there is no requirement for niche construction to result directly from genetic variation before it can influence the selection of genetic variation. For instance, niche construction can depend on learning, as is the case for

Odling-Smee, Laland, and Feldman (1996)

the British tits that may have changed their own selection pressures by opening foil milk-bottle tops, thereby gaining access to a new resource. The evolutionary consequences of this learned innovation are unknown, but it is possible that this new resource may now be selecting for some further change in these tits-for instance, for different digestive enzymes-or for improved learning ability (Fisher and Hinde 1949; Sherry and Galef 1984). Similarly, niche construction can also depend on culture, as happens when humans increase the frequency of the sickle-cell allele in their own populations by unwittingly increasing the prevalence of malaria in their environments through their agricultural practices (Durham 1991). The consequences for a gene at the second locus are the same, provided the effect of the niche construction on the environmental resources that constitute the source of selection is unchanged. The net result is an additional role for phenotypes in evolution. Phenotypes not only survive and reproduce differentially in the face of natural selection and chance but also modify some sources of selection in their environments by niche construction.

Relativization of Evolutionary Biology

On the basis of empirical evidence, Lewontin (1978, 1982, 1983) has argued that the "metaphor of adaptation" should be replaced by a "metaphor of construction." However, the acceptance of Lewontin's position demands more than just semantic adjustments to evolutionary theory. Niche construction changes the dynamic of the evolutionary process in fundamental ways because it precludes a description of evolutionary change relative only to autonomous environments. Instead, evolution now consists of endless cycles of natural selection and niche construction. Equally, it is no longer tenable to assume that the only way organisms can contribute to evolutionary descent is by passing on fit or unfit genes to their descendants relative to their environments, because they can also pass on modifications in those environments that are better or worse suited to their genes. Adaptation becomes a two-way street.

With the human species and the advent of material culture these effects are considerably amplified. There would be a reasonable argument that the Anthropocene is a synonym for niche construction enabled by technology.

Acknowledgments

We are grateful to R. C. Lewontin for his inspiration, support, and advice.

REFERENCES

Allen, T. F. H., and T. B. Starr. 1982. *Hierarchy: Perspectives for Ecological Complexity.* Chicago, IL: University of Chicago Press.

Boyd, R., and P. J. Richerson. 1985. *Culture and the Evolutionary Process.* Chicago, IL: University of Chicago Press.

Brown, J. L. 1983. "Cooperation: A Biologist's Dilemma." *Advances in the Study of Behavior* 13:1–37.

Connor, R. C. 1986. "Pseudo-Reciprocity: Investing in Mutualism." *Animal Behaviour* 34:1562–1584.

Cowley, D. E., and W. R. Atchley. 1992. "Quantitative Genetic Models for Development, Epigenetic Selection, and Phenotypic Evolution." *Evolution* 46:495–518.

Darwin, C. 1881. *The Formation of Vegetable Mold Through the Action of Works with Observations on Their Habits.* London, UK: J. Murray.

Durham, W. H. 1991. *Coevolution: Genes, Culture, and Human Diversity.* Stanford, CA: Stanford University Press.

Edmunds, M. 1974. *Defense in Animals.* New York, NY: Longman.

Feldman, M. W., and L. L. Cavalli-Sforza. 1976. "Cultural and Biological Evolutionary Processes: Selection for a Trait Under Complex Transmission." *Theoretical Population Biology* 9:238–259.

Fisher, J., and R. A. Hinde. 1949. "The Opening of Milk Bottles by Birds." *British Birds* 42:347–357.

Fox, D. L. 1979. *Biochromy: Natural Coloration of Living Things.* Berkeley, CA: University of California Press.

Futuyma, D. J., and M. Slatkin, eds. 1983. *Coevolution.* Sunderland, MA: Sinauer.

Hansell, M. H. 1984. *Animal Architecture and Building Behavior.* New York, NY: Longman.

Janzen, D. H. 1985. "The Natural History of Mutualisms." In *The Biology of Mutualisms,* edited by D. Boucher, 40–99. London, UK: Croom Helm.

Kirkpatrick, M., and R. Lande. 1989. "The Evolution of Maternal Characters." *Evolution* 43:485–503.

Krebs, J. R., and N. B. Davies. 1993. *An Introduction to Behavioural Ecology.* 3rd ed. Oxford, UK: Blackwell.

Lee, K. E. 1985. *Earthworms: Their Ecology and Relation with Soil and Land Use.* London, UK: Academic Press.

Lewontin, R. C. 1978. "Adaptation." *Scientific American* 239:156–169.

———. 1982. "Organism and Environment." In *Learning, Development and Culture,* edited by H.C. Plotkin, 151–172. New York, NY: Wiley.

———. 1983. "Gene, Organism, and Environment." In *Evolution from Molecules to Men*, edited by D. S. Bendall, 273–285. Cambridge, UK: Cambridge University Press.

Mathews, R. W., and J. R. Mathews. 1978. *Insect Behavior*. New York, NY: Wiley.

Mesterton-Gibbons, M., and L. A. Dugatkin. 1992. "Cooperation Among Unrelated Individuals: Evolutionary Factors." *Quarterly Review of Biology* 67:267–280.

Mount, A. B. 1964. "The Interdependence of The Eucalypts and Forest Fires in Southern Australia." *Australian Forestry* 28:166–172.

Odling-Smee, F. J. 1988. "Niche Constructing Phenotypes." In *The Role of Behavior in Evolution*, edited by H.C. Plotkin, 73–132. Cambridge, MA: MIT Press.

Rice, E. L. 1984. *Allelopathy*. 2nd ed. Orlando, FL: Academic Press.

Ricklefs, R. E. 1990. 3rd ed. New York, NY: Freeman.

Robertson, D. S. 1991. "Feedback Theory and Darwinian Evolution." *Journal of Theoretical Biology* 152:469–484.

Rothenbuhler, W. C. 1964. "Behavior Genetics of Nest Cleaning in Honey Bees. IV. Responses of F, and Backcross Generations to Disease-Killed Brood." *American Zoologist* 4:111–123.

Roughgarden, J. 1975. "Evolution of Marine Symbiosis: A Simple Cost-Benefit Model." *Ecology* 56:1201–1208.

Schluter, D., and L. Gustafsson. 1993. "Maternal Inheritance of Condition and Clutch Size in the Collared Flycatcher." *Evolution* 47:658–667.

Sherry, D. F., and B. G. Jr. Galef. 1984. "Cultural Transmission without Imitation–Milk Bottle Opening by Birds." *Animal Behaviour* 32:937–938.

Shukla, J., C. Nobra, and P. Sellars. 1990. "Amazon Deforestation and Climate Change." *Science* (Washington, DC) 247:1322–1325.

Spradbery, J. P. 1973. *Wasps*. London, UK: Sidgwick and Jackson.

Thompson, J. N. 1982. *Interaction and Coevolution*. New York, NY: Wiley.

van Valen, L. 1973. "A New Evolutionary Law." *Evolutionary Theory* 1:1–30.

von Frisch, K. 1975. *Animal Architecture*. London, UK: Hutchinson.

Waddington, C. H. 1959. "Evolutionary Systems; Animal and Human." *Nature* (London, UK) 183:1634–1638.

West, M. J., A. P. King, and A. A. Arberg. 1988. "The Inheritance of Niches: The Role of Ecological Legacies in Ontogeny." In *Handbook of Behavioral Neurobiology*, edited by E. Blass, 41-62. New York, NY: Plenum.

West-Eberhard, M. J. 1975. "The Evolution of Social Behavior by Kin Selection." *Quarterly Review of Biology* 50:1–33.

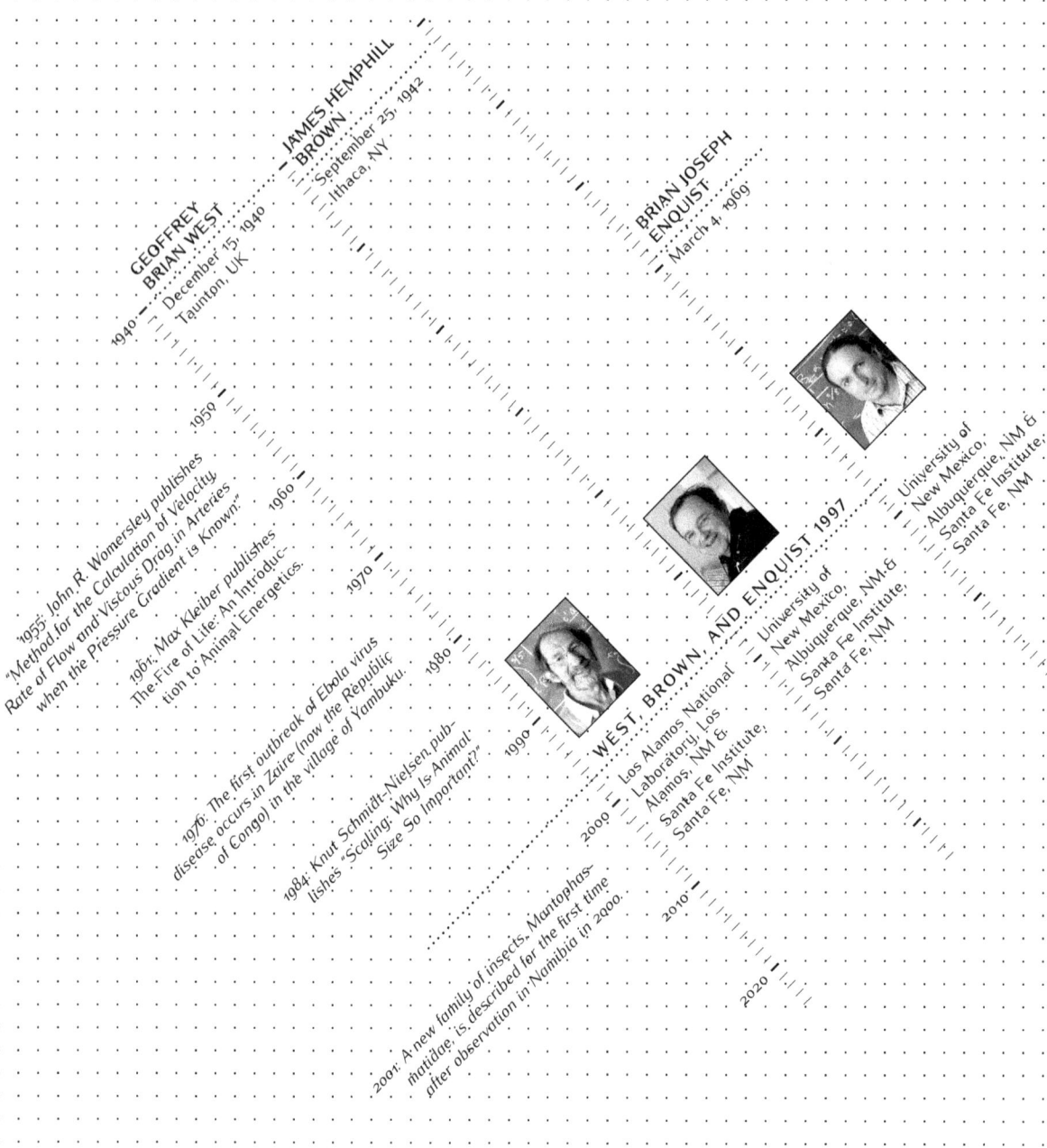

WEST, BROWN & ENQUIST

[83]

QUARTER-POWER ALLOMETRIC SCALING AND LIFE DYNAMICS

Pablo Marquet, Pontificia Universidad Católica de Chile and Santa Fe Institute

The concepts of scale, dimension, and geometry have a long and intertwined history at the root of both essential synthesis and revolutions in the history of science. From the "closed world of the ancients to the open world of the moderns" (Koyre 1957), they can be found in Galileo's geometric analysis of Dante's inferno (Peterson 2002), the design of ships and other structures, Einsteinian space-time, and Mandelbrot's fractal geometry (Mandelbrot 1975). A brilliant example in the science of complexity, this paper by Geoffrey West, James Brown, and Brian Enquist (WBE hereafter) falls into the same tradition, using scale, dimension, and geometry to provide a synthetic and holistic theory—in this case, a theory of life dynamics.

First, WBE solve a complex problem in biology related to the origin of allometric scaling laws of the form

$$B = b_0 W^\theta, \qquad (1)$$

where θ is the scaling exponent and b_0 a normalization constant. In explaining why θ follows a quarter-power scaling, and why it should be $\frac{3}{4}$ for the case of metabolism, WBE open a venue for understanding complex biological systems in general, from cells to societies. Metabolism is the transformation of energy and materials carried out by living entities in the process of keeping a far from equilibrium steady state characterized by using the energy and materials in their environment to grow, maintain their selves, and generate copies of themselves through reproduction. Thus, it is fair to say that, from cells to societies, metabolism (from the Greek word μεταβολή or change) encapsulated much of what life is. Metabolic scaling, as it turns out, has provided a solid foundation for understanding the rules of change or

G. B. West, J. H. Brown, and B. J. Enquist, "A General Model for the Origin of Allometric Scaling Laws in Biology," *Science* 276 (5309), 122–126 (1997).

Reprinted with permission from AAAS.

the dynamics of life and paved the way to understand biological scaling in general.

Metabolic scaling was one of the first scaling relationships that attracted scientific attention. It was initially formulated as a problem in heat emission by living entities in a time when heat as a form of energy was only beginning to be understood, thanks to the work of Carnot on heat engines in 1824. Thanks to Benjamin Thompson and James Prescott Joule, it later became apparent that work can be converted into heat, and both interact in determining the internal energy of a system as became established in the first law of thermodynamics. During the mid-nineteenth century, heat was a hot topic, and its emission by machines and living entities a puzzle. One of the first explanations of metabolic scaling, the surface law of metabolism, was postulated in 1838 by the mathematician Pierre Frédéric Sarrus and the medical physiologist Jean-François Rameaux. This seminal interdisciplinary collaboration used data on the respiration rate of two thousand human beings of different sizes (see Kleiber 1932). The logic behind the surface law is simple, appealing to the dimension of classical geometric bodies and the thermodynamics of heat dissipation (Fourier's law); the heat generated by the metabolism of a three-dimensional body is emitted through a two-dimensional surface, so as systems increase in size the heat production will relate to heat emission by an exponent of 3/2, which is Galileo's square-cube law, implying a formidable increase in the temperature of the system. Thus, to maintain a constant body temperature, assuming it has some selective value, metabolism should have been molded by natural selection to scale with system size with an exponent of 2/3. However, the 2/3 exponent was not empirically verified, leading scientists to search for additional explanations; as Kleiber (1947, p.525) pointed out, "there is no theoretical basis for the hypothesis that the metabolic rate of homeotherms should be exactly proportional to their particular surface area rather than to a more general function of body size." For Kleiber (1947, p. 538), progress requires integrating the heat transfer and blood circulation theory as follows: "In natural selection, those animals prove to be better fit whose rate of oxygen consumption is regulated so as to permit the more efficient temperature regulation as well as the more efficient transport of oxygen and nutrients." As more data became available, it became clear that the relationship between metabolism and size has an exponent of 3/4 instead of 2/3 (Kleiber 1932, 1947), but how this related to efficiency in distributing oxygen and nutrients was not apparent, notwithstanding developments on the theoretical analysis of physiological systems envisioned as efficient and economical (i.e., minimizing the energy costs of operation) and molded by natural selection, as introduced by Murray (1926).

After Kleiber's work there was a lull in theoretical explanations for metabolic scaling, but this did not impede empirical work, which accumulated in monographic treatment (see Peters 1983; Calder 1984; Schmidt-Nielsen 1984; Niklas 1994). These works showed that scaling relationships with exponents that are multiples of 1/4 are ubiquitous in biology and that there were no satisfactory explanations for them yet, notwithstanding several attempts to it (e.g. McMahon 1973). It is helpful to examine McMahon's work in some detail as it is an excellent example of the algebra of scaling relationships, how they can be combined, and the limitations associated with these analyses.

Building on D'Arcy Thompson's "science of form" (Thompson 1942), McMahon started by showing the existence of constraints to tree height and branch length due to buckling and bending, respectively. These so-called "elastic criteria" set the critical tree and branch length (l) to be a function of diameter (d) to the 2/3 power. As McMahon shows, and because of the elastic criteria, the length of limbs in animals scales as 2/3 of its diameter too. Thus, if

$$l \propto d^{2/3}, \qquad (2)$$

and since the total weight (W) of an organism goes as

$$W \propto ld^2, \qquad (3)$$

then from Eq. (2) it follows that $l^3 \propto d^2$. Replacing this into Eq. (3), we obtain that $l \propto W^{1/4}$, which implies that

$$d \propto W^{3/8}. \qquad (4)$$

Then McMahon assumed that the power output of a muscle and hypothetically "all the metabolic variables involved in maintaining the flow of energy to that muscle depend only on its cross-sectional area," which is proportional to d^2, hence

$$B \propto (W^{3/8})^2 \propto W^{3/4} \qquad (5)$$

where B is the maximum power output.

While this approach seems to explain the existence of a $\frac{3}{4}$ scaling law in metabolism, it may apply neither to microorganisms nor aquatic animals and plants, which, while not satisfying the elastic criteria, do follow a ¾ metabolic scaling law. Further, a typical organism is more than just muscles; at least in humans, the larger contribution to metabolism comes not from muscles but from other organs (heart, kidneys, brain, and liver) (Wang *et al.* 2010). Thus, a more holistic explanation seems in order.

WBE approach the problem of metabolic scaling as a particular case of a more general one (quarter-power scaling). Their model tackles the complexity and the geometry of living entities; living systems are neither Euclidean solid bodies nor homogeneous but composed of different integrated components or subunits with diverse chemical requirements and constraints, which is true from unicellular through multicellular organisms. In facing this complexity, WBE focus on a zeroth-order approximation to one of the major challenges for living organisms: how to distribute energy and essential materials to all their components (e.g., cells, tissues, and organs) according to their needs. For example, aerobic metabolism is fueled by oxygen dissolved at a constant concentration in the blood, which circulates through a network (i.e., the cardiovascular system), reaching all components of the living system. Thus, blood volume flow through the circulatory network can be used as a proxy for metabolic rate, and its scaling with organismal size would result from constraints on this network that affects volume flow. WBE think of this distribution network as composed of k levels that can fill the three-dimensional space formed by their interacting components so that it can reach out to each through successive branching. One way of doing this is by using a self-similar, space-filling fractal branching with area and volume preserving branching; the sum of the cross-sectional area of daughter branches at each level of branching, the same as the volume serviced by the network at each level, is the same and independent of the level k. Further, this branching ends at some point because the terminal units (e.g., the capillaries of the circulatory system) are invariant or independent of the size of the living system.

Once the network is defined, WBE solve for the hydrodynamics and elasticity equations for blood flow through the network, assuming that natural selection has selected for efficient networks that minimize energy dissipation, that is, the operational cost. This assumption translates into an optimal design that minimizes total resistance at the level of the whole network. It turns out that this requires that area-preserving branching is replaced by area-increasing branching in small vessels, slowing down blood flow and thus permitting for efficient diffusion of oxygen from capillaries to the group of cells they service (for discussion and further explanations, see West, Brown, and Enquist 1999, 2000; West and Brown 2005). It is this holistic, integrated approach that makes the WBE *sui generis*.

One of the salient features of the biological scaling theory initiated with this paper is the theory's efficiency; it makes many predictions with few free parameters (Marquet *et al.* 2014) and provides a solid foundation upon which to construct and expand the theory to account for fundamental processes in ecology, where it provides the first principles and foundations for the metabolic theory of ecology (Brown *et al.* 2004; Sibly, Brown, and Kodric-Brown 2012), in human systems (e.g., Bettencourt *et al.* 2007; Brown *et al.* 2011; Burger, Weinberger, and Marquet 2017; West 2017), and of phenomena such as sleep and cancer (Savage and West 2007; Brummer and Savage 2021).

All theories are incomplete. Like living systems, they change, expand, and metamorphose into different types. Biological scaling theory is a healthy and expanding field that has grown in many directions. Some areas, including evolution and paleobiology, have had limited growth, while others are more active, such as its integration with other ecological theories (stoichiometry, neutral theory, network theory, and the maximum entropy theory of ecology). Still others remain open challenges: a theory of information dynamics and integration in biological networks, or the relationship between information and metabolism. At the end of the day, life remains a triumvirate of matter, energy, and information fluxes.

REFERENCES

Bettencourt, L. M. A., J. Lobo, D. Helbing, C. Kühnert, and G. B. West. 2007. "Growth, Innovation, Scaling, and the Pace of Life in Cities." *Proceedings of the National Academy of Sciences* 104 (17): 7301–6. https://doi.org/10.1073/pnas.0610172104.

Brown, J. H., W. R. Burnside, A. D. Davidson, J. P. DeLong, W. C. Dunn, M. J. Hamilton, N. Mercado-Silva, *et al*. 2011. "Energetic Limits to Economic Growth." *BioScience* 61 (1): 19–26. https://doi.org/10.1525/bio.2011.61.1.7.

Brown, J. H., J. F. Gillooly, A. P. Allen, V. M. Savage, and G. B. West. 2004. "Toward a Metabolic Theory of Ecology." *Ecology* 85 (7): 1771–89. https://doi.org/10.1890/03-9000.

Brummer, Alexander B., and Van M. Savage. 2021. "Cancer as a Model System for Testing Metabolic Scaling Theory." *Frontiers in Ecology and Evolution* 9. https://www.frontiersin.org/article/10.3389/fevo.2021.691830.

Burger, J. R., V. P. Weinberger, and P. A. Marquet. 2017. "Extra-Metabolic Energy Use and the Rise in Human Hyper-Density." *Scientific Reports* 7 (March): 43869. https://doi.org/10.1038/srep43869.

Calder, W. A. 1984. *Size, Function, and Life History*. Cambridge, MA: Harvard University Press.

Kleiber, M. 1932. "Body Size and Metabolism." *Hilgardia. A Journal of Agricultural Science* 6 (11): 315–353. https://doi.org/10.3733/hilg.v06n11p315.

———. 1947. "Body Size and Metabolic Rate." *Physiological Reviews* 27 (4): 511–541. https://doi.org/10.1152/physrev.1947.27.4.511.

Koyre, A. 1957. *From the Closed World to the Infinite Universe*. Baltimore, MD: The Johns Hopkins Press.

Mandelbrot, B. B. 1975. *Les Objets Fractals: Forme, Hasard et Dimension*. Vol. 17. Paris: Flammarion.

Marquet, P. A., A. P. Allen, J. H. Brown, J. A. Dunne, B. J. Enquist, J. F. Gillooly, P. A. Gowaty, *et al*. 2014. "On Theory in Ecology." *BioScience* 64 (8): 701–710. https://doi.org/10.1093/biosci/biu098.

McMahon, T. 1973. "Size and Shape in Biology." *Science* 179 (4079): 1201–4. https://doi.org/10.1126/science.179.4079.1201.

Murray, C. D. 1926. "The Physiological Principle of Minimum Work: I. The Vascular System and the Cost of Blood Volume." *Proceedings of the National Academy of Sciences* 12 (3): 207–14. https://doi.org/10.1073/pnas.12.3.207.

Niklas, K. J. 1994. *Plant Allometry: The Scaling of Form and Process*. Chicago, IL: University of Chicago Press.

Peters, R. H. 1983. *The Ecological Implications of Body Size*. Cambridge, UK: Cambridge University Press.

Peterson, M. A. 2002. "Galileo's Discovery of Scaling Laws." *American Journal of Physics* 70 (6): 575–580. https://doi.org/10.1119/1.1475329.

Savage, V. M., and G. B. West. 2007. "A Quantitative, Theoretical Framework for Understanding Mammalian Sleep." *Proceedings of the National Academy of Sciences* 104 (3): 1051–1056. https://doi.org/10.1073/pnas.0610080104.

Schmidt-Nielsen, K. 1984. *Scaling. Why Is Animal Size so Important?* Cambridge, UK: Cambridge University Press.

Sibly, R. M., J. H. Brown, and A. Kodric-Brown. 2012. *Metabolic Ecology. A Scaling Approach.* Chichester, West Sussex, UK: Wiley-Blackwell.

Thompson, D. W. 1942. *On Growth and Form.* Cambridge, UK: Cambridge University Press.

Wang, Z., Z. Ying, A. Bosy-Westphal, J. Zhang, B. Schautz, W. Later, S. B. Heymsfield, and M. J. Müller. 2010. "Specific Metabolic Rates of Major Organs and Tissues across Adulthood: Evaluation by Mechanistic Model of Resting Energy Expenditure." *The American Journal of Clinical Nutrition* 92 (6): 1369–77. https://doi.org/10.3945/ajcn.2010.29885.

West, G. 2017. *Scale: The Universal Laws of Life, Growth, and Death in Organisms, Cities, and Companies.* New York, NY: Penguin.

West, G. B., and J. H. Brown. 2005. "The Origin of Allometric Scaling Laws in Biology from Genomes to Ecosystems: Towards a Quantitative Unifying Theory of Biological Structure and Organization." *Journal of Experimental Biology* 208 (9): 1575–92. https://doi.org/10.1242/jeb.01589.

West, G. B., J. H. Brown, and B. J. Enquist. 1999. "The Fourth Dimension of Life: Fractal Geometry and Allometric Scaling of Organisms." *Science* 284 (5420): 1677–1679. https://doi.org/10.1126/science.284.5420.1677.

———. 2000. *Scaling in Biology: Patterns and Processes, Causes and Consequences.* New York, NY: Oxford University Press.

A GENERAL MODEL FOR THE ORIGIN OF ALLOMETRIC SCALING LAWS IN BIOLOGY

Geoffrey B. West, Los Alamos National Laboratory and Santa Fe Institute,
James H. Brown, University of New Mexico and Santa Fe Institute, and
Brian J. Enquist, University of New Mexico and Santa Fe Institute

Abstract

Allometric scaling relations, including the 3/4 power law for metabolic rates, are characteristic of all organisms and are here derived from a general model that describes how essential materials are transported through space-filling fractal networks of branching tubes. The model assumes that the energy dissipated is minimized and that the terminal tubes do not vary with body size. It provides a complete analysis of scaling relations for mammalian circulatory systems that are in agreement with data. More generally, the model predicts structural and functional properties of vertebrate cardiovascular and respiratory systems, plant vascular systems, insect tracheal tubes, and other distribution networks.

Biological diversity is largely a matter of body size, which varies over 21 orders of magnitude (McMahon and Bonner 1983; Bonner 1983; Brown 1995). Size affects rates of all biological structures and processes from cellular metabolism to population dynamics (Schmidt-Nielsen 1984; Calder III 1984; Peters 1983). The dependence of a biological variable Y on body mass M is typically characterized by an allometric scaling law of the form

$$Y = Y_0 M^b \tag{1}$$

where b is the scaling exponent and Y_0 a constant that is characteristic of the kind of organism. If, as originally thought, these relations reflect geometric constraints, then b should be a simple multiple of one-third. However, most biological phenomena scale as quarter rather than third powers of body mass (Schmidt-Nielsen 1984; Calder III 1984; Peters 1983; Feldman and McMahon 1983): For example, metabolic rates B of entire organisms scale as $M^{3/4}$; rates of cellular metabolism, heartbeat, and maximal population growth scale as $M^{-1/4}$; and times of blood circulation, embryonic growth and development, and lifespan scale as $M^{1/4}$. Sizes of biological structures scale similarly: For example, the cross-sectional areas of mammalian aortas and of tree trunks scale as $M^{3/4}$. No general theory explains the origin of these laws. Current hypotheses, such as resistance to elastic buckling in terrestrial organisms (McMahon 1973) or diffusion of materials across hydrodynamic boundary layers in aquatic organisms (Patterson 1992), cannot explain why so many biological processes in nearly all kinds of animals (Schmidt-Nielsen 1984; Calder III 1984; Peters 1983), plants (Niklas 1994), and microbes (Hemmingsen 1950, 1960) exhibit quarter-power scaling.

We propose that a common mechanism underlies these laws: Living things are sustained by the transport of materials through linear networks that branch to supply all parts of the organism. We develop a quantitative model that explains the origin and ubiquity of quarter-power scaling; it predicts the essential features of transport systems, such as mammalian blood vessels and bronchial trees, plant vascular systems, and insect tracheal tubes. It is based on three unifying principles or assumptions: First, in order for the network to supply the entire volume of the organism, a space-filling fractal-like branching pattern (Mandelbrot 1977) is required. Second, the final branch of the network (such as the capillary in the circulatory system) is a size-invariant unit (Schmidt-Nielsen 1984; Calder III 1984). And third, the energy required to distribute resources is minimized (Thompson 1942); this final restriction is basically equivalent to minimizing the total hydrodynamic resistance of the system. Scaling laws arise from the interplay between physical and geometric constraints implicit in these three principles. The model presented here should be viewed as an idealized representation in that we ignore complications such as tapering of vessels, turbulence, and nonlinear effects. These play only a minor role in determining the dynamics of the entire network and could be incorporated in more detailed analyses of specific systems.

Most distribution systems can be described by a branching network in which the sizes of tubes regularly decrease (Fig. 1). One version is exhibited by vertebrate circulatory and respiratory systems, another by the "vessel–bundle" structure of multiple parallel tubes, characteristic

In this key passage, the major insight—that life is sustained by networks—becomes apparent, as well as the generality of the approach that goes well beyond metabolism to understand quarter-power scaling in general.

Subsequent work has expanded the model to different fractal-like network topologies.

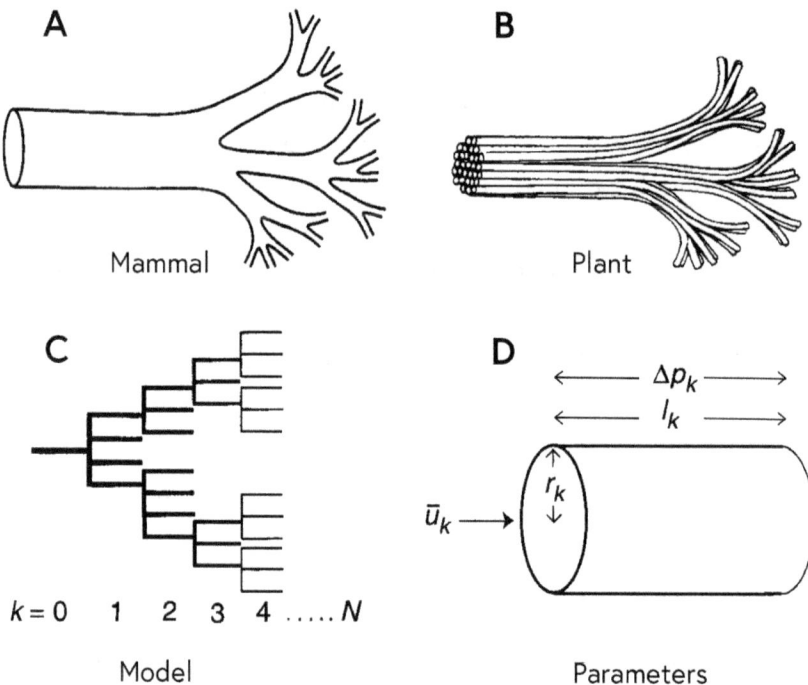

Figure 1. Diagrammatic examples of segments of biological distribution networks: (A) mammalian circulatory and respiratory systems composed of branching tubes; (B) plant vessel-bundle vascular system composed of diverging vessel elements; (C) topological representation of such networks, where k specifies the order of the level, beginning with the aorta ($k = 0$) and ending with the capillary ($k = N$); and (D) parameters of a typical tube at the kth level.

of plant vascular systems (Shinozaki *et al.* 1964a, 1964b; Zimmerman 1983; Tyree and Ewers 1991). Biological networks vary in the properties of the tube (elastic to rigid), the fluid transported (liquid to gas), and the nature of the pump (a pulsatile compression pump in the cardiovascular system, a pulsatile bellows pump in the respiratory system, diffusion in insect tracheae, and osmotic and vapor pressure in the plant vascular system). In spite of these differences, these networks exhibit essentially the same scaling laws.

For convenience we shall use the language of the cardiovascular system, namely, aorta, arteries, arterioles, and capillaries; the correspondence to other systems is straightforward. In the general case, the network is composed of N branchings from the aorta (level 0) to the capillaries (level N, denoted here by a subscript c) (Fig. 1C). A typical branch at some intermediate level k has length l_k, radius r_k, and pressure drop Δp_k (Fig. 1D).

The volume rate of flow is $Q_k = \pi r_k^2 \bar{u}_k$ where $\bar{u}_k$ is the flow velocity averaged over the cross section and, if necessary, over time.

Each tube branches into n_k smaller ones[1], so the total number of branches at level k is $N_k = n_0 n_1 \ldots n_k$. Because fluid is conserved as it flows through the system

$$\dot{Q}_0 = N_k \dot{Q}_k = N_k \pi r_k^2 \bar{u}_k = N_c \pi r_c^2 \bar{u}_c \tag{2}$$

which holds for any level k. We next introduce the important assumption, the second above, that the terminal units (capillaries) are invariant, so $r_c, l_c, \bar{u}_c$, and, consequently, Δp_c are independent of body size. Because the fluid transports oxygen and nutrients for metabolism, $\dot{Q}_0 \propto B$; thus, if $B \propto M^a$ (where a will later be determined to be $3/4$, then $\dot{Q}_0 \propto M^a$. Equation 2 therefore predicts that the total number of capillaries must scale as B, that is, $N_c \propto M^a$.

To characterize the branching, we introduce scale factors $\beta_k \equiv r_{k+1}/r_k$ and $\gamma_k \equiv l_{k+1}/l_k$. We shall prove that in order to minimize the energy dissipated in the system in the sense of the third principle above, the network must be a conventional self-similar fractal in that $\beta_k = \beta, \gamma_k = \gamma$, and $n_k = n$, all independent of k (an important exception is β_k in pulsatile systems). For a self-similar fractal, the number of branches increases in geometric proportion $\left(N_k = n^k\right)$ as their size geometrically decreases from level 0 to level N. Before proving self-similarity, we first examine some of its consequences.

Because $N_c = n^N$, the number of generations of branches scales only logarithmically with size

$$N = \frac{a \ln(M/M_\circ)}{\ln n} \tag{3}$$

where M_0 is a normalization scale for M [2]. Thus, a whale is 10^7 times heavier than a mouse but has only about 70% more branchings from aorta

---

[1] The branching of a vessel at level k into n_k smaller vessels (Fig. 1) is assumed to occur over some small, but finite, distance that is much smaller than either l_k or l_{k+1}. This relation is similar to that assumed in the Strahler method (Strahler 1953; Shinozaki et al. 1964a, 1964b; Zimmerman 1983; Tyree and Ewers 1991; Fung 1984). A generalization to nonuniform branching, where the radii and lengths at a given level may vary, is straightforward.

[2] Normalization factors, such as M_0, will generally be suppressed, as in Eq. 1. In general, all quantities should be expressed in dimensionless form; note, however, that this does not guarantee that they are size independent and scale as M°. For example, the Womersley number, α of Eq. 8, although dimensionless, scales as $M^{1/4}$.

to capillary. The total volume of fluid in the network ("blood" volume V_b) is

$$V_b = \sum_{k=0}^{N} N_k V_k = \sum_{k=0}^{N} \pi r_k^2 l_k n^k$$

$$= \frac{(n\gamma\beta^2)^{-(N+1)} - 1}{(n\gamma\beta^2)^{-1} - 1} n^N V_c \qquad (4)$$

where the last expression reflects the fractal nature of the system. As shown below, one can also prove from the energy minimization principle that $V_b \propto M$. Because $n\gamma\beta^2 < 1$ and $N \gg 1$, a good approximation to Eq. 4 is $V_b = V_0/(1 - n\gamma\beta^2) = V_c (\gamma\beta^2)^{-N}(1 - n\gamma\beta^2)$. From our assumption that capillaries are invariant units, it therefore follows that $(\gamma\beta^2)^{-N} \propto M$. Using this relation in Eq. 3 then gives

$$a = -\frac{\ln n}{\ln(\gamma\beta^2)} \qquad (5)$$

To make further progress requires knowledge of γ and β. We shall show how the former follows from the space-filling fractal requirement, and the latter, from the energy minimization principle.

A space-filling fractal is a natural structure for ensuring that all cells are serviced by capillaries. The network must branch so that a group of cells, referred to here as a "service volume," is supplied by each capillary. Because $r_k \ll l_k$ and the total number of branchings N is large, the volume supplied by the total network can be approximated by the sum of spheres whose diameters are that of a typical kth-level vessel, namely $4/3\pi (l_k/2)^3 N_k$. For large N, this estimate does not depend significantly on the specific level, although it is most accurate for large k. This condition, that the fractal be volume-preserving from one generation to the next, can therefore be expressed as $4/3\pi (l_k/2)^3 N_k \approx 4/3\pi (l_{k+1}/2)^3 N_{k+1}$. This equation relation gives $\gamma_k^3 \equiv (l_{k+1}/l_k)^3 \approx N_k/N_{k+1} = 1/n$, showing that $\gamma_k \approx n^{-1/3} \approx \gamma$ must be independent of k. This result for γ_k is a general property of all space-filling fractal systems that we consider.

The 3/4 power law arises in the simple case of the classic rigid-pipe model, where the branching is assumed to be area-preserving, that is, the sum of the cross-sectional areas of the daughter branches equals that of the parent, so $\pi r_k^2 = n\pi r_{k+1}^2$. Thus, $\beta_k \equiv r_{k+1}/r_k = n^{-1/2} = \beta$, independent of k When the area-preserving branching relation, $\beta =$

West, Brown, and Enquist (1997)

$n^{-1/2}$, is combined with the space-filling result for γ, Eq. 5 yields $a = 3/4$, so $B \propto M^{3/4}$. Many other scaling laws follow. For example, for the aorta, $r_0 = \beta_c^{-N_r} = N_c^{1/2} r_c$ and $l_0 = \gamma^{-N} r_c = N_c^{P/3} l_c$ yielding $r_0 \propto M^{3/8}$ and $l_0 \propto M^{1/4}$. This derivation of the $a = 3/4$ law is essentially a geometric one, strictly applying only to systems that exhibit area-preserving branching. This property has the further consequence, which follows from Eq. 2, that the fluid velocity must remain constant throughout the network and be independent of size. These features are a natural consequence of the idealized vessel-bundle structure of plant vascular systems (Fig. 1B), in which area-preserving branching arises automatically because each branch is assumed to be a bundle of n^{N-k} elementary vessels of the same radius (Shinozaki *et al.* 1964a, 1964b; Zimmerman 1983; Tyree and Ewers 1991). Pulsatile mammalian vascular systems, on the other hand, do not conform to this structure, so for them, we must look elsewhere for the origin of quarter-power scaling laws.

Some features of the simple pipe model remain valid for all networks: (i) The quantities γ and β play a dual scaling role: they determine not only how quantities scale from level 0 (aorta) to N (capillary) within a single organism of fixed size, but also how a given quantity scales when organisms of different masses are compared. (ii) The fractal nature of the entire system as expressed, for example, in the summation in Eq. 4 leads to a scaling different from that for a single tube, given by an individual term in the series. These network systems must therefore be treated as a complete integrated unit; they cannot realistically be modeled by a single or a few representative vessels. (iii) The scaling with M does not depend on the branching ratio n.

We next consider the dynamics of the network and examine the consequences of the energy minimization principle, which is particularly relevant to mammalian vascular systems. Pulsatile flow, which dominates the larger vessels (aorta and major arteries), must have area-preserving branching, so that $\beta = n^{-1/2}$, leading to quarter-power scaling. The smaller vessels, on the other hand, have the classic "cubic-law" branching (Thompson 1942), where $\beta = n^{-1/3}$, and play a relatively minor role in allometric scaling.

First consider the simpler problem of nonpulsatile flow. For steady laminar flow of a Newtonian fluid, the viscous resistance of a single tube

To model allometric scaling, and particularly metabolism, as resulting from properties of components was a tradition that this contribution terminates with.

2637

is given by the well-known Poiseuille formula $R_k = 8\mu l_k/\pi r_k^4$, where μ is the viscosity of the fluid. Ignoring small effects such as turbulence and nonlinearities at junctions, the resistance of the entire network is given by[3]

$$Z = \sum_{k=0}^{N} \frac{R_k}{N_k} = \sum_{k=0}^{N} \frac{8\mu l_k}{\pi r_k^4 n^k}$$
$$= \frac{\left[1 - \left(n\beta^4/\gamma\right)^{N+1}\right] R_c}{(1 - n\beta^4/\gamma) n^N} \quad (6)$$

Now, $n\beta^4/\gamma < 1$ and $N \gg 1$, so a good approximation is $Z = R_c/\left(1 - n\beta^4/\gamma\right) N_c$. Because R_c is invariant, $Z \propto N_c^{-1} \propto M$, which leads to two important scaling laws: blood pressure $\Delta p = \dot{Q}_0 Z$ must be independent of body size and the power dissipated in the system (cardiac output) $W = \dot{Q}_0 \Delta p \propto M^a$, so that the power expended by the heart in overcoming viscous forces is a size-independent fraction of the metabolic rate. Neither of these results depends on detailed knowledge of n, β, or γ, in contrast to results based on $V_b \propto M$, such as Eq. 5, $a = 3/4$, and $r_0 \propto M^{3/8}$. From Eq. 2, $\dot{Q}_0 = \pi r_0^2 \bar{u}_0$, which correctly predicts that the velocity of blood in the aorta $\bar{u}_0 \propto M^0$ (Schmidt-Nielsen 1984; Calder III 1984). However, an area-preserving scaling relation $\beta = n^{-1/2}$ also implies by means of Eq. 2 that $\bar{u}_k = \bar{u}_0$ for all k. This relation is valid for fluid flow in plant vessels (because of the vascular bundle structure)[4] and insect tracheae (because gas is driven by diffusion) (Krogh 1920); both therefore exhibit area-preserving branching, which leads to 3/4 power scaling of metabolic rate. Branching cannot be entirely area-preserving in mammalian circulatory systems because blood must slow down to allow materials to diffuse across capillary walls. However, the pulsatile nature of the mammalian cardiovascular system solves the problem.

Energy minimization constrains the network for the simpler nonpulsatile systems. Consider cardiac output as a function of all relevant variables: $W(r_k, l_k, n_k, M)$. To sustain a given metabolic rate in an organism

---

[3] This formula is not valid for plant vessel bundles because plants are composed of multiple parallel vessel elements. Their resistance is given by $Z = 8\mu l/N_c \pi r_c^4$, where l is the length of a single vessel element, r_c is its radius, and N_c is their total number.

[4] Shinozaki et al. (1964a, 1964b), Zimmerman (1983), and Tyree and Ewers (1991). This relation holds for plant vessels from the roots to the leaves, but not within leaves (Canney 1993).

of fixed mass M with a given volume of blood $V_b(r_k, l_k, n_k, M)$, the minimization principle requires that the cardiac output be minimized subject to a space-filling geometry. To enforce such a constraint, we use the standard method of Lagrange multipliers (λ, λ_k, and λ_M) and so need to minimize the auxiliary function

$$F(r_k, l_k, n) = W(r_k, l_k, n_k, M) + \lambda V_b(r_k, l_k, n_k, M) + \sum_{k=0}^{N} \lambda_k N_k l_k^3 + \lambda_M M \quad (7)$$

Because $B \propto Q_0$ and $W = \dot{Q}_0^2 Z$, this problem is tantamount to minimizing the impedance Z, which can therefore be used in Eq. 7 in place of W. First, consider the case where $n_k = n$, so that we can use Eqs. 4 and 6 for V_b and Z, respectively. For a fixed mass M, the auxiliary Lagrange function F, which incorporates the constraints, must be minimized with respect to all variables for the entire system (r_k, l_k, and n). This requires $\delta F/\delta l_k = \delta F/\delta r_k = \delta F/\delta n = 0$, which straightforwardly leads to $\beta_k = n^{-1/3}$. More generally, by considering variations with respect to n_k, one can show that $n_k = n$, independent of k. The result, $\beta_k = n^{-1/3}$, is a generalization of Murray's finding (Murray 1926), derived for a single branching, to the complete network. Now varying M and minimizing F in Eq. 7 ($\delta F/\delta M = 0$) leads to $V_b \propto M$, which is just the relation needed to derive Eq. 5. Although the result $\beta_k = n^{-1/3}$ is independent of k, it is not area-preserving and therefore does not give $a = 3/4$ when used in Eq. 5; instead, it gives $a = 1$. It does, however, solve the problem of slowing blood in the capillaries: Eq. 2 gives $\bar{u}_c/\bar{u}_0 = (n\beta^2)^{-N} = N_c^{-1/3}$. For humans, $N_c \sim 10^{10}$, so $\bar{u}_c/\bar{u}_0 \sim 10^{-3}$, in reasonable agreement with data (Caro et al. 1978). On the other hand, it leads to an incorrect scaling law for this ratio: $\bar{u}_c/\bar{u}_0 \propto M^{-1/4}$. Incorporating pulsatile flow not only solves these problems, giving the correct scaling relations ($a = 3/4$ and $\bar{u}_c/\bar{u}_0 \propto M^0$), but also gives the correct value for $\bar{u}_c/\bar{u}_0$.

A complete treatment of pulsatile flow is complicated; here, we present a simplified version that contains the essential features needed for the scaling problem. When an oscillatory pressure p of angular frequency ω is applied to an elastic (characterized by modulus E) vessel with wall thickness h, a damped traveling wave is created: $p = p_0 e^{i(\omega t - 2\pi z/\lambda)}$. Here, t is time, z is the distance along the tube, λ is the wavelength, and p_0 is

Here, it is clear that deviations from fractality may affect the value of the exponent, as has subsequently been shown by different authors.

the amplitude averaged over the radius; the wave velocity $c = 2\pi\omega\lambda$. Both the impedance Z and the dispersion relation that determines c are derived by solving the Navier-Stokes equation for the fluid coupled to the Navier equations for the vessel wall (Womersley 1955b, 1955a). In the linearized incompressible-fluid, thin-wall approximation, this problem can be solved analytically to give

$$\left(\frac{c}{c_0}\right)^2 \approx -\frac{J_2\left(i^{3/2}\alpha\right)}{J_0\left(i^{3/2}\alpha\right)} \quad and \quad Z \approx \frac{c_0^2 \rho}{\pi r^2 c} \qquad (8)$$

Here $\alpha \equiv (\omega\rho/\mu)^{1/2} r$ is the dimensionless Womersley number (see footnote 2), and $c_0 \equiv (Eh/2\rho r)^{1/2}$ is the Korteweg-Moens velocity. In general, both c and Z are complex functions of ω, so the wave is attenuated and disperses as it propagates. Consider the consequences of these formulas as the blood flows through progressively smaller tubes: For large tubes, α is large (in a typical human artery, $\alpha \approx 5$), and viscosity plays almost no role. Equation 8 then gives $c = c_0$ and $Z = \rho c_0 / \pi r^2$; because both of these are real quantities, the wave is neither attenuated nor dispersed. The r dependence of Z has changed from the nonpulsatile r^{-4} behavior to r^{-2}. Minimizing energy loss now gives h_k/r_k (and, therefore, c_k) independent of k and, most importantly, an area-preserving law at the junctions, so $\beta_k = n^{-1/2}$. This relation ensures that energy-carrying waves are not reflected back up the tubes at branch points and is the exact analog of impedance matching at the junctions of electrical transmission lines (Caro et al. 1978). As k increases, the sizes of tubes decrease, so $\alpha \to 0$ (in human arterioles, for example, $a \approx 0.05$), and the role of viscosity increases, eventually dominating the flow. Equation 8 then gives $c \approx i^{1/2}\alpha c_0/4 \to 0$, in agreement with observation (Caro et al. 1978). Because c and, consequently, λ now have imaginary parts, the traveling wave is heavily damped, leaving an almost steady oscillatory flow whose impedance is, from Eq. 8, given by the Poiseuille formula; that is, the r^{-4} behavior is restored. Thus, for large k, corresponding to small vessels, $\beta_k = n^{-1/3}$. We conclude that for pulsatile flow, β_k is not independent of k but rather has a steplike behavior (Fig. 2). This picture is well supported by empirical data.[5]

---

[5] Caro et al. (1978) and Fung (1984). See, for example, Iberall (1967) and Sherman (1981), which contain summaries of earlier data; also Zamir, Sinclair, and Wonnacott

The crossover from one behavior to the other occurs over the region where the wave and Poiseuille impedances are comparable in size. The approximate value of k where this occurs (say, $\bar{k}$) is given by $r_{\bar{k}}^2/l_{\bar{k}} \approx 8\mu/\rho c_0$, leading to $N - \bar{k} \equiv \bar{N} \approx \ln\left(8\mu l_c/\rho c_0 r_c^2\right)/\ln n$ independent of M. Thus, the number of generations where Poiseuille flow dominates should be independent of body size. On the other hand, the crossover point itself grows logarithmically: $\bar{k} \propto N \propto \ln M$. For humans, with $n = 3$ (Fung 1984), $\bar{N} \approx 15$ and $N \approx 22$ (assuming $N_c \approx 2 \times 10^{10}$), whereas with $n = 2$, $N \approx 24$ and $N \approx 34$. These values mean that in humans Poiseuille flow begins to compete with the pulse wave after just a few branchings, dominating after about seven. In a 3-g shrew, Poiseuille flow begins to dominate shortly beyond the aorta.

The derivation of scaling laws based on β_k derived from Eqs. 7 and 8 (Fig. 2) leads to the same results as before. For simplicity, assume that the crossover is sharp; using a gradual transition does not change the resulting scaling laws. So, for $k > \bar{k}$, define $\beta_k \equiv \beta_> = n^{-1/3}$ and, for $k < \bar{k}$, $\beta_k \equiv \beta_< = n^{-1/2}$. This predicts that area preservation only persists in the pulsatile region from the aorta through the large arteries, at most until $k \approx \bar{k}$. First consider the radius of the aorta r_0: its scaling behavior is now given by $r_0 \equiv r_c \beta_>^{\bar{k}-N} \beta_<^{\bar{k}} = r_c n^{1/3N + 1/6\bar{k}} = r_c n^{1/2N - 1/6\bar{N}}$ which gives $r_0 \propto M^{3/8}$ and, for humans, $r_0/r_c \approx 10^4$, in agreement with data (Schmidt-Nielsen 1984; Calder III 1984). Using Eq. 3 we obtain, for the ratio of fluid velocity in the aorta to that in the capillary, $\bar{u}_0/\bar{u}_c = N_c (r_c/r_0)^2 = n^{\bar{N}/3} \bar{u}_0/\bar{u}_c \approx 250$, independent of M, again in agreement with data. Because γ reflects the space-filling geometry, it remains unchanged, so we

---

(1992) and Li (1996). Care must be taken in comparing measurements with prediction, particularly if averages over many successive levels are used. For example, if $A_k \equiv \Sigma_k \pi r_k^2$ is the total cross-sectional area at level k, then for the aorta and major arteries, where $k < \bar{k}$ and the branching is area-preserving, we predict $A_0 = A_k$. Suppose, however, that the first K levels are grouped together. Then, if the resulting measurement gives $\bar{A}_K$, area-preserving predicts $\bar{A}_K = KA_0$ (but not $\bar{A}_K = A_0$). It also predicts $r_0^3 \approx n^{1/2} \sum N_k r_k^3$. Using results from LaBarbera (1990), who used data averaged over the first 160 vessels (approximately the first 4 levels), gives, for human beings, $A_0 \approx 4.90$ cm^2, $\bar{A}_K \approx 19.98$ cm^2, $r_0^3 \approx 1.95$ cm^3, and $\Sigma N_k r_k^3 \approx 1.27$ cm^3, in agreement with area preservation. LaBarbera, unfortunately, took the fact that $\bar{A}_K \neq A_0$ and $r_0^3 \approx \Sigma N_k r_k^3$ as evidence for cubic rather than area-preserving branching. For small vessels, where $k > \bar{k}$, convincing evidence for the cubic law can be found in the analysis of the arteriolar system by Ellsworth et al. (1987).

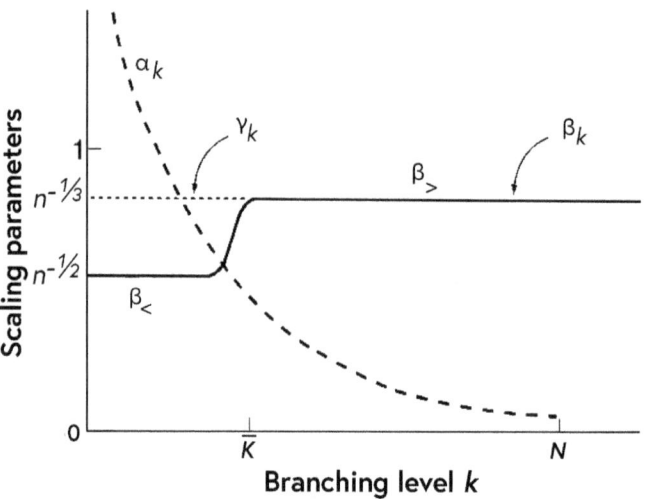

Figure 2. Schematic variation of the Womersley number α_k and the scaling parameters β_k and γ_k with level number (k) for pulsatile systems. Note the steplike change in β_k at $k = \bar{k}$ from area-preserving pulse-wave flow in major vessels to area-increasing Poiseuille-type flow in small vessels.

still have $l_0 \propto M^{1/4}$. Blood volume V_b, however, is more complicated

$$V_b = \frac{V_c}{(\beta_>^2\gamma)^N}\left\{\left(\frac{\beta_>}{\beta_<}\right)^{2\bar{k}}\frac{1-(n\beta_<^2\gamma)^{\bar{k}}}{1-(n\beta_<^2\gamma)}\right.$$
$$\left. + \left[\frac{1-(n\beta_>^2\gamma)^N}{1-(n\beta_>^2\gamma)} - \frac{1-(n\beta_>^2\gamma)^{\bar{k}}}{1-(n\beta_>^2\gamma)}\right]\right\} \tag{9}$$

This formula is a generalization of Eq. 4 and is dominated by the first term, which represents the contribution of the large tubes (aorta and arteries). Thus, $V_b \propto n^{N+1/3\bar{k}} \propto n^{4/3}N$, which, because it must scale as M, leads, as before, to $a = 3/4$. As size decreases, the second term, representing the cubic branching of small vessels, becomes increasingly important. This behavior predicts small deviations from quarter-power scaling ($a \geq 3/4$), observed in the smallest mammals(Schmidt-Nielsen 1984; Calder III 1984). An expression analogous to Eq. 9 can be derived for the total impedance of the system Z. It is dominated by the small vessels (arterioles and capillaries) and, as before, gives Δp and $\bar{u}_0 \propto M^0$. In order to understand allometric scaling, it is necessary to formulate an integrated model for the entire system. The present model should be viewed as an idealized zeroth-order approximation: it accounts for many of the features of distribution networks and can be used as a point of departure for more

More detailed analyses have indeed been carried out. Most significantly, the inclusion of temperature effects on metabolism, as well as food quality and stoichiometry, have been shown to be important.

detailed analyses and models. In addition, because it is quantitative, the coefficients, Y_0 of Eq. 1, can also, in principle, be derived. It accurately predicts the known scaling relations of the mammalian cardiovascular system (Table 1); data are needed to test other predictions. For example, the invariance of capillary parameters implies $N_c \propto M^{3/4}$ rather than the naive expectation $N_c \propto M$, so the volume serviced by each capillary must scale as $M^{1/4}$, and capillary density per cross-sectional area of tissue, as $M^{-1/12}$.

A minor variant of the model describes the mammalian respiratory system. Although pulse waves are irrelevant because the tubes are not elastic, the formula for Z is quite similar to Eq. 8. The fractal bronchial tree terminates in $N_A \propto M^{3/4}$ alveoli. The network is space-filling, and the alveoli play the role of the service volume accounting for most of the total volume of the lung, which scales as M. Thus, the volume of an alveolus $V_A \propto M^{1/4}$, its radius $r_A \propto M^{1/12}$, and its surface area $A_A \propto r_A^2 \propto M^{1/6}$, so the total surface area of the lung $A_L = N_A A_A \propto M^{11/12}$. This explains the paradox (Gehr et al. 1981) that A_A scales with an exponent closer to 1 than the 3/4 seemingly needed to supply oxygen. The rate of oxygen diffusion across an alveolus, which must be independent of M, is proportional to $\Delta p_{O_2} A_A / r_A$. Thus, $\Delta p_{O_2} \propto M^{-1/12}$, which must be compensated for by a similar scaling of the oxygen affinity of hemoglobin. Available data support these predictions (Table 1).

Our model provides a theoretical, mechanistic basis for understanding the central role of body size in all aspects of biology. Considering the many functionally interconnected parts of the organism that must obey the constraints, it is not surprising that the diversity of living and fossil organisms is based on the elaboration of a few successful designs. Given the need to redesign the entire system whenever body size changes, either during ontogeny or phylogenetic diversification, small deviations from quarter-power scaling sometimes occur (Peters 1983; Bennett and Harvey 1987; Bennett 1988; Harvey and Pagel 1991). However, when body sizes vary over many orders of magnitude, these scaling laws are obeyed with remarkable precision. Moreover, the predicted scaling properties do not depend on most details of system design, including the exact branching

This area has not been well explored. Much remains to be done.

In truth, the system should not necessarily go through redesigns when body size changes through evolution, but only through scale changes. Unless we are talking about macroevolutionary changes associated with the emergence of higher taxa whose internal networks are much different from the hierarchical fractal-like networks envisioned here.

Cardiovascular

Variable	Predicted Exponent	Observed
Aorta radius r_0	3/8 = 0.375	0.36
Aorta pressure Δp_0	0 = 0.00	0.032
Aorta blood velocity u_0	0 = 0.00	0.07
Blood volume V_b	1 = 1.00	1.00
Circulation time	1/4 = 0.25	0.25
Circulation distance l	1/4 = 0.25	ND
Cardiac stroke volume	1 = 1.00	1.03
Cardiac frequency ω	−1/4 = −0.25	−0.25
Cardiac output $\dot{E}$	3/4 = 0.75	0.74
Number of capillaries N_c	3/4 = 0.75	ND
Service volume radius	1/12 = 0.083	ND
Womersley number α	1/4 = 0.25	0.25
Density of capillaries	−1/12 = −0.083	−0.095
O_2 affinity of blood P_{50}	−1/12 = −0.083	−0.089
Total resistance Z	−3/4 = −0.75	−0.76
Metabolic rate B	3/4 = 0.75	0.75

Respiratory

Variable	Predicted Exponent	Observed
Tracheal radius	3/8 = 0.375	0.39
Interpleural pressure	0 = 0.00	0.004
Air velocity in trachea	0 = 0.00	0.02
Lung volume	1 = 1.00	1.05
Volume flow to lung	3/4 = 0.75	0.80
Volume of alveolus V_A	1/4 = 0.25	ND
Tidal volume	1 = 1.00	1.041
Respiratory frequency	−1/4 = −0.25	−0.26
Power dissipated	3/4 = 0.75	0.78
Number of alveoli N_A	3/4 = 0.75	ND
Radius of alveolus r_A	1/12 = 0.083	0.13
Area of alveolus A_A	1/6 = 0.083	ND
Area of lung A_L	11/12 = 0.92	0.95
O_2 diffusing capacity	1 = 1.00	0.99
Total resistance	−3/4 = −0.75	−0.70
O_2 consumption rate	3/4 = 0.75	0.76

Table 1. Values of allometric exponents for variables of the mammalian cardiovascular and respiratory systems predicted by the model compared with empirical observations. Observed values of exponents are taken from Schmidt-Nielsen (1984), Calder III (1984), and Peters (1983); ND denotes that no data are available.

pattern, provided it has a fractal structure[6]. Significantly, nonfractal systems, such as combustion engines and electric motors, exhibit geometric (third-power) rather than quarter-power scaling (McMahon and Bonner 1983). Because the fractal network must still fill the entire D-dimensional volume, our result generalizes to $a = D/(D+1)$. Organisms are three-dimensional, which explains the 3 in the numerator of the 3/4 power law, but it would be instructive to examine nearly two-dimensional organisms such as bryozoans and flatworms. The model can potentially explain how fundamental constraints at the level of individual organisms lead to corresponding quarter-power allometries at other levels. The constraints of body size on the rates at which resources can be taken up from the environment and transported and transformed within the body ramify to cause quarter-power scaling in such diverse phenomena as rate and duration of embryonic and postembryonic growth and development, interval between clutches, age of first reproduction, life span, home range and territory size, population density, and maximal population growth rate (McMahon and Bonner 1983; Schmidt-Nielsen 1984; Calder III 1984; Peters 1983). Because organisms of different body sizes have different requirements for resources and operate on different spatial and temporal scales, quarter-power allometric scaling is perhaps the single most pervasive theme underlying all biological diversity.

WBE elaborate on this topic in much more detail in their 1999 *Science* paper ("The Fourth Dimension of Life"). The basic idea is that the three-dimensional organism is embedded or functions as fourth-dimensional entity, thanks to the extra dimension provided by the space-filling network.

One of the theory's major successes has been its capacity to explain and predict processes at different level of biological organization, from cells to societies.

This is a fundamental insight. As it happens, in other ecological theories, such as in neutral theory, to be the same you have to be different (the same is different paradox). In this case, the pervasiveness of quarter-power allometric scaling implies some fundamental similarity and universality in life, the same yet different paradox.

Acknowledgments

J.H.B. is supported by NSF grant DEB-9318096, B.J.E. by NSF grant GER-9553623 and a Fulbright Fellowship, and G.B.W. by the Department of Energy.

REFERENCES

Bennett, A. F. 1988. "Structural and Functional Determinates of Metabolic Rate." *American Zoologist* 28 (2): 699–708.

Bennett, P. M., and P. H. Harvey. 1987. "Active and Resting Metabolism in Birds: Allometry, Phylogeny and Ecology." *Journal of Zoology* 213:327.

---

[6] This is reminiscent of the invariance of scaling exponents to details of the model that follow from renormalization group analyses, which can be viewed as a generalization of classical dimensional analysis.

Bonner, J. T. 1983. *The Evolution of Complexity by Means of Natural Selection*. Princeton, NJ: Princeton University Press.

Brown, J. H. 1995. *Macroecology*. Chicago, IL: University of Chicago Press.

Calder III, W. A. 1984. *Size, Function and Life History*. Cambridge, MA: Harvard University Press.

Canney, M. J. 1993. "The Transpiration Stream in the Leaf Apoplast: Water and Solutes." *Philosophical Transactions of the Royal Society B* 341 (87).

Caro, C. G., T. J. Pedley, R. C. Schroter, and W. A. Seed. 1978. *The Mechanics of Circulation*. Oxford, UK: Oxford University Press.

Ellsworth, M. L., A. Liu, B. Dawant, A. S. Popel, and R. N. Pittman. 1987. "Analysis of Vascular Pattern and Dimensions in Arteriolar Networks of the Retractor Muscle in Young Hamsters." *Microvascular Research* 34 (2): 168–183.

Feldman, H. A., and T. A. McMahon. 1983. "The 34 Mass Exponent for Energy Metabolism is not a Statistical Artifact." *Respiration Physiology* 52 (2): 149–163.

Fung, Y. C. 1984. *Biodynamics*. New York, NY: Springer-Verlag.

Gehr, P., D. K. Mwangi, A. Ammann, G. M. O. Maloiy, C. Richard Taylor, and Weibel E. R. 1981. "Design of the Mammalian Respiratory System. V. Scaling Morphometric Pulmonary Diffusing Capacity to Body Mass: Wild and Domestic Mammals." *Respiration Physiology* 44 (1): 61–86.

Harvey, P. H., and M. D. Pagel. 1991. *The Comparative Method in Evolutionary Biology*. Oxford, UK: Oxford University Press.

Hemmingsen, A. M. 1950. "The Relation of Standard (Basal) Energy Metabolism to Total Fresh Weight of Living Organisms." *Reports of the Steno Memorial Hospital* (Copenhagen, Denmark) 4:7–58.

———. 1960. "Energy Metabolism as Related to Body Size and Respiratory Surface, and its Evolution." *Report of Steno Memorial Hospital* (Copenhagen, Denmark) 9:1–110.

Iberall, A. S. 1967. "Anatomy and Steady Flow Characteristics of the Arterial System with an Introduction to its Pulsatile Characteristics." *Mathematical Biosciences* 1:375–395.

Krogh, A. 1920. "Pfluegers Arch. Gesamte Physiol." *Menschen Tiere* 179 (95).

LaBarbera, M. 1990. "Principles of Design of Fluid Transport Systems in Zoology." *Science* 249 (4972): 992–1000.

Li, J. K-J. 1996. *Comparative Cardiovascular Dynamics of Mammals*. Boca Raton, FL: CRC Press.

Mandelbrot, B. 1977. *The Fractal Geometry of Nature*. New York, NY: Freeman.

McMahon, T. A. 1973. "Size and Shape in Biology." *Science* 179:1201.

McMahon, T. A., and J. T. Bonner. 1983. *On Size and Life*. New York, NY: Scientific American Library.

Murray, C. D. 1926. *Proceedings of the National Academy of Sciences of the United States of America* 12 (207).

Niklas, K. J. 1994. *Plant Allometry: The Scaling of Form and Process*. Chicago, IL: University of Chicago Press.

Patterson, M. R. 1992. *Science* 255.

Peters, R. H. 1983. *The Ecological Implications of Body Size*. Cambridge, MA: Cambridge University Press.

Schmidt-Nielsen, K. 1984. *Scaling: Why Is Animal Size So Important?* Cambridge, UK: Cambridge University Press.

Sherman, T. F. 1981. "On Connecting Large Vessels to Small. The Meaning of Murray's Law." *The Journal of General Physiology* 78:431.

Shinozaki, K., Yoda K., Hozumi K., and Kira T. 1964a. "A Quantitative Analysis of Plant Form-the Pipe Model Theory I. Basic Analyses." *Japanese Journal of Ecology* 14 (3): 97–105.

———. 1964b. "A Quantitative Analysis of Plant Form-the Pipe Model Theory II." *Japanese Journal of Ecology* 14 (3): 133.

Strahler, A. N. 1953. "Revision of Hortons' Quantitative Factors in Erosional Terrain." *Transactions, American Geophysical Union* 34:345.

Thompson, D. W. 1942. *On Growth and Form*. Cambridge, UK: Cambridge University Press.

Tyree, M. T., and F. W. Ewers. 1991. "The Hydraulic Architecture of Trees and Other Woody Plants." *New Phytologist* 119:345–360.

Womersley, J. R. 1955a. "Method for the Calculation of Velocity, Rate of Flow and Viscous Drag in Arteries When the Pressure Gradient is Known." *The Journal of Physiology* 127:553.

———. 1955b. "Oscillatory Motion of a Viscous Liquid in a Thin-Walled Elastic Tube—I: The Linear Approximation for Long Waves." *The London, Edinburgh, and Dublin Philosophical Magazine and Journal of Science* 46 (373): 199–221.

Zamir, M., P. Sinclair, and T. H. Wonnacott. 1992. "Relation Between Diameter and Flow in Major Branches of the Arch of the Aorta." *Journal of Biomechanics,* 1303–10.

Zimmerman, M. H. 1983. *Xylem Structure and the Ascent of Sap*. Berlin, Germany: Springer-Verlag.

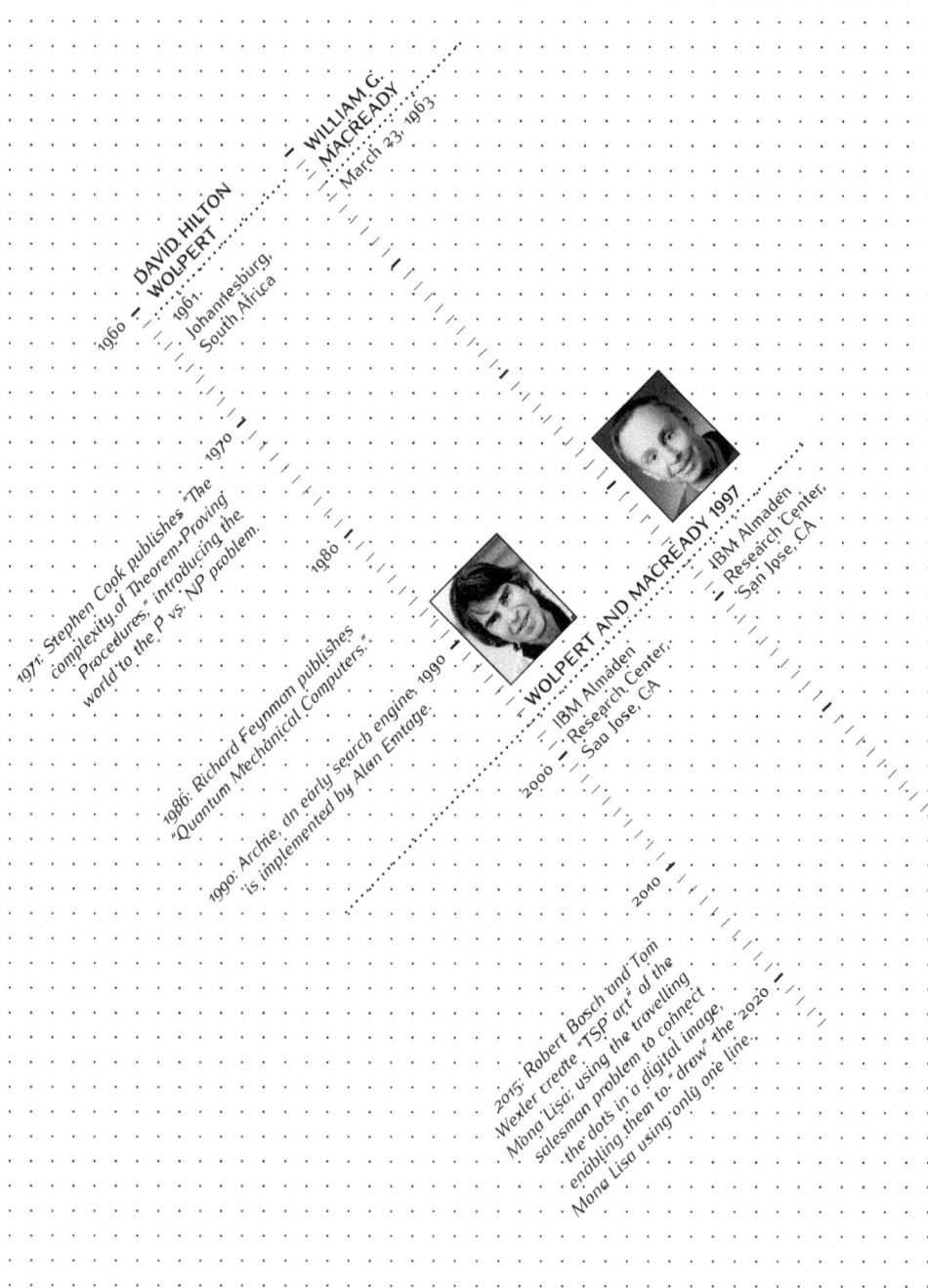

WOLPERT AND MACREADY

DAVID HILTON WOLPERT — 1961, Johannesburg, South Africa

WILLIAM G. MACREADY — March 23, 1963

1971: Stephen Cook publishes "The complexity of Theorem-Proving Procedures," introducing the world to the P vs. NP problem.

1986: Richard Feynman publishes "Quantum Mechanical Computers."

1990: Archie, an early search engine, is implemented by Alan Emtage.

WOLPERT AND MACREADY 1997 — IBM Almaden Research Center, San Jose, CA

IBM Almaden Research Center, San Jose, CA

2015: Robert Bosch and Tom Wexler create "TSP art" of the Mona Lisa, using the travelling salesman problem to connect the dots in a digital image, enabling them to "draw" the Mona Lisa using only one line.

[84]

NO FREE LUNCH

Charlie Strauss, Los Alamos National Laboratory, and Vijay Balasubramanian, University of Pennsylvania and Santa Fe Institute

Few theorems are as universal, poorly misunderstood, and absolutely infuriating as the series of no-free-lunch theorems. The theorems ultimately stem from the same basic insight but have been tailored to the specific cases of "search," "optimization," "generalization," and others with profound implications for those fields. I dare to say that many, many papers would never be published if all authors were required to answer how they were able to violate the iron-clad bounds of the no-free-lunch theorem.

In a nutshell, the theorem for search or optimization says that no search algorithm can find the global minimum of a finite search surface faster than any other on average. This has maddening implications, for whenever a paper debuts a new oh-so-clever search innovation and demonstrates it on a benchmark set where it outperforms the number of steps required to find the minimum compared to everything else there's an annoying corollary. If any algorithm is shown to outperform on a selected set of tests, it is equivalent to proving that it must underperform on average over all other problems outside the test set! That is, by claiming a better algorithm by empirical tests, you are actually proving it's worse.

Initially most people scoff. It seems insane that a hill-climbing search algorithm would on average find the minimum of surfaces as fast as a hill-descending algorithm. The answer to this paradox is counterintuitive. The hill climber never knows for sure that the next state it will select to test is necessarily uphill till it tries it; it's just following guidelines that tend to produce uphill behavior, like local slopes based on past states it tried. By random chance it can blunder into the lowest-energy state unexpectedly. What's counterintuitive is that, when averaged over all

D. H. Wolpert and W. G. Macready, "No Free Lunch Theorems for Optimization," *IEEE Transactions on Evolutionary Computation* 1 (1), 67–82 (1997).

©1997 IEEE. Reprinted with permission.

possible finite state potential energy surfaces, such accidents balance out the performance so that the hill climber will on average find the minimum as fast as the hill descender. Arggggg!

Initially, one is led to a state of despair. It is like working on perpetual motion and then learning about Boltzmann laws. Why bother, then? Well, to understand that one needs to first examine the underlying reason for the theorem. The framework for its derivation is to imagine a matrix representing points on a surface, each with an energy value. Imagine enumerating every possible combinatorial rearrangement of theses matrix values. One immediately sees that knowing the values of any set of these states in a given matrix tells you nothing predictive about the next state you will choose to interrogate. Thus there can be no clever method to pick the next state. A hill descender is just fooling itself into thinking it knows something about the future. Indeed, any search algorithm's trajectory on this surface is the same as some other on a permuted surface. Hence all averages need be the same.

But this leads us to a beautiful insight. Algorithms are better or worse than others only because they provide some other benefit besides the search. Here are a couple: Obviously the statement of the theorem should have included a caveat saying that one should never revisit the same state. Additionally, another performance measure is the number of states visited to find the global minimum, not the wall clock. But to not revisit requires having a memory of where you have been. And so an algorithm that revisits might require less memory storage than one that does not. Or, conversely, one that does store past information can perform better but only at that resource cost. Likewise an algorithm that chooses its next state more quickly, or perhaps evaluates a proxy potential energy function that is faster to compute might have better wall-clock performance.

But the really big insight is when you can say that you only want a search algorithm for a subclass of all possible problems. If this subset is not chosen randomly from all possible problems but has some underlying restriction, then perhaps then you could have an algorithm tuned to that restriction that can outperform. Thus, when publishing any new search and optimization algorithm, the author should be required to state what types of restrictions they would impose to define a class of problems and, if possible, explain why their algorithm aligns to that. However, it appears that in many cases defining that alignment may itself be NP-hard. So this insight requires deep thought, not simply empirical testing.

One such thing that nearly every person encountering this diabolical theorem leaps to is the idea that they are only interested in "smooth" surfaces, implying there is some bounded gradient across the matrix of values that forces neighbors to have some similarity to each other. It's the whole reason we think of things as "hill descending" or "climbing," so it's baked into our brains that this is how problems are. Not random matrices.

But it's not quite that simple. Imagine decomposing a smooth surface into the convolution of a smoothing function over an arbitrarily rough surface. You could even picture this as an underlying smaller matrix and interpolating new interstitial matrix elements by smoothing interpolation. In this view there is in effect a reduction on the size of the matrix. One could instead have run a search on the smaller but arbitrary matrix and gotten a speed up that way too rather than relying on gradients. The dimensionality reduction of "knowing" something is smooth may not be significant in very high-dimensional problems.

However, this can't be right, can it? After all, in almost every real-world problem domain optimizers work on, they find algorithms that empirically outperform others. So what is this saying? It's saying two possible things, I believe. It might be saying that, of all problems in the world, humans are only interested in the tractable ones. Ones where there seem to be some interesting principles creating it. For example, evolution may drive biological systems towards a subset of possible configurations that are somehow related. Aspects of physical laws (like, for example, no curve crossing in phase spaces) or equilibrium systems may have a similar effect of creating problems dominated by a small related subset. More arbitrary problems like the traveling salesman might therefore be the ones that defy easy search. The other possible explanation, or loophole, is that seldom do we actually care about finding the global minimum. Maybe we are interested in algorithms that tend to find a sufficiently lower energy state in the lowest amount of time on average. Of course, there are lots of problems, notably quantum computing, where it is critical to find the global minimum.

Thus one can see the beauty of this infuriating and annoying theorem. It truly spoils the fun of people who don't think about what the nature of the beast they are optimizing is. But, for those who embrace this, it raises the consciousness and drives further curiosity. We begin to wonder if perhaps by knowing which algorithms work best for certain classes of problems that maybe we would learn some hidden secrets of the subset that problem space lives in and thus understand nature better. At a minimum, we'd have to read fewer useless journal articles if authors were required to dwell on why their algorithms worked.

NO FREE LUNCH THEOREMS FOR OPTIMIZATION

*David H. Wolpert, IBM Almaden Research Center, and
William G. Macready, IBM Almaden Research Center
and Santa Fe Institute*

Abstract

A framework is developed to explore the connection between effective optimization algorithms and the problems they are solving. A number of "no free lunch" (NFL) theorems are presented which establish that for any algorithm, any elevated performance over one class of problems is offset by performance over another class. These theorems result in a geometric interpretation of what it means for an algorithm to be well suited to an optimization problem. Applications of the NFL theorems to information-theoretic aspects of optimization and benchmark measures of performance are also presented. Other issues addressed include time-varying optimization problems and *a priori* "head-to-head" minimax distinctions between optimization algorithms, distinctions that result despite the NFL theorems' enforcing of a type of uniformity over all algorithms.

Index Terms— Evolutionary algorithms, information theory, optimization.

A statement of the meaning of "no free lunch" theorems: better performance on some problems is necessarily offset by worse performance on others.

I. Introduction

The past few decades have seen an increased interest in general-purpose "black-box" optimization algorithms that exploit limited knowledge concerning the optimization problem on which they are run. In large part these algorithms have drawn inspiration from optimization processes that occur in nature. In particular, the two most popular black-box optimization strategies, evolutionary algorithms (Fogel, Owens, and Walsh 1966; Holland 1993; Schwefel 1995) and simulated annealing (Kirkpatrick, Gelatt, and Vecchi 1983), mimic processes in natural selection and statistical mechanics, respectively.

In light of this interest in general-purpose optimization algorithms, it has become important to understand the relationship between how well an algorithm a performs and the optimization problem f on which it is run. In this paper we present a formal analysis that contributes toward such an understanding by addressing questions like the following: given the abundance of black-box optimization algorithms and of optimization problems, how can we best match algorithms to problems (i.e., how best can we relax the black-box nature of the algorithms and have them exploit some knowledge concerning the optimization problem)? In particular, while serious optimization practitioners almost always perform such matching, it is usually on a heuristic basis; can such matching be formally analyzed? More generally, what is the underlying mathematical "skeleton" of optimization theory before the "flesh" of the probability distributions of a particular context and set of optimization problems are imposed? What can information theory and Bayesian analysis contribute to an understanding of these issues? How *a priori* generalizable are the performance results of a certain algorithm on a certain class of problems to its performance on other classes of problems? How should we even measure such generalization? How should we assess the performance of algorithms on problems so that we may programmatically compare those algorithms?

Broadly speaking, we take two approaches to these questions. First, we investigate what *a priori* restrictions there are on the performance of one or more algorithms as one runs over the set of all optimization problems. Our second approach is to instead focus on a particular problem and consider the effects of running over all algorithms. In the current paper we present results from both types of analyses but concentrate largely on the first approach. The reader is referred to the companion paper (Macready and Wolpert 1996) for more types of analysis involving the second approach.

We want to identify restrictions on performance by a fixed algorithm when run on all possible optimization problems.

We begin in Section II by introducing the necessary notation. Also discussed in this section is the model of computation we adopt, its limitations, and the reasons we chose it.

One might expect that there are pairs of search algorithms A and B such that A performs better than B on average, even if B sometimes outperforms A. As an example, one might expect that hill climbing usually outperforms hill descending if one's goal is to find a maximum of the cost function. One might also expect it would outperform a random search in such a context. One of the main results of this paper is that such expectations are incorrect. We prove two "no free lunch" (NFL) theorems in Section III that demonstrate this and more generally illuminate the

You might think that random search would on average be worse than a defined search algorithm. This turns out not to be true!

connection between algorithms and problems. Roughly speaking, we show that for both static and time-dependent optimization problems, the average performance of any pair of algorithms across all possible problems is identical. This means in particular that if some algorithm a_1's performance is superior to that of another algorithm a_2 over some set of optimization problems, then the reverse must be true over the set of all other optimization problems. (The reader is urged to read this section carefully for a precise statement of these theorems.) This is true even if one of the algorithms is random; any algorithm a_1 performs worse than randomly just as readily (over the set of all optimization problems) as it performs better than randomly. Possible objections to these results are addressed in Sections III-A and III-B.

In Section IV we present a geometric interpretation of the NFL theorems. In particular, we show that an algorithm's average performance is determined by how "aligned" it is with the underlying probability distribution over optimization problems on which it is run. This section is critical for an understanding of how the NFL results are consistent with the well-accepted fact that many search algorithms that do not take into account knowledge concerning the cost function work well in practice.

Section V-A demonstrates that the NFL theorems allow one to answer a number of what would otherwise seem to be intractable questions. The implications of these answers for measures of algorithm performance and of how best to compare optimization algorithms are explored in Section V-B.

In Section VI we discuss some of the ways in which, despite the NFL theorems, algorithms can have *a priori* distinctions that hold even if nothing is specified concerning the optimization problems. In particular, we show that there can be "head-to-head" minimax distinctions between a pair of algorithms, i.e., that when considering one function at a time, a pair of algorithms may be distinguishable, even if they are not when one looks over all functions.

In Section VII we present an introduction to the alternative approach to the formal analysis of optimization in which problems are held fixed and one looks at properties across the space of algorithms. Since these results hold in general, they hold for any and all optimization

Past performance turns out not to guarantee future returns.

Optimization algorithms work if they exploit some specific relation between the algorithm and the problem to be solved.

problems and thus are independent of the types of problems one is more or less likely to encounter in the real world. In particular, these results show that there is no *a priori* justification for using a search algorithm's observed behavior to date on a particular cost function to predict its future behavior on that function. In fact when choosing between algorithms based on their observed performance it does not suffice to make an assumption about the cost function; some (currently poorly understood) assumptions are also being made about how the algorithms in question are related to each other and to the cost function. In addition to presenting results not found in Macready and Wolpert (1996), this section serves as an introduction to the perspective adopted in Macready and Wolpert (1996).

We conclude in Section VIII with a brief discussion, a summary of results, and a short list of open problems.

We have confined all proofs to appendixes to facilitate the flow of the paper. A more detailed, and substantially longer, version of this paper, a version that also analyzes some issues not addressed in this paper, can be found in Wolpert and Macready (1995).

II. Preliminaries

We restrict attention to combinatorial optimization in which the search space $\mathcal{X}$, though perhaps quite large, is finite. We further assume that the space of possible "cost" values $\mathcal{Y}$ is also finite. These restrictions are automatically met for optimization algorithms run on digital computers where typically $\mathcal{Y}$ is some 32 or 64 bit representation of the real numbers.

The size of the spaces $\mathcal{X}$ and $\mathcal{Y}$ are indicated by $|\mathcal{X}|$ and $|\mathcal{Y}|$, respectively. An optimization problem f (sometimes called a "cost function" or an "objective function" or an "energy function") is represented as a mapping $f : \mathcal{X} \mapsto \mathcal{Y}$ and $\mathcal{F} = \mathcal{Y}^{\mathcal{X}}$ indicates the space of all possible problems. $\mathcal{F}$ is of size $|\mathcal{Y}|^{|\mathcal{X}'|}$—a large but finite number. In addition to static f, we are also interested in optimization problems that depend explicitly on time. The extra notation required for such time-dependent problems will be introduced as needed.

It is common in the optimization community to adopt an oracle-based view of computation. In this view, when assessing the performance of algorithms, results are stated in terms of the number of function evaluations required to find a given solution. Practically though, many optimization algorithms are wasteful of function evaluations. In particular, many algorithms do not remember where they have already searched and therefore often revisit the same points. Although any algorithm that is wasteful in this fashion can be made more efficient simply by remembering where it has been (cf. tabu search (Glover 1989, 1990)), many real-world algorithms elect not to employ this stratagem. From the point of view of the oracle-based performance measures, these revisits are "artifacts" distorting the apparent relationship between many such real-world algorithms.

Many optimization algorithms revisit the same points multiple times. This waste can be avoided by including memory as a resource.

This difficulty is exacerbated by the fact that the amount of revisiting that occurs is a complicated function of both the algorithm and the optimization problem and therefore cannot be simply "filtered out" of a mathematical analysis. Accordingly, we have elected to circumvent the problem entirely by comparing algorithms based on the number of distinct function evaluations they have performed. Note that this does not mean that we cannot compare algorithms that are wasteful of evaluations—it simply means that we compare algorithms by counting only their number of distinct calls to the oracle.

They will quantify "efficiency" by the number of distinct calls to an "oracle" that evaluates the cost function for free.

We call a time-ordered set of m distinct visited points a "sample" of size m. Samples are denoted by

$$d_m \equiv \{(d_m^x(1), d_m^y(1)), \cdots, (d_m^x(m), d_m^y(m))\}.$$

The points in a sample are ordered according to the time at which they were generated. Thus $d_m^x(i)$ indicates the $\mathcal{X}$ value of the ith successive element in a sample of size m and $d_m^y(i)$ is its associated cost or $\mathcal{Y}$ value. $d_m^y \equiv \{d_m^y(1), \cdots, d_m^y(m)\}$ will be used to indicate the ordered set of cost values. The space of all samples of size m is $\mathcal{D}_m = (\mathcal{X} \times \mathcal{Y})^m$ (so $d_m \in \mathcal{D}_m$) and the set of all possible samples of arbitrary size is $\mathcal{D} \equiv \cup_{m \geq 0} \mathcal{D}_m$.

As an important clarification of this definition, consider a hill-descending algorithm. This is the algorithm that examines a set of neighboring points in $\mathcal{X}$ and moves to the one having the lowest cost. The process is then iterated from the newly chosen point. (Often, implementations of hill descending stop when they reach a local minimum, but they can easily be extended to run longer by randomly jumping to a new unvisited point once the neighborhood of a local minimum has been exhausted.) The point to note is that because a sample contains all the previous points at which the oracle was consulted, it includes the $(\mathcal{X}, \mathcal{Y})$ values of *all* the neighbors of the current point, and not only the lowest cost one that the algorithm moves to. This must be taken into account when counting the value of m.

An optimization algorithm a is represented as a mapping from previously visited sets of points to a single new (i.e., previously unvisited) point in $\mathcal{X}$. Formally, $a : d \in \mathcal{D} \mapsto \{x \mid x \notin d^x\}$. Given our decision to only measure distinct function evaluations even if an algorithm revisits previously searched points, our definition of an algorithm includes all common black-box optimization techniques like simulated annealing and evolutionary algorithms. (Techniques like branch and bound (Lawler and Wood 1966) are not included since they rely explicitly on the cost structure of partial solutions.)

As defined above, a search algorithm is deterministic; every sample maps to a unique new point. Of course, essentially, all algorithms implemented on computers are deterministic,[1] and in this our definition is not restrictive. Nonetheless, it is worth noting that all of our results are extensible to nondeterministic algorithms, where the new point is chosen stochastically from the set of unvisited points. This point is returned to later.

Under the oracle-based model of computation any measure of the performance of an algorithm after m iterations is a function of the sample d_m^y. Such performance measures will be indicated by $\Phi(d_m^y)$. As an example, if we are trying to find a minimum of f, then a reasonable measure of the performance of a might be the value of the lowest $\mathcal{Y}$ value in $d_m^y : \Phi(d_m^y) = \min_i \{d_m^y(i) : i = 1 \cdots m\}$. Note that measures of performance based on factors other than d_m^y (e.g., wall clock time) are outside the scope of our results.

We shall cast all of our results in terms of probability theory. We do so for three reasons. First, it allows simple generalization of our results to stochastic algorithms. Second, even when the setting is deterministic, probability theory provides a simple consistent framework in which

---

[1] In particular, note that pseudorandom number generators are deterministic given a seed.

to carry out proofs. The third reason for using probability theory is perhaps the most interesting. A crucial factor in the probabilistic framework is the distribution $P(f) = P\left(f\left(x_1\right), \cdots, f\left(x_{|\mathcal{X}|}\right)\right)$. This distribution, defined over $\mathcal{F}$, gives the probability that each $f \in \mathcal{F}$ is the actual optimization problem at hand. An approach based on this distribution has the immediate advantage that often knowledge of a problem is statistical in nature and this information may be easily encodable in $P(f)$. For example, Markov or Gibbs random field descriptions (Kinderman and Snell 1980) of families of optimization problems express $P(f)$ exactly.

Exploiting $P(f)$, however, also has advantages even when we are presented with a single uniquely specified cost function. One such advantage is the fact that although it may be fully specified, many aspects of the cost function are *effectively* unknown (e.g., we certainly do not know the extrema of the function). It is in many ways most appropriate to have this effective ignorance reflected in the analysis as a probability distribution. More generally, optimization practitioners usually act as though the cost function is partially unknown, in that the same algorithm is used for all cost functions in a class of such functions (e.g.. in the class of all traveling salesman problems having certain characteristics). In so doing, the practitioner implicitly acknowledges that distinctions between the cost functions in that class are irrelevant or at least unexploitable. In this sense, even though we are presented with a single particular problem from that class, we act as though we are presented with a probability distribution over cost functions, a distribution that is nonzero only for members of that class of cost functions. $P(f)$ is thus a prior specification of the class of the optimization problem at hand, with different classes of problems corresponding to different choices of what algorithms we will use, and giving rise to different distributions $P(f)$.

Even if a cost function is fully specified, many things about it—for example, the extrema—may not be explicitly known.

The authors define a statistical measure of performance.

Given our choice to use probability theory, the performance of an algorithm a iterated m times on a cost function f is measured with $P\left(d_m^y \mid f, m, a\right)$. This is the conditional probability of obtaining a particular sample d_m under the stated conditions. From $P\left(d_m^y \mid f, m, a\right)$ performance measures $\Phi\left(d_m^y\right)$ can be found easily.

In the next section we analyze $P(d_m^y \mid f, m, a)$ and in particular how it varies with the algorithm a. Before proceeding with that analysis, however, it is worth briefly noting that there are other formal approaches to the issues investigated in this paper. Perhaps the most prominent of these is the field of computational complexity. Unlike the approach taken in this paper, computational complexity largely ignores the statistical nature of search and concentrates instead on computational issues. Much, though by no means all, of computational complexity is concerned with physically unrealizable computational devices (e.g., Turing machines) and the worst-case resource usage required to find optimal solutions. In contrast, the analysis in this paper does not concern itself with the computational engine used by the search algorithm, but rather concentrates exclusively on the underlying statistical nature of the search problem. The current probabilistic approach is complimentary to computational complexity. Future work involves combining our analysis of the statistical nature of search with practical concerns for computational resources.

III. The NFL Theorems

In this section we analyze the connection between algorithms and cost functions. We have dubbed the associated results NFL theorems because they demonstrate that if an algorithm performs well on a certain class of problems then it necessarily pays for that with degraded performance on the set of all remaining problems. Additionally, the name emphasizes a parallel with similar results in supervised learning (Wolpert 1996a, 1996b).

The precise question addressed in this section is: "How does the set of problems $F_1 \subset \mathcal{F}$ for which algorithm a_1 performs better than algorithm a_2 compare to the set $F_2 \subset \mathcal{F}$ for which the reverse is true?" To address this question we compare the sum over all f of $P(d_m^y \mid f, m, a_1)$ to the sum over all f of $P(d_m^y \mid f, m, a_2)$. This comparison constitutes a major result of this paper: $P(d_m^y \mid f, m, a)$ is independent of a when averaged over all cost functions.

Astonishingly, the performance averaged over cost functions is independent of the optimization algorithm!

Theorem 1. *For any pair of algorithms a_1 and a_2*

$$\sum_f P(d_m^y \mid f, m, a_1) = \sum_f P(d_m^y \mid f, m, a_2).$$

A proof of this result is found in Appendix A. An immediate corollary of this result is that for any performance measure $\Phi(d_m^y)$, the average over all f of $P(\Phi(d_m^y) \mid f, m, a)$ is independent of a. The precise way that the sample is mapped to a performance measure is unimportant.

This theorem explicitly demonstrates that what an algorithm gains in performance on one class of problems is necessarily offset by its performance on the remaining problems; that is the only way that all algorithms can have the same f-averaged performance.

A result analogous to Theorem 1 holds for a class of time-dependent cost functions. The time-dependent functions we consider begin with an initial cost function f_1 that is present at the sampling of the first $\mathcal{X}$ value. Before the beginning of each subsequent iteration of the optimization algorithm, the cost function is deformed to a new function, as specified by a mapping $T : \mathcal{F} \times \mathcal{N} \to \mathcal{F}$.[2] We indicate this mapping with the notation T_i. So the function present during the ith iteration is $f_{i+1} = T_i(f_i)$. T_i is assumed to be a (potentially i-dependent) bijection between $\mathcal{F}$ and $\mathcal{F}$. We impose bijectivity because if it did not hold, the evolution of cost functions could narrow in on a region of f's for which some algorithms may perform better than others. This would constitute an *a priori* bias in favor of those algorithms, a bias whose analysis we wish to defer to future work.

How best to assess the quality of an algorithm's performance on time-dependent cost functions is not clear. Here we consider two schemes based on manipulations of the definition of the sample. In scheme 1 the particular $\mathcal{Y}$ value in $d_m^y(j)$ corresponding to a particular x value $d_m^x(j)$ is given by the cost function that was present when $d_m^x(j)$ was sampled. In contrast, for scheme 2 we imagine a sample D_m^y given by the $\mathcal{Y}$ values from the present cost function for each of the x values in d_m^x. Formally if $d_m^x = \{d_m^x(1), \cdots, d_m^x(m)\}$, then in scheme 1 we have $d_m^y = \{f_1(d_m^x(1)), \cdots, T_{m-1}(f_{m-1})(d_m^x(m))\}$, and in scheme 2 we

Good performance on some problems is necessarily paid for by degradation on others.

---

[2] An obvious restriction would be to require that T does not vary with time, so that it is a mapping simply from $\mathcal{F}$ to $\mathcal{F}$. An analysis for T's limited in this way is beyond the scope of this paper.

have $D_m^y = \{f_m(d_m^x(1)), \cdots, f_m(d_m^x(m))\}$ where $f_m = T_{m-1}(f_{m-1})$ is the final cost function.

In some situations it may be that the members of the sample "live" for a long time, compared to the time scale of the dynamics of the cost function. In such situations it may be appropriate to judge the quality of the search algorithm by D_m^y; all those previous elements of the sample that are still "alive" at time m, and therefore their current cost is of interest. On the other hand, if members of the sample live for only a short time on the time scale of the dynamics of the cost function, one may instead be concerned with things like how well the "living" member(s) of the sample track the changing cost function. In such situations, it may make more sense to judge the quality of the algorithm with the d_m^y sample.

Results similar to Theorem 1 can be derived for both schemes. By analogy with that theorem, we average over all possible ways a cost function may be time dependent, i.e., we average over all T (rather than over all f). Thus we consider $\sum_T P(d_m^y \mid f_1, T, m, a)$ where f_1 is the initial cost function. Since T only takes effect for $m > 1$, and since f_1 is fixed, there are *a priori* distinctions between algorithms as far as the first member of the sample is concerned. After redefining samples, however, to only contain those elements added after the first iteration of the algorithm, we arrive at the following result, proven in Appendix B.

Theorem 2. *For all $d_m^y, D_m^y, m > 1$, algorithms a_1 and a_2, and initial cost functions f_1*

$$\sum_T P(d_m^y \mid f_1, T, m, a_1) = \sum_T P(d_m^y \mid f_1, T, m, a_2)$$

and

$$\sum_T P(D_m^y \mid f_1, T, m, a_1) = \sum_T P(D_m^y \mid f_1, T, m, a_2)$$

Performance averaged over time-dependent cost functions is independent of the algorithm.

So, in particular, if one algorithm outperforms another for certain kinds of cost function dynamics, then the reverse must be true on the set of all other cost function dynamics.

Although this particular result is similar to the NFL result for the static case, in general the time-dependent situation is more subtle. In particular, with time dependence there are situations in which there can be *a priori* distinctions between algorithms even for those members of the sample arising after the first. For example, in general there will be distinctions between algorithms when considering the quantity $\sum_f P(d_m^y \mid f, T, m, a)$. To see this, consider the case where $\mathcal{X}$ is a set of contiguous integers and for all iterations T is a shift operator, replacing $f(x)$ by $f(x-1)$ for all x [with $\min(x) - 1 \equiv \max(x)$]. For such a case we can construct algorithms which behave differently *a priori*. For example, take a to be the algorithm that first samples f at x_1, next at $x_1 + 1$, and so on, regardless of the values in the sample. Then for any f, d_m^y is always made up of identical $\mathcal{Y}$ values. Accordingly, $\sum_f P(d_m^y \mid f, T, m, a)$ is nonzero only for d_m^y for which all values $d_m^y(i)$ are identical. Other search algorithms, even for the same shift T, do not have this restriction on $\mathcal{Y}$ values. This constitutes an *a priori* distinction between algorithms.

A. IMPLICATIONS OF THE NFL THEOREMS

As emphasized above, the NFL theorems mean that if an algorithm does particularly well on average for one class of problems then it must do worse on average over the remaining problems. In particular, if an algorithm performs better than random search on some class of problems then it must perform *worse than random search* on the remaining problems. Thus comparisons reporting the performance of a particular algorithm with a particular parameter setting on a few sample problems are of limited utility. While such results do indicate behavior on the narrow range of problems considered, one should be very wary of trying to generalize those results to other problems.

Infuriatingly, performing better than random search on some problems requires the algorithm to perform worse on others.

Note, however, that the NFL theorems need not be viewed as a way of comparing function classes $\mathcal{F}_1$ and $\mathcal{F}_2$ (or classes of evolution operators T_1 and T_2, as the case might be). They can be viewed instead as a statement concerning any algorithm's performance when f is not fixed, under the uniform prior over cost functions, $P(f) = 1/|\mathcal{F}|$. If we wish instead to analyze performance where f is not fixed, as in this alternative interpretation of the NFL theorems, but in contrast with the NFL case f is now chosen from a nonuniform prior, then we must analyze

explicitly the sum

$$P(d_m^y \mid m, a) = \sum_f P(d_m^y \mid f, m, a) P(f) \qquad (1)$$

Since it is certainly true that any class of problems faced by a practitioner will not have a flat prior, what are the practical implications of the NFL theorems when viewed as a statement concerning an algorithm's performance for nonfixed f? This question is taken up in greater detail in Section IV but we offer a few comments here.

First, if the practitioner has knowledge of problem characteristics but does not incorporate them into the optimization algorithm, then $P(f)$ is effectively uniform. (Recall that $P(f)$ can be viewed as a statement concerning the practitioner's choice of optimization algorithms.) In such a case, the NFL theorems establish that there are no formal assurances that the algorithm chosen will be at all effective.

Second, while most classes of problems will certainly have some structure which, if known, might be exploitable, the simple existence of that structure does not justify choice of a particular algorithm; that structure must be known and reflected directly in the choice of algorithm to serve as such a justification. In other words, the simple existence of structure per se, absent a specification of that structure, cannot provide a basis for preferring one algorithm over another. Formally, this is established by the existence of NFL-type theorems in which rather than average over specific cost functions f, one averages over specific "kinds of structure," i.e., theorems in which one averages $P(d_m^y \mid m, a)$ over distributions $P(f)$. That such theorems hold when one averages over all $P(f)$ means that the indistinguishability of algorithms associated with uniform $P(f)$ is not some pathological, outlier case. Rather, uniform $P(f)$ is a "typical" distribution as far as indistinguishability of algorithms is concerned. The simple fact that the $P(f)$ at hand is nonuniform cannot serve to determine one's choice of optimization algorithm.

Finally, it is important to emphasize that even if one is considering the case where f is not fixed, performing the associated average according to a uniform $P(f)$ is not essential for NFL to hold. NFL can also be demonstrated for a range of nonuniform priors. For example, any prior of the form $\prod_{x \in \mathcal{X}} P'(f(x))$ (where $P'(y = f(x))$ is the

Some problems have a structure that can be exploited by an algorithm to achieve better performance. But the structure must be known, and the algorithm should be adapted to it. The mere existence of structure neither implies better performance nor justifies one or another algorithm.

distribution of $\mathcal{Y}$ values) will also give NFL theorems. The f-average can also enforce correlations between costs at different $\mathcal{X}$ values and NFL-like results will still be obtained. For example, if costs are rank ordered (with ties broken in some arbitrary way) and we sum only over all cost functions given by permutations of those orderings, then NFL remains valid.

The choice of uniform $P(f)$ was motivated more from theoretical rather than pragmatic concerns, as a way of analyzing the theoretical structure of optimization. Nevertheless, the cautionary observations presented above make clear that an analysis of the uniform $P(f)$ case has a number of ramifications for practitioners.

B. STOCHASTIC OPTIMIZATION ALGORITHMS

Thus far we have considered the case in which algorithms are deterministic. What is the situation for stochastic algorithms? As it turns out, NFL results hold even for these algorithms.

The results also apply to stochastic optimization algorithms like simulated annealing.

The proof is straightforward. Let σ be a stochastic "nonpotentially revisiting" algorithm. Formally, this means that σ is a mapping taking any sample d to a d-dependent distribution over $\mathcal{X}$ that equals zero for all $x \in d^x$. In this sense σ is what in statistics community is known as a "hyper-parameter," specifying the function $P\left(d_{m+1}^x(m+1) \mid d_m, \sigma\right)$ for all m and d. One can now reproduce the derivation of the NFL result for deterministic algorithms, only with a replaced by σ throughout. In so doing, all steps in the proof remain valid. This establishes that NFL results apply to stochastic algorithms as well as deterministic ones.

IV. A Geometric Perspective on the NFL Theorems

Intuitively, the NFL theorem illustrates that if knowledge of f, perhaps specified through $P(f)$, is not incorporated into a, then there are no formal assurances that a will be effective. Rather, in this case effective optimization relies on a fortuitous matching between f and a. This point is formally established by viewing the NFL theorem from a geometric perspective. Consider the space $\mathcal{F}$ of all possible cost functions. As previously discussed in regard to (1), the probability of obtaining some d_m^y is

$$P\left(d_m^y \mid m, a\right) = \sum_f P\left(d_m^y \mid m, a, f\right) P(f)$$

where $P(f)$ is the prior probability that the optimization problem at hand has cost function f. This sum over functions can be viewed as an inner product in $\mathcal{F}$. Defining the $\mathcal{F}$-space vectors $\vec{v}_{d_m^y,a,m}$ and $\vec{p}$ by their f components $\vec{v}_{d_m^y,a,m}(f) \equiv P(d_m^y \mid m,a,f)$ and $\vec{p}(f) \equiv P(f)$, respectively

$$P(d_m^y \mid m,a) = \vec{v}_{d_m^y,a,m} \cdot \vec{p}. \qquad (2)$$

This equation provides a geometric interpretation of the optimization process. d_m^y can be viewed as fixed to the sample that is desired, usually one with a low cost value, and m is a measure of the computational resources that can be afforded. Any knowledge of the properties of the cost function goes into the prior over cost functions $\vec{p}$. Then (2) says the performance of an algorithm is determined by the magnitude of its projection onto $\vec{p}$, i.e., by how aligned $\vec{v}_{d_m^y,a,m}$ is with the problems $\vec{p}$. Alternatively, by averaging over d_m^y, it is easy to see that $E(d_m^y \mid m,a)$ is an inner product between $\vec{p}$ and $E(d_m^y \mid m,a,f)$. The expectation of any performance measure $\Phi(d_m^y)$ can be written similarly.

In any of these cases, $P(f)$ or $\vec{p}$ must "match" or be aligned with a to get the desired behavior. This need for matching provides a new perspective on how certain algorithms can perform well in practice on specific kinds of problems. For example, it means that the years of research into the traveling salesman problem (TSP) have resulted in algorithms aligned with the (implicit) $\vec{p}$ describing traveling salesman problems of interest to TSP researchers.

Taking the geometric view, the NFL result that $\sum_f P(d_m^y \mid f,m,a)$ is independent of a has the interpretation that for any particular d_m^y and m, all algorithms a have the same projection onto the uniform $P(f)$, represented by the diagonal vector $\vec{1}$. Formally, $v_{d_m^y,a,m} \cdot \vec{1} = c(d_m^y,m)$ where c is some constant depending only upon d_m^y and m. For deterministic algorithms, the components of $v_{d_m^y,a,m}$ (i.e., the probabilities that algorithm a gives sample d_m^y on cost function f after m distinct cost evaluations) are all either zero or one, so NFL also implies that $\sum_f P^2(d_m^y \mid m,a,f) = c(d_m^y,m)$. Geometrically, this means that the length of $\vec{v}_{d_m^y,a,m}$ is independent of a. Different algorithms thus generate different vectors $\vec{v}_{d^y,a,m}$ all having the same length and lying on a cone with constant projection onto $\vec{1}$. A schematic of this situation is shown in Fig. 1 for the case where $\mathcal{F}$ is three dimensional. Because the

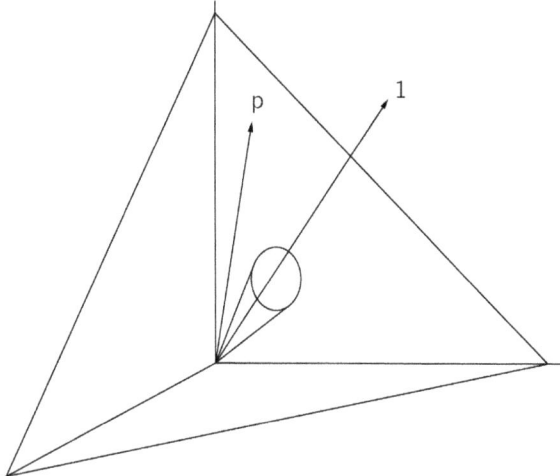

Figure 1. Schematic view of the situation in which function space $\mathcal{F}$ is three dimensional. The uniform prior over this space, $\vec{1}$, lies along the diagonal. Different algorithms a give different vectors v lying in the cone surrounding the diagonal. A particular problem is represented by its prior $\vec{p}$ lying on the simplex. The algorithm that will perform best will be the algorithm in the cone having the largest inner product with $\vec{p}$.

components of $\vec{v}_{c,a,m}$ are binary, we might equivalently view $\vec{v}_{d_m^y,a,m}$ as lying on the subset of Boolean hypercube vertices having the same hamming distance from $\vec{1}$.

Now restrict attention to algorithms having the same probability of some particular d_m^y. The algorithms in this set lie in the intersection of two cones—one about the diagonal, set by the NFL theorem, and one set by having the same probability for d_m^y. This is in general an $|\mathcal{F}| - 2$ dimensional manifold. Continuing, as we impose yet more d_m^y-based restrictions on a set of algorithms, we will continue to reduce the dimensionality of the manifold by focusing on intersections of more and more cones.

This geometric view of optimization also suggests measures for determining how "similar" two optimization algorithms are. Consider again (2). In that the algorithm only gives $\vec{v}_{d_m^y,a,m}$, perhaps the most straightforward way to compare two algorithms a_1 and a_2 would be by measuring how similar the vectors $\vec{v}_{d_m^y,a_1,m}$ and $\vec{v}_{d_m^y,a_2,m}$ are, perhaps by evaluating the dot product of those vectors. Those vectors, however, occur on the right-hand side of (2), whereas the performance of the algorithms—which is after all our ultimate concern—occurs on the lefthand side. This suggests measuring the similarity of two algorithms not directly in terms of their vectors $\vec{v}_{d_m^y,a,m}$ but rather in terms of the dot products of those vectors with $\vec{p}$. For example, it may be the case that algorithms behave very similarly for certain $P(f)$ but are quite different for other $P(f)$. In many respects, knowing this about two algorithms is of more interest than knowing how their vectors $\vec{v}_{d_m^y,a,m}$ compare.

As another example of a similarity measure suggested by the geometric perspective, we could measure similarity between algorithms based on similarities between $P(f)$'s. For example, for two different algorithms, one can imagine solving for the $P(f)$ that optimizes $P(d_m^y \mid m, a)$ for those algorithms, in some nontrivial sense.[3] We could then use some measure of distance between those two $P(f)$ distributions as a gauge of how similar the associated algorithms are.

Unfortunately, exploiting the inner product formula in practice, by going from a $P(f)$ to an algorithm optimal for that $P(f)$, appears to often be quite difficult. Indeed, even determining a plausible $P(f)$ for the situation at hand is often difficult. Consider, for example, TSP problems with N cities. To the degree that any practitioner attacks all N-city TSP cost functions with the same algorithm, he/she implicitly ignores distinctions between such cost functions. In this, that practitioner has implicitly agreed that the problem is one of how their fixed algorithm does across the set of all N-city TSP cost functions. But the detailed nature of the $P(f)$ that is uniform over this class of problems appears to be difficult to elucidate.

On the other hand, there is a growing body of work that does rely explicitly on enumeration of $P(f)$. For example, applications of Markov random fields (Kinderman and Snell 1980; Griffeath 1976) to cost landscapes yield $P(f)$ directly as a Gibbs distribution.

V. Calculational Applications of the NFL Theorems

In this section, we explore some of the applications of the NFL theorems for performing calculations concerning optimization. We will consider both calculations of practical and theoretical interest and begin with

---

[3] In particular, one may want to impose restrictions on $P(f)$. For instance, one may wish to only consider $P(f)$ that are invariant under at least partial relabeling of the elements in $\mathcal{X}$, to preclude there being an algorithm that will assuredly "luck out" and land on $\min_{x \in \mathcal{X}} f(x)$ on its very first query.

calculations of theoretical interest, in which information-theoretic quantities arise naturally.

A. INFORMATION-THEORETIC ASPECTS OF OPTIMIZATION

For expository purposes, we simplify the discussion slightly by considering only the histogram of number of instances of each possible cost value produced by a run of an algorithm, and not the temporal order in which those cost values were generated. Many real-world performance measures are independent of such temporal information. We indicate that histogram with the symbol $\vec{c}$; $\vec{c}$ has $\mathcal{Y}$ components $\left(c_{y_1}, c_{y_2}, \cdots, c_{y_{|y|}}\right)$, where c_i is the number of times cost value y_i occurs in the sample d_m^y.

Now consider any question like the following: "What fraction of cost functions give a particular histogram $\vec{c}$ of cost values after m distinct cost evaluations produced by using a particular instantiation of an evolutionary algorithm?"

At first glance this seems to be an intractable question, but the NFL theorem provides a way to answer it. This is because—according to the NFL theorem—the answer must be independent of the algorithm used to generate $\vec{c}$. Consequently, we can choose an algorithm for which the calculation is tractable.

Theorem 3. *For any algorithm, the fraction of cost functions that result in a particular histogram $\vec{c} = m\vec{\alpha}$ is*

$$\rho_f(\vec{\alpha}) = \frac{\left(c_1 c_2 \overset{m}{\cdots} c_{|\mathcal{Y}|}\right) |\mathcal{Y}|^{|\mathcal{X}|-m}}{|\mathcal{Y}|^{|\mathcal{X}|}} = \frac{\left(c_1 c_2 \overset{m}{\cdots} c_{|\mathcal{Y}|}\right)}{|\mathcal{Y}|^m}.$$

For large enough m, this can be approximated as

$$\rho_f(\vec{\alpha}) \cong C(m, |\mathcal{Y}|) \frac{\exp[mS(\vec{\alpha})]}{\prod_{i=1}^{|\mathcal{Y}|} \alpha_i^{1/2}}$$

where $S(\vec{\alpha})$ is the entropy of the distribution $\vec{\alpha}$, and $C(m, |\mathcal{Y}|)$ is a constant that does not depend on $\vec{\alpha}$.

This theorem is derived in Appendix C. If some of the $\vec{\alpha}_i$ are zero, the approximation still holds, only with $\mathcal{Y}$ redefined to exclude the y's corresponding to the zero-valued $\vec{\alpha}_i$. However, $\mathcal{Y}$ is defined and the normalization constant of (3) can be found by summing over all $\vec{\alpha}$ lying on the unit simplex (Strauss, Wolpert, and Wolf 1992).

A related question is the following: "For a given cost function, what is the fraction ρ_{alg} of all algorithms that give rise to a particular $\vec{c}$?" It turns out that the only feature of f relevant for this question is the histogram of its cost values formed by looking across all $\mathcal{X}$. Specify the fractional form of this histogram by $\vec{\beta}$ so that there are $N_i = \beta_i |\mathcal{X}|$ points in $\mathcal{X}$ for which $f(x)$ has the ith $\mathcal{Y}$ value.

In Appendix D it is shown that to leading order, $\rho_{\text{alg}}(\vec{\alpha}, \vec{\beta})$ depends on yet another information-theoretic quantity, the Kullback–Liebler distance (Cover and Thomas 1991) between $\vec{\alpha}$ and $\vec{\beta}$.

Theorem 4. *For a given f with histogram $\vec{N} = |\mathcal{X}|\vec{\beta}$, the fraction of algorithms that give rise to a histogram $\vec{c} = m\vec{\alpha}$ is given by*

$$\rho_{\text{alg}}(\vec{\alpha}, \vec{\beta}) = \frac{\prod_{i=1}^{|\mathcal{Y}|} \binom{N_i}{c_i}}{\binom{|\mathcal{X}|}{m}} \tag{3}$$

For large enough m, this can be written as

$$\rho_{\text{alg}}(\vec{\alpha}, \vec{\beta}) \cong C(m, |\mathcal{X}|, |\mathcal{Y}|) \frac{e^{-m D_{KL}(\vec{\alpha}, \vec{\beta})}}{\prod_{i=1}^{|\mathcal{Y}|} \alpha_i^{1/2}}$$

where $D_{KL}(\vec{\alpha}, \vec{\beta})$ is the Kullback-Liebler distance between the distributions $\vec{\alpha}$ and $\vec{\beta}$.

As before, C can be calculated by summing $\vec{\alpha}$ over the unit simplex.

B. MEASURES OF PERFORMANCE

We now show how to apply the NFL framework to calculate certain benchmark performance measures. These allow both the programmatic assessment of the efficacy of any individual optimization algorithm and principled comparisons between algorithms.

Without loss of generality, assume that the goal of the search process is finding a minimum. So we are interested in the ϵ dependence of $P(\min(\vec{c}) > \epsilon \mid f, m, a)$, by which we mean the probability that the minimum cost an algorithm a finds on problem f in m

distinct evaluations is larger than ϵ. At least three quantities related to this conditional probability can be used to gauge an algorithm's performance in a particular optimization run:

i) the uniform average of $P(\min(\vec{c}) > \epsilon \mid f, m, a)$ over all cost functions;

ii) the form $P(\min(\vec{c}) > \epsilon \mid f, m, a)$ takes for the random algorithm, which uses no information from the sample d_m;

iii) the fraction of algorithms which, for a particular f and m, result in a $\vec{c}$ whose minimum exceeds ϵ.

These measures give benchmarks which any algorithm run on a particular cost function should surpass if that algorithm is to be considered as having worked well for that cost function. Without loss of generality assume that the ith cost value (i.e., $\mathcal{Y}_i$) equals i. So cost values range from minimum of one to a maximum of $|\mathcal{Y}|$, in integer increments. The following results are derived in Appendix E.

Theorem 5.
$$\sum_f P(\min(\vec{c}) > \epsilon \mid f, m) = \omega^m(\epsilon)$$

where $\omega(\epsilon) \equiv 1 - \epsilon/|\mathcal{Y}|$ is the fraction of cost lying above ϵ. In the limit of $|\mathcal{Y}| \to \infty$, this distribution obeys the following relationship:

$$\frac{\sum_f E(\min(\vec{c}) \mid f, m)}{|\mathcal{Y}|} = \frac{1}{m+1}.$$

Unless one's algorithm has its best-cost-so-far drop faster than the drop associated with these results, one would be hard pressed indeed to claim that the algorithm is well suited to the cost function at hand. After all, for such a performance the algorithm is doing no better than one would expect it to when run on a randomly chosen cost function.

Unlike the preceding measure, the measures analyzed below take into account the actual cost function at hand. This is manifested in the dependence of the values of those measures on the vector $\vec{N} = |\mathcal{X}|\vec{\beta}$ given by the cost function's histogram.

Theorem 6. *For the random algorithm $\tilde{a}$*

$$P(\min(\vec{c}) \geq \epsilon \mid f, m, \tilde{a}) = \prod_{i=0}^{m-1} \frac{\Omega(\epsilon) - i/|\mathcal{X}|}{1 - i/|\mathcal{X}|} \tag{4}$$

where $\Omega(\epsilon) \equiv \sum_{i=\epsilon}^{|\mathcal{Y}|} N_i/|\mathcal{X}|$ is the fraction of points in $\mathcal{X}$ for which $f(x) \geq \epsilon$. To first order in $1/|\mathcal{X}|$

$$P(\min(\vec{c}) > \epsilon \mid f, m, \tilde{a}) = \\ \Omega^m(\epsilon)\left(1 - \frac{m(m-1)(1-\Omega(\epsilon))}{2\Omega(\epsilon)}\frac{1}{|\mathcal{X}|} + \cdots\right) \tag{5}$$

This result allows the calculation of other quantities of interest for measuring performance, for example the quantity

$$E(\min(\vec{c}) \mid f, m, \tilde{a}) = \sum_{\epsilon=1}^{|\mathcal{Y}|} \epsilon[P(\min(\vec{c}) \geq \epsilon \mid f, m, \tilde{a}) \\ - P(\min(\vec{c}) \geq \epsilon + 1 \mid f, m, \tilde{a})].$$

Note that for many cost functions of both practical and theoretical interest, cost values are approximately distributed Gaussianly. For such cases, we can use the Gaussian nature of the distribution to facilitate our calculations. In particular, if the mean and variance of the Gaussian are μ and σ^2, respectively, then we have $\Omega(\epsilon) = \text{erfc}((\epsilon - \mu)/\sqrt{2}\sigma)/2$, where erfc is the complimentary error function.

To calculate the third performance measure, note that for fixed f and m, for any (deterministic) algorithm a, $P(\vec{c} > \epsilon \mid f, m, a)$ is either one or zero. Therefore the fraction of algorithms that result in a $\vec{c}$ whose minimum exceeds ϵ is given by

$$\frac{\sum_a P(\min(\vec{c}) > \epsilon \mid f, m, a)}{\sum_a 1}.$$

Expanding in terms of $\vec{c}$, we can rewrite the numerator of this ratio as $\sum_{\vec{c}} P(\min(\vec{c}) > \epsilon \mid \vec{c}) \sum_a P(\vec{c} \mid f, m, a)$. The ratio of this quantity to $\sum_a 1$, however, is exactly what was calculated when we evaluated measure ii) [see the beginning of the argument deriving (4)]. This establishes the following theorem.

Theorem 7. *For fixed f and m, the fraction of algorithms which result in a $\vec{c}$ whose minimum exceeds ϵ is given by the quantity on the right-hand sides of (4) and (5).*

As a particular example of applying this result, consider measuring the value of $\min(\vec{c})$ produced in a particular run of an algorithm. Then imagine that when it is evaluated for equal to this value, the quantity given in (5) is less than $1/2$. In such a situation the algorithm in question has performed worse than over half of all search algorithms, for the f and m at hand, hardly a stirring endorsement.

None of the above discussion explicitly concerns the dynamics of an algorithm's performance as m increases. Many aspects of such dynamics may be of interest. As an example, let us consider whether, as m grows, there is any change in how well the algorithm's performance compares to that of the random algorithm.

To this end, let the sample generated by the algorithm a after m steps be d_m, and define $y' \equiv \min(d_m^y)$. Let k be the number of additional steps it takes the algorithm to find an x such that $f(x) < y'$. Now we can estimate the number of steps it would have taken the random search algorithm to search $\mathcal{X} - d_\mathcal{X}$ and find a point whose y was less than y'. The expected value of this number of steps is $1/z(d) - 1$, where $z(d)$ is the fraction of $\mathcal{X} - d_m^x$ for which $f(x) < y'$. Therefore $k + 1 - 1/z(d)$ is how much worse a did than the random algorithm, on average.

Next, imagine letting a run for many steps over some fitness function f and plotting how well a did in comparison to the random algorithm on that run, as m increased. Consider the step where a finds its nth new value of $\min(\vec{c})$. For that step, there is an associated k [the number of steps until the next $\min(d_m^y)$] and $z(d)$. Accordingly, indicate that step on our plot as the point $(n, k + 1 - 1/z(d))$. Put down as many points on our plot as there are successive values of $\min(\vec{c}(d))$ in the run of a over f.

If throughout the run a is always a better match to f than is the random search algorithm, then all the points in the plot will have their ordinate values lie below zero. If the random algorithm won for any of the comparisons however, that would give a point lying above zero. In general, even if the points all lie to one side of zero, one would expect that as the search progresses there would be a corresponding (perhaps systematic) variation in how far away from zero the points lie. That variation indicates when the algorithm is entering harder or easier parts of the search.

Note that even for a fixed f, by using different starting points for the algorithm one could generate many of these plots and then superimpose them. This allows a plot of the mean value of $k+1-1/z(d)$ as a function of n along with an associated error bar. Similarly, the single number $z(d)$ characterizing the random algorithm could be replaced with a full distribution over the number of required steps to find a new minimum. In these and similar ways, one can generate a more nuanced picture of an algorithm's performance than is provided by any of the single numbers given by the performance measure discussed above.

VI. Minimax Distinctions Between Algorithms

The NFL theorems do not directly address minimax properties of search. For example, say we are considering two deterministic algorithms a_1 and a_2. It may very well be that there exist cost functions f such that a_1's histogram is much better (according to some appropriate performance measure) than a_2's, but no cost functions for which the reverse is true. For the NFL theorem to be obeyed in such a scenario, it would have to be true that there are many more f for which a_2's histogram is better than a_1's than vice-versa, but it is only slightly better for all those f. For such a scenario, in a certain sense a_1 has better "head-to-head" minimax behavior than a_2; there are f for which a_1 beats a_2 badly, but none for which a_1 does substantially worse than a_2.

Formally, we say that there exists head-to-head minimax distinctions between two algorithms a_1 and a_2 iff there exists a k such that for at least one cost function f, the difference $E(\vec{c} \mid f, m, a_1) - E(\vec{c} \mid f, m, a_2) = k$, but there is no other f for which $E(\vec{c} \mid f, m, a_2) - E(\vec{c} \mid f, m, a_1) = k$. A similar definition can be used if one is instead interested in $\Phi(\vec{c})$ or d_m^y, rather than $\vec{c}$.

It appears that analyzing head-to-head minimax properties of algorithms is substantially more difficult than analyzing average behavior as in the NFL theorem. Presently, very little is known about minimax behavior involving stochastic algorithms. In particular, it is not known if there are any senses in which a stochastic version of a deterministic algorithm has better/worse minimax behavior than that deterministic algorithm. In fact, even if we stick completely to deterministic algorithms, only an extremely preliminary understanding of minimax issues has been reached.

What is known is the following. Consider the quantity

$$\sum_f P_{d^y_{m,1}, d^y_{m,2}}(z, z' \mid f, m, a_1, a_2)$$

for deterministic algorithms a_1 and a_2. (By $P_A(a)$ is meant the distribution of a random variable A evaluated at $A = a$.) For deterministic algorithms, this quantity is just the number of f such that it is both true that a_1 produces a sample with $\mathcal{Y}$ components z and that a_2 produces a sample with $\mathcal{Y}$ components z'.

In Appendix F, it is proven by example that this quantity need not be symmetric under interchange of z and z'.

Theorem 8. *In general*

$$\sum_f P_{d^y_{m,1}, d^y_{m,2}}(z, z' \mid f, m, a_1, a_2)$$
$$\neq \sum_f P_{d^y_{m,1}, d^y_{m,2}}(z', z \mid f, m, a_1, a_2).$$

This means that under certain circumstances, even knowing only the $\mathcal{Y}$ components of the samples produced by two algorithms run on the same unknown f, we can infer something concerning which algorithm produced each population.

Now consider the quantity

$$\sum_f P_{\vec{c}_1, \vec{c}_2}(z, z' \mid f, m, a_1, a_2)$$

again for deterministic algorithms a_1 and a_2. This quantity is just the number of f such that it is both true that a_1 produces a histogram z and that a_2 produces a histogram z'. It too need not be symmetric under interchange of z and z' (see Appendix F). This is a stronger statement than the asymmetry of d^y's statement, since any particular histogram corresponds to multiple samples.

It would seem that neither of these two results directly implies that there are algorithms a_1 and a_2 such that for some f a_1's histogram is much better than a_2's, but for no f's is the reverse is true. To investigate this problem involves looking over all pairs of histograms (one pair for each f) such that there is the same relationship between (the performances of the algorithms, as reflected in) the histograms. Simply having an inequality between the sums presented above does not seem to directly imply that the relative performances between the associated pair of histograms is asymmetric. (To formally establish this would involve creating scenarios in which there is an inequality between the sums, but no head-to-head minimax distinctions. Such an analysis is beyond the scope of this paper.)

On the other hand, having the sums be equal does carry obvious implications for whether there are head-to-head minimax distinctions. For example, if both algorithms are deterministic, then for any particular $f, P_{d^y_{m,1}, d^y_{m,2}}(z_1, z_2 \mid f, m, a_1, a_2)$ equals one for one (z_1, z_2) pair and zero for all others. In such a case, $\sum_f P_{d^y_{m,1}, d^y_{m,2}}(z_1, z_2 \mid f, m, a_1, a_2)$ is just the number of f that result in the pair (z_1, z_2). So $\sum_f P_{d^y_{m,1}, d^y_{m,2}}(z, z' \mid f, m, a_1, a_2) = \sum_f P_{d^y_{m,1}, d^y_{m,2}}(z', z \mid f, m, a_1, a_2)$ implies that there are no head-to-head minimax distinctions between a_1 and a_2. The converse, however, does not appear to hold.[4]

As a preliminary analysis of whether there can be head-to-head minimax distinctions, we can exploit the result in Appendix F, which concerns the case where $|\mathcal{X}| = |\mathcal{Y}| = 3$. First, define the following performance measures of two-element samples, $\Phi(d^y_2)$.

1. $\Phi(y_2, y_3) = \Phi(y_3, y_2) = 2$.

---

[4] Consider the grid of all (z, z') pairs. Assign to each grid point the number of f that result in that grid point's (z, z') pair. Then our constraints are i) by the hypothesis that there are no head-to-head minimax distinctions, if grid point (z_1, z_2) is assigned a nonzero number, then so is (z_2, z_1) and ii) by the no-free-lunch theorem, the sum of all numbers in row z equals the sum of all numbers in column z. These two constraints do not appear to imply that the distribution of numbers is symmetric under interchange of rows and columns. Although again, like before, to formally establish this point would involve explicitly creating search scenarios in which it holds.

2. $\Phi(y_1, y_2) = \Phi(y_2, y_1) = 0$.

3. Φ of any other argument $= 1$.

In Appendix F we show that for this scenario there exist pairs of algorithms a_1 and a_2 such that for one f a_1 generates the histogram $\{y_1, y_2\}$ and a_2 generates the histogram $\{y_2, y_3\}$, but there is no f for which the reverse occurs (i.e., there is no f such that a_1 generates the histogram $\{y_2, y_3\}$ and a_2 generates $\{y_1, y_2\}$).

So in this scenario, with our defined performance measure, there are minimax distinctions between a_1 and a_2. For one f the performance measures of algorithms a_1 and a_2 are, respectively, zero and two. The difference in the Φ values for the two algorithms is two for that f. There are no other f, however, for which the difference is -2. For this Φ then, algorithm a_2 is minimax superior to algorithm a_1.

It is not currently known what restrictions on $\Phi(d_m^y)$ are needed for there to be minimax distinctions between the algorithms. As an example, it may well be that for $\Phi(d_m^y) = \min_i \{d_m^y(i)\}$ there are no minimax distinctions between algorithms.

More generally, at present nothing is known about "how big a problem" these kinds of asymmetries are. All of the examples of asymmetry considered here arise when the set of X values a_1 has visited overlaps with those that a_2 has visited. Given such overlap, and certain properties of how the algorithms generated the overlap, asymmetry arises. A precise specification of those "certain properties" is not yet in hand. Nor is it known how generic they are, i.e., for what percentage of pairs of algorithms they arise. Although such issues are easy to state (see Appendix F), it is not at all clear how best to answer them.

Consider, however, the case where we are assured that, in m steps, the samples of two particular algorithms have not overlapped. Such assurances hold, for example, if we are comparing two hill-climbing algorithms that start far apart (on the scale of m) in $\mathcal{X}$. It turns out that given such assurances, there are no asymmetries between the two algorithms for m-element samples. To see this formally, go through the argument used to prove the NFL theorem, but apply that argument to the quantity $\sum_f P_{d^y_{m,1}, d^y_{m,2}}(z, z' \mid f, m, a_1, a_2)$ rather than $P(\vec{c} \mid f, m, a)$. Doing this establishes the following theorem.

Theorem 9. *If there is no overlap between $d^x_{m,1}$ and $d^x_{m,2}$, then*

$$\sum_f P_{d^y_{m,1}, d^y_{m,2}}(z, z' \mid f, m, a_1, a_2) \\ = \sum_f P_{d^y_{m,1}, d^y_{m,2}}(z', z \mid f, m, a_1, a_2).$$

An immediate consequence of this theorem is that under the no-overlap conditions, the quantity $\sum_f P_{C_1, C_2}(z, z' \mid f, m, a_1, a_2)$ is symmetric under interchange of z and z', as are all distributions determined from this one over C_1 and C_2 (e.g., the distribution over the difference between those C's extrema).

Note that with stochastic algorithms, if they give nonzero probability to all d^x_m, there is always overlap to consider. So there is always the possibility of asymmetry between algorithms if one of them is stochastic.

VII. $P(f)$-Independent Results

All work to this point has largely considered the behavior of various algorithms across a wide range of problems. In this section we introduce the kinds of results that can be obtained when we reverse roles and consider the properties of many algorithms on a single problem. More results of this type are found in Macready and Wolpert (1996). The results of this section, although less sweeping than the NFL results, hold no matter what the real world's distribution over cost functions is.

Here the authors consider different algorithms applied to the same problem.

Let a and a' be two search algorithms. Define a "choosing procedure" as a rule that examines the samples d_m and d'_m, produced by a and a', respectively, and based on those samples, decides to use either a or a' for the subsequent part of the search. As an example, one "rational" choosing procedure is to use a for the subsequent part of the search if and only if it has generated a lower cost value in its sample than has a'. Conversely we can consider an "irrational" choosing procedure that uses the algorithm that had *not* generated the sample with the lowest cost solution.

A formal definition of choosing between algorithms.

At the point that a choosing procedure takes effect, the cost function will have been sampled at $d_\cup \equiv d_m \cup d'_m$. Accordingly, if $d_{>m}$ refers to the samples of the cost function that come after using the choosing algorithm, then the user is interested in the remaining sample $d_{>m}$. As always, without loss of generality, it is assumed that the search algorithm selected by the choosing procedure does not return to any points in $d\cup$.[5]

The following theorem, proven in Appendix G, establishes that there is no *a priori* justification for using any particular choosing procedure. Loosely speaking, no matter what the cost function, without special consideration of the algorithm at hand, simply observing how well that algorithm has done so far tells us nothing *a priori* about how well it would do if we continue to use it on the same cost function. For simplicity, in stating the result we only consider deterministic algorithms.

Theorem 10. *Let d_m and d'_m be two fixed samples of size m, that are generated when the algorithms a and a', respectively, are run on the (arbitrary) cost function at hand. Let A and B be two different choosing procedures. Let k be the number of elements in $c_{>m}$. Then*

$$\sum_{a,a'} P\left(c_{>m} \mid f, d, d', k, a, a', A\right)$$
$$= \sum_{a,a'} P\left(c_{>m} \mid f, d, d', k, a, a', B\right).$$

Amazingly, given a fixed cost function, the performance of an algorithm on past samples does not tell us anything about future performance, at least without further assumptions or other *a priori* information.

Implicit in this result is the assumption that the sum excludes those algorithms a and a' that do not result in d and d' respectively when run on f.

In the precise form it is presented above, the result may appear misleading, since it treats all samples equally, when for any given f some samples will be more likely than others. Even if one weights samples according to their probability of occurrence, however, it is still true that,

---

[5] a can know to avoid the elements *it* has seen before. However *a priori*, a has no way to avoid the elements observed by a' has (and vice-versa). Rather than have the definition of a somehow depend on the elements in $d' - d$ (and similarly for a'), we deal with this problem by defining $c_{>m}$ to be set only by those elements in $d > m$ that lie outside of $d_\cup$. (This is similar to the convention we exploited above to deal with potentially retracing algorithms.) Formally, this means that the random variable $c_{>m}$ is a function of $d\cup$ as well as of $d_{>m}$. It also means there may be fewer elements in the histogram $c_{>m}$ than there are in the sample $d_{>m}$.

on average, the choosing procedure one uses has no effect on likely $c_{>m}$. This is established by the following result, proven in Appendix H.

Theorem 11. *Under the conditions given in the preceding theorem*

$$\sum_{a,a'} P(c_{>m} \mid f, m, k, a, a', \mathcal{A})\\ = \sum_{a,a'} P(c_{>m} \mid, f, m, k, a, a', \mathcal{B}).$$

These results show that no assumption for $P(f)$ alone justifies using some choosing procedure as far as subsequent search is concerned. To have an intelligent choosing procedure, one must take into account not only $P(f)$ but also the search algorithms one is choosing among. This conclusion may be surprising. In particular, note that it means that there is no intrinsic advantage to using a rational choosing procedure, which continues with the better of a and a', rather than using an irrational choosing procedure which does the opposite.

These results also have interesting implications for degenerate choosing procedures $\mathcal{A} \equiv$ {always use algorithm a} and $\mathcal{B} \equiv$ {always use algorithm a'}. As applied to this case, they mean that for fixed f_1 and f_2, if f_1 does better (on average) with the algorithms in some set $\mathcal{A}$, then f_2 does better (on average) with the algorithms in the set of all other algorithms. In particular, if for some favorite algorithms a certain "well-behaved" f results in better performance than does the random f, then that well-behaved f gives *worse than random* behavior on the set all remaining algorithms. In this sense, just as there are no universally efficacious search algorithms, there are no universally benign f which can be assured of resulting in better than random performance regardless of one's algorithm.

In fact, things may very well be worse than this. In supervised learning, there is a related result (Wolpert 1996b). Translated into the current context, that result suggests that if one restricts sums to only be over those algorithms that are a good match to $P(f)$, then it is often the case that "stupid" choosing procedures—like the irrational procedure of choosing the algorithm with the less desirable $\vec{c}$—outperform "intelligent" ones. What the set of algorithms summed over must be in order for a rational choosing procedure to be superior to an irrational procedure is not currently known.

VIII. Conclusions

A framework has been presented in which to compare general-purpose optimization algorithms. A number of NFL theorems were derived that demonstrate the danger of comparing algorithms by their performance on a small sample of problems. These same results also indicate the importance of incorporating problem-specific knowledge into the behavior of the algorithm. A geometric interpretation was given showing what it means for an algorithm to be well suited to solving a certain class of problems. The geometric perspective also suggests a number of measures to compare the similarity of various optimization algorithms.

More direct calculational applications of the NFL theorem were demonstrated by investigating certain information-theoretic aspects of search, as well as by developing a number of benchmark measures of algorithm performance. These benchmark measures should prove useful in practice.

We provided an analysis of the ways that algorithms can differ *a priori* despite the NFL theorems. We have also provided an introduction to a variant of the framework that focuses on the behavior of a range of algorithms on specific problems (rather than specific algorithms over a range of problems). This variant leads directly to reconsideration of many issues addressed by computational complexity, as detailed in Macready and Wolpert (1996).

Much future work clearly remains. Most important is the development of practical applications of these ideas. Can the geometric viewpoint be used to construct new optimization techniques in practice? We believe the answer to be yes. At a minimum, as Markov random field models of landscapes become more wide spread, the approach embodied in this paper should find wider applicability.

Appendix A
Proof for Static Cost Functions

We show that $\sum_f P(\vec{c} \mid f, m, a)$ has no dependence on a. Conceptually, the proof is quite simple but necessary bookkeeping complicates things, lengthening the proof considerably. The intuition behind the proof is straightforward: by summing over all f we ensure that the past performance of an algorithm has no bearing on its future performance. Accordingly, under such a sum, all algorithms perform equally.

The proof is by induction. The induction is based on $m = 1$, and the inductive step is based on breaking f into two independent parts, one for $x \in d_m^x$ and one for $x \notin d_m^x$. These are evaluated separately, giving the desired result.

For $m = 1$, we write the first sample as $d_1 = \{d_1^x, f(d_1^x)\}$ where d_1^x is set by a. The only possible value for d_1^y is $f(d_1^x)$, so we have

$$\sum_f P(d_1^y \mid f, m = 1, a) = \sum_f \delta(d_1^y, f(d_1^x))$$

where δ is the Kronecker delta function.

Summing over all possible cost functions, $\delta(d_1^y, f(d_1^x))$ is one only for those functions which have cost d_1^y at point d_1^x. Therefore that sum equals $|\mathcal{Y}|^{|\mathcal{X}|-1}$, independent of d_1^x

$$\sum_f P(d_1^y \mid f, m = 1, a) = |\mathcal{Y}|^{|\mathcal{X}|-1}$$

which is independent of a. This bases the induction.

The inductive step requires that if $\sum_f P(d_m^y \mid f, m, a)$ is independent of a for all d_m^y, then so also is $\sum_f P(d_{m+1}^y \mid f, m+1, a)$. Establishing this step completes the proof.

We begin by writing

$$\begin{aligned}P(d_{m+1}^y \mid f, m+1, a) &= P(\{d_{m+1}^y(1), \cdots, d_{m+1}^y(m)\}, \\ &\quad \cdot d_{m+1}^y(m+1) \mid f, m+1, a) \\ &= P(d_m^y, d_{m+1}^y(m+1) \mid f, m+1, a) \\ &= P(d_{m+1}^y(m+1) \mid d_m, f, m+1, a) \\ &\quad \cdot P(d_m^y \mid f, m+1, a)\end{aligned}$$

and thus

$$\begin{aligned}\sum_f P(d_{m+1}^y \mid f, m+1, a) = \sum_f P(d_{m+1}^y(m+1) \\ \mid d_m^y, f, m+1, a) \\ \cdot P(d_m^y \mid f, m+1, a).\end{aligned}$$

The new y value, $d^y_{m+1}(m+1)$, will depend on the new x value, f, and nothing else. So we expand over these possible x values, obtaining

$$\sum_f P\left(d^y_{m+1} \mid f, m+1, a\right) = \sum_{f,x} P\left(d^y_{m+1}(m+1) \mid f, x\right)$$
$$\cdot P\left(x \mid d^y_m, f, m+1, a\right)$$
$$\cdot P\left(d^y_m \mid f, m+1, a\right)$$
$$= \sum_{f,x} \delta\left(d^y_{m+1}(m+1), f(x)\right)$$
$$\cdot P\left(x \mid d^y_m, f, m+1, a\right)$$
$$\cdot P\left(d^y_m \mid f, m+1, a\right).$$

Next note that since $x = a(d^x_m, d^y_m)$, it does not depend directly on f. Consequently we expand in d^x_m to remove the f dependence in $P(x \mid d^y_m, f, m+1, a)$

$$\sum_f P\left(d^y_{m+1} \mid f, m+1, a\right) = \sum_{f,x,d^x_m} \delta\left(d^y_{m+1}(m+1), f(x)\right)$$
$$\cdot P\left(x \mid d_m, a\right)$$
$$\cdot P\left(d^x_m \mid d^y_m, f, m+1, a\right)$$
$$\cdot P\left(d^y_m \mid f, m+1, a\right)$$
$$= \sum_{f,d^x_m} \delta\left(d^y_{m+1}(m+1), f(a(d_m))\right)$$
$$\cdot P\left(d_m \mid f, m, a\right)$$

where use was made of the fact that $P(x \mid d_m, a) = \delta(x, a(d_m))$ and the fact that $P(d_m \mid f, m+1, a) = P(d_m \mid f, m, a)$.

The sum over cost functions f is done first. The cost function is defined both over those points restricted to d^x_m and those points outside of d^x_m. $P(d_m \mid f, m, a)$ will depend on the f values defined over points inside d^x_m while $\delta\left(d^y_{m+1}(m+1), f(a(d_m))\right)$ depends only on the f values defined over points outside d^x_m. (Recall that $a(d^x_m) \notin d^x_m$.) So we have

$$\sum_f P\left(d^y_{m+1} \mid f, m+1, a\right) = \sum_{d^x_m} \sum_{f(x \in d^x_m)} P(d_m \mid f, m, a) \qquad (6)$$
$$\sum_{f(x \notin d^x_m)} \delta\left(d^y_{m+1}(m+1), f(a(d_m))\right).$$

The sum $\sum_{f(x \notin d_m^x)}$ contributes a constant, $|\mathcal{Y}|^{|\mathcal{X}|-m-1}$, equal to the number of functions defined over points not in d_m^x passing through $(d_{m+1}^x(m+1), f(a(d_m)))$. So

$$\sum_f P(d_{m+1}^y \mid f, m+1, a) = |\mathcal{Y}|^{|\mathcal{X}|-m-1} \sum_{f(x \in d_m^x), d_m^x} P(d_m \mid f, m, a)$$

$$= \frac{1}{|\mathcal{Y}|} \sum_{f, d_m^x} P(d_m \mid f, m, a)$$

$$= \frac{1}{|\mathcal{Y}|} \sum_f P(d_m^y \mid f, m, a).$$

By hypothesis, the right-hand side of this equation is independent of a, so the left-hand side must also be. This completes the proof.

Appendix B
NFL Proof for Time-Dependent Cost Functions

In analogy with the proof of the static NFL theorem, the proof for the time-dependent case proceeds by establishing the a-independence of the sum $\sum_T P(c \mid f, T, m, a)$, where here c is either d_m^y or D_m^y.

To begin, replace each T in this sum with a set of cost functions, f_i, one for each iteration of the algorithm. To do this, we start with the following:

$$\sum_T P(c \mid f, T, m, a) = \sum_T \sum_{d_m^x} \sum_{f_2 \cdots f_m} P\left(c \mid \vec{f}, d_m^x, T, m, a\right)$$

$$\cdot P(f_2 \cdots f_m, d_m^x \mid f_1, T, m, a)$$

$$= \sum_{d_m^x} \sum_{f_2 \cdots f_m} P\left(\vec{c} \mid \vec{f}, d_m^x\right) P\left(d_m^x \mid \vec{f}, m, a\right)$$

$$\cdot \sum_T P(f_2 \cdots f_m \mid f_1, T, m, a)$$

where the sequence of cost functions, f_i, has been indicated by the vector $\vec{f} = (f_1, \cdots, f_m)$. In the next step, the sum over all possible T is decomposed into a series of sums. Each sum in the series is over the values T can take for one particular iteration of the algorithm. More formally, using $f_{i+1} = T_i(f_i)$, we write

$$\sum_T P(c \mid f, T, m, a) = \sum_{d_m^x} \sum_{f_2 \cdots f_m} P\left(\vec{c} \mid \vec{f}, d_m^x\right) P\left(d_m^x \mid \vec{f}, m, a\right)$$
$$\cdot \sum_{T_1} \delta(f_2, T_1(f_1)) \cdots$$
$$\cdot \sum_{T_{m-1}} \delta(f_m, T_{m-1}(T_{m-2}(\cdots T_1(f_1)))).$$

Note that $\sum_T P(c \mid f, T, m, a)$ is independent of the values of $T_{i>m-1}$, so those values can be absorbed into an overall a-independent proportionality constant.

Consider the innermost sum over T_{m-1}, for fixed values of the outer sum indexes $T_1 \cdots T_{m-2}$. For fixed values of the outer indexes, $T_{m-1}(T_{m-2}(\cdots T_1(f_1)))$ is just a particular fixed cost function. Accordingly, the innermost sum over T_{m-1} is simply the number of bijections of $\mathcal{F}$ that map that fixed cost function to f_m. This is the constant, $(|\mathcal{F}| - 1)!$. Consequently, evaluating the T_{m-1} sum yields

$$\sum_T P(c \mid f, T, m, a_1) \propto \sum_{d_m^x} \sum_{f_2 \cdots f_m} P\left(c \mid \vec{f}, d_m^x\right) P\left(d_m^x \mid \vec{f}, m, a\right)$$
$$\cdot \sum_{T_1} \delta(f_2, T_1(f_1)) \cdots$$
$$\cdot \sum_{T_{m-2}} \delta(f_{m-1}, T_{m-2} \cdot (T_{m-3}(\cdots T_1(f_1)))).$$

The sum over T_{m-2} can be accomplished in the same manner T_{m-1} is summed over. In fact, all the sums over all T_i can be done, leaving

$$\sum_T P(c \mid f, T, m, a_1) \propto \sum_{d_m^x} \sum_{f_2 \cdots f_m} P\left(D_m^y \mid \vec{f}, d_m^x\right)$$
$$\cdot P\left(d_m^x \mid \vec{f}, m, a\right) \quad (7)$$
$$= \sum_{d_m^x} \sum_{f_2 \cdots f_m} P\left(c \mid \vec{f}, d_m^x\right)$$
$$\cdot P(d_m^x \mid f_1 \cdots f_{m-1}, m, a).$$

In this last step, the statistical independence of c and f_m has been used.

Further progress depends on whether c represents d_m^y or D_m^y. We begin with analysis of the D_m^y case. For this case $P\left(c \mid \vec{f}, d_m^x\right) =$

$P\left(D_m^y \mid f_m, d_m^x\right)$, since D_m^y only reflects cost values from the last cost function, f_m. Using this result gives

$$\sum_T P\left(D_m^y \mid f, T, m, a_1\right) \propto \sum_{d_m^x} \sum_{f_2 \cdots f_{m-1}} P\left(d_m^x \mid f_1 \cdots f_{m-1}, m, a\right)$$
$$\cdot \sum_{f_m} P\left(D_m^y \mid f_m, d_m^x\right).$$

The final sum over f_m is a constant equal to the number of ways of generating the sample D_m^y from cost values drawn from f_m. The important point is that it is independent of the particular d_m^x. Because of this the sum over d_m^x can be evaluated eliminating the a dependence

$$\sum_T P\left(D_m^y \mid f, T, m, a\right)$$
$$\propto \sum_{f_2 \cdots f_{m-1}} \sum_{d_m^x} P\left(d_m^x \mid f_1 \cdots f_{m-1}, m, a\right) \propto 1.$$

This completes the proof of Theorem 2 for the case of D_m^y.

The proof of Theorem 2 is completed by turning to the d_m^y case. This is considerably more difficult since $P\left(\vec{c} \mid \vec{f}, d_m^x\right)$ cannot be simplified so that the sums over f_i cannot be decoupled. Nevertheless, the NFL result still holds. This is proven by expanding 7 over possible d_m^y values

$$\sum_T P\left(d_m^y \mid f, T, m, a\right) \propto \sum_{d_m^x} \sum_{f_2 \cdots f_m} \sum_{d_m^y} P\left(d_m^y \mid d_m^y\right)$$
$$\cdot P\left(d_m^y \mid \vec{f}, d_m^x\right)$$
$$\cdot P\left(d_m^x \mid f_1 \cdots f_{m-1}, m, a\right)$$
$$= \sum_{d_m^y} P\left(d_m^y \mid d_m^y\right) \qquad (8)$$
$$\cdot \sum_{d_m^x} \sum_{f_2 \cdots f_m} P\left(d_m^x \mid f_1 \cdots f_{m-1}, m, a\right)$$
$$\cdot \prod_{i=1}^m \delta\left(d_m^y(i), f_i\left(d_m^x(i)\right)\right).$$

The innermost sum over f_m only has an effect on the $\delta\left(d_m^y(i), f_i\left(d_m^x(i)\right)\right)$ term so it contributes $\sum_{f_m} \delta\left(d_m^y(m), f_m\left(d_m^x(m)\right)\right)$. This is a constant, equal to $|\mathcal{Y}|^{|\mathcal{X}|-1}$. This leaves

$$\sum_T P(d_m^y \mid f, T, m, a) \propto \sum_{d_m^y} P(d_m^y \mid d_m^y) \sum_{d_m^x} \sum_{f_2 \cdots f_{m-1}}$$
$$\cdot P(d_m^x \mid f_1 \cdots f_{m-1}, m, a)$$
$$\cdot \prod_{i=1}^{m-1} \delta(d_m^y(i), f_i(d_m^x(i))).$$

The sum over $d_m^x(m)$ is now simple

$$\sum_T P(d_m^y \mid f, T, m, a) \propto \sum_{d_m^y} P(d_m^y \mid d_m^y) \sum_{d_m^x(1)} \cdots \sum_{d_m^x(m-1)} \sum_{f_2 \cdots f_{m-1}}$$
$$\cdot P(d_{m-1}^x \mid f_1 \cdots f_{m-2}, m, a)$$
$$\cdot \prod_{i=1}^{m-1} \delta(d_m^y(i), f_i(d_m^x(i))).$$

The above equation is of the same form as (8), only with a remaining sample of size $m - 1$ rather than m. Consequently, in an analogous manner to the scheme used to evaluate the sums over f_m and $d_m^x(m)$ that existed in (8), the sums over f_{m-1} and $d_m^x(m-1)$ can be evaluated. Doing so simply generates more a-independent proportionality constants. Continuing in this manner, all sums over the f_i can be evaluated to find

$$\sum_T P(\vec{c} \mid f, T, m, a_1) \propto \sum_{d_m^y} P(\vec{c} \mid d_m^y) \sum_{d_m^x(1)} P(d_m^x(1) \mid m, a)$$
$$\cdot \delta(d_m^y(1), f_1(d_m^x(1))).$$

There is algorithm dependence in this result, but it is the trivial dependence discussed previously. It arises from how the algorithm selects the first x point in its sample, $d_m^x(1)$. Restricting interest to those points in the sample that are generated subsequent to the first, this result shows that there are no distinctions between algorithms. Alternatively, summing over the initial cost function f_1, all points in the sample could be considered while still retaining an NFL result.

Appendix C
Proof of ρ_f Result

As noted in the discussion leading up to Theorem 3, the fraction of functions giving a specified histogram $\vec{c} = m\vec{\alpha}$ is independent of the algorithm. Consequently, a simple algorithm is used to prove the theorem. The algorithm visits points in $\mathcal{X}$ in some canonical order, say $x_1, x_2, \cdots, x_m$. Recall that the histogram $\vec{c}$ is specified by giving the frequencies of occurrence, across the $x_1, x_2, \cdots, x_m$, for each of the $|\mathcal{Y}|$ possible cost values. The number of f's giving the desired histogram under this algorithm is just the multinomial giving the number of ways of distributing the cost values in $\vec{c}$. At the remaining $|\mathcal{X}| - m$ points in $\mathcal{X}$ the cost can assume any of the $|\mathcal{Y}|f$ values giving the first result of Theorem 3.

The expression of $\rho_f(\vec{\alpha})$ in terms of the entropy of $\vec{\alpha}$ follows from an application of Stirling's approximation to order $\mathcal{O}(1/m)$, which is valid when all of the c_i are large. In this case the multinomial is written

$$\ln \binom{m}{c_1 c_2 \cdots c_{|\mathcal{Y}|}} \cong m \ln m - \sum_{i=1}^{|\mathcal{Y}|} c_i \ln c_i$$

$$+ \frac{1}{2}\left[\ln m - \sum_{i=1}^{|\mathcal{Y}|} \ln c_i\right]$$

$$\cong m S(\vec{\alpha})$$

$$+ \frac{1}{2}\left[(1 - |\mathcal{Y}|)\ln m - \sum_{i=1}^{|\mathcal{Y}|} \ln \alpha_i\right]$$

from which the theorem follows by exponentiating this result.

Appendix D
Proof of ρ_{alg} Result

In this section the proportion of all algorithms that give a particular $\vec{c}$ for a particular f is calculated. The calculation proceeds in several steps

Since $\mathcal{X}$ is finite there are a finite number of different samples. Therefore any (deterministic) a is a huge, but finite, list indexed by all possible d's. Each entry in the list is the x that the a in question outputs for that d-index.

Consider any particular unordered set of $m(\mathcal{X}, \mathcal{Y})$ pairs where no two of the pairs share the same x value. Such a set is called an unordered path π. Without loss of generality, from now on we implicitly restrict the discussion to unordered paths of length m. A particular π is in or from a particular f if there is a unordered set of $m(x, f(x))$ pairs identical to π.

The numerator on the right-hand side of (3) is the number of unordered paths in the given f that give the desired $\vec{c}$.

The number of unordered paths in f that give the desired $\vec{c}$, the numerator on the right-hand side of (3), is proportional to the number of a's that give the desired $\vec{c}$ for f and the proof of this claim constitutes a proof of (3). Furthermore, the proportionality constant is independent of f and $\vec{c}$.

Proof: The proof is established by constructing a mapping $\phi : a \mapsto \pi$ taking in an a that gives the desired $\vec{c}$ for f, and producing a π that is in f and gives the desired $\vec{c}$. Showing that for any π the number of algorithms a such that $\phi(a) = \pi$ is a constant, independent of π, f, and $\vec{c}$. and that ϕ is single valued will complete the proof.

Recalling that every x value in an unordered path is distinct, any unordered path π gives a set of $m!$ different ordered paths. Each such ordered path π_{ord} in turn provides a set of m successive d's (if the empty d is included) and a following x. Indicate by $d\left(\pi_{ord}\right)$ this set of the first m d's provided by π_{ord}.

From any ordered path π_{ord} a "partial algorithm" can be constructed. This consists of the list of an a, but with only the $md\left(\pi_{ord}\right)$ entries in the list filled in, the remaining entries are blank. Since there are $m!$ distinct partial a's for each π (one for each ordered path corresponding to π), there are $m!$ such partially filled-in lists for each π. A partial algorithm may or may not be consistent with a particular full algorithm. This allows the definition of the inverse of ϕ: for any π that is in f and gives $\vec{c}$, $\phi^{-1}(\pi) \equiv$ (the set of all a that are consistent with at least one partial algorithm generated from π and that give $\vec{c}$ when run on f).

To complete the first part of the proof, it must be shown that for all π that are in f and give $\vec{c}$, $\phi^{-1}(\pi)$ contains the same number of elements, regardless of π, f, or c. To that end, first generate all ordered paths induced by π and then associate each such ordered path with a distinct m-element partial algorithm. Now how many full algorithm lists are consistent with at least one of these partial algorithm partial lists? How this question is answered is the core of this appendix. To answer this question, reorder the entries in each of the partial algorithm lists by permuting the indexes d of all the lists. Obviously such a reordering will not change the answer to our question.

Reordering is accomplished by interchanging pairs of d indexes. First, interchange any d index of the form $((d_m^x(1), d_m^y(1)), \cdots, (d_m^x(i \leq m), d_m^y(i \leq m)))$ whose entry is filled in any of our partial algorithm lists with $d'(d) \equiv ((d_m^x(1), z), \cdots, (d_m^x(i), z))$, where z is some arbitrary constant $\mathcal{Y}$ value and x_j refers to the jth element of $\mathcal{X}$. Next, create some arbitrary but fixed ordering of all $x \in \mathcal{X}$: $(x_1, \cdots, x_{|\mathcal{X}|})$. Then interchange any d' index of the form $((d_m^x(1), z), \cdots, (d_m^x(i \leq m), z))$ whose entry is filled in any of our (new) partial algorithm lists with $d''(d') \equiv ((x_1, z), \cdots, (x_m, z))$. Recall that all the $d_m^x(i)$ must be distinct. By construction, the resultant partial algorithm lists are independent of $\pi, \vec{c}$ and f, as is the number of such lists (it is $m!$). Therefore the number of algorithms consistent with at least one partial algorithm list in $\phi^{-1}(\pi)$ is independent of π, c and f. This completes the first part of the proof.

For the second part, first choose any two unordered paths that differ from one another, A and B. There is no ordered path A_{ord} constructed from A that equals an ordered path B_{ord} constructed from B. So choose any such A_{ord} and any such B_{ord}. If they disagree for the null d, then we know that there is no (deterministic) a that agrees with both of them. If they agree for the null d, then since they are sampled from the same f, they have the same single-element d. If they disagree for that d, then there is no a that agrees with both of them. If they agree for that d, then they have the same double-element d. Continue in this manner all the up to the $(m-1)$-element d. Since the two ordered paths differ, they must have disagreed at some point by now, and therefore there is no a that agrees with both of them. Since this is true for any A_{ord} from A and any B_{ord} from B, we see that there is no a in $\phi^{-1}(A)$ that is also in $\phi^{-1}(B)$. This completes the proof.

To show the relation to the Kullback-Liebler distance the product of binomials is expanded with the aid of Stirling's approximation when both N_i and c_i are large

$$\ln \prod_{i=1}^{|\mathcal{Y}|} \binom{N_i}{c_i} \cong \sum_{i=1}^{|\mathcal{Y}|} -\frac{1}{2} \ln 2\pi + N_i \ln N_i$$
$$- c_i \ln c_i - (N_i - c_i) \ln(N_i - c_i)$$
$$+ \frac{1}{2} (\ln N_i - \ln(N_i - c_i) - \ln c_i).$$

It has been assumed that $c_i/N_i \ll 1$, which is reasonable when $m \ll |\mathcal{X}|$. Expanding $\ln(1-z) = -z - z^2/2 - \cdots$, to second order gives

$$\ln \prod_{i=1}^{|\mathcal{Y}|} \binom{N_i}{c_i} \cong \sum_{i=1}^{|\mathcal{Y}|} c_i \ln\left(\frac{N_i}{c_i}\right) - \frac{1}{2} \ln c_i + c_i$$
$$- \frac{1}{2} \ln 2\pi - \frac{c_i}{2N_i}(c_i - 1 + \cdots).$$

Using $m/|\mathcal{X}| \ll 1$ then in terms of $\vec{\alpha}$ and $\vec{\beta}$ one finds

$$\ln \prod_{i=1}^{|\mathcal{Y}|} \binom{N_i}{c_i} \cong -mD_{KL}(\vec{\alpha}, \vec{\beta}) + m - m \ln\left(\frac{m}{|\mathcal{X}|}\right)$$
$$- \frac{|\mathcal{Y}|}{2} \ln 2\pi - \sum_{i=1}^{|\mathcal{Y}|} \frac{1}{2} \ln(\alpha_i m)$$
$$+ \frac{m}{2|\mathcal{X}|}\left(\frac{\alpha_i}{\beta_i}\right)(1 - \alpha_i m + \cdots)$$

where $D_{KL}(\vec{\alpha}, \vec{\beta}) \equiv \sum_i \alpha_i \ln(\beta_i/\alpha_i)$ is the Kullback-Liebler distance between the distributions $\vec{\alpha}$ and $\vec{\beta}$. Exponentiating this expression yields the second result in Theorem 4.

Appendix E
Benchmark Measures of Performance

The result for each benchmark measure is established in turn.

The first measure is $\sum_f P(\min(d_m^y) \mid f, m, a)$. Consider

$$\sum_f P(\min(d_m^y) \mid f, m, a) \qquad (9)$$

for which the summand equals zero or one for all f and deterministic a. It is one only if

i) $f(d_m^x(1)) = d_m^y(1)$

ii) $f(a[d_m(1)]) = d_m^y(2)$

iii) $f(a[d_m(1), d_m(2)]) = d_m^y(3)$

and so on. These restrictions will fix the value of $f(x)$ at m points while f remains free at all other points. Therefore

$$\sum_f P(d_m^y \mid f, m, a) = |\mathcal{Y}|^{|\mathcal{X}|-m}.$$

Using this result in 9 we find

$$\sum_f P(\min(d_m^y) > \epsilon \mid f, m) = \frac{1}{|\mathcal{Y}|^m} \sum_{d_m^y} P(\min(d_m^y) > \epsilon \mid d_m^y)$$

$$= \frac{1}{|\mathcal{Y}|^m} \sum_{d_m^y \ni \min(d_m^y) > \epsilon} 1$$

$$= \frac{1}{|\mathcal{Y}|^m} (|\mathcal{Y}| - \epsilon)^m$$

which is the result quoted in Theorem 5.

In the limit as $|\mathcal{Y}|$ gets large write

$$\sum_f E(\min(\vec{c}) \mid f, m) = \sum_{\epsilon=1}^{|\mathcal{Y}|} \epsilon [\omega^m(\epsilon - 1) - \omega^m(\epsilon)]$$

and substitute in for $\omega(\epsilon) = 1 - \epsilon/|\mathcal{Y}|$. Replacing ϵ with $\zeta + 1$ turns the sum into $\sum_{\zeta=0}^{|\mathcal{Y}|-1} [\zeta + 1] \left[\left(1 - \frac{\zeta}{|\mathcal{Y}|}\right)^m - \left(1 - \frac{\zeta+1}{|\mathcal{Y}|}\right)^m \right]$. Next, write $|\mathcal{Y}| = b/\Delta$ for some b and multiply and divide the summand by Δ. Since $|\mathcal{Y}| \to \infty$ then $\Delta \to 0$. To take the limit of $\Delta \to 0$, apply L'hopital's rule to the ratio in the summand. Next use the fact that Δ is going to zero to cancel terms in the summand. Carrying through the algebra and dividing by b/Δ, we get a Riemann sum of the form $\frac{m}{b^2} \int_0^b dx\, x(1 - x/b)^{m-1}$. Evaluating the integral gives the second result in Theorem 5.

The second benchmark concerns the behavior of the random algorithm. Summing over the $\mathcal{Y}$ values of different histograms $\vec{c}$, the performance of $\tilde{a}$ is

$$P(\min(\vec{c}) \geq \epsilon \mid f, m, \tilde{a}) = \sum_{\vec{c}} P(\min(\vec{c}) \geq \epsilon \mid \vec{c}) P(\vec{c} \mid f, m, \tilde{a})$$

Now $P(\vec{c} \mid f, m, \tilde{a})$ is the probability of obtaining histogram $\vec{c}$ in m random draws from the histogram $\vec{N}$ of the function f. This can be viewed as the definition of $\tilde{a}$. This probability has been calculated previously as $\prod_{i=1}^{|\mathcal{Y}|}\binom{N_i}{c_i}/\binom{|\mathcal{X}|}{m}$. So

$$P(\min(\vec{c}) \geq \epsilon \mid f, m, \tilde{a}) = \frac{1}{\binom{|\mathcal{X}|}{m}} \sum_{c_1=0}^{m} \cdots \sum_{c_{|\mathcal{Y}|}=0}^{m} \delta\left(\sum_{i=1}^{|\mathcal{Y}|} c_i, m\right)$$

$$\cdot P(\min(\vec{c}) \geq \epsilon \mid \vec{c}) \prod_{i=1}^{|\mathcal{Y}|} \binom{N_i}{c_i}$$

$$= \frac{1}{\binom{c\mathcal{X}'|}{m}} \sum_{c_c=0}^{m} \cdots \sum_{c_{Y\mathcal{I}}=0}^{m} \delta\left(\sum_{i=\epsilon}^{|\mathcal{Y}|} c_i, m\right)$$

$$\cdot \prod_{i=\epsilon}^{|\mathcal{Y}|} \binom{N_i}{c_i}$$

$$= \frac{\binom{\sum_{i=c^c}^{|\mathcal{Y}|} N_i}{m}}{\binom{|\mathcal{X}|}{m}} \equiv \frac{\binom{\Omega(\epsilon)|\mathcal{X}|}{m}}{\binom{|\mathcal{X}|}{m}}$$

which is (4) of Theorem 6.

Appendix F
Proof Related to Minimax Distinctions Between Algorithms

This proof is by example. Consider three points in $\mathcal{X}$, x_1, x_2, and x_3, and three points in $\mathcal{Y}$, y_1, y_2, and y_3.

1. Let the first point a_1 visits be x_1 and the first point a_2 visits be x_2.

2. If at its first point a_1 sees a y_1 or a y_2, it jumps to x_2. Otherwise it jumps to x_3.

3. If at its first point a_2 sees a y_1, it jumps to x_1. If it sees a y_2, it jumps to x_3.

Consider the cost function that has as the $\mathcal{Y}$ values for the three $\mathcal{X}$ values $\{y_1, y_2, y_3\}$, respectively.

For $m = 2$, a_1 will produce a sample (y_1, y_2) for this function, and a_2 will produce (y_2, y_3).

The proof is completed if we show that there is no cost function so that a_1 produces a sample containing y_2 and y_3 and such that a_2 produces a sample containing y_1 and y_2.

There are four possible pairs of samples to consider:

i) $[(y_2, y_3), (y_1, y_2)]$;

ii) $[(y_2, y_3), (y_2, y_1)]$;

iii) $[(y_3, y_2), (y_1, y_2)]$;

iv) $[(y_3, y_2), (y_2, y_1)]$.

Since if its first point is a y_2, a_1 jumps to x_2 which is where a_2 starts, when a_1's first point is a y_2 its second point must equal a_2's first point. This rules out possibilities i) and ii).

For possibilities iii) and iv), by a_1's sample we know that f must be of the form $\{y_3, s, y_2\}$, for some variable s. For case iii), s would need to equal y_1, due to the first point in a_2's sample. For that case, however, the second point a_2 sees would be the value at x_1, which is y_3, contrary to hypothesis. For case iv), we know that the s would have to equal y_2, due to the first point in a_2's sample. That would mean, however, that a_2 jumps to x_3 for its second point and would therefore see a y_2, contrary to hypothesis.

Accordingly, none of the four cases is possible. This is a case both where there is no symmetry under exchange of d^y's between a_1 and a_2, and no symmetry under exchange of histograms.

Appendix G
Fixed Cost Functions and Choosing Procedures

Since any deterministic search algorithm is a mapping from $d \subset \mathcal{D}$ to $x \subset \mathcal{X}$, any search algorithm is a vector in the space $\mathcal{X}^{\mathcal{D}}$. The components of such a vector are indexed by the possible samples, and the value for each component is the x that the algorithm produces given the associated sample. Consider now a particular sample d of size m. Given d, we can say whether or not any other sample of size greater than m has the (ordered) elements of d as its first m (ordered) elements. The set of those samples that do start with d this way defines a set of components of any algorithm vector a. Those components will be indicated by $a_{\supseteq d}$.

The remaining components of a are of two types. The first is given by those samples that are equivalent to the first $M < m$ elements in d for some M. The values of those components for the vector algorithm a will be indicated by $a_{\subset d}$. The second type consists of those components corresponding to all remaining samples. Intuitively, these are samples that are not compatible with d. Some examples of such samples are those that contain as one of their first m elements an element not found in d, and samples that re-order the elements found in d. The values of a for components of this second type will be indicated by $a_{\perp d}$.

Let $proc$ represent either A or B. We are interested in

$$\sum_{a,a'} P\left(c_{>m} \mid f, d_1, d_2, k, a, a', proc\right)$$
$$= \sum_{a_{\perp d}, a'_{\perp d'}} \sum_{a_{\subset d}, a_{\subset d'}} \sum_{a_{\supseteq d}, a'_{\supseteq d'}} P\left(c_{>m} \mid f, d, d', k, a, a', proc\right).$$

The summand is independent of the values of $a_{\perp d}$ and $a'_{\perp d}$ for either of our two d's. In addition, the number of such values is a constant. (It is given by the product, over all samples not consistent with d, of the number of possible x each such sample could be mapped to.) Therefore, up to an overall constant independent of d, d', f, and $proc$, the sum equals

$$\sum_{a_{\subset d}, a_{\subset d'}} \sum_{a_{\supseteq d}, a'_{\supseteq d'}} P\left(c_{>m} \mid f, d, d', a_{\supseteq d}, a'_{\supseteq d'}, a_{\subset d}, a'_{\subset d'}, proc\right).$$

By definition, we are implicitly restricting the sum to those a and a' so that our summand is defined. This means that we actually only allow one value for each component in $a_{\subset d}$ (namely, the value that gives the next x element in d) and similarly for $a'_{\subset d'}$. Therefore the sum reduces to

$$\sum_{a_{\supseteq d}, a'_{\supseteq d'}} P\left(c_{>m} \mid f, d, d', a_{\supseteq d}, a'_{\supseteq d'}, proc\right).$$

Note that no component of $a_{\supseteq d}$ lies in $d_\cup^x$. The same is true of $a'_{\supseteq d'}$. So the sum over $a_{\supseteq d}$ is over the same components of a as the sum over $a'_{\supseteq d'}$ is of a'. Now for fixed d and d', $proc$'s choice of a or a' is fixed. Accordingly, without loss of generality, the sum can be rewritten as

$$\sum_{a_{\supseteq d}} P\left(c_{>m} \mid f, d, d', a_{\supseteq d}\right)$$

with the implicit assumption that $c_{>m}$ is set by $a_{\supseteq d}$. This sum is independent of $proc$.

Appendix H
Proof of Theorem 11

Let *proc* refer to a choosing procedure. We are interested in

$$\sum_{a,a'} P\left(c_{>m} \mid f, m, k, a, a', proc\right)$$
$$= \sum_{a,a',d,d'} P\left(c_{>m} \mid f, d, d', k, a, a', proc\right)$$
$$\times P\left(d, d' \mid f, k, m, a, a', proc\right).$$

The sum over d and d' can be moved outside the sum over a and a'. Consider any term in that sum (i.e., any particular pair of values of d and d'). For that term, $P(d, d' \mid f, k, m, a, a', proc)$ is just one for those a and a' that result in d and d', respectively, when run on f, and zero otherwise. (Recall the assumption that a and a' are deterministic.) This means that the $P(d, d' \mid f, k, m, a, a', proc)$ factor simply restricts our sum over a and a' to the a and a' considered in our theorem.

Accordingly, our theorem tell us that the summand of the sum over d and d' is the same for choosing procedures A and B. Therefore the full sum is the same for both procedures.

Acknowledgment

The authors would like to thank R. Das, D. Fogel, T. Grossman, P. Helman, B. Levitan, U.-M. O'Reilly, and the reviewers for helpful comments and suggestions.

REFERENCES

Cover, T. M., and J. A. Thomas. 1991. *Elements of Information Theory*. New York, NY: Wiley.

Fogel, L. J., A. J. Owens, and M. J. Walsh. 1966. *Artificial Intelligence through Simulated Evolution*. New York, NY: Wiley.

Glover, F. 1989. "Tabu Search I." *ORSA Journal on Computing* 1 (3): 190–206.

———. 1990. "Tabu Search II." *ORSA Journal on Computing* 2:4–32.

Griffeath, D. 1976. "Introduction to Random Fields." In *Denumerable Markov Chains*, edited by J. G. Kemeny, J. L. Snell, and A. W. Knapp. New York, NY: Springer-Verlag.

Holland, J. H. 1993. *Adaptation in Natural and Artificial Systems.* Cambridge, MA: MIT Press.

Kinderman, R., and J. L. Snell. 1980. *Markov Random Fields and Their Applications.* Providence, RI: American Mathematical Society.

Kirkpatrick, S., D. C. Gelatt, and M. P. Vecchi. 1983. "Optimization by Simulated Annealing." *Science* 220:671–680.

Lawler, E. L., and D. E. Wood. 1966. "Branch-and-Bound Methods: A Survey." *Operations Research* 14 (4): 699–719.

Macready, W. G., and D. H. Wolpert. 1996. "What Makes an Optimization Problem Hard?" *Complexity* 5:40–46.

Schwefel, H.-P. 1995. *Evolution and Optimum Seeking.* New York, NY: Wiley.

Strauss, C. E. M., D. H. Wolpert, and D. R. Wolf. 1992. "Alpha, Evidence, and the Entropic Prior." *Maximum Entropy and Bayesian Methods* (Reading, MA), 113–120.

Wolpert, D. H. 1996a. "On Bias Plus Variance." *Neural Computation* 9:1271–1248.

———. 1996b. "The Lack of a Prior Distinctions Between Learning Algorithms." *Neural Computation* 8:1341–1390.

Wolpert, D. H., and W. G. Macready. 1995. *No Free Lunch Theorems for Search.* Technical report SFI-TR-05-010. Sante Fe, NM: Santa Fe Institute.

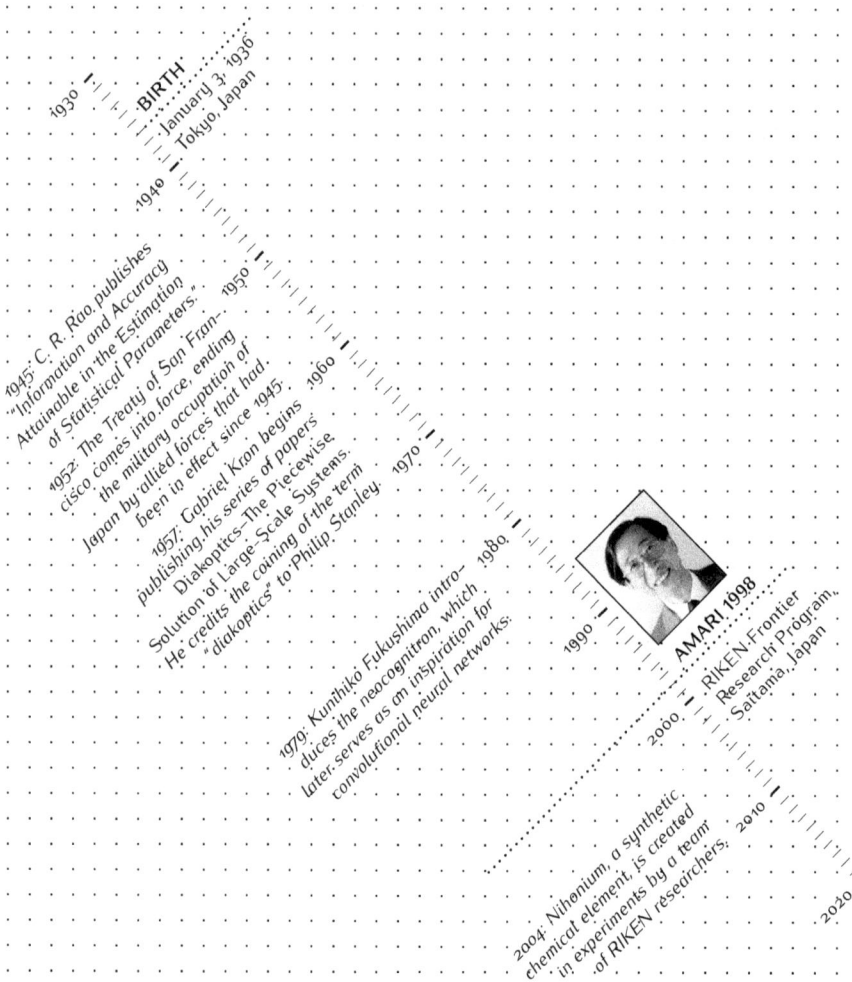

BIRTH
January 3, 1936
Tokyo, Japan

1945: C. R. Rao publishes "Information and Accuracy Attainable in the Estimation of Statistical Parameters."

1952: The Treaty of San Francisco comes into force, ending the military occupation of Japan by allied forces that had been in effect since 1945.

1957: Gabriel Kron begins publishing his series of papers Diakoptics—The Piecewise Solution of Large-Scale Systems. He credits the coining of the term "diakoptics" to Philip Stanley.

1979: Kunihiko Fukushima introduces the neocognitron, which later serves as an inspiration for convolutional neural networks.

AMARI 1998
RIKEN Frontier Research Program, Saitama, Japan

2004: Nihonium, a synthetic chemical element, is created in experiments by a team of RIKEN researchers.

SHUN-ICHI AMARI

[85]

FROM THE EUCLIDEAN TO THE NATURAL

Nihat Ay, Hamburg University of Technology

The idea of learning as an optimization process can be traced back to the early years of artificial neural networks. This idea has been very fruitful, ultimately leading to the recent successes of deep neural networks as learning machines. While being consistent with optimization, however, the first learning algorithms for neurons were inspired by neurophysiological and neuropsychological paradigms, most notably by the celebrated work of Donald Hebb (1949). Building on such paradigms, Frank Rosenblatt (1957) proposed an algorithm for training a simple neuronal model, which Warren McCulloch and Walter Pitts had introduced in their seminal article in 1943.[1] The convergence of this algorithm can be formally proved with elementary arguments from linear algebra (perceptron convergence theorem; see Novikoff 1962). The idea of learning as an optimization process, however, offers not only a unified conceptual foundation of learning, it also allows us to study learning from a rich mathematical perspective. In this context, the stochastic gradient descent method plays a fundamentally important role (Widrow 1963; Amari 1967; Rumelhart, Hinton, and Williams 1986). Nowadays, it represents the main instrument for training artificial neural networks, which brings us to Shun-Ichi Amari's article "Natural Gradient Works Efficiently in Learning." Let us unfold this title and thereby reveal the main insights of Amari's work.

What is a gradient? Before we discuss the notion of the natural gradient, let us first elaborate on the general notion of a gradient. Learning is interpreted as a process of optimization of some quantity, referred to as an objective function. The prediction error in supervised learning and the expected reward in reinforcement learning are

S.-I. Amari, "Natural Gradient Works Efficiently in Learning," *Neural Computation* 10 (2), 251–276 (1998).

Reprinted with permission from MIT Press.

---

[1] See *Foundational Papers in Complexity Science*, volume 1, chapter 5.

important instances of such an *objective function*, denoted by $\mathcal{L}$ in this introduction. Clearly, the learning should minimize the prediction error and maximize the expected reward. We can think of the corresponding optimization process as a search in a landscape, the search space, with valleys, hills and mountains where the height at a particular location is given by the value of the objective function $\mathcal{L}$ at that location. Say that we are initially standing at a point with low altitude and that the objective is to walk up to the highest location. Under ideal visibility conditions, where the optimal location can be seen very clearly, one can try to walk up to it in the most direct way. The problem with this strategy is that natural processes such as learning and evolution navigate in their respective landscapes almost blindly. Their view is very limited to a local neighborhood of their current location. Think of the difficulties that you face on a foggy day with visibility restricted to some maximal radius. All you can do is explore that visible neighborhood in order to decide in which direction to go. On the other hand, this is already enough to gradually walk up and thereby gain height. Here, the direction of maximal increase of the height is called the gradient direction. It is crucial to note that the neighborhood for probing the slope in various directions, and therefore the gradient itself, depends on the distance function (metric) defined on the search space for which we use the symbol $\mathcal{M}$ in what follows. If $\mathcal{M}$ carries a natural metric, this is then the one that has to be used in gradient-based optimization. With that natural metric, the gradient is then also called the natural gradient. Important results from information geometry are concerned with identifying such natural geometric structures, the Fisher–Rao metric and the Amari–Chentsov tensor being important instances of these (Amari and Nagaoka 2000; Amari 2016; Ay *et al.* 2017).

What is the *natural* gradient? In some sense, there is nothing special about the natural gradient. It is simply the gradient, as already described, based on a natural metric on $\mathcal{M}$. The "problem" arises when we express the elements of the search space with the help of a parametrization. In order to be more explicit, and without loss of generality, we restrict attention to a feed-forward neural network which represents input-output functions determined by the synaptic weights and biases of its neurons. Summarizing all parameters into one parameter vector $\theta = \theta_1, \ldots, \theta_n$, the functions in the search space are given in terms of a parametrization $\theta \mapsto f_\theta$, where f_θ is the function determined by the parameter vector θ. Thus, while being embedded in a high-dimensional space, the space of input–output functions, the search space $\mathcal{M}$ is constrained to all functions f_θ that the network can represent. Typically, $\mathcal{M}$ has a non-trivial geometric structure. It can be curved and have holes like a donut or a pretzel, as illustrated in figure 1. In the context of neural networks, it also has self-intersections and singularities. Now, a learning process in $\mathcal{M}$ is mediated by a corresponding process in the parameter domain. Two instances of such domains are shown as rectangles in figure 1. Interpreting the objective function $\mathcal{L}$, which is defined on the search space $\mathcal{M}$, as a function of the parameter vector θ, that is, $\mathcal{L}(\theta) := \mathcal{L}(f_\theta)$, we can compute the gradient in the parameter space as outlined above. Here, the parameter space carries the natural Euclidean geometry of $\mathbb{R}^n$, which implies that the neighborhood of a point, the ball of radius one centered around that point, looks completely symmetric. In the two parameter domains of figure 1, this is illustrated by a black dashed (lower left) and an outlined dashed circle (lower right), respectively. The actual neighborhood of $p \in \mathcal{M}$, shown as a symmetric black circle around p on $\mathcal{M}$ itself, however, is generically deformed into an ellipse in parameter space. This is why the probing of the slope of $\mathcal{L}(\theta)$ should be done on the respective ellipses, and not on the Euclidean circles, in order to be equivalent to the probing of the slope on the circle around p in $\mathcal{M}$.

Let us now be a bit more precise with the outlined idea of the natural gradient in order to connect it also formally to Amari's article. Basically, all aspects of the Euclidean geometry follow from the inner product $\langle v, w \rangle = \Sigma_{i=1}^n v_i w_i$. The slope of the function $\mathcal{L}$, considered as a function on the parameter space, in the direction of $v = (v_1, \ldots, v_n)$ is given by

$$\langle v, \nabla_\theta \mathcal{L} \rangle = \| v \| \| \nabla_\theta \mathcal{L} \| \cos(\alpha), \tag{1}$$

where $\nabla_\theta \mathcal{L} = \left(\frac{\partial \mathcal{L}}{\partial \theta_i}(\theta), \ldots, \frac{\partial \mathcal{L}}{\partial \theta_n}(\theta) \right)$ and α denotes the angle between v and $\nabla_\theta \mathcal{L}$. Thus, if we restrict attention to vectors v of length one, that is $\| v \| = \sqrt{\langle v, v \rangle} = 1$, the direction v of maximal slope (1) is proportional to

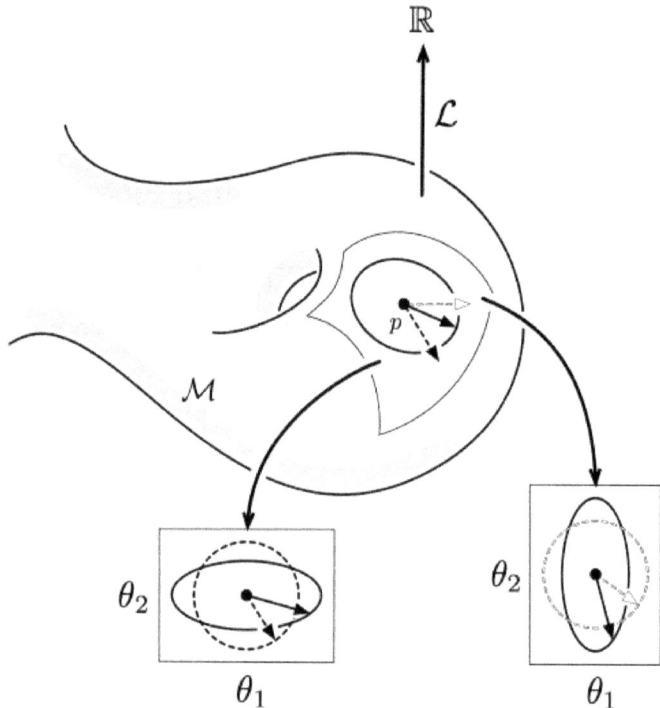

Figure 1. Illustration of the natural gradient of a function $\mathcal{L}$ on the search space $\mathcal{M}$.

$$\nabla_\theta \mathcal{L}, \qquad (2)$$

which is the gradient with respect to the natural geometry of the parameter space. On the other hand, the parameters are only a convenient instrument for implementing the learning. Given that they do not play the prime role here, the gradient that we compute with the help of parameters should be invariant with respect to parameter transformations. It is easy to see that the gradient (2) does not satisfy this condition. More precisely, if another parametrization is used to train the same neural network, say in terms of θ', then $\nabla_{\theta'}\mathcal{L}$ and $\nabla_\theta \mathcal{L}$ will typically correspond to completely different directions in the search space $\mathcal{M}$, and both of them will be different from the natural direction of change, the natural gradient, on $\mathcal{M}$.

The main idea of the natural gradient method is that a particular parametrization serves as a coordinate system for learning and has no special meaning. One can parametrize the actual search space in many equivalent ways. The geometry that we should use for optimization of the search is the one that is naturally defined on that space. As the change in the search space is mediated through a change in the parameter space, we have to correct the Euclidean gradient $\nabla_\theta \mathcal{L}$ so that it takes into account the geometry of $\mathcal{M}$ and is invariant with respect to the particular choice of the coordinate system. In order to do so, we associate with the natural basis $e_i, i = 1, \ldots, n$, of $\mathbb{R}^n$ vectors in the tangent space of $\mathcal{M}$, which we denote by $\partial_i, i = 1, \ldots, n$. In Amari's work, the geometry on $\mathcal{M}$ is given in terms of an inner product g, a so-called Riemannian metric, on $\mathcal{M}$. With this, we can define the matrix $G = (g_{ij})$, with

$$g_{ij} := g\left(\partial_i, \partial_j\right).$$

It turns out that the natural gradient, expressed in parameter space, is given by the formula

$$G^{-1}(\theta)\nabla_\theta \mathcal{L}. \tag{3}$$

The expression (3) is a correction of (2) that is invariant with respect to coordinate transformations. In theorem 1, Amari proves that (3) gives the direction of steepest ascent with respect to the inner product g on the search space $\mathcal{M}$. As the title of that paper suggests, respecting the geometry of the search space as outlined in this introduction greatly improves previously proposed gradient learning algorithms. In particular, Amari proves the efficiency of the natural gradient method based on the Cramér–Rao inequality. This result is presented as theorem 2, which can be considered as the main theorem of the article.

Since the publication of Amari's article, many authors have confirmed the strength of the natural gradient method in various contexts of learning, including reinforcement learning and robotics. More recent work addresses challenges within the field of deep neural networks (see Ay 2020, and references therein). The work of Amari has initiated fundamental directions of research within the field of neural networks and machine learning, where natural geometric structures are used to greatly improve learning algorithms.

REFERENCES

Amari, S. 1967. "A Theory of Adaptive Pattern Classifiers." *IEEE Transactions on Electronic Computers* EC-16 (3): 299–307. https://doi.org/10.1109/PGEC.1967.264666.

———. 2016. *Information Geometry and Its Applications.* Vol. 194. Applied Mathematical Sciences. Tokyo, Japan: Springer Tokyo. https://doi.org/10.1007/978-4-431-55978-8.

Amari, S., and H. Nagaoka. 2000. *Methods of Information Geometry.* Vol. 191. Translations of Mathematical Monographs. Oxford, UK: Oxford University Press.

Ay, N. 2020. "On the Locality of the Natural Gradient for Learning in Deep Bayesian Networks." *Information Geometry* 6:1–49. https://doi.org/10.1007/s41884-020-00038-y.

Ay, N., J. Jost, H. V. Lê, and L. Schwachhöfer. 2017. *Information Geometry.* Vol. 64. Ergebnisse der Mathematik und ihrer Grenzgebiete, 3. Folge / A Series of Modern Surveys in Mathematics. Cham, Switzerland: Springer. https://doi.org/10.1007/978-3-319-56478-4.

Hebb, D. O. 1949. *The Organization of Behaviour.* New York, NY: Wiley.

McCulloch, M., and W. Pitts. 1943. "A Logical Calculus of the Ideas Immanent in Nervous Activity." *Bulletin of Mathematical Biophysics* 5:115–133. https://doi.org/10.1007/BF02478259.

Novikoff, A. B. 1962. "On Convergence Proofs for Perceptrons." In *Symposium on the Mathematical Theory of Automata,* 12:615–622. Brooklyn, NY: Polytechnic Institute of Brooklyn.

Rosenblatt, F. 1957. "The Perceptron: A Probabilistic Model for Information Storage and Organization in the Brain." *Psychology Review* 65:386–407. https://doi.org/10.1037/h0042519.

Rumelhart, D. E., G. E. Hinton, and R. J. Williams. 1986. "Learning Internal Representations by Error Propagation." In *Parallel Distributed Processing,* 1:318–362. Cambridge, MA: MIT Press.

Widrow, B. 1963. *A Statistical Theory of Adaptation.* Oxford, UK: Pergamon Press.

NATURAL GRADIENT WORKS EFFICIENTLY IN LEARNING

Shun-ichi Amari

Abstract

When a parameter space has a certain underlying structure, the ordinary gradient of a function does not represent its steepest direction, but the natural gradient does. Information geometry is used for calculating the natural gradients in the parameter space of perceptrons, the space of matrices (for blind source separation), and the space of linear dynamical systems (for blind source deconvolution). The dynamical behavior of natural gradient online learning is analyzed and is proved to be Fisher efficient, implying that it has asymptotically the same performance as the optimal batch estimation of parameters. This suggests that the plateau phenomenon, which appears in the backpropagation learning algorithm of multilayer perceptrons, might disappear or might not be so serious when the natural gradient is used. An adaptive method of updating the learning rate is proposed and analyzed.

Introduction

The stochastic gradient method (Widrow 1963; Amari 1967; Tsypkin 1973; Rumelhart, Hinton, and Williams 1986) is a popular learning method in the general nonlinear optimization framework. The parameter space is not Euclidean but has a Riemannian metric structure in many cases. In these cases, the ordinary gradient does not give the steepest direction of a target function; rather, the steepest direction is given by the natural (or contravariant) gradient. The Riemannian metric structures are introduced by means of information geometry (Amari 1985; Murray and Rice 1993; Amari 1997; Amari, Kurata, and Nagaoka 1992). This article gives the natural gradients explicitly in the case of the space of perceptrons for neural learning, the space of matrices for blind source separation, and the space of linear dynamical systems for blind multichannel source deconvolution. This is an extended version of an earlier article (Amari 1996), including new results.

How good is natural gradient learning compared to conventional gradient learning? The asymptotic behavior of online natural gradient

Amari points out that what has been considered natural before, the Euclidean geometry of the parameter space, is not necessarily natural. Previously defined steepest descent algorithms do not actually point in the direction of steepest descent.

learning is studied for this purpose. Training examples can be used only once in online learning when they appear. Therefore, the asymptotic performance of online learning cannot be better than the optimal batch procedure where all the examples can be reused again and again. However, we prove that natural gradient online learning gives the Fisher-efficient estimator in the sense of asymptotic statistics when the loss function is differentiable, so that it is asymptotically equivalent to the optimal batch procedure (see also Amari 1995; Opper 1996). When the loss function is nondifferentiable, the accuracy of asymptotic online learning is worse than batch learning by a factor of 2 (see, for example, van den Broeck and Reimann 1996). It was shown in Amari et al. (1992) that the dynamic behavior of natural gradient in the Boltzmann machine is excellent.

It is not easy to calculate the natural gradient explicitly in multilayer perceptrons. However, a preliminary analysis (Yang and Amari 1997b), by using a simple model, shows that the performance of natural gradient learning is remarkably good, and it is sometimes free from being trapped in plateaus, which give rise to slow convergence of the backpropagation learning method (Saad and Solla 1995). This suggests that the Riemannian structure might eliminate such plateaus or might make them not so serious.

Online learning is flexible, because it can track slow fluctuations of the target. Such online dynamics were first analyzed in Amari (1967) and then by many researchers recently. Sompolinsky, Barkai, and Seung (1995), and Barkai, Seung, and Sompolinsky (1995) proposed an adaptive method of adjusting the learning rate (see also Amari 1967). We generalize their idea and evaluate its performance based on the Riemannian metric of errors.

The article is organized as follows. The natural gradient is defined in section 2. Section 3 formulates the natural gradient in various problems of stochastic descent learning. Section 4 gives the statistical analysis of efficiency of online learning, and section 5 is devoted to the problem of adaptive changes in the learning rate. Calculations of the Riemannian metric and explicit forms of the natural gradients are given in sections 6, 7, and 8.

2. Natural Gradient

Let $S = \{w \in R^n\}$ be a parameter space on which a function $L(w)$ is defined. When S is a Euclidean space with an orthonormal coordinate system w, the squared length of a small incremental vector dw connecting w and $w + dw$ is given by

$$|dw|^2 = \sum_{i=1}^{n} (dw_i)^2,$$

where dw_i are the components of dw. However, when the coordinate system is nonorthonormal, the squared length is given by the quadratic form

$$|dw|^2 = \sum_{i,j} g_{ij}(w) dw_i dw_j. \qquad (2.1)$$

Here, Amari presents a direct comparison of the Euclidean with the more general Riemannian geometry. The latter is stated in terms of equation (2.1) and is not further specified at this point. In section 3, Amari presents several instances of this geometry.

When S is a curved manifold, there is no orthonormal linear coordinates, and the length of dw is always written as in equation 2.1. Such a space is a Riemannian space. We show in later sections that parameter spaces of neural networks have the Riemannian character. The $n \times n$ matrix $G = (g_{ij})$ is called the Riemannian metric tensor, and it depends in general on w. It reduces to

$$g_{ij}(w) = \delta_{ij} = \begin{cases} 1, & i = j, \\ 0, & i \neq j \end{cases}$$

in the Euclidean orthonormal case, so that G is the unit matrix I in this case.

The steepest descent direction of a function $L(w)$ at w is defined by the vector dw that minimizes $L(w + dw)$ where $|dw|$ has a fixed length, that is, under the constraint

$$|dw|^2 = \varepsilon^2 \qquad (2.2)$$

for a sufficiently small constant ε.

This formula provides the actual direction of steepest descent, given by the Riemannian geometry rather than the Euclidean geometry of the parameter space.

Theorem 1. *The steepest descent direction of $L(w)$ in a Riemannian space is given by*

$$-\tilde{\nabla} L(w) = -G^{-1}(w) \nabla L(w) \qquad (2.3)$$

where $G^{-1} = (g^{ij})$ is the inverse of the metric $G = (g_{ij})$ and ∇L is the conventional gradient,

$$\nabla L(w) = \left(\frac{\partial}{\partial w_1} L(w), \ldots, \frac{\partial}{\partial w_n} L(w) \right)^T,$$

the superscript T denoting the transposition.

Proof. We put
$$dw = \varepsilon a,$$
and search for the a that minimizes
$$L(w + dw) = L(w) + \varepsilon \nabla L(w)^T a$$
under the constraint
$$|a|^2 = \sum g_{ij} a_i a_j = 1.$$
By the Lagrangean method, we have
$$\frac{\partial}{\partial a_i} \left\{ \nabla L(w)^T a - \lambda a^T G a \right\} = 0.$$
This gives
$$\nabla L(w) = 2\lambda G a$$
or
$$a = \frac{1}{2\lambda} G^{-1} \nabla L(w),$$
where λ is determined from the constraint.

We call
$$\tilde{\nabla} L(w) = G^{-1} \nabla L(w)$$
the natural gradient of L in the Riemannian space. Thus, $-\tilde{\nabla} L$ represents the steepest descent direction of L. (If we use the tensorial notation, this is nothing but the contravariant form of $-\nabla L$.) When the space is Euclidean and the coordinate system is orthonormal, we have
$$\tilde{\nabla} L = \nabla L. \tag{2.4}$$
This suggests the natural gradient descent algorithm of the form
$$w_{t+1} = w_t - \eta_t \tilde{\nabla} L(w_t), \tag{2.5}$$
where η_t is the learning rate that determines the step size.

> Once we have the correction of the usual Euclidean gradient by the natural one, we can use the well-known scheme for updating the parameters. Within this scheme, we go step by step in the direction of steepest descent of the objective function.

3. Natural Gradient Learning

Let us consider an information source that generates a sequence of independent random variables $z_1, z_2, \ldots, z_t, \ldots$, subject to the same probability distribution $q(z)$. The random signals z_t are processed by a processor (like a neural network) that has a set of adjustable parameters $\boldsymbol{w}$. Let $l(\boldsymbol{z}, \boldsymbol{w})$ be a loss function when signal z is processed by the processor whose parameter is w. Then the risk function or the average loss is

$$L(\boldsymbol{w}) = E[l(\boldsymbol{z}, \boldsymbol{w})], \qquad (3.1)$$

where E denotes the expectation with respect to $\boldsymbol{z}$. Learning is a procedure to search for the optimal $\boldsymbol{w}^*$ that minimizes $L(\boldsymbol{w})$.

The stochastic gradient descent learning method can be formulated in general as

$$\boldsymbol{w}_{t+1} = \boldsymbol{w}_t - \eta_t C(\boldsymbol{w}_t) \nabla l(\boldsymbol{z}_t, \boldsymbol{w}_t), \qquad (3.2)$$

where η_t is a learning rate that may depend on t and $C(\boldsymbol{w})$ is a suitably chosen positive definite matrix (see Amari 1967). In the natural gradient online learning method, it is proposed to put $C(\boldsymbol{w})$ equal to $G^{-1}(\boldsymbol{w})$ when the Riemannian structure is defined. We give a number of examples to be studied in more detail.

3.1 STATISTICAL ESTIMATION OF PROBABILITY DENSITY FUNCTION

In the case of statistical estimation, we assume a statistical model $\{p(\boldsymbol{z}, \boldsymbol{w})\}$, and the problem is to obtain the probability distribution $p(\boldsymbol{z}, \hat{\boldsymbol{w}})$ that approximates the unknown density function $q(\boldsymbol{z})$ in the best way—that is, to estimate the true $\boldsymbol{w}$ or to obtain the optimal approximation $\boldsymbol{w}$ from the observed data. A typical loss function is

$$l(\boldsymbol{z}, \boldsymbol{w}) = -\log p(\boldsymbol{z}, \boldsymbol{w}). \qquad (3.3)$$

The expected loss is then given by

$$L(\boldsymbol{w}) = -E[\log p(\boldsymbol{z}, \boldsymbol{w})]$$
$$= E_q \left[\log \frac{q(\boldsymbol{z})}{p(\boldsymbol{z}, \boldsymbol{w})} \right] + H_Z,$$

where H_Z is the entropy of $q(\boldsymbol{z})$ not depending on $\boldsymbol{w}$. Hence, minimizing L is equivalent to minimizing the Kullback–Leibler divergence

$$D[q(\boldsymbol{z}) : p(\boldsymbol{z}, \boldsymbol{w})] = \int q(\boldsymbol{z}) \log \frac{q(\boldsymbol{z})}{p(\boldsymbol{z}, \boldsymbol{w})} d\boldsymbol{z} \qquad (3.4)$$

of two probability distributions $q(\boldsymbol{z})$ and $p(\boldsymbol{z}, \boldsymbol{w})$. When the true distribution $q(\boldsymbol{z})$ is written as $q(\boldsymbol{z}) = p(\boldsymbol{z}, \boldsymbol{w}^*)$, this is equivalent to obtain the maximum likelihood estimator $\hat{\boldsymbol{w}}$.

This equation has to be contrasted with equation (2.5). While the parameter update rule (2.5) is based on the direction of steepest descent, the rule (3.2) takes into account that even that information is not fully available to the learning system. The learning is data-driven and can only use an estimate of the direction of steepest descent. This leads to the general stochastic gradient descent rule (3.2). Amari pioneered the method of a stochastic gradient descent in 1967, and nowadays it represents the most commonly used training method in the field of deep learning.

Amari presents a number of important instances of the outlined general theory in the following sections. The examples of sections 3.1 to 3.4 highlight different Riemannian structures for the individual natural gradient update rules.

The Riemannian structure of the parameter space of a statistical model is defined by the Fisher information (Rao 1945; Amari 1985)

$$g_{ij}(\boldsymbol{w}) = E\left[\frac{\partial \log p(\boldsymbol{x}, \boldsymbol{w})}{\partial \mathrm{w}_i} \frac{\partial \log p(\boldsymbol{x}, \boldsymbol{w})}{\partial \mathrm{w}_j}\right] \quad (3.5)$$

in the component form. This is the only invariant metric to be given to the statistical model (Chentsov 1972; Campbell 1985; Amari 1985). The learning equation (see equation 3.2) gives a sequential estimator $\hat{\boldsymbol{w}}_t$.

3.2 MULTILAYER NEURAL NETWORK

Let us consider a multilayer feedforward neural network specified by a vector parameter $\boldsymbol{w} = (\mathrm{w}_1, \ldots, \mathrm{w}_n)^T \in \boldsymbol{R}^n$. The parameter $\boldsymbol{w}$ is composed of modifiable connection weights and thresholds. When input $\boldsymbol{x}$ is applied, the network processes it and calculates the outputs $\boldsymbol{f}(\boldsymbol{x}, \boldsymbol{w})$. The input $\boldsymbol{x}$ is subject to an unknown probability distribution $q(\boldsymbol{x})$. Let us consider a teacher network that, by receiving $\boldsymbol{x}$, generates the corresponding output $\boldsymbol{y}$ subject to a conditional probability distribution $q(\boldsymbol{y} \mid \boldsymbol{x})$. The task is to obtain the optimal $\boldsymbol{w}^*$ from examples such that the student network approximates the behavior of the teacher.

Let us denote by $l(\boldsymbol{x}, \boldsymbol{w})$ a loss when input signal $\boldsymbol{x}$ is processed by a network having parameter $\boldsymbol{w}$. A typical loss is given,

$$l(\boldsymbol{x}, \boldsymbol{y}, \boldsymbol{w}) = \frac{1}{2}|\boldsymbol{y} - \boldsymbol{f}(\boldsymbol{x}, \boldsymbol{w})|^2, \quad (3.6)$$

where $\boldsymbol{y}$ is the output given by the teacher.

Let us consider a statistical model of neural networks such that its output $\boldsymbol{y}$ is given by a noisy version of $\boldsymbol{f}(\boldsymbol{x}, \boldsymbol{w})$,

$$\boldsymbol{y} = \boldsymbol{f}(\boldsymbol{x}, \boldsymbol{w}) + \boldsymbol{n}, \quad (3.7)$$

where $\boldsymbol{n}$ is a multivariate gaussian noise with zero mean and unit covariance matrix I. By putting $\boldsymbol{z} = (\boldsymbol{x}, \boldsymbol{y})$, which is an input-output pair, the model specifies the probability density of $\boldsymbol{z}$ as

$$p(\boldsymbol{z}, \boldsymbol{w}) = cq(\boldsymbol{x}) \exp\left\{-\frac{1}{2}|\boldsymbol{y} - \boldsymbol{f}(\boldsymbol{x}, \boldsymbol{w})|^2\right\}, \quad (3.8)$$

where c is a normalizing constant and the loss function (see equation 3.6) is rewritten as

$$l(\boldsymbol{z}, \boldsymbol{w}) = \text{const} + \log q(\boldsymbol{x}) - \log p(\boldsymbol{z}, \boldsymbol{w}). \quad (3.9)$$

Given a sequence of examples $(\boldsymbol{x}_1, \boldsymbol{y}_1), \ldots, (\boldsymbol{x}_t, \boldsymbol{y}_t), \ldots$, the natural gradient online learning algorithm is written as

$$\boldsymbol{w}_{t+1} = \boldsymbol{w}_t - \eta_t \tilde{\nabla} l(\boldsymbol{x}_t, \boldsymbol{y}_t, \boldsymbol{w}_t). \qquad (3.10)$$

Information geometry (Amari 1985) shows that the Riemannian structure is given to the parameter space of multilayer networks by the Fisher information matrix,

$$g_{ij}(\boldsymbol{w}) = E\left[\frac{\partial \log p(\boldsymbol{x}, \boldsymbol{y}; \boldsymbol{w})}{\partial \mathrm{w}_i} \frac{\partial p(\boldsymbol{x}, \boldsymbol{y}; \boldsymbol{w})}{\partial \mathrm{w}_j}\right]. \qquad (3.11)$$

We will show how to calculate $G = (g_{ij})$ and its inverse in a later section.

3.3 BLIND SEPARATION OF SOURCES

Let us consider m signal sources that produce m independent signals $s_i(t)$, $i = 1, \ldots, m$, at discrete times $t = 1, 2, \ldots$. We assume that $s_i(t)$ are independent at different times and that the expectations of s_i are 0. Let $r(s)$ be the joint probability density function of $\boldsymbol{s}$. Then it is written in the product form

$$r(\boldsymbol{s}) = \prod_{i=1}^{m} r_i(s_i). \qquad (3.12)$$

Consider the case where we cannot have direct access to the source signals $\boldsymbol{s}(t)$ but we can observe their m instantaneous mixtures $\boldsymbol{x}(t)$,

$$\boldsymbol{x}(t) = A\boldsymbol{s}(t) \qquad (3.13)$$

or

$$x_i(t) = \sum_{j=1}^{m} A_{ij} s_j(t),$$

where $A = (A_{ij})$ is an $m \times m$ nonsingular mixing matrix that does not depend on t, and $\boldsymbol{x} = (x_1, \ldots, x_m)^T$ is the observed mixtures.

Blind source separation is the problem of recovering the original signals $\boldsymbol{s}(t), t = 1, 2, \ldots$ from the observed signals $\boldsymbol{x}(t), t = 1, 2, \ldots$ (Jutten and Hérault 1991). If we know A, this is trivial, because we have

$$\boldsymbol{s}(t) = A^{-1}\boldsymbol{x}(t).$$

The "blind" implies that we do not know the mixing matrix A and the probability distribution densities $r_i(s_i)$.

A typical algorithm to solve the problem is to transform $\boldsymbol{x}(t)$ into

$$\boldsymbol{y}(t) = W_t \boldsymbol{x}(t), \tag{3.14}$$

where W_t is an estimate of A^{-1}. It is modified by the following learning equation:

$$W_{t+1} = W_t - \eta_t F(\boldsymbol{x}_t, W_t). \tag{3.15}$$

Here, $F(\boldsymbol{x}, W)$ is a special matrix function satisfying

$$E[F(\boldsymbol{x}, W)] = 0 \tag{3.16}$$

for any density functions $r(\boldsymbol{s})$ in equation 3.12 when $W = A^{-1}$. For W_t of equation 3.15 to converge to A^{-1}, equation 3.16 is necessary but not sufficient, because the stability of the equilibrium is not considered here.

Let $K(W)$ be an operator that maps a matrix to a matrix. Then

$$\tilde{F}(\boldsymbol{x}, W) = K(W) F(\boldsymbol{x}, W)$$

satisfies equation 3.16 when F does. The equilibrium of F and $\tilde{F}$ is the same, but their stability can be different. However, the natural gradient does not alter the stability of an equilibrium, because G^{-1} is positive-definite.

Let $l(\boldsymbol{x}, W)$ be a loss function whose expectation

$$L(W) = E[l(\boldsymbol{x}, W)]$$

is the target function minimized at $W = A^{-1}$. A typical function F is obtained by the gradient of l with respect to W,

$$F(\boldsymbol{x}, W) = \nabla l(\boldsymbol{x}, W). \tag{3.17}$$

Such an F is also obtained by heuristic arguments. Amari and Cardoso Amari and Cardoso (1997) gave the complete family of F satisfying equation 3.16 and elucidated the statistical efficiency of related algorithms.

From the statistical point of view, the problem is to estimate $W = A^{-1}$ from observed data $\boldsymbol{x}(1), \ldots, \boldsymbol{x}(t)$. However, the probability density function of $\boldsymbol{x}$ is written as

$$p_X(\boldsymbol{x}; W, r) = |W| r(W \boldsymbol{x}), \tag{3.18}$$

which is specified not only by W to be estimated but also by an unknown function r of the form 3.12. Such a statistical model is said to be semiparametric and is a difficult problem to solve (Bickel, Klassen, Ritov, & Wellner 1993), because it includes an unknown function of infinite degrees of freedom. However, we can apply the information-geometrical theory of estimating functions (Amari and Kawanabe 1997) to this problem.

When F is given by the gradient of a loss function (see equation 3.17), where ∇ is the gradient $\partial/\partial W$ with respect to a matrix, the natural gradient is given by

$$\tilde{\nabla} l = G^{-1} \circ \nabla l. \qquad (3.19)$$

Here, G is an operator transforming a matrix to a matrix so that it is an $m^2 \times m^2$ matrix. G is the metric given to the space $Gl(m)$ of all the nonsingular $m \times m$ matrices. We give its explicit form in a later section based on the Lie group structure. The inverse of G is also given explicitly. Another important problem is the stability of the equilibrium of the learning dynamics. This has recently been solved by using the Riemannian structure (Amari, Chen, & Chichocki 1997; see also Cardoso & Laheld 1996). The superefficiency of some algorithms has been also proved in Amari (1999) under certain conditions.

3.4 BLIND SOURCE DECONVOLUTION

When the original signals $s(t)$ are mixed not only instantaneously but also with past signals as well, the problem is called blind source deconvolution or equalization. By introducing the time delay operator z^{-1},

$$z^{-1} s(t) = s(t-1), \qquad (3.20)$$

we have a mixing matrix filter $\boldsymbol{A}$ denoted by

$$\boldsymbol{A}(z) = \sum_{k=0}^{\infty} A_k z^{-k}, \qquad (3.21)$$

where A_k are $m \times m$ matrices. The observed mixtures are

$$\boldsymbol{x}(t) = \boldsymbol{A}(z)\boldsymbol{s}(t) = \sum_k A_k \boldsymbol{s}(t-k). \qquad (3.22)$$

To recover the original independent sources, we use the finite impulse response model

$$\boldsymbol{W}(z) = \sum_{k=0}^{d} W_k z^{-1} \qquad (3.23)$$

of degree d. The original signals are recovered by

$$\boldsymbol{y}(t) = \boldsymbol{W}_t(z)\boldsymbol{x}(t), \tag{3.24}$$

where $\boldsymbol{W}_t$ is adaptively modified by

$$\boldsymbol{W}_{t+1}(z) = \boldsymbol{W}_t(z) - \eta_t \nabla l\left\{\boldsymbol{x}_t, \boldsymbol{x}_{t-1}, \ldots, \boldsymbol{W}_t(z)\right\}. \tag{3.25}$$

Here, $l\left(\boldsymbol{x}_t, \boldsymbol{x}_{t-1}, \ldots, \boldsymbol{W}\right)$ is a loss function that includes some past signals. We can summarize the past signals into a current state variable in the online learning algorithm. Such a loss function is obtained by the maximum entropy method (Bell and Sejnowski 1995), independent component analysis (Comon 1994), or the statistical likelihood method.

In order to obtain the natural gradient learning algorithm

$$\boldsymbol{W}_{t+1}(z) = \boldsymbol{W}_t(z) - \eta_t \tilde{\nabla} l\left(\boldsymbol{x}_t, \boldsymbol{x}_{t-1}, \ldots, \boldsymbol{W}_t\right),$$

we need to define the Riemannian metric in the space of all the matrix filters (multiterminal linear systems). Such a study was initiated by Amari (1987). It is possible to define G and to obtain G^{-1} explicitly (see section 8). A preliminary investigation into the performance of the natural gradient learning algorithm has been undertaken by Douglas, Cichocki, and Amari (1996) and Amari et al. (1997).

> This is a core section of the paper. The title "Natural Gradient Works Efficiently in Learning" actually has two interpretations. An obvious, and less precise, one is that the natural gradient is in some sense better than the usual Euclidean one. Interestingly, this interpretation of efficiency led to various confirmations of the statement that the natural gradient works efficiently. A second interpretation refers to a statistical notion of efficiency. This is clearly a more formal and precise interpretation. Thus, the title can be interpreted both by readers who know statistics and those who do not, and both interpretations are valid.

4. Natural Gradient Gives Fisher-Efficient Online Learning Algorithms

This section studies the accuracy of natural gradient learning from the statistical point of view. A statistical estimator that gives asymptotically the best result is said to be Fisher efficient. We prove that natural gradient learning attains Fisher efficiency.

Let us consider multilayer perceptrons as an example. We study the case of a realizable teacher, that is, the behavior of the teacher is given by $q(\boldsymbol{y} \mid \boldsymbol{x}) = p(\boldsymbol{y} \mid \boldsymbol{x}, \boldsymbol{w}^*)$. Let $D_T = \{(\boldsymbol{x}_1, \boldsymbol{y}_1), \ldots, (\boldsymbol{x}_T, \boldsymbol{y}_T)\}$ be T-independent input-output examples generated by the teacher network having parameter $\boldsymbol{w}^*$. Then, minimizing the log loss,

$$l(\boldsymbol{x}, \boldsymbol{y}; \boldsymbol{w}) = -\log p(\boldsymbol{x}, \boldsymbol{y}; \boldsymbol{w}),$$

over the training data D_T is to obtain $\hat{\boldsymbol{w}}_T$ that minimizes the training error

$$L_{\text{train}}(\boldsymbol{w}) = \frac{1}{T} \sum_{t=1}^{T} l\left(\boldsymbol{x}_t, \boldsymbol{y}_t; \boldsymbol{w}\right). \tag{4.1}$$

This is equivalent to maximizing the likelihood $\prod_{t=1}^{T} p(\boldsymbol{x}_t, \boldsymbol{y}_t; \boldsymbol{w})$. Hence $\hat{\boldsymbol{w}}_T$ is the maximum likelihood estimator. The Cramér-Rao theorem states that the expected squared error of an unbiased estimator satisfies

$$E\left[(\hat{\boldsymbol{w}}_T - \boldsymbol{w}^*)(\hat{\boldsymbol{w}}_T - \boldsymbol{w}^*)^T\right] \geq \frac{1}{T} G^{-1}, \quad (4.2)$$

where the inequality holds in the sense of positive definiteness of matrices. An estimator is said to be efficient or Fisher efficient when it satisfies equation 4.2 with equality for large T. The maximum likelihood estimator is Fisher efficient, implying that it is the best estimator attaining the Cramér–Rao bound asymptotically,

$$\lim_{T \to \infty} T E\left[(\hat{\boldsymbol{w}}_T - \boldsymbol{w}^*)(\hat{\boldsymbol{w}}_T - \boldsymbol{w}^*)^T\right] = G^{-1}, \quad (4.3)$$

where G^{-1} is the inverse of the Fisher information matrix $G = (g_{ij})$ defined by equation 3.11.

Examples $(\boldsymbol{x}_1, \boldsymbol{y}_1), (\boldsymbol{x}_2, \boldsymbol{y}_2) \ldots$ are given one at a time in the case of online learning. Let $\tilde{\boldsymbol{w}}_t$ be an online estimator at time t. At the next time, $t+1$, the estimator $\tilde{\boldsymbol{w}}_t$ is modified to give a new estimator $\tilde{\boldsymbol{w}}_{t+1}$ based on the current observation $(\boldsymbol{x}_t, \boldsymbol{y}_t)$. The old observations $(\boldsymbol{x}_1, \boldsymbol{y}_1), \ldots, (\boldsymbol{x}_{t-1}, \boldsymbol{y}_{t-1})$ cannot be reused to obtain $\tilde{\boldsymbol{w}}_{t+1}$, so the learning rule is written as

$$\tilde{\boldsymbol{w}}_{t+1} = \boldsymbol{m}(\boldsymbol{x}_t, \boldsymbol{y}_t, \tilde{\boldsymbol{w}}_t).$$

The process $\{\tilde{\boldsymbol{w}}_t\}$ is Markovian. Whatever learning rule $\boldsymbol{m}$ is chosen, the behavior of the estimator $\tilde{\boldsymbol{w}}_t$ is never better than that of the optimal batch estimator $\hat{\boldsymbol{w}}_t$ because of this restriction. The gradient online learning rule

$$\tilde{\boldsymbol{w}}_{t+1} = \tilde{\boldsymbol{w}}_t - \eta_t C \frac{\partial l(\boldsymbol{x}_t, \boldsymbol{y}_t; \tilde{\boldsymbol{w}}_t)}{\partial \boldsymbol{w}},$$

was proposed where C is a positive-definite matrix, and its dynamical behavior was studied by Amari (1967) when the learning constant $\eta_t = \eta$ is fixed. Heskes and Kappen (1991) obtained similar results, which ignited research into online learning. When η_t satisfies some condition, say, $\eta_t = c/t$, for a positive constant c, the stochastic approximation guarantees that $\tilde{\boldsymbol{w}}_t$ is a consistent estimator converging to $\boldsymbol{w}^*$. However, it is not Fisher efficient in general.

There arises a question of whether there exists a learning rule that gives an efficient estimator. If it exists, the asymptotic behavior of online

Statistical estimation is based on a stochastic data process so that the estimated parameter will always fluctuate around the optimal parameter. This fluctuation cannot be avoided, which is expressed by the Cramér-Rao inequality (4.2). The best strategy for estimation should then try to minimize the fluctuations so that the lower bound in (4.2) is reached, at least in the limit of a large set of data points. This is the statistical notion of efficiency, expressed by equation (4.3).

learning is equivalent to that of the best batch estimation method. This article answers the question affirmatively, by giving an efficient online learning rule (see Amari 1995; see also Opper 1996).

Let us consider the natural gradient learning rule,

$$\tilde{w}_{t+1} = \tilde{w}_t - \frac{1}{t}\tilde{\nabla} l(x_t, y_t, \tilde{w}_t). \tag{4.4}$$

Theorem 2. *Under the learning rule (see equation 4.4), the natural gradient online estimator $\tilde{w}_t$ is Fisher efficient.*

> This can be considered the main theorem of the article, which reflects the statement in the title in a precise statistical sense.

Proof. Let us denote the covariance matrix of estimator $\tilde{w}_t$ by

$$\tilde{V}_{t+1} = E\left[(\tilde{w}_{t+1} - w^*)(\tilde{w}_{t+1} - w^*)^T\right]. \tag{4.5}$$

This shows the expectation of the squared error. We expand

$$\frac{\partial l(x_t, y_t; \tilde{w}_t)}{\partial w} = \frac{\partial l(x_t, y_t; w^*)}{\partial w} + \frac{\partial^2 l(x_t, y_t; w^*)}{\partial w \partial w}(\tilde{w}_t - w^*) + O\left(|\tilde{w}_t - w^*|^2\right).$$

By subtracting w^* from the both sides of equation 4.4 and taking the expectation of the square of the both sides, we have

$$\tilde{V}_{t+1} = \tilde{V}_t - \frac{2}{t}\tilde{V}_t + \frac{1}{t^2}G^{-1} + O\left(\frac{1}{t^3}\right), \tag{4.6}$$

where we used

$$E\left[\frac{\partial l(x_t, y_t; w^*)}{\partial w}\right] = 0, \tag{4.7}$$

$$E\left[\frac{\partial^2 l(x_t, y_t; w^*)}{\partial w \partial w}\right] = G(w^*),$$

$$G(\tilde{w}_t) = G(w^*) + O\left(\frac{1}{t}\right), \tag{4.8}$$

because $\tilde{w}_t$ converges to w^* as guaranteed by stochastic approximation under certain conditions (see Kushner and Clark 1978). The solution of equation 4.6 is written asymptotically as

$$\tilde{V}_t = \frac{1}{t}G^{-1} + O\left(\frac{1}{t^2}\right),$$

proving the theorem.

The theory can be extended to be applicable to the unrealizable teacher case, where

$$K(\boldsymbol{w}) = E\left[\frac{\partial^2}{\partial \boldsymbol{w} \partial \boldsymbol{w}} l(\boldsymbol{x}, \boldsymbol{y}; \boldsymbol{w})\right] \qquad (4.9)$$

should be used instead of $G(\boldsymbol{w})$ in order to obtain the same efficient result as the optimal batch procedure. This is locally equivalent to the Newton–Raphson method. The results can be stated in terms of the generalization error instead of the covariance of the estimator, and we can obtain more universal results (see Amari 1993; Amari and Murata 1993).

Remark. In the cases of blind source separation and deconvolution, the models are semiparametric, including the unknown function r (see equation 3.18). In such cases, the Cramér–Rao bound does not necessarily hold. Therefore, Theorem 2 does not hold in these cases. It holds when we can estimate the true r of the source probability density functions and use it to define the loss function $l(\boldsymbol{x}, W)$. Otherwise equation 4.8 does not hold. The stability of the true solution is not necessarily guaranteed either. Amari, Chen, and Cichocki (1997) have analyzed this situation and proposed a universal method of attaining the stability of the equilibrium solution.

5. Adaptive Learning Constant

The dynamical behavior of the learning rule (see equation 3.2) was studied in Amari (1967) when η_t is a small constant η. In this case, $\boldsymbol{w}_t$ fluctuates around the (local) optimal value $\boldsymbol{w}^*$ for large t. The expected value and variance of $\boldsymbol{w}_t$ was studied, and the trade-off between the convergence speed and accuracy of convergence was demonstrated.

When the current $\boldsymbol{w}_t$ is far from the optimal $\boldsymbol{w}^*$, it is desirable to use a relatively large η to accelerate the convergence. When it is close to $\boldsymbol{w}^*$, a small η is preferred in order to eliminate fluctuations. An idea of an adaptive change of η was discussed in Amari (1967) and was called "learning of learning rules."

Sompolinsky et al. (1995) (see also Barkai et al., 1995) proposed a rule of adaptive change of η_t, which is applicable to the pattern classification problem where the expected loss $L(\boldsymbol{w})$ is not differentiable at $\boldsymbol{w}^*$. This article generalizes their idea to a more general case where $L(\boldsymbol{w})$ is differentiable and analyzes its behavior by using the Riemannian structure.

We propose the following learning scheme:

$$w_{t+1} = w_t - \eta_t \tilde{\nabla} l(x_t, y_t; \hat{w}_t) \tag{5.1}$$

$$\eta_{t+1} = \eta_t \exp\{\alpha [\beta l(x_t, y_t; \hat{w}_t) - \eta_t]\}, \tag{5.2}$$

where α and β are constants. We also assume that the training data are generated by a realizable deterministic teacher and that $L(w^*) = 0$ holds at the optimal value. (See Murata, Müller, Ziehe, and Amari 1996 for a more general case.) We try to analyze the dynamical behavior of learning by using the continuous version of the algorithm for the sake of simplicity,

$$\frac{d}{dt} w_t = -\eta_t G^{-1}(w_t) \frac{\partial}{\partial w} l(x_t, y_t; w_t), \tag{5.3}$$

$$\frac{d}{dt} \eta_t = \alpha \eta_t [\beta l(x_t, z_t; w_t) - \eta_t]. \tag{5.4}$$

In order to show the dynamical behavior of (w_t, η_t), we use the averaged version of equations 5.3 and 5.4 with respect to the current input-output pair (x_t, y_t). The averaged learning equation (Amari 1967, 1977) is written as

$$\frac{d}{dt} w_t = -\eta_t G^{-1}(w_t) \left\langle \frac{\partial}{\partial w} l(x, y; w_t) \right\rangle, \tag{5.5}$$

$$\frac{d}{dt} \eta_t = \alpha \eta_t \{\beta \langle l(x, y; w_t) \rangle - \eta_t\}, \tag{5.6}$$

where $\langle \ \rangle$ denotes the average over the current (x, y). We also use the asymptotic evaluations

$$\left\langle \frac{\partial}{\partial w} l(x, y; w_t) \right\rangle = \left\langle \frac{\partial}{\partial w} l(x, y; w^*) \right\rangle$$
$$+ \left\langle \frac{\partial^2}{\partial w \partial w} l(x, y; w^*)(w_t - w^*) \right\rangle$$
$$= G^*(w_t - w^*)$$
$$\langle l(x, y; w_t) \rangle = \frac{1}{2}(w_t - w^*)^T G^*(w_t - w^*),$$

where $G^* = G(w^*)$ and we used $L(w^*) = 0$. We then have

$$\frac{d}{dt} w_t = -\eta_t (w_t - w^*), \tag{5.7}$$

$$\frac{d}{dt} \eta_t = \alpha \eta_t \left\{ \frac{\beta}{2}(w_t - w^*)^T G^*(w_t - w^*) - \eta_t \right\}. \tag{5.8}$$

Now we introduce the squared error variable,

$$e_t = \frac{1}{2}(\boldsymbol{w}_t - \boldsymbol{w}^*)^T G^* (\boldsymbol{w}_t - \boldsymbol{w}^*), \quad (5.9)$$

where e_t is the Riemannian magnitude of $\boldsymbol{w}_t - \boldsymbol{w}^*$. It is easy to show

$$\frac{d}{dt}e_t = -2\eta_t e_t, \quad (5.10)$$

$$\frac{d}{dt}\eta_t = \alpha\beta\eta_t e_t - \alpha\eta_t^2. \quad (5.11)$$

The behavior of equations 5.10 and 5.11 is interesting. The origin $(0,0)$ is its attractor. However, the basin of attraction has a boundary of fractal structure. Anyway, starting from an adequate initial value, it has the solution of the form

$$e_t = \frac{a}{t},$$
$$\eta_t = \frac{b}{t}.$$

The coefficients a and b are determined from

$$a = 2ab$$
$$b = -\alpha\beta \text{ab} + \alpha b^2.$$

This gives

$$b = \frac{1}{2},$$
$$a = \frac{1}{\beta}\left(\frac{1}{2} - \frac{1}{\alpha}\right), \quad \alpha > 2.$$

This proves the $1/t$ convergence rate of the generalization error, that is, the optimal order for any estimator $\hat{\boldsymbol{w}}_t$ converging to $\boldsymbol{w}^*$. The adaptive η_t shows a nice characteristic when the target teacher is slowly fluctuating or changes suddenly.

6. Natural Gradient in the Space of Perceptrons

The Riemannian metric and its inverse are calculated in this section to obtain the natural gradient explicitly. We begin with an analog simple perceptron whose input-output behavior is given by

$$y = f(\boldsymbol{w} \cdot \boldsymbol{x}) + n, \quad (6.1)$$

While various Riemannian structures for the natural gradient were highlighted in sections 3.1 to 3.4, natural gradients are explicitly evaluated and studied to some extent in sections 6, 7, and 8. These sections also serve as an overview of Amari's work, which culminated in this milestone article. While the results of these sections are important in their respective application fields, the article's core message can be understood without a detailed study of these sections.

where n is a gaussian noise subject to $N(0, \sigma^2)$ and

$$f(u) = \frac{1 - e^{-u}}{1 + e^{-u}}. \tag{6.2}$$

The conditional probability density of y when $\boldsymbol{x}$ is applied is

$$p(y \mid \boldsymbol{x}; \boldsymbol{w}) = \frac{1}{\sqrt{2\pi}\sigma} \exp\left\{-\frac{1}{2\sigma^2}[y - f(\boldsymbol{w} \cdot \boldsymbol{x})]^2\right\}. \tag{6.3}$$

The distribution $q(\boldsymbol{x})$ of inputs $\boldsymbol{x}$ is assumed to be the normal distribution $N(0, I)$. The joint distribution of $(\boldsymbol{x}, y)$ is

$$p(y, \boldsymbol{x}; \boldsymbol{w}) = q(\boldsymbol{x}) p(y \mid \boldsymbol{x}; \boldsymbol{w}).$$

In order to calculate the metric G of equation 3.11 explicitly, let us put

$$\mathrm{w}^2 = |\boldsymbol{w}|^2 = \sum \mathrm{w}_i^2 \tag{6.4}$$

where $|\boldsymbol{w}|$ is the Euclidean norm. We then have the following theorem.

Theorem 3. *The Fisher information metric is*

$$G(\boldsymbol{w}) = \mathrm{w}^2 c_1(\mathrm{w}) I + \{c_2(\mathrm{w}) - c_1(\mathrm{w})\} \boldsymbol{w}\boldsymbol{w}^T, \tag{6.5}$$

where $c_1(\mathrm{w})$ and $c_2(\mathrm{w})$ are given by

$$c_1(\mathrm{w}) = \frac{1}{4\sqrt{2\pi}\sigma^2 \mathrm{w}^2} \int \{f^2(\mathrm{w}\varepsilon) - 1\}^2 \exp\left\{-\frac{1}{2}\varepsilon^2\right\} d\varepsilon,$$

$$c_2(\mathrm{w}) = \frac{1}{4\sqrt{2\pi}\sigma^2 \mathrm{w}^2} \int \{f^2(\mathrm{w}\varepsilon) - 1\}^2 \varepsilon^2 \exp\left\{-\frac{1}{2}\varepsilon^2\right\} d\varepsilon.$$

Proof. We have

$$\log p(y, \boldsymbol{x}; \boldsymbol{w}) = \log q(\boldsymbol{x}) - \log(\sqrt{2\pi}\sigma) - \frac{1}{2\sigma^2}[y - f(\boldsymbol{w} \cdot \boldsymbol{x})]^2.$$

Hence,

$$\frac{\partial}{\partial \mathrm{w}_i} \log p(y, \boldsymbol{x}; \boldsymbol{w}) = \frac{1}{\sigma^2}\{y - f(\boldsymbol{w} \cdot \boldsymbol{x})\} f'(\boldsymbol{w} \cdot \boldsymbol{x}) x_i$$

$$= \frac{1}{\sigma^2} n f'(\boldsymbol{w} \cdot \boldsymbol{x}) x_i.$$

The Fisher information matrix is given by

$$g_{ij}(\boldsymbol{w}) = E\left[\frac{\partial}{\partial \mathrm{w}_i} \log p \frac{\partial}{\partial \mathrm{w}_j} \log p\right]$$

$$= \frac{1}{\sigma^2} E\left[\{f'(\boldsymbol{w} \cdot \boldsymbol{x})\}^2 x_i x_j\right],$$

where $E[n^2] = \sigma^2$ is taken into account. This can be written, in the vector-matrix form, as

$$G(w) = \frac{1}{\sigma^2} E\left[(f')^2 xx^T\right].$$

In order to show equation 6.5, we calculate the quadratic form $r^T G(w) r$ for arbitrary r. When $r = w$,

$$w^T G w = \frac{1}{\sigma^2} E\left[\{f'(w \cdot x)\}^2 (w \cdot x)^2\right].$$

Since $u = w \cdot x$ is subject to $N(0, w^2)$, we put $u = w\varepsilon$, where ε is subject to $N(0, 1)$. Noting that

$$f'(u) = \frac{1}{2}\{1 - f^2(u)\},$$

we have,

$$w^T G(w) w = \frac{w^2}{4\sqrt{2\pi\sigma^2}} \int \varepsilon^2 \{f^2(w\varepsilon) - 1\}^2 \exp\left\{-\frac{\varepsilon^2}{2}\right\} d\varepsilon,$$

which confirms equation 6.5 when $r = w$. We next put $r = v$, where v is an arbitrary unit vector orthogonal to w (in the Euclidean sense). We then have

$$v^T G(w) v = \frac{1}{4\sigma^2} E\left[\{f^2(w \cdot x) - 1\}^2 (v \cdot x)^2\right].$$

Since $u = w \cdot x$ and $v = v \cdot x$ are independent, and v is subject to $N(0, 1)$, we have

$$v^T G(w) v = \frac{1}{4\sigma^2} E\left[(v \cdot x)^2\right] E\left[\{f^2(w \cdot x) - 1\}^2\right]$$

$$= \frac{1}{4\sqrt{2\pi\sigma^2}} \int \{f^2(w\varepsilon) - 1\}^2 \exp\left\{-\frac{\varepsilon^2}{2}\right\} d\varepsilon.$$

Since $G(w)$ in equation 6.5 is determined by the quadratic forms for n-independent w and v's, this proves equation 6.5.

To obtain the natural gradient, it is necessary to have an explicit form of G^{-1}. We can calculate $G^{-1}(w)$ explicitly in the perceptron case.

Theorem 4. *The inverse of the Fisher information metric is*

$$G^{-1}(w) = \frac{1}{w^2 c_1(w)} I + \frac{1}{w^4}\left(\frac{1}{c_2(w)} - \frac{1}{c_1(w)}\right) ww^T. \tag{6.6}$$

This can easily be proved by direct calculation of GG^{-1}. The natural gradient learning equation (3.10) is then given by

$$w_{t+1} = w_t + \eta_t \{y_t - f(w_t \cdot x_t)\} f'(w_t \cdot x_t)$$
$$\left[\frac{1}{w_t^2 c_1(w_t)} x_t + \frac{1}{w_t^4}\left(\frac{1}{c_2(w_t)} - \frac{1}{c_1(w_t)}\right)(w_t \cdot x_t) w_t\right]. \quad (6.7)$$

We now show some other geometrical characteristics of the parameter space of perceptrons. The volume V_n of the manifold of simple perceptrons is measured by

$$V_n = \int \sqrt{|G(w)|} dw \quad (6.8)$$

where $|G(w)|$ is the determinant of $G = (g_{ij})$, which represents the volume density by the Riemannian metric. It is interesting to see that the manifold of perceptrons has a finite volume.

Bayesian statistics considers that w is randomly chosen subject to a prion distribution $\pi(w)$. A choice of $\pi(w)$ is the Jeffrey prior or noninformative prior given by

$$\pi(w) = \frac{1}{V_n} \sqrt{|G(w)|}. \quad (6.9)$$

The Jeffrey prior is calculated as follows.

Theorem 5. *The Jeffrey prior and the volume of the manifold are given, respectively, by*

$$\sqrt{|G(w)|} = \frac{w}{V_n} \sqrt{c_2(w)\{c_1(w)\}^{n-1}}, \quad (6.10)$$

$$V_n = a_{n-1} \int \sqrt{c_2(w)\{c_1(w)\}^{n-1}} w^n dw, \quad (6.11)$$

respectively, where a_{n-1} is the area of the unit $(n-1)$-sphere.

The Fisher metric G can also be calculated for multilayer perceptrons. Let us consider a multilayer perceptron having m hidden units with sigmoidal activation functions and a linear output unit. The input-output relation is

$$y = \sum v_i f(\boldsymbol{w}_i \cdot \boldsymbol{x}) + n,$$

or the conditional probability is

$$p(y \mid \boldsymbol{x}; \boldsymbol{v}, \boldsymbol{w}_1, \ldots, \boldsymbol{w}_m) = c \exp\left[-\frac{1}{2}\left\{y - \sum v_i f(\boldsymbol{w}_i \cdot \boldsymbol{x})\right\}^2\right]. \tag{6.12}$$

The total parameter $\boldsymbol{w}$ consist of $\{\boldsymbol{v}, \boldsymbol{w}_1, \ldots, \boldsymbol{w}_m\}$. Let us calculate the Fisher information matrix G. It consists of $m+1$ blocks corresponding to these $\boldsymbol{w}_i$'s and $\boldsymbol{v}$.

From

$$\frac{\partial}{\partial \boldsymbol{w}_i} \log p(y \mid \boldsymbol{x}; \boldsymbol{w}) = n v_i f'(\boldsymbol{w}_i \cdot \boldsymbol{x}) \boldsymbol{x},$$

we easily obtain the block submatrix corresponding to $\boldsymbol{w}_i$ as

$$E\left[\frac{\partial}{\partial \boldsymbol{w}_i} \log p \frac{\partial}{\partial \boldsymbol{w}_i} \log p\right] = \frac{1}{\sigma^4} E[n^2] v_i^2 E\left[\{f'(\boldsymbol{w}_i \cdot \boldsymbol{x})\}^2 \boldsymbol{x}\boldsymbol{x}^T\right]$$

$$= \frac{1}{\sigma^2} v_i^2 E\left[\{f'(\boldsymbol{w}_i \cdot \boldsymbol{x})\}^2 \boldsymbol{x}\boldsymbol{x}^T\right].$$

This is exactly the same as the simple perceptron case except for a factor of $(v_i)^2$. For the off-diagonal block, we have

$$E\left[\frac{\partial}{\partial \boldsymbol{w}_i} \log p \frac{\partial}{\partial \boldsymbol{w}_j} \log p\right] = \frac{1}{\sigma^2} v_i v_j E\left[f'(\boldsymbol{w}_i \cdot \boldsymbol{x}) f'(\boldsymbol{w}_j \cdot \boldsymbol{x}) \boldsymbol{x}\boldsymbol{x}^T\right].$$

In this case, we have the following form,

$$G\boldsymbol{w}_i\boldsymbol{w}_j = c_{ij} I + d_{ii} \boldsymbol{w}_i \boldsymbol{w}_i^T + d_{ij} \boldsymbol{w}_i \boldsymbol{w}_j^T + d_{ji} \boldsymbol{w}_j \boldsymbol{w}_i^T + d_{jj} \boldsymbol{w}_j \boldsymbol{w}_j^T, \tag{6.13}$$

where the coefficients c_{ij} and d_{ij}'s are calculated explicitly by similar methods.

The $\boldsymbol{v}$ block and $\boldsymbol{v}$ and $\boldsymbol{w}_i$ block are also calculated similarly. However, the inversion of G is not easy except for simple cases. It requires inversion of a $2(m+1)$ dimensional matrix. However, this is much better than the direct inversion of the original $(n+1)m$-dimensional matrix of G. Yang and Amari (1997a) performed a preliminary study on the performance of the natural gradient learning algorithm for a simple multilayer perceptron. The result shows that natural gradient learning might be free from the plateau phenomenon. Once the learning trajectory is trapped in a plateau, it takes a long time to get out of it.

7. Natural Gradient in the Space of Matrices and Blind Source Separation

We now define a Riemannian structure to the space of all the $m \times m$ nonsingular matrices, which forms a Lie group denoted by $Gl(m)$, for the purpose of introducing the natural gradient learning rule to the blind source separation problem. Let dW be a small deviation of a matrix from W to $W + dW$. The tangent space T_W of $Gl(m)$ at W is a linear space spanned by all such small deviations dW_{ij}'s and is called the Lie algebra.

We need to introduce an inner product at W by defining the squared norm of dW

$$ds^2 = \langle dW, dW \rangle_W = \|dW\|^2.$$

By multiplying W^{-1} from the right, W is mapped to $WW^{-1} = I$, the unit matrix, and $W + dW$ is mapped to $(W + dW)W^{-1} = I + dX$, where

$$dX = dW W^{-1}. \tag{7.1}$$

This shows that a deviation dW at W is equivalent to the deviation dX at I by the correspondence given by multiplication of W^{-1}. The Lie group invariance requires that the metric is kept invariant under this correspondence, that is, the inner product of dW at W is equal to the inner product of dWY at WY for any Y,

$$\langle dW, dW \rangle_W = \langle dWY, dWY \rangle_{WY}. \tag{7.2}$$

When $Y = W^{-1}$, $WY = I$. This principle was used to derive the natural gradient in Amari, Cichocki, and Yang (1996); see also Yang and Amari (1997b) for detail. Here we give its analysis by using dX.

We define the inner product at I by

$$\langle dX, dX \rangle_I = \sum_{i,j} (dX_{ij})^2 = \operatorname{tr}\left(dX^T dX\right). \tag{7.3}$$

We then have the Riemannian metric structure at W as

$$\langle dW, dW \rangle_W = \operatorname{tr}\left\{\left(W^{-1}\right)^T dW^T dW W^{-1}\right\}. \tag{7.4}$$

We can write the metric tensor G in the component form. It is a quantity having four indices $G_{ij,kl}(W)$ such that

$$ds^2 = \sum G_{ij,kl}(W) dW_{ij} dW_{kl},$$
$$G_{ij,kl}(W) = \sum_m \delta_{ik} W^{-1}_{jm} W^{-1}_{lm}, \tag{7.5}$$

where W_{jm}^{-1} are the components of W^{-1}. While it may not appear to be straightforward to obtain the explicit form of G^{-1} and natural gradient $\tilde{\nabla} L$, in fact it can be calculated as shown below.

Theorem 6. *The natural gradient in the matrix space is given by*

$$\tilde{\nabla} L = (\nabla L) W^T W. \tag{7.6}$$

Proof. The metric is Euclidean at I, so that both $G(I)$ and its inverse, $G^{-1}(I)$, are the identity. Therefore, by mapping dW at W to dX at I, the natural gradient learning rule in terms of dX is written as

$$\frac{dX}{dt} = -\eta_t G^{-1}(I) \frac{\partial L}{\partial X} = -\eta_t \frac{\partial L}{\partial X}, \tag{7.7}$$

where the continuous time version is used. We have from equation 7.1

$$\frac{dX}{dt} = \frac{dW}{dt} W^{-1}. \tag{7.8}$$

The gradient $\partial L / \partial X$ is calculated as

$$\frac{\partial L}{\partial X} = \frac{\partial L(W)}{\partial W} \left(\frac{\partial W^T}{\partial X} \right) = \frac{\partial L}{\partial W} W^T.$$

Therefore, the natural gradient learning rule is

$$\frac{dW}{dt} = -\eta_t \frac{\partial L}{\partial W} W^T W,$$

which proves equation 7.6.

The $dX = dW W^{-1}$ forms a basis of the tangent space at W, but this is not integrable; that is, we cannot find any matrix function $X = X(W)$ that satisfies equation 7.1. Such a basis is called a nonholonomic basis. This is a locally defined basis but is convenient for our purpose. Let us calculate the natural gradient explicitly. To this end, we put

$$l(\boldsymbol{x}, W) = -\log \det |W| - \sum_{i=1}^{n} \log f_i(y_i), \tag{7.9}$$

where $\boldsymbol{y} = W\boldsymbol{x}$ and $f_i(y_i)$ is an adequate probability distribution. The expected loss is

$$L(W) = E[l(\boldsymbol{x}, W)],$$

which represents the entropy of the output $\boldsymbol{y}$ after a componentwise nonlinear transformation (Nadal and Parga 1994; Bell and Sejnowski

1995). The independent component analysis or the mutual information criterion also gives a similar loss function (Comon 1994; Amari et al. 1996; see also Oja & Karhunen 1995). When f_i is the true probability density function of the ith source, $l(\boldsymbol{x}, W)$ is the negative of the log likelihood.

The natural gradient of l is calculated as follows. We calculate the differential

$$dl = l(\boldsymbol{x}, W + dW) - l(\boldsymbol{x}, W) = -d\log\det|W| - \sum d\log f_i(y_i)$$

due to change dW. Then,

$$\begin{aligned} d\log\det|W| &= \log\det|W + dW| - \log\det|W| \\ &= \log\det\left|(W + dW)W^{-1}\right| = \log(\det|I + dX|) \\ &= \operatorname{tr} dX. \end{aligned}$$

Similarly, from $d\boldsymbol{y} = dW\boldsymbol{x}$,

$$\begin{aligned} \sum d\log f_i(y_i) &= -\varphi(\boldsymbol{y})^T dW \boldsymbol{x} \\ &= -\varphi(\boldsymbol{y})^T dX \boldsymbol{y}, \end{aligned}$$

where $\varphi(\boldsymbol{y})$ is the column vector

$$\varphi(\boldsymbol{y}) = [\varphi_1(y_1), \ldots, \varphi_m(y_m)],$$

$$\varphi_i(y_i) = -\frac{d}{dy}\log f_i(y_i).$$

This gives $\partial L/\partial X$, and the natural gradient learning equation is

$$\frac{dW}{dt} = \eta_t \left(I - \varphi(\boldsymbol{y})^T \boldsymbol{y}\right) W. \tag{7.11}$$

The efficiency of this equation is studied from the statistical and information geometrical point of view (Amari and Kawanabe 1997; Amari and Cardoso 1997). We further calculate the Hessian by using the natural frame dX,

$$d^2l = \boldsymbol{y}^T dX^T \dot{\varphi}(\boldsymbol{y}) dX \boldsymbol{y} + \varphi(\boldsymbol{y})^T dX dX \boldsymbol{y}, \tag{7.12}$$

where $\dot{\varphi}(\boldsymbol{y})$ is the diagonal matrix with diagonal entries $d\varphi_i(y_i)/dy_i$. Its expectation can be explicitly calculated (Amari, Chen, and Cichocki 1997). The Hessian is decomposed into diagonal elements and two-by-two diagonal blocks (see also Cardoso and Laheld 1996). Hence, the stability of the above learning rule is easily checked. Thus, in terms of dX, we can solve the two fundamental problems: the efficiency and the stability of learning algorithms of blind source separation (Amari and Cardoso 1997; Amari, Chen, and Cichocki 1997).

8. Natural Gradient in Systems Space

The problem is how to define the Riemannian structure in the parameter space $\{W(z)\}$ of systems, where z is the time-shift operator. This was given in Amari (1987) from the point of view of information geometry (Amari 1985, 1997; Murray & Rice 1993). We show here only ideas (see Douglas et al. 1996; Amari, Douglas, Cichocki, & Yang 1997, for preliminary studies).

In the case of multiterminal deconvolution, a typical loss function l is given by

$$l = -\log \det |W_0| - \sum_i \int p\{y_i; \boldsymbol{W}(z)\} \log f_i(y_i)\, dy_i, \quad (8.1)$$

where $p\{y_i; \boldsymbol{W}(z)\}$ is the marginal distribution of $\boldsymbol{y}(t)$ which is derived from the past sequence of $\boldsymbol{x}(t)$ by matrix convolution $\boldsymbol{W}(z)$ of equation 3.24. This type of loss function is obtained from maximization of entropy, independent component analysis, or maximum likelihood.

The gradient of l is given by

$$\nabla_m l = -\left(W_0^{-1}\right)^T \delta_{0m} + \varphi(\boldsymbol{y}_t)\, \boldsymbol{x}^T(t-m), \quad (8.2)$$

where

$$\nabla_m = \frac{\partial}{\partial W_m},$$

and

$$\nabla l = \sum_{m=0}^{d} (\nabla_m l)\, z^{-m}. \quad (8.3)$$

In order to calculate the natural gradient, we need to define the Riemannian metric G in the manifold of linear systems. The geometrical theory of the manifold of linear systems by Amari (1987) defines the Riemannian metric and a pair of dual affine connections in the space of linear systems.

Let

$$d\boldsymbol{W}(z) = \sum_m d\boldsymbol{W}_m z^{-m} \quad (8.4)$$

be a small deviation of $\boldsymbol{W}(z)$. We postulate that the inner product $\langle d\boldsymbol{W}(z), d\boldsymbol{W}(z)\rangle$ is invariant under the operation of any matrix filter $\boldsymbol{Y}(z)$,

$$\langle d\boldsymbol{W}(z), d\boldsymbol{W}(z)\rangle_{\boldsymbol{W}(z)} = \langle d\boldsymbol{W}(z)\boldsymbol{Y}(z), d\boldsymbol{W}(z)\boldsymbol{Y}(z)\rangle_{\boldsymbol{WY}}, \quad (8.5)$$

where $\boldsymbol{Y}(z)$ is any system matrix. If we put

$$\boldsymbol{Y}(z) = \{\boldsymbol{W}(z)\}^{-1},$$

which is a general system not necessarily belonging to FIR,

$$\boldsymbol{W}(z)\{\boldsymbol{W}(z)\}^{-1} = \boldsymbol{I}(z),$$

which is the identity system

$$\boldsymbol{I}(z) = I$$

not including any z^{-m} terms. The tangent vector $d\boldsymbol{W}(z)$ is mapped to

$$dX(z) = d\boldsymbol{W}(z)\{\boldsymbol{W}(z)\}^{-1}. \tag{8.6}$$

The inner product at I is defined by

$$\langle dX(z), dX(z) \rangle_I = \sum_{m,ij} (dX_{m,ij})^2, \tag{8.7}$$

where $dX_{m,ij}$ are the elements of matrix dX_m.

The natural gradient

$$\tilde{\nabla} l = G^{-1} \circ \nabla l$$

of the manifold of systems is given as follows.

Theorem 7. *The natural gradient of the manifold of systems is given by*

$$\tilde{\nabla} l = \nabla l(z) \boldsymbol{W}^T \left(z^{-1} \right) \boldsymbol{W}(z), \tag{8.8}$$

where operator z^{-1} should be operated adequately.

The proof is omitted. It should be remarked that $\tilde{\nabla} l$ does not belong to the class of FIR systems, nor does it satisfy the causality condition either. Hence, in order to obtain an online learning algorithm, we need to introduce time delay to map it to the space of causal FIR systems. This article shows only the principles involved; details will published in a separate article by Amari, Douglas, and Cichocki.

9. Conclusions

This article introduces the Riemannian structures to the parameter spaces of multilayer perceptrons, blind source separation, and blind source deconvolution by means of information geometry. The natural gradient learning method is then introduced and is shown to be statistically efficient. This implies that optimal online learning is as efficient as optimal batch learning when the Fisher information matrix exists. It is also suggested that natural gradient learning might be easier to get out of plateaus than conventional stochastic gradient learning.

Acknowledgments

I thank A. Cichocki, A. Back, and H. Yang at RIKEN Frontier Research Program for their discussions.

REFERENCES

Amari, S. 1967. "Theory of Adaptive Pattern Classifiers." *IEEE Transactions on Electronic Computers* EC-16 (3): 299–307.

———. 1977. "Neural Theory of Association and Concept-Formation." *Biological Cybernetics* 26:175–185.

———. 1985. *Differential-Geometrical Methods in Statistics.* Lecture Notes in Statistics 28. New York, NY: Springer-Verlag.

———. 1987. "Differential Geometry of a Parametric Family of Invertible Linear Systems—Riemannian Metric, Dual Affine Connections and Divergence." *Mathematical Systems Theory* 20:53–82.

———. 1993. "Universal Theorem on Learning Curves." *Neural Networks* 6:16–66.

———. 1995. "Learning and Statistical Inference." In *Handbook of Brain Theory and Neural Networks,* edited by M. A. Arbib, 52–26. Cambridge, MA: MIT Press.

———. 1996. "Neural Learning in Structured Parameter Spaces-Natural Riemannian Gradient." Edited by M. C. Mozer, M. I. Jordan, and Th. Petsche. *Advances in Neural Processing Systems* 9.

———. 1997. "Information Geometry." *Contemporary Mathematics* 203:81–95.

———. 1999. "Superefficiency in Blind Source Separation." *IEEE Transactions on Signal Processing* 47 (4): 936–944.

Amari, S., and J. F. Cardoso. 1997. "Blind Source Separation—Semiparametric Statistical Approach." *IEEE Transactions on Signal Processing* 45 (11): 2692–2700.

Amari, S., T. P. Chen, and A. Cichocki. 1997. "Stability Analysis of Learning Algorithms for Blind Source Separation." *Neural Networks* 10 (8): 1345–1351.

Amari, S., A. Cichocki, and H. H. Yang. 1996. "A New Learning Algorithm for Blind Signal Separation." *NIPS '95* 8.

Amari, S., S. C. Douglas, A. Cichocki, and H. H. Yang. 1997. "Multichannel Blind Deconvolution and Equalization Using The Natural Gradient." In *First IEEE Signal Processing Workshop on Signal Processing Advances in Wireless Communications,* 101–104. IEEE.

Amari, S., and M. Kawanabe. 1997. "Information Geometry of Estimating Functions in Semiparametric Statistical Models." *Bernoulli* 3 (1): 29–54.

Amari, S., K. Kurata, and H. Nagaoka. 1992. "Information Geometry of Boltzmann Machines." *IEEE Transactions on Neural Networks* 3 (2): 260–271.

Amari, S., and N. Murata. 1993. "Statistical Theory of Learning Curves under Entropic Loss Criterion." *Neural Computation* 5 (1): 140–153.

Barkai, N., H. S. Seung, and H. Sompolinsky. 1995. "Local and Global Convergence of On-Line Learning." *Physical Review Letters* 75:1415–1418.

Bell, A. J., and T. J. Sejnowski. 1995. "An Information-Maximization Approach to Blind Separation and Blind Deconvolution." *Neural Computation* 7:1129–1159.

Bickel, P. J., C. A. J. Klassen, Y. Ritov, and J. A. Wellner. 1993. *Efficient and Adaptive Estimation for Semiparametric Models.* Baltimore, MD: Johns Hopkins University Press.

Campbell, L. L. 1985. "The Relation Between Information Theory and the Differential-Geometric Approach to Statistics." *Information Sciences* 35:199–210.

Cardoso, J. F., and B. Laheld. 1996. "Equivariant Adaptive Source Separation." *IEEE Transactions on Signal Processing* 44:3017–3030.

Chentsov, N. N. 1972. "Statistical Decision Rules and Optimal Inference (in Russian)." Translated in English (1982), Rhode Island: AMS, *Nauka.*

Comon, P. 1994. "Independent Component Analysis, a New Concept?" *Signal Processing* 36:287–314.

Douglas, S. C., A. Cichocki, and S. Amari. 1996. "Fast Convergence Filtered Regressor Algorithms for Blind Equalization." *Electronics Letters* 32:2114–2115.

Heskes, T., and B. Kappen. 1991. "Learning Process in Neural Networks." *Physical Review* A44:2718–2762.

Jutten, C., and J. Hérault. 1991. "Blind Separation of Sources, an Adaptive Algorithm Based on Neuromimetic Architecture." *Signal Processing* 24 (1): 1–31.

Kushner, H. J., and D. S. Clark. 1978. *Stochastic Approximation Methods for Constrained and Unconstrained Systems.* Berlin, Germany: Springer-Verlag.

Murata, N., K. R. Müller, A. Ziehe, and S. Amari. 1996. "Adaptive On-Line Learning in Changing Environments." Edited by M. C. Mozer, M. I. Jordan, and Th. Petsche. *Advances in Neural Processing Systems* (Cambridge, MA) 9.

Murray, M. K., and J. W. Rice. 1993. *Differential Geometry and Statistics.* New York, NY: Chapman & Hall.

Nadal, J. P., and N. Parga. 1994. "Nonlinear Neurons in the Low Noise Limit—A Factorial Code Maximizes Information Transfer." *Network* 5:561–581.

Oja, E., and J. Karhunen. 1995. "Signal Separation by Nonlinear Hebbian Learning." *Computational Intelligence: A Dynamic System Perspective*, 83–97.

Opper, M. 1996. "Online Versus Offline Learning from Random Examples: General Results." *Physical Review Letters* 77:4671–4674.

Rao, C. R. 1945. "Information and Accuracy Attainable in the Estimation of Statistical Parameters." *Bulletin of the Calcutta Mathematical Society* 37:81–91.

Rumelhart, D. E., G. E. Hinton, and R. J. Williams. 1986. "Learning Internal Representations by Error Propagation." In *Parallel Distributed Processing*, 318–362. Cambridge, MA: MIT Press.

Saad, D., and S. A. Solla. 1995. "On-Line Learning in Soft Committee Machines." *Physical Review E* 52:4225–4243.

Sompolinsky, H., N. Barkai, and H. Seung. 1995. "On-Line Learning of Dichotomies." *Advances in Neural Information Processing Systems* 7.

Tsypkin, Ya. Z. 1973. *Foundation of the Theory of Learning Systems*. New York, NY: Academic Press.

van den Broeck, C., and P. Reimann. 1996. "Unsupervised Learning by Examples: On-Line Versus Off-Line." *Physical Review Letters* 76:2188–2191.

Widrow, B. 1963. *A Statistical Theory of Adaptation*. Oxford, UK: Pergamon Press.

Yang, H. H., and S. Amari. 1997a. "Adaptive On-Line Learning Algorithms for Blind Separation-Maximum Entropy and Minimal Mutual Information." *Neural Computation* 9 (7): 1457–1482.

———. 1997b. "Application of Natural Gradient in Training Multilayer Perceptrons." Unpublished manuscript.

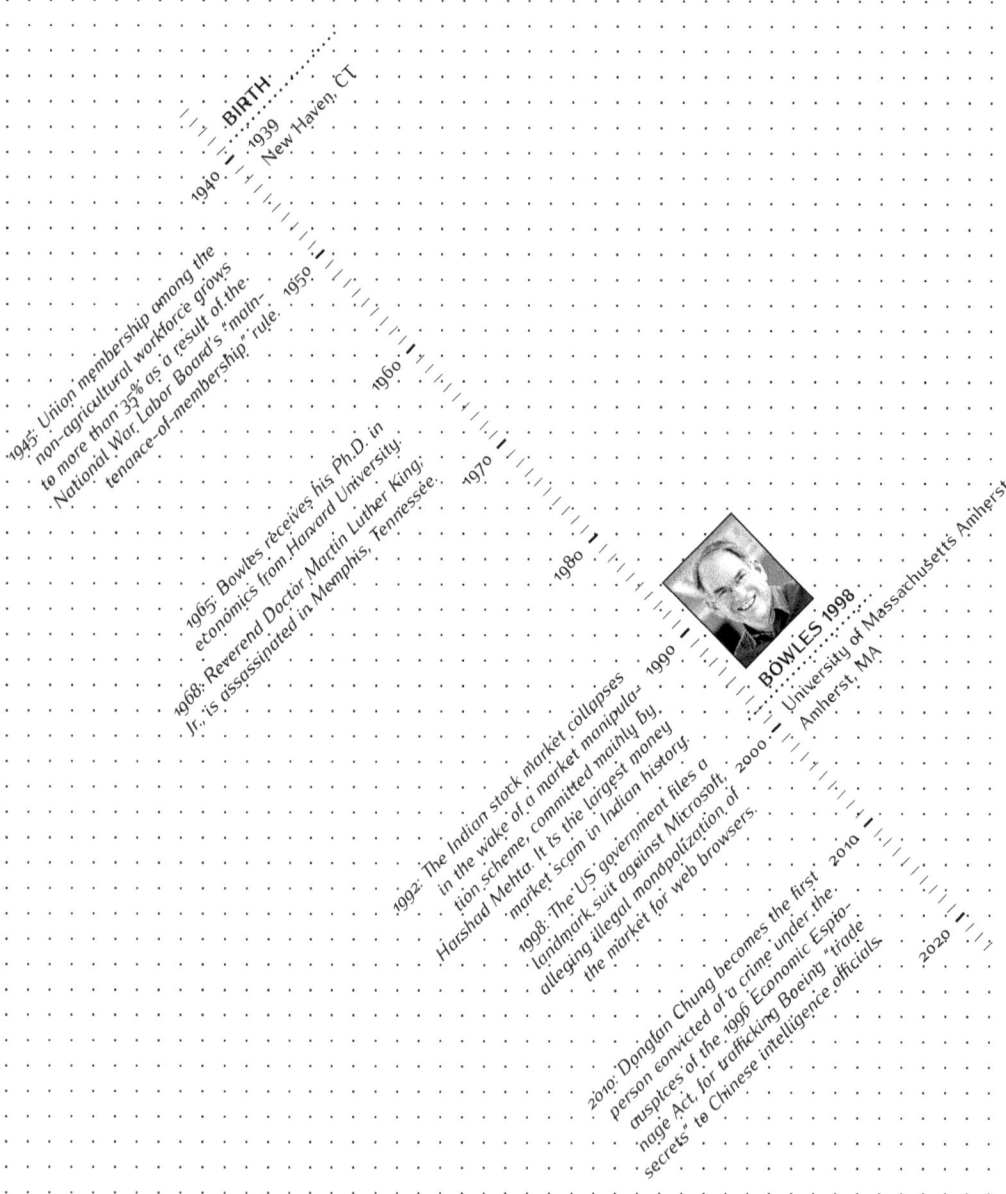

BIRTH: 1939, New Haven, CT

1945: Union membership among the non-agricultural workforce grows to more than 35% as a result of the National War Labor Board's "maintenance-of-membership" rule.

1965: Bowles receives his Ph.D. in economics from Harvard University.

1968: Reverend Doctor Martin Luther King, Jr., is assassinated in Memphis, Tennessee.

1992: The Indian stock market collapses in the wake of a market manipulation scheme, committed mainly by Harshad Mehta. It is the largest money market scam in Indian history.

1998: The US government files a landmark suit against Microsoft, alleging illegal monopolization of the market for web browsers.

BOWLES 1998, University of Massachusetts Amherst, Amherst, MA

2010: Donglan Chung becomes the first person convicted of a crime under the auspices of the 1996 Economic Espionage Act, for trafficking Boeing "trade secrets" to Chinese intelligence officials.

SAMUEL STEBBINS BOWLES

[86]

SAM BOWLES ON ENDOGENOUS PREFERENCES

Rajiv Sethi, Columbia University and Santa Fe Institute

In May 2021, the Governor of Ohio announced the launch of the state's Vax-a-Million initiative, under which any adult resident who had received at least one dose of a COVID-19 vaccine could enter a weekly lottery with a million-dollar prize. Youth aged twelve to seventeen could also enter the lottery and stood to win a full scholarship to a state school. There were five winners in each category over the course of the program, at a total cost to the state of $5.6 million. Similar initiatives were adopted in other states and in countries worldwide, often involving fixed payments to each vaccinated adult rather than lotteries.

Attempts to influence individual choices by offering financial incentives—conditional cash transfers or conditional cash lotteries—have become increasingly common. They have been used to influence a variety of behaviors and outcomes, including school attendance, academic achievement, smoking cessation, and regular exercise (Bastagli *et al.* 2019; Gneezy, Meier, and Rey-Biel 2011).

Sometimes the incentives work as anticipated, sometimes they fail to do so, and on occasion they backfire. Even in the case of Ohio's conditional cash lottery, there is some dispute about effectiveness: Barber and West (2022) and Brehm, Brehm, and Saavedra (2022) found the program to have significant benefits in boosting vaccination rates and lowering hospital admissions, while Lang, Esbenshade, and Willer (2023) found negligible and possibly even negative effects. Similarly, modest cash payments were found to be effective in raising vaccination rates in Sweden (Campos-Mercade *et al.* 2021), but a randomized controlled trial in the United States found negative effects for vaccine-hesitant individuals (Chang *et al.* 2021).

Understanding when and how such explicit incentives work, and why they sometimes fail, is now a vibrant area of research in economics.

S. Bowles, "Endogenous Preferences: The Cultural Consequences of Markets and Other Economic Institutions," *Journal of Economic Literature* 36, 75–111 (1998).

Copyright American Economic Association; reproduced with permission of the author and the Journal of Economic Literature.

But it was not always thus. Under traditional assumptions involving stable preferences and material self-interest, increasing the benefits of an activity should increase its incidence, and financial incentives should invariably work as anticipated. Empirical findings at odds with this prediction have forced economists to reconsider these basic assumptions.

Sam Bowles's 1998 paper on the cultural consequences of markets documented, shaped, and accelerated this process. This paper offered a wide-ranging and compelling case for thinking about preferences as endogenous to policy interventions, and indeed to social organization in general. Some aspects of the research agenda laid out in the paper have since been pursued vigorously, while other issues have yet to be systematically investigated.

For instance, much has been written about the crowding out of intrinsic motivation by the introduction of extrinsic rewards, following pioneering experimental findings by Gneezy and Rustichini (2000a, 2000b). Such effects can arise through psychological processes, responding to changes in framing and context. Paying a child to read may result in a loss of interest in reading for pleasure and lower the incidence of reading in later life. Similar effects can arise through informational channels, for instance by suggesting to decision makers that certain antisocial activities result in negligible harm to others. A small fine may be taken to imply that the activity is not especially harmful and result in greater incidence relative to the case of no fine at all.

In fact, once we allow for complex preferences, including a desire to signal one's own values and commitments to others, phenomena that appear to involve preference change may simply reflect changes in signaling ability. If one is paid for casting a vote in an election, the act of voting no longer signals public-spiritedness. Rewarding voting in this manner may therefore depress rather than boost turnout.

Such psychological and informational processes have received considerable attention over the past couple of decades. What has received less attention, and is in some respects the more challenging task, is the connection between social organization writ large and the preferences of individuals.

Bowles makes the point that the impersonal and ephemeral nature of market interactions changes the benefits and costs associated with the acquisition of various cultural traits. For instance, when contracts are explicit, complete, and externally enforced, there is little gain to possessing (and being seen to possess) socially valuable norms such as trust or reciprocity. These attributes may then decline in a population through a process of cultural transmission. And since these traits promote economic efficiency when transactions are not ephemeral or impersonal, or when contracts remain incomplete or implicit, such changes in preferences can be costly for a society (Ostrom 1990; Sethi and Somanathan 1996).

At the start of his paper, Bowles observes that "the axiom of exogenous preferences is as old as liberal political philosophy itself." This is true, but there also exists a literature—as old as liberal political philosophy itself—that was concerned precisely with the effects of institutional arrangements on preferences. In contrast with the focus on the erosion of socially beneficial values, however, this early work emphasized the suppression of socially harmful passions. Capitalism, it was argued, "would activate some benign human proclivities at the expense of some malignant ones… it would repress and perhaps atrophy the more destructive and disastrous components of human nature" (Hirschman 1977, 66). Montesquieu expressed this idea as follows: "commerce cures destructive prejudices . . . it polishes and softens barbarous mores" ([1748]1989, 338). Consistent with these views, later cross-cultural research by Bowles and others found that in small-scale societies of farmers, herders, and hunter-gatherers around the world, groups with more market exposure were more fair-minded in behavioral experiments (Henrich *et al.* 2001).

The fact that reasonable arguments can be made in support of two sharply opposing hypotheses illustrates the challenges of developing a coherent and testable theory of endogenous preferences. This is especially the case when we are thinking about the biggest of questions, such as the effects on norms and values of wholesale changes in social organization. But grapple with these questions we must, otherwise "the scope of economic inquiry is . . . truncated in ways which restrict its explanatory power, policy relevance, and ethical coherence (Bowles 1998, 75).

REFERENCES

Barber, A., and J. West. 2022. "Conditional Cash Lotteries Increase COVID-19 Vaccination Rates." *Journal of Health Economics* 81:102578. https://doi.org/10.1016/j.jhealeco.2021.102578.

Bastagli, F., J. Hagen-Zanker, L. Harman, V. Barca, G. Sturge, and T. Schmidt. 2019. "The Impact of Cash Transfers: A Review of the Evidence from Low- and Middle-Income Countries." *Journal of Social Policy* 48 (3): 569–594. https://doi.org/10.1017/S0047279418000715.

Bowles, S. 1998. "Endogenous Preferences: The Cultural Consequences of Markets and Other Economic Institutions." *Journal of Economic Literature* 36 (1): 75–111. https://www.jstor.org/stable/2564952.

Brehm, M. E., P. A. Brehm, and M. Saavedra. 2022. "The Ohio Vaccine Lottery and Starting Vaccination Rates." *American Journal of Health Economics* 8 (3). https://doi.org/10.1086/718512.

Campos-Mercade, P., A. N. Meier, F. H. Schneider, S. Meier, D. Pope, and E. Wengström. 2021. "Monetary Incentives Increase COVID-19 Vaccinations." *Science* 374 (6569): 879–882. https://doi.org/10.1126/science.abm0475.

Chang, T., M. Jacobson, M. Shah, R. Pramanik, and S. B. Shah. 2021. *Financial Incentives and Other Nudges Do Not Increase COVID-19 Vaccinations Among the Vaccine Hesitant*. Working Paper w29403. National Bureau of Economic Research. https://doi.org/10.3386/w29403.

Gneezy, U., S. Meier, and P. Rey-Biel. 2011. "When and Why Incentives (Don't) Work to Modify Behavior." *Journal of Economic Perspectives* 25 (4): 191–210. https://doi.org/10.1257/jep.25.4.191.

Gneezy, U., and A. Rustichini. 2000a. "A Fine Is a Price." *Journal of Legal Studies* 29 (1): 1–17.

———. 2000b. "Pay Enough or Don't Pay at All." *Quarterly Journal of Economics* 115 (3): 791–810.

Henrich, J., R. Boyd, S. Bowles, C. Camerer, E. Fehr, H. Gintis, and R. McElreath. 2001. "In Search of Homo Economicus: Behavioral Experiments in 15 Small-Scale Societies." *American Economic Review* 91 (2): 73–78.

Hirschman, A. O. 1977. *The Passions and the Interests: Political Arguments for Capitalism Before Its Triumph*. Princeton, NJ: Princeton University Press.

Lang, D., L. Esbenshade, and R. Willer. 2023. "Did Ohio's Vaccine Lottery Increase Vaccination? A Pre-Registered, Synthetic Control Study." *Journal of Experimental Political Science*, https://doi.org/10.1017/XPS.2021.32.

Montesquieu, C. [1748]1989. *The Spirit of the Laws*. Translated by A. M. Cohler, B. C. Miller, and H. S. Stone. Cambridge University Press.

Ostrom, E. 1990. *Governing the Commons: The Evolution of Institutions for Collective Action*. Cambridge, UK: Cambridge University Press.

Sethi, R., and E. Somanathan. 1996. "The Evolution of Social Norms in Common Property Resource Use." *American Economic Review* 86 (4): 766–88. http://www.jstor.org/stable/2118304.

ENDOGENOUS PREFERENCES: THE CULTURAL CONSEQUENCES OF MARKETS AND OTHER ECONOMIC INSTITUTIONS

Samuel Bowles, University of Massachusetts at Amherst

1. Hobbes' Fiction

[Let us]... return again to the state of nature, and consider men as if but even now sprung out of the earth, and suddenly (like mushrooms), come to full maturity, without any kind of engagement with each other. — Thomas Hobbes ([1651] 1949, p. 100)

If friends make gifts, gifts make friends.... Thus do primitive people transcend the Hobbesian chaos. — Marshall Sahlins (1972, p. 186)

Markets and other economic institutions do more than allocate goods and services: they also influence the evolution of values, tastes, and personalities. Economists have long assumed otherwise; the axiom of exogenous preferences is as old as liberal political philosophy itself. Thomas Hobbes' mushroom metaphor abstracts from the ways that society shapes the development of its members in favor of "taking individuals as they are." Reflecting this canon, most economists have not asked how we come to want and value the things we do.

Hobbes' fiction neatly elides the influence of social arrangements on the process of human development and thus greatly simplifies the task of economic theory. But the scope of economic inquiry is thereby truncated in ways which restrict its explanatory power, policy relevance, and ethical coherence. If preferences are affected by the policies or institutional arrangements we study, we can neither accurately predict nor coherently evaluate the likely consequences of new policies or institutions without taking account of preference endogeneity. In the pages which follow I review models and evidence concerning the impact of economic institutions on preferences, broadly construed, and comment on some implications for economic theory and policy analysis.[1]

The production and distribution of goods and services in any society is organized by a set of rules, among which are allocation by fiat in states, firms, and other organizations, patriarchal and other customary allocations based on gender, age, and kinship (as for example takes place within families), gift, theft, bargaining, and of course markets. Particular combinations of these rules give entire societies modifiers such as "capitalist," "traditional," "communist," "patriarchal," and "corporatist." These distinct allocation rules along with other institutions dictate what one must do or be to acquire one's livelihood. In so doing they impose characteristic patterns of interaction on the people who make up a society, affecting who meets whom, on what terms, to perform which tasks, and with what expectation of rewards.

One risks banality, not controversy, in suggesting that these allocation rules therefore influence the process of human development, affecting personality, habits, tastes, identities, and values. One cannot be too far out on a limb when in the company of Adam Smith as well as Edmund Burke, Alexis de Tocqueville and Karl Marx, John Stuart Mill and Frederick Hayek: all celebrated or lamented the effects of markets and other economic institutions on human development.[2] But consensus eludes any of the grand claims made concerning the nature of the effects or how they might be generated. The reason is that most writers have implicitly invoked a kind of functionalist correspondence between economic structures on the one hand and values, customs, and tastes on the other, without explaining the mechanisms by which

---

[1] I abstract from other forms of preference endogeneity such as the many variants of Harvey Leibenstein's (1950) "bandwagon" and "snob" effects or James Duesenberry's (1949) analysis of keeping up with the Jones' or Thorstein Veblen's (1934) emulation effects, or the interdependent preferences studied by Robert Pollak (1976). Rather, I here develop the research agenda suggested by Herbert Gintis' early (1971, 1972) investigations of the impact of economic institutions on preferences.

[2] In these pages and elsewhere, Albert Hirschman (1977, 1982) has catalogued early statements of the cultural effects of markets, obviating the need for more than passing mention here.

the former might affect the latter. Thoughtful works on the subject—Joseph Schumpeter (1950) on "the civilization of capitalism," Daniel Bell (1976) on "the cultural contradictions of capitalism," David Potter (1954) on the "people of plenty," Karl Polányi (1957) on "the great transformation," or Peter Laslett (1965) on "the world we have lost"—are surprisingly bereft of causal arguments.

Nonetheless, the argument that economic institutions influence motivations and values is plausible, and the amount of evidence consistent with the hypothesis is impressive. Many ethnographic and historical studies, for example, recount the impact of modern economic institutions on traditional or indigenous cultures.[3] The rapid rise of feminist values, the reduction in family size, and the transformation of sexual practices coincident with the extension of women's labor force participation likewise suggest that changes in economic organization may foster dramatic changes in value orientations.

Drawing on literatures from the other social sciences, history, and experimental economics, I have identified five effects of markets and other economic institutions on preferences. Few are supported by empirical evidence that will convince a confirmed skeptic, but most are plausible and consistent with substantial evidence.

Framing and situation construal: economic institutions are situations in the social psychological sense and thus have framing and other situation construal effects; people make different choices depending on whether the identical feasible set they face is generated by a market-like process or not (I address these issues in Section 4).

Intrinsic and extrinsic motivations: the ample scope of market choices and often extrinsic nature of market rewards may induce preference changes driven by individual desires for feelings of competence and self-determination; other institutions may have related effects (Section 5).

Effects on the evolution of norms: economic institutions influence the structure of social interactions and thus affect the evolution of norms by altering the returns to relationship-specific investments such as

---

[3] Among the more instructive not mentioned elsewhere in this essay are Kenelm Burridge (1969), Daniel Lerner (1958), Margaret J. Field (1960), T. Scarlett Epstein (1962), Michael Taussig (1980), and Jean Ensminger (1992).

reputation-building, affecting the kinds of sanctions that may be applied in interactions, and changing the likelihood of interaction for different types of people (Section 6).

Task performance effects: economic institutions structure the tasks people face and hence influence not only their capacities but their values and psychological functioning as well (Section 7).

Effects on the process of cultural transmission: in part for the above reasons, and in part independently, markets and other institutions affect the cultural learning process itself, altering the ways we acquire our values and desires, including child rearing and schooling, as well as informal learning rules such as conformism (Section 8).

Until recently, economic theory gave little guidance in understanding these effects, for it purposefully abstracted from what were considered to be the irrelevant sociological details of the exchange process. In the complete contracting world of Walrasian economics, for example, there is little reason for an economic actor to be concerned about his exchange partner's psychological makeup or moral commitments; moreover there is no way that these personal traits could be affected, if one were so concerned. Markets of this type, wrote Albert Hirschman in these pages (1982, p. 1473), are peopled by "large numbers of price taking anonymous buyers and sellers supplied with perfect information"... and "function without any prolonged human or social contact among or between the parties." Grocery markets approximate this ideal (a fact which may explain why fruit stands and fish markets figure so prominently in economics textbooks).

By contrast, now-standard microeconomic theories of labor, credit, and other markets as well as the contemporary theory of the firm treat economic interactions as personal, strategic, and durable connections among people whose identities matter for the outcomes. Aspects of social life once thought to be the province of psychology or sociology are thus seen to be essential to the explanation of the bread and butter of economics: prices and quantities. The theory of asymmetric information and incomplete contracting shows that markets may not clear in competitive equilibrium, leading to asymmetries between those on the short side of the market (able to secure all the transactions they desire) and those on the long side of the market, some of which may be unable to secure any transaction at all. An important consequence is the reappearance of complex, asymmetrically placed, opportunistic, and (especially) malleable economic actors more reminiscent of the flesh and blood dramatis personae of classical economics than the anemic and one-dimensional homo economicus of the standard textbooks.

Two aspects of exchanges with incomplete contracts account for this. First, where contracts are incompletely specified or costly to enforce, the ex post terms of an exchange may depend on the normative commitments and psychological makeup of the parties to the exchange; where the amount of work done on the job cannot be secured by a contract it will be influenced by the employee's work ethic or sense of alienation, for example.

Second, because of the durability of the exchange, one or both parties may have the capacity to structure the relationship so as to affect the preferences of their exchange partner (Bowles and Gintis 1993; Mulligan 1997). Paternalistic policies in lifetime employment firms are an example. Incomplete contracts thus provide both the motivation and the means for deliberate (as well as unwitting) preference modification in the exchange process.

Models of incomplete contracts not only dramatize the shortcomings of the exogenous preferences assumption, they also provide a basis for a more nuanced treatment of the effects of markets and other economic institutions on preferences. Walrasian grocery markets support personal interactions quite distinct from the long term relationship characteristic of a lifelong employment firm; and the differences in the structure of these exchanges appear to have effects on preferences, as we will see presently. Or, to take another example, there are significant differences in the personality effects on participants in markets which clear in equilibrium and those which do not, and in those markets which do not clear, for people on the short side of the market (whose advantageous positions may allow them to make take it or leave it offers) and those on the long side of the market, some of whom are simply excluded from the exchange process, while others fear losing the transaction they have secured. Thus the details of market structures—and in particular the ways in which social interactions are patterned—may be important.

I turn first to methodological issues. In the next section I ask what we mean by preferences and how they might be influenced by economic institutions; and in Section 3 I present a model illuminating the influence of economic institutions on the process of cultural evolution.

2. Social Interactions and the Evolution of Preferences

> *I do not know the fruit salesman personally; and I have no particular interest in his well-being. He reciprocates this attitude. I do not know, and I have no need to know whether he is in direst poverty, extremely wealthy, or somewhere in between... Yet the two of us are able to... transact exchanges efficiently because both parties agree on the property rights relevant to them.* —James Buchanan (1975, p. 17)

Preferences are reasons for behavior, that is, attributes of individuals that (along with their beliefs and capacities) account for the actions they take in a given situation. To explain why a person chose a point in a budget set, for example, one might make reference to her craving for the chosen goods, or to a religious prohibition against the excluded goods. Conceived this way, preferences go considerably beyond tastes, as an adequate account of individual actions would have to include values or what Amartya Sen (1977) terms commitments and John Harsanyi (1982) calls moral preferences (as distinct from personal preferences). Also included are the manner in which the individual construes the situation in which the choice is to be made (Lee Ross and Nisbett 1991), the way that the decision situation is framed (Amos Tversky and Kahneman 1986), compulsions, addictions, habits, and more broadly, psychological dispositions. Preferences may be strongly cognitively mediated—my enjoying ice cream may depend critically on my belief that ice cream does not make me fat—or they may be visceral reactions—like disgust or fear—evoking strong emotions but having only the most minimal cognitive aspects (Robert B. Zajonc 1980; David Laibson 1996; Loewenstein 1996; Rozin and Carol Nemeroff 1990).

The term "preferences" for these heterogeneous reasons for behavior is perhaps too narrow, and runs the risk of falsely suggesting that a single model of action is sufficient; Patrick H. Nowell-Smith's

(1954) "pro and con attitudes" or "reasons for choosing" are more descriptive, but unwieldy.[4]

For preferences to have explanatory power they must be sufficiently persistent to explain behaviors over time and across situations.[5] If preferences are endogenous with respect to economic institutions it will be important to distinguish between the effects of the incentives and constraints of an institutional setup (along with given preferences) on behaviors, and the effect of the institution on preferences per se. The key distinction is that where preferences (and not just behaviors) are endogenous they will have explanatory power in situations distinct from the institutional environments which account for their adoption. Thus, however acquired, preferences must be internalized, taking on the status of general motives or constraints on behavior. Values which become durable attributes of individuals—for example, the sense of one's own efficacy introduced below—may explain behaviors in novel situations, and hence are included in this broad concept of preferences.

We acquire preferences through genetic inheritance and learning. Very long lasting economic institutions, such as the social structures of the simple societies which predominated in the first 100,000 years (90 percent) of biologically modern human existence, could substantially affect gene distributions in a population and hence could provide part of a genetic explanation of preferences (Christopher Boehm 1993; Linnda Caporael et al. 1989; Feldman and Kevin Laland 1996; William Durham 1991). Nonetheless it seems likely that the more important effects of economic organization on preferences operate through cultural transmission, that is, learning. Drawing on the extensive literature on food tastes, Clark McCauley, Rozin, and Barry Schwartz (1994, p. 27) write:

---

[4]In order to account for an individual's actions preferences need not coincide with the reasons given by the particular individual, of course. Nor do preferences alone generally give a sufficient account of behaviors: my consumption of aspirin is accounted for by my aversion to pain plus my belief that aspirin will relieve the pain and that this little white object is indeed an aspirin, and so on.

[5]Benjamin Bloom (1964) documents stability over time of a range of measured personality traits. For particular psychological dimensions introduced below see Herman Witkin and John Berry (1975 p. 41; field independence), Paul Andrisani and Gilbert Nestel (1976 p. 161; internal-external locus of control), and Kohn and Carmi Schooler (1983 1983 p. 147; self-directedness).Ross and Nisbett (1991) provide a critical review of the evidence for intertemporal and cross situational consistency of behavior.

> *The human being comes into the world with certain likes and dislikes, such as innate dislike of pain, bitter tastes, and many types of strong stimulation, and an innate liking for certain types of touch or sweet tastes... Almost the entire adult ensemble of likes and dislikes is acquired, presumably in the process of enculturation.*

For this reason I will treat preferences as cultural traits, or learned influences on behavior: liking ice cream, or never lying, or reciprocating dinner invitations are cultural traits.[6]

We know surprisingly little about how we come to have the preferences we do; the theory of cultural evolution is thus similar to the theory of natural selection prior to its integration with Mendelian genetics. While it is comforting to recall that Darwin's contribution was possible even though he did not know how traits are passed on, this lacuna is nonetheless a major impediment to endogenizing preferences. We know that intentional motivations are sometimes involved; one learns to appreciate classical music because one notices that aficionados appear to enjoy it (Gintis 1972; Bowles and Gintis 1986; Gary Becker 1996). But instrumental motivations may be of limited importance compared to other influences such as mere exposure (Zajonc 1968), the unintended consequences of activities motivated by other ends (such as migration to a different culture in search of work), and conformism (Solomon Asch 1952; Muzafer Sherif 1937; Theodore Newcomb et al 1967). While the individual benefits accruing to those exhibiting particular cultural traits may affect these learning processes and hence the rate at which the traits are replicated, most preferences are not chosen in the usual sense of intentional action toward given ends. Rather, preferences are learned as an accent or a taste for a national cuisine is acquired, that is, by processes which may but need not be intentional.

This has the important implication that a convincing theory of preferences has to involve departures from standard economic methodology; preferences cannot simply be explained as equilibrium outcomes arising from choices by optimizing agents.

---

[6]The pioneering works in the formal theory of cultural evolution are Luigi Cavalli-Sforza and Feldman (1981) and Boyd and Peter Richerson 1985. Robert LeVine (1973) earlier developed what he termed a Darwinian "variation-selection model to culture-personality relations." Sarah Otto, Freddy Christiansen, and Feldman (1995), Robert Plomin and his collaborators (see Plomin and Denise Daniels 1987) as well as Rozin (1991). Rozin and T. A. Vollmecke (1986) provide evidence that food tastes, psychological functioning, and other traits are far from exhaustively determined by genetic inheritance.

However acquired, preferences are internalized: there is considerable evidence that preferences learned under one set of circumstances become generalized reasons for behavior. Thus economic institutions may induce specific behaviors—self-regarding, opportunistic, or cooperative, say—which then become part of the behavioral repertoire of the individual. The effects of mere exposure just mentioned provide a particularly transparent example: "likes" or habits initially induced by exposure or repetition become permanent reasons for behavior.

Learning by doing is another mechanism for the generalization of preferences: behaviors found successful in coping with the tasks defined by one sphere of life are generalized to other realms of life. Paul Breer and Edwin Locke (1965, p. 253) present substantial experimental evidence to this effect. They asked subjects to perform different sets of tasks and investigated changes in apparently unrelated values:

> *In a period of less than four hours and without a single verbal reference to family, fraternity, way of life, or any of the other areas measured, we succeeded in changing a wide variety of attitudes ranging from specific beliefs about the most effective way to organize a work group, to abstract values concerning the individual and society. This evidence was taken to mean that task experience is capable of exerting a very powerful influence on all sorts of beliefs, values, and preferences which, to the casual observer, appear to be only remotely related to the task itself.*

Finally, preferences may become generalized through a process which Leon Festinger (1957, p. 260) termed dissonance reduction:

> *the human organism tries to establish internal harmony, consistency or congruity among his opinions, attitudes, knowledge, and values. . . . there is a drive toward consonance among cognitions.*

The cognitive elements in dissonance could be one's values and a behavior, as when one is doing something which is inconsistent with one's values. Festinger (1957, pp. 271–73) frequently used this reasoning to explain "specific ideological changes or opinion changes subsequent to the change in a person's way of life" such as a:

> *sudden change in the job which a person does. A worker in a factory, for example may be promoted to the job of foreman. Suddenly he finds himself giving orders instead of receiving*

> them... these new actions will be dissonant in many instances with opinions and values which he acquired as a worker and still holds. In pursuit of dissonance reduction, one would expect this person to quite rapidly accept the opinions and values of other foreman, that is, opinions and values which are consonant with the things he now does.

Dissonance reduction thus provides another explanation for how economic circumstances may induce new preferences, and why the new preferences might become general reasons for behavior.

In contrast to the social interactions based approach taken here, many would emphasize the role of religious or political indoctrination or advertising in preference change. These intentional forms of inculcation are undoubtedly important, but where empirical studies are available, other influences appear as powerful if not more.[7] If I am right that acquiring preferences is akin to acquiring an accent, studies of language change may be illuminating. On the basis of intensive empirical study of linguistic change in Philadelphia, for example, William Labov concluded that

> linguistic traits are not transmitted across group boundaries simply by exposure in the mass media or in schools.... Our basic language system is not acquired from school teachers or from radio announcers, but from friends and competitors: those who we admire, and those who we have to be good enough to beat. (Labov 1983, p. 23)

The inference is not that institutions such as schools and churches are unimportant, but that understanding their role in the acquisition of cultural traits may be enhanced by seeing them—along with markets, firms, families, and governments—as distinct patterns of social

---

[7]Studies of preferences for brands of food, soap, movies, and other consumption items for which one would expect an important advertising effect indicate that personal contact is considerably more important than advertising in motivating brand changes (Elihu Katz and Paul Lazarsfeld 1955). Everett Rogers' (1962) classic study of diffusion of innovations found personal communication to be of substantial importance in the diffusion of both ideas and practices such as cooking methods. Some behavioral changes may be induced simply by providing information; in these cases media exposure appears to be effective. But where information alone is insufficient (changes in smoking behavior, e.g.) face to face contact appears to be more effective (June Flora, Nathan Maccoby, and John Farquhar 1989).

interaction affecting the diffusion of cultural traits in a population in ways often unrecognized by any of the participants.

3. Economic Institutions and Cultural Evolution

> *[The 17th century Salem "witches" and their defenders were] a group of people who were on the advancing edge of profound historical change. If from one angle they were diverging from an accepted norm of behavior, from another angle their values represented the "norm" of the future. In an age about to pass, the assertion of private will posed the direst possible threat to the stability of the community; in the age about to arrive it would form a central pillar on which that stability rested.* —Paul Boyer and Stephen Nissenbaum (1974, p. 109)

How might allocation rules affect the differential replication of cultural traits? The gist of an answer is given best by a concrete example. Erich Fromm and Michael Maccoby's (1970, p. 232) study of social character in a Mexican village led them to this conclusion:

> *In a relatively stable society (or class) with its typical social character there will always be deviant characters who are unsuccessful or even misfits under the traditional conditions. However in the process of socioeconomic change, new economic trends develop for which the traditional character is not well adapted, while a certain heretofore deviant character type can make optimal use of the new conditions. As a result the "ex deviants" become the most successful individuals or leaders of their society or class. They acquire the power to change laws, educational systems, and institutions in a way that . . . influences the character development of succeeding generations. . . . deviant and secondary trait personalities never fully disappear and hence . . . social changes always find the individuals and groups that can serve as the core for the new social order.*

The traits of the "ex deviants" in this example enjoy heightened replication propensities both directly (others may want to emulate the successful) and indirectly, because bearers of the traits become privileged cultural models, such as teachers.

The Fromm–Maccoby view is supported by the field research of LeVine in Nigeria. Using David McClelland's (1961) measures of achievement motivation and other value orientations, LeVine (1966) found significant differences among distinct cultural groups which were not explicable by religious or educational influences. However, the pattern of personality differences were consistent with the hypothesis that distinct motivations had evolved as adaptations to longstanding geographically determined differences in structures of competitive economic opportunity among the various cultural groups. Moreover as market-based and other competitive systems of advancement became more generalized during the course of the twentieth century, those exhibiting high levels of achievement orientation—some of them presumably "deviants" in the premodern compliance-based rather than achievement-oriented subcultures—gained positions of educational leadership, thus assuming roles as privileged cultural models.

Achievement orientation is one of many possible traits that may proliferate as market societies spread. Montesquieu, writing in 1748, argued that "commerce cures destructive prejudices . . . it polishes and softens barbarous mores."

These studies of Mexico and Nigeria suggest that new economic arrangements might affect cultural evolution in two ways: either by influencing the economic well-being of those exhibiting distinct traits or by altering the learning rules which make up the process of cultural transmission itself. The cultural transmission process translates economic well-being, exposure to role models, and other influences into replication of traits, and thus intervenes between payoffs and replication.

Evolutionary game theoretic models typically abstract entirely from the process of cultural transmission, representing payoffs associated with particular traits as if they were the only influences on the replication of traits.[8] By contrast, models of cultural evolution typically address what is known about the particulars of the process by which traits are acquired, distinguishing between vertical transmission from parents, oblique transmission from non-parental members of the previous generation (for example, teachers), and horizontal

---

[8]One could interpret the payoffs in evolutionary game theoretic models as the replication propensities themselves, but while thus formally accommodating analysis of the process of cultural transmission, this would add no insight to the distinct influences of the transmission process per se.

transmission from one's own cohort (as in the case of language change or fashion; Penelope Eckert 1988, 1982; Labov 1972, p. 304).[9] These models, as well as the studies of Mexico and Nigeria above, make it clear that a trait which is advantaged in the transmission process may diffuse in a population even if the economic benefits associated with the trait are inferior to the population average. Thus the effects of economic institutions on both payoffs to distinct traits and the cultural transmission process must be studied.

A particularly important example of how a trait may be advantaged in the transmission process is termed conformist transmission: the prevalence of a trait in a population may enhance the replication propensity of each representative of that trait, independently of the payoff to those exhibiting the trait. Under quite general conditions where learning is costly, conformist transmission may be efficient in the sense that an individual who sometimes adopts traits by simply copying what others are doing rather than on the basis of the payoffs associated with various actions will do better than those who always engage in costly investigation of the relevant payoffs (Boyd and Richerson 1985; Feldman, Kenichi Aoki, and Jochen Kumm 1996. Conformist transmission of preferences thus might have evolved under the influence of either genetic or cultural inheritance. Frequency dependent replication may also arise where groups that are numerically preponderant are disproportionately likely to occupy privileged roles as teachers or other cultural models. Persistent ethnic differences in food tastes, coupled with very low vertical (parental) transmission of these tastes (Rozin 1991) is a piece of evidence suggesting the importance of conformist transmission. Another, from empirical cross cultural psychological studies, is the importance of membership in particular tribes as a predictor of values orientations, independently of sources of livelihood, ecology, and other possible influences (Robert Edgerton 1971).

---

[9]Empirical studies in this tradition include Kuang-Ho Chen, Cavalli-Sforza, and Feldman (1982) and Cavalli-Sforza et al. (1982); Boyd and Richerson (1985) survey many empirical studies of these three transmission processes. Alberto Bisin and Thierry Verdier (1996) present a model of preference evolution integrating the cultural evolution and evolutionary game theory approaches.

The relationship between payoff-based and conformism or other frequency dependent influences on the replication of cultural traits and the ways that these may be influenced by economic institutions may be illustrated by means of a simple model based on Bowles (1996). The basic intuition is that the distribution of cultural traits in a population is determined as the equilibrium of a system whose exogenous elements are subject to long-term influence of markets and other economic institutions. Economic institutions affect the evolution of preferences by changing these exogenous determinants of the cultural equilibrium.

Suppose x and y are mutually exclusive cultural traits. Each member of a large population is a "cultural model" with replication propensities, r_x or r_y, defined as the number of copies of each model made at end of each period, possibly a generation. Agents implement the strategy dictated by their trait in a game which assigns benefits to each, following which the traits are replicated through an updating process described below, generating a new population frequency (one may think of the population as composed of single parents each with a single child, who in the process of growing up may or may not adopt the traits of the parent). *Cultural equilibrium* is defined as a frequency of traits which is stationary.

Members of the population are paired to play a two-person game, the payoffs of which are denoted $\pi(i,j)$, the payoff to playing trait i against a j-playing partner. The "game" may be one of the familiar interactions of the hawk–dove, prisoners' dilemma, or coordination game type. It might refer to an interaction as everyday as eating a meal together, or meeting in public, where the two traits might dictate matters of style ("wearing a tie") or taste ("wanting a drink"). Or it could refer to an exchange of goods or some more conventionally economic interaction. The payoff structure could be degenerate in the sense that my enjoyment of a beer may not depend at all on what you are eating or drinking; but the model is designed to address more interesting cases of emulation, social dilemmas and the like.

For any population frequency of the x trait, $p \in [0,1]$ let $\mu_{ij} = \mu_{ij}(p;\delta)$ be the probability of being paired with a j type conditional on being an i type, where $\delta \in [0,1]$ is a measure of the exogenously determined extent to which pairing is nonrandom.[10] If pairing is random $\delta = 0$ and the probability of meeting an x type is simply p, irrespective of one's own type: $\mu_x = \mu_{yx} = p$. But where residence is correlated with type, or where sorting by type takes place by means of social networks or other groups, the probability of meeting one's own type may exceed that given by the population frequency. The expected payoffs are thus

$$b_x(p;\delta) = \mu_{xx}\pi(x,x) + \mu_{xy}\pi(x,y)$$
$$b_y(p;\delta) = \mu_{yx}\pi(y,x) + \mu_{yy}\pi(y,y). \quad (1)$$

To take account of frequency dependent biases in cultural transmission suppose that conformist transmission is described by the conformist bias function $\sigma(p)$, which we write as $\sigma_x(p-k)$ and $\sigma_y(k-p)$ where for simplicity $\sigma_y = \sigma_x \equiv \sigma > 0$ and $k \in [0,1]$ is the value of p for which no bias operates. Further we define $\alpha \in [0,1]$, the degree of conformism, as the weight placed on $\sigma(p)$ as opposed to $b(p;\delta)$ in the transmission process. Thus we have the replication propensities:

$$r_x = \alpha\sigma(p-k) + (1-\alpha)(b_x(p;\delta) - b_y(p;\delta)) + 1$$
$$r_y = \alpha\sigma(k-p) + (1-\alpha)(b_y(p;\delta) - b_x(p;\delta)) + 1. \quad (2)$$

Where $p = k$, or if $\alpha = 0$ conformist transmission does not operate so replication depends solely on payoffs, as in conventional evolutionary game theoretic models. Equilibrium is defined by $dp/dt = 0$, which for $p \in (0,1)$ requires that the effects of conformist transmission offset the effects of unequal game outcomes so that $r_x = r_y$, or

$$\alpha\sigma(p-k)/(1-\alpha) = b_y(p;\delta) - b_x(p;\delta) \quad (3)$$

from which it can be seen that cultural equilibrium does not require equal payoffs. Figure 1 illustrates this equilibrium condition for the case of an interior stable equilibrium. This equilibrium can be seen to be stable because for $p > p^*$ the payoff advantage of the y trait (the righthand side of (3)) more than offsets its disadvantage due to

---

[10] The explicit relationship between the μ's and p is this: define δ as the degree of segmentation of the population, then for $p \in (0,1), \mu_{xx} = \delta + (1-\delta)p$ and $\mu_{xy} = (1-\delta)(1-p)$, from which it is clear that δ is a non-genetic analogue to the "degree of relatedness" in biological models (W. D. Hamilton 1975; Alan Grafen 1979).

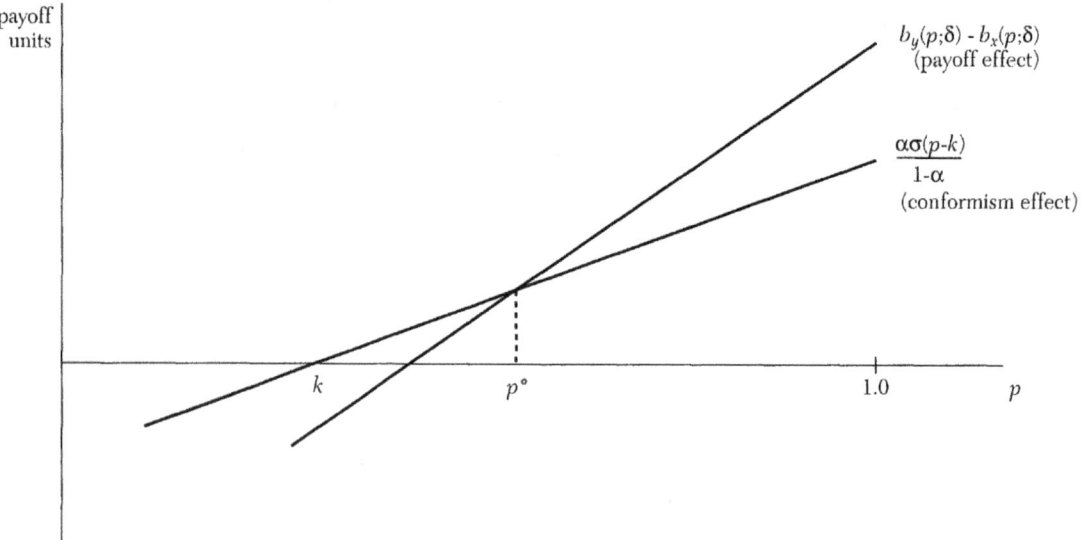

Figure 1. Cultural Equilibrium

conformist transmission (the lefthand side), as a result of which $r_y > r_x$ giving $dp/dt < 0$. So disturbances of p will be self correcting.

Markets and other economic institutions will affect the distribution of cultural traits in the population because they influence the determination of the exogenous variables in the above model:

i) the rules governing who interacts with whom (as indicated by the functions $\mu_{ij}(p, \delta)$ measuring the degree of segmentation into distinct social networks and other sources of nonrandom pairing);

ii) the payoffs $\pi(i, j)$ to any given interaction (determined by the frequency of interaction, ease of recognition of types, for example);

iii) the structure of the transmission process itself (in this case the nature and strength of conformism, σ, k, and α, including the assignment of some types as compulsory or otherwise advantaged models, such as teachers).

In more complex models, allowing for movement among population groups, cultural equilibria are influenced by migratory flows, which in turn are subject to the influence of economic institutions (Bowles and Gintis 1998).

To say that economic institutions have these effects is, of course, to compare markets, say, with some other allocation rule or to compare various types of markets. Allocation rules are differing mechanisms for coordinating the transfer of goods and services. Economists tend to focus on the relationships thereby established among the objects of exchange, relative prices, for example. But allocation rules also establish relationships among people, based on assignment to distinct positions with corresponding rights, status and obligations and patterns of interaction. Thus markets support interpersonal experiences distinct from other allocation rules. Robert Lane (1991), whose *The Market Experience* must be the starting point for any consideration of the psychology of markets, writes:

In spite of the variety of markets over time and across cultures, I believe that it is possible to conceive of a market experience that is typical, frequent, and paradigmatic for those who do market work for pay, use money and buy—rather than make, inherit or receive from government—the commodities with which they adorn their lives. (p. 4)
. . . Inevitably the market shapes how humans flourish, the development of their existences, their minds, and their dignity. (p. 17)

What is psychologically distinctive about markets as opposed to other allocation mechanisms? Max Weber ([1922]1978, p. 636) wrote "A market may be said to exist wherever there is competition for opportunities of exchange among a plurality of potential parties."[11] Markets structure social interactions "each of which is specifically ephemeral insofar as it ceases to exist with the act of exchanging the goods." As a result, according to Weber,

The market community as such is the most impersonal relationship of practical life into which humans can enter

---

[11] Georg Simmel (1990, p. 297) writes in similar vein that "money . . . is conducive to the removal of the personal element from human relationships through its indifferent and objective nature." Talcott Parsons (1949, p. 688) describes markets similarly: "When a man walks into a store in a strange city to make a purchase, his only relevant relation to the clerk behind the counter concerns matters of kind of good, price, etc. All other facts about both persons may be disregarded. Above all it is not necessary even to know whether the two have any further interests in common beyond the immediate transaction."

	ANONYMOUS	PERSONAL
EPHEMERAL	Ideal Markets	Ascriptive Markets
DURABLE	Bureaucracies	Communities

Figure 2. Allocation Rules as Learning Environments

with one another. This is not due to the potentiality of struggle among the interested parties which is inherent in the market relationship.... The reason for the impersonality of the market is its matter-of-factness, its orientation to the commodity and only to that.

In Weber's view, then, markets—at least ideally—are characterized by impersonality, ephemerality of contact, and ease of entry and exit.[12] This might be termed the economics textbook conception of competitive markets.

Contrast these arrangements with what Parsons (1967a, p. 507) calls the "two principal competitors" of the market: "requisitioning through the direct application of political power" and "non-political solidarities and communities." These allocation rules contrast with markets in at least one of the characteristics—impersonality and ephemerality—stressed by Weber. Centralized bureaucratic allocations are in some respects as impersonal as markets—at least ideally—but membership in the group defining the allocation is generally given, entry and exit costs are high (often involving a change in citizenship or at least residence), and contacts are far from ephemeral. In contrast to both bureaucratic and market allocation, kin-like directly interacting communities with stable membership exhibit neither ephemerality nor impersonality in their characteristic rules governing allocation.[13] Figure 2 presents these three ideal types along with a fourth— ephemeral and personal social interaction—which I have termed

---

[12]On the psychological dimensions of distinct economic arrangements, see Alan Page Fiske (1991, 1992). Among economists, Yoram Ben-Porath (1980, p. 4) provides the fullest description of the idealized interactions among market agents as "impersonal." See also Zelizer (1996).

[13]I have in mind allocation systems of the type described by Emile Durkheim's "organic solidarity" ([1933]1964) or Sahlins' (1972) "generalized reciprocity," or Ferdinand Tonnies' ([1887]1963) Gemeinschaft, or William Ouchi's "clans" (1980).

ascriptively ordered markets. Racially segmented spot labor markets are an example, as they are personal (the racial identities of the participants matter) but the contact among participants is not ongoing.

The contrast between personalized non-market transactions and the putative impersonality of market exchange is, of course, a matter of degree, particularly in markets characterized by the asymmetric information, incomplete contracts and hence importance of trust, ongoing interaction, and shared understandings of the type analyzed by the theory of social exchange.[14] Impersonality of contact and permeability of boundaries, while characteristic of all markets by comparison to other allocation rules, describe some markets more aptly than others. Thus in assessing the cultural effects of markets it will be necessary to distinguish not only between markets and other allocation rules such as bureaucracy and community, but among differing types of markets as well.

I turn now to evidence and reasoning concerning the five effects of economic institutions on preferences previewed in the introduction.

4. Markets, Situations, and Framing

> *Money is one of the shatteringly simplifying ideas of all time, and like any other new and compelling idea, it creates its own revolution.... [The] Tiv [of Nigeria] have tried to categorize money with other imported goods . . . to be ranked morally below subsistence. They have, of course, not been successful in so doing.* —Paul Bohannan (1959, pp. 500, 503)

Markets frame choices; we will see that a choice problem presented in a market environment may induce behaviors different from the identical problem in framed in a non-market way. Consider an example of market framing: paying a tax and receiving a governmental service differs relevantly from buying the identical service on a market. In the first case one may—as a citizen—feel entitled to the service (irrespective of the taxes paid) and may be unlikely to compare the value of the service to that of other goods and services whether traded or not; in the second

---

[14] Kenneth Arrow (1971), George Akerlof (1984) Heinz Hollander (1990), Rachel Kranton (1996), and Kreps (1990), have recently formalized some of Peter Blau's (1964) and George Homans' (1958) early reasoning concerning social exchange.

case one may feel that the good is acquired by dint of one's talents as an income earner, and may readily compare its price with other traded goods and services, the value of one's own labor time, and the like. This framing effect may thus be part of an account of why a particular action was taken; if different institutions induce different choices from an identical choice set institutions may affect preferences. This is because choices made under the influence of institutionally determined framing may later be repeated even in the absence of the framing effect if the effects of exposure to the object of choice, or dissonance reduction effects are strong; however, I am aware of no evidence to this effect.

Markets thus affect behavior in ways not fully captured by the fact that market-determined prices and endowments define the budget set: markets provide presumptive reasons why people possess the goods they do, and they prompt some comparisons while inhibiting others. I call these the construal effects of markets, borrowing the term from social psychology in preference to the familiar but narrower concept of framing effects. The construal effects of markets arise in large part because people appear to have what might be termed relational preferences: the terms on which they are willing to transact depends both on the perceived relationships among the exchanging parties, and on related concepts of fairness. Markets affect both.

There is considerable experimental evidence consistent with the importance of the construal effects of markets. Experimental markets and bargaining environments consistently yield discrepant results, with markets quickly converging to the competitive equilibria implied by self-regarding preferences, and bargaining games often yielding evidence consistent with other-regarding or relational preferences.[15] An example of the later is the comparative study of bargaining and market behavior in four cultures by Alvin Roth *et al.* (1991).

In their study both market and bargaining experiments were designed to have distributionally extreme equilibria, one player receiving all of the benefits. The market experiments quickly converged to this equilibrium in all four cultures. By contrast, proposers in

---

[15] Camerer and Richard Thaler (1995) is a recent survey of results for ultimatum and dictator experiments. Alternating offer bargaining experiments yield similar results (Jack Ochs and Roth 1989).

the ultimatum game (the bargaining situation) made much higher than equilibrium offers; and substantial positive offers were frequently rejected.[16] Positive offers are also common in dictator games. These results are consistent with a large body of experimental evidence by others, beginning with Werner Güth, Rolf Schmittenberger, and Bernd Schwarze (1982).

A considerable body of research has sought to explain this now well established aspect of ultimatum and dictator game play, with interest centering on the question (unrelated to my concern here) of whether the unexpected results were motivated by intrinsically generous preferences on the part of the proposers. But the subsequent experiments, particularly those by Elizabeth Hoffman *et al.* (1994), have yielded considerable insight on the construal effects of markets. Hoffman and her collaborators varied two aspects of the experimental environment for ultimatum and dictator games: proposers either won their position by doing well on a trivia quiz or were randomly assigned, and their relationship to their game partner was described either as an "exchange" (with prices elicited by the experimenter) or simply as "divide $10." The combined "earned status" plus "exchange" experimental condition approximates a market arrangement in that competitive success is not simply a matter of chance but is based on apparent accomplishment; and the exchange framing of the game structure is transparently more market-like than the "dividing the pie" framework.[17] Despite the fact that the experimental situation was otherwise identical, the two market like protocols yielded significantly smaller offers in both the ultimatum game and the dictator game.

---

[16] A high offer in an ultimatum game may not indicate generosity toward the other player, but rather the anticipation that low offers will be rejected. But the rejection of a positive offer is other regarding (perhaps motivated by spite or a commitment to reciprocal fairness) so I represent high offers by the proposer as evidence of other regarding behavior (either that of the proposer or the proposer's anticipation of the responder).

[17] If, as I have suggested, markets enhance the perception that one's possessions are acquired by merit rather than by chance, the extent of endowment effects—the unwillingness to part with goods in one's possession for prices considerably higher than the maximum price one would have paid to acquire them—may be subject to market effects. Experimental subjects reported by Loewenstein and Samuel Issacharoff (1994) were particularly unwilling to part with objects in their possession if they came to possess them as a result of winning an inconsequential contest. Thus market-acquired goods may be more subject to endowment effects than goods acquired as gifts or public transfers.

In other experiments, market-like anonymity generates behaviors differing from those induced by more personal settings. Communication or other conditions contributing to group identity or a reduction in social distance among experimental subjects increases contributions in public goods games (John Ledyard 1995; Robyn Dawes, Alphons van de Kragt, and John Orbell 1988) and induces cooperative play in prisoners' dilemma interactions (Peter Kollock 1998). Simon Gaechter and Fehr (1997) find that in a public goods interaction, even quite minimal (experimentally induced) social familiarity among subjects enhances the impact of social approval incentives (implemented by ex post revelation of subjects' identities and contributions); when familiarity and the public revelation of one's contributions is combined, a significant increase in participation results.

On the basis of experimental evidence from dictator and ultimatum games, Schotter, Avi Weiss, and Inigo Zapater (1996, p. 38) offer this conclusion:

The morality of economic agents embedded in a market context may . . . be quite different from their morality in isolation. While we are not claiming that people change their nature when they function in markets, it may be that the competition inherent in markets and the need to survive offers justifications for actions that in isolation would be unjustifiable.

A striking example illustrating this suggestion is found in Catherine Andre and Platteau's (1997, p. 32) study of the impact of the land market in Rwanda:

customary obligations attached to lineage lands, in particular obligations to redistribute land in favor of landscarce kith and kin, cease to apply when the lands are acquired through a purchase instead of being handed down within the lineage.

The experimental results might be summarized by saying that the more the experimental situation approximates a competitive (and complete contracts) market with many anonymous buyers and sellers, the less other-regarding behavior will be observed.[18] Because the above market and non-market experiments were conducted with the same subject pools, these

---

[18] On the differing outcomes of anonymous and face to face bargaining see Roth (1995) and Roy Radner and Schotter (1989).

results are consistent with the view that market-like situations induce self-regarding behavior, not by making people intrinsically selfish, but by evoking the self-regarding behaviors in their preference repertoires. Thus, the hypothesis that market situations induce self-regarding behavior does not imply that those living in non-market societies would be intrinsically less self-regarding.[19]

Where competitive markets approximate the law of the single price and where the extent of markets is such that few things do not have a price, markets have further construal effects: they facilitate comparison among disparate objects. The appropriate comparison is to settings (in families or in so-called "primitive exchange") in which goods may be transacted at vastly different exchange ratios depending on the social relationships among the parties to the exchange.[20] Reporting on a pre-market society in southeastern New Guinea, Raymond Firth (1958, p. 69) writes: "There is ... no final measure of the value of individual things, and no common medium whereby every type of good and service can be translated into terms of every other." Well working markets, by contrast, favor thinking of goods both abstractly (bananas in general, not this particular banana) and comparatively (objects seen as representing more or less market value, divorced from their particular uses or properties). Markets are thus powerful cognitive simplifiers, allowing radical reductions in the complexity with which one typically views an assortment of disparate goods.

A dramatic example is provided Bohannan's study (1959) of the extension of markets in an African subsistence economy, that of the Tiv in Nigeria.

> The most distinctive feature about the economy of the Tiv—and it is a feature they share with many, perhaps most, of the pre-monetary peoples—is what can be called a multicentric

What remains relatively mysterious, even today, is precisely how and why certain elements from a preference repertoire are selected in response to framing.

---

[19] In fact high ("other regarding") offers in the ultimatum games of Roth et al. were made in what would appear to be the most and least market-like societies in the sample—U.S. and the former Yugoslavia (by contrast to Israel and Japan). In the public goods experiments by Steven Kachelmeier and Mohamed Shehata (1997) subjects in Beijing acted no different than the Canadian subjects under conditions of anonymity but proved significantly less self regarding when the identity of the players was public knowledge.

[20] Sahlins' (1972) theory of primitive exchange is distinct from market exchange precisely in this deviation from the law of the single price.

> *economy ... in which a society's exchangeable goods fall into two or more mutually exclusive spheres, each marked by different institutionalization and different moral values.* (p. 492)

Among the Tiv, domestic goods, women, and prestige goods were all exchanged, and in the latter there was a monetary equivalent (brass rods) but "no one, save in the depths of extremity, ever paid brass rods for domestic goods" (p. 493), while "rights in women had no equivalent or 'price' in brass rods or in any other item save, of course, identical rights in another woman.... Exchanges within a category ... excite no moral judgements. Exchanges between categories, however, do excite a moral reaction" (p. 496). The extension of generalized markets, and with them money, eroded these arrangements:

> *General purpose money provides a common denominator among all the spheres, thus making the commodities within each expressible in terms of a single standard and hence immediately exchangeable.* (p. 500)

Among the Tiv, the set of permissible exchanges has expanded with the advent of markets and basic notions of what it means to have a well-ordered life have changed.

5. Markets and Motivation

> *In the realm of ends everything has either a price or dignity. Whatever has a price can be replaced by something else which is equivalent; whatever is above all price, and therefore has no equivalent, has dignity.* —Immanuel Kant ([1785]1949, p. 182)

This is perhaps not entirely true, especially in the information age, since summary statistics pertaining to past transactions are routinely recorded and published (customer ratings are an example).

The reward structures of markets may affect motivation independently of framing effects. The impersonality and ephemerality of contact which characterizes markets (by contrast to other allocation rules) imply that a market "transaction entails a full quid pro quo (with) no left-over business or outstanding balance" (Ben-Porath 1980, p. 4). By default, then, the incentives relevant to activities governed by markets frequently center on the quid pro quo and take the form of what social psychologists term extrinsic rather than intrinsic rewards or sanctions, namely rewards unrelated to the activities being motivated. On the basis of dozens of experiments by social psychologists over the past 30 years one may conclude that the salience of extrinsic reward in market activities will have effects on preferences.

A series of well-designed experiments show that the degree to which an activity is liked may be reduced by inducing subjects to engage in the activity as a means toward an extrinsic goal, such as being paid. The nature of the extrinsic reward is unimportant as long as it is clearly a quid pro quo. Mark Lepper, David Greene, and Nisbett (1973, p. 130) write "Contracting explicitly to engage in an activity for a reward (will) undermine interest in the activity, even when the reward is insubstantial or merely symbolic." Correspondingly when people are induced to engage in an activity with little or no extrinsic reward, they come to value the activity more highly, that is, they come to believe that their actions were intrinsically motivated (Lepper and Greene 1978; and Edward Deci and Richard Ryan 1985; Deci 1975).

Similar changes in evaluations induced by extrinsic rewards have been shown to affect subsequent behavior in non-experimental situations. Frey and Felix Oberholzer-Gee (1997) found that proposing financial compensation reduced Swiss citizens' willingness to host a nuclear waste facility. Richard Titmuss' (1971) claim that eliciting blood donations by monetary incentives had perverse effects on preferences lacked compelling evidence, as was pointed out by Arrow (1972) and Christopher Bliss (1972). However a field experiment by William Upton (1973) partially supports Titmuss' suggestion. Among 1,261 prospective blood donors in Kansas City and Denver, some were offered financial inducements, others not. Among those initially exhibiting strong motivations to contribute blood (as indicated by past donations), those offered financial inducements were substantially less likely to actually donate blood than those offered no financial reward. Among those expressing low intrinsic motivation, however, the prospect of financial reward had a (not statistically significant) positive effect on eventual donation.

The underlying psychological mechanism appears to be a fundamental desire for "feelings of competence and self determination" which are associated with intrinsically motivated behaviors (Deci 1975). Relatedly, a person's perceived degree of self-determination in making a choice influences the evaluation of the things over which the choice is being made. For example, risk imposed by others is weighed more negatively than risk chosen by the subject. (See Chauncy Starr 1969; see also Camerer and Howard Kunreuther 1989.)

While the evidence for extrinsic reward and other self-determination effects on preferences appears quite strong, the relevant data provide little support for the anti-market normative inferences sometimes thought to follow. First, the evidence does not implicate monetary rewards per se, but rather any extrinsic reward (including negative rewards such as punishments or admonitions). Moreover, distinctly non-market aspects of governance—close supervision, externally imposed time limits for work tasks for example—appear to have similar effects (Lepper and Greene 1978, p. 121).

Paying someone to perform a task which they might willingly have done without pay seems likely to undermine motivation; but this says little about the relative effectiveness of the various ways—pay, supervision, threat of job loss, etc.—to induce people to undertake tasks which they would rather *not* do. Second, while the extrinsic nature of market rewards may undermine motivations, the wide range of choices often afforded in market situations may support the sense of self-determination and thus induce positive motivational effects.

6. Markets, Reputations, and Norms

> *The real reason why all these economic obligations [among the Trobriand Islanders] are normally kept, and kept very scrupulously, is that failure to comply places a man in an intolerable position ... The honourable citizen is bound to carry out his duties, though his submission is not due to any instinct or intuitive impulse or mysterious "group sentiment," but to the detailed and elaborate working of a system, in which every act has its own place and must be performed without fail.... every one is well aware of its existence and in each concrete case he can foresee the consequences.* —Bronislaw Malinowski (1926, p. 40)

"Market-like arrangements" wrote Charles Schultze (1977, p. 18) "reduce the need for compassion, patriotism, brotherly love, and cultural solidarity." Minimizing the demand on what might now be termed social capital, plus the conviction that the market system is, as Hayek (1948, p. 11) wrote, "a system under which bad men can do least harm," are among the attractive features of markets. This is not to say, however, that markets make norms redundant; where contracts are incomplete or unenforceable, trustworthiness and other norms facilitate exchange. Arrow (1971, p. 22) writes:

> *In the absence of trust... opportunities for mutually beneficial cooperation would have to be foregone . . . norms of social behavior, including ethical and moral codes [may be] . . . reactions of society to compensate for market failures.*

But if, as Schultze and Hayek say, markets make fewer demands on people's elevated motivations, the impersonal and ephemeral nature of market interactions also affect the benefits and costs of acquiring cultural traits affecting socially valued behaviors. Markets thus affect not only the demand for, but also the supply of cultural traits. Among these are reputations for trustworthiness, generosity, and vengefulness.

In economics, little progress has been made in developing and testing a theory of preferences along these lines.

Where markets govern the exchange of well defined (meaning third party enforceable) property rights, reputations of any kind will tend to be both costly for people to acquire and of little benefit to those who do, and for these reasons unlikely to be favored by differential replication. A consequence is that markets lack the personal element of non-market connections, and as Ben-Porath (1980, p. 18) writes, with "[t]he development of markets . . . the benefits from a connection decline as identity becomes less important." Thus where markets approximate the ideal complete-contracting assumptions of the standard model, the adverse consequences of lack of trustworthiness or generosity may be attenuated; but at the same time markets may militate against the evolution of these traits. Thus markets may undermine the reproduction of traits necessary for efficient market transactions in the absence of complete contracting.

To see how this might be the case, I will consider a subset of norms which I call *nice traits*; these are behaviors which in social interactions confer benefits on others. Others would like to be paired with those exhibiting nice traits in an interaction. Included are such strategies as conditional or unconditional cooperation in a prisoners' dilemma game, contributing rather than withholding in a public goods game, or playing dove in a hawk–dove game.[21] As my subsequent examples will confirm,

---

[21] Bowles (1996) gives the following definition: in a population with traits x and x' the latter indicating all other traits, x is a nice trait if $\pi(x,x) > \pi(x,x')$,

it is not possible to generalize about the effect of markets on socially valued norms: "nice traits" may sustain collusion where competition would be more socially beneficial, for example (Rose-Ackerman 1997). Using a model similar to that presented above, I (1996) show that allocation rules which closely conform to idealized markets may support lower equilibrium population frequencies of nice traits, by comparison with alternative allocation rules which deviate from the market ideal. The intuition behind this result is that behaviors determined by nice traits affect others in non-contractible ways, and the regulation of non-contractible behaviors through market-like interactions generates analogues to familiar market failures which in many cases may be attenuated by deviations from the market ideal.

To see why this might be true consider the various ways-identified by biologists, students of cultural evolution, and evolutionary game theorists- that nice traits might flourish in a large population. All, I will suggest, may be weakened by the impersonality and ephemerality of contact that characterize markets. First, frequently repeated interaction of a given pair of individuals provides opportunities to sanction violations of norms and to reward nice traits. By contrast to other allocation rules, the ephemerality and anonymity of market interactions clearly militate against repeated pairings and hence against this mechanism for supporting nice traits. Second, frequent interaction of a limited number of people likewise lowers the cost of acquiring information about the recent behaviors of others, thus increasing the value of acquiring a reputation for being "nice."[22] The impersonality and ephemerality of contact in markets clearly militate against these mechanisms favoring nice traits.

$\pi(x', x) > \pi(x'x')$, and $\pi(x, x) > \pi(x', x')$. Other nice traits are "learn" rather than "imitate" in games of conformism and learning of the type studied by John Conlisk (1980) and Boyd and Richerson (1993).

[22] Bowles and Gintis (1997a) define a level of "optimal parochialism" based on the latter mechanism in a model of endogenous group formation in a large population. In a population paired to play one shot prisoners' dilemma games with the strategy set augmented by an opportunity to pay a cost to determine the type of the other agent and to cooperate if the other is either a cooperator or an "inspector," the fraction of defectors in a (stable, interior) equilibrium population distribution varies linearly with the cost of inspection.

Model	Effect favoring the replication of nice traits	Necessary structural characteristic	Relationship to idealized markets
RETALIATION: Taylor (1976), Drew Fudenberg–Eric Maskin (1986)	Punishment of antisocial behaviors	Frequent or long lasting interactions	Unlikely given ephemerality
REPUTATION: Kreps (1990), Carl Shapiro (1983), Bowles and Gintis (1997a)	Enhanced value of reputations for niceness	Low cost of information about others	Unlikely given impersonality, ephemerality
SEGMENTATION: Grafen (1979), Hamilton (1975), Axelrod-Hamilton (1981)	Advantageous pairing for those with nice traits	Non random pairing of agents	Unlikely given impersonality, low entry and exit costs
GROUP SELECTION: David Wilson and Elliot Sober (1994), Boyd and Richerson (1990)	Enhanced group selection pressures favor nice traits	Limited intergroup migration, non-random group formation	Unlikely given low entry and exit costs and impersonality

Figure 3. How Markets May Discourage the Evolution of Nice Traits

The third mechanism—segmentation—is less familiar to economists, having been introduced by biologists as "games among relatives."[23] Where, as in the model introduced above, populations are segmented so that individuals of a given "type" tend to interact disproportionately with one another, nice types will be favored. For example, if because of geographical or cultural segmentation, the probability of interacting with one's own type is greater than the population frequency of the trait, the equilibrium level of the nice trait will exceed the equilibrium distribution of traits in a population under random pairing. The reason is that segmentation partially internalizes the externality associated with the nice trait: the nonrandom pairing means that the benefits of niceness are disproportionately likely to be conferred on others bearing nice traits, thereby favoring the replication of nice traits. To the extent that the impersonality (and hence anonymity) of markets erodes the bases of segmentation, markets inhibit this mechanism that fosters the proliferation of nice traits.

---

[23] Grafen (1979) and Robert Axelrod and Hamilton (1981). Bowles (1996) and Bowles and Gintis (1998) study the effect of segmentation on the evolution of nice traits in a large population.

Finally, socially beneficial culturally transmitted traits may evolve if the pressure of cultural group selection is sufficiently strong. This occurs when the prevalence of nice traits in a subgroup enhances the average performance of the group sufficiently to allow the trait to proliferate even if it is disadvantaged in replication within each group. Group selection pressures vary with the extent of group differences in the distribution of traits among the subgroups in a large population, which in turn depends on the level of migration among groups and the extent to which the formation of new groups contributes to between group differences, for example by favoring the formation of groups which are more homogeneous than the population as a whole (Boyd and Richerson 1990). High entry and exit costs and other supports for population segmentation sustain the group differences which render the pressure of group selection effective. The low entry and exit costs typical of markets—by comparison to other allocation rules—weaken the pressures of group selection.

Figure 3 summarizes these four mechanisms supporting the replication of nice traits, and the manner in which market allocation rules may undermine them.[24] Because these conjectures predict differences between the distribution of cultural traits in whole populations, adequate testing would require comparative data in which entire populations governed by more or less market like arrangements are the units of analysis. As the following studies suggest, however, less demanding tests using experimental data on a common pool of subjects under varying institutional conditions suggest that deviations from idealized markets may induce "nice behaviors."

We have seen already that even in experimental markets characterized by complete contracting and distributionally extreme (unfair) equilibrium outcomes, the competitive equilibrium implied by self-regarding preferences is rapidly obtained in a wide variety of subject pools. This does not occur in experimental markets with incomplete contracts. In a series of experiments, Fehr and his co-authors have found that contractual incompleteness induces a pattern of reciprocity among subjects which has durable effects on competitive equilibrium (Fehr and Jean-Robert Tyran 1996; Fehr *et al.*

---

[24] The bibliographic references merely give examples of the relevant models; inferences concerning the relationship between economic institutions and the evolution of norms are my own.

1997; and Fehr, Gaechter, and Georg Kirchsteiger 1997). For example in an experimental labor market in which effort is selected by the "worker" after a wage offer is made by the "firm," the subgame perfect equilibrium based on self-regarding preferences in a one time interaction (offer the lowest wage, provide the lowest effort level) does not occur. Rather, "firms" offer wages higher than necessary and "workers" reciprocate by working harder than the minimum.

Relatedly, Kollock (1994, p. 341) investigated "the structural origins of trust in a system of exchange, rather than treating trust as an individual personality variable" with similar results. Using an experimental design based on the exchange of goods of variable quality, Kollock found that trust in and commitment to trading partners as well as a concern for one's own and others' reputations emerges when product quality is variable and non-contractible but not when it is contractible. These experimental results appear to capture some of the structure of actual exchanges. Ammar Siamwalla's (1978) study of marketing structures in Thailand contrasts the impersonal structure of the wholesale rice market—where the quality of the product is readily assayed by the buyer—with the personalized exchange based on trust in the raw rubber market—where quality is impossible to determine at the point of purchase.

These experimental results suggest that trust or reciprocity may depend on the form of the contract, contractual incompleteness leading to trusting and reciprocal behaviors, and conversely. Fehr, Gaechter, and Kirchsteiger (1997) present a surprising case of this in their experiments with "firms" and "workers." When they provided more complete contracting of labor effort through monitoring and the imposition of fines on workers in cases of verified shirking, worker effort significantly *declined*. Their interpretation is that explicit incentives may destroy trust- and reciprocity-based incentives.

Outside the experimenter's lab, of course, the degree of contractual incompleteness is not exogenous, and it may respond to the levels of trust and reciprocity exhibited by the relevant population of traders. For example, lower levels of trust and reciprocity would plausibly lead those designing contracts and the relevant enforcement environments to be willing to pay more for more complete contracts. Greif's (1994) analysis of the divergent cultural and institutional trajectories of the Genovese and Maghribi traders in the late medieval Mediterranean provides a well documented historical example. The individualism of the Genovese traders precluded the communitarian enforcement techniques of the Maghribi traders; but it also provided an impetus for the development and perfection of ultimately more successful third party enforcement of claims by the Genovese.

If levels of trust and reciprocity on the one hand and contractual incompleteness on the other are mutually determining one may define an equilibrium set of norms and contracts. If the nature of the mutual influences are as I have suggested, there may be any number of these equilibria, some with high levels of trust and relatively incomplete contracts (like the Maghribi traders) and others with the converse (like the Genovese). If this vastly oversimplified view captures something about the dynamics of cultural change, we might expect rapid shifts in both norms and contracts where exogenous events "tip" the society from the basin of attraction of one norm-contract equilibrium to another. Thus, it seems reasonable that some of the apparently profound cultural changes associated with the extension of markets in previously non-market systems might be explained by the structural characteristics in Figure 3 along with the increasingly contractual nature of transactions between people and the related incentives to reallocate time and effort away from human investments which are specific to a particular relationship (trust, ethnic or communal capital) and toward general investments (schooling).

Florencia Mallon's (1983) study of the growth of markets—particularly labor markets—and the erosion of community institutions in the central highlands of Peru during the early twentieth century suggests that some of the above mechanisms may have been at work. Central to the institutions of local solidarity among residents was the practice of contributing labor to road building, irrigation, and other communal projects: "Community membership itself, and access to village resources was defined in terms of a quota of labor time that households owed to the community as a whole." With the extension of labor markets, many found employment in distant mines for extended periods of time, eventually converting the labor dues they owed to the community to cash payments collected and sent home by migrants associations in the mining towns. But "migration, by commodifying relationships and separating them out from the intricately woven fabric of local life, was changing the very context within which community could be defined" (Mallon 1983, pp. 264–65).

Traditional institutions were further undermined by the sale of common lands (or charging fees for the use of the common lands) and the use of the proceeds to build schools and roads. Increased access of the richer peasants to distant markets for their produce freed them of dependence on the locality. The obligation to provide communal labor—or even money payments in their stead—thus became unenforceable, and the practice declined. The institutions which had directed community members' efforts and imagination toward common projects, and the dense network of social relationships sustaining this gave way to investments—schooling and transportation—whose returns were relatively independent of the community social fabric, and contributed little to it.[25]

The ethnographic literature on the environmental degradation of local commons provides numerous examples of similar processes (Jean-Marie Baland and Platteau 1995).

7. Markets, Firms, and Tasks

It is only our Western societies that quite recently turned man into an economic animal. But we are not yet all animals of the same species. —Marcel Mauss ([1925]1967, p. 74)

Learning by doing is a ubiquitous form of personal development; it applies to preferences no less than to skills. The activities we engage in and the tasks they present to us are not fully determined by technology; they depend as well on economic institutions. Thus economic institutions may shape preferences by influencing the tasks we perform.

We know from the experiments of Sherif (1937), Breer and Locke (1965), and others that task performance affects values. Relatedly, a substantial ethnographic literature suggests that differing modes of livelihood are associated with differing general attitudes and values. Edgerton's (1971) cross-cultural comparisons, for example, revealed a large and statistically significant relationship between the predominance of pastoral as opposed to farming livelihoods and the general cultural

---

[25] An analogous process, occurring in Botswana, is described by T. S. Zufferey (1986). See also Raul and Luis Garcia-Barrios (1990).

valuation of independence.[26] A plausible mechanism for both the Edgerton and the Breer and Locke findings is that strategies found successful in coping with the tasks defined by one sphere of life are generalized to other realms of life. Because markets and other economic institutions affect the kinds of tasks we confront and structure rewards and penalties to various behaviors, we may presume that they affect learning.

What, then do market tasks teach? Lane (1991, p. 11) reasons that the belief that one is effective in influencing his or her fate (called self-attribution):

> *is learned from experiences of acting and seeing the world respond, contingent responses. [Because] a transaction . . . requires mutually contingent responses . . . an economy based on transactions teaches self-attribution.*

I know of no evidence for or against this plausible conjecture; but if true, the effects on self-attribution may well depend on the structure of the relevant markets.

Consider, first, the following counter intuitive example. Market interactions are particularly likely to contribute to the sense of personal efficacy under conditions which allow what I term consumer sovereignty with teeth—that which occurs when the consumer's purchase confers a rent on the seller because price exceeds marginal cost. In this situation the consumer who switches to another supplier imposes a cost (the loss of the rent) on the seller (Gintis 1989). In this sense monopolistically competitive markets may provide at least as fertile a ground for the development of a sense of personal efficacy as do perfectly competitive markets: the latter will exhibit a larger number of potential suppliers, while the former will exhibit the stronger version of consumer sovereignty (because the consumer confers a rent on the seller whenever $p > mc$) albeit vis à vis a more limited array of suppliers.

If Lane is right that markets teach self-attribution or personal efficacy, the extent to which this is true appears to depend on one's success in market

---

[26]Independent minded people may become herders rather than farmers, of course, but Edgerton's results are robust even when the pastoralism/farming measure is not the actual means of livelihood (e.g., livestock ownership) but rather the geographical suitability of the relevant locale for each of these two pursuits. The relevant correlations are only slightly diminished when the underlying, and presumably exogenous, measure is used.

activities: income predicts self-attribution better than other demographic variables including level of schooling; whites are more self-attributing than African Americans; men are more self-attributing than women; and self-attribution rises with age until leveling off in middle age.[27]

What the market teaches depends not merely on the degree of success as measured by income, but also on the structural location in a market situation. That the inability to find suitable employment may undermine ones's sense of efficacy is unsurprising, but having a job can do the same. The relevant feature of the labor market is that it requires employees to relinquish (substantial, but not unlimited) authority over their actions to the employer (Herbert Simon 1951). The employer's authority can be effectively wielded because the labor market does not clear, and the employee is thus not indifferent to having the job or losing it (Bowles and Gintis 1992, 1993). Of course employees differ greatly in the degree to which they are subjected to hierarchical authority; and these differences appear to have psychological consequences.

Over a period of three decades Kohn and his collaborators have studied the relationship between one's position in an occupational hierarchy and the individual's valuation of self-direction and independence in their children, intellectual flexibility, and personal self-directedness, concluding that "the experience of occupational self direction has a profound effect on people's values, orientation, and cognitive functioning" (Kohn *et al.* (1990), p. 967; see also Kohn 1969, 1990). His collaborative study of Japan, the U.S. and Poland (1990) based on sample surveys of male employees (from the 1960s and 1970s) yielded cross culturally consistent findings: people who exercise self-direction on the job, also value self-direction more in other realms of their life (including child-rearing and leisure activities) and are less likely to exhibit the nexus of traits

---

[27] Gerald Gurin and Patricia Gurin (1976); James Birren, Walter Cunningham, and Koichi Yamamoto (1983). In the former study the normalized regression coefficient of income in a multiple regression predicting a measure of personal efficacy is twice as large as that for education and three times as large as that for race (p. 137). Longitudinal evidence suggests that internality (or self-attribution) and success are mutually determining; while an internal locus of control contributes to success, the reverse is also true (Andrisani and Nestel 1976). The strong relationship between income and self-attribution may thus arise because the more self-attributing people are also more successful (rather than the other way around); but this reasoning obviously does not apply to the substantial correlations with exogenous determinants of economic success such as race, gender, and age.

termed the authoritarian personality. (See Kohn and Schooler 1983, p. 142.)

These results do not arise because self-directed people select (or are selected into) jobs where occupational self-direction is substantial. In a series of related studies using longitudinal data, Kohn and his colleagues use two stage least squares estimation to address the question of reciprocal causation (personality dimensions as causes of job position); effects in both directions are found, but the job to personality causal effects are robust.[28] Compared to direct measures of occupational self-direction, covarying influences such as income, race, ethnicity, family structure, religion, and education were considerably less robust and consistent predictors of personality and tastes.

Kohn and his co-authors reason that "social structure affects individual psychological functioning mainly by affecting the conditions of people's own lives." Summarizing his earlier work on child rearing, Kohn (1969, p. 189) wrote:

> *Self direction, in short, requires opportunities and experiences that are much more available to people who are more favorably situated in the hierarchical order of society; conformity is the natural consequence of inadequate opportunity to be self-directed.*

But why should work experiences affect child-rearing values and leisure time preferences? Kohn concludes that:

> *The simple explanation that accounts for virtually all that is known about the effects of job on personality . . . is that the processes are direct: learning from the job and extending those lessons to off-the job realities.*[29]

---

[28] See Kohn and Carrie Schoenbach (1983). Jeylan Mortimer, Jon Lorence, and Donald Kumka (1986, p. 113) use similar methods to address the problem of endogeneity of occupational selection, and report a substantial causal effect of occupationally determined work autonomy on the sense of self-confidence.

[29] Kohn (1990, p. 59). Gabriel Almond and Sidney Verba (1963, pp. 180ff, 364ff) provide further evidence that work experiences are associated with generalized subjective orientations. Across all occupational types in five different countries, those who were consulted on the job scored significantly higher on a measure of subjective civic competence measuring the sense of personal efficacy in dealing with local and national government bodies. Differences (between those consulted and those not) in subjective competence scores within broad job types were larger than the differences between job types.

Thus, just as the wide range of choices and contingent reinforcement characteristic of consumer goods markets may promote personal efficacy, the surrender of authority to employers which characterizes the labor market appears to support far-reaching psychological effects, some of which undermine the sense of being in control of one's life.

Unlike Kohn's studies, most research on the relationship between job structures and personality do not adequately address the problem of mutual causation mentioned above. Robert Karasek (1978) however, was able to study the behavioral effects of exogenous *changes* in job structure (including both expert and self-reports of job characteristics) using panel data on the Swedish labor force over the years 1968–1974, a period of considerable experimentation with job redesign. He found that

> *workers whose jobs had become more passive also became passive in their leisure and political participation and workers with more active jobs became more active. These findings were significant in eight out of nine subpopulations controlled for education and family class background.*
>
> (Karasek 1990, pp. 53–54)

The effect of economic institutions on task performance and hence on personality may go beyond those stressed by Kohn and Lane. The seemingly desirable attribute of markets stressed by Schultze above—that they make few demands on our ethical reasoning—may have a negative counterpart in a reduced salience of moral concerns or capacity for moral reasoning. A recent public goods experiment suggests that these market effects may be important (Norman Frohlich and Joe Oppenheimer 1995). Subjects played five-person public goods games under two conditions: one group played the standard contribution game and the other played a modified game in which a randomized assignment of payoffs similar to the Rawlsian veil of ignorance made it optimal to contribute the maximal amount to the public good. Half of the subjects (in each treatment) were allowed to engage in discussion prior to each play (of course the discussion should have had no effect on the outcome of the standard game, as the dominant strategy is to contribute nothing). After eight rounds of play another seven rounds were conducted, this time with the same groups but with all playing the standard game. Among those who had been permitted discussion, those who had experienced the incentive

This is a fascinating observation, but it has also been argued that markets "activate some benign human proclivities at the expense of some malignant ones" and "repress and perhaps atrophy the more destructive and disastrous components of human nature" (Hirschman 1977, referencing David Hume 1752).

compatible modified game contributed significantly less in the final seven rounds than those whose only experience was the standard game, and (in subsequent questionnaires) revealed that their behavior was less guided by considerations of fairness.

The authors' explanation of this striking finding is that the incentive compatible mechanism rewarded those contributing to the public good, thus making self interest a good guide to action, while those experiencing the standard game succeeded only to the extent that they evoked considerations of fairness as a distinct motive. They conclude

> *The failure of the . . . (incentive compatible) mechanism to confront subjects with an ethical dilemma appears to lead to little or no learning in ethical behavior in the subsequent period. . . . It is an institution, like other incentive compatible devices, which can generate near optimal outcomes. . . . However from an ethical point of view it is not only unsuccessful as pertains to subsequent behavior; it appears to be actually pernicious. It undermines ethical reasoning and ethically motivated behavior.* (Frolich and Oppenheimer 1995, p. 44)

Thus far I have considered the direct effects of markets and other economic institutions on the evolution of preferences. But there are indirect effects as well.

8. Markets and the Process of Cultural Transmission

> *. . . the modern and the traditional consciousness of the [early 19th century] French peasant contended for mastery [in] . . . the form of an incessant struggle between the schoolmasters and the priests.* —Karl Marx ([1852]1963, p. 125)

Here I consider the influence of economic institutions on the structure of social interactions which make up the process of cultural transmission, that is on child-rearing practices, childhood and adolescent socialization, and the availability of and sometimes compulsory exposure to entirely new cultural models such as teachers and media figures. Some of these effects are the intentional result of people's attempt to acquire and to teach their children those traits required for adequate functioning in the social system; other effects are entirely unintended.

One avenue for the effects of economic organization on cultural transmission, the existence of a connection between forms of livelihood and patterns of child rearing, has been widely documented. Herbert Barry, Irvin Child, and Margaret Bacon (1959) categorized 79 mostly nonliterate societies according to the prevalent form of livelihood (animal husbandry, agricultural, hunting, and fishing) and the related ease of food storage or other forms of wealth accumulation, the latter being a well documented correlate of dimensions of social structure such as stratification. They combined these with evidence on the dominant forms of child rearing including obedience training, self-reliance, independence, and responsibility. They found large differences in the recorded child-rearing practices, concluding: "knowledge of the economy alone would enable one to predict with considerable accuracy whether a society's socialization pressures were primarily toward compliance or assertion."[30] Other studies have confirmed consistent relationships between these group differences in child-rearing practices and group differences in various measures of psychological functioning (Witkin and Berry 1975). For example, hunter gatherer societies stress in their child rearing (and achieve in their adults) greater independence, while more stratified agricultural societies stress (and achieve) greater obedience.

These results suggest economic structural effects on child rearing and thereby on personality, but do not shed light on the effects of modern economic institutions; indeed markets played a limited role in most of the societies studied by Barry, Bacon, and Child and Witkin and Berry. The expansion of markets may have had its largest impact in rendering inadequate the previously dominant family-based and heterogeneous forms of socialization studied by these authors. Ernest Gellner's (1983) account of the rise of nationalism is based on the transformation of socialization required by the spatial extension of the division of labor made possible by markets:

---

[30] The statistical relationships observed were not explainable by the covariation of child-rearing practices and type of livelihood with other measures of social structure such as unililnearity of descent, extent of polygyny, levels of participation of women in the predominant subsistence activity, size of population units, and the like. A societylevel rather than individual approach has been adopted in mnuch of the cross cultural literature on child rearing. See the work of Beatrice and John Whiting (Whiting 1963; J. and B. Whiting 1975).

> *In the closed local communities of the agrarian or tribal worlds, when it came to communication, context, tone, gesture, personality and situation were everything... Among intimates of a close community, explicitness would have been pedantic and offensive* [p. 33] ... *[but] the requirements of a modern economy obliges [us] to be able to communicate contextlessly and with precision with all comers in face to face ephemeral contacts.* (p. 140)

This requires

> *sustained frequent and precise communication between strangers involving a sharing of explicit meaning, transmitted in a standard idiom and in writing when required. For a number of converging reasons this society must be exo-educational: each individual is trained by specialists, not just by his own local group, if indeed he has one.* (p. 34)

As a result there emerged "a school transmitted culture not a folk transmitted one" (p. 36) in which children were "handed over by their kin groups to an educational machine" (p. 37). Universal schooling may be represented as a particular assignment of cultural models to children, one unprecedented in its divorce from family and degree of centralization.[31]

As a result the cultural transmission processes became markedly more conformist as cultural models were selected from (or by) dominant groups and a society-wide socialization system intruded into what was once an entirely local learning process.

We know strikingly little about the cultural impact of these historically novel forms of socialization; most studies of the impact of schooling and its relationship to the economy have stressed the contribution of schooling to cognitive functioning, not to values or personality. But the evidence that personality effects of schooling are important is substantial, if indirect. The substantial and apparently causal relationship between years of schooling attained by an individual and subsequent labor market earnings presents a puzzle, for available data suggests that a large part of the schooling-earnings relationship is not mediated by the effect of schooling on the level

---

[31] Robert Dreeben's (1968) book on the socialization tasks of schooling develops a similar argument based on "the liberating effect" of "the separation of the workplace from the household" (p. 129). See also Gintis (1971) and Bowles and Gintis (1997b).

of cognitive functioning. Schools make people smarter, and richer, but the latter effect—at least in the U.S.—is surprisingly independent of the former. The relevant evidence is this: the estimated effect of schooling on earnings is only modestly reduced if the individual level of cognitive skill is econometrically "held constant" by inclusion in an earnings function.[32]

Gintis and I (1997b) suggest that schooling may raise earnings through its contribution to the acquisition of such personality traits as a lower rate of time preference, a lower disutility of effort, or a cooperative relationship to authority figures, which are relevant to the work situation but which are not measured on the existing cognitive measures. We motivate this hypothesis using a standard principal agent model of the problem of labor discipline in an employment relationship characterized by incomplete contracts. If we are right, the structure of schooling would contribute to preparation for adult roles in a manner not dissimilar to that suggested by the Bacon, Child, and Barry study. But do schools produce these non-cognitive employment related traits?

To the best of my knowledge only one study has attempted to provide an answer; it does not provide a satisfactory basis for generalization, but it is nonetheless worth reviewing. The strategy of the study was to see if schools rewarded (and thus inferentially fostered the development of) people with the same personality traits that are valued by employers. In parallel investigations in distinct populations conducted during the early 1970s, Richard Edwards (1977) used peer-rated personality measures of employees in both private and public employment to predict supervisor ratings of these workers. Peter Meyer (1972) used the same peer rated personality variables to predict grade point averages of students in a high school, controlling for SAT (verbal and math) and IQ. Edwards found that employees judged by their workplace peers to be "perseverant," "dependable," "consistent," "punctual," "tactful," "identifies with work," and "empathizes" had significantly higher supervisor ratings, while those judged by their peers to be "creative" and "independent" were ranked poorly by supervisors. Meyer found virtually identical results for the high school students in his grading study: the exact traits predicting favorable supervisor ratings in the Edwards study, predicted good grades (holding

---

[32] Gintis (1971) first demonstrated this. Bowles and Gintis (1997b) reviews the large number of relevant estimates over a 40 year period.

constant cognitive scores). Teachers and employers in these samples reward the same personality traits.[33]

9. Conclusion

Political writers have established it as a maxim, that, in contriving any system of government ... every man ought to be supposed to be a knave and to have no other end, in all his actions, than his private interest. —David Hume ([1754]1898 p. 117)

Lawgivers make the citizen good by inculcating habits in them, and this is the aim of every lawgiver; if he does not succeed in doing that, his legislation is a failure. It is in this that a good constitution differs from a bad one. —Aristotle (1962, p. 1103)

This reflects a fundamental tradeoff in economics: to what extent should one abstract from some phenomena in order to generate insights into others? A central argument in this essay is that economists can't simply leave to other disciplines an understanding of preference formation.

Economists have followed Hume, rather than Aristotle, in positing a given and self-regarding individual as the appropriate behavioral foundation for considerations of governance and policy. The implicit premise that policies and constitutions do not affect preferences has much to recommend it: the premise provides a common if minimal analytical framework applicable to a wide range of issues of public concern, it expresses a prudent antipathy toward paternalistic attempts at social engineering of the psyche, it modestly acknowledges how little we know about the effects of economic structure and policy on preferences, and it erects a barrier both to ad hoc explanation and to the utopian thinking of those who invoke the mutability of human dispositions in order to sidestep difficult questions of scarcity and social choice. Realism, however, cannot be among the virtues invoked on behalf of the exogenous preferences premise.[34] Economic institutions, we have seen, may affect preferences through their direct influences on situational construal, forms of reward, the evolution

---

[33] We would like to know (but do not) if schools produce the traits they reward, and if traits valued by supervisors are rewarded by enhanced pay. The underlying studies are reported and compared in Bowles and Gintis (1976).

[34] Indeed Hume, immediately following the passage just quoted, muses that it is "strange that a maxim should be true in politics which is false in fact." While in academic settings most economists still adhere to the exogenous preferences canon and its "de gustibus non est disputandum" (George Stigler and Becker 1977) implication, many appear aware of its limitations when it comes to evaluating institutions and policies. Thus Becker (1995, p. 26) refers to "the effects of a free-market system on self-reliance, initiative, and other virtues" and referring to government transfers to the poor, claims that "the present system corrupts the values transmitted to children."

of norms, and task related learning as well as their indirect effects on the process of cultural transmission itself.

One hopes that the active research agenda now being pursued by economists, other social scientists and biologists in this area may soon allow more confidence in assessing the empirical magnitude and generality of these effects.[35] The following research priorities seem particularly important.

First, we know very little about the process of cultural transmission—who acquires what trait from whom, under what conditions, and why. Yet this information is critical to understanding how economic institutions may impact on preferences. Empirical studies of the relative importance of parents, other family members, friends, teachers, and others in cultural learning, and the interplay of cultural and genetic transmission would be very valuable.

Second, while we have evidence that traits acquired in one environment are then generalized to others (recall Kohn's studies of child rearing) we do not know how this takes place or how persistent the traits may be once the initiating environment is withdrawn.

Third, because imitation of prevalent traits, or enforced conformism may play an important role in the transmission of cultural traits, comparative studies of whole societies may provide insights not available in individual-based studies. An example may make this clear. Recall that Edgerton (1971, p. 195) found that pastoralists valued independent action more than farmers. But farmers in a predominantly pastoral tribe valued independence more (almost twice as much by his measure) than farmers in predominantly farming tribes, while pastoralists in predominantly farming tribes valued independence considerably less than did pastoralists in the pastoral tribe. Thus it appears that the predominant livelihood in a tribe may have cultural effects beyond the effects of the livelihood of the individual. Analysis of individual data within a single cultural group this may miss important effects of economic institutions operating on group differences. Comparative analysis of economic experiments implemented in the differing economic environments of distinct societies, including those with premodern economic institutions, would be illuminating.

---

[35] Avner Ben-Ner and Louis Putterman (1997) is a valuable collection of relevant work by economists.

I emphasize empirical studies because the proliferation of relevant theoretical models in economics has not been matched by empirical investigation. But important contributions could be made by two types of conceptual work.

Fourth, experiments in economics, sociology, and psychology have raised serious doubts about the behavioral accuracy of the minimalist conception of homo economicus: the individual actor with self-regarding and outcome-based preferences. Much of the impact of economic institutions on behavior may occur through the ways that particular institutional settings prompt individuals to draw one or another response—whether self-regarding, spiteful, generous, or other—from their varied behavioral repertoires. A concept of preferences more adequately grounded in the empirical study of behavior would assist in analyzing these processes.

Finally, an integration of the insights of the theory of cultural evolution with those of evolutionary game theory seems likely to be insightful, especially in view of the apparent importance of conformism in cultural transmission (and hence the needed modification of the concept of cultural equilibrium as suggested in Section 3).

Shortcomings of the existing empirical studies and the unsatisfactory "black box" nature of extant knowledge of social learning notwithstanding, the weight of both reason and evidence point strongly to the endogeneity of preferences. If preferences are indeed endogenous in the senses suggested here, four implications follow.

First, economics pays a heavy price for its self-imposed isolation from the other behavioral sciences. At its simplest, the conception underlying contemporary disciplinary boundaries is one of society marked by an implausible degree of specialization among institutions: families and religious institutions shape culture, governments govern, and economic institutions allocate resources. These disciplinary boundaries have favored the development of parochial, incompatible, and inadequate models of human behavior in the various disciplines, ranging from the oversocialized homo sociologicus to the undersocialized homo economicus (Mark Granovetter 1985). Recognition of the cultural effects of markets (and other economic institutions) may foster a more unified approach to the behavioral sciences, a benefit of which might be the more successful resolution of outstanding puzzles in economics.[36]

Second, the effectiveness of policies and their political viability may depend on the preferences they induce or evoke.[37] Hirschman (1985, p. 10) points out that economists typically assume otherwise and for this reason propose

> *to deal with unethical or antisocial behavior by raising the cost of that behavior rather than proclaiming standards and imposing prohibitions and sanctions. The reason is probably that they think of citizens as consumers with unchanging or arbitrarily changing tastes in matters civic as well as commodity-related behavior. . . . A principal purpose of publicly proclaimed laws and regulations is to stigmatize antisocial behavior and thereby to influence citizens' values and behavioral codes.*

Frohlich and Oppenheimer's and Fehr and Gaechter's experiments above suggest that raising the cost of an antisocial behavior and other incentive compatible devices may actually do harm. Moreover, the analysis in Section 6 of the evolution of nice traits suggests that approximating the market ideal by perfecting property rights may weaken non-market solutions to problems of social coordination. There is thus a norm-related analogue to the second best theorem of welfare economics: *where contracts are incomplete (and hence norms may be important in attenuating market failures), more closely approximating idealized complete contracting markets may exacerbate the underlying market failure (by undermining the reproduction of socially valuable*

---

[36] For example, given the poor empirical showing of most theories of wages (Truman Bewley 1995) an adequate understanding of wage setting institutions—including why employers do not generally charge job fees (H. Lorne Carmichael 1985) —would seem to require an account embracing effects of wages on such preferences as the disutility of labor and perceptions of just treatment, along lines suggested by Akerlof's (1984) analysis of gift exchange and Robert Solow's (1990) treatment of "labor markets as social institutions," as well as the work of Fehr and his coauthor mentioned above.

[37] Romer's (1996) account of the origins and evolution of the social security system addresses the ways that income transfer programs shape preferences; and Frey's (1997) econometric study of tax compliance in Switzerland explores the way that different constitutional arrangements affect a predisposition to tax avoidance. On the importance of considering the impact of environmental policy on environmental preferences see Cass Sunstein (1993).

norms such as trust or reciprocity) and result in a less efficient equilibrium allocation. An analogous caution applies to governmental, family based, or other solutions: for example, numerous experiments (as we have seen) suggest that "earning" a claim on a resource differs in psychologically important ways from simply receiving one, and an adequate understanding of public transfers would seem to require attention to these effects.

Third, preference endogeneity gives rise to a kind of market failure and suggests a reconsideration of some aspects of normative economics. The influence of our preferences on others is not even approximately captured by contracts: norms of generosity, non-aggression, or punishment of antisocial behaviors confer external benefits for example, while a taste (or addiction) for smoking confers external costs. Because our preferences have non-contractual effects on others, how we acquire them is a matter of public concern.

Just as the process of natural selection does not generally maximize average fitness, there is no reason to expect that the process of cultural transmission determining the equilibrium distribution of traits in the population will support a socially optimal outcome. The cultural equivalent of a market failure thus results; indeed the long-term persistence of socially and even individually disadvantageous norms is hardly open to question, extreme forms of blood revenge representing a particularly well documented example (Jon Elster 1989; Edgerton 1992; Boehm 1984). Because states, communities, and markets may influence the process of cultural evolution, any normative evaluation of the role and scope of these institutions must attempt to take their cultural effects into account.

Fourth, there thus may be a novel public interest in some types of economic arrangements which are commonly considered private. Uncoerced exchange among informed adults is often considered a private realm in which there is no public interest the absence of non-contractual effects on third parties. The philosopher David Gauthier (1986, pp. 95-96) writes "The operation of the market cannot in itself raise any evaluative issues . . . The presumption of free activity ensures that no one is subject to any form of compulsion or any type of limitation not already affecting her own actions as a solitary individual." But if preferences are shaped by markets and other economic institutions, both evaluative issues and a public interest may arise, for an individual's preferences induce actions imposing non-contractible costs and benefits on others. Thus part of the reasoning which conventionally establishes a public interest in the nature and amount of schooling —the socialization of children is to some extent a public good—would seemingly apply to the effects of economic institutions on preferences as well.

A broader concept of market failure is thus required, one encompassing the effects of economic policies and institutions on preferences and for this reason more adequate for the consideration of an appropriate mix of markets, communities, families, and states in economic governance.[38] Such a new welfare economics would of course have to confront the longstanding liberal philosophical reluctance to privilege some ends over others; that is, it would have to address the problem that Hobbes' mushroom fiction ellides.

Acknowledgments

Thanks to Eric Verhoogen, James Heintz, Stephanie Eckman, Nicole Huber, and Melissa Osborne for research assistance, to the University of Siena for providing an unparalleled environment for research and writing, to Robert Boyd, Colin Camerer, Robert Cialdini, Lilia Costabile, Joshua Cohen, Gerald Cohen, Martin Daly, Peter Dorman, Catherine Eckel, Marcus Feldman, Ernst Fehr, Nancy Folbre, Christina Fong, Bruno Frey, Herbert Gintis, Karla Hoff, Daniel Kahneman, Melvin Kohn, David Kreps, Timur Kuran, George Loewenstein, Elinor Ostrom, Bentley MacLeod, Paul Malherbe, Karl Ove Moene, Casey Mulligan, Richard Nisbett, Ugo Pagano, Jean-Philippe Platteau, John Roemer, Susan Rose-Ackerman, Paul Romer, Paul Rozin, Andrew Schotter, Paul Seabright, Gil Skillman, Peter Skott, Rohini Somanathan, Hillel Steiner, Philippe van Parijs, Burton Weisbrod, Elisabeth Wood, Erik Olin Wright, Viviana Zeliser, and two anonymous

---

[38] An example of the reasoning I am recommending is Michael Taylor's (1987) suggestion that the kinds of opportunism which the Hobbesian state is said to curb might be the consequence of living under a centralized authority, or more succinctly that the Hobbesian state produces Hobbesian man (and then more or less successfully curbs him). Analogous reasoning may apply to markets and homo economicus.

referees for guidance in the literatures covered here and comments on an earlier draft, and to the MacArthur Foundation for financial support.

REFERENCES

Akerlof, G. A. 1984. "A Theory of Social Custom, of Which Unemployment May Be One Consequence." In *An Economic Theorist's Book of Tales: Essays that Entertain the Consequences of New Assumptions in Economic Theory,* edited by G. A. Akerlof, 69–99. Cambridge, MA: Cambridge University Press.

Almond, G. A., and S. Verba. 1963. *The Civic Culture: Political Attitudes and Democracy in Five Nations.* Princeton, NJ: Princeton University Press.

Andre, C., and J. Platteau. 1997. "Land Relations Under Unbearable Stress: Rwanda Caught in the Malthusian Trap." *Journal of Economic Behavior and Organization* 34 (1): 1–55.

Andrisani, P. J., and G. Nestel. 1976. "Internal—External Control as Contributor and Outcome of Work Experience." *Journal of Applied Psychology* 61 (2): 156–65.

Aristotle. 1962. *Nicomachean Ethics.* Indianapolis, IN: Bobbs-Merrill.

Arrow, K. J. 1971. "Political and Economic Evaluation of Social Effects and Externalities." In *Frontiers of Quantitative Economics (Contributions to Economic Analysis Volume 71),* edited by M. D. Intriligator, 71:3–24. Contributions to Economic Analysis. North-Holland.

———. 1972. "Gifts and Exchanges." *Philosophy and Public Affairs* 1 (4): 343–62.

Asch, S. E. 1952. *Social Psychology.* Englewood Cliffs, NJ: Prentice-Hall, Inc.

Axelrod, R., and W. D. Hamilton. 1981. "The Evolution of Cooperation." *Science* 211 (27): 1390–96.

Baland, J.-M., and J.-P. Platteau. 1995. *Halting Degradation of Natural Resources: Is There a Role for Rural Communities?* Oxford, UK: Clarendon Press.

Barry III, H., I. L. Child, and M. K. Bacon. 1959. "Relation of Child Training to Subsistence Economy." *American Anthropologist* 61 (1): 51–63.

Becker, G. S. 1995. "The Best Reason to Get People off the Dole." *Business Week,* 26.

———. 1996. *Accounting For Tastes.* Cambridge, MA: Harvard University Press.

Bell, D. 1976. *The Cultural Contradictions of Capitalism.* New York, NY: Basic Books.

Ben-Ner, A., and L. Putterman, eds. 1997. *Economics, Values, and Organizations.* Cambridge, UK: Cambridge University Press.

Ben-Porath, Y. 1980. "The F-Connection: Families, Friends, and Firms and the Organization of Exchange." *Population and Development Review* 6 (1): 1–30.

Bewley, T. F. 1995. "A Depressed Labor Market as Explained by Participants." *The American Economic Review* 85 (2): 250–54.

Birren, J. E., W. R. Cunningham, and K. Yamamoto. 1983. "Psychology of Adult Development and Aging." *Annual Review of Psychology* 34:543–75.

Bisin, A., and T. Verdier. 1996. "The Economics of Cultural Transmission and the Dynamics of Preferences." Unpublished paper, Massachusetts Institute of Technology and CERAS.

Blau, P. M. 1964. *Exchange and Power in Social Life.* New York, NY: John Wiley & Sons.

Bliss, C. J. 1972. "Review of R. M. Titmuss, The Gift Relationship: From Human Blood to Social Policy." *Journal of Public Economics* 1:162–65.

Bloom, B. S. 1964. *Stability and Change in Human Characteristics.* New York, NY: John Wiley & Sons, Inc.

Boehm, C. 1984. *Blood Revenge: The Anthropology of Feuding in Montenegro and Other Tribal Societies.* University of Pennsylvania Press.

———. 1993. "Egalitarian Behavior and Reverse Dominance Hierarchy." *Current Anthropology* 34 (3): 227–54.

Bohannan, P. 1959. "The Impact of Money on an African Subsistence Economy." *Journal of Economic History* 19 (4): 491–503.

Bowles, S. 1996. *Markets as Cultural Institutions: Equilibrium Norms in Competitive Economies.* Technical report 1996--5. Dept. of Economics, University of Massachusetts.

Bowles, S., and H. Gintis. 1976. *Schooling in Capitalist America: Educational Reform and the Contradictions of Economic Life.* New York, NY: Basic Books, Inc.

———. 1986. *Democracy and Capitalism: Property, Community, and the Contradictions of Modern Social Thought.* London, UK: Routledge / Kegan Paul.

———. 1992. "Power and Wealth in a Competitive Capitalist Economy." *Philosophy & Public Affairs* 21 (4): 324–53.

———. 1993. "The Revenge of Homo Economicus: Contested Exchange and the Revival of Political Economy." *Journal of Economic Perspectives* 7 (1): 83–102.

———. 1998. "The Moral Economy of Communities: Structured Populations and the Evolution of Pro-Social Norms." *Evolution and Human Behavior* 19 (1).

———. 1997b. "Labor Discipline and the Returns to Schooling."

———. 1997a. "Optimal Parochialism: The Dynamics of Trust and Exclusion in Communities."

Boyd, R., and P. J. Richerson. 1985. *Culture and the Evolutionary Process.* Chicago, IL: University of Chicago Press.

———. 1990. "Group Selection among Alternative Evolutionarily Stable Strategies." *Journal of Theoretical Biology* 145:331–42.

———. 1993. "Rationality, Imitation, and Tradition." In *Nonlinear Dynamics and Evolutionary Economics,* edited by R. H. Day and Chen P., 131–51. New York, NY: Oxford University Press.

Boyer, P., and S. Nissenbaum. 1974. *Salem Possessed: The Social Origins of Witchcraft.* Cambridge, MA: Harvard University Press.

Breer, P. E., and E. A. Locke. 1965. *Task Experience as a Source of Attitudes.* Homewood, IL: The Dorsey Press.

Buchanan, J. M. 1975. *The Limits of Liberty.* Chicago, IL: University of Chicago Press.

Burridge, K. 1969. *New Heaven, New Earth: A Study of Millenarian Activities.* New York, NY: Schocken Books.

Camerer, C., and R. H. Thaler. 1995. "Anomalies: Ultimatums, Dictators and Manners." *Journal of Economic Perspectives* 9 (2): 209–19.

Camerer, C. F., and H. Kunreuther. 1989. "Decision Processes for Low Probability Events: Policy Implications." *Journal of Policy Analysis and Management* 8 (4): 565–92.

Caporael, L. R., R. M. Dawes, J. M. Orbell, and A. J. C. Van De Kragt. 1989. "Selfishness Examined: Cooperation in the Absence of Egoistic Incentives." *Behavioral and Brain Sciences* 12:683–99.

Carmichael, H. L. 1985. "Can Unemployment Be Involuntary? Comment." *American Economic Review* 75 (5): 1213–14.

Cavalli-Sforza, L. L., and M. W. Feldman. 1981. *Cultural Transmission and Evolution: A Quantitative Approach.* Princeton, NJ: Princeton University Press.

Cavalli-Sforza, L. L., M. W. Feldman, K. H. Chen, and S. M. Dornbusch. 1982. "Theory and Observation in Cultural Transmission." *Science* 218:19–27.

Chen, K., L. L. Cavalli-Sforza, and M. W. Feldman. 1982. "A Study of Cultural Transmission in Taiwan." *Human Ecology* 10 (3): 365–82.

Conlisk, J. 1980. "Costly Optimizers Versus Cheap Imitators." *Journal of Economic Behavior and Organization* 1 (3): 275–93.

Dawes, R. M., A. J. C. Van De Kragt, and J.M. Orbell. 1988. "Not Me or Thee but We: The Importance of Group Identity in Eliciting Cooperation in Dilemma Situations: Experimental Manipulations." *Acta Psychologica* 68:83–97.

Deci, E. L. 1975. *Intrinsic Motivation.* New York, NY: Plenum Press.

Deci, E. L., and R. M. Ryan. 1985. *Intrinsic Motivation and Self-Determination in Human Behavior.* New York, NY: Plenum Press.

Dreeben, R. 1968. *On What is Learned in School.* Reading, CA: Addison-Wesley.

Duesenberry, J. S. 1949. *Income, Saving and the Theory of Consumer Behavior.* Cambridge, MA: Harvard University Press.

Durham, W. H. 1991. *Coevolution: Genes, Culture, and Human Diversity.* Stanford, CA: Stanford University Press.

Durkheim, E. [1933]1964. *The Division of Labor in Society.* New York, NY: Free Press.

Eckert, P. 1982. "Clothing and Geography in a Suburban High School." Edited by P. C. Kottak. *Researching American Culture* (Ann Arbor, MI), 139–44.

———. 1988. "Adolescent Social Structure and the Spread of Linguistic Change." *Language in Society* 17 (2): 183–207.

Edgerton, R. B. 1971. *The Individual in Cultural Adaptation.* Berkeley, CA: University of California Press.

———. 1992. *Sick Societies: Challenging the Myth of Primitive Harmony.* New York, NY: Free Press.

Edwards, R. C. 1977. "Personal Traits and 'Success' in Schooling and Work." *Educational and Psychological Measurement* 37:125–38.

Elster, J. 1989. "Social Norms and Economic Theory." *Journal of Economic Perspectives* 3 (4): 99–117.

Ensminger, J. 1992. *Making a Market: The Institutional Transformation of an African Society.* Cambridge, UK: Cambridge University Press.

Epstein, T. S. 1962. *Economic Development and Social Change in South India.* Manchester, UK: Manchester University Press.

Fehr, E., Kirchler. E., A. Weichbold, and S. Gächter. 1997. "When Social Norms Overpower Competition: Gift Exchange in Exerimental Labor Markets."

Fehr, E., S. Gaechter, and G. Kirchsteiger. 1997. "Reciprocity as a Contract Enforcement Device: Experimental Evidence." *Econometrica* 65 (4): 833–60.

Fehr, E., and J. Tyran. 1996. "Institutions and Reciprocal Fairness." *Nordic Journal of Political Economy* 23 (2): 133–44.

Feldman, M. W., K. Aoki, and J. Kumm. 1996. *Individual Versus Social Learning: Evolutionary Analysis in a Fluctuating Environment.* Technical report 96-05—030. Santa Fe, NM: Santa Fe Institute.

Feldman, M. W., and K. N. Laland. 1996. *Gene-Culture Coevolutionary Theory.* Technical report 96-05-033. Santa Fe, NM: Santa Fe Institute.

Festinger, L. 1957. *A Theory of Cognitive Dissonance.* Stanford, CA: Stanford University Press.

Field, M. J. 1960. *Search for Security: An Ethno-Psychiatric Study of Rural Ghana.* Evanston, IL: Northwestern University Press.

Firth, R. W. 1958. *Work and Wealth of Primitive Communities,* 62–81. New York, NY: Mentor Books.

Fiske, A. P. 1991. *Structures of Social Life: The Four Elementary Forms of Human Relations.* New York, NY: Free Press.

———. 1992. "The Four Elementary Forms of Sociality: Framework for a Unified Theory of Social Relations." *Psychological Review* 99 (4): 689–723.

Flora, J., N. Maccoby, and J. W. Farquhar. 1989. "Communication Campaigns to Prevent Cardiovascular Disease: The Stanford Community Studies." In *Public Communication Campaigns,* edited by Ronald E. R. and Charles K. A., 233–52. Newbury Park, CA: Sage.

Frey, B. S. 1997. "A Constitution for Knaves Crowds Out Civic Virtue." *The Economic Journal* 107 (443): 1043–53.

Frey, B. S., and F. Oberholzer-Gee. 1997. "The Cost of Price Incentives: An Empirical Analysis of Motivation Crowding-out." *The American Economic Review* 87 (4): 746–55.

Frolich, N., and J. A. Oppenheimer. 1995. "The Incompatibility of Incentive Compatible Devices and Ethical Behavior: Some Experimental Results and Insights." *Public Choice Studies* 25:24–51.

Fromm, E., and M. Maccoby. 1970. *Social Character in a Mexican Village: A Sociopsychoanalytic Study.* Englewood Cliffs, NJ: Prentice—Hall.

Fundenberg, D., and E. Maskin. 1986. "Folk Theorem in Repeated Games with Discounting or with Incomplete Information." *Econometrica* 54 (3): 533–54.

Gaechter, S., and E. Fehr. 1997. "Collective Action as a Partial Social Exchange." Unpublished paper.

Garcia-Barrios, R., and L. Garcia-Barrios. 1990. "Environmental and Technological Degradation in Peasant Agriculture: A Consequence of Development in Mexico." *World Development* 18 (11): 1569–85.

Gauthier, D. P. 1986. *Morals by Agreement.* Oxford, UK: Clarendon Press.

Gellner, E. 1983. *Nations and Nationalism.* Ithaca, NY: Cornell University Press.

Gintis, H. 1971. "Education, Technology, and the Characteristics of Worker Productivity." *American Economic Review* 61 (2): 266–79.

———. 1972. "A Radical Analysis of Welfare Economics and Individual Development." *The Quarterly Journal of Economics* 86 (4): 572–99.

———. 1989. "The Power to Switch: On the Political Economy of Consumer Sovereignty." In *Unconventional Wisdom: Essays on Economics in Honor of John Kenneth Galbraith.* Boston, MA: Houghton Mifflin.

Grafen, A. 1979. "The Hawk–Dove Game Played Between Relatives." *Animal Behavior* 27 (3): 905–07.

Granovetter, M. S. 1985. "Economic Action and Social Structure: The Problem of Embeddedness." *The American Journal of Sociology* 91 (3): 481–510.

Greif, A. 1994. "Cultural Beliefs and the Organization of Society: A Historical and Theoretical Reflection on Collectivist and Individualist Societies." *Journal of Political Economy* 102 (5): 912–50.

Gurin, G., and P. Gurin. 1976. "Personal Efficacy and the Ideology of Individual Responsibility." In *Economic Means for Human Needs,* edited by B. Strumpel, 131–57. Ann Arbor, MI: Institute for Social Research.

Güth, W., R. Schmittberger, and B. Schwarze. 1982. "An Experimental Analysis of Ultimatum Bargaining." *Journal of Economic Behavior and Organization* 3 (4): 367–88.

Hamilton, W. D. 1975. "Innate Social Aptitudes of Man: an Approach from Evolutionary Genetics." In *Biosocial Anthropology,* edited by R. Fox, 115–32. New York, NY: John Wiley.

Harsanyi, J. C. 1982. "Morality and the Theory of Rational Behavior." In *Utilitarianism and Beyond,* edited by A. Sen and B. Williams, 39–62. Cambridge, UK: Cambridge University Press.

Hayek, F. A. von. 1948. *Individualism and Economic Order.* Chicago, IL: University of Chicago Press.

Hirschman, A. O. 1977. *The Passions and the Interests: Political Arguments for Capitalism Before Its Triumph.* Princeton, NJ: Princeton University Press.

———. 1982. "Rival Interpretations of Market Society: Civilizing, Destructive, or Feeble?" *Journal of Economic Literature* 20 (4): 1463–84.

———. 1985. "Against Parsimony: Three Easy Ways of Complicating Some Categories of Economic Discourse." *Economics & Philosophy* 1 (1): 7–21.

Hobbes, T. 1949. *De Cive or The Citizen.* New York, NY: Appleton-Century-Crofts, Inc.

Hoffman, E., K. McCabe, K. Shachat, and V. Smith. 1994. "Preferences, Property Rights, and Anonymity in Bargaining Games." *Games and Economic Behavior* 7 (3): 346–80.

Hollander, H. 1990. "A Social Exchange Approach to Voluntary Cooperation." *American Economic Review* 80 (5): 1157–67.

Homans, G. C. 1958. "Social Behavior as Exchange." *American Journal of Sociology* 63 (6): 597–606.

Hume, D. [1754]1898. *Essays: Moral, Political and Literary.* London, UK: Longmans, Green and Co.

Intriligator, M. D. 1971. *Frontiers of Quantitative Economics.* Amsterdam, Netherlands: North-Holland Publishing Company.

Kachelmeier, S. J., and M. Shehata. 1997. "Internal Auditing and Voluntary Cooperation in Firms: A Cross-Cultural Experiment." *Accounting Review* 72 (3): 407–31.

Kant, I. [1785]1949. *The Philosophy of Kant: Immanuel Kant's Moral and Political Writings.* Edited by C. J. Friedrich. New York, NY: Modern Library.

Karasek, R. 1978. *Job Socialization: A Longitudinal Study of Work, Political and Leisure Activity.* Technical report 59.

Karasek, R., and T. Theorell. 1990. *Healthy Work: Stress, Productivity, and the Reconstruction of Working Life.* New York, NY: Basic Books.

Katz, E., and P. F. Lazarsfeld. 1955. *Personal Influence: The Part Played by People in the Flow of Mass Communications.* Glencoe, IL: Free Press.

Kohn, M., A. Naoi, C. Schoenbach, C. Schooler, and K. Slomczynski. 1990. "Position in the Class Structure and Psychological Functioning in the United States, Japan, and Poland." *American Journal of Sociology* 95 (4): 964–1008.

Kohn, M. L. 1969. *Class and Conformity: A Study In Values.* Homewood, IL: Dorsey Press.

———. 1990. "Unresolved Issues in the Relationship Between Work and Personality." In *The Nature of Work: Sociological perspectives,* 36–68. New Haven, CT: Yale University Press.

Kohn, M. L., and C. Schoenbach. 1983. "Class, Stratification, and Psychological Functioning." In *Work and Personality: An Inquiry Into the Impact of Social Stratification,* 154–89. Ablex.

Kohn, M. L., and C. Schooler. 1983. *Work and Personality: An Inquiry Into the Impact of Social Stratification.* Norwood, NJ: Ablex.

Kollock, P. 1994. "The Emergence of Exchange Structures: An Experimental Study of Uncertainty, Commitment, and Trust." *American Journal of Sociology* 100 (2): 313–45.

———. 1998. "Transforming Social Dilemmas: Group Identity and Cooperation." In *Modeling Rationality, Morality, and Evolution,* edited by P. Danielson, 185–209. New York, NY: Oxford University Press.

Kranton, R. E. 1996. "Reciprocal Exchange: A Self-Sustaining System." *American Economic Review* 86 (4): 830–51.

Kreps, D. M. 1990. "Corporate Culture and Economnic Theory." In *Perspectives on Positive Political Economy,* 90–143. Cambridge, UK: Cambridge University Press.

Labov, W. 1972. *Sociolinguistic Patterns.* Philadelphia, PA: University of Pennsylvania Press.

———. 1983. *De Facto Segregation of Black and White Vernaculars.* Technical report. Project on Linguistic Change and Variation. Lingiustics Laboratory, University of Pennsylvania.

Laibson, D. 1996. "A Cue-Theory of Consumption."

Lane, R. E. 1991. *The Market Experience.* Cambridge, UK: Camnbridge University Press.

Laslett, P. 1965. *The World We Have Lost.* London, UK: Methuen.

Ledyard, J. O. 1995. "Public Goods: A Survey of Experimental Research." In *The Handbook of Experimental Economics,* edited by J. H. Kagel and A. E Roth, 111–94. Princeton, NJ: Princeton University Press.

Leibenstein, H. 1950. "Bandwagon, Snob, and Veblen Effects in the Theory of Consumers' Demand." *The Quarterly Journal of Economics* 64:183–207.

Lepper, M. R., and D. Greene, eds. 1978. *The Hidden Costs of Reward: New Perspectives on the Psychology of Human Motivation.* Hillsdale, NJ: Lawrence Erlbaum.

Lepper, M. R., D. Greene, and R. E. Nisbett. 1973. "Undermining Children's Intrinsic Interest With Extrinsic Reward: A Test of the 'Overjustification' Hypothesis." *Journal of Personality and Social Psychology* 28 (1): 129–37.

Lerner, D. 1958. *The Passing of Traditional Society: Modernizing the Middle East.* New York, NY: Free Press.

LeVine, R. A. 1966. *Dreams and Deeds: Achievement Motivation in Nigeria.* Chicago, IL: University of Chicago Press.

———. 1973. *Culture, Behavior, and Personality.* London: Hutchinson & Co.

Loewenstein, G. 1996. "Out of Control: Visceral Influences on Behavior." *Organizational Behavior and Human Decision Processes* 65:272–92.

Loewenstein, G., and S. Issacharoff. 1994. "Source Dependence in the Valuation of Objects." *Journal of Behavioral Decision Making* 7:157–68.

Malinowski, B. 1926. *Crime in Savage Society.* London, UK: Routledge & Kegan Paul.

Mallon, F. E. 1983. *The Defense of Community in Peru's Central Highlands: Peasant Struggle and Capitalist Transition 1860-1940.* Princeton, NJ: Princeton University Press.

Marx, K. [1852]1963. *The Eighteenth Bruimaire of Louis Bonaparte.* New York, NY: International Publishers.

Mauss, M. [1925]1967. *The Gift: Forms and Functions of Exchange in Archaic Societies.* New York, NY: W. W. Norton.

Mccauley, C., P. Rozin, and B. Schwartz. 1994. "On the Origin and Nature of Preference and Values."

McClelland, D. 1961. *The Achieving Society.* Princeton, NJ: Van Nostrand.

Meyer, P. J. 1972. "Schooling and the Reproduction of the Social Division of Labor." Unpublished honors thesis.

Mortimer, J. T., J. Lorence, and D. S. Kumka. 1986. *Work, Family, and Personality: Transition to Adulthood.* Norwood, MA: Ablex Publishing.

Mulligan, C. B. 1997. *Parental Priorities and Economic Inequality.* Chicago, IL: University of Chicago Press.

Newcomb, T. M. 1967. *Persistence and Change: Bennington College and Its Students After Twenty-Five Years.* New York, NY: John Wiley & Sons, Inc.

Nowell-Smith, P. H. 1954. *Ethics.* London, UK: Penguin.

Ochs, J., and A. E. Roth. 1989. "An Experimental Study of Sequential Bargaining." *American Economic Review* 79 (3): 355–84.

Otto, S. P., F.B. Christiansen, and M. W. Feldman. 1995. *Genetic and Cultural Inheritance of Continuous Traits*. Technical report 0064. Morrison Institute for Population and Resource Studies, Stanford University.

Ouchi, W. G. 1980. "Markets, Bureaucracies, and Clans." *Administrative Science Quarterly* 25 (1): 129–41.

Parsons, T. 1949. *The Structure of Social Action: A Study in Social Theory with Special Reference to a Group of Recent European Writers*. 2nd ed. Glencoe, IL: Free Press.

———. 1967a. "On the Concept of Political Power." In *Sociological Theory and Modern Society*, edited by T. Parsons, 232–262. New York, NY: Free Press.

———, ed. 1967b. *Sociological Theory and Modern Society*. New York, NY: Free Press.

Plomin, R., and D. Daniels. 1987. "Why Are Children in the Same Family So Different from One Another." *Behavioral and Brain Sciences* 10 (1): 1–60.

Polányi, K. 1957. *The Great Transformation: The Political and Economic Origins of Our Time*. Beacon Hill, MA: Beacon Press.

Pollak, R. A. 1976. "Interdependent Preferences." *The American Economic Review* 66 (3): 309–20.

Potter, D. M. 1954. *People of Plenty: Economic Abundance and the American Character*. Chicago, IL: University of Chicago Press.

Radner, R., and A. Schotter. 1989. "The Sealed Bid Mechanism: An Experimental Study." *Journal of Economic Theory* 48 (1): 179–220.

Rogers, E. M. 1962. *Diffusion of Innovations*. New York, NY.

Romer, P. 1996. "Preferences, Promises and the Politics of Entitlement." In *Individual and Social Responsibility: Child Care, Education, Medical Care and Long-Term Care in America*, edited by V. Fuchs, 195–228. Chicago, IL: University of Chicago Press.

Rose-Ackerman, S. 1997. "Gifts and Bribe." In *Economics, Values and Organization*. Cambridge, UK: Cambridge University Press.

Ross, L., and R. E. Nisbett. 1991. *The Person and the Situation*. Philadelphia, PA: Temple University Press.

Roth, A. E. 1995. "Bargaining Experiments." In *The Handbook of Experimental Economics*, edited by J. H. Kagel and A. E. Roth, 253–346. Princeton, NJ: Princeton University Press.

Roth, A. E., V. Prasnikar, M. Okuno-Fujiwara, and S. Zamir. 1991. "Bargaining and Market Behavior in Jerusalem, Ljubljana, Pittsburgh, and Tokyo: An Experimental Study." *American Economic Review* 81 (5): 1068–95.

Rozin, P. 1991. "Family Resemblance in Food and Other Domains: The Family Paradox and the Role of Parental Congruence." *Appetite* 16 (2): 93–102.

Rozin, P., and C. Nemeroff. 1990. "The Laws of Sympathetic Magic: a Psychological Analysis of Similarity and Contagion." In *Cultural Psychology: Essays on Comparative Human Development*, edited by J.W. Stigler, Shweder R.A., and G. Herdt, 205–32. Cambridge, UK: Cambridge University Press.

Rozin, P., and T. A. Vollmecke. 1986. "Food Likes and Dislikes." *Annual Review of Nutrition* 6:433–56.

Sahlins, M. P. 1972. *Stone Age Economics*. Chicago, IL: Aldine Publishing Company.

Schotter, A., A. Weiss, and I Zapater. 1996. "Fairness and Survival in Ultimatum and Dictatorship Games." *Journal of Economic Behavior & Organization* 31 (1): 37–56.

Schultze, C. L. 1977. *The Public Use of Private Interest.* Washington, DC: Brookings Institution Press.

Schumpeter, J. A. 1950. *Capitalism, Socialism and Democracy.* 3rd ed. New York, NY: Harper & Row.

Sen, A. 1977. "Rational Fools: A Critique of the Behavioral Foundations of Economic Theory." *Philosophy and Public Affairs* 6 (4): 317–44.

Shapiro, C. 1983. "Premiums for High Quality Products as Returns to Reputations." *The Quarterly Journal of Economics* 98 (4): 659–79.

Sherif, M. 1937. "An Experimental Approach to the Study of Attitudes." *Sociometry* 1:90–98.

Siamwalla, A. 1978. "Farmers and Middlemen: Aspects of Agricultural Marketing in Thailand." *Economic Bulletin for Asia and the Pacific* 29 (1): 38–50.

Simmel, G. 1990. *The Philosophy of Money.* London, UK: Routledge.

Simon, H. A. 1951. "A Formal Theory of the Employment Relationship." *Econometrica* 19:293–305.

Solow, R. M. 1990. *The Labor Market as a Social Institution.* Cambridge, MA: Blackwell.

Starr, C. 1969. "Social Benefits Versus Technological Risk." *Science* 165:1232–38.

Stigler, G. J., and G. S. Becker. 1977. "De Gustibus Non Est Disputandum." *The American Economic Review* 67 (2): 76–90.

Sunstein, C. R. 1993. "Endogenous Preferences, Environmental Law." *The Journal of Legal Studies* 22 (2): 217–54.

Taussig, M. T. 1980. *The Devil and Commodity Fetishism in South America.* Chapel Hill, NC: University of North Carolina Press.

Taylor, M. 1976. *Anarchy and Cooperation.* Cambridge, UK: Cambridge University Press.

———. 1987. *The Possibility of Cooperation.* Cambridge, UK: Cambridge University Press.

Titmuss, R. M. 1971. *The Gift Relationship: From Human Blood to Social Policy.* Pantheon Books.

Tonnies, F. 1963. *Community & Society.* New York, NY: Harper & Row.

Tversky, A., and D. Kahneman. 1986. "Rational Choice and the Framing of Decisions." *The Journal of Business* 59 (4): 251–78.

Upton III, W. E. 1973. "Altruism, Attribution, and Intrinsic Motivation in the Recruitment of Blood Donors." PhD diss., Cornell University.

Veblen, T. 1934. *The Theory of the Leisure Class.* New York, NY: Modern Library.

Weber, M. [1922]1978. *Economy and Society: An Outline of Interpretive Sociology, Vols. I and II.* Edited by G. Roth and C. Wittich. Berkeley, CA: University of California Press.

Whiting, B. B., ed. 1963. *Six Cultures: Studies of Child Rearing.* New York: John Wiley and Sons.

Whiting, B. B., and J. W. M. Whiting. 1975. *Children of Six Cultures: A Psycho-Cultural Analysis.* Cambridge, MA: Harvard University Press.

Wilson, D. S., and E. Sober. 1994. "Reintroducing Group Selection to the Human Behavioral Sciences." *Behavioral and Brain Sciences* 17:585–654.

Witkin, H. A., and J. W. Berry. 1975. "P. Differentiation in Cross-Cultural Perspective." *Journal of Cross-Cultural Psychology* 6 (1): 4–87.

Zajonc, R. B. 1968. "Attitudinal Effects of Mere Exposure." Monograph Supplement, *Journal of Personality and Social Psychology* 9 (2): 1–27.

———. 1980. "Feeling and Thinking: Preferences Need No Inferences." *American Psychologist* 35 (2): 151–75.

Zelizer, V. A. 1996. "Payments and Social Ties." *Sociological Forum* 11 (3): 481–95.

Zufferey, F. S. 1986. *A Study of Local Institutions and Resource Management Inquiry in Eastern Central District.* Land Tenure Center, University of Wisconsin-Madison.

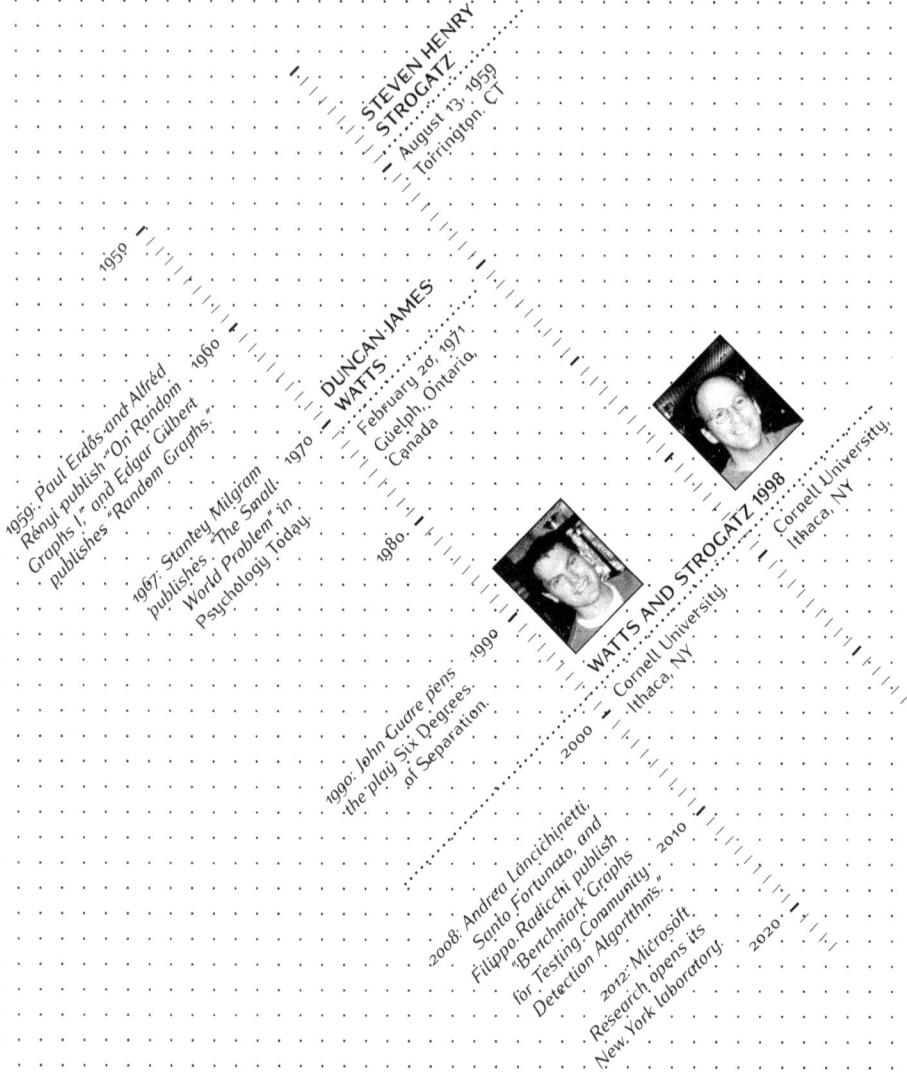

WATTS & STROGATZ

[87]

HOW THE SMALL-WORLD MODEL TRANSFORMED HOW WE THINK ABOUT CONNECTIVITY

Michelle Girvan, University of Maryland and Santa Fe Institute

The field of network science as we know it today is widely thought to have been ignited by Duncan Watts and Steve Strogatz's 1998 paper "The Collective Dynamics of 'Small-World' Networks." Network science is a highly interdisciplinary field that focuses on how the pattern of connectivity between a system's components is critical to its function. Applications range from brain networks to power grids to epidemics. Researchers working in network science have a wide diversity of backgrounds, including physical, mathematical, social, biological, and computational sciences. In trying to explain complex phenomena by probing interaction patterns, network science is also closely tied to complex-systems science.

While network science as a field is relatively young, researchers were, of course, studying networks long before Watts and Strogatz introduced their "small-world" network model. The difference between research within the sphere of what we now call network science and those earlier studies (e.g., graph theory proofs by mathematicians, network algorithms by computer scientists, social network analysis by sociologists) is that the aim of network science, like that of complex-systems science, is to unravel and understand the common drivers of complexity that appear across a wide range of applications.

A major contribution of the Watts–Strogatz "small-world" network model was to demonstrate that, by tuning between regularity and randomness, networks could simultaneously have both high clustering and short average path length. In the context of a social network, high clustering means that two of your friends are likely to be friends with one another. And short average path length is associated with the notion of "six degrees

D. J. Watts and S. H. Strogatz, "Collective Dynamics of 'Small-World' Networks," *Nature* 393, 440–442 (1998).

Reprinted with permission of the authors and Springer Nature.

of separation"—the property that the average number of links between two individuals in the network is relatively short.

Well-studied networks within the physics, applied mathematics, and complex systems communities exhibited one or the other but not both of these properties. Abstract mathematical models of physical, biological, and technological systems tended to assume one of two extremes for the connectivity pattern: completely regular lattice structures, with strong local structure and potential for high clustering, or completely random networks, with short average path lengths.

The small-world network model of Watts and Strogatz interpolated between these extremes, probing the regime between the perfectly regular and the perfectly random—where we expect that real-world networks reside. To accomplish this, the model begins with nodes placed uniformly around a ring, and then each node is connected to its k nearest neighbors on either side. Next, with probability p, each link is rewired by choosing one of its ends at random and connecting it to a random node. At $p = 0$, the network is a highly clustered one-dimensional lattice with long average path length. At $p = 1$, the network is random with low clustering and short average path length. In between, they find a significant region over which the resulting networks simultaneously have high clustering and short average path length, incorporating elements of both regularity and randomness. Importantly, they show that adding only a small amount of randomness to the regular network is enough to make the path lengths drop to near the level of the fully random network.

A critical feature of the Watts–Strogatz paper was that it not only contained an elegant abstract model, but it also connected that model to data. Watts and Strogatz reported clustering and average path length values for three different networks: a biological network (the neural network of the nematode *C. elegans*), a proxy for a social network (the network of actors connected through movies), and a technological network (the power grid of the western United States and Canada). In all three cases they showed that, as compared with a random graph with the same number of links, the clustering values of the real networks were far higher, while the average path lengths were almost as short. In other words, the empirical networks showed structural signatures of both regularity and randomness. Several studies have revealed this trend to be ubiquitous across diverse datasets (for a review and compilation of statistics across a wide range of networks see Newman 2018).

By connecting their model with data, Watts and Strogatz paved the way for network science to serve as a bridge between abstract complex systems models and data-driven applications. Like another foundational paper in network science—Barabási and Albert's (1999) "preferential-attachment" model of networks—Watts and Strogatz's paper was firmly rooted in the complex systems community by beginning with a simple model meant to explain the origin of complex organization across many applications. Unlike many earlier simple models of complex systems, these two foundational papers in network science generated tremendous excitement from application area experts by demonstrating how they might be directly relevant to important questions in empirical research.

While the paper by Watts and Strogatz may have sparked the field of network science, there were certainly other factors that contributed to its explosive growth. The most important of these was arguably the near-simultaneous emergence of the era of big data, which meant that large-scale networks were being mapped out in contexts ranging from the internet to protein interaction networks to social media networks. But more data cannot advance our understanding of the world without models to help us unravel its complexities, especially penetrating and insightful models like the one introduced by Watts and Strogatz.

REFERENCES

Barabási, A.-L., and R. Albert. 1999. "Emergence of Scaling in Random Networks." *Science* 286 (5439): 509–512. https://doi.org/10.1126/science.286.5439.5.

Newman, M. 2018. *Networks: An Introduction*. Oxford, UK: Oxford University Press. https://doi.org/10.1093/acprof:oso/9780199206650.001.0001.

COLLECTIVE DYNAMICS OF 'SMALL-WORLD' NETWORKS

Duncan J. Watts, Cornell University
and Steven H. Strogatz, Cornell University

Abstract

> This paper is best known for the simple model of network structure it introduced, which came to be known as the Watts–Strogatz model of "small-world" networks. However, it's worth noting that the authors were motivated by the interplay between structure and dynamics right from the start, as evidenced by the inclusion of "dynamics" in the title.

Networks of coupled dynamical systems have been used to model biological oscillators (Winfree 1980; Kuramoto 1984; Strogatz and Stewart 1993; Bressloff, Coombes, and De Souza 1997), Josephson junction arrays (Braiman, Lindner, and Ditto 1995; Wiesenfeld 1996), excitable media (Gerhardt, Schuster, and Tyson 1990), neural networks (Collins, Chow, and Imhoff 1995; Hopfield and Herz 1995; Abbott and van Vreeswijk 1993), spatial games (Nowak and May 1992), genetic control networks (Kauffman 1969) and many other self-organizing systems. Ordinarily, the connection topology is assumed to be either completely regular or completely random. But many biological, technological and social networks lie somewhere between these two extremes. Here we explore simple models of networks that can be tuned through this middle ground: regular networks 'rewired' to introduce increasing amounts of disorder. We find that these systems can be highly clustered, like regular lattices, yet have small characteristic path lengths, like random graphs. We call them 'small-world' networks, by analogy with the small-world phenomenon (Milgram 1967; Kochen 1989) (popularly known as six degrees of separation; Guare 1990). The neural network of the worm *Caenorhabditis elegans*, the power grid of the western United States, and the collaboration graph of film actors are shown to be small-world networks. Models of dynamical systems with small-world coupling display enhanced signal-propagation speed, computational power, and synchronizability. In particular, infectious diseases spread more easily in small-world networks than in regular lattices.

To interpolate between regular and random networks, we consider the following random rewiring procedure (Fig. 1). Starting from a ring lattice with n vertices and k edges per vertex, we rewire each edge at random with probability p. This construction allows us to 'tune' the graph between regularity ($p = 0$) and disorder ($p = 1$), and thereby to probe the intermediate region $0 < p < 1$, about which little is known.

We quantify the structural properties of these graphs by their characteristic path length $L(p)$ and clustering coefficient $C(p)$, as defined in Fig. 2 legend. Here $L(p)$ measures the typical separation between two vertices in the graph (a global property), whereas $C(p)$ measures the cliquishness of a typical neighbourhood (a local property). The networks of interest to us have many vertices with sparse connections, but not so sparse that the graph is in danger of becoming disconnected. Specifically, we require $n \gg k \gg \ln(n) \gg 1$, where $k \gg \ln(n)$ guarantees that a random graph will be connected (Bollabás 1985). In this regime, we find that $L \sim n/2k \gg 1$ and $C \sim 3/4$ as $p \to 0$, while $L \approx L_{\text{random}} \sim \ln(n)/\ln(k)$ and $C \approx C_{\text{random}} \sim k/n \ll 1$ as $p \to 1$. Thus the regular lattice at $p = 0$ is a highly clustered, large world where L grows linearly with n, whereas the random network at $p = 1$ is a poorly clustered, small world where L grows only logarithmically with n. These limiting cases might lead one to suspect that large C is always associated with large L, and small C with small L.

On the contrary, Fig. 2 reveals that there is a broad interval of p over which $L(p)$ is almost as small as L_{random} yet $C(p) \gg C_{\text{random}}$. These small-world networks result from the immediate drop in $L(p)$ caused by the introduction of a few long-range edges. Such 'short cuts' connect vertices that would otherwise be much farther apart than L_{random}. For small p, each short cut has a highly nonlinear effect on L, contracting the distance not just between the pair of vertices that it connects, but between their immediate neighbourhoods, neighbourhoods of neighbourhoods and so on. By contrast, an edge removed from a clustered neighbourhood to make a short cut has, at most, a linear effect on C; hence $C(p)$ remains practically unchanged for small p even though $L(p)$ drops rapidly. The important implication here is that at the local level (as reflected by $C(p)$), the transition to a small world is almost undetectable. To check the robustness of these results, we have tested many different types of initial regular graphs, as well as different algorithms for random rewiring, and all give qualitatively similar results. The only requirement is that the rewired edges must typically connect vertices that would otherwise be much farther apart than L_{random}.

The idealized construction above reveals the key role of short cuts. It suggests that the small-world phenomenon might be common in sparse networks with many vertices, as even a tiny fraction of short cuts would suffice. To test this idea, we have computed L and C

In the network science literature, "small-world property" is sometimes used to mean only that a network has a short average path length. Others use "small-world networks" in the way Watts and Strogatz introduced in the abstract and in their model, to refer to networks that exhibit both short average path lengths and high clustering. But note that they contribute to the confusion of the terminology by saying here that a random network is a "poorly clustered small world."

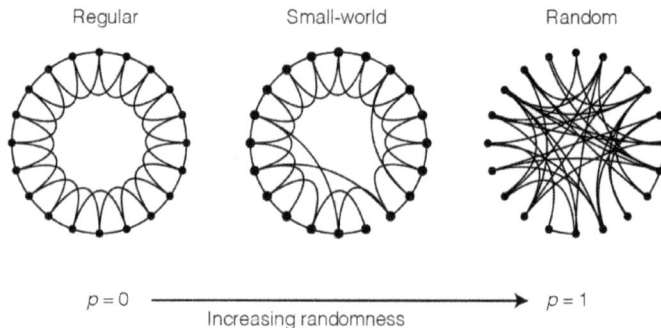

Figure 1. Random rewiring procedure for interpolating between a regular ring lattice and a random network, without altering the number of vertices or edges in the graph. We start with a ring of n vertices, each connected to its k nearest neighbours by undirected edges. (For clarity, $n = 20$ and $k = 4$ in the schematic examples shown here, but much larger n and k are used in the rest of this Letter.) We choose a vertex and the edge that connects it to its nearest neighbour in a clockwise sense. With probability p, we reconnect this edge to a vertex chosen uniformly at random over the entire ring, with duplicate edges forbidden; otherwise we leave the edge in place. We repeat this process by moving clockwise around the ring, considering each vertex in turn until one lap is completed. Next, we consider the edges that connect vertices to their second-nearest neighbours clockwise. As before, we randomly rewire each of these edges with probability p, and continue this process, circulating around the ring and proceeding outward to more distant neighbours after each lap, until each edge in the original lattice has been considered once. (As there are $nk/2$ edges in the entire graph, the rewiring process stops after $k/2$ laps.) Three realizations of this process are shown, for different values of p. For $p = 0$, the original ring is unchanged; as p increases, the graph becomes increasingly disordered until for $p = 1$, all edges are rewired randomly. One of our main results is that for intermediate values of p, the graph is a small-world network: highly clustered like a regular graph, yet with small characteristic path length, like a random graph. (See Fig. 2.)

for the collaboration graph of actors in feature films (generated from data available at http://us.imdb.com), the electrical power grid of the western United States, and the neural network of the nematode worm *C. elegans* (Achacoso and Yamamoto 1992). All three graphs are of scientific interest. The graph of film actors is a surrogate for a social network (Wasserman and Faust 1994), with the advantage of being much more easily specified. It is also akin to the graph of mathematical collaborations centred, traditionally, on P. Erdös (partial data available at http://www.acs.oakland.edu/~grossman/erdoshp.html). The graph of the power grid is relevant to the efficiency and robustness of power networks (Phadke and Thorp 1988). And *C. elegans* is the sole example of a completely mapped neural network.

Table 1 shows that all three graphs are small-world networks. These examples were not hand-picked; they were chosen because of their inherent interest and because complete wiring diagrams were available.

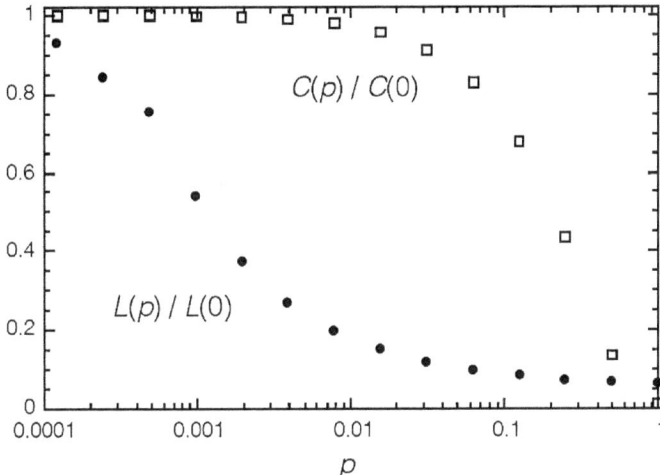

Figure 2. Characteristic path length $L(p)$ and clustering coefficient $C(p)$ for the family of randomly rewired graphs described in Fig. 1. Here L is defined as the number of edges in the shortest path between two vertices, averaged over all pairs of vertices. The clustering coefficient $C(p)$ is defined as follows. Suppose that a vertex v has k_v neighbours; then at most $k_v(k_v - 1)/2$ edges can exist between them (this occurs when every neighbour of v is connected to every other neighbour of v). Let C_v denote the fraction of these allowable edges that actually exist. Define C as the average of C_v over all v. For friendship networks, these statistics have intuitive meanings: L is the average number of friendships in the shortest chain connecting two people; C_v reflects the extent to which friends of v are also friends of each other; and thus C measures the cliquishness of a typical friendship circle. The data shown in the figure are averages over 20 random realizations of the rewiring process described in Fig. 1, and have been normalized by the values $L(0), C(0)$ for a regular lattice. All the graphs have $n = 1,000$ vertices and an average degree of $k = 10$ edges per vertex. We note that a logarithmic horizontal scale has been used to resolve the rapid drop in $L(p)$, corresponding to the onset of the small-world phenomenon. During this drop, $C(p)$ remains almost constant at its value for the regular lattice, indicating that the transition to a small world is almost undetectable at the local level.

EMPIRICAL EXAMPLES OF SMALL-WORLD NETWORKS

	L_{actual}	L_{random}	C_{actual}	C_{random}
Film actors	3.65	2.99	0.79	0.00027
Power grid	18.7	12.4	0.080	0.005
C. elegans	2.65	2.25	0.28	0.05

Table 1. Characteristic path length L and clustering coefficient C for three real networks, compared to random graphs with the same number of vertices (n) and average number of edges per vertex (k). (Actors: $n = 225,226, k = 61$. Power grid: $n = 4,941, k = 2.67$. *C. elegans*: $n = 282, k = 14$.) The graphs are defined as follows. Two actors are joined by an edge if they have acted in a film together. We restrict attention to the giant connected component (Bollabás 1985) of this graph, which includes $\sim 90\%$ of all actors listed in the Internet Movie Database (available at http://us.imdb.com), as of April 1997. For the power grid, vertices represent generators, transformers and substations, and edges represent high-voltage transmission lines between them. For *C. elegans*, an edge joins two neurons if they are connected by either a synapse or a gap junction. We treat all edges as undirected and unweighted, and all vertices as identical, recognizing that these are crude approximations. All three networks show the small-world phenomenon: $L \geqslant L_{\text{random}}$ but $C \gg C_{\text{random}}$.

Since this paper was published, a wide variety of real-world networks have been shown to exhibit similar properties of low average path length and high clustering. Examples include coauthorship networks, word co-occurence networks, metabolic networks, and food webs (Newman 2018).

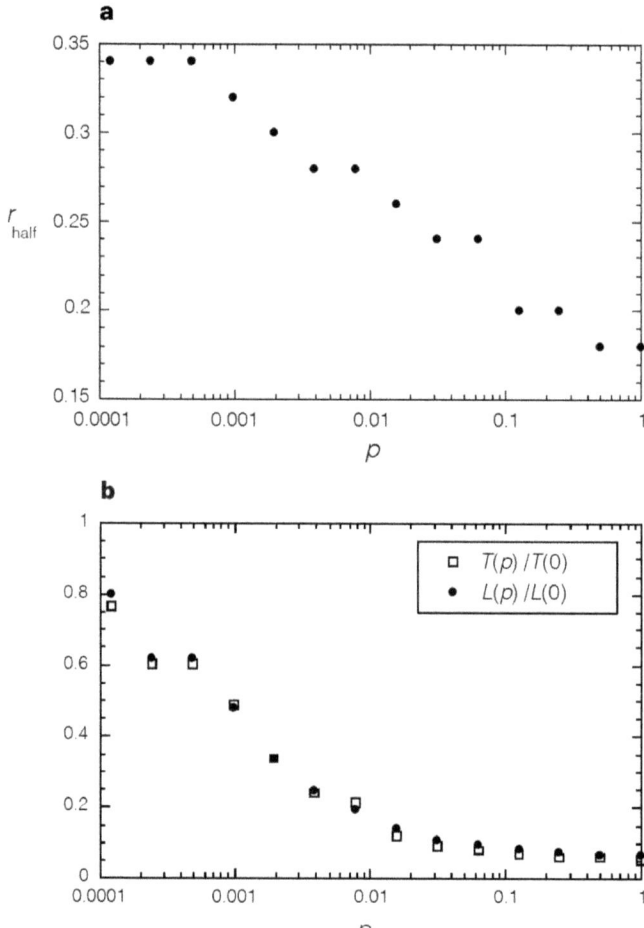

Figure 3. Simulation results for a simple model of disease spreading. The community structure is given by one realization of the family of randomly rewired graphs used in Fig. 1. **a**, Critical infectiousness r_{half}, at which the disease infects half the population, decreases with p. **b**, The time $T(p)$ required for a maximally infectious disease ($r = 1$) to spread throughout the entire population has essentially the same functional form as the characteristic path length $L(p)$. Even if only a few per cent of the edges in the original lattice are randomly rewired, the time to global infection is nearly as short as for a random graph.

Thus the small-world phenomenon is not merely a curiosity of social networks (Milgram 1967; Kochen 1989) nor an artefact of an idealized model—it is probably generic for many large, sparse networks found in nature.

We now investigate the functional significance of small-world connectivity for dynamical systems. Our test case is a deliberately simplified model for the spread of an infectious disease. The population structure is modelled by the family of graphs described in Fig. 1. At time $t = 0$, a single infective individual is introduced into an otherwise healthy population. Infective individuals are removed permanently (by immunity or death) after a period of sickness that lasts one unit of dimensionless time. During this time, each infective individual can infect each of its healthy neighbours with probability r. On subsequent time steps, the disease spreads along the edges of the graph until it either infects the entire population, or it dies out, having infected some fraction of the population in the process.

Two results emerge. First, the critical infectiousness r_{half}, at which the disease infects half the population, decreases rapidly for small p (Fig. 3a). Second, for a disease that is sufficiently infectious to infect the entire population regardless of its structure, the time $T(p)$ required for global infection resembles the $L(p)$ curve (Fig. 3b). Thus, infectious diseases are predicted to spread much more easily and quickly in a small world; the alarming and less obvious point is how few short cuts are needed to make the world small.

Our model differs in some significant ways from other network models of disease spreading (Sattenspiel and Simon 1988; Longini Jr. 1988; Hess 1996; Blythe, Castillo-Chavez, and Palmer 1991; Kretschmar and Morris 1996). All the models indicate that network structure influences the speed and extent of disease transmission, but our model illuminates the dynamics as an explicit function of structure (Fig. 3), rather than for a few particular topologies, such as random graphs, stars and chains (Sattenspiel and Simon 1988; Longini Jr. 1988; Hess 1996; Blythe, Castillo-Chavez, and Palmer 1991). In the work closest to ours, Kretschmar and Morris (1996) have shown that increases in the number of concurrent partnerships can significantly accelerate the propagation of a sexually-transmitted disease that spreads along

Interestingly, while the connection with dynamics was a strong motivation for this work, most references to this paper focus solely on the structural model. The second part of the paper, which focuses on dynamics, gets much less attention.

The crucial insight here is that a very small amount of randomness is enough to deliver the advantages or disadvantages of a small world.

the edges of a graph. All their graphs are disconnected because they fix the average number of partners per person at $k = 1$. An increase in the number of concurrent partnerships causes faster spreading by increasing the number of vertices in the graph's largest connected component. In contrast, all our graphs are connected; hence the predicted changes in the spreading dynamics are due to more subtle structural features than changes in connectedness. Moreover, changes in the number of concurrent partners are obvious to an individual, whereas transitions leading to a smaller world are not.

We have also examined the effect of small-world connectivity on three other dynamical systems. In each case, the elements were coupled according to the family of graphs described in Fig. 1. (1) For cellular automata charged with the computational task of density classification (Das, Mitchell, and Crutchfield 1994), we find that a simple 'majority-rule' running on a small-world graph can outperform all known human and genetic algorithm-generated rules running on a ring lattice. (2) For the iterated, multi-player 'Prisoner's dilemma' (Nowak and May 1992) played on a graph, we find that as the fraction of short cuts increases, cooperation is less likely to emerge in a population of players using a generalized 'tit-for-tat' (Axelrod 1984) strategy. The likelihood of cooperative strategies evolving out of an initial cooperative/non-cooperative mix also decreases with increasing p. (3) Small-world networks of coupled phase oscillators synchronize almost as readily as in the mean-field model (Kuramoto 1984), despite having orders of magnitude fewer edges. This result may be relevant to the observed synchronization of widely separated neurons in the visual cortex (Gray *et al.* 1989) if, as seems plausible, the brain has a small-world architecture.

As of May 2024, this paper had over 54,000 citations according to Google Scholar. In hoping to stimulate further studies, Watts and Strogatz couldn't have imagined how influential their work would become.

We hope that our work will stimulate further studies of small-world networks. Their distinctive combination of high clustering with short characteristic path length cannot be captured by traditional approximations such as those based on regular lattices or random graphs. Although small-world architecture has not received much attention, we suggest that it will probably turn out to be widespread in biological, social and man-made systems, often with important dynamical consequences.

Acknowledgements

We thank B. Tjaden for providing the film actor data, and J. Thorp and K. Bae for the Western States Power Grid data. This work was supported by the US National Science Foundation (Division of Mathematical Sciences).

REFERENCES

Abbott, L. F., and C. van Vreeswijk. 1993. "Asynchronous States in Neural Networks of Pulse-Coupled Oscillators." *Physical Review E* 48 (2): 1483–1490.

Achacoso, T. B., and W. S. Yamamoto. 1992. *AY's Neuroanatomy of C. Elegans for Computation*. Boca Raton, FL: CRC Press.

Axelrod, R. 1984. *The Evolution of Cooperation*. New York, NY: Basic Books.

Blythe, S. P., C. Castillo-Chavez, and J. S. Palmer. 1991. "Toward a Unified Theory of Sexual Mixing and Pair Formation." *Mathematical Biosciences* 107:379–405.

Bollabás, B. 1985. *Random Graphs*. London, UK: Academic.

Braiman, Y., J. F. Lindner, and W. L. Ditto. 1995. "Taming Spatiotemporal Chaos with Disorder." *Nature* 378:465–467.

Bressloff, P. C., S. Coombes, and B. De Souza. 1997. "Dynamics of a Ring of Pulse-Coupled Oscillators: A Group Theoretic Approach." *Physical Review Letters* 79:2791–2794.

Collins, J. J., C. C. Chow, and T. T. Imhoff. 1995. "Stochastic Resonance without Tuning." *Nature* 376:236–238.

Das, R., M. Mitchell, and J. P. Crutchfield. 1994. In *Parallel Problem Solving from Nature,* edited by Y. Davido, H.-P. Schwefel, and R. Männer, 866:344–353. Lecture Notes in Computer Science. Berlin, Germany: Springer.

Gerhardt, M., H. Schuster, and J. J. Tyson. 1990. "A Cellular Automaton Model of Excitable Media Including Curvature and Dispersion." *Science* 247:1563–1566.

Gray, C. M., P. König, A. K. Engel, and W. Singer. 1989. "Oscillatory Responses in Cat Visual Cortex Exhibit Intercolumnar Synchronization which Reflects Global Stimulus Properties." *Nature* 338:334–337.

Guare, J. 1990. *Six Degrees of Separation: A Play*. New York, NY: Vintage Books.

Hess, G. 1996. "Disease in Metapopulation Models: Implications for Conservation." *Ecology* 77:1617–1632.

Hopfield, J. J., and A. V. M. Herz. 1995. "Rapid Local Synchronization of Action Potentials: Toward Computation with Coupled Integrate-and-Fire Neurons." *Proceedings of the National Academy of Sciences,* 6655–6662.

Kauffman, S. A. 1969. "Metabolic Stability and Epigenesis in Randomly Constructed Genetic Nets." *Journal of Theoretical Biology* 22:437–467.

Kochen, M., ed. 1989. *The Small World.* Norwood, NJ: Ablex.

Kretschmar, M., and M. Morris. 1996. "Measures of Concurrency in Networks and the Spread of Infectious Disease." *Mathematical Biosciences* 133:165–195.

Kuramoto, Y. 1984. *Chemical Oscillations, Waves, and Turbulence.* Berlin, Germany: Springer.

Longini Jr., I. M. 1988. "A Mathematical Model for Predicting the Geographic Spread of New Infectious Agents." *Mathematical Biosciences* 90:367–383.

Milgram, S. 1967. "The Small World Problem." *The Psychology Today* 2:60–67.

Nowak, M. A., and R. M. May. 1992. "Evolutionary Games and Spatial Chaos." *Nature,* 826–829.

Phadke, A. G., and J. S. Thorp. 1988. *Computer Relaying for Power Systems.* New York, NY: Wiley.

Sattenspiel, L., and C. P. Simon. 1988. "The Spread and Persistence of Infectious Diseases in Structured Populations." *Mathematical Biosciences* 90:341–366.

Strogatz, S. H., and I. Stewart. 1993. "Coupled Oscillators and Biological Synchronization." *Scientific American* 269 (6): 102–109.

Wasserman, S., and K. Faust. 1994. *Social Network Analysis: Methods and Applications.* Cambridge, UK: Cambridge University Press.

Wiesenfeld, K. 1996. "New Results on Frequency-Locking Dynamics of Disordered Josephson Arrays." *Physica B* 222:315–319.

Winfree, A. T. 1980. *The Geometry of Biological Time.* New York, NY: Springer.

DAVID PINES
June 8, 1924
Kansas City, MO

PETER GUY WOLYNES
April 27, 1953
Chicago, IL

BRANKO P. STOJKOVIĆ
January 21, 1967
Belgrade, Serbia

JÖRG SCHMALIAN
April, 1965
Halberstadt, East Germany

ROBERT BETTS LAUGHLIN
November 1, 1950
Visalia, CA

1940

1950 — 1959: Richard Feynman presents his lecture "There's Plenty of Room at the Bottom" for the first time.

1960

1970

1980 — 1987: Marc Mezard, Giorgio Parisi, and Miguel Angel Virasoro publish "Spin Glass Theory and Beyond."

1990

LAUGHLIN, PINES, SCHMALIAN, STOJKOVIĆ, AND WOLYNES 2000 — 2000: Venki Ramakrishnan's laboratory completely describes the full molecular structure of the 30S subunit of a ribosome.

Stanford University, Palo Alto, CA

University of California, LANL, and University of Illinois Urbana–Champaign, Urbana, IL

Iowa State University, Ames, IA

Los Alamos National Laboratory, Los Alamos, NM

University of Illinois Urbana–Champaign, Urbana, IL

DEATH
May 3, 2018
Urbana, IL
cancer

2010 — 2003: Samsung introduces Silver Nano, their antibacterial coating, composed of silver nanoparticles, which they use in various appliances to inhibit bacterial cell growth.

2020

STOJKOVIĆ PHOTO © LARRY LAMSA (CREATIVE COMMONS DEED - ATTRIBUTION 4.0 INTERNATIONAL)

LAUGHLIN, PINES, SCHMALIAN, STOJKOVIĆ & WOLYNES

[88]

MESOCOSMOS: IN SEARCH OF UNIVERSAL LAWS OF LIVING MATERIAL

Christopher P. Kempes, Santa Fe Institute

Cells occupy an unusual position in the grand scale of things, not truly atomic, but not truly macroscopic either. For example, the smallest bacterial cells are composed of millions of atoms, yet certain macromolecules, such as the ribosome, exist in very small quantities (e.g., there are $<$ 100 ribosomes for the smallest bacteria) (Kempes *et al.* 2017). This discreteness has downstream consequences, for example, as the intrinsic noise associated with randomly splitting a small pool of molecules into two daughter cells. In describing the physics of cells we are caught between the relatively easy physics of single atoms, where quantum mechanics dominates, and the classical physics of macroscopic systems such as hydraulics. This is exactly the point of Robert Laughlin, David Pines, Joerg Schmalian, Branko P. Stojković, and Peter Wolynes in "The Middle Way." They aren't just interested in cells but in all "mesoscopic" matter, with a particular interest in biological matter. The advances in condensed matter physics and material science associated with mesoscopic phenomena have been immense since the time that "The Middle Way" was published (e.g., Fish, Wagner, and Keten 2021), but here I want to focus on the biological aspects of this foundational paper, where the mesoscopic is relevant to everything from protein folding to the structure of tissues in multicellular organisms. Indeed, the mesoscopic is still at the heart of the ongoing revolution occurring in biological matter, both from the perspective of detailed dynamics and in terms of broad organizing principles and concepts.

"The Middle Way" did a fantastic job of identifying many of the key ideas and challenges for the following twenty years of biology. Namely, that (1) mesoscopic phenomena would come to dominate much of the thinking in biology, (2) that principles beyond evolution are needed to truly understand life, (3) that mere genetic cataloging would fall short

R. B. Laughlin, D. Pines, J. Schmalian, B. P. Stojković, and P. Wolynes, "The Middle Way," *PNAS* 97 (1), 32–37 (2000).

Copyright (2000) National Academy of Sciences, USA. Reprinted with permission.

of creating a complete understanding of biology, (4) that free energy principles come to organize our understanding of protein folding, and (5) the importance of phases in cells. Below I focus on a few of these ideas in more detail in order to highlight the importance of finding the "middle way" in biology.

A Shift in Perspective

An important argument in this paper is that biology needs other organizational principles beyond evolution by natural selection, which the authors point out is a perspective held by most of the field. This is a critical point for moving our understanding of life forward, especially for questions addressing the origins of life, alternate evolutionary histories, the search for life beyond Earth, and the systematization of organismal diversity into "laws." However, it is also important to note that there is a long history of people calling for such principles, a notable early example being D'Arcy Thompson (1917), who famously, and a bit defensively, stated,

> *But I maintain that it is no less an exaggeration if we tend to neglect these direct physical and mechanical modes of causation altogether, and to see in the characters of a bone merely the results of variation and heredity...*

The utility of searching for principles in biology remains one of the most important ongoing conversations in biology, with a variety of recent successes and continued tension. Laughlin *et al.* critique "bioinformatics as librarianship" and argue that such a focus naturally follows from the assumption that there are no organizing principles and that only exhaustive cataloging of physiology can explain how cells function. The authors were spot on in forecasting that the complete enumeration of various *omics* would still fall short of explaining problems of key interest in biology, such as the genotype-phenotype map, development in multicellular organisms, and the complete set of organismal physiology and dynamics. The authors are then honest about the fact that they don't really know what the organizing principles of mesoscopic systems are, or even if such principles exist. Over the last twenty-five years there have been successes in understanding such mesoscopic living materials, which we will come to later, but I also want to point out that the authors may be overly focused on thinking of life purely as the emergent properties of mesoscopic material. That is, the bookending perspectives are that either life is an emergent property of certain scales and physics or that it is a set of broad principles and functions *contending* with the constraints of particular scales and physics. The authors are more focused on the former end of this spectrum and are in search of "protected properties" of the mesoscopic, which they define as generic behavior that is conserved across systems independent of the details. Here they want to find properties like ferromagnetism that are part of a universality class and transcend the details of the system. Such properties likely exist for life, but there is also utility in turning this problem around a bit in order to see life as finding general properties or solutions through evolution under the constraints of mesoscopic physics (Kempes and Krakauer 2021). For example, the error threshold sets hard limits on the fidelity with which the information storage system of any living system must be replicated. At different physical scales, and using different materials, this will set limitations on how information storage and replication is achieved. Understanding the details of the DNA replication system depends on knowing both the physical and material constraints at the mesoscale and the more abstract constraint of the error threshold. Perhaps life's materials can only be understood in terms of an evolutionary co-optimization of physical and functional constraints (Kempes and Krakauer 2021).

Similarly, metabolic scaling theory has emerged as one of the great biological syntheses over the last twenty-five years (e.g., West 2017). Here many organism features are known to systematically change following a power law of system size. Such scaling behavior has a natural connection with the types of universality classes that Laughlin *et al.* are looking for, and indeed exist for the smallest cells (Kempes, Dutkiewicz, and Follows 2011; Kempes *et al.* 2017), but, again, are the product of an evolution optimization of particular biological functions under physical constraints. They are not necessarily protected properties of mesoscopic matter. It may be the case that such protected properties are discovered for living systems in the future, and indeed such a search should continue as these would revolutionize a general understanding of life, but those properties may exist at a more abstract or functional level, with a variety of diverse implementations in particular materials.

The Strange Materials of Life

The authors make repeated calls for new understanding of mesoscopic dynamics and experiments to probe this physical scale, and the last twenty years has seen a revolution in experimental techniques and the dynamics and material properties that these techniques have uncovered. Here the range of physical dynamics observed is awe-inspiring. These include recognizing that diffusion is anomalous in cells (Banks and Fradin 2005), the complicated construction dynamics of amazing molecular machines like the flagellum system (Minamino and Namba 2004), and the impressive range of phenomena associated with "active matter" such as transport along cytoskeleton networks or induced fluid flows around such networks (Pantaloni, Clainche, and Carlier 2001; Qu *et al.* 2021). In many cases, there is still a huge amount of work to be done in organizing these processes into principles, but one emerging synthesis is that of phase separation, which is perhaps most strongly connected with the proposals of "The Middle Way." Recent work has connected the appearance of separated liquid phases within cells with important physiological consequences such as separating reactions, increasing local concentrations, reducing noise, and allowing for signal transport (Hyman, Weber, and Jülicher 2014; Klosin *et al.* 2020; McSwiggen *et al.* 2019). People are still calling for general organizing principles to understand the importance of multiple phases at mesoscopic scales within cells.

A Non-Equilibrium World

It is important to note that within biophysics a key way to understand life that has recently emerged is the perspectives of non-equilibrium thermodynamics and stochastic thermodynamics. While the authors of "The Middle Way" discuss non-ergodicity, they underemphasize the transient and non-equilibrium nature of cells. That cells are driven systems transitioning across a set of states toward division, where distinct physiology is expressed in time (e.g., the onset of genome replication) or in response to environmental conditions (e.g., the nonreplicating survival state), is less emphasized by the authors. The non-equilibrium and stochastic properties of cellular dynamics is emerging as one of the most important considerations for the mesoscopic. Here cells must achieve function in the face of stochasticity and where fundamental limitations, such as the Landauer bound, matter for biological materials. Non-equilibrium thermodynamic considerations of information processing and erasure, sensing and signaling, protein folding, cell replication, and chemical reaction networks is emerging as a central consideration in biology (e.g., Laughlin, de Ruyter van Steveninck, and Anderson 1998; England 2013; Kempes *et al.* 2017; Kolchinsky and Wolpert 2018; Mehta and Schwab 2012; Ouldridge and Rein ten Wolde 2017; Rao and Esposito 2016).

In summary, "The Middle Way" is an impressively prescient article that foreshadowed the next two decades of biophysics. What this paper calls us to do—to search for unifying principles in mesoscopic biological material—would still ring true as a central challenge if it were written today.

REFERENCES

Banks, D. S., and C. Fradin. 2005. "Anomalous Diffusion of Proteins Due to Molecular Crowding." *Biophysical Journal* 89 (5): 2960–71. https://doi.org/10.1529/biophysj.104.051078.

England, J. L. 2013. "Statistical Physics of Self-Replication." *The Journal of Chemical Physics* 139 (12). https://doi.org/10.1063/1.4818538.

Fish, J., G. J. Wagner, and S. Keten. 2021. "Mesoscopic and Multiscale Modelling in Materials." *Nature Materials* 20:774–786. https://doi.org/10.1038/s41563-020-00913-0.

Hyman, A. A., C. A. Weber, and F. Jülicher. 2014. "Liquid-Liquid Phase Separation in Biology." *Annual Review of Cell and Developmental Biology* 30:39–58. https://doi.org/10.1146/annurev-cellbio-100913-013325.

Kempes, C. P., S. Dutkiewicz, and M. J. Follows. 2011. "Growth, Metabolic Partitioning, and the Size of Microorganisms." *Proceedings of the National Academy of Sciences* 109 (2): 495–500. https://doi.org/10.1073/pnas.1115585109.

Kempes, C. P., and D. C. Krakauer. 2021. "The Multiple Paths to Multiple Life." *Journal of Molecular Evolution* 89:415–426. https://doi.org/10.1007/s00239-021-10016-2.

Kempes, C. P., D. H. Wolpert, Z. Cohen, and J. Pérez-Mercader. 2017. "The Thermodynamic Efficiency of Computations Made in Cells Across the Range of Life." *Philosophical Transactions of the Royal Society A: Mathematical, Physical and Engineering Sciences* 375 (2109): 20160343. https://doi.org/10.1098/rsta.2016.0343.

Klosin, A., F. Oltsch, T. Harmon, A. Honigmann, F. Jülicher, A. A. Hyman, and C. Zechner. 2020. "Phase Separation Provides a Mechanism to Reduce Noise in Cells." *Science* 367:464–8. https://doi.org/10.1126/science.aav6691.

Kolchinsky, A., and D. H. Wolpert. 2018. "Semantic Information, Autonomous Agency and Non-Equilibrium Statistical Physics." *Interface Focus* 8 (6): 20180041. https://doi.org/10.1098/rsfs.2018.0041.

Laughlin, S. B., R. de Ruyter van Steveninck, and J. C. Anderson. 1998. "The Metabolic Cost of Neural Information." *Nature Neuroscience* 1:36–41. https://doi.org/10.1038/236.

McSwiggen, D. T., M. Mir, X. Darzacq, and R. Tjian. 2019. "Evaluating Phase Separation in Live Cells: Diagnosis, Caveats, and Functional Consequences." *Genes & Development* 33 (23-24): 1619–34. https://doi.org/10.1101/gad.331520.119.

Mehta, P., and D. J. Schwab. 2012. "Energetic Costs of Cellular Computation." *Proceedings of the National Academy of Sciences* 109 (44): 17978–82. https://doi.org/10.1073/pnas.1207814109.

Minamino, T., and K. Namba. 2004. "Self-Assembly and Type III Protein Export of the Bacterial Flagellum." *Microbial Physiology* 7 (1-2): 5–17. https://doi.org/10.1159/000077865.

Ouldridge, T. E., and Pieter Rein ten Wolde. 2017. "Fundamental Costs in the Production and Destruction of Persistent Polymer Copies." *Physical Review Letters* 118 (15): 158103.

Pantaloni, D., C. L. Clainche, and M.-F. Carlier. 2001. "Mechanism of Actin-Based Motility." *Science* 292 (5521): 1502–1506. https://doi.org/10.1126/science.1059975.

Qu, Z., D. Schildknecht, S. Shadkhoo, E. Amaya, J. Jiang, H. J. Lee, D. Larios, F. Yang, R. Phillips, and M. Thomson. 2021. "Persistent Fluid Flows Defined by Active Matter Boundaries." *Communications Physics* 4 (198): 1–9.

Rao, R., and M. Esposito. 2016. "Nonequilibrium Thermodynamics of Chemical Reaction Networks: Wisdom from Stochastic Thermodynamics." *Physical Review X* 6 (4): 041064. https://doi.org/10.1103/PhysRevX.6.041064.

Thompson, D'A. W. 1917. *On Growth and Form*. Cambridge, UK: Cambridge University Press.

West, G. 2017. *Scale: The Universal Laws of Life, Growth, and Death in Organisms, Cities, and Companies*. New York, NY: Penguin.

THE MIDDLE WAY

R. B. Laughlin, Stanford University,
David Pines, Institute for Complex Adaptive Matter,
Joerg Schmalian, Los Alamos National Laboratory,
Branko P. Sojković, Iowa State University, and
Peter Wolynes, University of Illinois

Abstract

Mesoscopic organization in soft, hard, and biological matter is examined in the context of our present understanding of the principles responsible for emergent organized behavior (crystallinity, ferromagnetism, superconductivity, etc.) at long wavelengths in very large aggregations of particles. Particular attention is paid to the possibility that as-yet-undiscovered organizing principles might be at work at the mesoscopic scale, intermediate between atomic and macroscopic dimensions, and the implications of their discovery for biology and the physical sciences. The search for the existence and universality of such rules, the proof or disproof of organizing principles appropriate to the mesoscopic domain, is called the middle way.

Limits of Understanding

Seeing is the beginning of understanding. This may seem an obvious truism, yet it conflicts with a dogma central to much of science, that knowledge of the underlying physical laws alone is sufficient for us to understand all things, even ones that cannot be seen. But the conflict is only apparent, for the dogma is false. Although behavior of atoms and small molecules can be predicted with reasonable accuracy starting from the underlying laws of quantum mechanics, the behavior of large ones cannot, for the errors always eventually run out of control as the number of atoms increases because of exponentially increasing computer requirements. At the same time, however, very large aggregations of particles have some astonishing properties, such as the ability to levitate magnets when they are cooled to cryogenic temperatures, that are commonly acknowledged to be "understood." How can this be? The answer is that these properties are actually caused by collective organizing principles that formally grow out of the microscopic rules but are in a real sense independent of them.

We say that superfluidity, ferromagnetism, metallic conduction, hydrodynamics, and so forth are "protected" properties of matter—generic behavior that is reliably the same one system to the next, regardless of details (Laughlin and Pines 2000). There are more sophisticated ways of articulating this idea, such as stable fixed point of the renormalization group, but these all boil down to descriptions of behavior that emerges spontaneously and is stable against small perturbations of the underlying equations of motion. Unfortunately, the observational tools with which these principles were discovered work only at long wavelengths. Furthermore, the mathematical tools that have been used to justify the existence of protected properties from the theoretical view have focused on reaching asymptopia, the existence of a thermodynamic limit of a nearly infinite number of particles. More is clearly different (Anderson 1972). But we also must ask is plenty nearly enough? One could debate whether the existence of protected behavior on the macroscopic level is a fundamental truth because of quantum mechanics or is a historical accident because that is where we have had the tools to discover protectorates. However, the fact is that the length scale between atoms and small molecules on the one hand and macroscopic matter on the other is a regime into which we cannot presently see and about which we therefore know very little. This state of affairs would not be of much concern if there were a desert of physical phenomena between the very large and the very small. But, as we all know, there is life in the desert.

The miracles of nature revealed by modern molecular biology are no less astonishing than those found by physicists in macroscopic matter. Their existence leads one to question whether as-yet-undiscovered organizing principles might be at work at the mesoscopic scale, at least in living things. This is by any measure a central philosophical controversy of modern science, for a commonly held view is that there are no principles in biology except for Darwinian evolution. But what if this view is just a consequence of our inability to see? Indeed the rules of self-organization at macroscopic length scales were not self-evident at the time of their discovery and were accepted as true only after repeated

This is the authors' essential goal: can we find properties of biological materials that transcend the specifics or details of a particular system? Such properties would open new windows for bioengineering, synthetic biology, the origins of life, understanding biological diversity, and the search for life beyond Earth. However, it is useful to note that it is still an open question whether such properties emerge out of material physics or exist at a more abstract level of biological function.

The addition of other organizing principles to Darwinian evolution is a critical step for biology. It is indeed a commonly held view that no principles beyond evolution organize biology.

confrontations with experiment left no alternative. The existence of similar rules at the mesoscopic scale would have profound implications for all of science, not just biology, for noncrystalline matter often has curious and poorly understood behavior suggestive of mesoscopic organization. It is thus a question worth asking. We call the search for the existence of mesoscopic protectorates—the proof or disproof of organizing principles appropriate to the mesoscopic domain—the middle way.

Life in the Desert

Twentieth-century science has uncovered the fact that there are numerous large molecules that carry out the processes of life. Although the functions carried out by these molecules are still very incompletely understood, they are amazing to an extent rarely appreciated by physical scientists and engineers. Proteins can catalyze a vast number of unrelated chemical reactions. They can pick out one substrate from thousands of chemically similar ones. They can act like computers executing a sequence of instructions. They can alter their activity through the presence of specific affector molecules in their environments. They can function as signals or receptors for these signals. They can be poisons. They can assemble together spontaneously to form mechanical structures like the cytoskeleton or viruses. The precedent of life allows no other conclusion than that mesoscopic objects organize themselves and function in ways unlike anything we know at very large or very small scales.

Nonbiological systems also have interesting mesoscopic behavior, although it is not as well understood. Glasses, for example, which have structure on this scale, exhibit a strange low temperature-specific heat, and at higher temperatures, memory effect, and nonergodicity, behavior also seen in protein crystals. They are unstable and age, i.e., interconvert their structures slowly over time while showing no significant changes in x-ray scattering, in contrast to the stability and time-independence of crystalline solids (Ediger, Angell, and Nagel 1996). They also exhibit a wide range of time scales of motion, including indications that entire mesoscopic regions reconfigure themselves cooperatively. All of these phenomena are organizational, in that the atomic constituents of glasses and interactions are well known, but how they cooperate to yield the observed behavior is not.

This is an important point and particularly relevant to the origins of life: certain structures are the equilibrium state in specific contexts. Thus, this type of physics explains why and how certain types of structure can be obtained "for free," that is, from the physics alone without much prior biological evolution.

Some kinds of inanimate mesoscopic self-organization can be easily visualized, and perhaps not coincidentally are identified as understood. For example, a variety of mesoscopic structures, some of which are aptly analogous to the cellular membrane, can be formed by assembling artificial polymers in solution or amphiphiles in water-oil mixtures (deGennes and Taupin 1982; Gelbart and Ben Shaul 1996). There are also spherical micelles, self-assembled droplets of surfactant, and interpenetrating networks of water and lipids closely related to structures within the Golgi apparatus (Alberts *et al.* 1994). Such amphiphillic assemblies exhibit dynamics at a range of long time scales similar to the relaxation seen in glasses. Another instance of visible selforganization is the organogel, a simple monomer that does not crystallize easily out of solution but instead forms fibrous webs with complex internal substructure similar to those found in organic gelatins (Geiger *et al.* 1999).

Mesoscopic organization also occurs as a purely electronic phenomenon in systems with relatively defect-free atomic lattices. For example, electrons in semiconductors engineered to the mesoscopic scale show a wealth of incipient ordering phenomena that continue to surprise. There are spin glasses, systems that exhibit remanence, hysteresis, memory, and so forth but consist only of unpaired spins on impurity sites communicating through conventional exchange (Fischer and Hertz 1991; Mezard, Parisi, and Virasoro 1987). There is the class of strongly correlated electronic materials, including heavy-fermion metals, high-T_c and organic superconductors, and colossal magnetoresistive manganites, which exhibit many strange behaviors at the mesoscopic scale that have thus far defied description. Among these behaviors are dynamic magnetic domains (stripes) (Tranquada, Ichikawa, and Uchida 1999), and anomalous low-frequency spin fluctuations (Curro *et al.* 1999) in the cuprate superconductors, large low temperature-specific heats in the heavy electron systems (Fisk *et al.* 1988), and extreme impurity sensitivity. These latter effects have not been conclusively identified as mesoscopic, but their failure to disappear as the sample quality improves is highly suggestive. Ideas about mesoscopic organization in correlated-electron materials are particularly relevant to the larger issue of measurement because they are so obviously prejudiced by the lack of mesoscopic eyes.

Conflicts of Principle

The existence or nonexistence of mesoscopic organizing principles has become an issue of deeply held belief, rarely discussed in public yet informing much of what we do. Whether this situation is the result of intrinsic limitations on measurement capability is perhaps debatable, but its effect on science is unmistakable. For example, our experience with macroscopic physics argues strongly for the fundamental impossibility of proceeding from sequence to structure to function in biology by means of computer modeling unless there are principles that protect the calculations and make them predictive. Thus this agenda of the computational biologist tacitly acknowledges the existence of principles, even at the same time that some of its adherents forcefully disavow the idea. Similarly our experience with macroscopic organization tells us that rules that are dreamt up without the benefit of physical insight are nearly always wrong, for correct rules are really natural phenomena and therefore must be discovered, not invented. The widely held view of bioinformatics as librarianship effectively proceeds from the assumption that there are no principles, for otherwise the ad hoc organizations of data would be seen as theories without physical basis and therefore meaningless. But the way forward in science begins with understanding what one doesn't understand—identifying which parts of one's world view are informed and which parts are prejudice. Are there organizing principles in mesoscopic systems? The truth is that we do not know one way or the other. The experimental record has not yet spoken. But it is clear that the question is sufficiently important that it cannot be evaded much longer. Whether we want to or not, we are now forced to take a stand.

In the world of biology, that at least some simple rules operate at the mesoscopic scale is demonstrated by the fact that some amino acid sequences fold and others do not. This distinction, which is quite sharp for large proteins, is arguably attributed to energy landscapes that funnel the molecule through a sequence of configurations that are virtually never metastable, so that the folded state can be reached by

This is a critical point: unless we understand the general principles of biological materials every mapping from the microscale physics to macroscale physiology will be particular and hard to understand *a priori*.

Indeed, cataloging alone has been less effective than originally hoped by bioinformatics.

any one of a large number of paths (Wolynes, Onuchic, and Thirumalai 1995; Onuchic, Luthey-Schulten, and Wolynes 1997). Even more persuasive is the observation of the nonuniqueness of the sequence that folds into a protein with a particular structure, say that of myoglobin. This happens reliably for sequences that almost appear randomly related to each other, so it would appear that small perturbations of the underlying system still preserve myoglobiness, which could then be regarded as an emergent collective property. How this occurs is only partially understood.

There is also evidence that not only the final structure but also the average properties of the structures that form on the routes to the folded state are largely shared by members of a family of folds. Statistics of the partially folded state vary only weakly with sequence, but strongly depend on topology (Alm and Baker 1999). This robustness of folding behavior makes the empirical case for some protected behavior of mesoscopic biological matter. Does there then exist a funnel protectorate?

Not only structure, but also some aspects of biomolecular function appear to be protected. This protection is most elegantly seen in the polymorphism of enzymes: in the same individual slightly different sequence versions of the same enzyme catalyze appropriate reactions (Xue and Yeung 1995). Single molecule experiments on enzymes show that biological catalysts sometimes have highly fluctuating rates from copy to copy (Edman *et al.* 1999). Yet the organism lives. This is a hint that protection in biology may arise from the evolutionary necessity of tolerating diversity. But is that the only cause?

Outside the biological world there is circumstantial evidence for protection at mesoscopic scales. Glasses often are thought of as just very slow liquids. Explaining their dynamics then would be just a question of getting the local molecular interactions right and studying the movement of the atoms on a computer. Arguing against this is the well-known correlation between transport properties and configurational entropy, known since the 1940s to occur across a wide range of substances (Kauzmann 1948). This correlation is sufficiently good that it can be used to engineer the properties of glassy polymers via addition of plasticizers. There are experimental hints from neutron scattering

(Mezei, Knaak, and Farrago 1989) and NMR (Tracht *et al.* 1998) experiments that glassy dynamics involves motion on mesoscopic length scales. But we are truly stymied at getting more details at these length scales by the lack of better tools for ferreting out organization at this size range.

While the transition from liquid to glass lies in a regime where classical statistical mechanics probably holds sway, in the low-temperature quantum regime there is evidence for protected behavior related to structures we cannot see. All amorphous substances show a linear-specific heat, a result found experimentally and a shock to theorists brought up on Debye's continuum description of solids at low temperature. Although theorists cleverly resolved the problem by pointing out the existence of two level tunneling systems (Anderson, Halperin, and Varma 1972), it has later surprised them to find universal characteristics of the density and scattering properties of these two level systems in a wide range of chemically distinct substances (Yu and Leggett 1988). Despite recent progress (Strehlow, Enss, and Hunklinger 1998; Enss and Hunklinger 1997) no entirely convincing microscopic identification of what is actually tunneling has yet been made by experiment.

Struggling to Overcome Large and Small Prejudices

The success of the sciences of the small and large has been based on some simple general guidelines. One of these guidelines is the expectation that systems possess a unique favored state and that the important motions of the system can be described as combinations of excitations that involve structures in some sense close to that favored state (Landau 1959). These excitations may scatter off each other, but primarily retain their integrity during their motions (Pines 1963). Occasionally, a second state can emerge through a phase transition. In this case also, there is considerable understanding of how structures self similar on all length scales can emerge near a continuous phase transition. But these principles are no longer sufficient in the mesoscopic realm. The phenomena of nonexponential dynamics and aging suggest that many states, each potentially very long lived, can be found for systems with mesoscopic organization. In some situations no single one dominates.

The distinctions here for mesoscopic systems are critical, and in fact, much recent work has come to show the importance of phase separation in cells.

Not all motions can be simply described as fluctuations near one of these states. Transitions between states are also important (Sherrington 1997).

In one-component systems, either classical or quantum mechanical, a candidate principle for understanding the breakdown of the elementary excitation picture and the emergence of mesoscopic organization is nonlinear feedback. This is illustrated in one approach to strongly correlated electron systems in which the interaction between electrons plays the dominant role in determining system behavior. Feedback occurs because the interaction between charge carriers that can dramatically alter the nature of the excitations is itself determined by the excitations it alters (Monthoux and Pines 1995). This feedback has, of course, been known since Debye's theory of electrolytes was pushed outside the dilute unit, but its quantum mechanical consequences are more subtle, because the speed at which an environmental disturbance disappears can determine the nature of an interaction. An example is the emergence of an effective dynamically attractive interaction between the essential repulsive helium atoms in 3He, an attraction that leads to Cooper pairs. Dynamical feedback often will just renormalize the excitations, which is clearly the case when the associated feedback is negative, in which case the system tends to stay in its existing state. If it is positive, however, it can give rise to a transition or crossover to another state, one that may possess organization on the mesoscopic scale. Such feedback is believed by many to be responsible for the remarkable behavior found in the normal state of the underdoped cuprate superconductors, where, as shown in Fig. 1, mesoscopic organization may be present in one or more of the three distinct phases of matter found as one lowers the temperature in the normal state before the system finally makes its transition to the superconducting state (Pines 1998).

Mesoscopic organization induced by feedback may not be confined to strongly correlated electron systems. The layered structure argued by theorists to exist at densities just below nuclear matter density in the neutron-rich crust of a neutron star (Pethick and Ravenhall 1991) represents an additional example from the quantum domain. The well-known mode-mode coupling theory (Götze 1989) for classical fluids also represents an attempt to use dynamical feedback to account for the nascent mesoscopic organization found in a system of strongly correlated atoms moving in a liquid. Dynamical feedback in liquids may be visualized as a cage effect. The slow motions of the neighbors of a given molecule allow them to provide a frictional cage on a central molecule's motion, slowing it. Because, in the democratic tradition, these neighbors would have their own cages, they must slow, too. These equations predict a transition to a nonergodic state in which molecules remain localized near their initial locations. Some of the predictions of this feedback theory are borne out in neutron scattering studies of liquids (Mezei, Knaak, and Farrago 1989), but others are not. It now seems that this theory indicates a kind of stability limit for the usual picture of a liquid as merely a dense gas, all of whose motions occur on the natural microscopic time scale of intermolecular collisions. The predicted nonergodicity signals the need to describe more complex motions involving transitions between widely different configurations (Wolynes 1992).

One concept to describe this complexity of classical liquids, glasses and proteins, is the energy landscape (Frauenfelder, Sugar, and Wolynes 1991; Stillinger and Weber 1984). Energy landscapes try to capture the idea that, although any many body system has myriad microscopic states, these can be organized into a collection of basins. These basins are robust to small external perturbations. Motions within these basins can be described much as for the simpler systems and occur on the natural microscopic time scale. On the other hand, the experimental clues suggest that unlike the simpler systems, here there are a large number of structurally distinct basins. Many of these are distant from each other, but have comparable energies. The arrangements of these often are pictured as low dimensional plots. These caricatures of energy landscapes are meant to capture the idea of the diversity of the basins and the nature of the bottlenecks and energy barriers in configurations that prevent the system from rapidly moving from one state to another. The difficulty with these pictures is that the only fully accurate picture of an energy landscape would have an extremely high dimension. One can ascribe a coordinate system locally to any one basin and perhaps a few similar neighboring ones, but this does not apply throughout the configuration space. An analogous, but much simpler situation arises when making flat maps of the spherical Earth, where the topology of the sphere makes the position of the pole on a two-dimensional plot ambiguous. Quantitative treatments of thermodynamics and dynamics of energy landscapes currently try to use only statistical information about landscape topography. One prototype landscape is very rugged. On such a landscape, explicitly found for some statistical models, you can find configurations of comparable and rather low energies that are quite different in appearance. In Fig. 2 two different configurations of holes in simple model of a transition metal oxide are shown, along with their energies. Although these states can interconvert, they do so in a very complex way, involving large-scale rearrangements of structure and correspondingly large activation energies. A similar situation would be found most of the time for the energy landscape of a polymer of amino acids, if one chooses its sequence at random. The property of having such a set of low-energy states is connected with the idea of replica symmetry breaking (Edwards and Anderson 1975; Mezard and Parisi 1996; Kirkpatrick and Wolynes 1987): different copies of the same system may well fall into different long-lived states through accidents of detailed molecular motion.

A huge amount of progress has been made since in understanding how reliable polymer folding is connected with energy landscapes, and free-energy based models have become predictive for certain coarse-grained folding outcomes.

Another prototype landscape for mesoscopic systems is not so rugged but has one dominant basin of attraction. This so-called funnel landscape, shown in Fig. 3, is not typically found for most polymers of amino acids, but seems to describe the important special case of the proteins of nature that evolve to fold into a small set of related states. Out of all possible sequences, funnel landscapes are exponentially rare compared with rugged landscapes.

Although the complexity of the energy landscape is probably a fact of life at these mesoscopic scales, its origin often has been pictured as caused by frustration (Toulouse 1977), examples of which are depicted in Fig. 4. Frustration is an anthropomorphic and therefore perhaps provisional candidate concept. To explain the concept, we imagine the energies governing the motions of the system can be partitioned into competing parts. Of course the system does not know how we divide its energy up. Sometimes, however, the division seems very natural to us. For example, in a magnetic alloy some impurity spins will be directly

Laughlin et al. (2000)

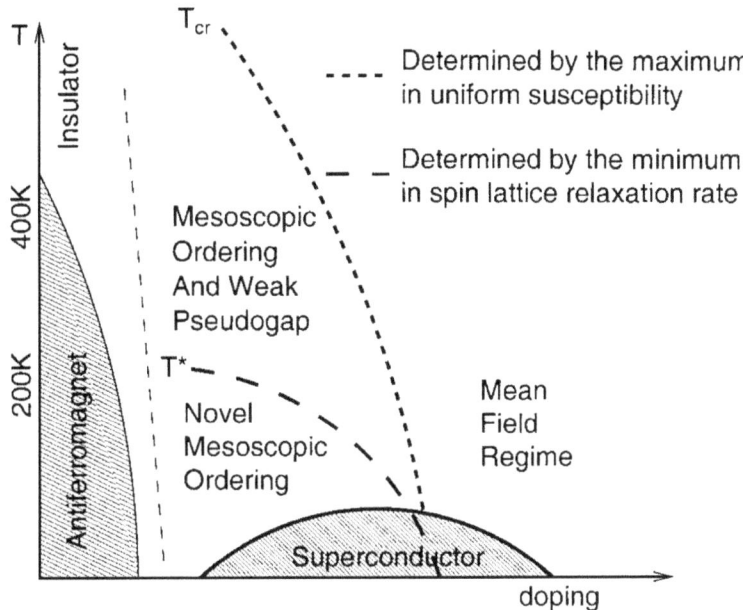

Figure 1. Generic phase diagram of high temperature superconducting cuprates. The true thermodynamic phases (antiferromagnetic at low doping and superconducting at higher doping are depicted by the shaded regions. The remaining lines correspond to crossovers, visible in a variety of experiments.

coupled in such a way to favor their becoming parallel to each other whereas others at different separations will be coupled so as to favor an antiparallel arrangement. The tendency of these individual parts of the energy to produce local order cannot be simultaneously satisfied in any given individual system configuration. This frustration suggests the possibility that quite different states can be stable and compete with each other, giving rise to the diversity of the landscape. The common frustration of the interactions between different pairs of amino acids is the cause of the random polypeptide's rugged energy landscape. The funnel landscape emerges only for those special sequences for which there is a structure in which nearly all the different interactions are simultaneously minimized; i.e., biological proteins are only minimally frustrated.

Both the amphiphile systems and the correlated electron systems also have been described by using the concept of frustration. In the case of amphiphiles the conflict arises between the tendency of the hydrophobic forces to separate lipid and water. The head group of the

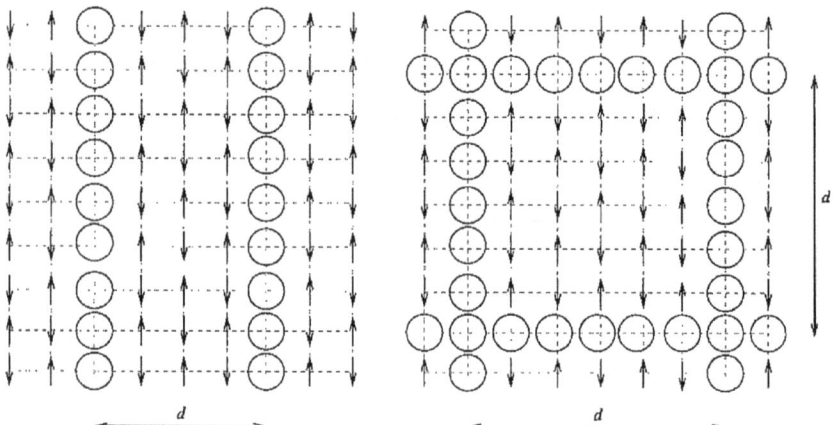

Figure 2. Stripe and grid configurations of holes (circles) in an antiferromagnetic background (arrows, indicating spin direction) at the same hole concentration level (d denotes the distance between stripe or grid lines). Note the change in sign of the local magnetization in the magnetic domains (p phase shift), which makes motion of the holes across line segments energetically inexpensive. Both configurations correspond to low energy states, with a large activation energy for a transition between them, caused by the Coulomb repulsion between holes.

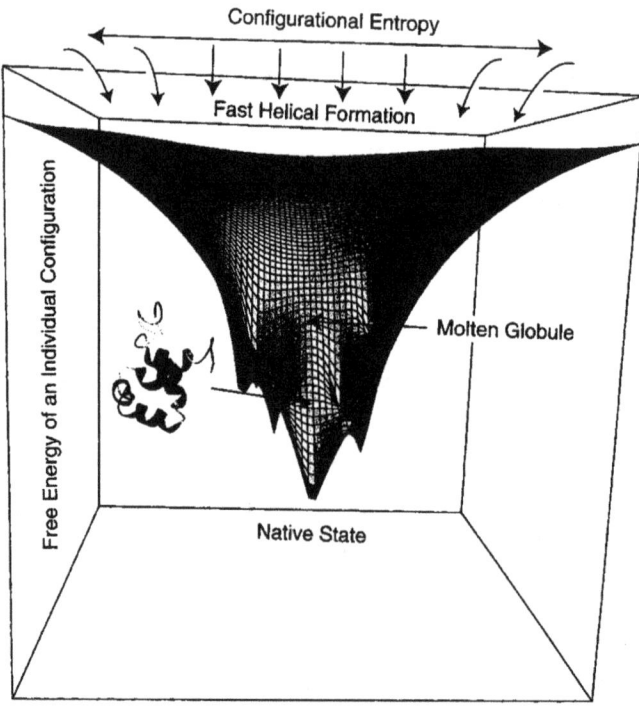

Figure 3. Sketch of a funnel landscape found in certain protein structures.

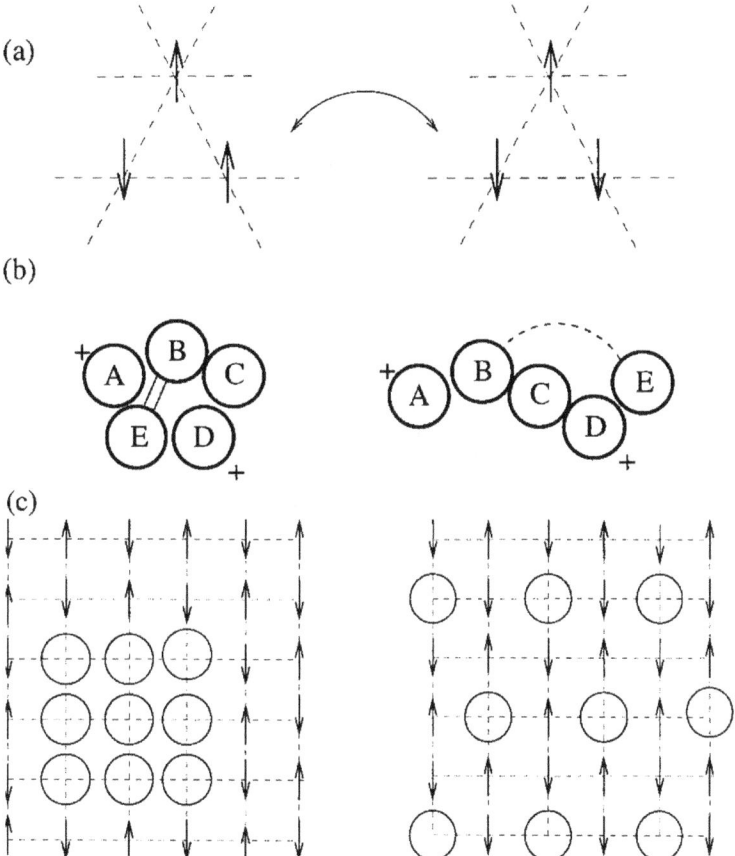

Figure 4. Examples of frustration. (a) For antiferromagnetically (AF) coupled Ising spins on a triangular lattice one of the AF bonds is always broken. (b) Folding of heteropolymers can be frustrated by the competing, e.g., bonding (indicated by solid lines) and Coulomb interactions (indicated by 1) between different constituents (A–E). (c) AF interactions in doped transition metal oxides energetically favor a phase-separated state, which is unfavorable for the Coulomb interaction, whereas the Coulomb interaction favors a Wigner crystal state that is unfavorable for the AF interactions; the result of the competition (frustration) yields formation of patterns, such as those shown in Fig. 2.

amphiphile has a tendency to remain in the water phase and the tail group in the lipid. Head and tail must remain connected, however. This example is rather analogous to the origin of frustration in the protein situation. Various views exist on the source of frustration in correlated electron systems. One idea is that positively charged vacancies, induced by chemical doping, inhibit any intrinsic magnetic order but cannot completely avoid magnetic regions. To do this they would have to group together, which in turn leads to a large Coulomb interaction between them. Thus magnetic order and electrostatic interactions frustrate each other and are in conflict, which is surprisingly similar to the amphiphile problem. In a sense the vacancies act like a surfactant.

Energy landscape pictures have a hard time indicating how distant parts of a system communicate with each other. It is thought that generally mesoscopic parts of the system with frustration will break up into domains or droplets. This idea started by McMillan has been extended to describe a large number of systems, including glasses and proteins (McMillan 1984; Bray and Moore 1986; Fisher and Huse 1986, 1988; Wolynes 1997). The interface energy between the droplets is scale dependent. Scale dependence of the interaction energy may be a more objective way of quantifying the concept of frustration. Droplet excitations for large systems with mesoscopic correlations may replace the concept of collective modes or quasiparticles relevant to simple systems with a single preferred state. One interesting question is whether the droplets should be thought of themselves as simple objects or have within them a complex energy landscape. Also the existence and nature of droplet configurations remains controversial because no mesoscopic probe has yet been devised to clearly visualize them.

Although the feedback idea has been applied to both quantum and classical systems, this rapid survey of candidate concepts largely borrowed from macroscopic systems for use in mesoscopically organized systems shows that the situation in quantum mechanics is much less developed than even for the classical systems. For example, frustration often is used to argue that a correlated electron system will not order. The resulting state then is often described as a resonance hybrid, but the properties of these quantum mechanical superpositions of different states are hard to deduce. What takes the place of energy landscapes? The dynamics and role of droplet excitations for highly quantum systems is still quite murky, which is unfortunate because attempts to build quantum computers will doubtless require this sort of understanding.

The Mesoscopic Frontier

To many people the world of mesoscale phenomena would seem to be intrinsically confined, but we do not believe this is so. First, the richness of experimental phenomena in the field shows that the subject is still in its infancy. In the short run even the basics have to be more firmly established. The nascent theoretical concepts are sketchy because they have been informed primarily by experiments on the wrong length scales. Indeed an argument can be made that the lack of appropriate probes for characterizing mesoscopic order is not the result of lack of scientific attention or inadequate funds but may represent intrinsic physical limitations. Our own ability to see is based on sensing the multiple correlations intrinsic in the complex shape of an object. The constraints of quantum mechanics limit the complexity of correlations that can be measured with light or particles when they have a wavelength sufficiently short to resolve an individual mesoscopic object. We may have to destroy an object if we want to study it.

Experiments have now been carried out at the appropriate length scales that do indeed reveal a huge diversity of interesting behavior.

Still, there is considerable hope for progress on the experimental side. Scanning tunneling microscopy and atomic force microscopy allow us to measure mesoscale phenomena, albeit only on the surface of complex objects. The time scale for such measurements also requires improvement. Other techniques probing small mesoscale regions in three dimensions are likely to be developed. For example, time-dependent x-ray spectroscopy and x-ray speckle dynamics (Dierker *et al.* 1995) using synchrotron radiation will allow probes of structure and dynamics beyond the currently available simple static diffraction pattern. Nonlinear and fluctuation spectroscopies (Weissman 1997), including improved neutron scattering and single molecule techniques (Rigler and Wolynes 1999), also should help. The scientific community needs to support these efforts to establish the experimental basis for the development of scientific principles of mesoscopic organization. Clearly, experiment alone will not be enough and theorists will have to work hard to keep up with the onslaught of new information.

The discovery of physical principles at mesoscale will reinforce the attack by biologists on the mysteries of cellular function. But, beyond this, a framework for understanding mesoscopic organization will be an extraordinary help in the effort to create an entirely artificial system with the complex adaptive behavior characteristic of life. Such artificial systems should be capable of a variety of functions that present biological systems cannot perform.

In any event, the applicability of the science of mesoscale organization that we believe can be developed will not be limited to the world between angstroms and centimeters. Organization following similar principles may well be manifested in astrophysics. As we have noted, complex structures already have been proposed for the exotic matter expected in neutron stars, while ideas developed to explain mescoscopic organization on Earth may be useful in explaining the origin of large-scale structure in the universe.

Acknowledgments

We thank fellow participants in the workshop Mesoscopic Organization in Matter, sponsored by the Institute for Complex Adaptive Matter, and held at Los Alamos August 24–28, 1999, for their contributions, both formal and informal, to that workshop, which have helped shape our thinking on this topic. This work was supported primarily by the Department of Energy and by the National Science Foundation under Grant DMR-9813899. Additional support was provided by the Science and Technology Center for Superconductivity under National Science Foundation Grant DMR 91-2000.

REFERENCES

Alberts, B., D. Bray, J. Lewis, M. Raff, K. Roberts, and J. D. Watson. 1994. *Molecular Biology of the Cell*. New York, NY: Garland.

Alm, E., and D. Baker. 1999. "Matching Theory and Experiment in Protein Folding." *Current Opinion in Structural Biology* 9:189–196.

Anderson, P. W. 1972. "More Is Different: Broken Symmetry and the Nature of the Hierarchical Structure of Science." *Science* 177:393.

Anderson, P. W., B. Halperin, and C. Varma. 1972. "Anomalous Low-Temperature Thermal Properties of Glasses and Spin Glasses." *Philosophical Magazine* 25:1–9.

Bray, A. J., and M. A. Moore. 1986. "Scaling Theory of the Ordered Phase of Spin Glasses." In *Heidelberg Colloquium on Glassy Dynamics: Lecture Notes in Physics,* edited by J. L. van Hemmen and I. Morgenstern, 121. Berlin, Germany: Springer.

Curro, N., B. J. Suh, P. C. Hammel, M. Hücker, B. Büchner, U. Ammerahl, and A. Revcolevschi. 1999. *Inhomogeneous Low Frequency Spin Dynamics in $La_{1.65}Eu_{0.2}Sr_{0.15}CuO_4$*. Technical report LA-UR-99-5452. Los Alamos National Laboratory.

deGennes, P. G., and C. Taupin. 1982. "Microemulsions and the Flexibility of Oil/Water Interfaces." *The Journal of Physical Chemistry* 86:2294–2304.

Dierker, S. B., R. Pimdak, R. M. Fleming, I. K. Robinson, and L. Berman. 1995. "X-Ray Photon Correlation Spectroscopy Study of Brownian Motion of Gold Colloids in Glycerol." *Physical Review Letters* 75:449–452.

Ediger, M. D., C. A. Angell, and S. R. Nagel. 1996. "Supercooled Liquids and Glasses." *The Journal of Physical Chemistry* 100:13200–13212.

Edman, L., Z. Földes-Papp, S. Wennmalm, and R. Rigler. 1999. "The Fluctuating Enzyme: A Single Molecule Approach." *Chemical Physics* 247:11–22.

Edwards, S. F., and P. W. Anderson. 1975. "Theory of Spin Glasses." *Journal of Physics F: Metal Physics* 5:965–974.

Enss, C., and S. Hunklinger. 1997. "Incoherent Tunneling in Glasses at Very Low Temperatures." *Physical Review Letters* 79:2831–2834.

Fischer, K. H., and J. A. Hertz. 1991. *Spin Glasses*. Cambridge, UK: Cambridge University Press.

Fisher, D. S., and D. A. Huse. 1986. "Ordered Phase of Short-Range Ising Spin-Glasses." *Physical Review Letters* 56:1601–1604.

———. 1988. "Equilibrium Behavior of the Spin-Glass Ordered Phase." *Physical Review B* 38:386–411.

Fisk, Z., D. Hess, C. J. Pethick, D. Pines, J. Smith, J. Thompson, and J. Willis. 1988. "Heavy-Electron Metals: New Highly Correlated States of Matter." *Science* 239:33–42.

Frauenfelder, H., S. Sugar, and P. G. Wolynes. 1991. "The Energy Landscapes and Motions of Proteins." *Science* 254:1598–1663.

Geiger, C., M. Stanesch, L. H. Chen, and D. G. Whitten. 1999. "Organogels Resulting from Competing Self-Assembly Units in the Gelator: Structure, Dynamics, and Photophysical Behavior of Gels Formed from Cholesterol–Stilbene and Cholesterol–Squaraine Gelators." *Langmuir* 15:2241–2245.

Gelbart, W. M., and A. Ben Shaul. 1996. "The "New" Science of "Complex Fluids"." *The Journal of Physical Chemistry* 100:13169–13189.

Götze, W. 1989. "Aspects of Structural Glass Transitions." In *Liquids, Freezing, and Glass Transitions,* edited by J. P. Hansen, D. Levesque, and J. Zinn-Justin, 287. Amsterdam, Netherlands: North-Holland.

Kauzmann, W. 1948. "The Nature of the Glassy State and the Behavior of Liquids at Low Temperatures." *Chemical Reviews* 43 (2): 219–256.

Kirkpatrick, T. R., and P. G. Wolynes. 1987. "Stable and Metastable States in Mean-Field Potts and Structural Glasses." *Physical Review B* 36:8552–8564.

Landau, L. D. 1959. "On the Theory of the Fermi Liquid." *Soviet Physics JETP* 35:70–76.

Laughlin, R. B., and D. Pines. 2000. "The Theory of Everything." *Proceedings of the National Academy of Sciences* 97 (1): 28–31.

McMillan, W. L. 1984. "Cluster-Quench Simulation of the Two-Dimensional Random Ising Model." *Journal of Physics C* 47:3189–3193.

Mezard, M., and G. Parisi. 1996. "A Tentative Replica Study of the Glass Transition." *Journal of Physics A* 29:6515–6524.

Mezard, M., G. Parisi, and M. A. Virasoro. 1987. *Spin Glass Theory and Beyond*. Vol. 9. World Scientific Lecture Notes in Physics. Teaneck, NJ: World Scientific.

Mezei, F., W. Knaak, and B. Farrago. 1989. "Glass Transition and Proton Dynamics in Ca(NO3)$_2$·4H$_2$O: A Neutron Spin Echo Study." *Physica B* 156:182–184.

Monthoux, P., and D. Pines. 1995. "$d_{x^2-y^2}$ Pairing and Spin Fluctuations in the Cuprate Superconductors: A Progress Report." *Journal of Physics and Chemistry of Solids* 56:1651–1658.

Onuchic, J. N., Z. Luthey-Schulten, and P. G. Wolynes. 1997. "Theory of Protein Folding: The Energy Landscape Perspective." *Annual Review of Physical Chemistry* 48:545–600.

Pethick, C. J., and D. Ravenhall. 1991. *Neutron Stars: Theory and Observation*. Edited by D. Pines and J. Ventura. 3–20. Dordrecht, Netherlands: Kluwer Academic Publishers.

Pines, D. 1963. *Elementary Excitations in Solids*. New York, NY: Benjamin.

———. 1998. "The Spin Fluctuation Model for High Temperature Superconductivity: Progress and Prospects." In *The Gap Symmetry and Fluctuations in High-Tc Superconductors,* edited by J. Bok, G. Deutscher, D. Pavuna, and S.A. Wolf, 111–142. New York, NY: Plenum.

Rigler, R., and P. G. Wolynes. 1999. "Preface." *Chemical Physics* 247 (1): vii–viii.

Sherrington, D. 1997. "Introduction and Overview." In *Landscape Paradigms in Physics and Biology: Concepts, Structures and Dynamics,* edited by H. Frauenfelder, A.R. Bishop, A. Garcia, A. S. Perelson, P. Schuster, D. Sherrington, and P.J. Swart, 117. Amsterdam, Netherlands: North-Holland.

Stillinger, F. H., and T. Weber. 1984. "Packing Structures and Transitions in Liquids and Solids." *Science* 225:983–989.

Strehlow, P., C. Enss, and S. Hunklinger. 1998. "Evidence for a Phase Transition in Glasses at Very Low Temperature: A Macroscopic Quantum State of Tunneling Systems?" *Physical Review Letters* 80:5361–5364.

Toulouse, G. 1977. "Theory of the Frustration Effect in Spin Glasses: I." *Communications Physics* 2:115–119.

Tracht, W., M. Wilhelm, A. Heuer, H. Feng, K. Schmidt-Rohr, and H.W. Spiess. 1998. "Length Scale of Dynamic Heterogeneities at the Glass Transition Determined by Multidimensional Nuclear Magnetic Resonance." *Physical Review Letters* 81:2727–2730.

Tranquada, J., N. Ichikawa, and S. Uchida. 1999. "Glassy Nature of Stripe Ordering in $La_{1.6-x}Nd_{0.4}Sr_xCuO_4$." *Physical Review B* 59:14712–14722.

Weissman, M. B. 1997. "Sorting Out Landscapes by Mesoscopic Noise." *Physica D* 107:421–429.

Wolynes, P. G. 1992. "Randomness and Complexity in Chemical Physics." *Accounts of Chemical Research* 25:513–519.

———. 1997. "Folding Funnels and Energy Landscapes of Larger Proteins Within the Capillarity Approximation." *Proceedings of the National Academy of Sciences* 94:6170–6174.

Wolynes, P. G., J. N. Onuchic, and D. Thirumalai. 1995. "Navigating the Folding Routes." *Science* 267:1619–1620.

Xue, Q., and E. Yeung. 1995. "Differences in the Chemical Reactivity of Individual Molecules of an Enzyme." *Nature* (London, UK) 373:681–683.

Yu, C., and A. J. Leggett. 1988. "Low Temperature Properties of Amorphous Materials: Through a Glass Darkly." *Comments on Condensed Matter Physics* 14:231–251.

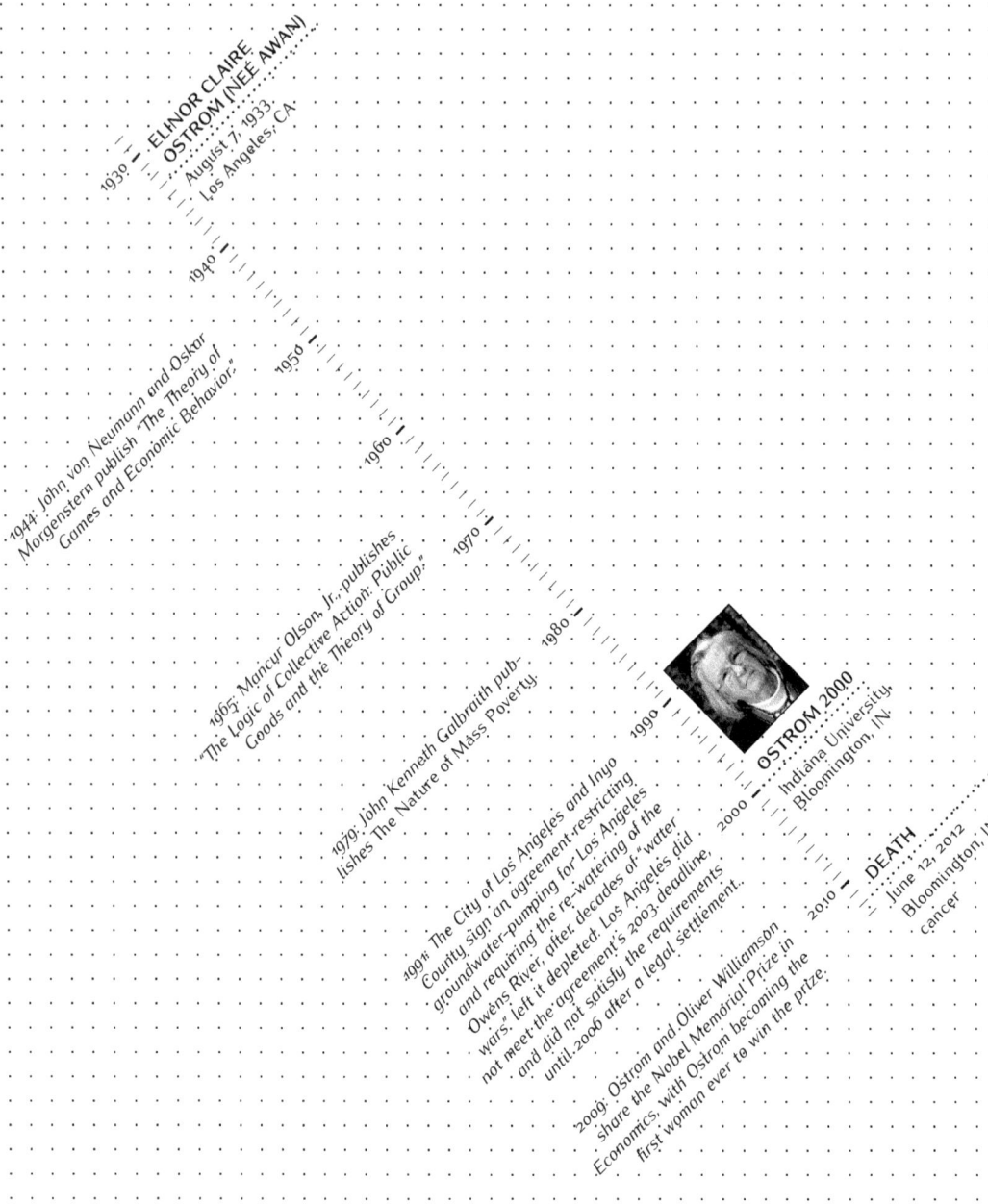

1930 — **ELINOR CLAIRE OSTROM (NEÉ AWAN)**
August 7, 1933, Los Angeles, CA.

1944: John von Neumann and Oskar Morgenstern publish "The Theory of Games and Economic Behavior."

1965: Mancur Olson, Jr., publishes "The Logic of Collective Action: Public Goods and the Theory of Groups."

1979: John Kenneth Galbraith publishes The Nature of Mass Poverty.

1991: The City of Los Angeles and Inyo County sign an agreement restricting groundwater-pumping for Los Angeles and requiring the re-watering of the Owens River, after decades of "water wars," left it depleted. Los Angeles did not meet the agreement's 2003 deadline, and did not satisfy the requirements until 2006 after a legal settlement.

OSTROM 2000
Indiana University, Bloomington, IN.

2009: Ostrom and Oliver Williamson share the Nobel Memorial Prize in Economics, with Ostrom becoming the first woman ever to win the prize.

DEATH
June 12, 2012
Bloomington, IN
cancer

ELINOR CLAIRE OSTROM (NEÉ AWAN)

[89]
COLLECTIVE ACTION &
THE DYNAMICS OF INSTITUTIONS

John M. Anderies, Arizona State University

In a series of books and papers spanning the 1980s and 1990s, Elinor Ostrom challenged dominant notions of how groups of people could solve problems related to common-pool resources. Her work was motivated by debates spurred in part by Garrett Hardin's essay "Tragedy of the Commons" (which should be titled "Tragedy of Open Access," as "the commons" was historically an effective property regime for the use of shared land resources) regarding environmental policy responses to commons problems in the decades preceding the publication of *Governing the Commons* in 1990. Specifically, policy recommendations occupied two ends of the institutional spectrum: totally centralized, top-down regulation by a benevolent social planner and fully decentralized, market-based solutions founded on secure, clearly defined property rights—the preferred solution in economics. In the context of evolutionary theories of cooperation, these two ends of the spectrum represent a case where coordination is forced by a benevolent dictator or coordination is generated endogenously via the market mechanism, conditioned by institutional arrangements that define and enforce property rights in such a way that generates the correct prices so that perfectly rational agents acting in their self-interest will converge to a market equilibrium that maximizes the social value (prevents over-exploitation) of a shared resource. Ostrom's work in the field suggested there were many alternatives to these two extremes (Ostrom 1990).

This paper is an evolution of Ostrom's work on trying to understand how institutions emerge in groups, and it takes a decidedly complex-systems lens. Much of the work on complex systems in the social domain has been focused on the question of how cooperation emerges in groups if we assume that agents are selfish and rational—that is,

E. Ostrom, "Collective Action and the Evolution of Social Norms," *Journal of Economic Perspectives* 14 (3), 137–158 (2000).

Reprinted with permission of the Indiana University Foundation, Inc.

are "rational egoists," as Ostrom refers to them in this paper. In these studies, the lens typically focuses on the individual: how can changes in the way the game is played induce selfish actors to cooperate, rooted in the evolutionary paradigm from biology and human ecology? Ostrom's work adds another layer to this question from a political-science angle: how can groups of people generate rules and norms (institutions) that they agree to adhere to in order facilitate collective action (cooperation)? The difference here is between a purely emergent (self-organizing) feature in a complex system where individual agents interact and one that is partly designed (the institutions—the rules of the game) and partly self-organizing, that is, emergent patterns of interaction among agents who play according to the rules. Of course, the creation of the rules is itself a problem of cooperation—of providing a public good—a second-order social dilemma that needs to be solved in order to solve a first-order commons dilemma (preventing the overuse of a common-pool resource).

Although Ostrom spends quite a bit of time in this article on what had been learned from the laboratory about contextual factors that affect collective action and the role of agent diversity in facilitating cooperation, it is her discussion of collective action outside the lab that may be the more important contribution of this paper. In the section on the evolution of rules and norms in the field, Ostrom leverages findings from experiments that support the existence of multiple types of agents (conditional cooperators and altruistic punishers) and their role in promoting cooperation to lay out elements of a recipe for "crafting rules through a dynamic process." Human behavior is context-dependent, and humans co-create context. This is very much a complex systems problem with its emphasis on sensitivity to initial conditions, multiple basins of attraction, stability (what Ostrom often referred to as "long enduring institutions"), and the fragility of long-run attractors in dynamical systems.

Based on Ostrom's extensive field work on communities that manage irrigation systems, groundwater basins, fisheries, and forests, she extracted eight institutional design principles that characterize the existing rules and norms she observed were associated with long-enduring institutions and how those rules were changed (Ostrom 1990;

Ostrom, Gardner, and Walker 1994). These cases were, however, static pictures of extant institutions and say little about the process by which communities developed the capacity to create and maintain these rules. This paper provides the link. The indirect evolutionary approach presents a mechanism by which actors may develop norms through cognitive processes activated by, for example, learning, face-to-face communication, and self-determination. Norms form the basis of formalized institutions, (i.e., rules) as communities extend and codify them in a self-reinforcing way. One of the overarching factors enabling the creation and adherence to formal institutions suggested by the indirect evolutionary approach Ostrom discusses in this paper is trust. Trust, and the factors that build or erode it, has subsequently become a very intense area of collective-action research.

With this and several related articles, Ostrom influenced subsequent research that developed experimental contexts both in the lab and the field. Her work resulted in deeper explorations of the factors that impact a group's capacity to create institutions and identify which institutional configurations will endure. This work also spurred a large literature testing whether the institutional design principles really do promote successful management of shared resources (e.g., Cox, Arnold, and Tomás 2010; Baggio *et al.* 2016). This article in particular, and Ostrom's work more broadly, has catalyzed research connecting evolutionary science, political science, economics, and systems science through the complexity lens. For example, it has been instrumental in the development of theories of cultural group selection (e.g., Waring, Goff, and Smaldino 2017), viewing social-ecological systems as complex adaptive systems (e.g., Preiser *et al.* 2018), and leveraging work in control theory to study the robustness of institutions (e.g., Cifdaloz *et al.* 2010). These continue to be fertile areas of research with many open questions that can leverage complex systems science ideas and techniques.

In addition to contributing widely to the theory of how groups generate institutions to coordinate actions for the common good, Ostrom's work also informs questions of practical importance for human societies. If a single broad principle emerges from Ostrom's work, it may be this: institutions developed through inclusive self-organizing governance based on processes that build trust tend to work better than those designed and imposed by outside groups. However, such institutions are very difficult to scale up. In the present age of globalized societies facing planetary common-pool resource problems in the form of climate change and biodiversity loss, this article provides much to think about—and a clear set of principles to inform the design of global institutions as they respond to these complex challenges.

REFERENCES

Baggio, J. A., A. J. Barnett, I. Perez-Ibarra, U. Brady, E. Ratajczyk, N. Rollins, C. Rubiños, et al. 2016. "Explaining Success and Failure in the Commons: The Configural Nature of Ostrom's Institutional Design Principles." *International Journal of the Commons* 10 (2): 417–439. https://doi.org/10.18352/ijc.634.

Cifdaloz, O., A. Regmi, J. M. Anderies, and A. A. Rodriguez. 2010. "Robustness, Vulnerability, and Adaptive Capacity in Small-Scale Social-Ecological Systems: The Pumpa Irrigation System in Nepal." *Ecology and Society* 15 (3): 39. https://doi.org/10.5751/ES-03462-150339.

Cox, M., G. Arnold, and S. V. Tomás. 2010. "A Review of Design Principles for Community-Based Natural Resource Management." *Ecology and Society* 15 (4): 38. https://doi.org/10.5751/ES-03704-150438.

Hardin, G. 1968. "The Tragedy of the Commons." *Science* 162 (3859): 1243–1248. https://doi.org/10.1126/science.162.3859.1243.

Ostrom, E. 1990. *Governing the Commons: The Evolution of Institutions for Collective Action.* Cambridge, UK: Cambridge University Press.

Ostrom, E., R. Gardner, and J. Walker. 1994. *Rules, Games, and Common-Pool Resources.* Ann Arbor, MI: University of Michigan Press.

Preiser, R., R. Biggs, A. De Vos, and C. Folke. 2018. "Social-Ecological Systems as Complex Adaptive Systems." *Ecology and Society* 23 (4). https://doi.org/10.5751/ES-10558-230446.

Waring, T. M., S. H. Goff, and P. E. Smaldino. 2017. "The Coevolution of Economic Institutions and Sustainable Consumption via Cultural Group Selection." *Ecological Economics* 131:524–532. https://doi.org/10.1016/j.ecolecon.2016.09.022.

COLLECTIVE ACTION AND THE EVOLUTION OF SOCIAL NORMS

Elinor Ostrom, Indiana University, Bloomington

With the publication of *The Logic of Collective Action* in 1965, Mancur Olson challenged a cherished foundation of modern democratic thought that groups would tend to form and take collective action whenever members jointly benefitted. Instead, Olson (1965, p. 2) offered the provocative assertion that no self-interested person would contribute to the production of a public good: "[U]nless the number of individuals in a group is quite small, or unless there is coercion or some other special device to make individuals act in their common interest, *rational, self-interested individuals will not act to achieve their common or group interests.*" This argument soon became known as the "zero contribution thesis."

The idea that rational agents were not likely to cooperate in certain settings, even when such cooperation would be to their mutual benefit, was also soon shown to have the structure of an n-person prisoner's dilemma game (Hardin 1971, 1982). Indeed, the prisoner's dilemma game, along with other social dilemmas, has come to be viewed as the canonical representation of collective action problems (Lichbach 1996). The zero contribution thesis underpins the presumption in policy textbooks (and many contemporary public policies) that individuals cannot overcome collective action problems and need to have externally enforced rules to achieve their own long-term self-interest.

The zero contribution thesis, however, contradicts observations of everyday life. After all, many people vote, do not cheat on their taxes, and contribute effort to voluntary associations. Extensive fieldwork has by now established that individuals in all walks of life and all parts of the world voluntarily organize themselves so as to gain the benefits of trade, to provide mutual protection against risk, and to

> This view has changed considerably, based in part on ideas in this article. For example, this paper argues that context matters. It has since been pointed out in the literature that the prisoner's dilemma has a very specific institutional context designed to promote defection: communication is not allowed, prisoners are held in separate cells, etc. As such, the prisoner's dilemma is not a good metaphor for common-pool resource problems. Lack of institutions to promote cooperation is not analogous to the existence of institutions to prevent it.

create and enforce rules that protect natural resources.[1] Solid empirical evidence is mounting that governmental policy can frustrate, rather than facilitate, the private provision of public goods (Montgomery and Bean 1999). Field research also confirms that the temptation to free ride on the provision of collective benefits is a universal problem. In all known self-organized resource governance regimes that have survived for multiple generations, participants invest resources in monitoring and sanctioning the actions of each other so as to reduce the probability of free riding (Ostrom 1990).

While these empirical studies have posed a severe challenge to the zero contribution theory, these findings have not yet been well integrated into an accepted, revised theory of collective action. A substantial gap exists between the theoretical prediction that self-interested individuals will have extreme difficulty in coordinating collective action and the reality that such cooperative behavior is widespread, although far from inevitable.

Modern theories of collective action do attempt to integrate these factors based on research spurred by this and related work of Ostrom. While much progress has been made, there are significant data challenges—it is very costly and time consuming to characterize social-ecological systems over time and track "successful collective action."

Both theorists and empirical researchers are trying to bridge this gap. Recent work in game theory—often in a symbiotic relationship with evidence from experimental studies—has set out to provide an alternative micro theory of individual behavior that begins to explain anomalous findings (McCabe, Rassenti, and Smith 1996; Rabin 1993; Fehr and Schmidt 1999; Selten 1991; Bowles 1998). On the empirical side, considerable effort has gone into trying to identify the key factors that affect the likelihood of successful collective action (Feeny *et al.* 1990; Baland and Platteau 1996; Ostrom 2001).

This paper will describe both avenues of research on the underpinnings of collective action, first focusing on the experimental evidence and potential theoretical explanations, and then on the real-world empirical evidence. This two-pronged approach to the problem has been a vibrant area of research that is yielding many insights. A central finding is that the world contains multiple types of individuals, some more willing than others to initiate reciprocity to achieve the benefits of collective action. Thus, a core question is how potential

---

[1] See Milgrom and Weingast (1990) and Bromley *et al.* (1992). An extensive bibliography by Hess (1999) on diverse institutions for dealing with common pool resources can be searched on the web at http://www.indiana.edu/;workshop/wsl/wsl.html or obtained on a CD-ROM disk.

cooperators signal one another and design institutions that reinforce rather than destroy conditional cooperation. While no full-blown theory of collective action yet exists, evolutionary theories appear most able to explain the diverse findings from the lab and the field and to carry the nucleus of an overarching theory.

Laboratory Evidence on Rational Choice in Collective Action Situations

Most studies by political economists assume a standard model of rational individual action—what I will call a rational egoist. A wide range of economic experiments have found that the rational egoist assumption works well in predicting the outcome in auctions and competitive market situations (Kagel and Roth 1995). While subjects do not arrive at the predicted equilibrium in the first round of market experiments, behavior closely approximates the predicted equilibrium by the end of the first five rounds in these experiments. One of the major successes of experimental economics is to demonstrate the robustness of microeconomic theory for explaining market behavior.

In regard to collective action situations, on the other hand, the results are entirely different. Linear public good experiments are widely used for examining the willingness of individuals to overcome collective action problems. In a linear public good experiment, each individual is endowed with a fixed set of assets and must decide how many of these assets to contribute to a public good. When an individual makes a contribution of, say, 10 units to the public good, each of the participants in the group, including that individual, receive a benefit of, say, five units apiece. In this setting, the optimal outcome for the group of players as a whole is for everyone to contribute all of their endowments to provide the public good (if a group of 10 people, each individual contribution of 10 will have a social payoff of 50!). However, the unique equilibrium for rational egoists in a single-shot game is that everyone contributes zero, since each individual has access to benefits of the public good funded by the contributions of others, without paying any costs.[2]

If the public goods game is played for a finite number of rounds, zero is also the predicted equilibrium for every round. Rational egoists will reason that zero contribution is the equilibrium in the last round, and because they expect everyone to contribute zero in the last round, they also expect everyone to contribute zero in the second-to-last round, and eventually by backward induction they will work their way to the decision not to contribute to the public good in the present. Of course, these predictions are based on the assumptions that all players are fully rational and interested only in their own immediate financial payoff, that all players understand the structure of the game fully and believe that all other players are fully rational, and that no external actor can enforce agreements between the players.

Since the first public good experiments were undertaken by Dawes, McTavish, and Shaklee (1977), a truly huge number of such experiments has been undertaken under various conditions (see Davis and Holt 1993; Ledyard 1995; Offerman 1997, for an overview). By now seven general findings have been replicated so frequently that these can be considered the core facts that theory needs to explain.

1) Subjects contribute between 40 and 60 percent of their endowments to the public good in a one-shot game as well as in the first round of finitely repeated games.

2) After the first round, contribution levels tend to decay downward, but remain well above zero. A repeated finding is that over 70 percent of subjects contribute nothing in the announced last round of a finitely repeated sequence.

---

[2]In a linear public good game, utility is a linear function of individual earnings,

$$U_i = U_i \left[(E - x_i) + A \cdot P(\sum x_i) \right],$$

where E is an individual endowment of assets, x_i is the amount of this endowment contributed to provide the good, A is the allocation formula used to distribute the group benefit to individual players, and P is the production function. In a linear public good game, A is specified as $1/N$ and $0 < 1/N < P < 1$ (but both of these functions vary in other types of collective action). So long as $P < 1$, contributing to the collective good is never an optimal strategy for a fully self-interested player.

3) Those who believe others will cooperate in social dilemmas are more likely to cooperate themselves. A rational egoist in a public good game, however, should not in any way be affected by a belief regarding the contribution levels of others. The dominant strategy is a zero contribution no matter what others do.

4) In general, learning the game better tends to lead to more cooperation, not less. In a clear test of an earlier speculation that it just took time for subjects to learn the predicted equilibrium strategy in public good games, Isaac, Walker, and Williams (1994) repeated the same game for 10 rounds, 40 rounds, and 60 rounds with experienced subjects who were specifically told the end period of each design. They found that the rate of decay is inversely related to the number of decision rounds. In other words, instead of learning not to cooperate, subjects learn how to cooperate at a moderate level for ever-longer periods of time!

5) 5) Face-to-face communication in a public good game—as well as in other types of social dilemmas—produces substantial increases in cooperation that are sustained across all periods including the last period (Ostrom and Walker 1997).[3] The strong effect of communication is not consistent with currently accepted theory, because verbal agreements in these experiments are not enforced. Thus, communication is only "cheap talk" and makes no difference in predicted outcomes in social dilemmas. But instead of using this opportunity to fool others into cooperating, subjects use the time to discuss the optimal joint strategy, to extract promises from one another, and to give verbal tongue-lashings when aggregate contributions fall below promised levels. Interestingly, when communication is implemented by allowing subjects to signal promises to cooperate through their computer terminals, much less cooperation occurs than in experiments allowing face-to-face communication.

---

[3] Even more startling, Bohnet and Frey (1999) find that simply allowing subjects to see the other persons with whom they are playing greatly increases cooperation as contrasted to completely anonymous situations. Further, Frank, Gilovich, and Regan (1993) find that allowing subjects to have a face-to-face discussion enables them to predict who will play cooperatively at a rate significantly better than chance.

6) When the structure of the game allows it, subjects will expend personal resources to punish those who make below-average contributions to a collective benefit, including the last period of a finitely repeated game. No rational egoist is predicted to spend anything to punish others, since the positive impact of such an action is shared equally with others whether or not they also spend resources on punishing. Indeed, experiments conducted in the United States, Switzerland, and Japan show that individuals who are initially the least trusting are more willing to contribute to sanctioning systems and are likely to be transformed into strong cooperators by the availability of a sanctioning mechanism (Fehr and Gächter 2000). The finding that face-to-face communication is more efficacious than computerized signaling is probably due to the richer language structure available and the added intrinsic costs involved in hearing the intonation and seeing the body language of those who are genuinely angry at free riders (Ostrom 1998a).

7) The rate of contribution to a public good is affected by various contextual factors including the framing of the situation and the rules used for assigning participants, increasing competition among them, allowing communication, authorizing sanctioning mechanisms, or allocating benefits.

These facts are hard to explain using the standard theory that all individuals who face the same objective game structure evaluate decisions in the same way![4] We cannot simply resort to the easy criticism that undergraduate students are erratic. Increasing the size of the payoffs offered in experiments does not appear to change the broad patterns of empirical results obtained.[5] I believe that one is forced by these

---

[4] Although the discussion here focuses on collective action and public good games in particular, a broader range of experiments exists in which the rational egoist's prediction pans out badly. These include the ultimatum game, the dictator game, the trust game, and common-pool resources games with communication.

[5] Most of these experiments involve ultimatum games but the findings are quite relevant. Cameron (1995), for example, conducted ultimatum experiments in Indonesia and thereby was able to use sums that amounted to three months' wages. In this extremely tempting situation, she still found that 56 percent of the Proposers allocated between 40 and 50 percent of this very substantial sum to the Responder.

well-substantiated facts to adopt a more eclectic (and classical) view of human behavior.

Building a Theory of Collective Action with Multiple Types of Players

From the experimental findings, one can begin to put together some of the key assumptions that need to be included in a revised theory of collective action. Assuming the existence of two types of "norm-using" players—"conditional cooperators" and "willing punishers"—in addition to rational egoists, enables one to start making more coherent sense out of the findings of the laboratory experiments on contributions to public goods.

Conditional cooperators are individuals who are willing to initiate cooperative action when they estimate others will reciprocate and to repeat these actions as long as a sufficient proportion of the others involved reciprocate. Conditional cooperators are the source of the relatively high levels of contributions in one-shot or initial rounds of prisoner's dilemma and public good games. Their initial contributions may encourage some rational egoists to contribute as well, so as to obtain higher returns in the early rounds of the game (Kreps *et al.* 1982). Conditional cooperators will tend to trust others and be trustworthy in sequential prisoner's dilemma games as long as the proportion of others who return trust is relatively high. Conditional cooperators tend to vary, however, in their tolerance for free riding. Some are easily disappointed if others do not contribute, so they begin to reduce their own contributions. As they reduce their contributions, they discourage other conditional cooperators from further contributions. Without communication or institutional mechanisms to stop the downward cascade, eventually only the most determined conditional cooperators continue to make positive contributions in the final rounds.

The first four findings are consistent with an assumption that conditional cooperators are involved in most collective action situations. Conditional cooperators are apparently a substantial proportion of the population, given the large number of one-shot and finitely repeated experiments with initial cooperation rates ranging from 40 to 60 percent. Estimating that others are likely to cooperate should increase their willingness to cooperate. Further, knowing the number of repetitions will be relatively long, conditional cooperators can restrain their disappointment with free riders and keep moderate levels of cooperation (and joint payoffs) going for ever-longer periods of time.

The fifth and sixth findings depend on the presence of a third type of player who is willing, if given an opportunity, to punish presumed free riders through verbal rebukes or to use costly material payoffs when available. Willing punishers may also become willing rewarders if the circle of relationships allows them to reward those who have contributed more than the minimal level. Some conditional cooperators may also be willing punishers. Together, conditional cooperators and willing punishers create a more robust opening for collective action and a mechanism for helping it grow. When allowed to communicate on a face-to-face basis, willing punishers convey a considerable level of scorn and anger toward others who have not fully reciprocated their trust and give substantial positive encouragement when cooperation rates are high. Even more important for the long-term sustainability of collective action is the willingness of some to pay a cost to sanction others. The presence of these norm-using types of players is hard to dispute given the empirical evidence. The key question now is: How could these norm-using types of players have emerged and survived in a world of rational egoists?

These findings provide a basis for building an essential feature of cooperation: trust. Trust is a central example of an emergent feature in a complex system that is essential for the functioning of human societies as detailed by Francis Fukuyama in his book of that name.

EMERGENCE AND SURVIVAL OF MULTIPLE TYPES OF PLAYERS IN EVOLUTIONARY PROCESSES

Evolutionary theories provide useful ways of modeling the emergence and survival of multiple types of players in a population. In a strict evolutionary model, individuals inherit strategies and do not change strategies in their lifetime. In this approach, those carrying the more successful strategies for an environment reproduce at a higher rate. After many iterations the more successful strategies come to prominence in the population (Axelrod 1986). Such models are a useful starting point for thinking about competition and relative survival rates among different strategies.[6]

[6]For examples of strict evolutionary models involving collective action, see Nowak and Sigmund (1998), Sethi and Somanathan (1996), and Epstein and Axtell (1996).

Human evolution occurred mostly during the long Pleistocene era that lasted for about 3 million years, up to about 10,000 years ago. During this era, humans roamed the earth in small bands of hunter-gatherers who were dependent on each other for mutual protection, sharing food, and providing for the young. Survival was dependent not only on aggressively seeking individual returns but also on solving many day-to-day collective action problems. Those of our ancestors who solved these problems most effectively, and learned how to recognize who was deceitful and who was a trustworthy reciprocator, had a selective advantage over those who did not (Barkow, Cosmides, and Tooby 1992).

Evolutionary psychologists who study the cognitive structure of the human brain conclude that humans do not develop general analytical skills that are then applied to a variety of specific problems. Humans are not terribly skilled at general logical problem solving (as any scholar who has taught probability theory to undergraduates can attest). Rather, the human brain appears to have evolved a domain-specific, human-reasoning architecture (Clark and Karmiloff-Smith 1991). For example, humans use a different approach to reasoning about deontic relationships—what is forbidden, obligated, or permitted—as contrasted to reasoning about what is true and false. When reasoning about deontic relationships, humans tend to check for violations, or cheaters (Manktelow and Over 1991). When reasoning about whether empirical relationships are true, they tend to use a confirmation strategy (Oaksford and Chater 1994). This deontic effect in human reasoning has repeatedly been detected even in children as young as three years old and is not associated with overall intelligence or educational level of the subject (Cummins 1996).

Thus, recent developments in evolutionary theory and supporting empirical research provide strong support for the assumption that modern humans have inherited a propensity to learn social norms, similar to our inherited propensity to learn grammatical rules (Pinker 1994). Social norms are shared understandings about actions that are obligatory, permitted, or forbidden (Crawford and Ostrom 1995). Which norms are learned, however, varies from one culture to another, across families, and with exposure to diverse social norms expressed

The paper cited here laid the foundation for the Institutional Grammar Tool that has since been extended and elaborated. The institutional grammar tool parses institutional statements into building blocks that dictate how the formal rule would be instantiated in practice and captures the linguistic structure of how different cultural groups translate experiences into formal statements. Understanding this connection is critical for a theory of the evolution of institutions.

within various types of situations. The intrinsic cost or anguish that an individual suffers from failing to use a social norm, such as telling the truth or keeping a promise, is referred to as guilt, if entirely self-inflicted, or as shame, when the knowledge of the failure is known by others (Posner and Rasmusen 1999).

THE INDIRECT EVOLUTIONARY APPROACH TO ADAPTATION THROUGH EXPERIENCE

Recent work on an indirect evolutionary approach to the study of human behavior offers a rigorous theoretical approach for understanding how preferences—including those associated with social norms—evolve or adapt (Güth 1995; Güth and Yaari 1992). In an indirect evolutionary model, players receive objective payoffs, but make decisions based on the transformation of these material rewards into intrinsic preferences. Those who value reciprocity, fairness, and being trustworthy add a subjective change parameter to actions (of themselves or others) that are consistent or not consistent with their norms. This approach allows individuals to start with a predisposition to act in a certain way—thus, they are not rational egoists who only look forward—but it also allows those preferences to adapt in a relatively short number of iterations given the objective payoffs they receive and their intrinsic preferences about those payoffs.

Given the potentially high transaction cost of implementing formal institutions to guide individual actions toward better social outcomes, leveraging such intrinsic preferences has since become an active area of research to guide policy. The work of Richard Thaler and Cass Sunstein and their successful popular book *Nudge* (2008) is a notable example.

Social dilemmas associated with games of trust, like sequential prisoner's dilemma games, are particularly useful games for discussing the indirect evolutionary approach. In such games, if two players trust each other and cooperate, they can both receive a moderately high payoff. However, if one player cooperates and the other does not, then the one who did not cooperate receives an even higher payoff, while the other receives little or nothing. For a rational egoist playing this game, the choice is not to trust, because the expectation is that the other player will not trust, either. As a result, both players will end up with lower payoffs than if they had been able to trust and cooperate. When considering such games, it is useful to remember that most contractual relationships—whether for private or public goods—have at least an element of this basic structure of trying to assure mutual trust. An indirect evolutionary approach explains how a mixture of norm-users and rational egoists would emerge in settings where standard rational choice theory assumes the presence of rational egoists alone.

In this approach, social norms may lead individuals to behave differently in the same objective situation depending on how strongly they value conformance with (or deviance from) a norm. Rational egoists can be thought of as having intrinsic payoffs that are the same as objective payoffs, since they do not value the social norm of reciprocity. Conditional cooperators (to take only one additional type of player for now) would be modeled as being trustworthy types and would have an additional parameter that adds value to the objective payoffs when reciprocating trust with trustworthiness. By their behavior and resulting interaction, however, different types of players are likely to gain differential objective returns. In a game of trust where players are chosen from a population that initially contains some proportion of rational egoists and conditional cooperators, the level of information about player types affects the relative proportion of rational egoists and conditional cooperators over time. With complete information regarding types, conditional cooperators playing a trustworthy strategy will more frequently receive the higher payoff, while rational egoists will consistently receive a lower payoff, since others will not trust them.

Only the trustworthy type would survive in an evolutionary process with complete information (Güth and Kliemt 1998, p. 386). Viewed as a cultural evolutionary process, new entrants to the population would be more likely to adopt the preference ordering of those who obtained the higher material payoffs in the immediate past (Boyd and Richerson 1985). Those who were less successful would tend to learn the values of those who had achieved higher material rewards (Börgers and Sarin 1997).[7] Where a player's type is common knowledge, rational egoists would not survive. Full and accurate information about all players' types, however, is a very strong assumption and unlikely to be met in most real world settings.

---

[7] Eshel, Samuelson, and Shaked (1998) develop a learning model where a population of Altruists who adopt a strategy of providing a local public good interacts in a local neighborhood with a population of Egoists who free ride. In this local interaction setting, Altruists' strategies are imitated sufficiently often in a Markovian learning process to become one of the absorbing states. Altruists interacting with Egoists outside a circular local neighborhood are not so likely to survive.

If there is no information about player types for a relatively large population, preferences will evolve so that only rational egoists survive.[8] If information about the proportion of a population that is trustworthy is known, and no information is known about the type of a specific player, Güth and Kliemt (1998) show that first players will trust second players as long as the expected return of meeting trustworthy players and receiving the higher payoff exceeds the payoff obtained when neither player trusts the other. In such a setting, however, the share of the population held by the norm-using types is bound to decline. On the other hand, if there is a noisy signal about a player's type that is at least more accurate than random, trustworthy types will survive as a substantial proportion of the population. Noisy signals may result from seeing one another, face-to-face communication, and various mechanisms that humans have designed to monitor each other's behavior.

EVIDENCE TESTING THE INDIRECT EVOLUTIONARY APPROACH

An indirect evolutionary approach is able to explain how a mixture of contingent cooperators and rational egoists would emerge in settings where traditional game theory predicts that only rational egoists should prevail. The first six of the seven core findings summarized above were in part the stimulus for the development of the indirect evolutionary theory and the seventh is not inconsistent (see below for further discussion of it). Given the recent development of this approach, direct tests of this theory are not extensive. From the viewpoint of an indirect evolutionary process, participants in a collective action problem would start with differential, intrinsic preferences over outcomes due to their predispositions toward norms such as reciprocity and trust. Participants would learn about the likely behavior of others and shift their behavior in light of the experience and the objective payoffs they

---

[8]This implies that, in a game where players know only their own payoffs and not the payoffs of others, they are more likely to behave like rational egoists. McCabe and Smith (1999) show that players tend to evolve toward the predicted, subgame perfect outcomes in experiments where they have only private information of their own payoffs and to cooperative outcomes when they have information about payoffs and the moves made by other players (see also McCabe, Rassenti, and Smith 1996).

have received. Several recent experiments provide evidence of these kinds of contingent behaviors and behavioral shifts.[9]

In a one-shot, sequential, double-blind prisoner's dilemma experiment, for example, the players were asked to rank their preferences over the final outcomes after they had made their own choice, but before they knew their partner's decision. Forty percent of a pool of 136 subjects ranked the cooperative outcome (C,C) higher than the outcome if they defect while the other cooperates (D,C), and 27 percent were indifferent between these outcomes, even though their individual payoff was substantially higher for them in the latter outcome (Ahn *et al.* 1998).[10] This finding confirms that not all players enter a collective action situation as pure forward-looking rational egoists who make decisions based solely on individual outcomes. Some bring with them a set of norms and values that can support cooperation.

On the other hand, preferences based on these norms can be altered by bad experiences. After 72 subjects had played 12 rounds of a finitely repeated prisoner's dilemma game where partners were randomly matched each round, rates of cooperation were very low and many players had experienced multiple instances where partners had declined to cooperate, only 19 percent of the respondents ranked (C,C) above (D,C), while 17 percent were indifferent (Ahn *et al.* 1999). In this setting, the norms supporting cooperation and reciprocity were diminished, but not eliminated, by experience.

In another version of the prisoner's dilemma game, Cain (1998) first had players participate in a "dictator game"—in which one player divides a sum of money and the other player must accept the division, whatever it is—and then a prisoner's dilemma game. Stingy players, defined as those who retained at least 70 percent of their endowment in the earlier dictator game, tended to predict that all players would defect in the

---

[9] Further, Kikuchi, Watanabe, and Yamagishi (1996) have found that those who express a high degree of trust are able to predict others' behavior more accurately than those with low levels of trust.

[10] To examine the frequency of nonrational egoist preferences, a group of 181 undergraduates was given a questionnaire containing a similar payoff structure on the first day of classes at Indiana University in January 1999. They were asked to rank their preferences. In this nondecision setting, 52 percent reflected preferences that were not consistent with being rational egoists; specifically, 27 percent ranked the outcome (C,C) over (D,C) and 25 percent were indifferent.

prisoner's dilemma game. Nice players, defined as those that gave away at least 30 percent of their endowment, tended to predict that other nice players would cooperate and stingy players would defect. Before playing the prisoner's dilemma game, players were told whether their opponent had been "stingy" or "nice" in the dictator game. Nice players chose cooperation in the prisoner's dilemma game 69 percent of the time when they were paired with other nice players and 39 percent of the time when they were paired with stingy players.

Finally, interesting experimental (as well as field) evidence has accumulated that externally imposed rules tend to "crowd out" endogenous cooperative behavior (Frey 1994). For example, consider some paradoxical findings of Frohlich and Oppenheimer (1996) from a prisoner's dilemma game. One set of groups played a regular prisoner's dilemma game, some with communication and some without. A second set of groups used an externally imposed, incentive-compatible mechanism designed to enhance cooperative choices. In the first phase of the experiment, the second set gained higher monetary returns than the control groups, as expected. In the second phase of the experiment, both groups played a regular prisoner's dilemma game. To the surprise of the experimenters, a higher level of cooperation occurred in the control groups that played the regular prisoner's dilemma in both phases, especially for those who communicated on a face-to-face basis. The greater cooperation that had occurred due to the exogenously created incentive-compatible mechanism appeared to be transient. As the authors put it (p. 180), the removal of the external mechanism "seemed to undermine subsequent cooperation and leave the group worse off than those in the control group who had played a regular... prisoner's dilemma."

Several other recent experimental studies have confirmed the notion that external rules and monitoring can crowd out cooperative behavior.[11] These studies typically find that a social norm, especially

---

[11] Bohnet, Frey, and Huck (1999) set up a sequential prisoner's dilemma, but add a regulatory regime where a "litigation process" is initiated if there is a breach of performance. Cardenas, Stranlund, and Willis (2000) describe an experiment based on harvesting from a common-pool resource conducted in three rural villages in Columbia where exogenous but imperfect rule enforcement generated less cooperation than allowing face-to-face communication.

in a setting where there is communication between the parties, can work as well or nearly as well at generating cooperative behavior as an externally imposed set of rules and system of monitoring and sanctioning. Moreover, norms seem to have a certain staying power in encouraging a growth of the desire for cooperative behavior over time, while cooperation enforced by externally imposed rules can disappear very quickly. Finally, the worst of all worlds may be one where external authorities impose rules but are only able to achieve weak monitoring and sanctioning. In a world of strong external monitoring and sanctioning, cooperation is enforced without any need for internal norms to develop. In a world of no external rules or monitoring, norms can evolve to support cooperation. But in an in-between case, the mild degree of external monitoring discourages the formation of social norms, while also making it attractive for some players to deceive and defect and take the relatively low risk of being caught.

The Evolution of Rules and Norms in the Field

Field studies of collective action problems are extensive and generally find that cooperation levels vary from extremely high to extremely low across different settings. (As discussed above, the seventh core finding from experimental research is that contextual factors affect the rate of contribution to public goods.) An immense number of contextual variables are also identified by field researchers as conducive or detrimental to endogenous collective action. Among those proposed are: the type of production and allocation functions; the predictability of resource flows; the relative scarcity of the good; the size of the group involved; the heterogeneity of the group; the dependence of the group on the good; common understanding of the group; the size of the total collective benefit; the marginal contribution by one person to the collective good; the size of the temptation to free ride; the loss to cooperators when others do not cooperate; having a choice of participating or not; the presence of leadership; past experience and level of social capital; the autonomy to make binding rules; and a wide diversity of rules that are used to change the structure of the situation (see literature cited in Ostrom 2001).

Some consistent findings are emerging from empirical field research. A frequent finding is that when the users of a common-pool resource organize themselves to devise and enforce some of their own basic rules, they tend to manage local resources more sustainably than when rules are externally imposed on them (for example, Tang 1992; Blomquist 1992; Baland and Platteau 1996; Wade 1994). Common-pool resources are natural or humanly created systems that generate a finite flow of benefits where it is costly to exclude beneficiaries and one person's consumption subtracts from the amount of benefits available to others (Ostrom, Gardner, and Walker 1994). The users of a common-pool resource face a first-level dilemma that each individual would prefer that others control their use of the resource while each is able to use the resource freely. An effort to change these rules is a second-level dilemma, since the new rules that they share are a public good. Thus, users face a collective action problem, similar in many respects to the experiments discussed above, of how to cooperate when their immediate best-response strategies lead to suboptimal outcomes for all. A key question now is: How does evolutionary theory help us understand the well-established finding that many groups of individuals overcome both dilemmas? Further, how can we understand how self-organized resource regimes, that rarely rely on external third-party enforcement, frequently outperform government-owned resource regimes that rely on externally enforced, formal rules?

THE EMERGENCE OF SELF-ORGANIZED COLLECTIVE ACTION

From evolutionary theory, we should expect some individuals to have an initial propensity to follow a norm of reciprocity and to be willing to restrict their own use of a common pool resource so long as almost everyone reciprocates. If a small core group of users identify each other, they can begin a process of cooperation without having to devise a full-blown organization with all of the rules that they might eventually need to sustain cooperation over time. The presence of a leader or entrepreneur, who articulates different ways of organizing to improve joint outcomes, is frequently an important initial stimulus (Frohlich, Oppenheimer, and Young 1971; Varughese 1999).[12]

---

[12]Empirical studies of civil rights movements, where contributions can be very costly, find that organizers search for ways to assure potential participants of the importance

If a group of users can determine its own membership—including those who agree to use the resource according to their agreed-upon rules and excluding those who do not agree to these rules—the group has made an important first step toward the development of greater trust and reciprocity. Group boundaries are frequently marked by well-understood criteria, like everyone who lives in a particular community or has joined a specific local cooperative. Membership may also be marked by symbolic boundaries and involve complex rituals and beliefs that help solidify individual beliefs about the trustworthiness of others.

DESIGN PRINCIPLES OF LONG-SURVIVING, SELF-ORGANIZED RESOURCE REGIMES

Successful self-organized resource regimes can initially draw upon locally evolved norms of reciprocity and trustworthiness and the likely presence of local leaders in most community settings. More important, however, for explaining their long-term survival and comparative effectiveness, resource regimes that have flourished over multiple generations tend to be characterized by a set of design principles. These design principles are extensively discussed in Ostrom (1990) and have been subjected to extensive empirical testing.[13] Evolutionary theory helps to explain how these design principles work to help groups sustain and build their cooperation over long periods of time.

We have already discussed the first design principle—the presence of clear boundary rules. Using this principle enables participants to know who is in and who is out of a defined set of relationships and thus with whom to cooperate. The second design principle is that the local rules-in-use restrict the amount, timing, and technology of harvesting the resource; allocate benefits proportional to required inputs; and are crafted to take local conditions into account. If a group of users is going to harvest from

---

of shared internal norms and that many others will also participate (Chong 1991). Membership in churches and other groups that jointly commit themselves to protests and other forms of collective action is also an important factor (Opp, Voss, and Gern 1995).

[13] The design principles that characterize long-standing common-pool resource regimes have now been subject to considerable further empirical studies since they were first articulated (Ostrom 1990). While minor modifications have been offered to express the design principles somewhat differently, no empirical study has challenged their validity, to my knowledge (Morrow and Hull 1996; Asquith 1999; Bardhan 1999; Lam 1998).

a resource over the long run, they must devise rules related to how much, when, and how different products are to be harvested, and they need to assess the costs on users of operating a system. Well-tailored rules help to account for the perseverance of the resource itself. How to relate user inputs to the benefits they obtain is a crucial element of establishing a fair system (Trawick 1999). If some users get all the benefits and pay few of the costs, others become unwilling to follow rules over time.

In long-surviving irrigation systems, for example, subtly different rules are used in each system for assessing water fees used to pay for maintenance activities, but water tends to be allocated proportional to fees or other required inputs (Bardhan 1999). Sometimes water and responsibilities for resource inputs are distributed on a share basis, sometimes on the order in which water is taken, and sometimes strictly on the amount of land irrigated. No single set of rules defined for all irrigation systems in a region would satisfy the particular problems in managing each of these broadly similar, but distinctly different, systems (Tang 1992; Lam 1998).

The third design principle is that most of the individuals affected by a resource regime can participate in making and modifying their rules. Resource regimes that use this principle are both able to tailor better rules to local circumstances and to devise rules that are considered fair by participants. The Chisasibi Cree, for example, have devised a complex set of entry and authority rules related to the fish stocks of James Bay as well as the beaver stock located in their defined hunting territory. Berkes (1987, p. 87) explains that these resource systems and the rules used to regulate them have survived and prospered for so long because effective "social mechanisms ensure adherence to rules which exist by virtue of mutual consent within the community. People who violate these rules suffer not only a loss of favor from the animals (important in the Cree ideology of hunting) but also social disgrace." Fair rules of distribution help to build trusting relationships, since more individuals are willing to abide by these rules because they participated in their design and also because they meet shared concepts of fairness (Bowles 1998).

In a study of 48 irrigation systems in India, Bardhan (1999) finds that the quality of maintenance of irrigation canals is significantly lower on those systems where farmers perceive the rules to be made by a local elite. On the other hand, those farmers (of the 480 interviewed) who responded that the rules have been crafted by most of the farmers, as contrasted to the elite or the government, have a more positive attitude about the water allocation rules and the rule compliance of other farmers. Further, in all of the villages where a government agency decides how water is to be allocated and distributed, frequent rule violations are reported and farmers tend to contribute less to the local village fund. Consistent with this is the finding by Ray and Williams (1999) that the deadweight loss from upstream farmers stealing water on government-owned irrigation systems in Maharashtra, India, approaches one-fourth of the revenues that could be earned in an efficient water allocation and pricing regime.

Few long-surviving resource regimes rely only on endogenous levels of trust and reciprocity. The fourth design principle is that most long-surviving resource regimes select their own monitors, who are accountable to the users or are users themselves and who keep an eye on resource conditions as well as on user behavior. Further, the fifth design principle points out that these resource regimes use graduated sanctions that depend on the seriousness and context of the offense. By creating official positions for local monitors, a resource regime does not have to rely only on willing punishers to impose personal costs on those who break a rule. The community legitimates a position. In some systems, users rotate into this position so everyone has a chance to be a participant as well as a monitor. In other systems, all participants contribute resources and they hire monitors jointly. With local monitors, conditional cooperators are assured that someone is generally checking on the conformance of others to local rules. Thus, they can continue their own cooperation without constant fear that others are taking advantage of them.

On the other hand, the initial sanctions that are imposed are often so low as to have no impact on an expected benefit-cost ratio of breaking local rules (given the substantial temptations frequently involved). Rather, the initial sanction needs to be considered more as information both to the person who is "caught" and to others in the community. Everyone can make an error or can face difficult problems leading them to break a rule. Rule infractions, however, can generate a downward cascade of cooperation in a group that relies only on conditional cooperation and has no capacity to sanction (for example Kikuchi *et al.* 1998). In a regime that uses graduated punishments, however, a person who purposely or by error breaks a rule is notified that others notice the infraction (thereby increasing the individual's confidence that others would also be caught). Further, the individual learns that others basically continue to extend their trust and want only a small token to convey a recognition that the mishap occurred. Self-organized regimes rely more on what Margaret Levi calls "quasi-voluntary" cooperation than either strictly voluntary or coerced cooperation (Levi 1988). A real threat to the continuance of self-organized regimes occurs, however, if some participants break rules repeatedly. The capability to escalate sanctions enables such a regime to warn members that if they do not conform they will have to pay ever-higher sanctions and may eventually be forced to leave the community.

The design principles have subsequently been tested extensively. In general, there is a correlation between the presence of the design principles and the "success" of collective action although success is difficult to define in social-ecological systems. These studies also suggest that success can be more closely related to the biophysical and technological characteristics of the system rather than institutional characteristics.

Let me summarize my argument to this point. When the users of a resource design their own rules (Design Principle 3) that are enforced by local users or accountable to them (Design Principle 4) using graduated sanctions (Design Principle 5) that define who has rights to withdraw from the resource (Design Principle 1) and that effectively assign costs proportionate to benefits (Design Principle 2), collective action and monitoring problems are solved in a reinforcing manner (Agrawal 1999).

Individuals who think a set of rules will be effective in producing higher joint benefits and that monitoring (including their own) will protect them against being a sucker are willing to undertake conditional cooperation. Once some users have made contingent self-commitments, they are then motivated to monitor other people's behavior, at least from time to time, to assure themselves that others are following the rules most of the time. Conditional cooperation and mutual monitoring reinforce one another, especially in regimes where the rules are designed to reduce monitoring costs. Over time, further adherence to shared norms evolves and high levels of cooperation are achieved without the need to engage in very close and costly monitoring to enforce rule conformance.

The operation of these principles is then bolstered by the sixth design principle that points to the importance of access to rapid, low-cost, local arenas to resolve conflict among users or between users and officials. Rules, unlike physical constraints, have to be understood to be effective. There are always situations in which participants can interpret a rule that they have jointly made in different ways. By devising simple, local mechanisms to get conflicts aired immediately and resolutions that are generally known in the community, the number of conflicts that reduce trust can be reduced. If individuals are going to follow rules over a long period of time, some mechanism for discussing and resolving what constitutes a rule infraction is necessary to the continuance of rule conformance itself.

The capability of local users to develop an ever-more effective regime over time is affected by whether they have minimal recognition of the right to organize by a national or local government. This is the seventh design principle. While some resource regimes have operated for relatively long times without such rights (Ghate 2000), participants have had to rely almost entirely on unanimity as the rule used to change rules. (Otherwise, any temporarily disgruntled participant who voted against a rule change could go to the external authorities to threaten the regime itself!) Unanimity as a decision rule for changing rules imposes high transaction costs and prevents a group from searching for better matched rules at relatively lower costs.

Users frequently devise their own rules without creating formal, governmental jurisdictions for this purpose. In many in-shore fisheries, for example, local fishers devise extensive rules defining who can use a fishing ground and what kind of equipment can be used (Acheson 1988; Schlager 1994). As long as external governmental officials give at least minimal recognition to the legitimacy of such rules, the fishers themselves may be able to enforce the rules. But if external governmental officials presume that only they can make authoritative rules, then it is difficult for local users to sustain a self-organized regime (Johnson and Libecap 1982).

When common pool resources are somewhat larger, an eighth design principle tends to characterize successful systems—the presence of governance activities organized in multiple layers of nested enterprises. The rules appropriate for allocating water among major branches of an irrigation system, for example, may not be appropriate for allocating water among farmers along a single distributory channel. Consequently, among long-enduring self-governed regimes, smaller-scale organizations tend to be nested in ever-larger organizations. It is not unusual to find a large, farmer-governed irrigation system, for example, with five layers of organization each with its own distinct set of rules (Yoder 1992).

THREATS TO SUSTAINED COLLECTIVE ACTION

All economic and political organizations are vulnerable to threats, and self-organized resource-governance regimes are no exception. Both exogenous and endogenous factors challenge their long-term viability. Here we will concentrate on those factors that affect the distribution of types of participants within a regime and the strength of the norms of trust and reciprocity held by participants. Major migration (out of or into an area) is always a threat that may or may not be countered effectively. Out-migration may change the economic viability of a regime due to loss of those who contribute needed resources. In-migration may bring new participants who do not trust others and do not rapidly learn social norms that have been established over a long period of time. Since collective action is largely based on mutual trust, some self-organized resource regimes that are in areas of rapid settlement have disintegrated within relatively short times (Baland and Platteau 1996).

In addition to rapid shifts in population due to market changes or land distribution policies, several more exogenous and endogenous threats have been identified in the empirical literature (Sengupta 1991; Bates 1987; and literature cited in Ostrom 1998b; Britt 2000). These include: 1) efforts by national governments to impose a single set of rules on all governance units in a region; 2) rapid changes in technology, in factor availability, and in reliance on monetary transactions; 3) transmission failures from one generation to the next of the operational principles on which self-organized governance is based; 4) turning to external sources of help too frequently; 5) international aid that does not take account of indigenous knowledge and institutions; 6) growth of corruption and other forms of opportunistic behavior; and 7) a lack of large-scale institutional arrangements that provide fair and low-cost resolution mechanisms for conflicts that arise among local regimes, educational and extension facilities, and insurance mechanisms to help when natural disasters strike at a local level.

Contextual variables are thus essential for understanding the initial growth and sustainability of collective action as well as the challenges that long-surviving, self-organized regimes must try to overcome. Simply saying that context matters is not, however, a satisfactory theoretical approach. Adopting an evolutionary approach is the first step toward a more general theoretical synthesis that addresses the question of how context matters. In particular, we need to address how context affects the presence or absence of conditional cooperators and willing punishers and the likelihood that the norms held by these participants are adopted and strengthened by others in a relevant population.

Conclusion

Both laboratory experiments and field studies confirm that a substantial number of collective action situations are resolved successfully, at least in part. The old-style notion, pre-Mancur Olson, that groups would find ways to act in their own collective interest was not entirely misguided. Indeed, recent developments in evolutionary theory—including the study of cultural evolution—have begun to provide genetic and adaptive underpinnings for the propensity to cooperate based on the development and growth of social norms. Given the frequency and diversity of collective action situations in all modern economies, this represents a more optimistic view than the zero contribution hypothesis. Instead of pure pessimism or pure optimism, however, the picture requires further work to explain why some contextual variables enhance cooperation while others discourage it.

Empirical and theoretical work in the future needs to ask how a large array of contextual variables affects the processes of teaching and evoking social norms; of informing participants about the behavior of others and their adherence to social norms; and of rewarding those who use social norms, such as reciprocity, trust, and fairness. We need to understand how institutional, cultural, and biophysical contexts affect the types of individuals who are recruited into and leave particular types of collective action situations, the kind of information that is made available about past actions, and how individuals can themselves change structural variables so as to enhance the probabilities of norm-using types being involved and growing in strength over time.

Further developments along these lines are essential for the development of public policies that enhance socially beneficial, cooperative behavior based in part on social norms. It is possible that past policy initiatives to encourage collective action that were based primarily on externally changing payoff structures for rational egoists may have been misdirected—and perhaps even crowded out the formation of social norms that might have enhanced cooperative behavior in their own way. Increasing the authority of individuals to devise their own rules may well result in processes that allow social norms to evolve and thereby increase the probability of individuals better solving collective action problems.

REFERENCES

Acheson, J. M. 1988. *The Lobster Gangs of Maine.* Hanover, NH: University Press of New England.

Agrawal, A. 1999. *Greener Pastures: Politics, Markets, and Community among a Migrant Pastoral People.* Durham, NC: Duke University Press.

Ahn, T-K., E. Ostrom, D. Schmidt, and J. Walker. 1998. *Trust and Reciprocity: Experimental Evidence from PD Games.* Technical report. Workshop in Political Theory and Policy Analysis, working paper. Bloomington, IN: Indiana University.

———. 1999. *Dilemma Games: Game Parameters and Matching Protocols.* Technical report. Workshop in Political Theory and Policy Analysis, working paper. Bloomington, IN: Indiana University.

Asquith, N. M. 1999. *How Should the World Bank Encourage Private Sector Investment in Biodiversity Conservation? A Report Prepared for Kathy MacKinnon, Biodiversity Specialist, The World Bank.* Technical report. Washington, DC and Durham, NC: Sanford Institute of Public Policy ard Duke University.

Axelrod, R. 1986. "An Evolutionary Approach to Norms." *American Political Science Review* 80 (4): 1095–1111.

Baland, J.-M., and J.-P. Platteau. 1996. *Halting Degradation of Natural Resources: Is There a Role for Rural Communities?* Oxford, UK: Clarendon Press.

Bardhan, P. 1999. *Water Community: An Empirical Analysis of Cooperation on Irrigation in South India.* Technical report. Working paper. Berkeley, CA.

Barkow, J. H., L. Cosmides, and J. Tooby, eds. 1992. *The Adapted Mind: Evolutionary Psychology and the Generation of Culture.* Oxford, UK: Oxford University Press.

Bates, R. H. 1987. *Essays on the Political Economy of Rural Africa.* Edited by J. H. Barkow, L. Cosmides, and J. Tooby. Berkeley, CA: University of California Press.

Berkes, F. 1987. "Common Property Resource Management and Cree Indian Fisheries in Subarctic Canada." In *The Question of the Commons: The Culture and Ecology of Communal Resources.* Edited by B. J. McCay and J. Acheson, 66–91. Tucson, AZ: University of Arizona Press.

Blomquist, W. 1992. *Dividing the Waters: Governing Groundwater in Southern California.* San Francisco, CA: ICS Press.

Bohnet, I., and B. S. Frey. 1999. "The Sound of Silence in Prisoner's Dilemma and Dictator Games." *Journal of Economic Behavior and Organization* 38 (1): 43–58.

Bohnet, I., B. S. Frey, and S. Huck. 1999. *More Order with Less Law: On Contract Enforcement, Trust, and Crowding.* Technical report. Working paper. Cambridge, MA: Harvard University.

Börgers, T., and R. Sarin. 1997. "Learning Through Reinforcement and Replicator Dynamics." *Journal of Economic Theory* 77:1–14.

Bowles, S. 1998. "Endogenous Preferences: The Cultural Consequences of Markets and Other Economic Institutions." *Journal of Economic Literature* 36:75–111.

Boyd, R., and P. J. Richerson. 1985. *Culture and the Evolutionary Process.* Chicago, IL: University of Chicago Press.

Britt, C. 2000. *Forestry and Forest Policies.* Technical report. Workshop in Political Theory and Policy Analysis, working paper. Bloomington, IN: Indiana University.

Bromley, D. W., D. Feeny, M. A. McKean, P. Peters, J. L. Gilles, R. J. Oakerson, C. F. Runge, and J. T. Thomson, eds. 1992. *Making the Commons Work: Theory, Practice, and Policy.* San Francisco, CA: ICS Press.

Cain, M. 1998. "An Experimental Investigation of Motives and Information in the Prisoner's Dilemma Game." In *Advances in Group Processes,* edited by J. Skvoretz and J. Szmatka, 15:133–60. Greenwich, CT: JAI Press.

Cameron, L. 1995. *Raising the Stakes in the Ultimatum Game: Experimental Evidence from Indonesia.* Technical report. Discussion paper. Princeton, NJ: Princeton University.

Cardenas, J.-C., J. K. Stranlund, and C. E. Willis. 2000. "Local Environmental Control and Institutional Crowding-Out." *World Development* 28 (10): 1719–1733.

Chong, D. 1991. *Collective Action and the Civil Rights Movement.* Chicago, IL: University of Chicago Press.

Clark, A., and A. Karmiloff-Smith. 1991. "The Cognizer's Innards: A Psychological and Philosophical Perspective on the Development of Thought." *Mind and Language* 8 (4): 487–519.

Crawford, S. E. S., and E. Ostrom. 1995. "A Grammar of Institutions." *American Political Science Review* 89 (3): 582–600.

Cummins, D. D. 1996. "Evidence of Deontic Reasoning in 3- and 4-Year-Olds." *Memory and Cognition* 24:823–29.

Davis, D. D., and C. A. Holt. 1993. *Experimental Economics.* Princeton, NJ: Princeton University Press.

Dawes, R. M., J. McTavish, and H. Shaklee. 1977. "Behavior, Communication, and Assumptions about Other People's Behavior in a Commons Dilemma Situation." *Journal of Personality and Social Psychology* 35 (1): 1–11.

Epstein, J. M., and R. L. Axtell. 1996. *Growing Artificial Societies: Social Science from the Bottom Up.* Cambridge, MA: MIT Press.

Eshel, I., L. Samuelson, and A. Shaked. 1998. "Altruists, Egoists, and Hooligans in a Local Interaction Model." *American Economic Review* 88 (1): 157–79.

Feeny, D., F. Berkes, B. J. McCay, and J. M. Acheson. 1990. "The Tragedy of the Commons: Twenty-Two Years Later." *Human Ecology* 18 (1): 1–19.

Fehr, E., and S. Gächter. 2000. "Cooperation and Punishment in Public Goods Experiments." *American Economic Review* 90 (4): 980–994.

Fehr, E., and K. Schmidt. 1999. "A Theory of Fairness, Competition, and Cooperation." *Quarterly Journal of Economics* 114 (3): 817–68.

Frank, R. H., T. Gilovich, and D. T. Regan. 1993. "The Evolution of One-Shot Cooperation: An Experiment." *Ethology and Sociobiology* 14 (4): 247–56.

Frey, B. S. 1994. "How Intrinsic Motivation is Crowded Out and In Rationality and Society." *Rationality and Society* 6 (3): 334–52.

Frohlich, N., and J. A. Oppenheimer. 1996. "Experiencing Impartiality to Invoke Fairness in the n-PD: Some Experimental Results." *Public Choice* 86:117–35.

Frohlich, N., J. A. Oppenheimer, and O. Young. 1971. *Political Leadership and Collective Goods.* Princeton, NJ: Princeton University Press.

Ghate, R. 2000. *The Role of Autonomy in Self-Organizing Process: A Case Study of Local Forest Management in India.* Technical report. Workshop in Political Theory and Policy Analysis, working paper. Bloomington, IN.

Güth, W. 1995. "An Evolutionary Approach to Explaining Cooperative Behavior by Reciprocal Incentives." *International Journal of Game Theory* 24:323–44.

Güth, W., and H. Kliemt. 1998. "The Indirect Evolutionary Approach: Bridging the Gap between Rationality and Adaptation." *Rationality and Society* 10 (3): 377–99.

Güth, W., and M. Yaari. 1992. "An Evolutionary Approach to Explaining Reciprocal Behavior in a Simple Strategic Game," edited by U. Witt, 23–34. Ann Arbor, MI: University of Michigan Press.

Hardin, R. 1971. "Collective Action as an Agreeable n-Prisoners' Dilemma." *Science* 16 (5): 472–81.

———. 1982. *Collective Action.* Baltimore, MD: Johns Hopkins University Press.

Hess, C. 1999. *A Comprehensive Bibliography of Common Pool Resources.* Workshop in Political Theory and Policy Analysis. Indiana University, Bloomington, IN.

Isaac, R. M., J. Walker, and A. W. Williams. 1994. "Group Size and the Voluntary Provision of Public Goods: Experimental Evidence Utilizing Large Groups." *Journal of Public Economics* 54 (1): 1–36.

Johnson, R. N., and G. D. Libecap. 1982. "Contracting Problems and Regulation: The Case of the Fishery." *American Economic Review* 72 (5): 1005–1022.

Kagel, J., and A. Roth, eds. 1995. *The Handbook of Experimental Economics.* Princeton, NJ: Princeton University Press.

Kikuchi, M., M. Fujita, E. Marciano, and Y. Hayami. 1998. *State and Community in the Deterioration of a National Irrigation System.* Paper presented at the World Bank-EDI Conference on "Norms and Evolution in the Grassroots of Asia," Stanford University, February 6-7.

Kikuchi, M., Y. Watanabe, and T. Yamagishi. 1996. "Accuracy in the Prediction of Others' Trustworthiness and General Trust: An Experimental Study." *Japanese Journal of Experimental Social Psychology* 37 (1): 23–36.

Kreps, D. M., P. Milgrom, J. Roberts, and R. Wilson. 1982. "Rational Cooperation in the Finitely Repeated Prisoner's Dilemma." *Journal of Economic Theory* 27 (2): 245–52.

Lam, W. F. 1998. *Governing Irrigation Systems in Nepal: Institutions, Infrastructure, and Collective Action.* Oakland, CA: ICS Press.

Ledyard, J. 1995. "Public Goods: A Survey of Experimental Research." In *The Handbook of Experimental Economics,* edited by J. Kagel and A. Roth, 111–94. Princeton, NJ: Princeton University Press.

Levi, M. 1988. *Of Rule and Revenue.* Berkeley, CA: University of California Press.

Lichbach, M. I. 1996. "The Cooperator's Dilemma." (Ann Arbor, MI).

Manktelow, K. I., and D. E. Over. 1991. "Social Roles and Utilities in Reasoning with Deontic Conditionals." *Cognition* 39:85–105.

McCabe, K. A., S. J. Rassenti, and V. L. Smith. 1996. "Game Theory and Reciprocity in Some Extensive Form Experimental Games." *Proceedings of the National Academy of Sciences* 93 (23): 13421–13428.

McCabe, K. A., and V. L. Smith. 1999. *Strategic Analysis by Players in Games: What Information Do They Use.* Technical report. Working paper. Tucson, AZ: University of Arizona Economic Research Laboratory.

Milgrom, D. C., P. R. North, and B. R. Weingast. 1990. "The Role of Institutions in the Revival of Trade: The Law Merchant, Private Judges, and the Champagne Fairs." *Economics and Politics* 2 (1): 1–23.

Montgomery, M. R., and R. Bean. 1999. "Market Failure, Government Failure, and the Private Supply of Public Goods: The Case of Climate-Controlled Walkway Networks." *Public Choice* 99 (3/4): 403–37.

Morrow, C. E., and R. W. Hull. 1996. "Donor-Initiated Common Pool Resource Institutions: The Case of the Yanesha Forestry Cooperative." *World Development* 24 (10): 1641–1657.

Nowak, M. A., and K. Sigmund. 1998. "Evolution of Indirect Reciprocity by Image Scoring." *Nature* 393 (6685): 573–77.

Oaksford, M., and N. Chater. 1994. "A Rational Analysis of the Selection Task as Optimal Data Selection." *Psychological Review* 101 (4): 608–31.

Offerman, T. 1997. *Beliefs and Decision Rules in Public Goods Games: Theory and Experiments.* Dordrecht, Netherlands: Kluwer Academic Publishers.

Olson, M. 1965. *The Logic of Collective Action: Public Goods and the Theory of Groups.* Cambridge, MA: Harvard University Press.

Opp, K.-D., P. Voss, and C. Gern. 1995. *Origins of Spontaneous Revolution.* Ann Arbor, MI: University of Michigan Press.

Ostrom, E. 1990. *Governing the Commons: The Evolution of Institutions for Collective Action.* New York, NY: Cambridge University Press.

———. 1998a. "A Behavioral Approach to the Rational Choice Theory of Collective Action." *American Political Science Review* 92 (1): 1–22.

———. 1998b. "Institutional Analysis, Design Principles, and Threats to Sustainable Community Governance and Management of Commons." In *Law and the Governance of Renewable Resources: Studies from Northern Europe and Africa,* edited by E. Berge and N. C. Stenseth, 27–53. Oakland, CA: ICS Press.

———. 2001. "Reformulating the Commons." In *Protecting the Commons: A Framework For Resource Management In The Americas,* edited by J. Burger, E. Ostrom, R. Norgaard, D. Policansky, and B. Goldstein. Washington, DC: Island Press.

Ostrom, E., R. Gardner, and J. Walker. 1994. *Rules, Games, and Common-Pool Resources.* Ann Arbor, MI: University of Michigan Press.

Ostrom, E., and J. Walker. 1997. "Neither Markets Nor States: Linking Transformation Processes in Collective Action Arenas." In *Perspectives on Public Choice: A Handbook,* edited by D. C. Mueller, 35–72. Cambridge, UK: Cambridge University Press.

Pinker, S. 1994. *The Language Instinct.* New York, NY: William Morrow.

Posner, R. A., and E. B. Rasmusen. 1999. "Creating and Enforcing Norms, with Special Reference to Sanctions." *International Review of Law and Economics* 19 (3): 369–82.

Rabin, M. 1993. "Incorporating Fairness into Game Theory and Economics." *American Economic Review* 83:1281–1302.

Ray, I., and J. Williams. 1999. "Evaluation of Price Policy in the Presence of Water Theft." *American Journal of Agricultural Economics* 81:928–41.

Schlager, E. 1994. "Fishers' Institutional Responses to Common-Pool Resource Dilemmas." In *Rules, Games, and Common-Pool Resources,* edited by E. Ostrom, R. Gardner, and J. Walker, 247–65. Ann Arbor, MI: University of Michigan Press.

Selten, R. 1991. "Evolution, Learning, and Economic Behavior." *Games and Economic Behavior* 3 (1): 3–24.

Sengupta, N. 1991. *Managing Common Property: Irrigation in India and the Philippines.* New Delhi, India: Sage.

Sethi, R., and E. Somanathan. 1996. "The Evolution of Social Norms in Common Property Resource Use." *American Economic Review* 86 (4): 766–88.

Tang, S. Y. 1992. *Institutions and Collective Action: Self Governance in Irrigation.* San Francisco, CA: ICS Press.

Trawick, P. 1999. *The Moral Economy of Water: 'Comedy' and 'Tragedy' in the Andean Commons.* Technical report. Department of Anthropology, working paper. Lexington, KY: University of Kentucky.

Varughese, G. 1999. "Villagers, Bureaucrats, and Forests in Nepal: Designing Governance for a Complex Resource." PhD diss., Indiana University.

Wade, R. 1994. *Village Republics: Economic Conditions for Collective Action in South India.* San Francisco, CA: ICS Press.

Yoder, R. D. 1992. *Performance of the Chhattis Mauja Irrigation System, a Thirty-Five Hundred Hectare System Built and Managed by Farmers in Nepal.* Colombo, Sri Lanka: International Irrigation Management Institute.

BIOGRAPHICAL ODDITIES, VOLUME 4*

WILLIAM BRIAN ARTHUR | July 31, 1945 – | *c.f. Chapters 67, 77*

Arthur left Belfast shortly after the British army occupied the city as part of Operation Banner. He initially began a PhD at the University of Michigan because, when given an alphabetized list of prospective schools by a supervisor, the Ann Arbor address was listed first.

When making his now-famous model for the El Farol Bar Problem, Arthur changed the day on which live Irish music was actually performed from Wednesday to Thursday, so as to bury the lede and further prevent the overcrowding that he was modeling.

ALAN STUART PERELSON | April 11, 1947 – | *c.f. Chapters 68, 79*

In the 1950 US census, Perelson's name is spelled wrong and then awkwardly crossed out.

Perelson is an experienced woodworker.

In 2014, Thomson Reuters ranked Perelson as one of the "World's Most Influential Scientific Minds" of the previous decade.

JOHN ARCHIBALD WHEELER | July 9, 1911 – April 13, 2008 | *c.f. Chapter 70*

Allegedly, a student in 1967 suggested the term "black hole" to Wheeler, which he adopted. This is one origin story for the coining of the term.

At his summer home in Maine, Wheeler liked to fire beer cans from an old cannon on the property for fun.

In 1953, while traveling by train, Wheeler misplaced classified political documents and plans pertaining to the H-bomb. He insisted that he wedged the file behind a toilet and accidentally left it, but ran back moments later to find the envelope empty. Wheeler was reprimanded, but others lost their jobs over the incident, including Joint Committee on Atomic Energy director William Borden, who had prepared the documents for Wheeler. Borden subsequently turned his attention to J. Robert Oppenheimer, based on information he prepared for Wheeler's file, alleging that Oppenheimer was a Soviet spy, which ultimately led to the security hearing that removed Oppenheimer's Q clearance. The file was never found.

* Editor's note: In life we retain control over (some of) the details of our existence. Dates and details provided here reflect the information authors are willing to share broadly.

WILLIAM SAMUEL BIALEK | Aug. 14, 1960 – | *c.f. Chapter 71*

Bialek's parents were born in 1918 and 1919, in areas of Russia that would soon become part of Poland. His father was born on the day the Treaty of Versailles was ratified, and so he often joked to his son that when he was born he was Russian in the morning and Polish by the evening.

Bialek credits the inspiration behind his initial drive to study physics to a biographical profile of Isidor Rabi, written by Jeremy Bernstein in 1975, which changed his teenage view of what a physicist could be and do.

FREDERICK MARTIN RIEKE | 1966 – | *c.f. Chapter 71*

Rieke is the principal investigator for his own laboratory at the University of Washington. Among the list of Rieke Lab associates provided on the lab's website, Rieke lists his two children as honorary "former members" of the lab.

ROBERT DE RUYTER VAN STEVENINCK | May 31, 1954 – | *c.f. Chapter 71*

At the University of California, Berkeley, after painting the walls of physics classroom 397 floor-to-ceiling with blackboard paint, van Steveninck, Warland, Bialek, and Rieke covered the room with equations and figures and together conceived of the ideas for their book, *Spikes: Exploring the Neural Code*, in chalk.

DAVID KARSTEN WARLAND | April 8, 1963 – | *c.f. Chapter 71*

He is the son of composer and conductor Dale Warland, who is one of the only choral conductors ever to be inducted into the American Classical Music Hall of Fame.

Warland owns and operates a small farm, plays the cello, and is a talented woodworker who once made a wooden model of the famous "rainbow bridge" of China.

JOHN H. MILLER | Sept. 8, 1959 – | *c.f. Chapter 72*

Born into the fourth generation of a family of cattle ranchers, Miller grew up ranching and learned to bring in cattle on horseback. Several of his family members preferred to bring the cattle in with Cadillacs rather than horses or ATVs.

He was the Santa Fe Institute's first postdoc and has remained involved with the Institute ever since.

Miller learned to scuba dive in his home state of Colorado. In high school and college he worked as a professional diver in Colorado as well as in the British Overseas Territories in the West Indies.

FOUNDATIONAL PAPERS IN COMPLEXITY SCIENCE

KRISTIAN LINDGREN | Aug. 30, 1960 – | c.f. Chapter 73

At the time of publication, Lindgren is currently hard at work designing and building his very own space observatory on an island called Flatön off of the West coast of Sweden.

Lindgren is one of the founders of the master's program in complex adaptive systems at his alma mater, Chalmers University of Technology.

JOHN STEPHEN LANSING | May 19, 1950 – | c.f. Chapter 75

In 2018, Lansing took a team of researchers into dense jungle in Indonesian Borneo to meet with a group of nomadic, traditionally cave-dwelling hunter–gatherers. This small group, the Punan Batu, can trace their genetics back some 7,000 years, largely independent of other local indigenous groups and Austronesian peoples. After Lansing and his team agreed to help the Punan Batu defend their rights to keep their ancestral homesites safe from deforestation, evidence of the group's genetic descent, as well as their unique song-language, convinced the Regency of Bulungan to support the Punan Batu in their conservation claims.

JAMES NEVIN KREMER | July 19, 1945 – | c.f. Chapter 75

Kremer met Lansing by complete happenstance when both were surfing at the same beach in Southern California, shortly after Lansing had realized the project in Bali could really benefit from bringing an ecologist on board...

Nineteenth-century American composer and pianist Ethelbert Nevin is a distant relation of Kremer's and the source of his middle name.

MELANIE MITCHELL | 1958 – | c.f. Chapter 76

After reading Douglas Hofstadter's *Gödel, Escher, Bach*, Mitchell was determined to work with Hofstadter, and wrote him a letter to this effect. It recevied no reply. A year later, she spoke to him after he gave a lecture at MIT and called several times, but he never got back to her. Finally, Mitchell called Hofstadter after hours at 11:00 pm, and he answered the phone. Impressed by Mitchell's ideas and persistance, Hofstadter agreed to supervise her work on the analogy-making program CopyCat, and she began her PhD under his supervision shortly thereafter.

Biographical Oddities

PETER THOMAS HRABER | 1967 – | *c.f. Chapter 76*

Hraber plays the bagpipes and has been a member of Albuquerque & Four Corners Pipes & Drums, a Los Alamos-based bagpipe band.

Among the dedications in his PhD dissertation, Hraber thanks his two terriers for providing emotional support during his grad school years.

STEPHANIE FORREST | 1958 – | *c.f. Chapter 79*

In the early years of the Santa Fe Institute, space was at a premium and the Institute went through several major relocations. At one juncture, Forrest shared an office with SFI founder George Cowan, who enjoyed babysitting Forrest's infant child when she occasionally brought her to the office.

Forrest is an avid horseback rider and has trained several horses herself.

RAJESH CHERUKURI | 1970 – | *c.f. Chapter 79*

Cherukuri's undergraduate alma mater, Osmania University, is the seventh oldest in India.

According to Cherukuri, during his tenure as Vice President of Engineering for Bomgar Corporation, his leadership and changes to the department increased company revenue from $20 million to $100 million.

WALTER FONTANA | Dec. 3, 1960 – | *c.f. Chapter 80*

Fontana is a retired single-engine airplane pilot and paraglider pilot with a P4 rating, the second highest rating in paragliding, which requires upwards of 200 logged flights and over 75 hours of flight time.

Born in the South Tyrol region of Italy, Fontana's birthplace is formally registered as Merano/Meran, as local law requires both Italian and German designations for placenames, due to the region's proximity to Germany.

PETER FLORIAN STADLER | Dec. 24, 1965 – | *c.f. Chapter 80*

In 1997, Stadler was drafted into mandatory military service, an assignment which brought him no joy. On his personal website, as of the publication of this volume, he describes his time in the military with one hand-typed emoticon: :-(

Stadler is a connoisseur of chile peppers and New Mexican recipes. He is also an aficionado of Austrian and German beer varieties.

IVO LUDWIG HOFACKER | Nov. 23, 1964 – | c.f. Chapter 80

In the acknowledgements to the publication of ViennaRNA 2.0, a follow-up to the widely-successful ViennaRNA package on which Hofacker also worked, the authors, including Hofacker, dedicate the new package to Peter Schuster in honor of his seventieth birthday.

MURRAY GELL-MANN | Sept. 15, 1929 – May 24, 2019 | c.f. Chapter 81

A lifelong birder, Gell-Mann especially enjoyed birding with his older brother, Ben.

Gell-Mann and his first wife, Margaret, owned and bred Siberian huskies, and also Burmese cats, for many years. While Carl Sagan was housesitting for the Gell-Manns, their beloved show cat Shadow escaped the house and went missing. Sagan forgot to write the couple a letter to let them know the cat was lost until a month afterwards.

He was an avid collector of art, coins, stamps, neckties, and restaurant menus.

SETH LLOYD | Aug. 2, 1960 – | c.f. Chapter 81

In 2022, Lloyd starred in a sci-fi short film called "Steeplechase," the first of an ongoing mini-series following Lloyd and fellow physicist Michele Reilly on violent paradoxical excursions through time and reality, encountering alternate versions of themselves. Lloyd will also star in the next installment in the series, "Stag Hunt."

F. JOHN ODLING-SMEE | Oct. 10, 1935 – | c.f. Chapter 82

Born in the United Kingdom, Odling-Smee joined the Royal Navy at age eighteen and became a pilot in training in the Fleet Air Arm at the height of the Cold War. When told that there was the slightest possibility that he might be instructed to drop and H-bomb on a Soviet territory, he left the Navy.

For a time, Odling-Smee was a journalist at two major newspapers in Bristol, where he worked with then-unknown playwright Sir Tom Stoppard, who encouraged him to apply to undergraduate programs at University College London. Odling-Smee took the advice and began his research career.

KEVIN N. LALA | Oct. 5, 1962 – | *c.f. Chapter 82*

When he was born, Lala's family name was Laland. Of Parsi Indian heritage, Lala's family had chosen to anglicize their last name in an effort to mitigate difficulties their children faced, directly brought on by racism. In 2023, to celebrate his heritage and reconnect to his roots, he officially changed his name to its original form, Lala.

Lala has studied social learning and imitation his entire career. He has passed this habit of imitation onto the next generation, he discovered; while mowing the lawn, his son follows along with a fake lawnmower. Just like dad!

MARCUS FELDMAN | Nov. 14, 1942 – | *c.f. Chapter 82*

While Feldman's father was an engineer, he encouraged his son to take up any career he wanted. However, Feldman sr. had a talent for math, a trait his son strove to mirror, leading to the younger Feldman's love of mathematics. While biology was not in his plan for his longterm career, Feldman says the only other path he once considered, besides math, was psychiatry.

GEOFFREY BRIAN WEST | Dec. 15, 1940 – | *c.f. Chapter 83*

In May of 2006, West was named one of the Time 100 for the year by *Time Magazine*.

In 2012, West was a consultant for the popular Science channel documentary series "Through the Wormhole," on an episode dealing with the fine-tuned universe and questions about the driving forces behind the existence of the universe as we know it.

JAMES HEMPHILL BROWN | Sept. 25, 1942 – | *c.f. Chapter 83*

Brown has had a lifelong interest in natural history, beginning at a young age, which he attributes to the influences of two Cornell women: Sally Spofford, an administrative assistant at the Cornell Laboratory of Ornithology, and Brown's own mother, Catherine Brown, who earned her master's in zoology at Cornell.

BRIAN JOSEPH ENQUIST | March 4, 1969 – | c.f. Chapter 83

In high school, Enquist was the recorded voice for the Wilmington, Delaware morning and afternoon automated call-in weather forecasts.

Enquist has climbed some 40% of Colorado's "fourteeners," the fifty-three mountain peaks in the state of Colorado that exceed 14,000 feet of elevation.

In college, Enquist and a few friends decided to start a business—"Three Guys and a Truck—Moving & Landscaping"—to break into both aforementioned industries at once, but the venture struggled to take off, so Enquist returned to his previous college job, bagging groceries.

DAVID HILTON WOLPERT | 1961 – | c.f. Chapter 84

Wolpert is a writer of illustrative prose and free-form poetry, often presented online alongside photographs and experiences of his.

A quote of Wolpert's, which heads up a section of his website, is "I am not smart enough to be a mathematician, nor careful enough to be an experimentalist. Everything else is fair game."

WILLIAM G. MACREADY | March 23, 1963 – | c.f. Chapter 84

Macready looks back fondly on his time as a postdoctorate researcher at the Santa Fe Institute, during which time this paper with Wolpert was published. His best ideas, he believes, often came to him while hiking in the mountains in Santa Fe with coworkers and friends.

SHUN-ICHI AMARI | Jan. 3, 1936 – | c.f. Chapter 85

Amari is an avid player of the Japanese strategy game Go. He laments that computer programs like DeepMind's AlphaGo Zero are capable of beating many of the world's top Go players. In his retirement, Amari happily spends his time playing Go with other humans instead.

SAMUEL STEBBINS BOWLES | 1939 – | c.f. Chapter 86

In the 1940 US census, Bowles is listed among his family as the youngest "daughter," a clerical error.

Chester Bowles, Bowles's father, was a diplomat who served in his career as a congressman, the governor of Connecticut, an Under Secretary of State, and twice served as US Ambassador to India. On his first assignment to India, he moved his family there with him, including then-twelve-year-old Samuel.

In 1967, the Reverend Dr. Martin Luther King asked Bowles to contribute a paper as background material for his Poor People's March. Driving King to a canvassing event in Cambridge, Massachusetts, Bowles mentioned his boyhood interest in Buddhism and nonviolence, stemming from his childhood years in India. During this hurried drive, Bowles narrowly avoided colliding with a bus, after which King remarked "you certainly don't drive nonviolently, young man."

DUNCAN JAMES WATTS | Feb. 20, 1971 – | c.f. Chapter 87

Watts is an excellent rock climber. In 1996, he and two friends climbed the Nose of El Capitan in Yosmite National Park. A harrowing 3,000 ft. ascent, the climb took them five days to complete.

Born in Canada, Watts was raised in his parents' home countries of Australia and Scotland. At 16, he joined the Royal Australian Navy and went into officer training while simultaneously studying physics. During this time, he read James Gleick's *Chaos*, which inspired his interest in nonlinear dynamics.

STEVEN HENRY STROGATZ | Aug. 13, 1959 – | c.f. Chapter 87

At age nine, Strogatz won the Punt, Pass, and Kick football contest in his hometown of Torrington, Connecticut. He refers to this as the high point of his football career.

Strogatz is a world-class chess player. In 2006, the Cornell Chess Club invited American chess grandmaster Larry Christiansen to a simultaneous exhibition, and Strogatz signed up to play on a whim. Strogatz beat the grandmaster after several hours of play, the only competitor in the group of 35 to do so.

FOUNDATIONAL PAPERS IN COMPLEXITY SCIENCE

ROBERT BETTS LAUGHLIN | Nov. 1, 1950 – | *c.f. Chapter 88*

When Laughlin received the call announcing his Nobel Prize win in 1998, he had recently "fixed" the phone in his bedroom, unintentionally rendering it silent. The landline connected the call instead to his young son's bedroom, where it came through on his Mickey Mouse-themed telephone. Laughlin jokes that he must be the only person to ever receive their Nobel news via Mickey Mouse phone.

The premise of Laughlin's paper, "Pumped Thermal Grid Storage with Heat Exchange," became the inspiration for what is now the product of Malta, Inc., a thermoelectric energy storage system which, on its own or when combined with any currently viable source of electricity generation—power plants, solar arrays, wind farms, etc.—can store energy for upwards of a week and reconvert it to electricity as needed without using hazardous materials.

DAVID PINES | June 8, 1924 – May 3, 2018 | *c.f. Chapter 88*

In 1944, shortly after beginning graduate school in Berkeley, twenty-year-old Pines was drafted into the Navy to serve in WWII. After two years of service, inspired by previous mentorship he had received from J. Robert Oppenheimer, he transferred to Princeton and completed his PhD there.

On at least three occasions, Pines was a collaborator on projects that later won their researchers the Nobel Prize in Physics, but by happenstance he was either involved too early on in the project, or he was the fourth collaborator in the group, which left him out of the running every time (a Nobel cannot be shared more than thrice).

JÖRG SCHMALIAN | April, 1965 – | *c.f. Chapter 88*

In the late 1980s, Schmalian spent one summer working in construction in Russia on a Soviet factory near the Ural Mountains. He received a commendation for his excellent work in a formal ceremony commemorating the commissioned factory, and this was his first ever public recognition for his work.

What was that work? Brick-laying for the factory bathrooms.

His favorite pasttimes include road-biking and hiking in the Black Forest in Southwest Germany, and rowing on the river Rhine.

BRANKO STOJKOVIĆ | 1967 – | *c.f. Chapter 88*

Stojković attended gymnasium in Belgrade in his native Serbia before moving to the United States for college.

He is a strong swimmer, a fan of Aretha Franklin, and has channeled his physics degrees into success as a professional in the financial sector.

PETER GUY WOLYNES | April 21, 1953 – | *c.f. Chapter 88*

Wolynes has likened newly made protein molecules to strings of spaghetti, splattering all over the place.

He has conducted extensive research into the architecture and function of genomes, including the reproduction of exact physical traits among family members. Once, among family heirlooms, he found a framed photograph of a woman in her twenties who looked exactly like his daughter, only to find that it was actually his grandmother. He hung the photograph on his wall.

Protein structure prediciton software that Wolynes pioneered bears the name AWSEM-MD as an homage to a quip from a student who, seeing the software's full name, created the acronym AWSEM from its title. When the student said this was a joke, Wolynes decided it should be more than a joke, and the name stuck.

ELINOR CLAIRE OSTROM | Aug. 7, 1933 – June 12, 2012 | *c.f. Chapter 89*

Growing up after the Great Depression and during WWII, Ostrom witnessed scarcity on major and local levels, including the water wars over Los Angeles-area pumping rights, happening right in her backyard. When she was assigned to a research team looking at groundwater basins in the area, she was impressed by one basin whose jurisdictions found a way to cooperate effectively, leading to the subject of her dissertation and the themes of her life's work.

When Ostrom won her Nobel, shared with Oliver E. Williamson, many economists were not yet familiar with her work and assumed that the award had actually gone to economist Bengt Holmstrom.

Ostrom's high school did not allow her to take trigonometry. As a result, Ostrom could not enroll in UCLA's economics programs, and received all her degrees in political science instead. She is one of only three women to have won the Nobel Memorial Prize in Economic Sciences, and one of only a handful of laureates to have done so with degree backgrounds from outside the field.

EDITOR

DAVID C. KRAKAUER is the President and William H. Miller Professor of Complex Systems at the Santa Fe Institute. His research explores the evolution of intelligence and stupidity on Earth. This includes studying the evolution of genetic, neural, linguistic, social, and cultural mechanisms supporting memory and information processing, and exploring their shared properties. He served as the founding director of the Wisconsin Institutes for Discovery, codirector of the Center for Complexity and Collective Computation, and professor of mathematical genetics, all at the University of Wisconsin, Madison. He has been a visiting fellow at the Genomics Frontiers Institute at the University of Pennsylvania, a Sage Fellow at the Sage Center for the Study of the Mind at the University of California, Santa Barbara, a longterm fellow of the Institute for Advanced Study, and visiting professor of evolution at Princeton University. In 2012, he was included in the *Wired Magazine* Smart List: Fifty People Who Will Change the World. In 2016, he was included in *Entrepreneur Magazine*'s list of visionary leaders advancing global research and business.

Krakauer was previously chair of faculty and a resident professor and external professor at the Santa Fe Institute. A graduate of the University of London where he earned degrees in biology and computer science, he received his D.Phil. in evolutionary theory from Oxford University in 1995 and continued there as a postdoctoral fellow.

THE SANTA FE INSTITUTE PRESS

The SFI Press endeavors to communicate the best of complexity science and to capture a sense of the diversity, range, breadth, excitement, and ambition of research at the Santa Fe Institute; To provide a distillation of work at the frontiers of complex-systems science across a range of influential and nascent topics;

To change the way we think.

SEMINAR SERIES
New findings emerging from the Institute's ongoing working groups and research projects, for an audience of interdisciplinary scholars and practitioners.

ARCHIVE SERIES
Fresh editions of classic texts from the complexity canon, spanning SFI's four decades of advancing the field.

COMPASS SERIES
Provocative, exploratory volumes aiming to build complexity literacy in the humanities, industry, and the curious public.

SCHOLARS SERIES
Texts featuring foundational ideas, systems of knowledge, emerging methodologies, and areas of application in the complex-systems science world.

— MORE FROM SFI PRESS —

Foundational Papers in Complexity Science, Volumes 1, 2 & 3
David C. Krakauer, ed.

The Complex World: An Introduction to the Foundations of Complexity Science
David C. Krakauer

For additional titles, inquiries, or news about the Press, visit us at
WWW.SFIPRESS.ORG.

SFI PRESS BOARD OF ADVISORS

Sam Bowles
Professor, University of Siena;
SFI Resident Faculty

Sean Carroll
Homewood Professor of Natural
Philosophy, Johns Hopkins University;
SFI Fractal Faculty & External Faculty

Katherine Collins
Head of Sustainable Investing,
Putnam Investments; SFI Trustee
& Chair of the Board

Jennifer Dunne
SFI Resident Faculty;
SFI Vice President for Science

Mirta Galesic
SFI Resident Faculty

Ricardo Hausmann
Director, Center for International
Development & Professor, Harvard
University; SFI Science Board
Co-chair; SFI External Faculty

Chris Kempes
SFI Resident Faculty

Manfred Laublicher
President's Professor, Arizona State University;
SFI External Faculty

Simon Levin
James S. McDonnell Distinguished University
Professor in Ecology and Evolutionary
Biology at Princeton University

Ian McKinnon
Founding Partner, Sandia
Holdings LLC; SFI Trustee

John Miller
Professor, Carnegie Mellon University;
SFI Science Steering Committee Chair

William H. Miller
Chairman & CEO, Miller
Value Partners; SFI Trustee
& Chairman Emeritus

Melanie Mitchell
SFI Resident Faculty
SFI Davis Professor of Complexity,
SFI Science Board co-chair

Cristopher Moore
SFI Resident Faculty

Sidney Redner
SFI Resident Faculty

Geoffrey West
SFI Resident Faculty;
SFI Toby Shannan Professor of
Complex Systems & Past President

David Wolpert
SFI Resident Faculty

EDITORIAL

David C. Krakauer
Publisher/Editor-in-Chief

Sienna Latham
Managing Editor

Zato Hebbert
Production Assistant

Laura Egley Taylor
Advisor/Designer

Ellis Wylie
Post-Production
Coordinator

Additional editorial support provided by Katie Mast, Bronwynn Woodsworth, and Shafaq Zia.

ABOUT THE SANTA FE INSTITUTE

The Santa Fe Institute is the world headquarters for complexity science, operated as an independent, nonprofit research and education center located in Santa Fe, New Mexico. Our researchers endeavor to understand and unify the underlying, shared patterns in complex physical, biological, social, cultural, technological, and even possible astrobiological worlds. Our global research network of scholars spans borders, departments, and disciplines, bringing together curious minds steeped in rigorous logical, mathematical, and computational reasoning. As we reveal the unseen mechanisms and processes that shape these evolving worlds, we seek to use this understanding to promote the well-being of humankind and of life on Earth.

COLOPHON

The body copy for this book was set in EB Garamond, a typeface designed by Georg Duffner after the Ebenolff-Berner type specimen of 1592. Headings are in Kurier, created by Janusz M. Nowacki, based on typefaces by the Polish typographer Małgorzata Budyta, and Cochin, a typeface produced in 1912 by Georges Peignot and based on the copperplate engravings of French 17th century artist Nicolas Cochin, for whom the typeface is named. For footnotes and captions, we have used CMU Bright, a sans serif variant of Computer Modern, created by Donald Knuth for use in TeX, the typesetting program he developed in 1978.

The SFI Press complexity glyphs used throughout this book were designed by Brian Crandall Williams.

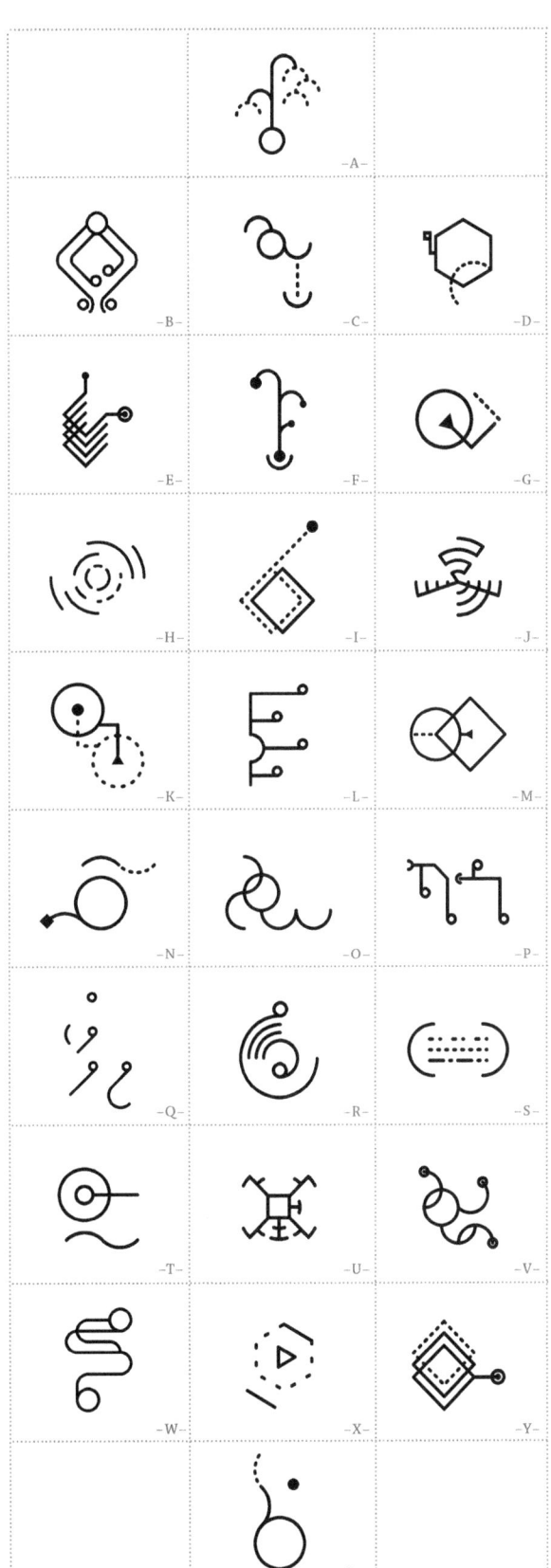

SANTA FE INSTITUTE
COMPLEXITY GLYPHS

SCHOLARS SERIES

www.ingramcontent.com/pod-product-compliance
Lightning Source LLC
LaVergne TN
LVHW061650050325
805089LV00006B/103